AF601892

ENVIRONMENTAL RESOURCE MANAGEMENT
Critical Issues

ENVIRONMENTAL RESOURCE MANAGEMENT
Critical Issues

Editor
Professor (Dr.) Arvind Kumar
FLS (London), FASc (Swiss), FISEC, FSESc, FAZ, FZA (Gold Medalist)
Pro-vice Chancellor
S.K.M. University, Dumka – 814 101 (Jharkhand)
&
Dr. Pashupati Kumar Roy
Reader & Head, Department of Political Science
A.S. College, Deoghar

2021
Daya Publishing House®
A Division of
Astral International Pvt. Ltd.
New Delhi – 110 002

Reprinted, 2021
ISBN: 9789351240297 (INT. EDN.)

Publisher's Note:

Every possible effort has been made to ensure that the information contained in this book is accurate at the time of going to press, and the publisher and author cannot accept responsibility for any errors or omissions, however caused. No responsibility for loss or damage occasioned to any person acting, or refraining from action, as a result of the material in this publication can be accepted by the editor, the publisher or the author. The Publisher is not associated with any product or vendor mentioned in the book. The contents of this work are intended to further general scientific research, understanding and discussion only. Readers should consult with a specialist where appropriate.

Every effort has been made to trace the owners of copyright material used in this book, if any. The author and the publisher will be grateful for any omission brought to their notice for acknowledgement in the future editions of the book.

Published by : **Daya Publishing House®**
A Division of
Astral International Pvt. Ltd.
– ISO 9001:2015 Certified Company –
4736/23, Ansari Road, Darya Ganj
New Delhi-110 002
Ph. 011-43549197, 23278134
e-mail: info@astralint.com
Website: www.astralint.com

Preface

Our nation which was unable to form a single needle before independence, now is marching side by side with Europe. One day will come; India may become superpower of the world. But increasing developmental process and accompanying environmental issues are the two aspects of a single coin. Developed countries like U.S.A., Germany, France, China and Japan vomit huge toxic waste products without proper caring of life on this watery planet. As a result of this, threats of ecological imbalance are hovering on. Strange it may sound but times are testimony to strategic use of environment for global supremacy. Current environmental issues and their management, ecological security, socio-economic constraints, environmental education and awakening and futuristic planning are the motto of this compiled book.

Realizing the imminent dangers of ignoring environmental issues has been some kind of new awakening and a new world order has addressed itself alive to environmental problems. How best can one maintain, upgrade and improve the environment with judicious utilization of this natural treasure for benefit of humanity is a trenchant question? Environmental issues are not merely technical, scientific or academic problem, but are intermeshed with social, economic and political fabric and ethos of human beings.

So, in order to save the nature, the slogan of "Lab to Land" should be enforced with great vigour and zeal, various management strategies should be formulated and mass awakening among all strata of society right from the common people to top executives should be done, otherwise millions of people will be forced to live as Ecological Refugees. Therefore, ecological security is the crying need of the hour. Keeping above facts in view, the present book has been edited but this is just my beginning in right direction, but not the end to stall the environmental catastrophe.

This book entitled "*Environmental Resource Management: Critical Issues*" is the unique compilation of the most recent and informative research articles of internationally acknowledged experts in the field of Environmental sciences with the intention of providing a sufficient depth of the subject to satisfy the needs at a level which will be comprehensive and interesting. This book will be useful to the students, research scholars, scientists in the field of Environmental management and ecoplanners, politicians and other people with similar interest.

I consider myself extremely fortunate in having got the invaluable support and erudite suggestions of eminent persons like Prof. (Dr.) Christian Ulrichs of Germany, Dr. M.A. Kabir Chowdhury of Malaysia, Prof. Dr. Ahmed H. Al-Harbi of Saudi Arabia, Prof. Tej Kumar Shrestha of Nepal, Prof. P.V. Dehadrai, Former Director, CIFRI, Dr. Dilip Kumar, Director, CIFE, Mumbai, Prof. K.C. Pandey, Former V.C. (Lucknow), Prof. A.R. Yousuf of University of Kashmir, Srinagar (J & K), Professor N.C. Datta of Calcutta University, Professor S.K. Konar of Kalyani University, Professor D.K. Belsare of Bhopal University, Professor U.C. Goswami of Gauhati University, Professor P.C. Mishra of Sambalpur University, Professor P. Natarajan of Kerala University, Professor P.S. Murthy of Bangalore University, Professor Ajit Varma of JNU, Professor A.L. Bhatia of Jaipur University, Professor A.K. Mittal of BHU, Professor S.P. Hosmani of Mysore University, Professor K. Kapoor of Udaipur University, Professor K.B. Reddy of Nagarjuna University, Professor M. Vikram Reddy of Pondicherry University, Professor B.K. Tiwari of NEHU, Professor G. Tripathi of Jodhpur University, Professor R. Ramalingam of Annamalai University, Professor Sharif U. Ahmad of Nagaland University, Professor G.K. Kulkarni of Aurangabad University, Professor S.U. Mehram of Nagpur University, Professor S.K. Battish of PAU, Professor B.M. Sharma of Manipur University, Professor B.D. Joshi of Hardwar University, Professor G.C. Pandey of Faizabad University, Professor K.C. Sharma of Ajmer University, Professor M. Raziuddin of Hazaribag University, Professor U.S. Bagde of Mumbai University, Professor Gurdeep Singh of I.S.M., Dhanbad, Dr. P.K. Goel of Karad, Professor S.P. Roy and Shri Tribhuwan Poddar of Bhagalpur University, Bhagalpur for encouragements.

I also express my deep sense of gratitude to my parents whose blessings have always prompted me to pursue academic activities deeply. I am also thankful to my wife, *Kumari Bimla* and my two lovely sons, *Kumar Pallav Shivshankaran* and *Kumar Prasun Ramakrishnan* whose natural smiles extended to me relief all through this tiresome endeavour.

Last but not the least, I am also thankful to Mr. Anil Mittal, Proprietor, Daya Publishing House, New Delhi for taking keen interest in bringing out of this book. Finally, I will always remain a debtor to all my well-wishers for their blessings, without which this book would not have come into existence.

Dumka

Professor A. Kumar

Pashupati Kumar Roy

Contents

Chapter 1

Content and Meaning of Environmental Management in India

I. Sundar[1], Sivakumar[1], P.K. Roy[2] and A. Kumar[3]
[1]Department of Economics, Directorate of Distance Education, Annamalai University, Annamalainagar – 608 002, Tamil Nadu, India
[2]Department of Political Science, A.S. College, Deoghar, S.K.M. University, Dumka, Jharkhand
[3]Environmental Science Research Unit, Post Graduate Department of Zoology, S.K.M. University, Dumka – 814 101, Jharkhand, India

ABSTRACT

The study and management of biodiversity and of nature's services are the key research foci of the discipline of Ecosystem Management. Ecosystem Management simply means the manipulation of land and waterscapes for human ends. Accordingly, academic staff works in a broad range of environments with varied resource uses including nature conservation, mining, forestry and agriculture. The biodiversity that is the focus of management is composed of both indigenous and exotic biota. Most importantly, as a 'management' discipline, staff recognizes the importance and legitimacy of social, economic and political concerns as well as ecological considerations, and the need to encompass a broad range of spatial and temporal scales in devising socio-economic and biophysical solutions to natural resource management problems. This paper deals with Ecosystem Management and its related components. The content themes discussed in multiple area and principle and constituents of ecosystem management have been discussed in this paper. It outlines the possible ways of managing biodiversity focusing on holistic resources management principles.

Keywords: *Ecosystem management, Principle of ecosystem management, Biodiversity management.*

Introduction

"Ecosystem management focuses on the conditions of the ecosystem, with goals of maintaining soil productivity, gene conservation, biodiversity, landscape patterns, and the array of ecological processes." It is a process that integrates scientific knowledge of ecological relationships within a complex socio-political and values framework toward the general goal of protecting native ecosystem integrity over the long run." Further it is said to be resource management system designed to maintain or enhance ecosystem health and productivity while producing essential commodities and other values to meet human needs and desires within the limits of socially, biologically and economically acceptable risk."

Concept Theme

"Ecosystem management is the integration of ecological, economic, and social principles to manage biological and physical systems in a manner safeguarding the long-term ecological sustainability, natural diversity, and productivity of the landscape. The primary goal of ecosystem management is to develop and implement management that conserves, restores, and maintains the ecological integrity, productivity, and biological diversity of public lands. Sustainable ecosystems provide many benefits for wildlife and humans such as habitat for fish and wildlife, clean drinking water for communities, wood, fiber, forage, recreational, and economic opportunities" (US Bureau of Land Management, 1994).

Ecosystem management is the integration of ecological, economic, and social factors in order to maintain and enhance the quality of the environment to meet current and future needs. It is a holistic approach to natural resource management. Applying ecosystem management involves collaboration of partners, land managers, and scientists, with inclusion and consent of the public, particularly on public land. "Ecosystem management, then, is a concept of management that is more inclusive of the variables that impinge on management, occurring at larger scales, and over more extended time horizons than has been common in the past" (Thomas, 1996).

Common Elements within Ecosystem Management Definitions

Merely discussing ecosystem management themes and definitions is unsatisfactory for practical use. Incorporating ecosystem management principles into land management is a process that must evolve over time. Part of the complexity is due to the scientific uncertainty about what comprises an ecosystem management plan. Even at the small individual forest unit we are still learning about the effect of different management practices on forest ecological processes and species. Given our limited knowledge at the stand level, how can we plan for cumulative effects at the larger landscape level? A famous ecologist once said that ecosystems are not only more complex than we think - they are more complex than we can think. Therefore, perhaps the most important aspect of Ecosystem Management is to view it as a learning process, where landowners and forest managers adapt their plans to changing information. Landowners should set clear goals about what they would like to achieve, and be flexible in addressing changing information and needs.

Five Main Ecosystem Management Goals

Maintain viable populations of all native species in situ. Represent, within protected areas, all native ecosystem types across their natural range of variation. Maintain evolutionary and ecological processes (*i.e.*, disturbance regimes, hydrological processes, nutrient cycles, etc.). Manage over period of time long enough to maintain the evolutionary potential of species and ecosystems. Accommodate human use and occupancy within these constraints. (Grumbine, 1994). Further ecosystem Management includes the following elements:

Sustainability

Ecosystem Management does not focus primarily on "deliverables" but rather regards intergenerational sustainability as a precondition. Goals. Ecosystem Management establishes measurable goals that specify future processes and outcomes necessary for sustainability. Sound Ecological Models and Understanding. Ecosystem Management relies on research performed at all levels of ecological organization. Complexity and Connectedness. Ecosystem Management recognizes that biological diversity and structural complexity strengthen ecosystems against disturbance and supply the genetic resources necessary to adapt to long-term change. The Dynamic Character of Ecosystems Recognizing that change and evolution are inherent in ecosystem sustainability. Context and Scale: Ecosystem processes operate over a wide range of spatial and temporal scales, and their behavior at any given location is greatly affected by surrounding systems. Thus, there is no single appropriate scale or time frame for management. Humans as Ecosystem Components. Ecosystem Management values the active role of humans in achieving sustainable management goals. Adaptability and Accountability. Ecosystem Management acknowledges that current knowledge and paradigms of ecosystem function are provisional, incomplete, and subject to change. Management approaches must be viewed as hypotheses to be tested by research and monitoring programmes."

Principles of Ecosystem Management

The ecosystem concept can be applied to all lands, not just those managed for natural values. Four biological and social system properties of ecosystems underlie successful environmental assessments and have implications for management:

1. Ecological systems are continually changing.
2. There may be substantial heterogeneity in impacts from a particular action.
3. Systems may exhibit several levels of stable behaviour; and
4. There is an organized connection between parts, but everything is not connected to everything else.

Given the political boundaries of park and wilderness ecosystems and unavoidable and uncertain human influence, change will occur in these systems. Given this reality, a four-step approach can be used as a process to meet the ecosystem goals:

1. Define goals and measurable targets for ecosystem condition.
2. Define the ecosystem boundaries for the primary components.
3. Develop management strategies to achieve goals that transcend political boundaries; and
4. Establish a programme to assess the effectiveness of the management strategies in achieving the identified goals (Agee and Johnson, 1988).

Ecosystem management is management driven by explicit goals, executed by policies, protocols, and practices, and made adaptable by monitoring and research based on our best understanding of the ecological interactions and processes necessary to sustain ecosystem composition, structure, and function. The following are fundamental scientific precepts for ecosystem management:

1. Spatial and temporal scale are critical.
2. Ecosystem function depends on its structure, diversity, and integrity;

3. Ecosystems are dynamic in space and time; and
4. Uncertainty, surprise, and limits to knowledge.

Ecosystem management is not a rejection of an anthropogenic for a totally biocentric worldview. Rather, it is management that acknowledges the importance of human needs while at the same time confronting the reality that the capacity of our world to meet those needs in perpetuity has limits and depends on the functioning of ecosystems (Christensen and Bartuska, 1996).

The transition to ecosystem management in parks and wilderness will involve a gradual shift in management thinking and behavior. Top priority issues in ecosystem management are:

1. What is the important research needs associated with ecosystem management?
2. What is the important general management, planning, and communications issues associated with ecosystem management?
3. What are the important challenges to ecosystem management in the areas of conflict resolution and cooperation?
4. What are the important limits and constraints to ecosystem management?

Management can be defined as human intervention in the dynamic processes, which determine the composition of plant and animal communities, so as to maintain a particular desired pattern or series of processes.

Conservation

Is here taken to be a wider process of sustainable management of renewable natural resources, as an essential foundation for the human future on this planet. Although the temperate communities are not great reservoirs of biological diversity, the scientific management of temperate communities for conservation is critical since the sustainable development of the temperate zones is essential for the human future, and the preservation of such diversity as they exhibit is important. Moreover, because these ecosystems are relatively well known, and their management receives greater resources than in most tropical areas, work here can be a valuable source of knowledge and methodology. In particular, such work can emphasize:

The value of detailed scientific insights obtained through the study of key elements in temperate ecosystems and species. The value of temperate communities in their own right as components of global diversity and as valued features of the countries in which they occur. The value of demonstrating in the temperate regions how human populations and demands can be brought into balance with ecosystems. The use of temperate examples to demonstrate the management techniques available to implement strategies for conservation and sustainable development and the utility of ecology as a basis for both conservation for human use and conservation of the diversity of nature in its own right.

However, it is crucial that temperate zone ecologists are outward looking and involve themselves in the daunting problems of the tropical zones (Holdgate, 1989).

Management of Biodiversity

Tropical forests are very rich in biological diversity and form an important economic and ecological resource. This biodiversity is of great value for communities living in or near these forests as a ready source of subsistence and cash income, and for the world at large as a source of tropical timber and non-timber products and a repository of genetic and chemical information.

However, this biological complexity is diminishing rapidly. The main human (anthropogenic) actions causing loss of tropical forest biodiversity along with the strategies of management for forest biodiversity are analyzed. Not only is the biophysical component important in management for biodiversity, but the active participation and support of local people, the national government, and international cooperation as a whole, are essential for an effective and sustainable development of tropical forests (Khasa and Dancik, 1997).

The perspectives and constraints that need to be addressed to establish biodiversity as a new scientific field. The perspectives include biodiversity as:

1. The unifying factor within the science of biology.
2. The cornerstone for agriculture, animal husbandry, forestry and aquaculture utilization.
3. The source of information for understanding the diversity of landscapes and the diversity of populations necessary for land use and regional development.
4. Fundamental in the new era of industrial applications.
5. The tool to establish a much-needed bridge within the social and cultural role; and
6. A pillar of human development where a new synthesis between globalization and diversity is achieved (di Castri and Younes, 1996).

The biodiversity management requires holistic management Holistic Resource Management (HRM) is a process of goal setting, decision making and monitoring which integrates social, ecological, and economic factors. Biodiversity enhancement is a fundamental principle in HRM and students are taught that biodiversity is the foundation of sustainable profit. In the HRM process, practitioners develop a holistic goal which includes:

1. Quality of life values.
2. Forms of production to support those values; and
3. Landscape planning, which should protect and enhance biodiversity and support ecosystem processes of succession, energy flow, hydrological and nutrient cycling (Stinner *et al.*, 1997).

In this context the following points can be considered.

Use a regional perspective when considering biological diversity. Think beyond the boundaries of specific ownership when planning and Managing. Plan and manage over large areas rather than using a stand-by-stand approach. Consider the cumulative impact of individual projects on regional population and resources.

Emphasize multispecies and ecosystem management instead of single species and tree management. Provide habitat sufficient to maintain species of concern (*e.g.* large wide ranging mammals.), not just sufficient habitat to attract immigrants from more productive sources. Maintain or create spatial patterns. Conduct ecological surveys and inventories. Monitor problem species and inventories (Probst and Crow, 1991).

It could be seen clearly from the above discussion that ecosystem and biodiversity management involves various issues and components. The issues and approaches discussed in the paper bring to insight in the area of ecosystem management principle.

References

Agee, J.K. and Johnson, D.R., 1988. Introduction to ecosystem management. In: *Ecosystem Management for Parks and Wilderness*, (Eds.) J.K. Agee and D.R. Johnson. University of Washington Press, Seattle and London, pp. 3–14.

Christensen, N., Bartuska, A., Brown, J., Carpenter, S., Antonio, C.D., Francis, R., Franklin, J., MacMahon, J., Noss, R., Parsons, D., Peterson, C., Turner, M. and Woodmansee, R., 1996. The Report of the Ecological Society of America Committee on the Scientific Basis for Ecosystem Management. *Ecological Applications*, 6: 665–691.

Di Castri, F. and Younes, T., 1996. Introduction: Biodiversity, the emergence of a new scientific field, its perspectives and constraints. In: *Biodiversity, Science and Development, Towards a New Partnership*, (Eds.) F. di Castri and T. Younes. CAB International, Wallingford, UK.

Grumbine, R.E., 1994. What is ecosystem management? *Conservation Biology*, 8: 27–39.

Holdgate, M., 1989. Conservation in a world context. In: The scientific management of temperate communities for conservation. The 31st symposium of the British Ecological Society, South Hampton, (Eds.) I. Spellerberg, F. Goldsmith and M. Morris. Blackwell Scientific Publications. Oxford, U.K.

Khasa, P. and Dancik, B., 1997. Sustaining tropical forest biodiversity. In: *Sustainable Forests: Global Challenges and Local Solutions*, (Eds.) O. Bourman and D. Brand. The Haworth Press.

Probst, J. and Crow, T., 1991. Integrating biological diversity and resource management: An essential approach to productive, sustainable ecosystems. *Journal of Forestry*, 89: 12–17.

Stinner, D., Stinner, B. and Martslof, E., 1997. Biodiversity as an organizing principle in agroecosystem management: Case studies of holistic resource management practitioners in the USA. *Agriculture, Ecosystems and Environment*, 62: 199–213.

Thomas, J.W., 1996. Forest Service perspective on ecosystem management. *Ecological Applications*, 6: 703–705.

U.S. Department of the Interior, 1994. Bureau of Land Management: Ecosystem management in the BLM: From concept to commitment. Gov. Pub. BLM/SC/Gi–94/005+1736.

Chapter 2

Waterlogged Area as New Horizon for Aquaculture Development: Golden Dream to the Rural Communities of Bangladesh

M. Shahadat Hossain* and Nani Gopal Das
Institute of Marine Sciences and Fisheries, University of Chittagong, Chittagong – 4331, Bangladesh

ABSTRACT

Begumgonj situated under the district of Noakhali, Bangladesh is suitable for rice cultivation during dry season, but unsuitable during rainy season due to deep waterlogged situation. Recently, the availability of waterlogged paddy lands and the potentials of prawn/fish aquaculture development have addressed to stimulate income generation, employment opportunity and supply nutritive food for sustainable rural livelihoods through earning of high-valued products. Landsat TM image and thematic information were analyzed using ENVI and ArcView GIS software and recorded 22,186 ha seasonal waterlogged area occupied by 294 polygons that covers about 52 per cent of total study area. Aquaculture suitability was identified on the basis of water depth and water logging duration for nursing and culture of fish/prawn as well as indigenous brood fish rearing through the establishment of local fish sanctuary. Rapid Participatory Rural Appraisal was carried out with different stakeholder groups to find out the available resources of waterlogged areas as well as their importance for community livelihoods. The paddy cum prawn culture as rotational basis may be economically viable and ecologically sound in this region. This is the prime way to increase rural economy through the development

---

* Tel. 88-031-710347, Mobile: 01711-720950, Fax: 88-031-713109 E-mail: hossainms@yahoo.com.

of primary production sector in the waterlogged paddy lands. As a result, employment opportunity of waterlogged-area-based villagers could secure their livelihoods during rainy season and discourage the on-going migration tendency of rural people. The study advocates that the local community in the waterlogged area requires the maximum extension and training support to utilize the dormant waterlogged resources for the improvement of socio-economic status of the rural people by promoting sustainable aquaculture.

Keywords: *Waterlogged area, Paddy-prawn rotation, Livelihood, Satellite image, and Suitability classification.*

Introduction

Aquaculture activities have been improved significantly in the recent years aiming the increased production target. The socio-economic benefits derived from aquaculture expansion provide the provision of nutritive foods contributing improved life style to the poor, income generation and employment opportunity, diversification of fish production and create scope for foreign exchange earning through export of high-valued products. Aquaculture is also treated as potential input to compensate for the low growth rate of capture fisheries. Sustainable aquaculture can contribute to the prevention and control of aquatic pollution since it relies on a good-quality water resource.

A major feature of the Begumgonj thana situated under Noakhali district, Bangladesh is the rice cultivation during dry season (November to March), but becomes waterlogged during rainy season. Following acute shortage of steam corridor, canal, tributaries, creeks and river coupled with heavy rain during monsoon season lead to create abnormal situation of water logging condition in the vast areas that creates a major cause of low agricultural productivity. The regular water logging means that in absence of alternative livelihood opportunities due to water logging situation poor people have been forced to move out of their villages to seek alternative employment as day labourers or in such occupations as ricksha/van pulling. It is perceived that the waterlogged areas could often contribute a much larger potential for aquaculture development, especially through prawn culture. Karim (1989) and Alam (2001) have recognized the potential for prawn farming in the waterlogged paddy lands of Noakhali. Chowdhury *et al.* (2003) mentioned the possible environmental impact of stocking prawn on biodiversity in the waterlogged paddy lands and social impact on poor households using the aquatic resources as a major source of livelihood during the monsoon season. The present study has commissioned to elucidate further on these issues.

Satellite remote sensing data are useful to updating of existing map or generating new map in land use and watershed areas during both dry and wet seasons. During observations, many earth features of interest have already identified, mapped and studied on the basis of their spectral characteristics (Lillisand and Kiefer, 2000). GIS and satellite remote sensing are very essential tools for planning of aquaculture development (Burrough, 1986; Setiato, 1993). GIS also serve as an analytical and predicting tool for aquaculture development and to test the consequence of various development decisions before their use in the landscape (Aguilar-Manjarrez and Ross, 1995). Other uses of GIS are in storing, managing and analyzing of spatial and non-spatial data in efficient manner (Kapetsky *et al.*, 1987; Burrough and McDonnel, 1998). Similar observations were made by Vibulsresth *et al.* (1993) and Venkataratnam *et al.* (1997). In recent years, significant advances have been made in the field of remote sensing of Bangladesh, specially in shrimp farming development in the coastal zone (Hossain *et al.*, 2001), change detection of Sunderbans mangrove forest (Islam *et al.*, 1997), mapping shrimp farming and agricultural areas (Shahid *et al.*, 1992), mapping suitable areas for saltpan development (Hossain *et al.*, 2003a) and mapping suitable areas for mangrove afforestation (Hossain *et al.*, 2003b).

The Begumgonj thana possesses an extensive seasonal waterlogged area of various depths. The water types support multitudes of aquatic flora and fauna, where fisheries contribute a major source of employment as well as main dietary protein for the rural poor. The reproduction, breeding and spawning of open inland fisheries are tuned and adjusted to the rhythm and amplitude of monsoon flooding. In the mean time the waterlogged paddy lands become inter-logged and integrated into a single biological production system, where the brood stocks and newborn young species undertake migrations in the nearby canals, steam corridors and the flood plains (Das and Hossain, 2005).

The objectives of this study are to make a full inventory of the waterlogged paddy lands and map their spatial location with suitability classification for community-based sustainable aquaculture development. The results of this research will be very essential in formulation of comprehensive and effective utilization of waterlogged paddy lands of Begumgonj thana.

Description of the Study Area

The geographical location of Begumgonj thana is in between latitude 22°50′ and 23°05′ N and longitude 90°55′ and 91°10′ E (Figure 2.1). Rain is the main sources of water in the Begumgonj thana. The general availability of water, however, remains high in most of the area. Despite this fact, irrigated agriculture during the dry season draws heavily on groundwater resources. The early rainfall fill up of water retention areas, ponds, ditches, and low lying areas and thus additional rainfall during the on-going rainy season are just over flowing or keep the areas waterlogged for about 6 months (May to October). Begumgonj thana is low elevated in comparison to the surrounding areas and thus water cannot be drained out. As a result, water stagnation is a common phenomenon during heavy rainfall. In Noakhali district, temperature varies from 12 to 34°C. The lowest temperature was recorded in December–February and the highest in April–June. Rainfall is abundant but seasonal. The highest rainfall was found in June–August while the lowest was in November–January. The maximum sunshine hours were found in October–November and February–May, where the minimum sunshine hours were recorded in June–July (Table 2.1).

Table 2.1: Monthly Temperature, Rainfall and Sunshine of Noakhali District

Month	*Temperature (°C)*		*Rainfall (mm)*	*Bright Sunshine (hours)*
	Min	*Max*		
January	12.86	23.56	Nil	197.3
February	16.82	28.68	67.2	238.4
March	17.40	29.71	85.4	254.4
April	24.48	33.87	100.6	254.6
May	25.34	34.18	200.4	223.9
June	25.43	30.50	949.0	87.1
July	26.41	29.25	478.2	168.8
August	26.40	32.14	435.2	203.3
September	26.22	32.14	267.0	163.0
October	25.08	32.25	298.6	252.5
November	20.06	29.66	Nil	275.9
December	16.89	25.66	33.4	–

Source: Meteorology Department, Maijdee (Personnel communication).

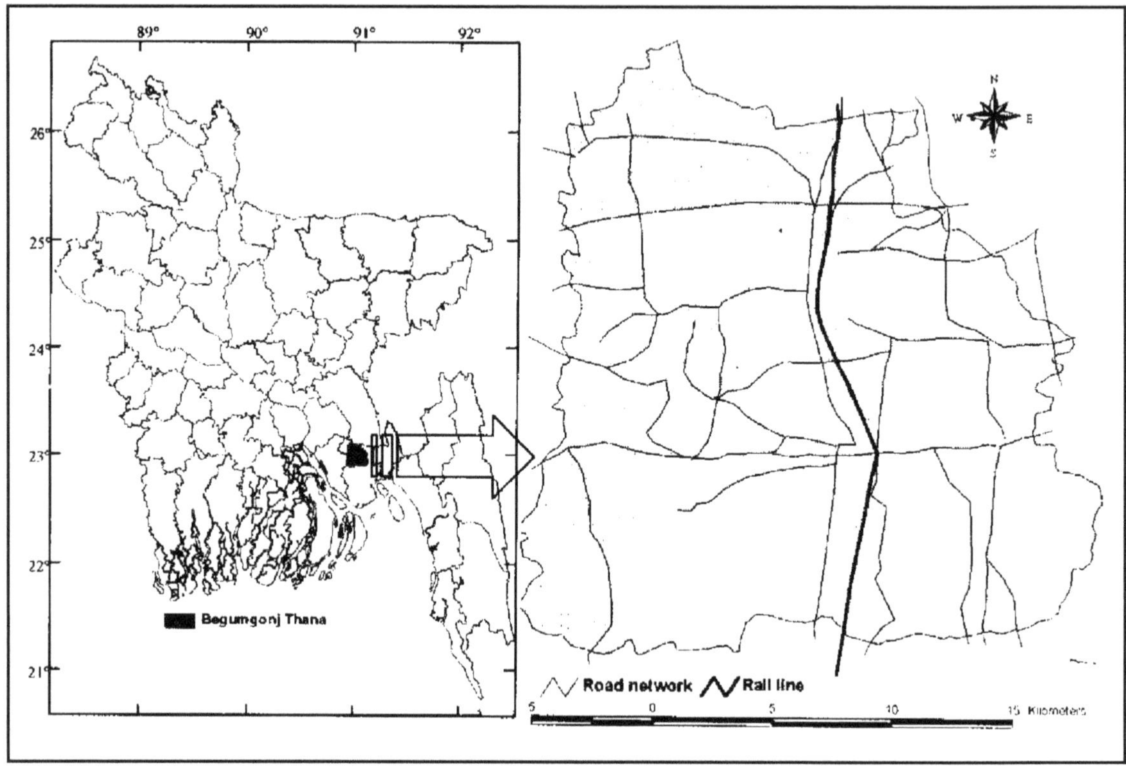

Figure 2.1: Geographical Location of the Begumgonj Thana, Noakhali

Materials and Methods

Data Collection

Landsat TM image acquired on 14th January 2001 was used for the present study, which was provided by SPARRSO (Bangladesh Space Research and Remote Sensing Organization). Topographic maps at 1 : 10,000 scale, published by the Survey of Bangladesh and 1 : 50,000 scale, published by LGED (Local Government and Engineering Department) were used for geometric correction. The Thana maps with union boundary of Begumganj thana were digitized using Arc View GIS software. Geometric correction was performed with fourth order polynomial (Research Systems Inc., 2000a) and the error was (0.56) controlled less than one pixel (30 m) for 50 ground control points (GCP). Road joint, rail line, cross-point of small canal and road, and prominent features (*e.g.* bazar, school, college campus) were selected as the main GCPs in the image (Das and Hossain, 2005). Image processor for the analysis was ENVI (The Environment for Visualizing Image). Ground data were obtained by field investigation of about 85 plots, including some permanent ones. All other required data such as climatic condition, land use, thana and union boundary, demography, etc. were collected from the respective departments. The collected data were synthesized and the useful data were extracted for this study.

Satellite Image Classification

Map to image registration was used to create imge data covering the whole area. ISODATA unsupervised classification (use information from the image itself to identify spectral clusters, which are interpreted as classes) was performed considering minimum and maximum classes of 5–10, 10–15 and 15–20, where the 10–15 classes turned out to be useful. Supervised classifications were carried out on the basis of Region of Interest (ROIs), where the ground truth or so-called training areas (collected during field investigation) were regions of terrain with known properties or characteristics (Research Systems Inc., 2000b). Parallelopiped and maximum likelihood classification strategy were applies but maximum likelihood classification was found to be most useful for discriminating the category of interest. After finishing the image processing, the reference points along with the study area were chosen for ground verification. The reference points were surveyed for the ground truthing for collecting data and comparing the preliminary map to real world. The preliminary maps of the waterlogged areas were revised and finalized by using Arc View GIS software.

Participatory Approach

Participatory appraisal involves a series of qualitative multidisciplinary approaches to learning about local-level conditions and local peoples' perspectives. Rapid Participatory Rural Appraisal (RRA/PRA) was carried out using field observations and community level group meetings with different stakeholder groups on the waterlogged areas from May to August 2004 to gather primary information following the approaches of Pido (1995), Pido *et al.* (1996), Townsley (1996), IIRR (1998) and Hossain *et al.* (2004) to know the available resources of the waterlogged areas as well as their importance for community livelihoods. Direct observation prevents rapid appraisal from being misled by myth (Chambers, 1980) and it often provides more valid and less costly information than other research methods (KKU, 1987). Group meetings with local communities are the important way of learning about local conditions and resources (Pelto and Pelto, 1978).

Three PRA tools such as, the seasonal calendar, daily activities and family food analysis were chiefly employed. The seasonal calendar technique was helpful for documenting regular cyclical periods (*i.e.*, seasonal) and significant events that occur during a year and influence the life of the community (Tripp and Woolley, 1989; IIRR, 1998; Townsley, 1996 and Pido *et al.*, 1996). Community members were asked questions in group meeting regarding the duration of the rainy and dry season, environmental conditions, land use patterns, and their activities. Groupmeetings had several advantages, including access to a large body of knowledge and mutual checking. There was a tendency of self-correcting mechanism within the group whereby if one person put across an over-favourable picture of his/her own or group's behaviour, a peer may give a more realistic observation. In cross checking among different groups, a high degree of uniformity was maintained. Some couples were requested to participate in a time allocation study and they were asked to identify their 24 hours activities. This type of task give an idea how they spend their time and find out if they have spare time for additional livelihood activities. Nutritional adequacy of the foods and beverages consumed by the family members were assessed. It provides an opportunity to gather information on food sources, allocation, diversity and their security. The research team interviewed ten families of different unions to record the schedule and number of meals taken each day as well as the foodstuff to describe the family diet for household nutritional information. Observations were recorded through transects across the area defined by using maps and satellite images. Photographs were taken as evidence of certain facts before interpretation. To produce useable outputs, observations were recorded as drawings and notes. Stakeholder selection and analysis were required to find out the appropriate groups for collecting information and to reveal the relationship between their activities. A checklist of topics was used to aid the memory.

Results and Discussion

Land Use Pattern

Most of the people are engaged in agricultural activities of rice and vegetables. About 20 to 25 years before the people used to practice Aush (seedling in dry season and harvesting in rainy season) and Amon (seedling in rainy season and harvesting in dry season) cultivation as their main crop with lower production rate (8–10 kg/decimal). In the recent years, the high yield verities of IRRI have become the main crop with higher production rate (25–30 kg/decimal) instead of Aush and Amon. The upland areas, roadside of village and homestead surroundings have been used for seasonal vegetable cultivation round the year. During winter months, some elevated lands also used for vegetable production but most of the waterlogged paddy lands remain barren during rainy season.

Inventory of Waterlogged Area

The Begumgonj thana consists of one Pourashava and 26 unions. It is the largest Thana of Bangladesh in terms of union number. The rainwater follows a course along inland depressions and these parts get waterlogged during rainy season. The digitized map (based on the analysis of satellite image) recorded 22,186 ha seasonal waterlogged area spreading in 294 polygons (Figure 2.2) that occupies about 52 per cent of the total area. Among 294 polygons (locally called hators), the largest and smallest were recorded as 314.70 and 11.88 ha respectively. Most of the hators are within the range of 12 to 100 ha, where 108 are of 12–50 ha, 115 of 50–100 ha and 62 of 100–200 ha and the remaining 9 hators are more than 200 ha (Figure 2.3). The waterlogged area was classified into three categories on the basis of water depth and water logging duration. Under the union of Kadirpur, Kutubpur, Chhayani, Rajgonj, Alayerpur, Sonaimuri, Ambarnagar and Jirtali 35 polygons were recorded where rainwater retains less than three months having water depth less than three feet. Most of the polygons (217) retain the water for 3–5 months with water depth of 3–4 ft and the remaining 42 polygons retain rainwater for more than 5 months having water depth more than 4 feet.

Classification of Waterlogged Area

The satellite image and digitized map of Begumgonj thana including union and pourashava boundary were used in this study. The team decided to classify the waterlogged areas into three categories on the basis of water depth and water logging duration. The water depth interval were < 3 feet, 3–4 feet and > 4 feet, where the interval of water logging duration were < 3 months, 3–5 months and > 5 months (Table 2.2). The spatial development has clearly indicated the location and extent of the waterlogged area. The diagnostic factors considered for suitability assessment of community-based aquaculture development are water depth and water logging duration. The suitability map of seasonal waterlogged area was prepared to identify the suitable zone for nursing the fry and rearing the cultivable species or brood fish/prawn (Table 2.3). Hossain *et al.* (2001) made similar observation for suitable shrimp farming area selection in the Cox's Bazar coastal zone of Bangladesh. The suitable area for nursing the fish/prawn fry was calculated about 2,000 ha in 35 hators (Class 1). Most of the waterlogged areas of Chhayani, Rajgonj, Alayerpur, Jirtal and Sonaimuri union were found to be located in Class 1 of the suitability classification. Class 2 represents about 16,000 ha in 216 hators, most of which lie through out the entire thana. About 4,000 ha in 43 hators were measured to represent class 3, which mainly locates in Kadirpur, Rasulpur, Hajipur, Nateshwar, and Nadana union.

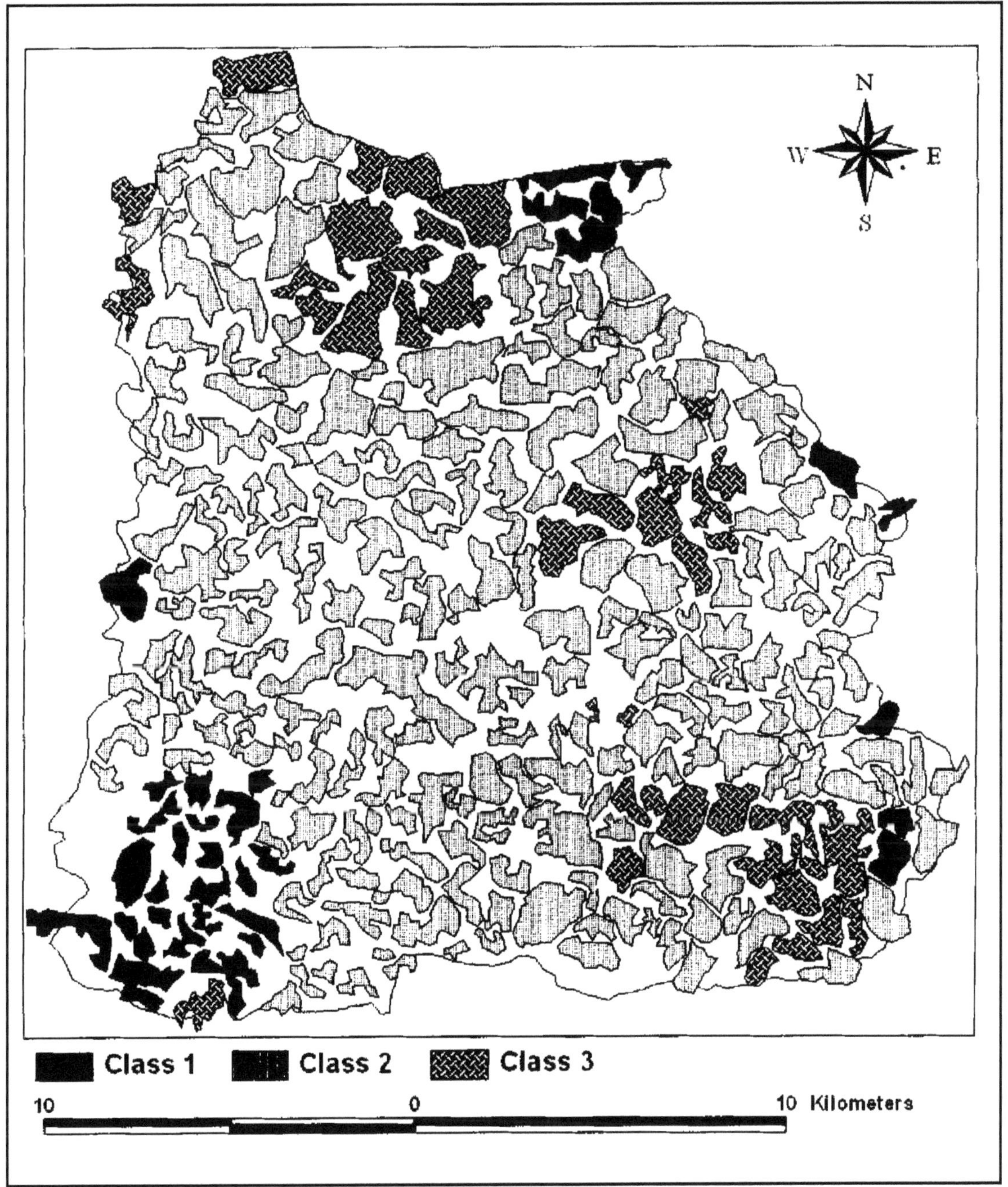

Figure 2.2: Waterlogged Area of Begumganj Thana with Suitability Classification for Aquaculture Development

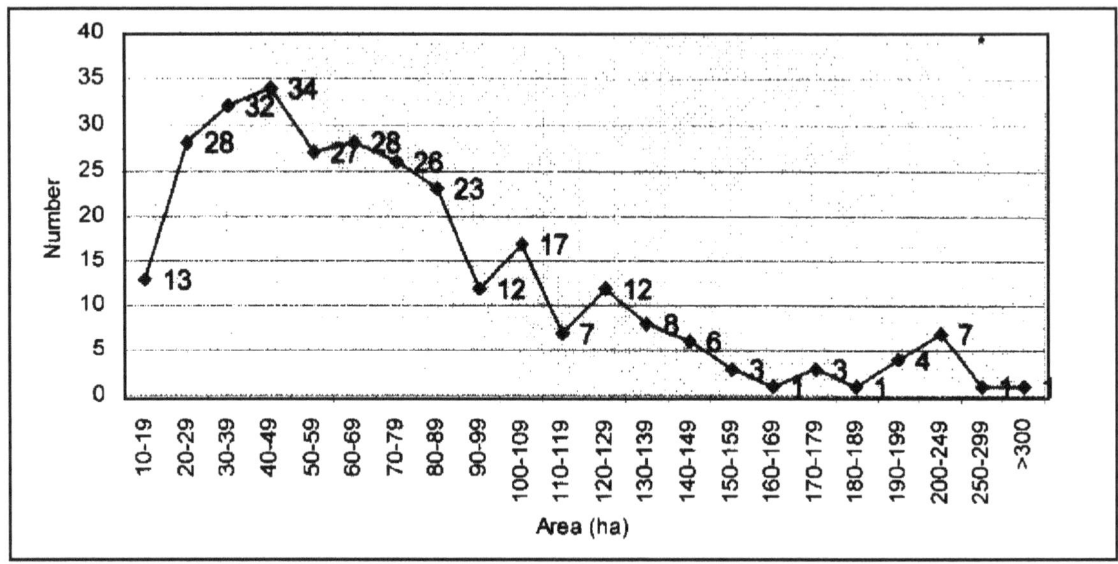

Figure 2.3: Size Variation of Waterlogged Paddy Lands of Begumganj Thana

Table 2.2: Criteria of Different Classes and their Prospective Mode of Utilization

Class	*Criteria*	*Mode of utilization*
Class I	Water depth < 3 ft and water logging duration < 3 months or Water depth < 3 ft and water logging duration 3–5 months or Water depth < 3 ft and water logging duration> 5 months or Water depth 3–4 ft and water logging duration < 3 months or Water depth> 4 ft and water logging duration < 3 months	Suitable for nursing of fish/prawn fry/PL
Class 2	Water depth 3–4 ft and water logging duration 3–5 months or Water depth 3–4 ft and water logging duration> 5 months or Water depth> 4 ft and water logging duration 3–5 months	Suitable for prawn, tilapia and silver barb culture
Class 3	Water depth >4 ft and water logging duration >5 months	Suitable for prawn, carp, tilapia and silver barb as well as brood fish rearing

Table 2.3: Suitability Classes of Waterlogged Area for Prawn/Fish Culture at Begumgonj Thana

Suitability Classes	*Number of Polygon (Locally Called Hator)*	*Area (ha)*
Class 1	35	2.009
Class 2	216	16.273
Class 3	43	3.904
Total	294	22,186

Fisheries Resources for Community Livelihood

Almost all the waterlogged areas are private land, owned by surrounding people and clearly demarcated into small pieces by iels (dikes) that are used as paddy lands. During rainy season the demarcated iles become invisible and the entire waterlogged area seems to be a single water body (Figure 2.4). Aquatic macrophytes and water hyacinth grow in most of the waterlogged areas and these are used as feed for the cattle. Indigenous small-sized fishes, such as clupeids, snakehead, minnows and barbs, eels, gobies, etc (Haque *et al.*, 1999) are commonly available in the waterlogged area (Figure 2.5). The poor strata of the community, those who have no income generating option in rainy days, usually depend on the fisheries and other resources of the waterlogged area. They catch fishes from nearby waterlogged area for their own consumption and sell the surplus to the local hat, which help them to solve their livelihood during lean period. Different local gears are used to catch the indigenous fishes; some notable are jaijal (cast net), koijal (fixed net), bori (hook), thelajal (push net), koach or teda, and anta (bamboo basket trap), which are also used in oxbow lakes (Middendorp *et al.*, 1996). Each gear is used for operation in specific water depth to catch target species such as anta used in 20–30 cm water depth to catch small indigenous fishes (Figure 2.5). It was reported by the people that the maximum fish catch occurs in the early morning and the peak time of buy/sell in local hat (market) is in between 7.00–9.00 am. Second catch occurs from noon to afternoon and brings to local hat at 5.00–7.00 pm. It was reported that during heavy rain, most of the household ponds are flooded and cultured species (mainly carps) are escaped to the nearby waterlogged area. Thus, Rohu (*Labeo rohita*), Catla (*Catla catla*), Mrigal (*Cirrhinus mrigala*), Grass carp (*Ctenopharyngodon idella*), Common carp (*Cyprinus carpio*), Tilapia, etc., are additionally available during harvest in the waterlogged area. During field survey it was observed that some individuals and CBO (Community Based Organizations) cooperatives have used bamboo fence to demarcate the waterlogged area into small bamboo-pan for carp stocking, which never exceeds 2 per cent of the total waterlogged area. Average yield has found 50 kg/ha in six months (June–November) in 2004, which are lower than the oxbow lake yields of 689 kg/ha/year (Haque *et al.*, 1999). No agricultural activities were reported in the waterlogged area

Figure 2.4: Single Water Body of a Typical Waterlogged Area of Begumganj Thana

Figure 2.5: Anta (Bamboo Basket Trap) to Catch Small Indigenous Fish of Begumganj Thana

during rainy season due to high water and dense aquatic macrophytes. When water depth decrease (October–November) the local community drain-out/pump-out the remaining water to catch the fishes (Figure 2.6) and then prepare the land for agricultural activities.

Seasonal Activity

Usually the rearing of poultry and livestock has been practiced year round though duckery depends solely on water logging period *i.e.*, May to November. Paddy (IRRI) and vegetable are the main cultivable items during winter season (November–March). Some rabi crops (green chili, peanuts, radish, bean, arum, tomato, etc.) are also reported to be cultivated in the elevated land, where no irrigation option for paddy cultivation. Fishing is done in waterlogged paddy fields during rainy months. Fish culture has been practiced in waterlogged areas for 6–7 months that covers around 2 per cent of the total waterlogged areas. Some people are engaged in rice mill year-round and some other works in brickfield from November–April. The rainy season paralyzes most of the activities of the poor community and thus the poor strata of the community face miserable condition in their daily life. Moreover, different types of human diseases, mainly diarrhea spread over the community in rainy season (Figure 2.7).

Family Food Analysis

The rural communities usually eat chapatti (locally called *roti*) at breakfast but sometimes they also eat watered-rice (keep cooked rice in water over night, locally called *hani bhat* or *panta bhat*) with chili and onion. Rice is the main food during lunch and dinner. In the daily food menu, the community

Figure 2.6: Pump-out the Water and Catch Even the Last Individual of Fish

Activities	Months											
	Jan	Feb	Mar	Apr	May	Jun	Jul	Aug	Sep	Oct	Nov	Dec
Paddy (IRRI)	■	■	■									■
Vegetable	■	■	■	■							■	■
Raining					■	■	■	■	■	■		
Water logging					■	■	■	■	■	■	■	
Fishing							■	■	■	■	■	■
Poultry rearing	■	■	■	■	■	■	■	■	■	■	■	■
Duck rearing					■	■	■	■	■	■	■	
Livestock	■	■	■	■	■	■	■	■	■	■	■	■
Fish culture					■	■	■	■	■	■	■	
Daily labour	■	■	■	■	■						■	■
Rice mill worker	■	■	■	■	■	■	■	■	■	■	■	■
Brick field worker	■	■	■	■							■	■
Service	■	■	■	■	■	■	■	■	■	■	■	■
Hardship						■	■	■	■	■		
Diseases					■	■	■	■				

Figure 2.7: Seasonal Calendar Showing Different Activities in the Waterlogged Area of Shibpur Village of Rasulpur Union, Begumgonj Thana

prefers fish than beef and tries to arrange fish and beef at every alternative day. In Bangladesh per capita fish consumption increased from 13.1 kg in 1997–98 to 14.4 kg in 2000–01 (BBS, 2004). On the other hand, the world average per caput fish consumption has been increasing over time, from 8.9 kg in 1961 to 15.8 kg in 1996 (Ye, 1999). The smell of potato and vegetable (either curry or fry) indicate the completion of dining arrangement, particularly in lunch and dinner. It was reported that they eat seasonal vegetable with fruits almost every day, as these are easily available in their own homestead garden (Table 2.4).

Table 2.4: Food Menu of Janab Amin Ullah's Family, Shibour Village of Rasulpur Union, Begumgonj Thana

Food items	*Consumption frequency*
Chapatti (locally called ruti)	Most of the morning
Watered-rice (locally called *hani bhat* or *panta bhat*)	Occasionally in the morning
Rice	Two times (lunch and dinner)
Fish	3–4 days/week
Beef	2–3 days/week
Potato	4–5 days/week
Vegetable (red amaranth, Spinach, Snake gourd, Bitter gourd, Ridge gourd, Pumpkin, Ladies finger, Turnip, Cauliflower, Radish, Eggplants, Cucumber, Lima Bean, etc.)	Seasonal
Fruits (Jackfruit, mango, water melon, papaya, palm, pineapple, wood apple, citron, banana, etc.)	Seasonal

Daily Activities

The major activities of the inhabitants are agriculture, fishing, trading, either engaged as daily labourer or as owners of such household activities. One person may be engaged in two or more different occupations *i.e.*, one family may have agricultural land, tractor for plough and shop in the locality. As many occupations are seasonal, so a person can take up different activities in different seasons. The daily activities of men are connected with intensive labour for income generation of their family, while the women's activities are solely dependent on family affairs.

Men usually pass their time in agricultural work with tea break and relax. Occasionally they are engaged in husbandry of cattle. During the rainy months (May–October) they catch fish from the surrounding water body for family consumption as well as for selling. Most of the morning they visit village hat either to buy or sell fresh fish and vegetable. They spend their evening time for trading, gossiping, purchase of household needs, meet with friend / relative at village hat or nearby shop. They also pass their leisure time with family members, neighbours and relatives. Women in rural communities do not participate directly in income generating activities. Usually they look after their families. These include the daily chore of childcare, collecting water, cooking, washing, chicken and duck rearing, homestead gardening, sewing cloths, making handicrafts, child education, and gossiping with neighbours. However, they have spare time for other livelihood activities. The dark, soundless village night gives them opportunity for sound sleeping and relaxation.

The daily activities of Janab Amin Ullah have shown as a typical example. He catches fish in the last half of the night for own consumption and also for selling. Amin Ullah has trading with small shop at the village and sometimes work n the field. Most of the evening he visits village hat for selling-buying of household items, gossiping with friends/relatives, watching TV, exchange national and international news, etc. Mrs. Amin Ullah is responsible for housework, childcare, cooking, etc. (Figure 2.8).

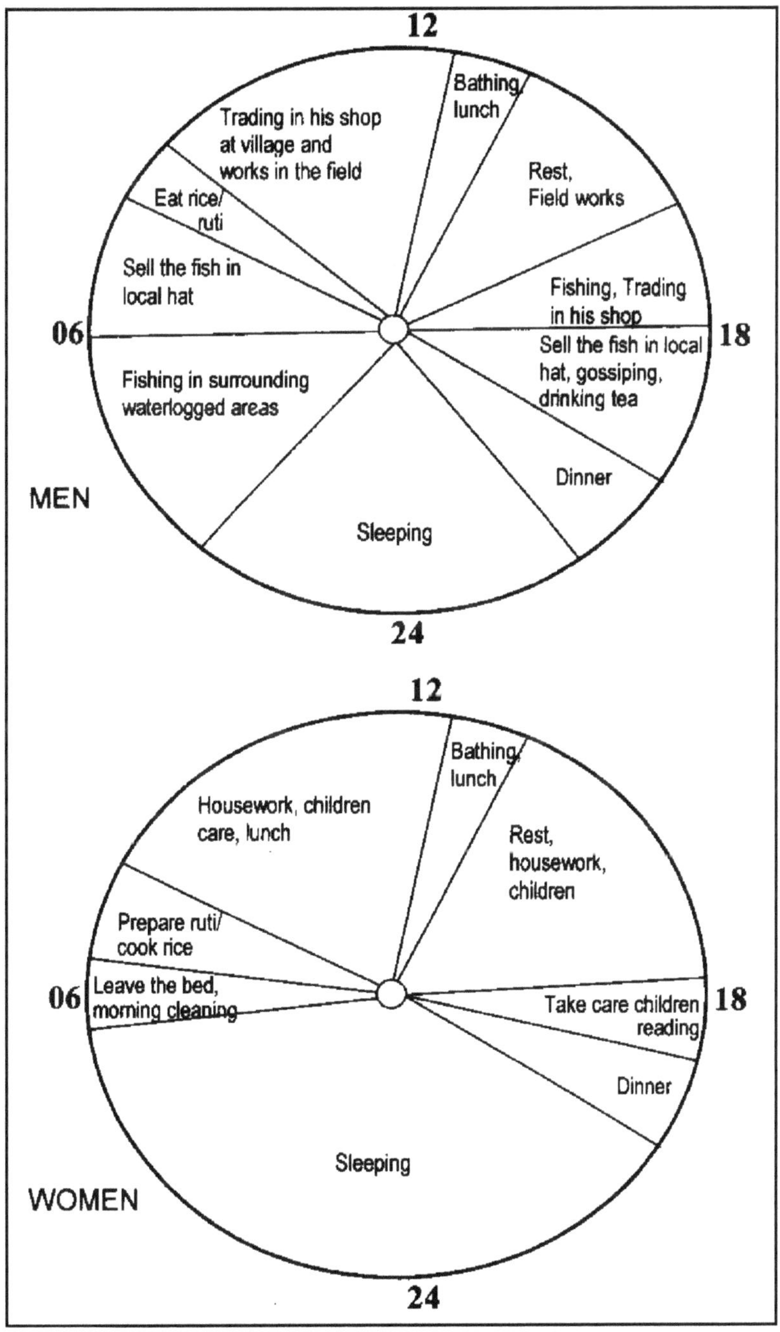

Figure 2.8: Daily Activity Charts of Janab Amin Ullah and Mrs. Amin Ullah in Shibpur Village, Begumgonj Thana

Management Framework

The waterlogged areas of Begumgonj remain unutilised with no economic returns. About 22,000 ha waterlogged paddy lands are identified as suitable for nursing and culture of fish/prawn as well as indigenous brood fish rearing through the establishment of local level fish sanctuary for biodiversity conservation. Different types of integrated programs with special stress to prawn/fish culture may be undertaken for overall economic upliftment of the local community by properly utilizing the vast untapped waterlogged areas in this region. Since wild growth of different fish species including prawn is prevalent in chronic waterlogged area, it is clear that community based prawn/fish culture in these waterlogged areas may be an excellent example to stimulate income generation, employment opportunity and supply nutritive food for sustainable rural livelihood. Rural communities need to be empowered economically, personally, educationally and politically to ensure their participation in planning for their future upliftment. The social, economic and political empowerment of the people through participation will help improve sustainable development in the waterlogged areas. An excellent, comprehensive account of ways to improve prawn farming may be provided in the framework that deals with some important processes involved in the waterlogged area of Begumgonj (Figure 2.9).

Conclusions and Recommendations

In the thrust for rapid aquaculture development, management of waterlogged areas in the Begumgonj thana has not been paid adequate attention in the planning process due to lack of requisite data as well as resource crunch. Community based aquaculture is a suitable allied activity for small-scale producers to augment their income and to promote ecological soundness farms in waterlogged area. The paddy and prawn culture in rotational basis may be economically viable and ecologically sound in this region. Species selection is important for aquaculture development in seasonal waterlogged paddy land. The polyculture of silver carp, grass carp and prawn in waterlogged area can ensure sustainable utilization of the available natural foods that ensure good production. Excess aquatic plants may hamper the growth of prawn that consumed by grass carp and keep the aquatic environment favourable to prawn. Thus, the farmers can earn substantial return from both grass carp and prawn. Taking advantage of this situation it is easy to retrieve this waterlogged area for community based aquaculture.

Local community need to inform about the availability of waterlogged paddy lands and make them aware the potential utilization as alternative income generating options. The prospects of fish/prawn farming need proper dissemination among the local community to ensure men and women participation, as the waterlogged area exists surrounding their house. They can take care of the fish/prawn culture, even from their yard or house, without hampering on going daily activities. Thus, traditional men-cantered activities will shift to integrated men-women participated activities that will ensure social empowerment and equity, as well as sharing of responsibility through good understanding. Therefore, it is the task of the people to sustain the available water resources to future generation. The World Commission on Environment and Development (WCED, 1987) defined the objective of sustainable development as "to meet the needs of the present generation without jeopardising the possibilities of future generation to meet their own needs". This is the only way to increase rural economy through the development of primary production sector in the waterlogged paddy lands. As a result, waterlogged-area-based village level employment opportunity will increase by discouraging the on going rural-urban migration tendency.

Management issues
- Dormant waterlogged area
- Suitable for aquaculture
- Locally available feed ingredients
- Seasonal livelihood options
- Rural-urban migration tendency
- Seasonal flooding
- Biodiversity conservation
- SIS brood bank establishment

Natural resources
- Waterlogged paddy land
- Stagnant rainwater
- Indigenous fisheries
- Aquatic macrophytes
- Water hyacinth
- Water lily

Opportunities
- Soil and water quality
- Fry availability
- Communication facility
- Marketing channel
- Indigenous fishing trap
- Available feed ingredients
- Fish culture in bamboo pen
- Meteorological parameter
- Duckery

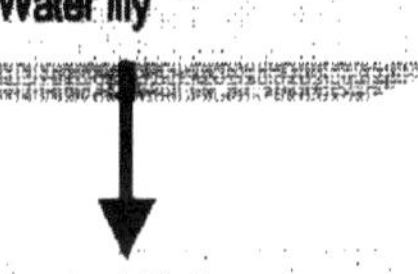

Management Approach
- Waterlogged area zoning
  - Suitable area allocation
- Paddy-prawn rotation
- Community mobilization
  - Community meeting
  - Beneficiary selection
- Community training on
  - Hatchery, nursery and culture pond
- Homestead hatchery development
- On-farm feed production
  - Locally produced soybean
  - Dried Chewa fish from Hatiya Island
  - Locally produced rice bran
- Carrying capacity
  - Environmental CC
  - Culture area intensity
  - Economic CC
- Post-stocking management
  - Partial harvesting
  - Full harvesting
  - SIS conservation
- Traceability
- Licensing
- Code of conduct

Integrated management requirements among
- Stakeholders viz. farmer, researcher, investor, trader, donor
- Sectors viz. fisheries, agriculture, environment, land revenue
- Levels of government viz. local, thana, district, national
- Disciplines viz. natural sciences, social sciences, engineering
- Available resources viz. personnel, funds, materials, equipments
- Responsibilities viz. community, administrator, government
- Gender viz. men, women
- Research tools viz. remote sensing, GIS, PRA, RRA, CBFM
- Research approaches viz. bottom-up, top-down, vertical, horizontal

Figure 2.9: Integrated Framework for Aquaculture Development in Waterlogged Area of Begumgonj, Noakhali, Bangladesh

Acknowledgements

The authors are grateful to Danish International Development Assistance (DANIDA) for providing funds to make a full inventory of the waterlogged areas and map their spatial location with suitability classification. They are also grateful to Dr. Harvey Demaine, Mr. Rashed Alam and Masud Rana of GNAEP (Greater Noakhali Aquaculture Extension Project) for their encouragement.

References

Aguilar-Manjarrez, J. and Ross, L.G., 1995. Geographical Information System (GIS), environmental models for aquaculture development in Sinaloa state, Mexico. *Aquaculture International*, 3(2): 103–115.

Alam, R., 2001. Feasibility study on freshwater prawn (*Macrobrachium rosenbergii*). Prospect of fresh water prawn at GNAEP and PBAEP areas. The report of feasibility study prepared for GNAEP and PBAEP, Ministry of Fisheries and Livestock and Danida.

BBS (Bangladesh Bureau of Statistics), 2004. *Statistical Pocket Book*. Bangladesh Bureau of Statistics, Ministry of Planning, Government of the People's Republic of Bangladesh.

Burrough, P.A., 1986. *Principles of Geographic Information System for Land Resources Assessment*. Oxford University Press, New York.

Burrough, P.A. and McDonnell, R.A., 1998. *Principles of Geographical Information Systems*. Oxford University Press, New York, 333 p.

Chambers, R., 1980. *Rapid Rural Appraisal: Rationale and Repertoire*. Discussion paper. Institute of Development Studies, Sussex, UK.

Chowdhury, A., Uddin, K., Halder, S., Colavito, L. and Collis, W., 2003. Environmental impact assessment of GNAEC freshwater prawn (*Macrobrachium rosengergii*) farming promotion. GOB/DANIDA Fisheries Support Unit, Winrock International, Dhaka, Bangladesh.

Das, N.G. and Hossain, M.S., 2005. Livelihood and resource assessment for aquaculture development in waterlogged paddy lands: Remote sensing, GIS and participatory approach. GNAEP of GOB–DANIDA and University of Chittagong, Bangladesh, 129 p.

Das, N.G. and Hossain, M.S., 2004. Resource assessment and livelihoods analysis of waterlogged paddy lands at greater Noakhali: Remote Sensing, GIS and participatory approach. Final report Submitted to GNAEP of GOB–DANIDA, 86 p.

Haque, A.K.M., Middendorp, H.AJ. and Hasan, M.R., 1999. Impact of carp stocking on the abundance and biodiversity of non-stocked indigenous fish species in culture based fisheries in oxbow lakes. In: *Sustainable Inland Fisheries Management in Bangladesh*, (Eds.) Middendorp, H.A.J., Thompson, P.M. and R.S. Pomeroy. ICLARM Conference Proceedings, Manila, Philippines, 58: 141–148.

Hossain, M.S., Lin, C.K., Demaine, D., Tokunaga, M. and Hussain, M.Z., 2001. Integrated GIS and remote sensing approaches for suitable shrimp farming area selection in the coastal zone of Bangladesh. *Asia–Pacific Remote Sensing and GIS Journal*, 14: 33–39.

Hossain, M.S., Lin, C.K, Demaine, H., Tokunaga, M. and Hussain, M.Z., 2003a. Land use zoning for solar salt production in Cox's Bazar Coast of Bangladesh: A remote sensing and GIS analysis. *Asian Journal of Geoinformatics*, 3(4): 69–77.

Hossain, M.S., Lin, C.K., Tokunaga, M. and Hussain, M.Z., 2003b. Remote Sensing and GIS application for suitable mangrove afforestation area selection in the coastal zone of Bangladesh. *Geocarto International*, 18(1): 61–65.

Hossain, M.S., Khan, Y.S.A., Chowdhury, S.R., Saifullah, S.M., Kashem, M.B. and Jabbar, S.M.A., 2004. Environment and socio-economic aspects: A community based approach from Chittagong coast, Bangladesh. *Jahangirnagar University Journal of Science*, 27: 155–176.

Islam, M.J., Alam, M.S. and Elahi, K.M., 1997. Remote sensing for change detection in the Sunderbans, Bangladesh. *Geocarto International*, 12(3): 91–100.

IIRR, 1998. *Participatory Methods in Community-based Coastal Resource Management in 3 Vols.* International Institute of Rural Reconstruction, Silang, Cavite, Philippines.

Kapetsky, J.M., McGregor, L. and Nanne, L.H., 1987. A geographical information system and satellite remote sensing to plan for aquaculture development: A FAO-UNESCO/GRID cooperative study in Costa Rica. *FAO Fisheries Technical Paper*, 287: 51.

Karim, M., 1989. Present status, scopes and constraints of *Macrobrachium rosenbergii* culture in the greater Noakhali district. Technical Findings of a Danida–financed mission on identification of the socio-economic feasibility of freshwater shrimp culture in old Noakhali district, October–November.

KKU (Khon Kaen University), 1987. Rural systems research and farming systems research projects, Thailand. In: *Proceedings of the 1985 International Conference on Rapid Rural Appraisal.*

Lillisand, T.M. and Kiefer, R.W., 2000. *Remote Sensing and Image Interpretation*, 4th edn. John Wiley and Sons, Inc., New York, 724 pp.

Middendorp, H.A.J., Hasan, M.R. and Apu, N.A., 1996. Community fisheries management of freshwater lakes in Bangladesh. *NAGA, ICLARM Q.*, 19: 4– 8.

Pido, M.D., 1995. The application of rapid rural appraisal techniques in coastal resources planning: Experience in Malampaya Sound, Philippines. *Ocean and Coastal Management*, 26(1): 57–72.

Pido, M. D., Pomeroy, R.S., Carlos, M.B. and Garces, L.R., 1996. *A Handbook for Rapid Appraisal of Fisheries Management Systems* (Version I). ICLARM, Manila, Philippines.

Pelto, P. and Pelto, G., 1978. *Anthropological Research: The Structure of Inquiry*. Cambridge University Press, Cambridge.

Research Systems Inc., 2000a. *ENVI User's Guide.* ENVI version 3.4, Research Systems Inc., USA.

Research Systems Inc., 2000b. *Exploring ENVI, Training Course Manual.* Better Solutions Consulting Limited, Liability Company, USA.

Setiato, 1993. *Planning of Coastal Aquaculture in Lampung Province, Indonesia.* AIT Thesis No. AE–93–31. Asian Institute of Technology, Bangkok, Thailand.

Shahid, M.A., Pramanik, M.A.H, Jabbar, M.A. and Ali, S., 1992. Remote sensing application to study the coastal shrimp farming area in Bangladesh. *Geocarto International*, 2: 5–13.

Townsley, P., 1996. Rapid rural appraisal, participatory rural appraisal and aquaculture. *FAO Fisheries Technical Paper*, Rome, Italy, 358: 109.

Venkataratnam, L., Thammappa, S.S., Sankar, T.R. and Anis, S., 1997. Mapping and monitoring prawn farming areas through remote sensing techniques. *Geocarto International*, 12(2): 23–29.

Vibulsresth, S., Ratanasermpong, S., Dowreang, D. and Silapathong, C., 1993. Mangrove monitoring using satellite data. *TRSC Newsletter*, 10(1): 1–2.

WCED (World Commission on Environment and Development), 1987. *Our Common Future.* Oxford University Press, Oxford.

Ye, Y., 1999. Historical consumption and future demand for fish and fishery products: Exploratory calculations for the years 20 15/2030. *FAO Fish. Circ.*, FAO, Rome, Italy, 946: 32.

Chapter 3

Reconnaissance of Wildlife Status and Conservation Attempts in Kashmir Valley, India

M.A. Khan
Division of Environmental Sciences,
S.K. University of Agricultural Sciences and Technology of Kashmir,
Shalimar, G.P.O. Box No. 726, Srinagar – 190 001, Kashmir, India
E-mail: ma_khan16@yahoo.co.in

ABSTRACT

This article deals the status of wildlife in hilly region of Kashmir valley. The diversity of wildlife intimately inter-linked with other bioresources including forests, ranges and agriculture has determining impact on the environmental sustainability. The environmental causes of depletion of wildlife are highlighted. The long-term ecological security warrants wildlife conservation (esp. *Hangul,* a rare Kashmir stag) on robust scientific management. Public awareness/education and environmental legislation shall go a long way in arresting the wildlife decline in the Kashmir region.

Keywords: *Wildlife status, Threats, Depletion causes, Conservation, Kashmir.*

Introduction

The State of Jammu and Kashmir enjoys enviable position in view of its rich natural resources and wildlife. Wildlife encompasses all uncultivated flora and fauna. Every species has the right to live and every threatened species must be protected from its extinction. Biological resources including forests, ranges, agriculture, fishery and wildlife have important role in the environment. Water, wilderness and wildlife are irrevocably interlinked. Forests provide long-term ecological security in protecting wildlife and conserve our life-support system.

Wildlife management is an important component of applied ecology that ranks very high in public interest and is attracting global attention. Conservation of wildlife includes the preservation of all species, the enhancement of wildlife habitat, the control of wildlife problems and the consumptive use of wildlife from the rapid extinction. However, mounting agriculture, industrial and demographic pressures have caused drastic shrinkage of wilderness areas which are the richest repositories of wildlife and biodiversity. Their continued existence is crucial for the long-term survival of the biodiversity and the supporting ecosystem.

The present communication provides a reconnaissance of the status of wildlife and conservation measures in the Kashmir Valley.

Status of Wildlife

The wildlife of J&K state is mainly represented by myriad of animals such as mammals, avifauna, reptiles, fish, amphibian fauna including some unique species. About 16 per cent of Indian mammals, birds, reptiles, amphibians and butterflies are reportedly present in the State (Reza and Choudary, 2003). Table 3.1 gives the number of species reported (WWF, 2001) from the region.

Table 3.1: Biodiversity of Jammu and Kashmir

Species	Number
Mammal	75
Bird	358
Reptile	68
Amphibian	14
Fish	44
Butterflies	158
Insects	225
Flora	3054
Medicinal plants	100

Threats to Wildlife in Kashmir

The survival of wildlife is inextricably linked to the ecosystem-health supporting the populations. However, human-greed and apathy of the concerned agencies is causing a great deal of concern to wildlife environmentalists. Trade and commerce (Khan, 2004) constitute main potential threats to the decline of wildlife throughout the world. The decimation of many natural animal and plant species occurs as a result of habitat-loss through clearing of vast areas for urban and industrial development, agriculture, grazing and also to meet the fuel-wood requirement. During the past 50–60, years the socio economic pressures have been building up in Kashmir at a very rapid pace resulting, among other things, in greater exploitation of natural resources and large-scale deforestation aimed at providing more and more land for agriculture and building purposes. With the cutting of forests and poor management of wetlands, the wildlife habitat goes on shrinking unabatedly. The populations were gradually pushed on to comparatively unproductive and unsafe patches of their habitat where they not only face the scarcity of food and water, but get exposed to poachers causing considerable decline in their number.

The increase in human population and associated activities exert tremendous pressure on a variety of wildlife and their products. Due to denudation in the upper catchment areas a lot of silt has accelerated 'filling-in' of water-bodies which has aggravated the ecological problems of the waterfowl populations in their habitat (Khan, 2004). Large number of people living around the wetlands have been encroaching these areas which has been drained for agriculture, urban expansion and other purposes. Over the past few years, a number of trees were felled in Srinagar City Forest for the creation/development of a Golf Course on an area already notified as a Wildlife habitat. Such an action of the then Jammu and Kashmir Government evoked a spate of media outcry and viewed as latest man-induced interference leading to wildlife habitat deterioration.

Table 3.2: Status of Mammals and Important Bird Species of Kashmir

Sl.No.	Common Name	Scientific Name	Status*
1.	Mammal**		
	Hangul (Kashmiri stag)	*Cervus elephus hanglu*	Sch. I
	Himalayan Brown Bear	*Ursus arctos*	Sch. I
	Himalayan Black Bear	*Ursus torquatus*	Sch. II
	Tibetan Lynx	*Lynx lynx isabellinus*	Sch.I
	Leopard	*Panthera pardus*	Sch. I
	Snow-Leopard	*Pantlzera uncia*	Sch. I
	Wild Boar	*Sus cristatus*	Sch. III
	Himalayan Ibex	*Capra ibex*	Sch. I
	Markhor	*Capra falconeri*	Sch. I
	Serow	*Capricornis sumatraensis humei*	Sch. I
	Goral	*Nemorhaedus goral*	Sch. III
	Musk Deer	*Muschus muschiefrous*	Sch. I
	Tahr	*Hemitragus jemlahicus*	Sch.I
	Langur	*Presby tis entellus*	Sch. II
	Himalayan Yellow Throated-Pine	*Martes flavigula*	Sch. II
	Red Fox	*Vulpes vulpes*	Sch. II
	Jackal	*Canis aureus*	Sch. II
2.	Avifauna***		
[Kashmir region comprises of 126 species of birds out of which the main resident game birds are threatened to extinction are as below]			
	Western Tragopan	*Tragopan melanocephalus*	Sch. I
	Cheer Pheasant	*Catreus wallichi*	Sch. I
	Monal Pheasant	*Lophophorus impejanus*	Sch. I
	Common Koklas	*Purcrasia macroloph*	Sch. II
	Himalayan Snow-Cock	*Tetroagallus himalayensis*	Sch. I

*: In Schedules of W(P)A.

**: *Source*: Anon (2002).

***: *Source*: Singh and Singh (1995).

The presence of the Government Sheep Breeding Farm has been recorded as an interference in the ecology of Dachigam National Park (ANON, 1985). Large quantities of grass are cut from within the national park for winter fodder (Gruisen, 1983). Shahtoosh is a classic case which constitutes flourishing trade in shawls and other garments; the product has its source in animal, *Chiru* which is facing the threat of extinction causing a worldwide concern for its survival. It is a believed that *Chiru*, the Tibetan Antelope (*Pantholps hodgsoni*) is killed for procuring the wool. Thus, the depletion of wildlife is attributed to following main factors:

1. Environmental changes arising from alteration, degradation and destruction of the natural habitats.
2. Animal-killing for flesh, feathers, skin and antler, often illicitly
3. Deforestation, which otherwise serve wildlife refugees
4. Agricultural expansion and other industrial or urbanization activities
5. Uncontrolled and unlimited grazing of domestic livestock indiscriminately
6. Disturbance in balance of nature (*e.g.* predator-prey relationship)
7. Poaching
8. Commercial exploitation without ensuring species regeneration
9. Indiscriminate use of chemical pesticides, herbicides and insecticide
10. Other factors include forest fires, road constructions etc.

As a result of major factors responsible for depletion of wildlife fauna, the important species like *Hangul*, Snow Leopard, Markhor and Musk Deer have been included in the list of animals facing grave dangers of extinction. The Jackal has been regarded as a vermin and is falling victim to commercial exploitation for fur trade in Kashmir. The Snow Leopard is killed for the skin. Likewise, Ibex, Himalayan Musk Deer, the Blue Bull and migratory birds are subject to hunting and smuggling. The Musk Deer is exploited primarily for its musk pod from which musk is extracted and is used in perfume industry. The musk is believed to fetch around 40,000 to 59,000 U.S. dollars per kg in the international market.

Poaching and hunting of rare species for obtaining horns, teeth, antlers etc., which sell at exorbitant cost reduce their number to the great extent and lead them to extinction.

Like other environmental problems more than a decade-long political upheaval in Kashmir also caused considerable pressure on wildlife species. *Hangul*, a rare Kashmiri stag, is believed to be on the verge of extinction with its population registering alarming fluctuation in Dachigam National Park. Due to poaching, the *Hangul* population declined drastically from a crude estimate of 2000 in 1947 (Gee, 1966) to below 200 in 1965 (Schaller, 1969). More recent information provided by Santoshi (2004) reports (Table 3.3) that the *Hangul* is on the verge of extinction with its population showing drastic reduction during 2004. The data indicate that as against 815 in 1988 the census conducted in March reveals the number about 209–243 only.

Table 3.3: Population Changes of *Hangul* in Dachigam National Park, Kashmir

Year	*Population*
1980	415
1982	470
1983	525
1984	605
1986	725
1988	815
1992	120
1994	175
1995	290
1996	338
1997	373
1998	360
1999	325
2000	470
2001	483
2004	About 209–243

The factors responsible for its declining population include the scarcity of adequate grazing ground, habitat-interference, increased biotic competition, lower breeding rate and poaching. Estimated population (140) of *Hangul* in 1970 hiked to 815 in 1988. Later, a long spell of turmoil in Kashmir took a heavy toll of the animal and drastically declined to a mere number of 120 during 1992. Such a scenario has developed despite the launch of "Project *Hangul*" during 1970 in Kashmir valley.

Attempts at Conservation

The Government of Jammu and Kashmir, from time to time, made some attempts, often half-hearted, to conserve the wildlife fauna and their habitat. However there is an urgent need to have an in-depth look at the formal and informal conservation efforts in the entire State. This should include a study of relationship between loss of biodiversity including wildlife due to major development projects like road building, hydro-electric dams, large-scale housing, agriculture, fruit growing and other development schemes.

Afforestation programmes on war-footing in degraded forest areas and sustainable eco-development in the catchment of water resources should be actively pursued.

Public awareness and their support is crucial for management of national parks, sanctuaries and water-bodies. The State government should also address the conservation and management of varied wildlife fauna. Wetland ecosystems which provide congenial habitat for resident and migratory avifauna deserve proper attention.

Collection of information related to present and past status and distribution of wildlife species gains paramount importance. The scientific research and development in the field of biological diversity and environmental problems must be promoted. Suggesting remedial measures for restoration of habitat and conservation of endangered wildlife species is need of the hour.

Over-exploitation of forest and wildlife have already played a havoc with the ecology of the region. Whole-hearted dedication to the task of rehabilitating the denuded forests to protect wildlife and, wherever possible, to arrest the grazing of mountain pastures by cattle shall go a long way in conserving the wildlife.

For conservation of endangered species like *Hangul, Markhor, Musk Deer,* Dachigam has got the status of national park in 1981. The area was a hunting reserve or *'rakh'* of the Maharaja of J&K from about 1910 until 1947 when its management has reverted to Forest Department. The area of Dachigam National Park is about 14,100 ha. After independence, the responsibility for the area reverted to the J&K State Government and its administration passed successively to Fisheries Department (1947–54), Forest Department (1954–60), Tawaza Entertainment Department (1972–1977) and finally back to Forest Department (Directorate of Game Preservation) in 1978.

The number of wildlife sanctuaries in Kashmir region has increased for conservation of endangered species. The wildlife sanctuaries in the Kashmir are Lachipora Wildlife Sanctuary (WLS), Limber WLS, Gulmarg, Overa WLS, Rajparian WLS (Daksum), Hirpora WLS and Thajwas (Baltal) WLS. Moreover, the conservation reserves like Harwan, Nishat, Cheshmashahi, Pahalgam, Hokersar, Dal Lake, Anchar in Kashmir also harbour a large number of bird species and other animals.

The State of Jammu and Kashmir, has several laws/rules governing the protection wildlife and forests (see Anon., 1981) The survival of Wildlife is inextricably linked to their habitability in ecosystems. The Wildlife (Protection) Act, 1972 and the Forest Act, 1927 are aimed at preservation and protection of environment through the protection of biodiversity for the conservation of wildlife habitat. Some other major laws/rules include rules for the Demarcation of Forests, 1914; the Cattle Traspass Act, 1977 Samvat (1920 AD); The Gulmarg Forest Rules, 1999 Samvat (1932 AD); The Pahalgam Forest Rules, 1999 Samvat (1932 AD); The Jammu and Kashmir Game Preservation Act, 1998 Sanvat (1942 AD) and Notifications/Rules regarding this Act; Rules for the Management of undemarcated Forests, 1947; The Jammu and Kashmir Prevention of Specified Trees Act, 1969 and its amendments in 1972; The Jammu and Kashmir Land Improvement Scheme Act, 1972 and its Rules (1973); The Jammu and Kashmir State Forest Corporation Act, 1978. Thus, the management of wildlife is governed and facilitated by a set of policies/legislations and if implemented effectively should support wildlife conservation and their sustainable utilization.

References

Anonymous, 1981. *Forest Law Manual of Jammu and Kashmir State,* Vols. I and II. Jammu and Kashmir Forest Department, Srinagar.

Anonymous, 1985. *Ecological-cum-Management Plan for Dachigam National Park, Jammu and Kashmir State 1985–90*. Department of Wildlife Protection, Srinagar, 56 pp.

Anonymous, 2002. *An Introduction to the Wildlife of Jammu and Kashmir*. Department of Wildlife Protection, Srinagar.

Gee, E.P., 1966. Report on the status of the Kashmir Stag, October, 1965. *J. Bomb. Nat. Hist. Soc.,* p. 62–115.

Gruisen, J.V., 1983. The *Hangul*: Dachigam's endangered deer. *Sanctuary (Asia),* pp. 114–131.

Khan, M.A., 2004. *Assessment of Anthropogenic Pressures on the Ecology of Hokersar Wetland (Kashmir) with Special Reference to Conservation.* Final Tech. Rep., Ministry of Environment and Forest, New Delhi, pp. 130.

Reza, A. and Chaudhury, H., 2003. Biodiversity of Jammu and Kashmir. *J. Environ. Artc. Arch. Biodiver.*

Santoshi, N., 2004. *Hangul* on verge of extinction. *Hindustan Times,* 13 May, 2005.

Schaller, R.G., 1969. Observations on *Hangul* or Kashmiri stag, *Cervus elaphus hanglu. J. Bomb. Nat. Hist. Soc.,* 66: 1–7

Singh, R. and Singh, S., 1995. *A Pocket Book of Indian Pheasants*. Wildlife Institute of India, Dehradun.

Chapter 4

Integrated Groundwater Prospects Analysis Using Remote Sensing and GIS Techniques

S.S. Asadi[1], Padmaja Vuppala[2], P. Srilatha[3] and M. Anji Reddy[4]

[1]Lecturer, [2]Junior Research Fellow, [3]Project Fellow, [4]Director, Centre for Environment, Institute of Science and Technology, Jawaharlal Nehru Technological University, Hyderabad – 500 072, A.P., India

ABSTRACT

Water plays a vital role in the development of any activity in the area and the availability of surface and groundwater governs the process of planning and development. The surface water resources are inadequate to fulfill the water demand (Amaresh *et al.*, 2002). Productivity through groundwater is quite high as compared to surface water, but groundwater resources have not yet been properly developed through exploration. Keeping this in view, the present study attempts to prepare groundwater prospects map and delineate areas with high groundwater potential using an integrated approach of remote sensing and GIS.

The study area chosen is bounded by 78°5′ and 78°30′ E longitude and 17°15′ and 17°30′ N latitude and covers SOI toposheet 56K/7 which includes parts of Medak and Hyderabad districts of A.P. The present study utilizes IRS-ID PAN and LISS-III merged geocoded data on 1 : 50000 scale to delineate groundwater prospect zones. The information on base, drainage, lithology, structure, land use/land cover, slope, geomorphology and hydrology were generated and integrated to obtain a final groundwater prospects map for the study area. Land use/Land cover, geomorphology and structures maps have been prepared using IRS-ID PAN and LISS-III merged satellite data by visual interpretation. Topographic information has been collected from SOI toposheet at 1 : 50000 scale.

The groundwater potentiality of the area has been assessed through integration of relevant layers, which include hydrogeomorphology, structures, lineament, slope, etc. in Arc/Info environment. Four categories of groundwater potential zones *viz.*, very good-good, good moderate, poor-nil and nil are delineated. In the final map, artificial recharge structures to increase groundwater levels in areas with less groundwater potential are also recommended. The study highlights the utility of Remote Sensing data and use of GIS to prepare database on different layers, analyze their relationship and prepare an integrated map even in a complex terrain. This type of integrated study clearly demonstrates the capabilities of remote sensing and GIS techniques in demarcation of the different groundwater potential zones and is very useful for the decision makers to protect the natural resources.

Keywords: *Groundwater prospects, Remote Sensing, Geographical Information System (GIS), Integrated study.*

Introduction

Water is a natural, vital and a key resource in all-economic activities ranging from agriculture to industry. Only a tiny fraction (3 per cent) of the planet's abundant water is available to us as fresh water while the remaining 97 per cent is found in the oceans and is too salty for drinking, irrigation, or industry (Anjaneyulu, 2004). About 2.997 per cent of the available fresh water is locked up in ice caps or glaciers or is buried so deep that it costs too much to extract.

Groundwater constitutes an important source of water supply for various purposes, such as domestic, industries and agriculture needs (Khairul Anam Musa *et al.*, 2000). The resource can be optimally used and sustained only when the quantity and quality of groundwater is assessed. It has been observed that the lack of standardization of methodology in estimating the groundwater, and improper tools for handling the same, leads to miscalculation of estimation of groundwater. It is essential to maintain a proper balance between the groundwater quantity and its exploitation. Otherwise it leads to the large scale decline in groundwater levels, which ultimately cause a serious problem for sustainable agricultural production (Anand Kumar and Sanjay Tomar, 2002).

Exploitation and utilization of available groundwater requires proper understanding of its origin, occurrence and movement, which are directly or indirectly controlled by terrain characteristics. Remote Sensing, with its advantages of spatial, spectral and temporal availability of data covering large and inaccessible areas within short time, has become a very effective tool in assessing, monitoring and conserving groundwater resources (Anji Reddy, 2001). Satellite data provides quick and useful baseline information on the parameters controlling the occurrence and movement of groundwater such as geology/lithology, structure/lineaments geomorphology, soils, land use/land cover and hydrological parameters etc. These parameters have to be integrated to assess groundwater. However, the conventional techniques have the limitation to study these parameters together because of the non-availability of data, integration tools and modeling techniques. The advent of GIS has added new vistas in the field of groundwater resource mapping and management (Rao Toleti *et al.*, 2000). The various thematic layers generated using remote sensing data like Geology/lithology, structure/lineament, geo-morphology, land use/land cover could be integrated with slope, drainage density and other collateral data in a GIS framework. This could be further analysed in GIS domain using logical conditions to derive groundwater zones as well as artificial recharge sites. This concept of integrated remote sensing and GIS has proved to be an efficient tool in groundwater studies (Krishnamurthy *et al.*, 1996; Saraf and Chaudhary, 1998; Khan and Mohrana, 2002). Keeping this in view, the present study aims at groundwater prospects mapping to increase the groundwater levels in the arid and semiarid area integrating Remote Sensing and GIS.

Description of Study Area

The area chosen for the study is bounded by 78°15′ and 78°30′ E longitude and 17°15′ and 17°30′ N latitude and covers SOI toposheet 56 K/7 which includes parts of Medak and Hyderabad districts of A.P. The mean minimum temperature in the study area is 13°C to 17°C in December and January, but it rises to 26°C to 29°C in March with an average annual rainfall of 884.8 mm (Census of India, 1991a and 1991b). In Hyderabad district more than 75 per cent of the rainfall is received during the southwest monsoon season, *i.e.*, from June to September. The soils of Medak district are mainly red earths comprising loamy sands, sandy loams and sandy clay loams, while in Hyderabad the soils are mainly brownish sandy and Ready loamy.

Objectives

1. To prepare a hydrogeomorphology map of the study area on 1 : 50,000 scale using IRS-ID PAN and LISS-III merged satellite data.
2. To prepare other thematic layers *viz.*, land use/land cover, drainage, watershed, slope, geology, structures and infiltration rate using satellite data and survey of India toposheets by visual interpretation technique.
2. To collect collateral information on population, rainfall pattern etc. for creation of attribute database.
3. To prepare an integrated groundwater prospects map and delineate groundwater potential zones by assessing geomorphological, lithological, structural and hydrological details.
4. To identify sites for the construction of artificial recharge structures in order to increase the groundwater potential in the study area.

Methodology

Data Collection

Different data products required for the study include Survey of India (SOI) 56 K/7 toposheet on 1 : 50,000 scale and IRS-I D PAN and LISS-III satellite imagery collected from National Remote Sensing Agency (NRSA). Collateral and demographic details are collected from Government organizations like the Bureau of Economics and Statistics.

Satellite Data Processing

The raw IRS-ID PAN and LISS-III satellite imageries collected from NRSA are geo-referenced using the ground control points with SOI toposheets as a reference. These two imageries are further merged to obtain a fused, high resolution (5.8 m of PAN) and colored (R, G, B bands of LISS-III) output using ERDAS Image processing software. The study area is then identified, delineated and subsetted from this fused data based on the latitude and longitude values and a final hard copy output is prepared for the generation of thematic maps using visual interpretation technique.

Generation of Thematic Maps

The raster based thematic maps are prepared from satellite imagery and toposheets using visual interpretation technique. The procedure consists of a set of image elements, which help in the recognition or interpretation of various land use/land cover features systematically on the enhanced satellite imagery during the classification of features (Lillesand, 2000). The procedure for thematic mapping used in this study is the system pioneered by United States Geological Survey (USGS), which is modified by National Remote Sensing Agency (NRSA) according to Indian conditions. These maps are converted to vector format by scanning using an AO flatbed deskjet scanner and digitization using AutoCAD software. The digital outputs are prepared using the editing and analyzing functions of Arc/Info and ARCVIEW GIS software (Siva Sankar, 2003). The GIS digital database consists of thematic maps like land use/land cover, drainage, soil, physiography, slope, road network, geomorphology, geology, structures and finally preparation of groundwater prospects map integrating all the above thematic details. The overall methodology employed for the present study is given in Figure 4.1.

Results and Discussion

Base Map

The base information drawn from toposheets and updated using satellite imagery can be used as a baseline data of physical features of the other thematic maps. The features included in the base map are the major water bodies *viz.*, rivers and tanks, major settlements, major roads, railways and other towns. The major settlements in the present study area are Koti, Himayathnagar, Malakpet, Balanagar and Kukatpally of Hyderabad district and Jamwada, Peddamagalaram, and Shaikpet of Medak district. The major water bodies extracted are Osman sagar, Himayathsagar, Hussainsagar, Miralam cheruvu and Musi river. The South Central Railway line and the NH-7, NH-9 connecting different state highways along with many metalled and unmetalled roads pass through the study area.

Drainage Map

All the rivers, their tributaries and small stream channels along with major and minor tanks are extracted from the toposheet to form the drainage map. Based on the drainage network, the weathering profile which controls the infiltration, groundwater table and discharge of surface water along the major streams and rivers is studied. A proper understanding of the major faults and the influence of groundwater flow are understood from drainage system and its controls. The study of drainage pattern, density and individual stream pattern provides clues to the distribution and attitude of the underlying rock formations and geologic structures. The drainage patterns observed in the study area are mainly of dendritic and deranged pattern. Dense dendritic drainage pattern is observed for the entire study area except in NE part. Dense drainage density is observed in NW and SW part, the medium drainage in SE part and coarse drainage density in NE part of the study area. Coarse textured drainage density indicates high infiltration, low runoff, low relief and permeable stratum (Seshagiri Rao, 2003).

Watershed Map

The watershed map is prepared in accordance with the National Watershed Atlas, 1990. The present study area comes under Region-4 which has a total area of 1130.48 lakh hectares and sub-divided into 8 basins. Within this region, the study area falls partially under basin-O *i.e.* the Krishna basin that has a total area of 272.03 lakh hectares and partially under basin-E. Within the Basin-O, the present study area is located in Catchment-l in the lowermost part of the basin below the Nagarjunasagar dam and is further divided into A, B, C, D and E sub-catchments. A total of five watersheds are covered partially in the study area *viz.*, 4E6D1, 4D1E5, 4D1E6, 4D1E7 and 4D1E8.

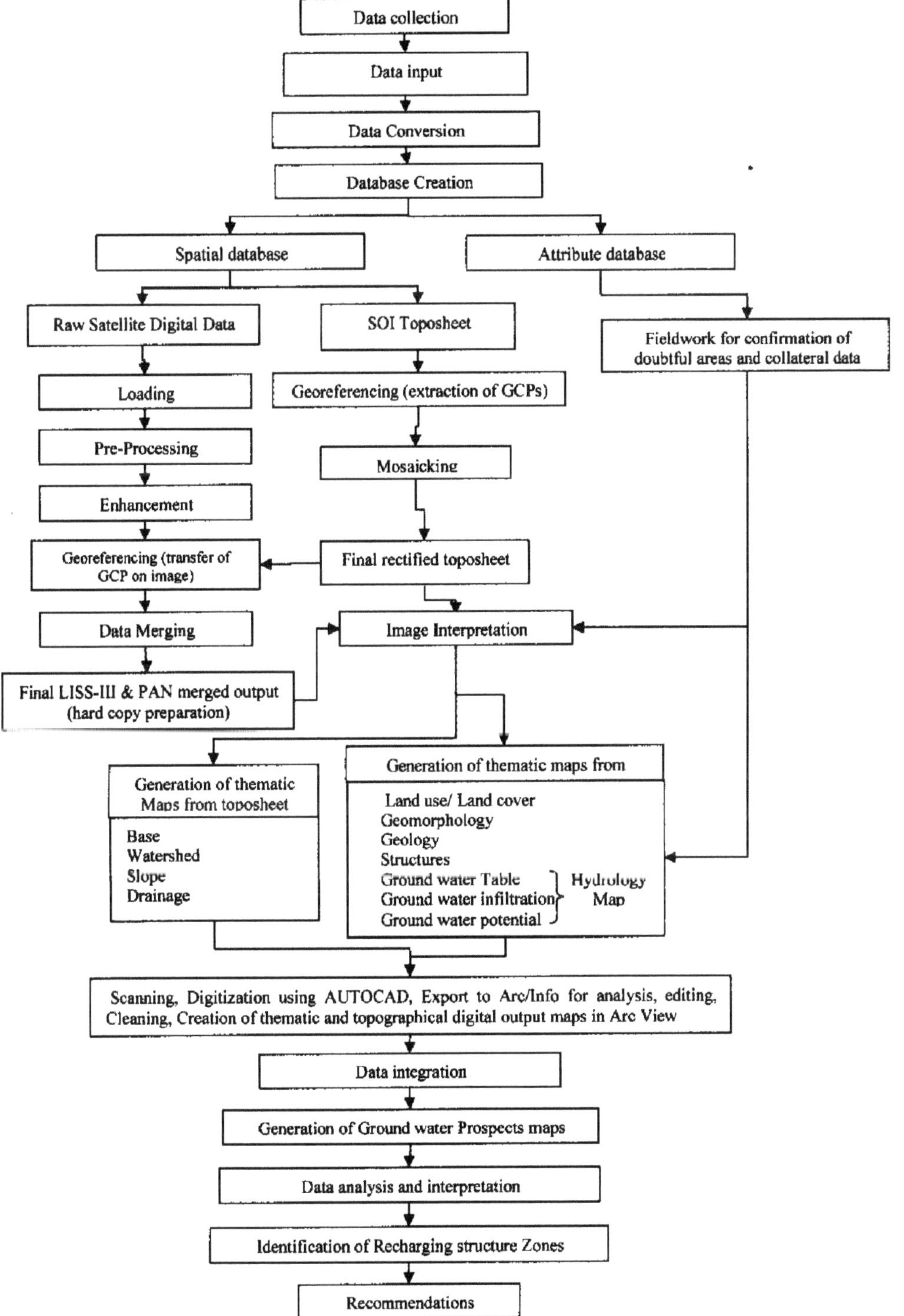

Figure 4.1: Flow Chart Showing the Methodology Adopted for the Present Study

Slope Map

The slope map of the study area is prepared from the contours and spot heights observed on the toposheet. Slope classes 1, 2, 3 and 4 are observed in the study area. Nearly level slope class covers 31 per cent of the total study area and is observed in the NE, NW and SW parts. 60 per cent of the study area in the SE part is under very gently sloping class and 91 per cent of study area in the central and SE region is under gently sloping class. Only 0.11 per cent of the total area is moderately sloping.

Land Use/Land Cover Map

The land use/land cover categories such as built-up land, agriculture, forest, water body and wastelands have been identified and mapped from the study area (Figure 4.2). About 27.86 per cent of the study area is under built-up land, which includes Hyderabad city, the residential area (84.53 per cent), industrial (9.81 per cent) and villages (5.59 per cent). The agriculture area occupying 36.05 per cent of total area are clearly delineated into single crop, double crop, fallow land and plantations from the satellite imagery. Single crop has been observed mostly in southern part of the study area and double crop on banks of the river Musi and other major streams. About 8.69 per cent of the total study area is occupied by water bodies such as Hussain Sagar, Miralam Tank, Himayath Sagar, Osman Sagar, Musi, Singur and Manjira rivers. About 2.01 per cent of the study area is under scrub forest. Wastelands occupy 27.85 per cent of the study area which is further classified into land with scrub (93.97 per cent), land without scrub (1.98 per cent), barren sheet rock area (3.8 per cent), and gullied/ ravinous land (0.18 per cent).

Geomorphology Map

The geomorphological classes observed in the study area are Pediplain with moderate weathering (PPM) occupying 25.20 per cent of total area, pediplain with shallow weathering (PPS) occupying 74.79 per cent, pediment (PD) occupying 9 per cent, pediment inselberg complex (PIC) occupying 15 per cent and dyke occupying 0.19 per cent. Based on the geomorphology, the weathering thickness of the soil is estimated to be 10–12 m in PPM and 0–10 m in PPS. In PD and PIC classes the soil weathering thickness is very less. The infiltration rates are moderate to good in PPM. Moderate to good yields of groundwater are observed in PPS, while in PO and PIC classes poor to nil yields are observed (Figure 4.3).

Structural Map

Structural features found in the study area are two types of lineaments namely, conformed lineament and inferred lineaments. Lineaments are linear fractures commonly associated with dislocation and deformations. They provide the pathways for groundwater movement and are hydrogeologically very important (Sankar *et al.*, 1996). The conformed lineaments are observed in the central part and in SW corner of the study area. The inferred lineaments are observed in NW, SW and SE corners of the study area.

Geology Map

The study area is undulating with scattered denudation hills and ridges in the granite gneiss terrain. The altitude varies from 500 mts in the west to 400 mts in the east with maximum elevation of 650 mts in the SW region. The study area constitutes mainly a granitic terrain (pink-grey) exposing a variety of archaean granitorides of peniusular gneissic complex (PGC) and schistoic (older metamorphic) rocks (Figure 4.4). They are intruded by basic dykes (Proterozoic) and covered locally by the deccan traps (upper cretaceous to lower Eocene). The granitorides (PGC) of the study area are

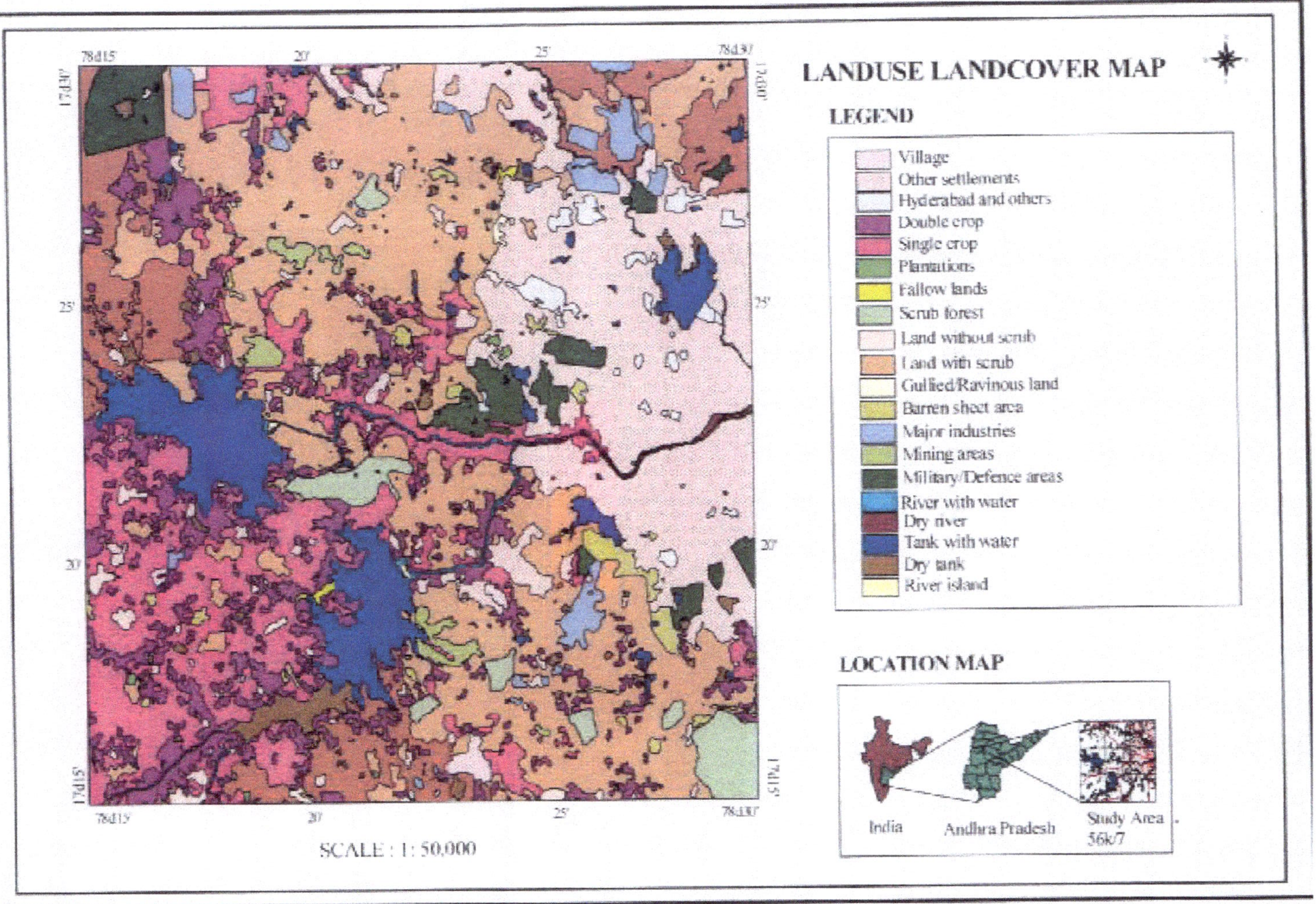

Figure 4.2: Land Use/Land Cover Map

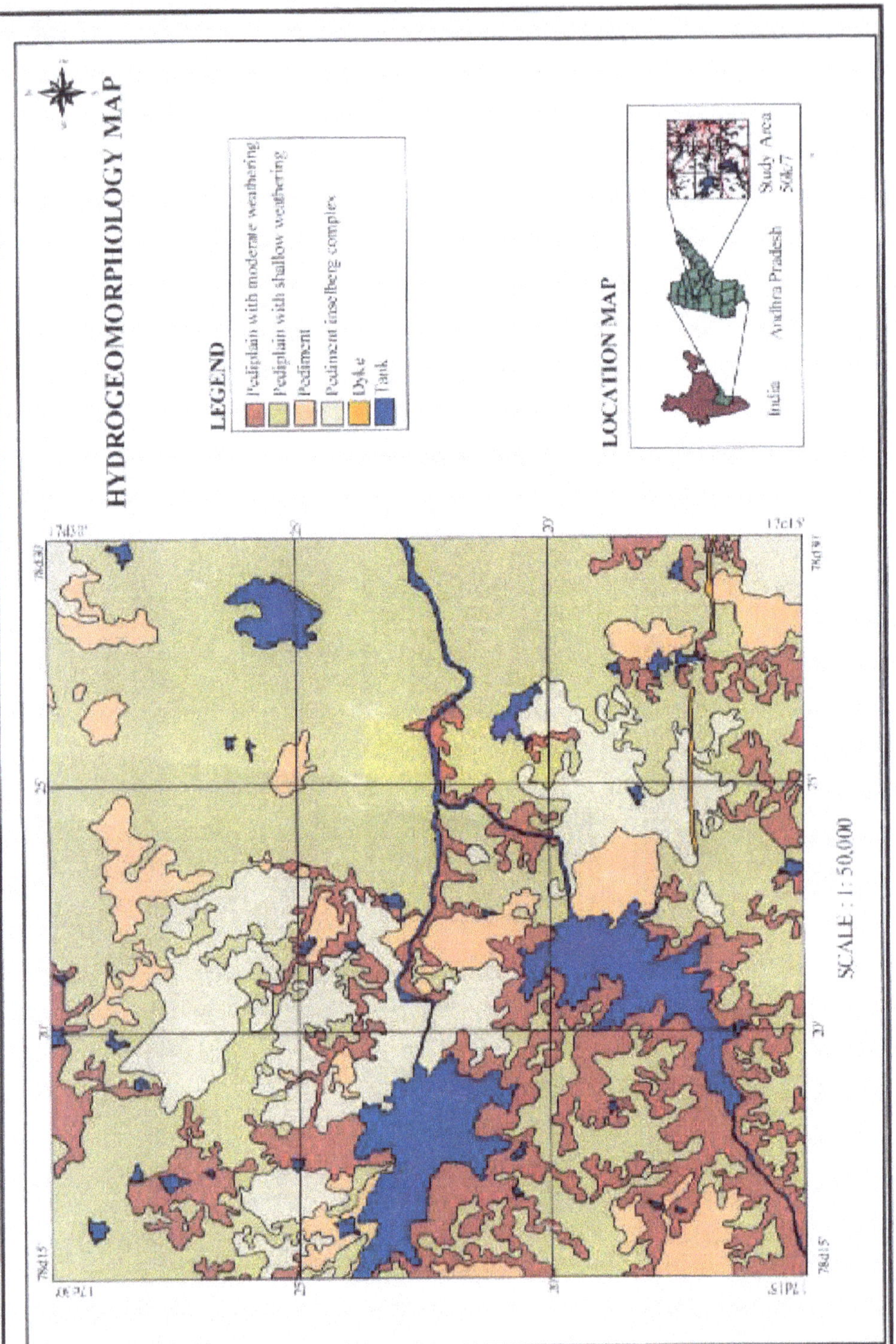

Figure 4.3: Hydrogeomorphology Map

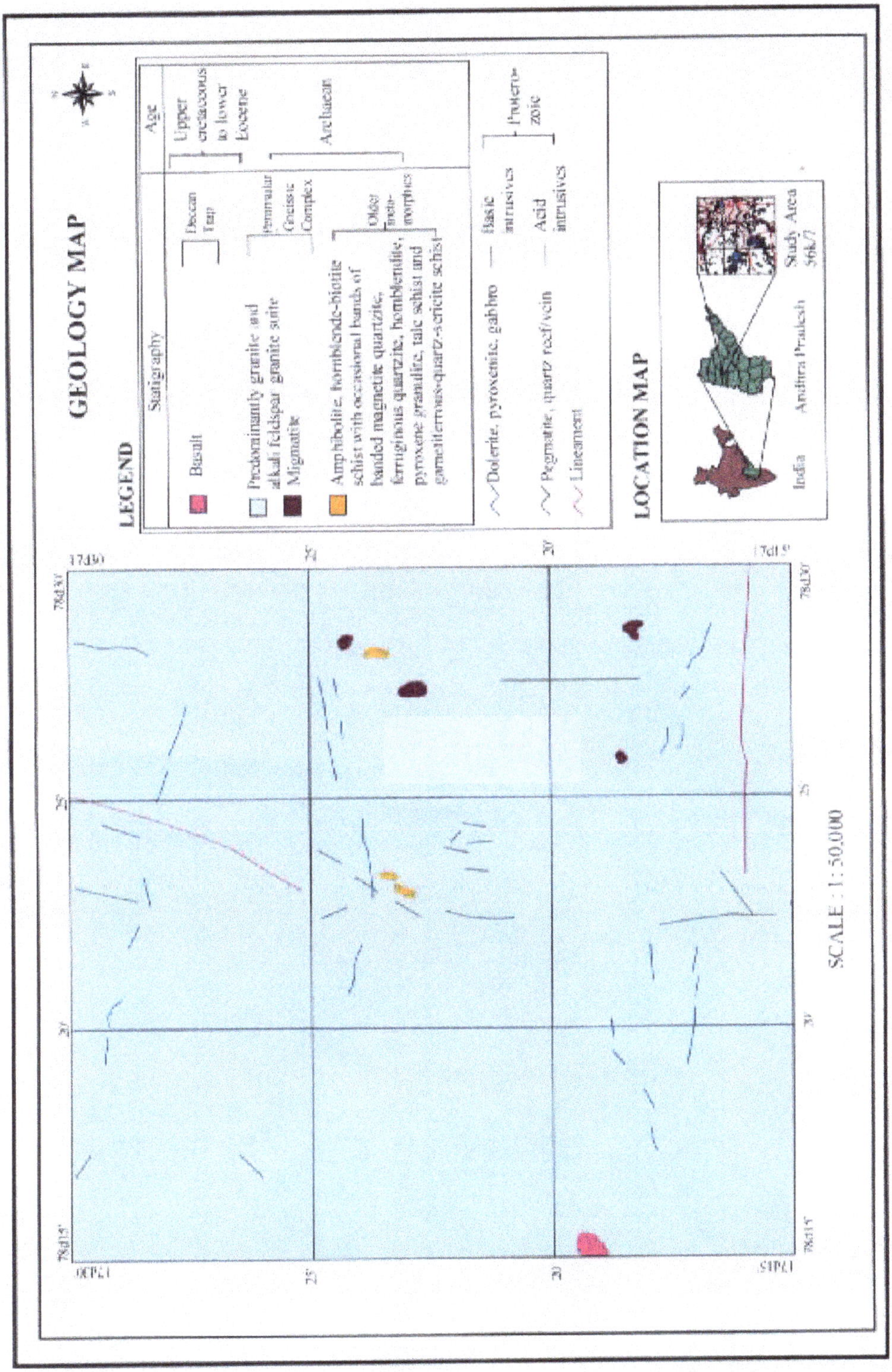

Figure 4.4: Geology Map

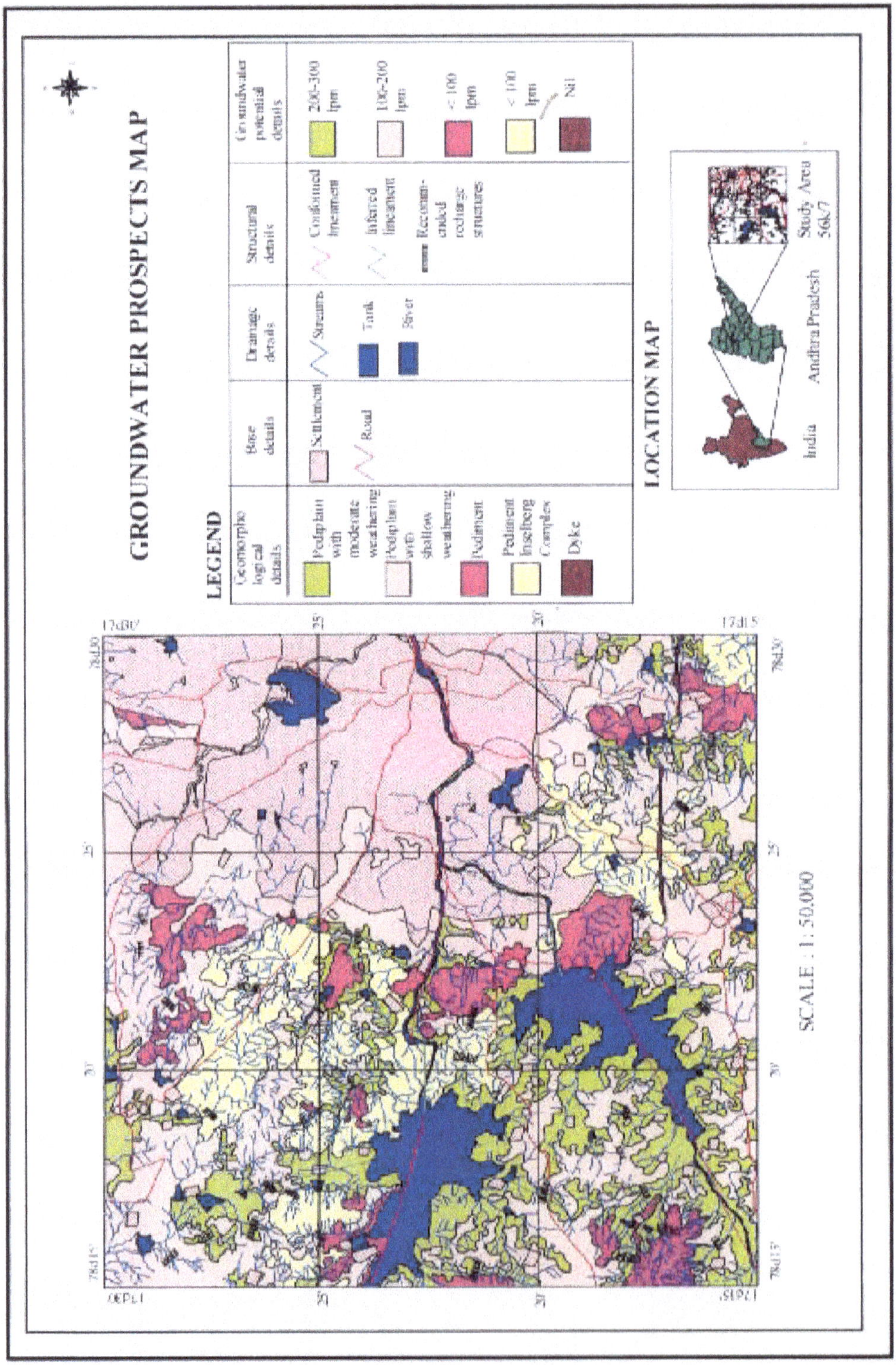

Figure 4.5: Groundwater Prospects Map

classified into three main suites based on the field relation and petro-chemical characteristics *viz.*, Tonalite-troudhjemite quartzchlorite (IT), Adanellite-granodiorite-quartzdiorite (AG) and Granite alkelifeldspar granite (Mary Hoffman, 1993). The geological categories observed in the study area are mainly granite (99.62 per cent), migmatite (0.172 per cent), basalt (0.118 per cent), amphibolite (0.082 per cent), and some lineaments, dolerites and pegmatites.

Groundwater Infiltration Map

The infiltration map the study area prepared according to the guidelines given by CPCB and is divided into five zones, *i.e.*, high, medium, medium low, low and nil zones. In 19 per cent of the total area along Musi and other surface water bodies the infiltration rates are high. More than half of the study area *i.e.* 56 per cent is under the medium zone, 9 per cent is under medium-low, 14 per cent under low infiltration zone and 2 per cent is under nil zone. In the high infiltration zone the weathering thickness of the soil is observed to be 20 mts, in medium zone it is 10 to 20 mts, and in medium to nil zone weathering thickness is observed to be less than 10 mts.

Groundwater Table Map

Based on the CPCB guidelines, the groundwater table in the study area is divided into three zones *viz.*, high (< 5 m bgl), medium (5 to 15 m bgl) and low (> 15 m bgl). 82 per cent of the total study area is under the medium zone, 18 per cent is under low zone and 0.2 per cent is under high zone area.

Groundwater Potential Map

The groundwater potential map of the study area is classified into four zones namely, very good to good potential (300–500 lpm) occupying 19 per cent of the total area, good to moderate potential zone (200–300 lpm), moderate to poor (200–100 lpm) occupying 23 per cent of the total area and poor to nil zone (0–50 lpm) occupying only 2 per cent of total area. Based on this map it is concluded that the groundwater potential in the entire study area is moderate to poor.

Groundwater Prospects Map

The groundwater prospects map is prepared integrating the geological, geomorphological, structural, hydrological and other collateral data information of the study area (Figure 4.5). The final integrated map exhibiting the details of the above maps not only aids in identification of areas with low groundwater potential but also helps in increasing the groundwater potential in these areas through identification of sites for construction of artificial recharge structures like check dams and percolation tanks. These sites are recommended based on the topographic condition and hydrological capacity of the terrain.

Conclusions and Recommendations

The final groundwater prospects map is prepared by integrating the thematic details of geomorphology, geology, structures, watershed, drainage, groundwater table etc. of the study area. The present study gives detail information of the terrain condition along with the groundwater status. It also helps in identifying areas for construction of artificial recharge structures in order to improve the groundwater levels in the perspective of groundwater use for future generations. This type of study is useful for the administrative officials to take quick decisions and implement them to meet the needs of future generations. The following recommendations are made to increase the groundwater levels in problematic areas along with construction of artificial recharge structures:

1. All the residents, whether dependent on surface water or groundwater should be made conscious of the value of fresh water. Awareness campaign for conservation and augmentation of groundwater should be taken up on a large scale not only to motivate but also to mobilize people in undertaking different artificial recharge methods.
2. To improve the groundwater level and to avoid further depletion of groundwater levels construction of artificial recharge structures like roof top/road top rainwater harvesting structures, pits and scavenger wells are recommended in houses/apartments and in most of the vacant areas such as Govt. office premises, public parks, schools, colleges and universities.
3. Hyderabad has no separate storm drainage system. Since it is mixed with the sewerage system, large quantities of storm runoff are lost, which could be used for recharge. It is therefore suggested that wherever possible storm water be separated and used for recharge with adequate planning.
4. Intense road network and development activities with less vacant land between dwelling units have reduced surface area. Such localities are to be prioritized and separate planning has to be taken up for recharge.
5. Regulation and control of groundwater development is to be made operational for not only protecting the environment but also to ensure equity in sharing groundwater.
6. The density of groundwater observation network may be increased for monitoring water levels and quality at close intervals for more intense planning and development.

References

Anjaneyulu, Y., 2004. *Introduction to Environmental Science*. B.S Publications, Hyderabad.

Anji Reddy, M., 2001. *Text book of Remote Sensing and Geographical Information Systems*, 2nd edn. B.S Publications, Hyderabad.

Census of India, 1991a. Series 2, Part 12(A and B), *District Census Handbook*, Ranga Reddy, Published by the Government of Andhra Pradesh.

Census of India, 1991b. Series 2, Part 12(A and B), *District Census Handbook*, Hyderabad, Published by the Government of Andhra Pradesh.

Hoffman, Mary, 1993. *Dictionary of Geology*, Special Indian edition. GOYL Saab Publishers and Distributors, Delhi.

Khan, M.A. and Moharana, P.C., 2002. Use of remote sensing and GIS in the delineation and characterization of groundwater prospect zones. *Journal of Indian Society of Remote Sensing*, 30(3): 131–141.

Krishnamurthy, J., Venkataesa Kumar, N., Jayraman, V. and Manivel, M., 1996. An approach to demarcate groundwater potential zones through remote sensing and GIS. *International Journal of Remote Sensing*, 17(10): 1867–1884.

Kumar, Anand and Tomar, Sanjay, 2002. Application of remote Sensing and GIS for groundwater Assessment. *Development Alternatives Newsletter*, p. 12.

Lillesand, Thomas M. and Keifer, Ralph W., 2000. *Remote Sensing and Image Interpretation*. John Wiley and Sons, New York.

Musa, Khairul Anam, Akhir, Juhari Mat and Abdullah, Ibrahim, 2000. Groundwater prediction potential zone in Langat basin using the integration of remote sensing and GIS. *ACRS Proceedings.*

Rao Toleti, B.V.M., Chaudhary, B.S., Mothi Kumar, K.E., Saroha, G.P., Yadav, Manoj, Singh, Ajeet Sharma, M.P., Pandey, A.C. and Singh, P.K., 2000. Integrated groundwater resources mapping in Gurgaon district, (Haryana), India using remote sensing and GIS techniques. *ACRS Proceedings.*

Sankar, K., Jegatheesan, M.S. and Balasubramanian, A., 1996. Geoelectrical resistivity studies in the Kanyakumari district, Tamil Nadu. *Journal of Applied Hydrology*, 9(1 and 2): 83–90.

Saraf, A.K. and Chaudhary, P.R., 1998. Integrated remote sensing and GIS for groundwater exploration and identification of artificial recharges sites. *International Journal of Remote Sensing*, 19(10): 1825–1841.

Seshagiri Rao, K.V., 2003. *Watersheds: Comprehensive Development*. B.S. Publications, Hyderabad, India.

Singh, Amaresh Kr., Raviprakash, S., Mishra, D. and Singh, Samarendra, 2002. Groundwater potential modelling in Chandraprabha subwatershed, V.P. using Remote Sensing, geoelectrical and GIS. *Map India.*

Siva Sankar, A., 2003. Evaluation of impacts of landuse changes on groundwater quality of a part of Hyderabad city using Remote Sensing, GIS and *In situ* studies. Ph.D Thesis, Hyderabad.

Chapter 5

Antimicrobial Properties of Selected Coastal Plants and Marine Algae from East Coast of Tamil Nadu

C. Rathika[1], S.M. Fazeela Mahaboob Begum[2] and K. Balakrishnan[1]
[1]Department of Biotechnology, Bharathidasan University, Tiruchirappalli – 620 024
[2]Department of Microbiology, Sri Vellaichamy Nadar College, Madurai – 625 021

ABSTRACT

Six algae and four coastal plants were studied for their antibacterial activity against eighteen and nine pathogenic and opportunistic microorganism respectively. Alcoholic extracts were tested for their antibacterial activity by using Kirby-Bauer disk diffusion method at a sample volume of 20 μl. Extracts were added either as raw sample or as airdried filter papers impregnated with the extracts. In the present study, two marine algal species showed efficient bacteriostatic activity for a wide range of microorganisms. Marine algae such as like *Acanthopora* and *Sargassum* revealed significant antibacterial activity. The maximum zone of inhibition was observed in *Dictiyota* against *Salmonella typi* (34 mm) and *Zymomonas moblis* and (22 mm) respectively. The coastal plant extracts also showed significant antibacterial activity. The extracts from coastal plants *such as Suaeda* and *Sesuvium* showed antibacterial activity against eight microbes out of nine tested. Thus the present investigation revealed that the extracts of coastal plants and algae, exhibited remarkable bacteriostatic activity to selected pathogenic and opportunistic microorganisms.

Introduction

Marine ecosystems are a rich source of medicinal plants and for miracle drugs. The ocean is a store house of valuable chemicals and natural resources like marine algae. More than 150,000 macro algae or seaweed species are found in the oceans of the globe but only few of them were identified (Harvey, 1988). Marine algae have been traditionally used as food and medicine. Marine algae contain the essential amino acids and polyunsaturated fatty acids, necessary vitamins and minerals and larger amount of dietary fibers. These contain a variety of biologically active substances, which possess antibacterial (Liao *et al.*, 2003) antiviral (Hudson *et al.*, 1993). Marine algal polysaccharides and proteinaceous substances have valuable functions in immune modulation and stimulation (Hori *et al.*, 1988; Otterlei *et al.*, 1991; Yoshizawa *et al.*, 1993; Shan *et al.*, 1999; Yoshizawa *et al.*, 1993; Son *et al.*, 2001). Secondary or primary metabolites from these organisms may be potential bioactive compounds of interest for the pharmacological industry (Attaway and Zeborsky, 1993). As a consequence of an increasing demand for the biodiversity in screening programmes, seeking therapeutics drugs from natural products there is now a greater interest in the marine organisms especially marine algae (Faulkner 1993; Schever, 1987). Special attention has been reported for antibacterial antiviral and/or antifungal activities related to marine algae against several pathogens (Ballesteros *et al.*, 1992; Bhakuni *et al.*, 1990, 1992; Caccamese and Azzolina 1979; Deig *et al.*, 1974; Perez *et al.*, 1990; Richards *et al.*, 1978; Selvaraj *et al.*, 1989; Siddhanta *et al.*, 1997). The search for antimicrobial agents has taken a definite direction in developed countries. While antiviral drugs are high on priority, mycobacterium and gram negative bacterial infections are still among the most common, serious and pathogenic diseases worldwide (Neu Harold, 1992; Bloom and Murray, 1992). Many bioactive and pharmacologically active substances have been isolated from algae. Sachithnanthan and Sivapalan (1975), Srinivasa Rao and Parekh (1981) showed that crude extracts of selected Indian seaweeds were active against gram positive bacteria. Marine algae have received comparatively less bioactive attention with reference to the screening for bioactive compounds. There are a number of seaweeds with economic potential (Critchley, *et al.*, 1998). Sea weeds are low in fats but contain vitamins and bioactive compounds like terpenoids and sulpahated polysaccharides (a potential natural antioxidant which are not found in land plants) (Lahaye and Kaffer, 1997). Reports show that the *Sargasum* spp. is found to have the highest free radical scavenging property (Matsukawa, 1997; Yan *et al.*, 1998).

Materials and Methods

Collection of Plant Materials

Six algae namely *Acanthapora spicifera* (Red), *Sargassum wightii* (Brown), *Dictyota dichotoma. Caulerpa racemosa (Green), Hydrucclathrus clathratus* (Brown), *Hormophysa triquetra* (Brown) and four coastal plants such as *Salicornia brachiata, Halophila thouars* (Sea grass), *Sesuvium portulacastrum, Suaeda maritima* were collected from CMFRI as a gift (Central marine fisheries and research institute, Mandapam) and from coastal areas of Devipatinam, Ramnad district, Tamil Nadu.

Preparation of Plant Extracts

The plant parts were cleaned and shade dried for one month. After proper drying the dried plant/algal samples were powdered. 150 mg of this powder was used to extract separately in soxhelt apparatus with 80 per cent ethyl alcohol as a solvent.

Microorganisms

Antibacterial activity of alcohol extracts of six algal and four coastal plant samples were evaluated against 18 different microorganisms. The coastal plants were tested against nine different microorganisms. The bacterial species used for the antimicrobial assay includes *Proteus* spp., *Escherichia coli, Lactobacillus* spp., *Enterobacter* spp., *Pseudomonas* spp., *Pseudomonas fluorescence, Streptococcus* spp., *Staphylococcus aureus, Bacillus thuringiensis, Rabitonia eutropha, Bravibacterium, Salmonella typhi, Klebsiella* spp., *Zymomonas mobilis, Xanthomonas campestris, Streptococcus marcesens, Vibrio cholera, Shigella flexneri.* The bacterial strains were maintained on nutrient agar at 4° centigrade and re-cultured every two weeks.

Antibacterial Susceptibility Test

Antibacterial activity was performed by slight modification of the method originally described by Bauer *et al.* (1966), which has been widely used for the antibacterial susceptibility testing (Barry and Thornsberry, 1985). Sterile Muller-Hinton agar plates were prepared and a loopful of each test bacteria was taken from the stock culture and spread on the surface of the solidified agar plate (6–12 hours old culture). After 5–10 minutes, the solvent extract (impregnated) sterile disc was placed on the agar surface by using a sterile forceps. The plates were incubated at 37°C for 24 hours. The discs were observed for the formation of clear zone of inhibition around the discs.

Table 5.1: Antibacterial Activity of Marine Algae Against Various Microorganisms

Sl. No.	*Name of Bacterial Species*	*Zone of Inhibition mm*					
		Acanthapora spicifera (Red)	*Sargasum wightii (Brown)*	*Dictyota dichotoma*	*Caulerpa racemosa (Green)*	*Hydrucclathrus clathratus (Brown)*	*Hormophysa triquetra (Brown)*
1.	*Escherichia coli*	8	8	–	–	–	–
2.	*Proteus* spp.	6	8	2	–	–	2
3.	*Pseudomonas* spp.	14	17	–	–	–	4
4.	*Pseudomonas fluorescence*	6	14	–	–	–	6
5.	*Entero bacter* spp.	6	16	–	–	–	8
6.	*Lacto bacillus* spp.	6	–	–	–	–	8
7.	*Bacillus thuringenesis*	6	6	–	–	–	2
8.	*Staphylococcus aureus*	14	8	–	2	–	4
9.	*Streptococcus* spp.	6	6	–	–	2	–
10.	*Rabitonia eutropha*	–	8	6	–	–	–
11.	*Bravi bacterium* spp.	12	–	8	–	–	4
12.	*Salmonella typhi*	–	4	34	12	–	–
13.	*Klebsiella* spp.	8	10	6	–	–	4
14.	*Zvmomonas mobilis*	–	–	22	–	–	2
15.	*Xanthomonas campestris*	–	2	10	–	–	2
17.	*Vibrio cholerae*	2	–	8	–	–	2
18.	*Shigella flexneri*	–	10	8	–	–	–

Results and Discussion

Results of preliminary antimicrobial screening of the extracts of marine algae and coastal plants were summarized in Tables 5.1 and 5.2). The algal extracts showed different degrees of antibacterial

Table 5.2: Antibacterial Activity of the Extracts of Coastal Plant on Selected Microorganisms

Sl.No.	Name of the Coastal Plant	Name of Bacterial Species — Zone of Inhibition (mm)																	
		Proteus spp.		*Escherichia coli*		*Lacto-bacillus spp.*		*Entero-bacter spp.*		*Pseudomonas spp.*		*Pseudo-monas flouresence*		*Staphylo-coccus spp.*		*Strepto-coccus*		*Bacillus thuringiensis*	
		1		2		3		4		5		6		7		8		9	
		S	A	S	A	S	A	S	A	S	A	S	A	S	A	S	A	S	A
1.	*Salicornia brachiata*	–	–	9	8	10	2	10	5	2	–	8	5	2	2	12	11	–	–
2.	*Holophila thouars*	–	–	7	11	10	8	7	2	2	2	–	–	2	2	10	2	–	–
3.	*Sessuvium portulacastrum*	–	–	8	2	10	8	8	8	10	11	9	6	11	2	10	8	10	9
4.	*Suaeda maritime*	8	7	2	2	29	8	8	8	7	9	8	7	10	8	8	2	–	–

S: Aqueous extract; AL: Extract added and air dried.

The marine algae *Acanthopora spicifera* and *Sargassum* showed antibacterial activity against 13 microorganisms tested. An efficient antibacterial activity was observed when the extract was dissolved in solvents and applied to the discs, whereas an air dried samples did not show sufficient antibacterial activity (except for *Hydrucclathrus clathratus* against *Rabitonia eutropha*– results not shown). This result can be related to volatile antimicrobial compounds in the samples, such as hydrogen peroxide, terpenoid, and bromo-ether compounds (Masuda *et al.*, 1997; Rosell and Srivastava, 1987). Another reason might be the loss of active materials that may be present in algae, like volatile fatty acids, during the drying process. An organic solvent provides a higher efficiency in extracting compounds for antimicrobial activities compared to water-based methods (Uma-Filno *et al.*, 2002; Masuda *et al.*, 1997).

The maximum diameter of zone of inhibition was observed against for *Acanthapora spicifera* (*Red*) algae (14 mm), against *Pseudomonas* spp., *Sargasum wightii* (*Brown*) algae against *Enterobacter* spp. (16 mm) and for *Dictiyota dycottoma* against *Salmonella typhi* and *Zymomonas mobilis* (34 mm and 22 mm respectively). *Dictiyota* showed much elevated antibacterial activity than *Caulerpa racemosa* (*Green*) (34 mm vs. 12 mm respectively). Several different organic solvents have been used to screening algae for antibacterial activity. Olessen *et al.* (1963) identified antibacterial activity in the chloroform and acetone extracts of *Falkenbergia billebrandii* against *S. aureus.* Sastry *et al.* (1994) showed antibacterial activity against Gram-positive and Gram negative pathogenic strains after successive extraction with benzene, chloroform and methanol. Likewise, Mahasneh *et al.* (1995) have shown antibiotic activity in organic extracts of six species of marine algae against multi-antibiotic resistant bacteria. Antimicrobial activity depends on both algal species and efficiency on extraction of their active principles (Vieira *et al.*, 1971). Out of the four, three plants *Salicornia, Sesuvium* and *Suaeda maritima* were showed antibacterial activity against eight microorganisms. The *Halophila thouars* (sea grass) showed antibacterial activity against six microbes. It did not show any bacteriostatic activity against *Pseudomonas fluorescence, Proteus* spp. and *Bacillus thuringiensis.* The present findings proved that some of these marine algal species possess compounds that can kill and inhibit the pathogenic disease causing microorganisms. Thus from the above results it is concluded that the coastal plants and marine algae can be exploited as promising candidates for drug discovery. Further studies are required to understand and to use these products as an alternative to synthetic antibiotics.

References

Attaway, D.H., and Zaborsky, O.R., 1993. Marine biotechnology. *I. Pharm. Bioac. Nat. Prod.*

Ballesteros, E., Martin, D. and Uriz, M.J., 1992. Biological activity of extracts from some Mediterranean macrophytes. *Bot Marina*, 35: 481–485.

Barry, A.L. and Thornsberry, C., 1985. Susceptibility tests: Disc diffusion test procedures. In: *Manual of Clinical Microbiology*, (Eds.) Lennette Balows, E.H., Hauster, W.J. and H.J. Shadomy. American Society for Microbiology, Washington, DC, pp. 978–987.

Bauer, A.W., Kirby, W.M.M., Sherris, J.S. and Turk, M., 1966. Antibiotic susceptibility testing by a Standard single disc method. *American Journal of Clinical Pathology*, 36: 493–496.

Bhakuni, D.S., Goel, A.K., Jain, S., Mehrotra, B.N., Srimal, R.C. and Srinivastava, M.N., 1992. Bioactivity of marine organisms, Part VI: Screening of some marine flora from Indian coasts. *Indian Journal of Experimental Biology*, 30(6): 512–517.

Bhakuni, D.S., Goel, A.K., Jain, S., Mehrotra, B.N. and Srimal, R.C., 1990. Screening of Indian plants for biological activity, Part XIV. *Indian Journal of Experimental Biology*, 28(7): 619–637.

Bloom, B.R. and Murray, C.J.L., 1992. Tuberculosis: Commentary on a reemerging killer. *Science*, 257: 1055–1064.

Caccamese, S. and Azzolina, R., 1979. Screening for antimicrobial activities in marine algae from Eastern Sicily. *Planta Med.*, 37: 333–339.

Critchley, A.T., Gillespie, R.D. and Rotman, K.W.G., 1998. In: *Several Resources of the World*, (Eds.) Critchley, M., and A.T. Ohno. Japan International Cooperation Agency, Japan, p. 413–425.

Deig, E.F., Ehresmann, D.W., Hatch, M.T. and Riedlinger, D.J., 1974. Inhibition of herpesvirus replication by marine algae extracts. *Antimicrob. Agents Chemother.*, 6(4): 524–525.

Faulkner, D.J., 1993. Marine natural products chemistry: Introduction. *Chem. Rev.*, 93: 1671–1673.

Harvey, W., 1988. *Biotechnology*, 6: 488–492.

Hori, K., Ikegami, S., Miyazawe, K. and Ito, K., 1988. Mitogenic and antineoplastic isoaggutinins from the red alga *Solieria robusta*. *Physiochemistry*, 27: 2063–2067.

Hudson, J.B., Kim, J.H., Lee, M.K., Dewreede, R.E., and Hong, Y.K., 1999. Antiviral compounds in extracts of Korean seaweeds, Evidence for multiple activities. *Journal of Applied Phycology*, 10: 427–434.

Lahaye, M. and Kaffer, B., 1997. Seaweed dietary fibres structure physicochemical and biological properties relevant to intestinal Physiology. *Sci. Aliments*, 17: 563–564.

Liao, W.R., Lin, J.Y., Shieh, W.Y. and Huang, R., 2003. Antibiotic activity of lectins from marine algae against marine *vibrios*. *Journal of Industrial Microbiology and Biotechnology*, 30: 433–439.

Mahasneh, J., Jamal, M., Kashasneh, M. and Ziodeh, M., 1995. Antibiotic activity of marine algae against multiantibiotic resistant bacteria. *Microbial.*, 83: 23–26.

Matsukawa, R., 1997. A comparison of screening methods for antioxidant activity in seaweeds. *J. Appl. Phycol.*, p. 929–935.

Masuda, M., Abe, T. and Seto, S., 1997. Diversity of halogenated secondary metabolites in the red alga *Laurencia nipponica* (Rhodomelaceae, ceramiales). *J. Phyco1.*, 33: 196–208.

Neu Harold, 1992. The crisis in antibiotic resistance. *Science*, pp. 257.

Olessen, P.E., Maretzki, A. and Almodovar, L.A., 1963. An investigation of antimicrobial substances from marine algae. *Bot. Marina*, 6: 226–232.

Otterlei, M., Ostgaard, K., Skjak-Braek, G., Smidsrod, O., Soon-Shiong, P. and Espevik, T., 1991. Induction of cytokine production from human monocytes Stimulated with alginate. *Journal of Immunotherapy*, 10: 286–291.

Perez, R.M., Avila, J.G., Perez, S., Martinez, A. and Martinez, G., 1990. Antimicrobial activity of some American algae. *J. Ethanophramacol.*, 29(1): 111–116.

Richards, J.T., Kern, E.R., Glasgow, L.A., Overall, J.C., Deign, E.F. and Hatch, M.T., 1978. Antiviral activity of extracts from marine algae. *Antimicrob. Agents Chemother.*, 14(1): 24–30.

Rosell, K.G. and Srinivastava, L.M., 1987. Fatty acids as antimicrobial substances in brown algae. *Hydrobiologia*, 151/152: 471–475.

Sachithnanathan, K. and Sivapalan, A., 1975. Antibacterial properties of some marine algae of Sri Lanka. *Bulletin of Fisheries Research Station, Sri Lanka*, 26: 5–9.

Sastry, V.M.V.S. and Rao, G.R.K., 1994. Antibacterial substances from marine algae: Successive extraction using benzene, choloroform and methanol. *Bot. Marina*, 37: 357–360.

Scheuer, P.J., 1997. *Bioorganic Marine Chemistry, Vol. 3.* Springer, Berlin Heidelberg, New York.

Selvaraj, R., Manivasagam, S., Purushothaman, A. and Subramanian, A., 1989. Preliminary investigation on antibacterial activity of some marine diatoms. *Indian J. Med. Res.*, 89: 198–200.

Shan, B., Yoshida, Y., Kuroda, E. and Yarnashita, U., 1999. Immunomodulating activity of seaweed extraction human lymphocytes *in vitro. International Journal of Immunopharmacology,* 21: 59–70.

Son, E., Moon, E., Rhee, D. and Pyo, S.2001. Stimulation of various functions in murine peritoneal macrophages by high manuronicacid-contain in galginate (HMA) exposure *in vivo. International Journal of Immunopharmacology,* 1: 147–154.

Siddhanta, A.K., Mody, K.H., Ramavat, B.K., Chawhan, V.D., Garg, H.S., Goel, A.K., Jinandra Doss M., Srinivastava, M.N., Patnaik, G.K. and Kamboj, V.P., 1997. Bioactivity of marine organisms, Part 8: Screening of some marine flora of western coast of India. *Indian Journal of Experimental Biology,* 35: 638–643.

Srinivasa Rao, P. and Parekh, K.S., 1981. Antibacterial activity of Indian Seaweed extracts. *Botanica Marina,* 24: 577–582

Uma-Filho, J.V.M., Carvalho, A.F.F.U. and Freitas, S.M. *et al.*, 2002. Antibacterial activity of extracts of six macro algae from the northeastern Brazilian coast. *Brazilian Journal of Microbiology,* 33: 311–313.

Vieira, F.P. and Caland-Noronha, M.C., 1971. Atividade antibiotica de algum as marin has do estado do ceara. *Arquir. Cienc. Mar,* 11: 91–93.

Yan, X., Nagata, T. and Fan, X., 1998. Antioxidant activities in some common seaweed. *Plants, Foods, Hum. Nutr.*, 52: 253–262

Yoshizawa, Y., Enomoto, A., Todho, H., Ametani, A. and Kaminogawa, S., 1993. Activation of marine macrophages by polysaccharide fractions from marine algae (Porphyrayezoensis). *Bioscience Biotechnology Biochemistry*, 57: 1862–1866.

Chapter 6

Correlation Coefficient (r) Value Between LCC Vs SPAD as Influenced by Hybrids and Nitrogen Management Practices

S. Ramesh, S. Ravi and B. Chandrasekaran
Department of Agronomy, Tamil Nadu Agricultural University, Coimbatore – 641 003, Tamil Nadu

ABSTRACT

Field experiments were conducted at wetland research farm of Tamil Nadu Agricultural University, Coimbatore during *Kharif* 2000 and 2001 to study the relationship between the LCC values and SPAD reading, and the LCC values Vs Grain yield as influenced by nitrogen management practices in rice hybrids. Treatments consisted of two hybrids in the main plot and seven levels of N management practices during *Kharif* 2000 and eight sub-plot treatments during *Kharif* 2001. Application of green manure @ 6.25 tha^{-1} combined with LCC cv5 based N application (N_4) registered, higher level of LCC and SPAD values than the other treatments. LCC Vs SPAD values had relatively stronger relationship in both the years at all the stages of crop growth and also confirmed the existence of strong positive correlation co-efficient between LCC Vs grain yield at all the stages of crop except at 14 days after transplanting.

Keywords: Hybrids, Nitrogen management, Leaf colour chart, SPAD.

Introduction

To sustain the self-sufficiency in the coming decades the rice production needs to be geared up every year by almost 2 million tones. This is a daunting task, in view of plateauing trend observed in yield potential of high yielding varieties, and decreasing and declining natural resource base. Among the many genetic approaches being explored to break the yield barrier in rice, hybrid rice technology appears to be the most feasible and readily adoptable one. Swaminathan (1998) emphasized precision farming to reduce the cost of production and improve the productivity on an ecologically sustainable basis. Matching the time course N demand of the rice crop with N supply leading to maximum utilization is the precision N management. Assessing the leaf N concentration is one of the techniques to maintain specific N concentration at different physiological stages by which the yield potential can be exploited. This is essentially because N concentration in leaves is highly correlated with concentration of the CO_2 fixing enzyme in photosynthesis (Greenwood *et al.*, 1991) and hence, the potential maximum rate of photosynthesis could be linearly related to the N concentration in the leaf (Black, 1993). One of the recently introduced N management approach for estimating the leaf N concentration was by the measurement of leaf greenness. Leaf green colour was calibrated very precisely to the leaf N content (Tanno, 1988). Among the different tools available to measure the leaf greenness, the non destructive measurement of leaf green colour intensity using leaf green colour charts (LCC) and the instrument namely, chlorophyll meter (SPAD-502) are gaining much importance. Hence in this context, it becomes imperative to conduct this experiment to study the relative accuracy of LCC to assess the leaf N status and it was compared and correlated with chlorophyll meter reading (IRRI, 1996) which is highly essential to determine the time of N topdressing for specific cultivar groups such as hybrids.

Materials and Methods

Field experiments to study the effect of green manuring and fertilizer N application through LCC grades on rice hybrids were conducted during *Kharif* 2000 and 2001 in wetland research farm of Tamil Nadu Agricultural University, Coimbatore. The soil of the experimental field was moderately drained, deep clay loam, classified taxonomically as *Typic haplustalf*. The soils were low in available soil N, medium in available soil P and high in available soil K respectively. The experiment was laid out in split plot design with three replications. The main plot treatments consisted of two hybrids PA6201 and CoRH2 and the sub-plots with seven levels of N management practices during *Kharif* 2000 whereas during *Kharif* 2001 one more sub-plot treatment was included which is the existing recommended practice under Integrated nitrogen management practices (N_8). The rice hybrid PA6201 was released in 2000 by Central Release Committee (CVRC). It is a medium duration (125–130 days) semi dwarf, erect type having long panicle with long slender grain. The rice hybrid $CoRH_2$ released during 1998 by Paddy Breeding Station, Tamil Nadu Agricultural University, Coimbatore is also a medium duration semi-dwarf type having compact panicle with medium grain type.

The details of the treatments and notations used are as under:

A. Main plot: Hybrid rice

H_1: PA6201

H_2: CoRH2

B. Sub-Plot: Nitrogen management

N_1: Basal + 3 splits of top dressing recommended dose (200 $kgha^{-1}$)

N_2: Green manure @6.25 tha^{-1} + N application as per LCC cv 3

N_3: Green manure 6.25 tha^{-1} + N application as per LCC cv 4

N_4: Green manure 6.25 tha^{-1} + N application as per LCC cv 5

N_5: 20 kg Nha^{-1} as basal + N as per LCC cv 3

N_6: 20 kg Nha^{-1} as basal + N as per LCC cv 4

N_7: 20 kg Nha^{-1} as basal + N as per LCC cv 5

N_8: Green manure @ 6.25 tha^{-1} + 3 splits of recommended dose (200 kgha^{-1})

An absolute control was maintained separately in the experimental field to find out the efficiency of soil applied nitrogen. The above ground biomass of *Sesbania aculeata* at the age of 40 days was harvested, weighed and applied to the respective plots as per the treatment schedule. The green manure was spread uniformly and incorporated one week before transplanting. The fertilizer nitrogen was applied to the rice crop as per the treatments. In LCC based N management treatments, the LCC values were recorded as per the standard procedure (IRRI, 1996) at weekly intervals starting from 14 DAT to flowering. Leaf colour chart consists of six strips from light yellowish green (No. 1) to dark green (No. 6). The rice leaf colour was compared with standard LCC under the same environmental condition. The colour of the single leaf was measured by holding the leaf, colour chart vertically and placing the middle part of leaf, 1 cm in front of colour strip for comparison (IRRI, 1996). Ten readings were taken for each plot and average was computed by rounding off to the nearest 0.5 to determine the need for N top dressing. Whenever the LCC values fall below the fixed critical level the amount of N to be applied at different growth stages was as per the schedule given below (IRRI, 1996).

Crop Growth Stages	*N to be Applied (kgha^{-1})*
Early growth stage (14–21 DAT)	20
Rapid growth stage (28–42 DAT)	30
Late growth stage (45-Flowering)	20

Table 6.1: Leaf Colour Chart Based N Application at Weekly Intervals-*Kharif* 2000

Time of Fertilizer Application (DAT)		H_1N_1	H_1N_2	H_1N_3	H_1N_4	H_1N_5	H_1N_6	H_1N_7	H_2N_1	H_2N_2	H_2N_3	H_2N_4	H_2N_5	H_2N_6	H_2N_7
Basal		50	–	–	–	20	20	20	50	–	–	–	20	20	20
Early growth stages (EGS)	14	–	–	20	20	–	20	20	–	–	20	20	–	20	20
	21	50	20	–	20	20	–	20	50	20	–	20	20	–	20
Rapid growth stages (RGS)	28	–	–	30	–	–	30	–	–	–	30	–	–	30	–
	35	–	30	–	30	30	–	30	–	30	–	30	30	–	30
Late growth stages (LGS)	42	50	–	30	–	–	30	–	50	–	30	–	–	30	–
	49	–	20	–	20	20	–	20	–	20	–	20	20	–	20
	56	–	–	20	20	–	20	20	–	–	20	20	–	20	20
	63	50	20	–	20	20	–	20	50	20	–	20	20	–	20
	70	–	–	20	20	–	20	20	–	–	20	20	–	20	20
Total kgha^{-1}		**200**	**90**	**120**	**150**	**110**	**140**	**170**	**200**	**90**	**120**	**150**	**110**	**140**	**170**

The total quantity of N applied as per the treatments are furnished in Tables 6.1 and 6.2. All the treatments including control received a uniform dose of 50 kg P_2O_5 and 50 kg K_2Oha^{-1}. The entire dose of P as Single Superphosphate (16 per cent P_2O_5) was applied as basal before transplanting. Potassium in the form of Muriate of Potash (60 per cent K_2O) was applied in three equal splits *viz.*, 50 per cent as basal and 25 per cent each at active tillering and panicle initiation stages.

Chlorophyll meter (SPAD 502) readings were recorded from 14 DAT in all the treatments at 7 days interval until flowering stage. The top most fully opened leaf was chosen for SPAD measurement and SPAD readings were taken on one side of the midrib of the leaf blade, mid way between leaf base and leaf tip. At early growth stages, measurements were taken near the tip of the leaf and even on the midrib, since the leaf blade was too narrow for SPAD measurements. Ten readings were taken at random for each plot and average was arrived. The correlation coefficient studies were made between LCC Vs SPAD values and between LCC V s grain yield.

Results and Discussion

Leaf Colour Chart (LCC) Values

Leaf colour chart (LCC) readings were recorded from 14 DAT to flowering stage in all the treatments at an interval at seven days (Tables 6.3 and 6.4). The two rice hybrids *viz.*, PA 6201 and CoRH2 did not exhibited any distinct colour difference among themselves at every point of LCC reading in both the seasons. This is evidenced by the observed LCC values for H_1 and H_2, their corresponding number of split doses and the total quantity of N applied to each treatment. However, variations in the LCC values were observed due to N management practices. The number of N splits across N_1 to N_7 was 4, 5, 7, 5, 6 and 7 and their corresponding quantity of N was 200, 90, 120, 150, 110, 140 and 170 kgha $^{-1}$ for both PA6201 and CoRH2 hybrids in both the seasons. Considering the growth and yield performance of the hybrids, the N_4 treatment *i.e.*, green manure @ 6.25 tha^{-1} followed by inorganic N at LCC critical value 5 based N applications was significantly superior to all other treatments. Therefore, N_4 treatment is found to be the optimum N management practice which requires seven splits of top dressing, amounting to 150 kg Nha^{-1}. This treatment resulted in a saving of 50 kg Nha^{-1} as compared to the recommended dose of 200 kg Nha^{-1} alone (N_1) besides the yield advantage.

SPAD Values

The SPAD measurements were taken in the same leaves used for LCC measurements at the same time and intervals. The observations were made across the hybrids and N application practices. The recorded SPAD values were used for the correlation studies with LCC values (Tables 6.5 and 6.6).

Correlation Coefficient Between LCC Vs SPAD Values

The correlation coefficient studies were made between LCC Vs SPAD values and between LCC Vs grain yield. The relationship between LCC Vs SPAD values revealed the existence of a strong positive correlation in both the seasons at all the stages of crop growth (Table 6.7). The r values ranged from 0.53 and 0.59 at 14 DAT to 0.95 at 49 DAT respectively for *Kharif* 2000 and 2001. Also it indicated the closeness during 28, 35, 42 and 49 DAT in both the years. However, the relationship between LCC Vs SPAD values at 14 DAT was exhibited significantly. With regard to correlation coefficient between LCC Vs Grain yield, there was a significant positive correlation coefficient (0.37 to 0.82) at all the time of observations except at 14 DAT (Table 6.8). The relationship between LCC Vs Grain yield was almost similar to that of LCC Vs SPAD in both the seasons. In case of SPAD Vs Grain yield (0.86 to 0.96) higher positive correlation was registered in all the stages of observations. Johnkutty and Palaniappan (1996) reported that the higher positive correlation coefficient between SPAD Vs Grain yield (0.75 to 0.90) in lowland rice at Coimbatore, Tamil Nadu.

Table 6.2: Leaf Colour Chart Based N Application at Weekly Intervals-*Kharif* 2001

Time of Fertilizer Application (DAT)		H_1N_1	H_1N_2	H_1N_3	H_1N_4	H_1N_5	H_1N_6	H_1N_7	H_1N_8	H_2N_1	H_2N_2	H_2N_3	H_2N_4	H_2N_5	H_2N_6	H_2N_7	H_2N_8
Basal		50	–	–	–	20	20	20	–	50	–	–	–	20	20	20	–
Early growth	14	–	–	20	20	–	20	20	–	–	–	20	20	–	20	20	–
stages (EGS)	21	50	20	–	20	20	–	20	50	50	20	–	20	20	–	20	50
Rapid growth	28	–	–	30	–	–	30	–	–	–	–	30	–	–	30	–	–
stages (RGS)	35	–	30	–	30	30	–	30	–	–	30	–	30	30	–	30	–
	42	50	–	30	–	–	30	–	50	50	–	30	–	–	30	–	50
Late growth	49	–	20	–	20	20	–	20	–	–	20	–	20	20	–	20	–
stages (LGS)	56	–	–	20	20	–	20	20	–	–	–	20	20	–	20	20	–
	63	50	20	–	20	20	–	20	50	50	20	–	20	–	–	20	50
	70	–	–	20	20	–	20	20	–	–	–	20	20	20	20	20	–
Total kgha^{-1}		**200**	**90**	**120**	**150**	**110**	**140**	**170**	**150**	**200**	**90**	**120**	**150**	**110**	**140**	**170**	**15**

Table 6.3: LCC Values Recorded at Different Days After Transplanting in the Two Hybrids as Influenced by Different N Regimes. Mean of 10 Observations in *Kharif* 2000

Hybrid/ N-Regime	*Leaf Colour Chart (LCC) Values at Different Days After Transplanting*								
	14	*21*	*28*	*35*	*42*	*49*	*56*	*63*	*70*
H_1N_1	3.5	4.0	5.0	4.5	4.0	4.5	4.5	4.5	4.5
H_1N_2	3.5	3.0	4.0	3.0	3.0	4.0	4.0	3.0	3.5
H_1N_3	3.5	4.5	3.5	4.5	4.0	4.5	4.0	4.5	4.0
H_1N_4	3.5	4.5	5.0	4.5	5.0	4.5	4.5	4.5	4.5
H_1N_5	3.0	3.0	3.5	3.0	4.0	3.0	3.5	3.0	3.5
H_1N_6	3.0	3.5	3.5	4.0	3.5	4.0	3.5	4.0	3.5
H_1N_7	3.0	3.5	5.0	4.5	5.0	4.5	4.0	4.5	4.5
H_2N_1	3.5	3.5	5.0	4.5	4.0	4.5	4.5	4.0	4.5
H_2N_2	3.5	3.0	4.0	3.0	4.0	3.0	4.0	3.0	3.5
H_2N_3	3.5	4.5	4.0	4.5	4.0	4.5	4.0	4.5	4.0
H_2N_4	3.5	4.5	5.0	4.5	5.0	4.5	4.5	4.5	4.5
H_2N_5	3.0	3.0	3.5	3.0	3.5	3.0	3.5	3.0	3.5
H_2N_6	3.0	3.5	3.5	4.0	3.5	4.0	3.5	4.0	3.5
H_2N_7	3.0	4.0	5.0	4.5	5.0	4.5	4.0	4.5	4.5

Table 6.4: LCC Values Recorded at Different Days After Transplanting in the Two Hybrids as Influenced by Different N Regimes. Mean of 10 Observations in *Kharif* 2001

Hybrid/ N-Regime	*Leaf Colour Chart (LCC) Values at Different Days After Transplanting*								
	14	*21*	*28*	*35*	*42*	*49*	*56*	*63*	*70*
H_1N_1	3.5	4.0	5.0	4.5	4.0	4.5	4.5	4.5	4.5
H_1N_2	3.5	3.0	4.0	3.0	4.0	3.0	4.0	3.0	3.5
H_1N_3	3.5	4.5	4.0	4.5	4.0	4.5	4.0	4.5	4.0
H_1N_4	3.5	4.5	5.0	4.5	5.0	4.5	4.5	4.5	4.5
H_1N_5	3.0	3.0	3.5	3.0	4.0	3.0	3.5	3.0	3.5
H_1N_6	3.0	3.5	3.5	4.0	3.5	4.0	3.5	4.0	3.5
H_1N_7	3.0	4.0	5.0	4.5	5.0	4.5	4.0	4.5	4.5
H_1N_8	3.5	4.5	5.0	4.5	5.0	4.5	4.5	4.0	4.5
H_2N_1	3.5	4.0	5.0	4.5	4.0	4.5	4.5	4.0	4.5
H_2N_2	3.5	3.0	4.0	3.0	4.0	3.0	4.0	3.0	3.5
H_2N_3	3.5	4.5	4.0	4.5	4.0	4.5	4.0	4.5	4.0

Table 6.5: SPAD Values Recorded at Different Days After Transplanting in the Two Hybrids as Influenced by Different N Regimes. Mean of 10 Observations in *Kharif* 2000

Hybrid/ N-Regime	*Chlorophyll Meter (SPAD) Values at Different Days After Transplanting*								
	14	*21*	*28*	*35*	*42*	*49*	*56*	*63*	*70*
H_1N_1	38.0	42.5	46.1	43.7	42.3	44.0	43.4	42.9	40.9
H_1N_2	36.5	36.0	38.3	36.0	38.1	35.6	37.7	35.0	36.8
H_1N_3	37.0	41.3	39.6	41.8	40.2	42.1	40.0	42.3	41.0
H_1N_4	39.5	44.0	48.4	45.0	46.5	45.1	44.8	44.0	43.2
H_1N_5	35.1	33.8	37.3	34.0	36.3	34.0	35.5	34.7	36.0
H_1N_6	35.5	39.9	38.5	39.1	37.4	40.2	38.6	41.1	38.8
H_1N_7	37.9	40.2	44.0	41.5	42.7	42.0	40.5	42.5	40.5
H_2N_1	37.2	41.0	44.5	42.3	41.2	42.9	42.1	41.8	39.5
H_2N_2	35.0	35.9	37.3	35.4	37.9	43.1	36.8	34.6	36.0
H_2N_3	36.1	40.0	38.7	40.5	39.1	41.4	39.2	41.1	40.0
H_2N_4	38.4	43.1	46.8	44.3	45.3	44.6	43.5	42.8	42.0
H_2N_5	34.2	32.7	35.6	33.7	35.0	33.8	34.6	34.0	34.0
H_2N_6	34.5	37.8	37.0	40.0	36.3	39.1	37.0	40.1	37.9
H_2N_7	36.4	39.5	42.0	40.8	41.5,	40.2	39.7	41.0	39.0

Table 6.6: SPAD Values Recorded at Different Days After Transplanting in the Two Hybrids as Influenced by Different N Regimes. Mean of 10 Observations in *Kharif* 2001

Hybrid/ N-Regime	*Chlorophyll Meter (SPAD) Values at Different Days After Transplanting*								
	14	*21*	*28*	*35*	*42*	*49*	*56*	*63*	*70*
H_1N_1	39.1	43.0	47.3	44.0	43.1	45.0	44.0	43.3	41.7
H_1N_2	36.9	35.7	39.9	37.1	39.0	36.0	38.2	35.6	37.2
H_1N_3	38.9	42.0	40.9	42.4	41.3	43.0	40.0	43.5	42.5
H_1N_4	40.0	45.2	49.1	46.7	47.0	46.0	45.0	44.5	44.0
H_1N_5	35.9	34.4	38.4	35.2	37.1	35.0	36.0	35.3	36.4
H_1N_6	36.6	40.3	39.0	40.5	38.3	41.8	39.3	42.2	39.3
H_1N_7	38.0	41.7	45.8	42.3	43.6	43.0	41.0	43.7	41.2
H_1N_8	39.6	44.4	48.1	45.6	45.4	44.4	43.9	43.8	43.1
H_2N_1	39.5	41.5	45.0	43.0	42.3	43.6	42.9	42.5	40.4
H_2N_2	35.9	36.3	38.6	36.0	38.8	36.0	37.5	35.1	36.8
H_2N_3	37.0	40.9	39.4	42.2	40.7	42.3	40.0	42.4	41.6
H_2N_4	39.1	44.6	47.9	45.0	46.2	45.5	44.4	43.6	42.5
H_2N_5	35.0	33.3	36.8	34.6	35.8	34.7	35.0	34.3	35.7
H_2N_6	35.4	38.9	38.0	40.7	37.1	40.0	38.1	41.7	38.4
H_2N_7	37.2	40.0	43.4	41.6	42.6	41.1	40.7	42.0	39.8
H_2N_8	38.2	42.7	46.5	43.9	44.1	43.4	41.8	43.4	41.9

Table 6.7: Correlation Coefficient (r) Between LCC Vs SPAD

Time of Observation (Days after transplanting)	*No. of Observation*	*Kharif 2000*	*Kharif 2001*
14	30	0.53*	0.59**
21	30	0.83**	0.92**
28	30	0.92**	0.94**
35	30	0.93**	0.93**
42	30	0.84**	0.86**
49	30	0.95**	0.95**
56	30	0.89**	0.88**
63	30	0.66**	0.94**
70	30	0.79**	0.82**

Table 6.8: Correlation Coefficient (r) Between LCC Vs Grain Yield

Time of Observation (Days after transplanting)	*No. of Observation*	*Kharif 2000*	*Kharif 2001*
14	30	0.37	0.44
21	30	0.70**	0.79**
28	30	0.73**	0.75**
35	30	0.81**	0.80**
42	30	0.62**	0.67**
49	30	0.81**	0.80**
56	30	0.71**	0.74**
63	30	0.82**	0.75**
70	30	0.79**	0.80**

*: Significant at 5 per cent level ($p = 0.05$); **: Significant at 1 per cent level ($p = 0.01$).

References

Black, C.A., 1993. *Soil Fertility Evaluation and Control*. Lewis Publishers, Boca Raton, pp. 171–175.

Greenwood, D.J., Gasta, G., Lemaire, G., Draycott, A., Millard, P. and Neeteson, J.J., 1991. Growth rate and per cent N of field grown crops: Theory and Experiments. *Ann. Bot.*, 67: 181–190.

IRRI, 1996. *Use of Leaf Colour Chart (LCC) for N Management in Rice*. Int. Rice Res. Inst., Manila 1099, Philippines.

Johnkutty, I. and Palaniappan, S.P., 1996. Nitrogen utilization growth and yield of rice as affected by green manuring and timing of isotope and non-isotope N. *Ph.D. Thesis*, Tamil Nadu Agric. Univ., Coimbatore.

Swaminathan, M.S., 1998. Rice and national food and livelihood security. Directorate of Rice Research, Silver Jubilee endowment Lecture, April 16, 1998. Punjab Agricultural University, Ludhiana.

Tanno, F., 1988. Forecasting and diagnosis methods on the plant growth for cultivars, koshihikari and sasanishiki, using multiple parameters. *Jpn. J. Soil Sci. and Plant Nutr.*, 59: 423–428.

Chapter 7

A Review of Biofertilizers and Biocides: A Best Alternative of Chemical Fertilizers and Pesticides

Deepali and Kamal K. Gangwar
Department of Zoology and Environmental Sciences,
Gurukul Kangri University, Haridwar – 249 404

In India, due to growing human population it is necessary to increase crop production and land productivity to fulfill their food requirements. It is estimated that up to 2020, total production required by country will become 321 million tones of food grains and for their proper growth nutrient requirement will become 28.8 million tones but only 21.6 million tone nutrient will be available from a deficit about 7.2 million tones. To increase crop production use of chemical fertilizers and plant protection materials is also increased. Increasing use of chemical fertilizers in agriculture make country self sufficient in food production but it pollute environment and cause slow deterioration of living beings (Saxena and Joshi, 2002a).

Plant requires essential nutrients like nitrogen (N), phosphorus (P), potassium (K), and several other minerals for their growth and receives them from soil (Arya, 2000). Nitrogen and phosphate relationship is very important and shows a direct impact on productivity of soil (Hatchinson and Richards, 1921). N,P,K fertilizers used for the crop improvement are not completely utilized by crops but excess of fertilizers washed away from land by rainwater and causes water pollution. A large amount of nutrients that are washed away in water bodies during rainy season can cause eutrophication and the whole stretch of water many become chocked.

The liberal use of chemical fertilizers especially in paddy fields may affect the growth inhabiting microorganisms (Bishara, 1978, 79; Konar and Sarkar, 1983). These fertilizers, reach into water bodies, also affect fishes (Palanichamy, 1985) and other organisms living their. In agriculture, chemical fertilizers are used extensively but they are costly and also have various adverse effects on soils *i.e.* depletes water holding capacity, soil fertility and disparity in soil nutrients. Due to insufficient uptake of these fertilizers by plants results in the leaching away from soil. Hence, it is necessary to develop some low cost fertilizers which work without disturbing nature.

Microbial Fertilizers/Biofertilizers/Biomanure

From the 1950s to 1970s considerable number of nitrogen fixing bacteria were found to be associated with crop. Several soil microbiologists suggests that nitrogen fixing bacteria associated with the plants may be the source of agronomically significant nitrogen inputs to the sugarcane crop in Brazil (Ray and Handerson, 2001). A number of microorganisms (bacteria, fungi and algae) are considered as beneficial for agriculture and used as biofertilizers. Microbial consortia are inoculated in the field for the improvement and supply of nitrogen, phosphorus, potassium and other essential elements which are necessary for the proper growth of plants. Microorganisms produce a range of extra cellular enzyme which has the potential to mediate utilization of organic sources of nitrogen and phosphorus in soil (Saxena and Joshi, 2002b).

Biofertilizers are supposed to be a safe alternative to chemical fertilizers to minimize the ecological disturbance. These are natural, organic, non-pollutant, cheap products that are required in a small dose (Gahukar, 2005–06). Microorganisms can fix atmospheric nitrogen, and solubilize insoluble phosphates, produce growth promoting substances for plants *e.g.* vitamins and hormones (Yojana, 1992). Many workers have reported that the uses of biofertilizers are beneficial for soil, as well as for crops. It is the safest method to maintain soil fertility. Biofertilizers are also called as "Microbial inoculants" (Subha Rao, 1982) due to use of many nitrogen fixing bacteria and cyanobacteria. Microbial inoculants are considered as new feature of agricultural system that facilitate or enhance the microbial process in the soil. Common biofertilizers such as *azotobacter, azospirillum, rhizobium,* blue green algae etc. are good nitrogen fixer and *Bacillus megatherium var phosphaticum, Aspergillus awamori, Penicillium digitatum* etc. are good phosphate solubilizers.

Nitrogen fixing organisms such as *azotobacter, Azotomonas, Azotococcus, Biejerinckia, cyanobacteria* (*Anabaena* spp.) can fix nitrogen under aerobic condition and can be utilized for nitrogen deficient soil, while some facultative anaerobes such as *Bacillus, Klebsiella, Rhodopseudomonas* and some species of the anaerobic genus *Clostridia* fix nitrogen under anaerobic condition (Emerich and Wall, 1985). Nitrogen fixation by nitrogen fixing microorganism is catalyzed by the enzyme 'nitrogenase' which is inhibited by free oxygen (Emerich and Wall, 1985). The reduction of atmospheric nitrogen to ammonia by the nitrogenase enzymes is expressed as follows:

$$N_2 + 8H^+ + 6e^- \rightarrow 2NH_4^+$$

Microbes are effective in inducing plant growth as they secrets plant growth promoters (auxins, abscisic acid, gibberellic acid, cytokinins, ethylene) and affects seed germination and root growth (Ramarethinum *et al.*, 2005). They also play considerable role in decomposition of organic materials and enrichment of compost.

Biofertilizers have 75 per cent moisture and it could be applied to the field directly. Biofertilizers contained 3.5 per cent–4 per cent nitrogen, 2 per cent–2.5 per cent phosphorus and 1.5 per cent potassium. In terms of N : P : K, it was found to be superior to farmyard manure and other type of manure (Mukhopadhyay, 2006).

Table 7.1: N : P : K Composition of Different Field Manure

Manure Type	Nitrogen per cent	Phosphorus As P_2O_5, per cent	Potassium as K_2O per cent
Farmyard manure	0.40	0.20	0.40
Urban compost	0.60	0.50	0.60
Green manure	0.600.70	0.100.20	1.25
Biofertilizer (50 per cent day)	1.802.40	1.001.20	0.600.80

Biofertilizers can be produced in the anaerobic digester used for production of biogas as a co-product of biogas. It consists of remaining residual solids after draining bioliquid. Excess sludge can perhaps serve as a fertilizers additive due to nitrogen fixing capability (Kargi and Ozmihci, 2004).

At present, annual production of biofertilizers is estimated as around 7000 tones from nearly 70 units and expected consumption of biofertilizers is approximately 6000 tones. In India nearly 3.5 lakh small and large biogas units have been installed during 1981–1991 (Makhopadhyay, 2006). Now Central Government also provides some financial assistance for setting up biofertilizer units.

Biofertilizers are an alternative to the conventional approach as they have lower cost than the chemical fertilizers and when they are required in bulk can be generated at the farm itself hence; these are economically attractive for the farmers (Venkatramani, 1996). The extent of cultivation depends on the market type and its proximity to the processing facility.

Transportation costs and distance to markets will affect product value and its potential use (Saxena and Joshi, 2002b).

Nitrogen Fixing Bacteria

Rhizobia

Legume plants have root nodules, where atmospheric nitrogen fixation is done by bacteria belonging to genera, *Rhizobium* (fast growing rhizobia), *Bradyshzodium* (slow growing rhizobia), *Sinorhizobium*, *Azorhizobium* and *Mesorhizobium* collectively called as rhizobia (Jordan, 1984; Chimote and Kashyap, 2001).

Rhizobium is a free living, gram negative, non-sporulating, aerobic, motile and rod shaped bacterium which occur in soil. It occurs in the roots of leguminous plants and forms nodules where it fixes nitrogen in the presence of leghaemoglobin. Leghaemoglobin promotes O_2 utilization in bacteroids and favours nitrogen fixation and converts atmospheric nitrogen into ammonia in presence of phosphorus and molybdenum (Gahukar, 2001). In the absence of leghaemoglobin nitrogen fixation can not takes place. When rhizobial culture is inoculated in field, pulse crops yield can be increased due to rhizobial symbiosis (Dubey, 2001). *Rhizobium* resides in roots of beans, grams, groundnut and soyabean etc. *Rhizobium* is a crop specific inocultant. It can fix 15–20 kg N/ha and increase crop yield upto 20 per cent. Higher Mg content due to inoculation of *Rhizobium* was observed by Kumudha (2005). Bhaskar and Kashyap (2004) reported that wild type strain of *R. ciceri* 18–7 mutant M126 and complemented mutant M126 (C_4) were characterized for symbiotic properties (*viz.*, acetylene reduction assay, total nitrogen content, nodule number and fresh and dry weight of the injected plants) and nitrogenase activity).

Azorhizobium is a stem nodule forming bacteria and for fixes nitrogen symbionts of the stem nodule also produce large amount of IAA that promotes plant growth (Ghose and Basu, 1997). They have isolated *Azorhizobium caulinodans* from the stem nodule of the leguminous emergent hydrophyte *Aeschynomene aspera* and their purpose of study was to check the ability of *Azorhizobium* spp. for EPS (extracellular polysaccharides) production for its importance in nodule symbiosis and in industry. EPS acts as determinants of host plant specificity and playa role in the first step of root hair infection in N_2 fixation (Olivares *et al.,* 1984). Ghosh and Basu (2001) also reported the plant *Aeschynomene aspera* possess many sessile stem nodule containing high amount of IAA.

Bradyrhizobium is also reported us good nitrogen fixer by various workers. Mathew *et al.* (2003) reported that when Mucunna seed were applied with an effective isolate of *Bradyrhizobium* in rubber (*Hevea brasiliensis*) Plantation, prior to souring it increased plant growth and consequently plant biomass, reduction in the weed population and increased soil microbial population were reported. It was also reported that *Bradyrhizobiuin* strain inoculation causes improvement in total organic carbon, N_2, available phosphorus and potassium in the soil.

Diazotrophs

These are aerobic chemolithotrophs and anaerobic photoautotrophs. These are non nodule forming bacteria. They include members of the families:

Azotobacteracae *e.g. Azotobacter*

They are the free living aerobic, photoautotrophic, non-symbiotic bacteria. They secretes vitamin-B complex, gibberellins, napthalene, acetic acid and other substances (Jain, 1998 and Gohukar, 2001) that inhibit certain root pathogens and improves root growth and uptake of plant nutrients. It was also reported that Azotobacter inoculation is effective in soil only in the presence of a native *Azotobacter* population. It was also reported that *Azotobacter* inoculation is effective in soil only in the presence of a native *Azotobacter* population. It occurs in the roots of *Paspalum notatum* (tropical grasses) and other spp. of this genus or other genera (Dobereiner, 1970). It adds 15–93 kg N/ha/annum on *P. notatum* roots (Dobereiner *et al.,* 1973). *Azotobacter indicunl* occurs in acidic soil in sugarcane plant roots. It can apply in cereals, millets, vegetables and flowers through seed, seedlings soil treatment.

Spirillaceae–*e.g. Azospirillum, Herbaspirillum*

These are gram negative, free living, associative symbiotic and non-nodule forming, aerobic bacteria. *Azospirillum* is a wide spread bacterium, occurs in the roots of dicots and mono cot plants *i.e.* corn, sorghum, wheat etc. (Tarrand *et al.,* 1978 and Elmerich, 1984), and responsible for nitrogen fixation in association with several cereals (Balandreau, 1983 and Vose, 1983).

It is easy to culture and identify. *Azospirillum* is found to be very effective in increasing 10–15 per cent yield of cereal crops and fixes N_2 upto 20–40 per cent kg/ha. Different *A. brasillense* strains inoculation in the wheat seed causes increase in seed germination, plumule and radicle length (Tien *et al.,* 1979; Gunasekaran and Purushothaman, 1980). *Azospirillum* inoculations cause plant growth. It may be due to either by nitrogenase activity or by its ability to produce plant growth promoters (Okon and Labandera-Gonzalez, 1994). Ray *et al.* (2004) have conducted a field experiment to assess the response of winter and autumn rice varieties to *Azospirillim brasillense,* strain Sp7 and *Azospirillum lipoferum,* strain C_2 with and without inorganic nitrogen. They have observed that there was an increase in the grain yield of all the 4 varieties as well as dry matter. They have also reported that *Azospirillim brasillense* (Sp7) inoculation gave higher per cent grain and straw yield than *A. Lipoferum* (C_2).

Herbaspirillum species occurs in roots, stems and leaves of sugarcane and rice. They produce growth promoters (IAA, Gibberillins, Cytokinins) That promotes root development and enhances uptake of plant nutrients (N, P, K). In legume plants, *Azospirillum* fixes atmospheric nitrogen.

Acetobacter Diazotrophicus

Another diazotroph is *Acetobacter diazotrophicus* occurs in roots, stem and leaves of sugarcane and sugar beat crops as nitrogen fixer and applied through soil treatment. It also produces growth promoters *e.g.* IAA. That helps in nutrients uptake, seed germination, and root growth. This bacterium fixes nitrogen upto 15 kg/ha/year and enhance upto 0.5–1 per cent crop yield (Gahukar, 2005–06). It releases various organic acids (Mahesh Kumar, 1999), succinate, tarterate, citrate and gluconate etc. which generated H^+ ion that dissolve mineral phosphate and make it easily available to plants. Mowade and Bhattacharya (2000) have reported that organism have the capacity to solubilising insoluble phosphate in Pikovskay's medium and Sperber's medium and tested for its response to different antibiosis under in vitro condition.

Cyanobacteria (Blue Green Algae)

Nostoc, Anabaena, Oscillatoria, Aulosira, Lyngbya etc. are the prokaryotic organisms and phototropic in nature. They play an important role in enriching paddy field soil by fixing atmospheric nitrogen and supply vitamin B complex and growth promoting substance (Sharma, 1986) which makes the plant grow vigorously. In paddy field there is no need of cyanobacterial inoculation. They also convert insoluble phosphorus into soluble form by excreting organic acids. Cyanobacteria fixes 20–30 kg/N/ha and increase 10–15 per cent crop yield when applied at 10 kg/ha. Cyanobacteria oxygenate the water impounded in the field. The cyanobacterial mat in paddy fields also reduces loss of moisture from the soil. Blue green algae can be used in the form of flakes or these flakes can be powdered and mixed with farmyard manure and soil then can be applied in the field. Algal flakes are dried and mixed at the rate of 10^{15} kg/ha, after one week rice is planted (Arya, 2000).

Azolla-Anabaena Symbiosis

It is a free floating, aquatic fern found on water surface having a cyanobacterial symbiont *Anabaena azollae* (heterocystous) in their leaves. Azolla fixes atmospheric nitrogen in association with nitrogen fixing cyanobacteria in paddy field and excrete organic nitrogen in water during its growth and also immediately upon trampling. Azolla is used as biofertilizer in India, USA, Sri Lanka, China, Indonesia, Bangladesh and many other countries. Azolla fixes 40, 60 kg N/ha in a month and increases 10–20 per cent yield cif paddy crops (Kumar, 2004).

Azolla contributes nitrogen, phosphorus (15–20 kg/ha/month), potassium (20–25 kg/ha/month) and organic carbon etc. and when applied in rice field also suppresses weed growth. It is also susceptible to high temperature (>40°C) and scarcity of water (Gahukar, 2005–06). Azolla also absorbs traces of potassium from irrigation water. Azolla can be used as green manure before rice planting. Azolla spp. are metal tolerant hence, can be applied near heavy metal polluted areas.

Phosphate Solubilising Bacteria

Pseudomonas fluorescens, Bacillus megatherium var. phosphaticum, Acrobacter acrogens, Nitrobacter spp., *Escherichia freundii, Serratia* spp., *Pseudomonas striata, Bacillus polymyxa* are the bacteria have phosphate solubilising ability.

'Phosphobacterin' are the bacterial fertilizers containing cells of *Bacillus megatherium var. phosphaticum* prepared firstly by USSR scientists. They increased about 10 to 20 per cent crop yield (Cooper, 1959).

It can be applied for low land and upland rice. They dissolve soil phosphate and make it available for crops. A large number of soil microorganisms have the ability to solubilize inorganic phosphates through their metabolic activity directly or indirectly (Arya, 2000). They produces plant growth promoting hormones *e.g.* IAA, GA and various organic acids *e.g.* lactic acid, citric acid, fumaric acid, succinic acid etc. These organic acids help in phosphate solubilising activity of soil.

Phosphate Solubilizing Fungi

Some fungi also have phosphate dissolving ability *e.g. Aspergillus niger, Aspergillus awamori, Penicillium digitatum* etc.

Plant Growth Promoting Rhizobacteria (PGPR)

Recently ability of plant growth promoting rhizobacteria is also studies for microbial control. They are also called as microbial pesticides *e.g. Bacillus* spp. and *Pseudomonas fluorescens*. Sipirin (2000) reported that the growth and survival or vetiver is possible without nitrogen and phosphorus application especially in the infertile soil with the help of diazotrophs including the genera of *Pseudomonas*. Chakraborti *et al.* (2003) reported that four bacterial spp. *Serratia marcescens* TR10, *Ochrobactrum ocnthropi* TR9, *Bacillus pumilus* TR24 and Bacillus spp. TR16, isolated from rhizosphere of tea plant have antagonistic ability against the root pathogens of the tea plant. They have also reported that *Srratia* spp. and *Ochrobactrum* spp. are able to promote growth of plants. It was founded by Paul *et al.* (2003) that *Pseudomonas fluorescens* application to the black pepper rhizophere resulted in easy mobilisation of the essential nutrients in the rhizosphere microcosm and resulted in enhanced uptake of nutrients, which reflected in increased plant biomass.

A study was carried out by Joseph *et al.* (2003) in the Kerala from 10 rubber growing areas and isolated 17 fluorescent rhizobacterial strains and founded that all the 17 isolates shows antagonism against 5 pathogens of rubber. They also observed that these organism also improve the growth of *H. brasiliensis* and associated crop cover *i.e.* shoot weight, root weight, nodulation, nitrogenase activity.

Mycorrhiza

Mycorrhizas are developed due to the symbiosis between some specific root inhabiting fungi and plant roots. This means these are fungi grow on roots of the plants. There are specific fungi for vegetables, fodder crops, flowers, trees etc. Mycorrhizas are used as biofertilizers (Kumudha, 2005) and biocides. They absorb nutrients such as manganese, phosphorus, iron, sulphur, zinc etc. from the soil and pass it on to the plant. They also show higher tolerance to high soil temperature. Mycorrhizal fungi increase the yield of crop fields by 30–40 per cent. Mycorrhiza also produces plant growth promoting substances. It was also reported that seedlings having mycorrhizal fungi growing on their roots grow faster after inoculation. Mycorrhiza occurs in low land and upland rice and it mobilize phosphorus required by rice.

Commonly mycorrhizas are of two types: Ectomycorrhiza and Endomycorrhiza. Both are different in structure and systematic position of fungi.

Ectomycorrhiza

They occur in the roots of the higher plants *e.g.* gymnosperms and angiosperms. These are the fungi requires pH 5–6 for their proper growth. Excessive use of inorganic fertilizers and shading supresses the development of ectomycorrhizal fungi. The poor development of mycorrhizal fungi in the roots of plants causes stunted growth and chlorosis of leaves.

Endomycorrhiza (VAM Fungi)

They occur commonly in the roots of crop plants. These fungi penetrate the cortical cells and get established them intracellularly by secreting extracellular enzymes. VAM fungal hyphae enhance the uptake of phosphorus and other nutrients that is a responsible for plant growth stimulation including roots and shoot length (Hayman, 1980). VAM also enhances the growth of black pepper, protects from *Phytophthora capsid, Radopholus similis* and *Melvidogyne incognita* (Anandraj *et al.,* 2001). It was reported by Kumudha (2005) that VAM inoculation was found to be effective in chlorophyll pigmentation. Trappe and Fogel (1977) reported that the increased nitrogen uptake in VAM might also be due to the increased phosphorus uptake which in turn might enhances the activity of NAD dependent enzyme which might contribute to nitrate reductase activity. VAM fungi enhance water uptake in plants and also provide heavy metals tolerance to plants.

Benefits from Biofertilizers

They are environmentally sound, pollution free and of low cost. They increase crop yield upto 10–40 per cent and fix nitrogen upto 40–50 kg. After using 3–4 years continuously, there is no need of application of biofertilizers because parental inoculums are sufficient for growth and multiplication. They improve soil texture, pH, and other properties of soil. They produces plant growth promoting substances *e.g.* IAA amino acids, vitamins etc.

Precautions

1. They are living organism hence, handling should be careful.
2. Biofertilizers should be used before expiry date.
3. They are species specific hence, particular biofertilizer should be used for a particular crop plant.
4. Recommended dose of biofertilizer should be applied.
5. They should not be exposed to direct sunlight.

References

Arya, A., 2000. Biofertilizer for sustainable plant development. *Everyman's Science,* 35(1): 19–25.

Anandraj, M., Venugopal, M.N., Veena, S.S., Kumar, A. and Sarma, Y.R., 2001. Eco-friendly management of disease of spices. *Indian Species,* 38(3): 28–31.

Bhaskar, V.V. and Kashyap, L.R., 2004. Azide resistance in *Rhizobium ciceri* linked with superior symbiotic nitrogen fixation. *Ind. J. Exp. Biol.,* 42: 1177–2285.

Bishara, N.F., 1978. Fertilizing fish ponds II growth of *Mugil cephla* in Egypt by pond fertilization and feeding. *Aquaculture,* 13: 361–367.

Bishara, N.F., 1979. Fertilizing fish ponds III growth of *Mugil apito* in Egypt by pond fertilization and feeding. *Aquaculture,* 16: 47–55.

Balandreau, J., 1983. Microbiology of the association. *Can. J. Microbiol.,* 29: 851–859.

Chakraborty, V., Chakraborty, B.N., Roy Chawdhary, N., Tongden, C. and Basnet, M., 2003. Investigations on plant growth promoting rhizobacteria of tea rhizosphere. *6th International PGPR Workshop,* 5–10 October, Calicut, India.

Chimote, V. and Kashyap, L.R., 2001. Lipochito oligosaccharides and legume rhizobium symbiosis: A new concept. *Indian J. Exp. Biol.,* 39: 401–409.

Cooper, 1959. *Soil Fertilizers*, 22: 227 – 233.

Dobereiner, J., 1970. *Zent. Bakteriol* II. 124: 224–230.

Dobereiner, J., Day, J.M. and Dart, P.J., 1973. *Perg. Agr Pe. Bras*, 8: 153–157.

Dubey, R.C., 2001. *A Textbook of Biotechnology*. S. Chand and Company Ltd. New Delhi.

Elmerich, C., 1984. *Biotechnology*, 2: 967–978.

Emerich, D.W. and Wall, J.D., 1985. Nitrogen fixation In: *Comprehensive Biotechnology*, Vol. 4, (Ed.) M. Moo-young. Pergamon Press, N.Y., pp. 73–106.

Gahukar, R.T., 2001. *Kisan World*, 28(3): 25–27.

Gahukar, R.T., 2005–06. Potential and use of biofertilizers in India. *Everyman's Science*, 40(5): 354–361.

Ghosh, A.C. and Basu, P.S., 1997. Culture growth and IAA production by a microbial diazotropic symbiont of stem-nodule of the legume *Aeschynomene aspera. Folia. Microbiol.*, 42: 595.

Ghosh, A.C. and Basu, P.S., 2001. Extracellular polysaccharide production by *Azorhizobium caulinodans* from stem nodules of Leguminous emergent hydrophyte, *Aeschynomene aspera. Ind. J. Exp. Biol.*, 39: 155–159.

Gunasekaran, S. and Purushothaman, D., 1980. Nitrogen fixation by Azopirillum in the rhizophere of cotton. In: *National Symposium on Biological Nitrogen Fixation in Relation to Crop Production*, T.N. Agric University, Coimbtore, pp. 27.

Hayman, D.S., 1980. Mycorrhiza and crop production. *Nature*, London, 287: 487–488.

Hutchinson, H.B. and Richards, E.H., 1921. *J. Ministry of Agriculture*, 28: 398.

Jain, R., 1998. *Millennium Biofertilizer Guide*. Nargi Printer, New Delhi, pp. 52.

Jordan, D.C., 1984. *Bergeys Manual of Systematic Bacteriology*, pp. 234.

Joseph, K., Sunju, Y., George, J., Mathew, J. and Kunivilla Jacob, C., 2003. Plant growth promoting Rhizobacteria in Rubber (*Hevea brasiliensis*) plantations. In: *6th International PGPR Workshop*, 5–10 October, Calicut, India.

Konar, S.K. and Sarkar, S.K., 1983. Acute toxicity of agricultural fertilizers to fish. *Geobios*, 10: 6–9.

Kargi, F. and Ozmihci, S., 2004. Batch biological treatment of nitrogen deficient synthetic wastewater using Azotobacter supplemented activated sludge. *Bioresource Technology*, 94: 113–117.

Kumar, N., 2004. *Indian Farmer's Digest*, 37(2): 7–9.

Kumudha, P., 2005. Studies on the effect of biofertilizers on the germination of *Acacia nilotica* Linn. *Seeds Ad. Plant Sci.*, 18(11): 679–684.

Mahesh Kumar, K.S. *et al.*, 1999. *Curr. Sci.*, 76: 874–875.

Mathew, J., Joseph, K., Lakshmanan, R., Jose, G., Kethandaraman, R. and Kuruvilla Jacob, C., 2003. Effect of Bradyrhizobium inoculation of *Mucuna bracteala* and its impact on the properties of soil under Hevea. In: *6th International PGPR Workshop*, 5–10 October, Calicut, India.

Mowade, S. and Bhattacharya, P., 2000. Resistance of P-solubilosing Acetobacter diazotrophicus to antibiotics. *Curr. Sci*, 79(II).

Mukhopadhyay, S.N., 2006. Eco-friendly products through process biotechnology in the provision of biotechnology economy: Recent advances. *Technorama*, A. Supplement to IEI News, March.

Okon, Y. and Labandera-Gonzalez, C.A., 1994. Agronomic application of Azospirillum: An evaluation of 20 years worldwide field inoculation. *Soil Biol., Biochem.*, 26: 1591–1601.

Olivares, J., Bedman, E.J. and Martinez, M.E., 1984. Infertility of *Rhizohium melilotias* affected by extracellular polysaccharides. *J. Appl. Bacteriol.*, 56: 389.

Palanichamy, S., Seeniammal, K. and Arunachalam, S., 1985. Effects of agricultural fertilizers complex on food consumption and growth in the fish *Sarotherodon mossambicum* (Trewaves). *J. Env. Biol.*, 6(2): 71–76.

Paul, D., Srinivasan, V., Anandraj, M. and Sarma, Y.R., 2003. *Pseudomonas fluorescens* mediated nutrient flux in the black pepper Rhizosphere microcosm and enhanced plant growth. In: 6th *International PGPR Workshop*, 5–10 October, Calicut, India.

Ramarethinam, S., Murugusen, N.V. and Rajalakshmi, N., 2005. *Pestology*, 20(4): 12–14.

Ray, N., Baruah, R. and Goswami, S.R., 2004. Effect of seedling bacterization of wetland rice with Azospirillum. *Ind. J. Environ. and Ecoplan.*, 8(3): 843–848.

Ray, R.N. and Handerson, G., 2001. Endophytic nitrogen fixation in sugarcane: Present knowledge and future applications, Technical expert meeting on increasing the use of biological nitrogen fixation (BNF) in agriculture, FAO, Rome.

Saxena, P. and Joshi, N., 2002a. A comparative study of compost formation by the *Mucor* sps. and *Penicillium* sps. on the soil. *Him. J. Env. Zool.*, 16(1): 83–96.

Saxena, P. and Joshi, N., 2002b. In: Role of microorganisms in the decomposition of organic wastes matter. *Ph.D. Thesis*, Gurukul Kangri University, Haridwar.

Sipirin, S., Thinathorn, A., Pintarak, A. and Aibcharoen, P., 2000. Effect of associative nitrogen fixing bacterial inoculation on growth of vetiver grass. A poster paper presented at ICV–2.

Subha Rao, N.S., 1982. In: Biofertilizers. In: *Advances in Agriculture Microbiology* (Ed.). Oxford and IBH Publication, New Delhi, pp. 219–242.

Sharma, V.K., 1986. A review of recent work on pesticide studies on the nitrogen fixing algae. *J. Env. Biol.*, 7(3): 171–176.

Tarrand, J.J., Kreig, N.R. and Dobereiner, J., 1978. A taxonomic study of the spirillum lipoferum groups, with description of a new genus, *Azospirillum gen. novo* and two sps. *Azosperillum brasiliense* sp novo. *Can. J. Microbiol.*, 24: 967–980.

Tien, T.M., Gaskins, M.H. and Hubbeli, D.H., 1979. Plant growth substances produced by *Azospirillim brasilense* and their effect on the growth of pearl millet (*Penniselum americanum* L.) *Appl. Environ. Microbiol.*, 37: 1016–1024.

Trappe, J.M. and Fogel, R.S., 1977. In: Ecosystem function of mycorrhizae in the below ground ecosystem: A synthesis of plant associated process, (Ed.) J.K. Marshall. Colarado State University, Range, Sci, Deptt. Sci. Services, No. 26 Fur Collins.

Venkatramani, G., 1996. In: *Hindus Survey of Indian Agriculture*, 29.

Vose, P.B., 1983. Development in non legume N_2 fixing systems. *Can. J. Microbial.*, 29: 837–850.

Yojana, 1992. In: Biofertilizers. In: *Advances in Agricultural Microbiology*, (Ed.) N.S. Subba Rao. Oxford and IBH Pub. Co., New Delhi, pp. 219–242.

Chapter 8

Effect of Enriched Municipal Solid Waste Compost Application on Soil Available Micronutrients

R. Kavitha* and P. Subramanian
Department of Environmental Sciences,
Tamil Nadu Agricultural University, Coimbatore – 641 003, Tamil Nadu, India

ABSTRACT

A field experiment was conducted during Kharif 2004 to investigate the influence of enriched municipal solid waste compost application on available micro-nutrient status of the soil in rice crop. There were seven treatment combinations in this experiment. The DTPA extractable cationic micronutrients (Cu, Zn, Mn and Fe) were significantly increased with use of manures and fertilizers. Micronutrient availability was highest in treatment of 100 per cent N through Enriched Municipal Solid Waste Compost (EMSWC) applied plot.

Keywords: *Enriched municipal solid waste compost, Micronutrients.*

Introduction

In urban areas, especially in the rapid urbanizing cities of the developing world, problems and issues of Municipal Solid Waste Management (MSWM) are of immediate importance. Reuse of organic waste material, often contributing to more than 50 per cent of the total waste amount, is still fairly limited but often has great recovery potential. Composting is the simplest yet best process for solid

* Corresponding Author: E-mail: kavithaenviro2004@yahoo.com.

waste management for our condition. City compost produced at mechanical composting plants throughout the Asia-Pacific region (India, Nepal, Pakistan, Philippines, Indonesia and Thailand) are generally low in plant nutrients and therefore their acceptability by farmers has been poor (Gaur and Singh, 1993). Hence enrichment is necessary for improving nutrient status and quality of compost. Enrichment of compost by nitrogen fixing bacteria and P solubilizing fungi is one of the ways of improving nutrient content of the final product. In recent past, chemical fertilizers have played very significant role in providing nutrients for intensive cropping, using high yielding varieties, which heralded the green revolution in the country. The continuous use of high analysis fertilizers in an unbalanced manner has resulted in additional problems of soil fertility. This progressively impoverishment the soils of their native nutrient reserves, leading to the appearance of the multiple nutrient deficiencies, in particular micronutrient deficiencies. So proper blending of chemical fertilizers with organic manure prepared from available materials will not only improve soil health but also help to maximize sustainable production (Gopal Reddy and Suryanarayana Reddy, 1999). Hence, the present study was undertaken to evaluate the changes in available micronutrient under enriched municipal solid waste compost and inorganic fertilizer combinations.

Materials and Methdos

The Compost utilized for the experiment was provided by the mechanized municipal solid waste composting plant. The Compost was enriched with poultry litter (10 per cent), spent wash (10 per cent), rock phosphate (0.5 per cent) and microbial consortium (0.5 per cent) consist of *Azotobacter*, Phosphobacteria and *Pseudomonas*. The chemical properties of the enriched and unemiched compost material are shown in the Table 8.1.

Table 8.1: Chemical Components in the MSW Compost

Sl.No.	Parameters	Concentration	
		Unenriched MSW Compost	Enriched MSW Compost
1.	pH	8.10	7.88
2.	EC (dS m^{-1})	2.23	4.88
3.	Organic Carbon (per cent)	11.25	10.44
4.	Nitrogen (per cent)	0.64	1.75
5.	Phosphorus (per cent)	0.56	1.16
6.	Potassium (per cent)	0.70	1.83
7.	Calcium (per cent)	2.32	4.18
8.	Magnesium (per cent)	0.65	0.97
9.	Sodium (per cent)	0.66	0.75
10.	Copper (mg kg^{-1})	317.00	223
11.	Zinc (mg kg^{-1})	1303.00	2110
12.	Iron (mg kg^{-1})	12659.00	16923.75
13.	Manganese (mg kg^{-1})	301.00	345.05

A field experiment was conducted during kharif 2004 at Wetland farm of Tamil Nadu Agricultural University, Coimbatore that is geographically situated in the North Western part of Tamil Nadu at 11° N latitude and 77° E longitude at an altitude of 426.72 m above mean sea level (MSL). The soil of the experimental field was deep, moderately drained, clay loam and taxonomically classified as Typic haplustalf. The soil nutrient status was low, medium and high for available nitrogen, phosphorous and potassium respectively.

The composite soil samples collected prior to the experiment were analysed. Table 8.2 shows selected physical and physio-chemical properties of the soil. The experimental field was irrigated with good quality water from a bore-well situated near by the field. The field experiment was conducted using rice variety ADT 43 with duration of 110 days. The experiment was laid out in a Randomized Block Design with three replications. All the treatments were allotted at random to plots within each replication. The experimental layout was kept undisturbed throughout the course of investigation. The treatments were applied with compost just before the transplanting of the crop. The treatment details are given below.

Table 8.2: Physical and Physio-chemical Properties of the Soil of the Experimental Field

Sl.No.	Parameters	Values
1.	Bulk density (g cc^{-1})	1.11
2.	Particle density (g cc^{-1})	1.81
3.	Porosity (per cent)	44.44
4.	pH	8.15
5.	EC (dS m^{-1})	0.56
6.	Organic carbon (per cent)	0.63

Treatment Details

T_1: Recommended dose of NPK through Inorganic fertilizers (120 : 38 : 38 kg ha^{-1} of N,P,K)

T_2: 100 per cent N through Municipal Solid Waste Compost (MSWC)

T_3: 100 per cent N through Enriched Municipal Solid Waste Compost (EMSWC)

T_4: 75 per cent N through Inorganic fertilizers + 25 per cent N through MSWC

T_5: 75 per cent N through Inorganic fertilizers + 25 per cent N through EMSWC

T_6: 5 t of EMSWC + 25 per cent N through Inorganic fertilizers

T_7: 5 t of Vermicompost + 25 per cent N through Inorganic fertilizers.

Compost Analysis

Total nitrogen content of the sample was estimated using the Microkjeldahl's method, as explained by Humphries (1956) and Phosphorus content by Vanadomolybdate method (Jackson, 1973). Total potassium was estimated (Jackson, 1973) by feeding the triacid extract in the flame photometer and the readings were recorded. Total sodium content was estimated through same procedure adopted for estimation of potassium. Total calcium and magnesium were determined using versenate titration method (Tandon, 1995). The estimation of micronutrients *viz.*, Cu, Zn, Mn and Fe was done by feeding the tri acid extract in the Atomic Absorption Spectrophotometer by using appropriate cathode lamp and slit width (APHA, 1989).

Soil Analysis

Pre-planting composite soil samples were taken from the experimental field from a depth of 0–20 cm. Similarly plot wise soil samples were collected from 0–20 cm depth at 30 DAT (Days after transplanting), 60 DAT and at harvest, dried under shade, sieved through 2 mm sieve and analyzed for chemical properties.

DTPA Extractable Micronutrients

For the estimation of available micronutrients (Cu, Zn, Mn, Fe), 10 g of air-dried soil was weighed and taken in a 150 ml shaking bottle and 20 ml of 0.5 M DTPA extract was added and kept for 2 hours shaking. The contents were then filtered using Whatman No. 42 filter paper and the extract was directly fed into the Atomic Absorption Spectrophotometer using appropriate cathode lamp and slit width (Lindsay and Norvell, 1978). The readings (X) were taken and the available micronutrient content was calculated. Analysis of variance was used to test significance ($P < 0.05$) of treatment effects.

Results and Discussion

Available Copper

The available copper content of soil in the treatments followed an increasing trend up to 60 DAT and then declined at harvest stage. The copper content ranged from 3.09 to 4.25, 2.93 to 4.85 and 2.00 to 3.27 mg kg^{-1} at 30 DAT, 60 DAT and at harvest stages respectively (Table 8.3). The highest value was observed on 60th day after planting (4.85 mg kg^{-1}) in T_3 (100 per cent N through EPGOM) treatment and the lowest value was observed at harvest stage (2.0 mg kg^{-1}) in the T_1 (absolute control) treatment. Among the different treatments the highest mean copper content of 4.12 mg kg^{-1} was recorded in T_3 (100 per cent N through EPGOM), which was significantly differed all other treatments.

Table 8.3: Effect of Enriched POABS Green Organic Manure on DTPA Extractable Copper and Zinc (mg kg^{-1})

Treatments	*Copper*				*Zinc*			
	30 DAT	*60 DAT*	*At harvest*	*Mean*	*30 DAT*	*60 DA T*	*At harvest*	*Mean*
T_1	3.09	2.93	2.00	2.67	4.33	3.27	2.23	3.28
T_2	4.06	4.18	3.08	3.77	6.08	6.43	6.70	6.40
T_3	4.25	4.85	3.27	4.12	6.23	6.50	7.03	6.59
T_4	3.26	3.80	3.12	3.39	4.39	5.13	5.24	4.92
T_5	3.10	3.87	3.17	3.38	5.48	5.77	5.99	5.75
T_6	3.13	4.06	3.14	3.44	4.06	5.58	6.09	5.24
T_7	2.14	3.63	2.15	2.64	4.64	5.00	5.39	5.01
Mean	3.29	3.90	2.85	3.35	5.03	5.38	5.52	5.31
	SEd			*CD (0.05)*	*SEd*			*CD (0.05)*
T	0.043			0.088	0.098			0.198
D	0.029			0.058	0.064			0.129
T × D	0.075			0.153	0.170			0.344

T_1: Recommended dose of NPK through Inorganic fertilizers (120 : 38 : 38 kg ha^{-1} of N, P, K); T_2: 100 per cent N through Municipal Solid Waste Compost (MSWC); T_3: 100 per cent N through Enriched Municipal Solid Waste Compost (EMSWC), T_4: 75 per cent N through Inorganic fertilizers + 25 per cent N through MSWC; T_5: 75 per cent N through Inorganic fertilizers + 25 per cent N through EMSWC; T_6: 5 t of EMSWC + 25 per cent N through Inorganic fertilizers; T_7: 5 t of Vermicompost + 25 per cent N through Inorganic fertilizers.

Available Zinc

The available zinc content of rhizosphere soil over various crop growth stages ranged from 4.06 to 6.23, 3.27 to 6.50 and 2.23 to 7.03 at 30 DAT, 60 DAT and at harvest respectively (Table 8.3). The content appeared to show an increasing trend with the advancement of crop growth stages. The individual and combined application of compost with inorganic fertilizer recorded higher zinc content over application of inorganic fertilizer alone at all crop growth stages. The mean zinc content was highest in the T_3 (100 per cent N through EPGOM) treatment with a value of 6.59 mg kg^{-1}, which was significantly different from all other treatments. The T_1 (absolute control) treatment had the lowest mean value of zinc (3.28 mg kg^{-1}).

Available Manganese

The changes on soil available manganese during the crop growth stages are furnished in Table 8.4. It ranged from 12.57 to 23.17, 10.20 to 16.60 and 8.03 to 19.03 mg kg^{-1} at 30 DAT, 60 DAT and at harvest stages respectively. The individual and combined application of compost could register significantly higher manganese content at all stages of crop growth. At 30 DAT, T_3 (100 per cent N through EPGOM) recorded the highest value of 23.17 mg kg^{-1} and this was followed by T_2 (100 per cent N through PGOM). The lowest value of 12.57 mg kg^{-1} was observed in T_1 (absolute control). Similar trend was noticed at later stages of crop growth.

Table 8.4: Effect of Enriched POABS Green Organic Manure on DTPA Extractable Manganese and Iron (mg kg^{-1})

Treatments	*Manganese*				*Iron*			
	30 DAT	*60 DAT*	*At harvest*	*Mean*	*30 DAT*	*60 DA T*	*At harvest*	*Mean*
T_1	12.57	10.20	8.03	10.27	19.17	17.03	16.97	17.72
T_2	20.87	15.60	17.80	18.09	22.27	31.00	22.97	25.41
T_3	23.17	16.60	19.03	19.60	23.40	32.90	23.17	26.49
T_4	14.00	11.33	13.73	13.02	20.53	25.63	18.00	21.39
T_5	17.83	14.23	15.20	15.75	21.30	26.57	21.30	23.06
T_6	18.20	14.10	15.87	16.06	21.87	27.80	22.33	24.00
T_7	15.17	12.40	13.90	13.82	20.80	25.60	20.33	22.24
Mean	17.40	13.49	14.79	15.23	21.33	26.65	20.72	22.90
	SEd			*CD (0.05)*	*SEd*			*CD (0.05)*
T	0.339			0.6847	0.422			0.852
D	0.2217			0.4482	0.276			0.558
T × D	0.5867			1.1859	0.730			1.476

T_1: Recommended dose of NPK through Inorganic fertilizers (120 : 38 : 38 kg ha^{-1} of N, P, K); T_2: 100 per cent N through Municipal Solid Waste Compost (MSWC); T_3: 100 per cent N through Enriched Municipal Solid Waste Compost (EMSWC), T_4: 75 per cent N through Inorganic fertilizers + 25 per cent N through MSWC; T_5: 75 per cent N through Inorganic fertilizers + 25 per cent N through EMSWC; T_6: 5 t of EMSWC + 25 per cent N through Inorganic fertilizers; T_7: 5 t of Vermicompost + 25 per cent N through Inorganic fertilizers.

Available Iron

The available iron content ranged from 19.17 to 23.40, 17.03 to 32.90 and 16.97 to 23.17 mg kg $^{-1}$ at 30 DAT, 60 DAT and harvest stages respectively (Table 8.4). The iron content did not show much variation at initial crop growth stages, however it shot up conspicuously at 60 DAT with a progressive decline thereafter. The effect of compost application in increasing the iron content as against inorganic fertilizer was prominent in the treatments. The iron content was maximum in T_3 (100 per cent N through EPGOM) and minimum in T_1 (absolute control) at all stages of crop growth *viz.*, 30 DAT, 60 DAT and at harvest stages respectively.

There was significant increase in the amount of iron, copper, manganese and zinc extractable from soils after compost application. The increase might be attributed due to the addition of enriched compost, which is having higher micronutrient level. Similar results were reported by Cottrell (1975) and Mohr (1979) with respect to zinc and copper and by Giordano *et al.* (1975) with respect to manganese and iron. In accordance with the above result, Gallardo *et al.* (1984) found that increasing application of town refuse compost linearly increased the residual extractable zinc in two soils of different fertility, while similar effects for iron were recorded in the low fertility soil, but not in the other soil.

Conclusion

City waste Compost contains appreciable amounts of available copper, zinc, manganese and iron. Micronutrients along with microorganisms playa very important role in supplying adequate amount of copper, zinc, manganese and iron to soil and plant ecosystem. Therefore the application of enriched compost alone and in combination with inorganic fertilizer has prominently increased the level of micronutrient in the soil.

References

APHA (American Public Health Association), 1989. *Standard Methods for the Examination of Water and Wastewater*, 19th edn. Washington, D.C.

Tandon, H.L.S., 1995. *Methods of Analysis of Soils, Plants, Waters and Fertilizers*. Fertilizer Development and Consultation Organization, New Delhi, pp. 143.

Cottrell, N.M., 1975. Disposal of municipal wastes on sandy soil: Effect of plant nutrient uptake. *M.Sc. Thesis*, Oregon State University, Corvallis, Oregon.

Gallardo-Lara, F., Robles, J., Esteban, E., Azcon, M. and Nogales, R., 1984. Poder fertilizante de un compost de basura urbana. II. Efecto directo y residual sobre la asimilabilidad de Fe y Zn. In: *Proceedings of I Congreso Nacional de la Cienica del Suelo*, Vol. 1., SECS, Madrid, p. 393–403.

Giordano, P.M., Mortvedt, J.J. and Mays, D.A., 1975. Effect of municipal wastes on crop yields and uptake of heavy metals. *J. Environ. Qual.*, 4: 394–399.

Gaur, A.C. and Singh, G., 1982. Influences of Azotobacter and rock phosphate on enriching mechanized compost. In: *Recycling of Crop, Animal and Industrial Waste in Agriculture*, (Ed.) H.L.S. Tandon, pp. 12.

Gopal Reddy, B. and Reddy, M. Suryanarayana, 1999. Effect of integrated nutrient management on soil available micronutrients in maize-soyabean cropping sequence. *J. Res. ANGRAU*, 27(3): 24–28.

Humphries, E.C., 1956. Mineral components and ash analysis. *Modem Method of Plant Analysis*, Springer-Verlag, Berlin, p. 468–502.

Jackson, M.L., 1973. *Soil Chemical Analysis*. Prentice-Hall of India Private Limited, New Delhi, pp. 498.

Lindsay, W.L. and Norvell, W.A., 1978. Development of DTPA soil test for zinc, iron manganese and copper. *Soil Sci. Soc. Am. J.*, 42: 421–428.

Mohr, H.D., 1979. Effect of garbage sewage-sludge compost on the heavy metal content of vineyard soils, grapevines organs and must. *Weinberg Keller*, 26(8): 333–344.

Chapter 9

Ecological Enumeration of Tree Vegetation in and Around Hirakud Township and Strategy for Future Plantation Programme

P.C. Mishra[1], S.P. Mishra[1], A.S.P. Mishra[2] and Niranjan Behera[1]
[1]Department of Environmental Sciences, School of Life Sciences, Sambalpur University, Jyoti Vihar - 768 019, Orissa
[2]Hindalco, Hirakud, Orissa

ABSTRACT

A detailed study was undertaken in Hirakud industrial town of Orissa to enlist the tree species, both naturally grown and planted and to find out frequency density, girth analysis and canopy size. *Cassia spectabilis* was the dominant species. Other co-dominant species are *Acacia auriculoformis, Zizyphus jujuba, Graviella robusta, Azadiracta indica* and *Mangifera indica* is the dominant canopy bearing species in the area. Based on the relative density and canopy cover combined, important value index (IVI) for different species were calculated. The values showed a range of 0.9–24.56.*Casia spectabilis* is the dominant species in the area with a IVI of 24.56. Total canopy coverage of different sites was estimated to be 1895439 m^2 over a total plantation area of 62 ha. The total CO_2 uptake by the green vegetation in a day amounted to 225.18 tons per day or 82189.85 tons per annum. The dust collecting capacity of the entire vegetation of Hirakud complex was calculated to be 69.75 tonnes. Assuming that the turn over rate of dust collection by leaves is 2–4 times a year because of weather parameters,the actual dust collection may vary from 140–280 tons per annum in the complex. The various dust tolerant and air pollution tolerant species have been recommended for plantation in the town.

Keywords: *Tree Vegetation, Carbon sequestration, Dust collecting capacity.*

Introduction

The concept of improved quality of life in urban habitat was earlier confined to proper solid waste disposal, wastewater treatment, water supply, roads and transport, access to institutional, commercial and trading facilities. But presently, the need for integrating environmental management by way of plantation, setting up of parks and gardens and maintenance of urban wetlands is getting adequate attention (Juwarkar *et al.*, 2006). A green city is not merely for a visual, aesthetic pleasure but to give the citizens a chance to breath clear air, desist serious cardio-pulmonary diseases and help citizens to appreciate nature and biodiversity. It also reduces the impact of air pollution on human health. Additionally, with urban greening a significant shelter for indicators of biodiversity can be created *viz.*, birds and butterflies. Urban greening can also help in controlling microclimate and recharging of groundwater and above all it can have immense bio-aesthetic value to augment quality of life. Therefore, the urban/industrial greening programme has to combine the need for meeting the demand of reducing pollution load and increasing aesthetic values. Re-vegetation of reclaimed lands present an excellence opportunity to optimize the carbon sequestration and dust collection efficiencies on these lands. With this background, a detailed ecological enumeration of tree vegetation of Hirakud industrial town was undertaken to assess its potential for carbon sequestration and dust collecting capacity and recommend strategies for future plantation programme.

Methodology

For the study, Hirakud Industrial Complex was subjected to preliminary survey with respect to the existence of the vegetation cover. Preliminary survey revealed that vegetation covers in the area do exist but as discontinuous localized patches. The following localities were selected for subsequent samplings:

1. Plantation area near main gate (close to river side colony).
2. Orchard near works Manager Bunglow.
3. Plantation on hillock near rickshaw para.
4. K plantation site near Christian para.
5. Plantation area on ash mound.
6. Plant species found on the hillock near rerolling mills colony.
7. Plantation on sludge pond area.
8. Plant species on the roadside on the way to ash mound.
9. Plant species in the colony area.
10. Roadside plants in riverside colony.
11. Vegetation in Hira Cable premises.
12. Roadside plants approaching Hirakud.
13. Plant cover close to Executive Bunglow and Guest House.

Each of the localities were frequently surveyed for different vegetation analysis between November 2005–January 2006.

During the survey in each site, various existing tree and shrub species were recorded and subsequently identified through botanical examination. For the taxonomic identification flora books of Haines (1961) and Saxena and Brahman (1995) were referred. Ecological enumeration of tree vegetation was made through quadrate sampling technique (Cottam and Curtis, 1956) with respect to following parameters.

1. Frequency
2. Density
3. Girth analysis, and
4. Tree canopy size.

Size of quadrate laid for the purpose was usually 10 × 10 m, except the road side plantation study where it was 1 × 10 m. Tree canopy was assessed by measuring the diameter of ground area covered by the canopy of the plant. Assuming the canopy area to be a circular one, the canopy cover is calculated by Πr^2 formulae (Misra, 1968). Total canopy coverage of a species was calculated by multiplying canopy area figure with total number of individuals.

Results

The following section presents the result of the study with data reflected in the tables and photographs have been provided of the respective study sites.

Plantation Area Near Main Gate (Close to River Side Colony)

The plantation covers an area of approximately 0.53 ha. In the site 6 tree species were recorded (Table 9.1). *Delonix regia* was the dominant species showing 90 per cent frequency with average girth size (cbh) of 90 cm/tree and density of 15 individuals/100 m^2. It was followed by *Azadirachta indica,* which exhibited a density of 4/100 m^2 and average girth of 72 cm/tree. Total canopy coverage by different species in the site varied from 19 m^2 to 395 m^2/100 m^2 with the maximum coverage by *Dolonix regia* and the minimum by *Pongamia pinnata.* Contribution of different plant species to the canopy coverage in the site has been illustrated in Figure 9.1. As revealed from the figure, *Delonix regia* contributed maximum of 67 per cent to the canopy coverage and *Pongamia pinnata* and *Eucalyptus maculatus* contributed as low as 3 per cent to the canopy coverage. It was further calculated that the total canopy coverage of the site amounted to 62580 m^2.

Table 9.1: List of Plant Species and their Ecological Enumeration in Plantation Area Near Main Gate Close to River Side Colony

Sl.No.	Species	Local Name	Frequency (per cent)	Density (No/ 100 m^2)	Girth Size Range (cm)	Average Girth Size (cm)	Canopy Coverage of Individual Tree (m^2)	Total Canopy Coverage (m^2/100 m^2)
1.	*Delonix regia* (DR)	Kishna Chuda	90	5	52–120	90	79	395
2.	*Azadirachta indica* (AI)	Neem	10	4	60–80	72	20	80
3.	*Polyalthia longifolia* (PL)	Deodar	05	2	40–60	48	20	40
4.	*Gmelina arborea* (GA)	Gamhari	05	1	40–60	52	19	19
5.	*Eucalyptus maculata* (EM)	Potash	05	1	80–120	95	19	19
6.	*Pongamia pinnata* (PP)	Karanja	05	1	20–60	45	38	38

Total canopy coverage: 591 m^2/100 m^2.

Total canopy coverage of the site: 31323 m^2.

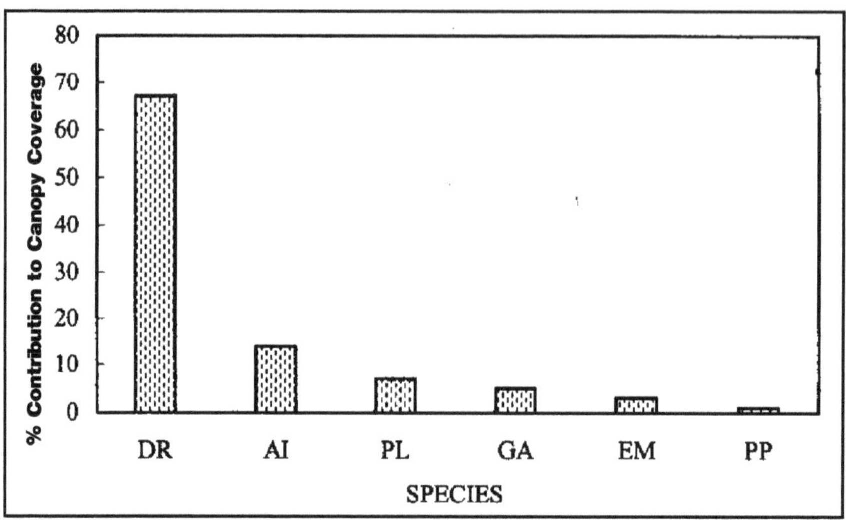

Figure 9.1: Per cent Contribution of Different Species to Total Canopy Coverage in the Site Near Main Gate

Orchard Near Works Manager Bunglow

The orchard is located close to the works manager bung low in the riverside colony. It covers an area of 2.42 ha. In the site, 12 tree species were recorded during the survey (Table 9.2). Out of the 12 species, most frequently encountered species are *Delonix regia* and *Lagerostroemia flosregineae* and these two species exhibited a density of 3 and 2 per 100 m^2 respectively. *Cassia spectabilis* (Chakunda) showed the maximum density of 4/100 m^2. Girth size (circumference at breast height: CBH) of different plants varied from 30 cm to a maximum of 250 cm (in case of *Ficus benghalensis*). Canopy coverage of individual tree varied from 3 to 79 m^2. From this, total canopy coverage of 12 different species were calculated. It showed a range of 9–79 m^2. *Ficus benghalensis* showed the maximum canopy coverage of

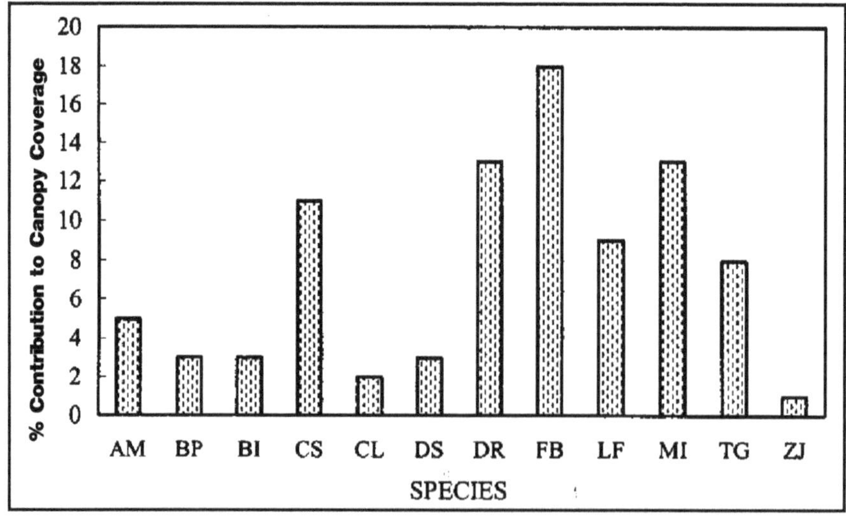

Figure 9.2: Per cent Contribution of Different Species to Total Canopy Coverage in the Site Near to Works Manager Bunglow

79 m^2 followed by *Delonix regia* and *Mangifera indica*. Figure 9.2 illustrates the percent contribution of different plant species to the total canopy coverage of the site. As evident from the figure, *Cassia spectabilis* contributed a maximum of 14 per cent to the total canopy coverage in the site. Total canopy coverage of the site was estimated to be 10,8900 m^2.

Table 9.2: List of Plant Species and their Ecological Enumeration in Orchard Near Works Manager Bunglow

Sl.No.	*Species*	*Local Name*	*Frequency (per cent)*	*Density (No/ 100 m²)*	*Girth Size Range (cm)*	*Average Girth Size (cm)*	*Canopy Coverage of Individual Tree (m²)*	*Total Canopy Coverage (m²/100 m²)*
1.	*Aegle marmelos* (AM)	Bela	05	2	65–80	75	12.5	25
2.	*Bahunia purpurea* (BP)	Kuliari	05	2	30–50	45	07	14
3.	*Butea monosperma* (BM)	Palash	20	2	80–95	90	07	14
4.	*Cassia spectabilis* (CS)	Chakunda	10	4	80–110	85	13	50
5.	*Citrus lemon* (CL)	Lemon	15	3	30–50	35	03	09
6.	*Dolbergia sisoo* (OS)	Sisoo	05	2	80–120	90	07	14
7.	*Delonix regia* (DR)	Gulmohar	25	3	60–100	75	20	59
8.	*Ficus benghalensis* (FB)	Bata	05	1	250	250	79	79
9.	*Lagerostroemia flosregineae* (LF)	Patali	25	2	40–50	45	20	39
10.	*Mangifera indica* (MI)	Mango	05	3	80–120	90	20	59
11.	*Tectona grandis* (TG)	Teak	10	3	70–90	80	13	38
12.	*Zizyphus jujuba* (ZI)	Barkoli	10	2	60–80	65	13	50

Total canopy coverage: 450 m^2/100 m^2

Total canopy coverage of the site: 108900 m^2.

Plantation on Hillock Near Rickshaw Para

The plantation covers an area of approximately 1 ha and is on a hillock. The site is also close to the proposed rerodding plant side. In the area, plantation has been done during 1998 and exclusively two plant species were recorded. These are *Acacia auriculiformis* and *Pongamia pinnata*. Table 9.3 presents data about their ecological distribution.

Table 9.3: List of Plant Species and their Ecological Enumeration on a Slope of Hillock Near Rickshaw Para

Sl.No.	*Species*	*Local Name*	*Frequency (per cent)*	*Density (No/ 100 m²)*	*Girth Size Range (cm)*	*Average Girth Size (cm)*	*Canopy Coverage of Individual Tree (m²)*	*Total Canopy Coverage (m²/100 m²)*
1.	*Acacia auriculiformis* (AA)	Acacia	80	24	30–70	40	6	144
2.	*Pongamia pinnata* (PP)	Karanja	40	4	30–40	35	8	32

Total canopy coverage: 176 m^2/100 m^2.

Total canopy coverage of the site: 17600 m^2.

In the site, out of 2 planted species *Acacia* sp. is the dominant plant species, showing higher frequency, density, girth size and canopy coverage. Total canopy coverage extended by *Acacia* was calculated to be 144 $m^2/100\ m^2$, against that of 32 $m^2/100\ m^2$ by *Pongamia* sp. Per cent contribution to the total canopy of the site has been illustrated in Figure 9.3.

As per the figure, contribution to canopy coverage in the site by *Acacia* was calculated to be 82 per cent. Total canopy coverage of the site was estimated to be 17600 m^2.

K_2 Plantation Site Near Christian Para

The plantation site covers an area of 51 acre out of which 5 acre (roughly) found to be vacant fallow land and rest of the area have been developed into the plantation site by the Hindalco. In the area 10 plant species were recorded (Table 9.4). Out of these, *Cassia spectabilis* exhibited the maximum

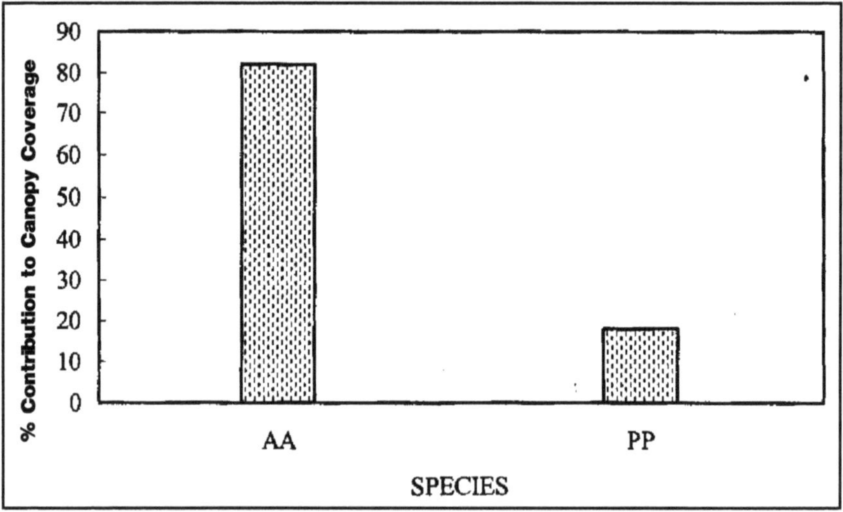

Figure 9.3: Per cent Contribution of Different Species to Total Canopy Coverage in the Site Near Rickshaw Para

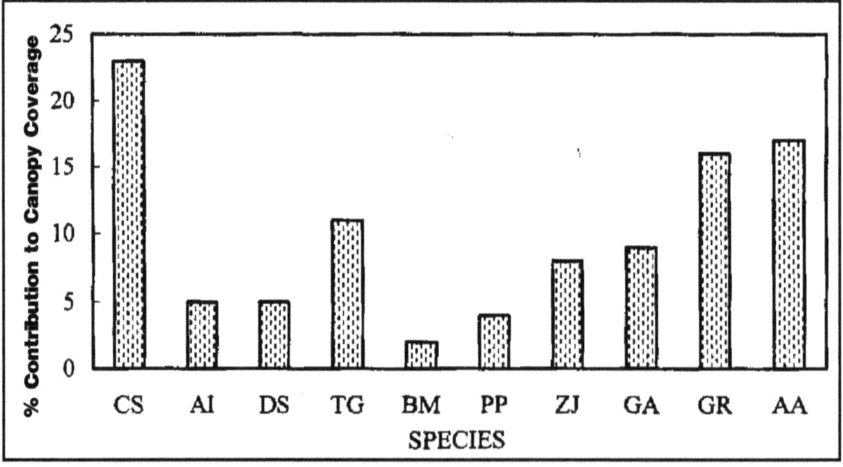

Figure 9.4: Per cent Contribution of Different Species to Total Canopy Coverage in the K_2 Plantation Site Near Christian Para

frequency of 70 per cent and the maximum density of 7 individuals per 100 m^2. Girth size of the tree individuals varied from 20–90 cm. *Acacia auriculiformis* exhibited the maximum average girth of 80 cm. Canopy coverage by individual tree varied from 2 to 4.5 m^2 with maximum canopy by *Pongamia pinnata*. However, total canopy coverage was observed to be maximum (28 m^2) in case of *Cassia spectabilis*. Percent contribution of different species to the total canopy coverage has been illustrated in Figure 9.4. Total canopy coverage of the site was estimated to be 226187 m^2.

Table 9.4: List of Plant Species and their Ecological Enumeration in Plantation Area in K$_2$ Plantation Site

Sl.No.	Species	Local Name	Frequency (per cent)	Density (No/ 100 m^2)	Girth Size Range (cm)	Average Girth Size (cm)	Canopy Coverage of Individual Tree (m^2)	Total Canopy Coverage (m^2/100 m^2)
1.	*Cassia spectabilis* (CS)	Chakunda	70	7	60–70	75	4.0	28
2.	*Azadirachta indica* (AI)	Neem	05	2	30–50	45	3.0	06
3.	*Dalbergia sisoo* (OS)	Sisoo	20	3	10–30	25	3.5	07
4.	*Tectona grandis* (TG)	Teak	15	4	20–40,	30	3.0	12
5.	*Butea monosperma* (BM)	Palash	10	1	20–40	30	2.5	2.5
6.	*Pongamia pinnata* (PP)	Karanja	20	1	20–40	35	4.5	4.5
7.	*Zizyphus jujuba* (ZI)	Barakoli	10	5	20–30	25	2.0	10
8.	*Gmelina arborea* (GA)	Gamhari	10	3	20–40	30	3.5	10.5
9.	*Grevillea robusta* (GR)	Exotic	15	5	40–50	45	4.0	20
10.	*Acacia auriculiformis* (AA)	Acacia	20	6	70–90	80	3.5	21

Total canopy coverage: 121.5 m^2/100 m^2.

Total canopy coverage of the site: 226187 m^2.

Plantation in Ash Mound

The plantation on ash mund covers an area of 33 acre. In the area, 10 tree species are found to be planted (Table 9.5). On the basis of the frequency, *Acacia auriculiformis* showed the maximum frequency of 70 per cent followed by *Delonix regia* (60 per cent). However, *Delonix regia* showed maximum density. Girth size of tree species varied from 5 to 220 cm with maximum girth of bamboo bush. Canopy coverage varied from 2 to 35 m^2 per individual tree with maximum canopy by *Delonix regia*. Also *Delonix regia* exhibited maximum canopy coverage on area basis. Figure 9.5 illustrates per cent contribution of different species to the total canopy coverage in the site. Total canopy coverage in the site was estimated to be 575605 m^2.

Plant Species Found on the Hillock Near Rerolling Mills Colony

The site covers an approximate area of 1 ha and is on a hillock. In the site, 12 species were recorded during the survey. Of these, *Cassia spectabilis* exhibited the maximum frequency of 70 per cent and maximum density of 5 per 100 m^2. Girth size of tree individuals ranged from 20 to 200 cm. *Albizzia lebbeck* exhibited the maximum average girth size (175 cm). Canopy coverage of individual trees varied from 3 m^2 to 80 m^2 with *Mangifera indica* exhibiting maximum girth. However, among all the 12 species,

Cassia spectabilis exhibited maximum total canopy coverage (Table 9.6), followed by *Azadirachta indica.* Least canopy coverage was contributed by the *Citrus lemon* in the site. Percent contribution to the canopy coverage of the site by different species have been presented in Figure 9.6. As per the figure, *Cassia spectabilis* contributed maximum of 27 per cent to the total canopy coverage in the site. Total canopy coverage of the site was estimated to be 145600 m^2.

Table 9.5: List of Plant Species and their Ecological Enumeration in Plantation Area on Ash Mound

Sl.No.	*Species*	*Local Name*	*Frequency (per cent)*	*Density (No/ 100 m²)*	*Girth Size Range (cm)*	*Average Girth Size (cm)*	*Canopy Coverage of Individual Tree (m²)*	*Total Canopy Coverage (m²/100 m²)*
1.	*Acacia auriculiformis* (AA)	Acacia	70	4	30–35	33	12	48
2.	*Acacia glauca* (AG)	Acacia	30	2	20–30	25	7	14
3.	*Albizia lebbeck* (AL)	Sirish	20	2	20–40	30	8	16
4.	*Bombusa flurndinaceae* (BA)	Bamboo	20	2	200–220	210	5	10
5.	*Cassia spectabilis* (CS)	Chalrunda	50	2	15–25	20	11	22
6.	*Dalbergia sisoo* (DS)	Sisoo	40	2	10–30	20	4	08
7.	*Delonix regia* (DR)	Gulmohar	60	8	35–50	40	35	280
8.	*Pongamia pinnata* (PP)	Karanja	30	3	25–45	35	7	21
9.	*Zizyphus jujuba* (ZI)	Barakoli	10	4	5–10	07	2	08
10.	*Peltophorum pterocarpum* (PPt)		20	1	20–30	25	4	04

Total canopy coverage: 431 m^2/100m^2.

Total canopy coverage of the site: 575605 m^2.

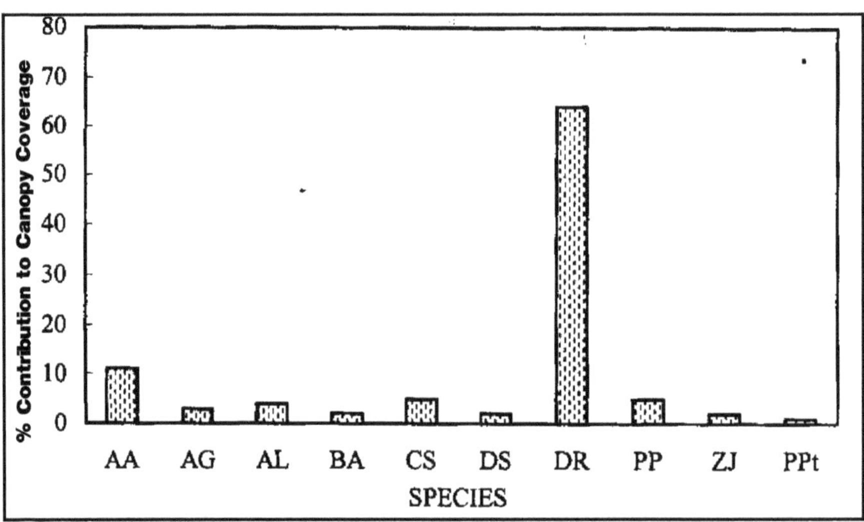

Figure 9.5: Per cent Contribution of Different Species to Total Canopy Coverage in the Site at Ash Mound

Table 9.6: List of Plant Species and their Ecological Enumeration on a Hillock Near Rerolling Mills Colony

Sl.No.	Species	Local Name	Frequency (per cent)	Density (No/ 100 m²)	Girth Size Range (cm)	Average Girth Size (cm)	Canopy Coverage of Individual Tree (m²)	Total Canopy Coverage (m²/100 m²)
1.	*Cassia spectabilis* (CS)	Bilati Chakunda	70	5	100–120	90	79	395
2.	*Mangifera indica* (MI)	Mango	40	2	70–100	80	80	160
3.	*Casia fistula* (CF)	Sunari	10	1	50–70	60	50	50
4.	*Azadirachta indica* (AI)	Neem	20	3	30–40	30	78	234
5.	*Anacardium accidentalis* (AAC)	Kaju	20	2	30–50	40	20	40
6.	*Artocarpus heterophyllus* (AH)	Panas	30	3	70–90	80	50	150
7.	*Albizzia lebbeck* (AL)	Sirish	25	2	450–200	175	77	154
8.	*Acacia auriculiformis* (AA)	Acacia	40	2	50–80	60	76	152
9.	*Psidium guajava* (PG)	Pijuli	10	1	20–30	25	20	20
10.	*Eucalyptus maculata* (EM)	Patash	10	2	30–60	50	18	36
11.	*Syzigium cumuni* (SC)	Jamuna	30	2	110–130	120	21	42
12.	*Citrus lemon* (CL)	Lemon	10	1	20–30	25	03	03

Total canopy coverage: 1456 m²/100 m².

Total canopy coverage of the site: 145600 m².

Plantation on Sludge Pond Area

The site covers an area of 17 acre. In the site, 6 plant species are recorded (Table 9.7). *Acacia auriculiformis* exhibited the maximum frequency and density in the site followed by *Dalbergia sisoo.*

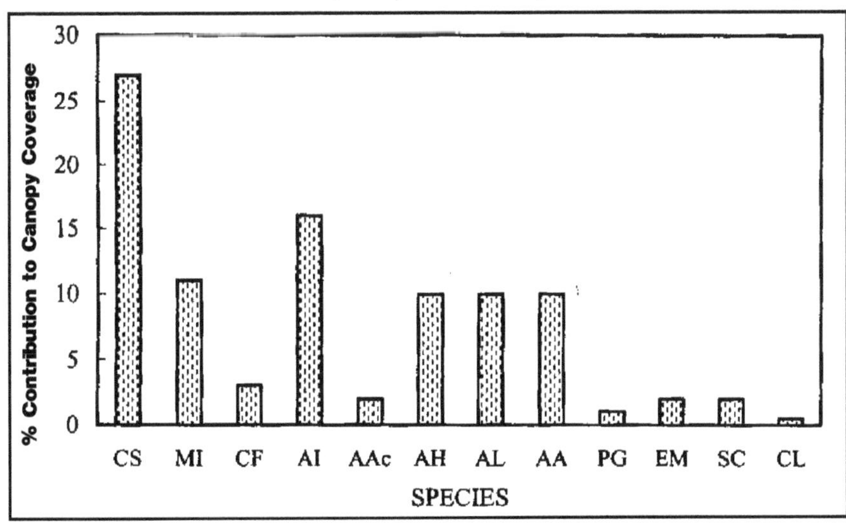

Figure 9.6: Per cent Contribution of Different Species to Total Canopy Coverage in the Site at Near Rerolling Mill Colony

Girth size of different individuals ranged from 10–50 cm. Average girth size of 40 cm was noted both for *Gmelina arborea* and *Cassia spectabilis*, which was found to be maximum among all the six species in the site. Average canopy coverage of individual tree varied from a minimum of 3 m^2 (*Gmelina arborea*) to a maximum of 12 m^2 (for both *Acacia auriculiformis* and *Cassia spectabilis*). Total canopy coverage per 100 m^2 was found to be the maximum (96 m^2/100 m^2) in case of *Acacia auriculiformis*. Total canopy coverage of the site was calculated to be 117,646 m^2. Percent contribution of different species to the total canopy coverage of the site has been illustrated in the Figure 9.7.

Table 9.7: List of Plant Species and their Ecological Enumeration in Sludge Pond Area

Sl.No.	Species	Local Name	Frequency (per cent)	Density (No/ 100 m^2)	Girth Size Range (cm)	Average Girth Size (cm)	Canopy Coverage of Individual Tree (m^2)	Total Canopy Coverage (m^2/100 m^2)
1.	*Dalbergia sisoo* (DS)	Sisoo	60	6	10–30	20	4	24
2.	*Acacia auriculiformis* (AA)	Acacia	70	8	15–35	20	12	96
3.	*Gmelina arborea* (GA)	Gambhari	20	2	30–50	40	3	06
4.	*Gaviella robusta* (GR)	Exotic sp.	30	2	25–45	30	7	14
5.	*Casia spectabilis* (CS)	Bada Chakunda	20	2	30–50	40	12	24
6.	*Azadiracta indica* (AI)	Neem	10	1	20–40	30	7	07

Total canopy coverage: 171 m^2/100 m^2.

Total canopy coverage of the site: 117646 m^2.

Plant Species on the Roadside on the Way to Ash Mund

Plantation on the left side of the road on the way to the ash mund covers an area of 5 ha (1 km length × 50 m breadth). In the area, 12 plant species were recorded. *Acacia glauca* was found to be the

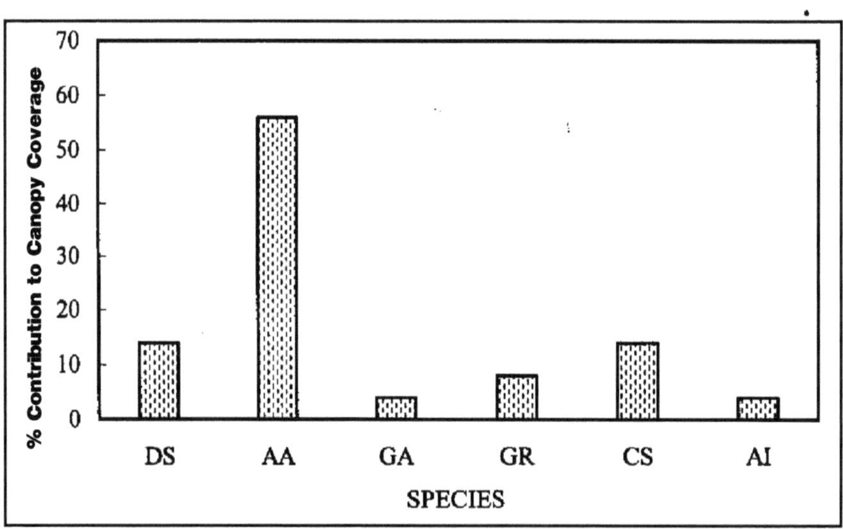

Figure 9.7: Per cent Contribution of Different Species to Total Canopy Coverage in the Site at Sludge Pond Area

most frequent species in the area (Table 9.8). However, *Mangifora inidca* exhibited the maximum density of 8 per 100 m^2 area. Girth size range of individuals of different species was noted to be 20–120 cm. On average basis, *Bambusa arundinaceae* exhibited maximum girth size of 110 cm, followed by *Ficus religiosa* (85 cm). Individuals of *Bambusa arundinaceae* showed the maximum average girth size of 110 cm and it was followed by *Ficus benghalensis* (75 cm). Canopy coverage of individual tree was estimated to be highest in *Ficus benghalensis* (42 m^2). However, total canopy coverage was observed to be the maximum in case of *Mangifera indica* (232 m^2/100 m^2). Percent contribution of different species to the total canopy coverage of the site has been illustrated in Figure 9.8. As per the figure, *Mangifera indica* contributed as much as 38 per cent to the canopy coverage of the site. Total canopy coverage of the site was calculated to be 309000 m^2.

Table 9.8: List of Plant Species and their Ecological Enumeration on Roadside on the Way to Ash Mound

Sl.No.	Species	Local Name	Frequency (per cent)	Density (No/ 100 m^2)	Girth Size Range (cm)	Average Girth Size (cm)	Canopy Coverage of Individual Tree (m^2)	Total Canopy Coverage (m^2/100 m^2)
1.	*Ficus benghalensis* (FB)	Bata	5	1	70–80	75	42	42
2.	*Ficus religiosa* (FR)	Pipal	5	1	80–90	85	30	30
3.	*Phoenix regia* (PR)	Khajuria	10	1	20–30	25	2	2
4.	*Pongamia pinnata* (PP)	Karanja	50	5	30–40	35	2	10
5.	*Acacia auriculiformis* (AA)	Acacia	5	2	30–40	35	20	40
6.	*Casurina equisitifolia* (CE)	Jhaun	5	1	40–60	50	13	13
7.	*Tamarindus indica* (TI)	Tentuli	5	1	30–40	35	20	20
8.	*Mangifera indica* (MI)	Mango	50	8	60–80	70	29	232
9.	*Casia spectabilis* (CS)	Bada Chakunda	5	3	60–70	65	28	84
10.	*Dalbergia sisoo* (DS)	Sisoo	20	3	30–50	40	8	24
11.	*Bambusa arundinacea* (BA)	Bamboo	10	3	100–120	110	7	21
12.	*Acacia glauca* (AG)	Acacia	70	4	30–60	50	25	100

Total canopy coverage: 618 m^2/100 m^2.

Total canopy coverage of the site: 309000 m^2.

Plant Species in the Colony Area

Survey of colony area (spreading over approximately 8 ha) revealed the presence of 25 tree species (Table 9.9). Some of the frequently encountered species in the area are *Mangifora indica, Mimosups elongi, Pettophorum pterocarpum* and *Ficus benghalensis.* Among the 25 species, density was found to be maximum in case of *Cassia spectabilis.* Girth size range of the individuals of different species varied from 20 cm to 210 cm. Average girth size data indicated maximum girth for both *Ficus religiosa* and *Ficus benghalensis.* Canopy area of tree individuals showed a range of 5–286 m^2 with maximum area in case of *Ficus benghalensis.* Calculation of total canopy coverage for different species also revealed the maximum value in case of *Ficus benghalensis.* Contribution of different species to the canopy coverage of the site has been illustrated in Figure 9.9. As per the figure, species like *Ficus benghalensis, Pongamia*

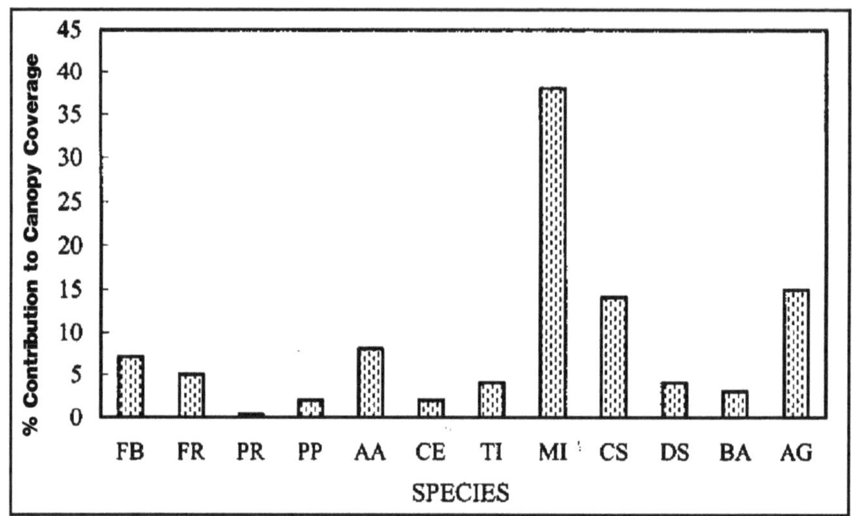

Figure 9.8: Per cent Contribution of Different Species to Total Canopy Coverage on the Roadside on the Way to Ash Mound

pinnata, Ficus religiosa are the dominant contributors to the canopy coverage of the site. Total canopy coverage of the site was calculated to be 149600 m^2.

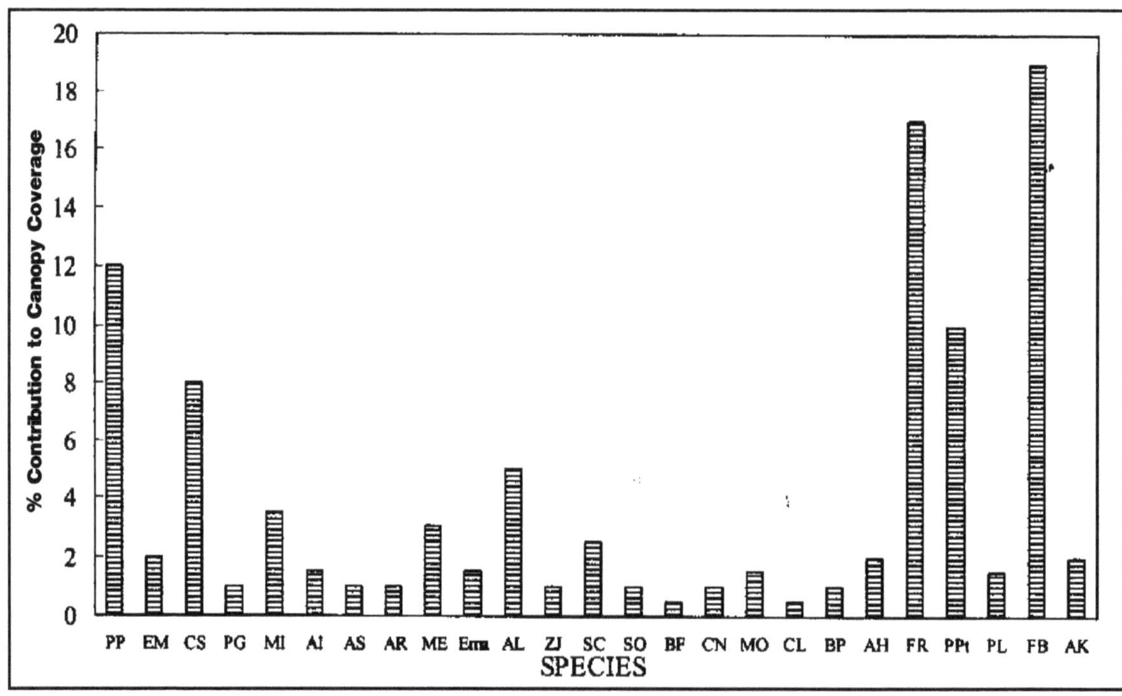

Figure 9.9: Per cent Contribution of Different Species to Total Canopy Coverage in the Colony Area

Table 9.9: List of Plant Species and their Ecological Enumeration in Other Colony Area

Sl.No.	Species	Local Name	Frequency (per cent)	Density (No/ 100 m^2)	Girth Size Range (cm)	Average Girth Size (cm)	Canopy Coverage of Individual Tree (m^2)	Total Canopy Coverage (m^2/100 m^2)
1.	*Pongamia pinnata* (PP)	Karanja	30	2	80–95	75	11.5	2.3
2.	*Eucalyptus maculata* (EM)	Potash	20	1	100–120	110	3.8	3.8
3.	*Casia spectabilis* (CS)	Chakunda	40	4	40–60	50	3.5	14.0
4.	*Psidum guajava* (PG)	Pears	20	1	40–50	45	2.3	2.3
5.	*Mangifera indica* (MI)	Mango	60	2	40–60	50	3.6	7.3
6.	*Azadirachta indica* (AI)	Neem	40	2	40–60	50	1.5	3.0
7.	*Annona squamosa* (AS)	Ata	10	1	20–40	30	2.0	2.0
8.	*A nnona reticulata* (AR)	Sita ata	10	1	20–40	30	2.0	2.0
9.	*Mimosups elongi* (ME)	Baul	60	2	30–50	40	3.1	6.2
10.	*Eagle mannelos* (EMa)	Bel	10	1	30–50	40	2.5	2.5
11.	*Albizia lebbek* (AL)	Sirish	20	1	60–80	70	9.1	9.1
12.	*Zizyphus jujuba* (ZJ)	Barokoli	20	1	20–40.	30	1.75	1.75
13.	*Syzigium cumuni* (SC)	Jamun	20	2	40–60	50	2.25	4.5
14	*Schleichera oleosa* (SO)	Kusum	10	1	40–60	50	2.0	2.0
15.	*Borassus flabellifer* (BF)	Tala	10	1	70–90	80	0.9	0.9
16.	*Cocos nucifera* (CN)	Coconut	20	1	70–80	75	2.25	2.25
17.	*Moringa oleifera* (MO)	Sajana	10	1	50–70	60	2.9	2.9
18.	*Citrus lemon* (CL)	Lemon	20	1	20–30	25	0.6	0.6
19.	*Bahunia purpurea* (BP)	Kuler	20	1	40–50	45	1.4	1.4
20.	*Artocarpus* (AH) *heterophyllus*	Panash	10	1	45–55	50	3.6	3.6
21.	*Ficus religiosa* (FR)	Peepal	40	1	190–210	200	31.5	31.5
22.	*Peltophorum pterocarpum* (PPt)	Radhachuda	60	2	60–80	70	9.0	18.0
23.	*Polyalthia longifolia* (PL)	Deodar	30	2	30–50	40	1.5	3.0
24.	*Ficus benghalensis* (FB)	Bara	60	1	190–210	200	35.8	35.8
25.	*Anthocephalus kadamba* (AK)	Kadam	40	1	30–50	40	3.6	3.6

Total canopy coverage: 187 m^2/100 m^2.

Total canopy coverage of the site: 149600 m^2.

Road Side Plants in Riverside Colony

Road in riverside colony stretches approximately 780 meter. While surveying vegetation of the road, 7 species were recorded, out of which most frequently recorded species was *Mimosups elongi.* Its density was also noted to be the maximum among the 7 recorded species (Table 9.10). All plants exhibited girth size range of 30–90 cm. *Delonix regia* showed the maximum average girth of 70 cm and the maximum canopy coverage. Contribution of different species to the canopy coverage of the site has been illustrated in Figure 9.10. Figure shows that in the site, *Delonix regia* contributed a maximum of 34 per cent to the total canopy coverage of the site. It was followed by *Mimosups elongi,* which contributed 30 per cent to the canopy coverage. Least contribution of only 0.5 per cent came from the species like *Atrocarpus heterophylla.* In the entire stretch of road, canopy coverage was estimated to be 8081 m².

Table 9.10: List of Plant Species and their Ecological Enumeration in Riverside Colony Road

Sl.No.	Species	Local Name	Frequency (per cent)	Density (No/ 100 m²)	Girth Size Range (cm)	Average Girth Size (cm)	Canopy Coverage of Individual Tree (m²)	Total Canopy Coverage (m²/100 m²)
1.	*Mimosups elongi* (ME)	Baula	70	4	50–70	60	78	312
2.	*Delonix regia* (DR)	Gulmohar	30	2	60–90	70	176	352
3.	*Acadirachta indica* (AI)	Neem	20	2	40–60	50	50	100
4.	*Pongamia pinnata* (PP)	Karanja	30	2	40–60	50	113	226
5.	*Polyalthia longifolia* (PL)	Deodara	30	3	30–60	45	07	21
6.	*Emblica officinalis* (EO)	Anla	05	1	30–60	45	12	12
7.	*Artocarpus heterophylla* (AH)	Panash	05	1	40–70	55	05	05

Total canopy coverage: 1028 m²/100 m².

Total canopy coverage of the site: 8081 m².

Vegetation in Hira Cable Premises

In the Hira Cable premises, vegetation/plantation was confined to approximately 2 acres of land. Survey of vegetation revealed presence of 8 plant species. Out of this, *Cocos nucifera* and *Mangifera*

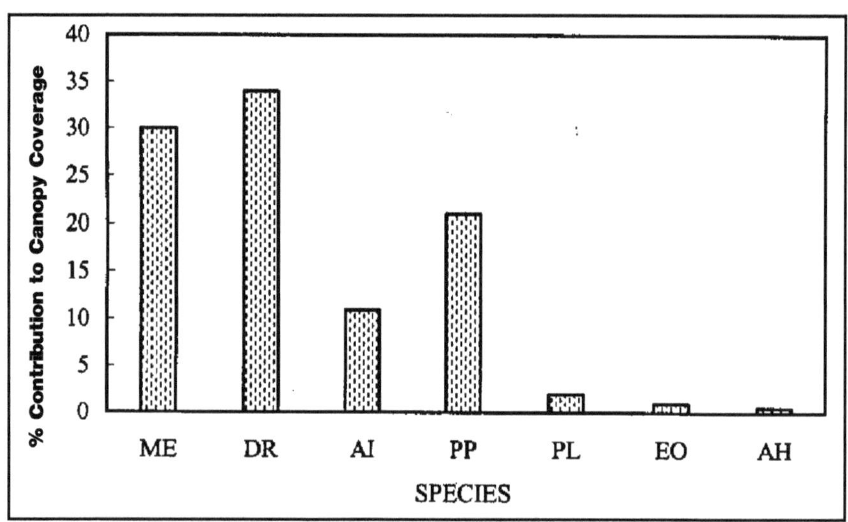

Figure 9.10: Per cent Contribution of Different Species to Total Canopy Coverage on the Riverside Colony Area

indica showed maximum frequency of 60 per cent. On the basis of density, *Delonix regia* was the dominant species showing a maximum density of 4/100 m^2. Individuals of different plant species exhibited a girth size range of 40–130 cm with maximum average girth size in case of *Mangifera indica* (Table 9.11). Canopy coverage by individual tree varied from 7 to 50 m^2/tree. Maximum canopy coverage per individual tree was noted in case of *Mangifera indica* and *Ficus glomerata*. Total canopy coverage by different species showed a range of 14–152 m^2/100 m^2 with maximum coverage by *Delonix regia*. Contribution to total canopy coverage by different species in the site has been illustrated in Figure 9.11. As per the figure, maximum contribution came from *Delonix regia*. It was followed by *Mangifera indica* and *Cassia spectabilis*. Total canopy coverage of the site was calculated to be 105400 m^2.

Table 9.11: List of Plant Species and their Ecological Enumeration in Hira Cable Premises

Sl.No.	*Species*	*Local Name*	*Frequency (per cent)*	*Density (No/ 100 m^2)*	*Girth Size Range (cm)*	*Average Girth Size (cm)*	*Canopy Coverage of Individual Tree (m^2)*	*Total Canopy Coverage (m^2/100 m^2)*
1.	*Delonix regia* (DR)	Gulmohar	40	4	60–70	65	38	152
2.	*Cassia spectabilis* (CS)	Chakunda	40	3	70–80	75	28	84
3.	*Pongamia pinnata* (PP)	Karanja	40	2	60–80	75	28	56
4.	*Cocos nucifera* (CN)	Coconut	60	2	60–80	70	07	14
5.	*Mangifera indica* (MI)	Mango	60	2	90–130	120	50	100
6.	*Ficus glomerata* (FG)	Dimri	20	1	40–60	50	50	50
7.	*Lagerostroemia flosreginae* (LF)	Patali	20	1	80–120	90	38	38
8.	*Cassia alata* (CA)	Bilati Chakunda	20	1	40–50	45	28	28

Total canopy coverage: 527 m^2/100 m^2

Total canopy coverage of the site: 105400 m^2

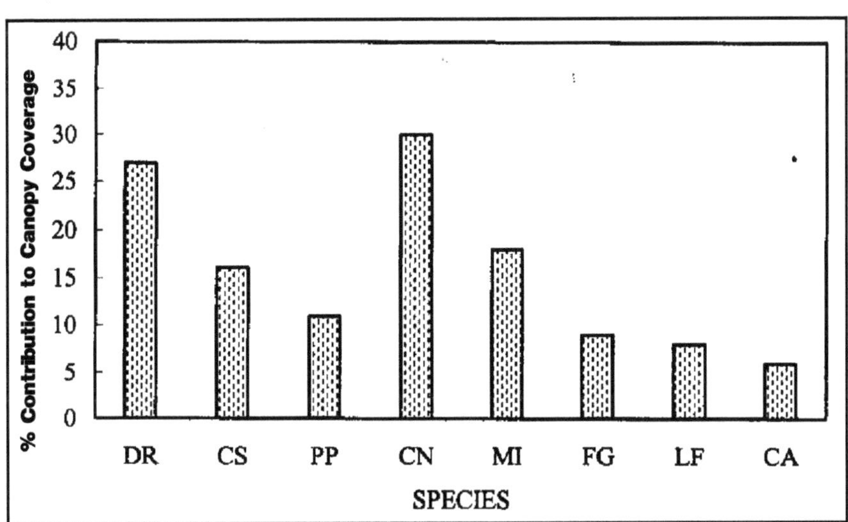

Figure 9.11: Per cent Contribution of Different Species to Total Canopy Coverage in the Hira Cable Premises

Road Side Plants Approaching Hirakud

The road approaching Hirakud stretches approximately 1.7 km. For this survey sampling began with the point close to FCI till the road enters into the Hirakud market area. During the survey, 14 species of plants are recorded. *Cassia siamea* and *Cassia spectabilis* both were found to maximally encountered frequent species on the roadside (Table 9.12). Both the species also shared highest density. Girth size range of different plants varied from 20 cm to a maximum of 250 cm. *Ficus benghalensis* exhibited the maximum average girth of 225 cm at breast height. Also it exhibited the maximum canopy coverage and was followed by *Ficus religiosa*. Contribution of different species to the total canopy coverage of the roadside has been illustrated in Figure 9.12. *Ficus benghalensis* contributed maximum *i.e.* 53 per cent of the total canopy coverage. Total canopy coverage of the site was calculated to be 21197 m^2.

Table 9.12: List of Plant Species and their Ecological Enumeration on Roadsides Approaching Hirakud

Sl.No.	Species	Local Name	Frequency (per cent)	Density (No/ 100 m^2)	Girth Size Range (cm)	Average Girth Size (cm)	Canopy Coverage of Individual Tree (m^2)	Total Canopy Coverage (m^2/100 m^2)
1.	*Acacia aureculiformis* (AA)	Acacia	50	3	40–50	45	12	36
2.	*Alstonia scholaris* (AS)	Chhatim	40	3	30–40	35	10	30
3.	*Cassia simea* (CSi)	Chakunda	60	4	40–60	50	13	52
4.	*Cassia spectabilis* (CS)	Bilati Chakunda	60	4	50–60	55	11	44
5.	*Spathodea companulatae* (SC)	Panituri	05	1	50–60	55	12	12
6.	*Peltolorum pterocarpum* (PPt)	Radha Chuda	20	1	40–50	45	12	12
7.	*Pongamia pinnata* (PP)	Karanja	30	1	40–50	45	15	15
8.	*Mangifera indica* (MI)	Mango	20	2	40–70	60	16	32
9.	*Palbargia sisoo* (OS)	Sisoo	20	2	20–40	30	08	16
10.	*Tectona grandia* (TG)	Teak	20	2	30–50	40	08	16
11.	*Ficus benghalensis* (FB)	Bata	10	2	200–250	225	330	660
12.	*Ficus religiosa* (FR)	Pipal	10	1	210–220	215	280	280
13.	*Azadirachta indica* (AI)	Neem	10	2	60–80	70	14	28
14.	*Ficus glomerata* (FG)	Dimri	20	1	40–60	50	12	12

Total canopy coverage: 1243 m^2/100 m^2

Total canopy coverage of the site: 21197 m^2

Plant Cover Close to Executive Bunglow and Guest House

Survey of the area around executive bunglows and guest house revealed the presence of 20 tree species (Table 9.13). Analysis of their frequency indicated the dominance of *Polyalthia longifolia*. It also showed the maximum density. Girth size range of different species varied from 30 cm to 300 cm with highest average girth in case of *Ficus benghalensis*. Canopy coverage per individual trees of different

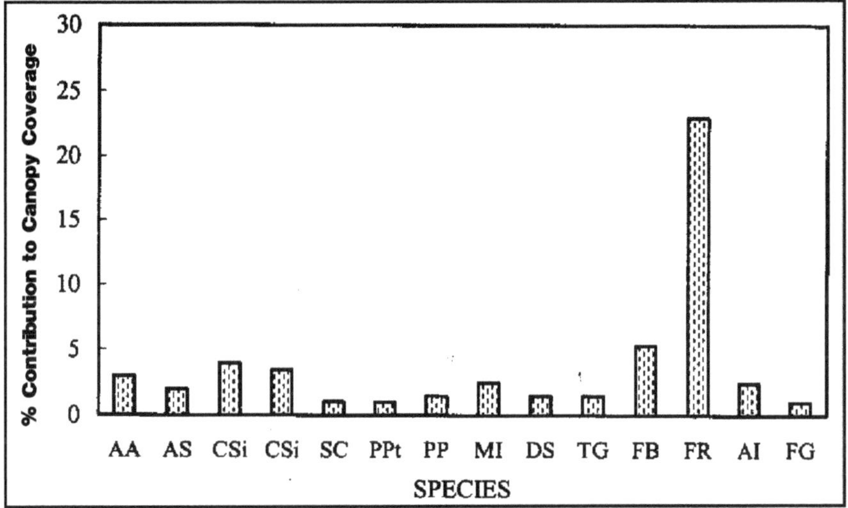

Figure 9.12: Per cent Contribution of Different Species to Total Canopy Coverage on the Road Side Approaching Hirakud

species showed a range of 3 m^2 (in case of *Borassus flabellifera*) to 314 m^2 (in case of *Ficus benghalensis*). Total canopy coverage per 100 m^2 area was calculated to be 793 m^2. Contribution of different species to total canopy coverage has been illustrated in Figure 9.13. From the figure, it was evident that *Ficus benghalensis* maximally contributed to the total canopy coverage (40 per cent). Total canopy coverage of the site was estimated to be 79300 m^2.

Discussion

Ecological Enumeration

The analysis of data presented in Table 9.14 reveals that on the basis of the relative density, *Cassia spectabilis* was the dominant species. Other co-dominant species are *Acacia auriculoformis, Zizyphus*

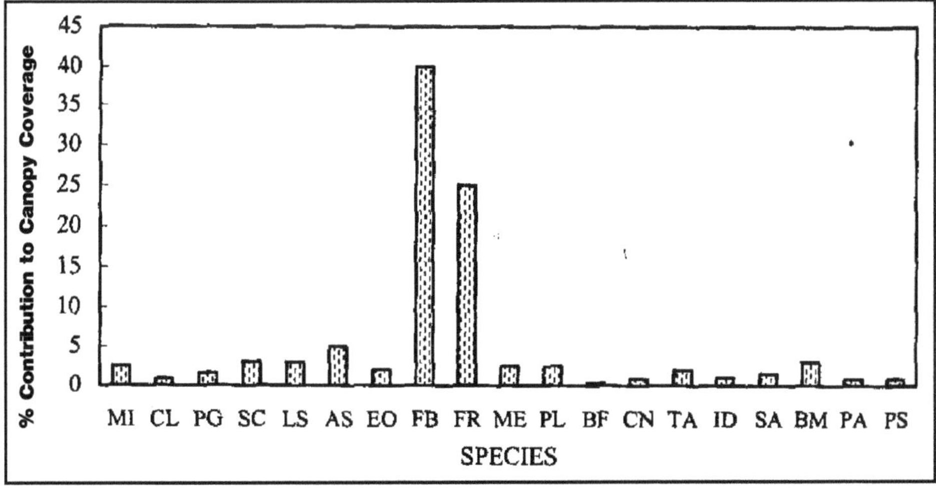

Figure 9.13: Per cent Contribution of Different Species to Total Canopy Coverage Near to Executive Bunglow and Guest House

jujuba, Graviella robusta, Azadiracta indica and *Mangifera indica.* Rest of the species exhibited relative density values below 5 (in 0–100 scale). On the basis of the canopy coverage, *Ficus benghalensis* is the dominant canopy bearing species in the area. Some other species like *Casia spectabilis, Acacia auriculiformis, Delonix regia* and *Ficus religiosa* are the important canopy bearing species. Rest of the species showed a relative canopy cover value of below 7 (in 0–100 scale). Based on the relative density and canopy cover combined, important value index (IVI) for different species were calculated. The values shoed a range of 0.9–24.56. *Casia spectabilis* is the dominant species in the area with a IVI of 24.56. Other codominant associates are *Acacia auriculiformis, Ficus benghalensis, mangifera indica* and *Graviela robusta.* Rest of the species shoed IVI less than 9 in 0–200 scale).

Table 9.13: List of Plant Species and their Ecological Enumeration Near Executive Bunglow and Guest House

Sl.No.	*Species*	*Local Name*	*Frequency (per cent)*	*Density (No/ 100 m²)*	*Girth Size Range (cm)*	*Average Girth Size (cm)*	*Canopy Coverage of Individual Tree (m²)*	*Total Canopy Coverage (m²/100 m²)*
1.	*Mangifera indica* (MI)	Mango	10	1	80–12	100	20	20
2.	*Citrus lemon* (CL)	Lemon	10	2	30–60	40	03	6
3.	*Psidium guajava* (PG)	Pears	10	1	40–60	50	13	13
4.	*Moringa oleifera* (MO)	Sajana	10	1	60–90	80	13	13
5.	*Syzigium cumini* (SC)	Jamu	15	1	90–120	100	28	28
6.	*Litchi Chinensis* (LC)	Litchu	10	1	80–120	95	28	28
7.	*A chras sapota* (AS)		10	1	80–100	90	38	38
8.	*Emblica officinalis* (EO)	Anla	5	1	40–60	50	13	13
9.	*Ficus benghalensis* (FB)	Bata	5	1	300	300	314	314
10.	*Ficus religiosa* (FR)	Pepal	5	1	250	250	201	201
11.	*Mimosops elongi* (ME)	Baula	10	2	70–90	85	10	20
12.	*Polyalthia longifolia* (PL)	Deodam	20	3	60–80	70	07	21
13.	*Borassus flabellifera* (BF)	Palm	5	1	90	90	03	3
14.	*Cocos nucifera* (CN)	Cocoanut	5	1	100–120	110	06	6
15.	*Terminalia arjuna* (TA)	Aljuna	10	1	140–180	150	13	13
16.	*Inga dulcis* (10)		5	1	40–60	50	08	8
17.	*Saraca ashoka* (SA)	Ashoka	5	1	30–40	35	12	12
18.	*Butea monosporma* (BM)	Palash	10	2	60–80	70	12	24
19.	*Plumeria alba* (PA)		10	1	60–90	80	06	6
20.	*Ptetocarpus santalimm* (PS)		5	1	30–50	45	06	6

Total canopy coverage: 793 m²/100 m²

Total canopy coverage of the site: 29 m²

Total Canopy Coverage of the Hirakud Industrial Complex

Total canopy coverage of the Hirakud Industrial complex has been presented in Table 9.14. Total canopy coverage of different sites was estimated to be 1790039 m² over a total plantation area of 62 ha.

Data in the table indicates that the plantation on ash mund area supports maximum canopy which contributes maximally (32 per cent) to the total canopy coverage. Other prominent contributors are Roadside plantation to ash mund (17 per cent), K_2 plantation site (12.6 per cent) and colony area plantation(8 per cent). Contribution from Rickshaw para plantation colony and Roadside plantation to Hirakud township are only 1 and 1.2 per cent respectively. Per hactre basis, canopy coverage was maximum in hillock near rerolling mills colony and minimum in roadside plantation to Hirakud township are only 1 and 1.2 respectively. Per hactre basis, canopy coverage was maximum in hillock near rerolling mills colony and minimum in roadside plantation in Hirakud. In K_2 plantation site, the canopy coverage per hactre is towards lower range and this suggest that there is ample scope to raise canopy coverage in these area.

Table 9.14: List of Plants Found in and Around Hirakud

Sl.No.	*Species*	*Relative Density Value*	*Relative Canopy Coverage Value*	*Important Value Index*
1.	*Acacia auriculiformis*	13	8.28	21.28
2.	*Acacia glauca*	1	2.61	3.61
3.	*Achras sapota*	0.6	0.18	0.78
4.	*Aegle marmalos*	1	0.57	1.57
5.	*Albizzia labbeck*	0.7	1.23	1.93
6.	*Alstonia scholaris*	0.2	0.18	0.38
6.	*Anacardium occidentalis*	0.1	0.18	0.28
7.	*Annona reticulata*	0.4	0.06	0.46
8.	*Annona squamosa*	0.4	0.06	0.46
9.	*Anthocephalus kadamba*	0.4	0.09	0.49
10.	*Artocarpus heterophyllus*	0.6	1.23	1.83
11.	*Azadirachta indica*	5.0	3.63	5.63
12.	*Bahunia purpurea*	0.7	0.27	0.97
13.	*Bambusa aurundinaceae*	1.0	';0.6	1.6
14.	*Borassus flabellifer*	0.4	0.09	0.49
15.	*Butea monosperma*	1.5	0.57	2.07
16.	*Cassa fistula*	0.06	0.48	0.54
17.	*Cassia alata*	5.0	1.4	6.4
18	*Citrus lemon*	1.1	0.24	1.34
19	*Cassia siamea*	0.2	0.27	0.47
20	*Cassia spectabilis*	14.6	9.96	24.56
21.	*Casurina equisetifolia*	0.3	0.33	0.63
22	*Cocos nucifera*	1.0	0.42	1.42
23.	*Dalbergia sisso*	1.0	0.18	1.18
24.	*Delonix regia*	4;0	7.74	4.74
26.	*Emblica officinalis*	0.07	0.12	0.19
27.	*Eucalypatus maculata*	0.6	0.92	1.52

Contd...

Table 9.14–Contd...

Sl.No.	Species	Relative Density Value	Relative Canopy Coverage Value	Important Value Index
28.	*Ficus benghalensis*	1.2	17.1	18.3
29.	*Ficus glomerata*	0.7	2.28	2.98
30.	*Ficus religiosa*	1.0	7.2	8.2
31.	*Gmelina arborea*	4.3	1.23	5.53
32	*Graviella robusta*	6.6	2.55	9.15
33.	*Inga dulcis*	0.06	0.03	0.09
34.	*Lagerostroemia flosreginae*	0.7	2.1	2.8
35.	*Litchi chinensis*	0.06	0.15	0.21
36.	*Mangifera indica*	5.0	10.5	15.5
37	*Mimosups elongi*	0.8	1.8	2.6
38	*Moringa oleifera*	0.5	0.12	0.62
39	*Peltophorum pterocarpum*	1.3	0.84	2.14
40	*Phoenix regia*	0.3	0.06	0.36
41.	*Pterocarpus sartalinus*	0.06	0.03	0.09
42.	*Plumeria alba*	0.06	0.03	0.09
43.	*Polyalthia longifolia*	1.4	4.23	5.63
44.	*Pongamia pinnata*	5.0	4.05	9.05
45.	*Psidium guajava*	0.6	0.27	0.87
46.	*Saraca ashoka*	0.06	0.03	0.09
47.	*Schleichera oleosa*	0.5	0.06	0.56
48	*Spathodea companulatae*	0.2	0.24	0.44
49.	*Syzigium cumuni*	1.2	0.06	1.26
50.	*Tamarindus indica*	0.3	0.54	0.84
51.	*Tectona grandis*	6.0	2.25	8.25
52.	*Terminalia arjuna*	0.06	0.1	0.16
53.	*Zizyphus jujuba*	7.0	1.68	8.68

Role of Vegetation in Hirakud Industrial Complex in Carbon Sequestration

Green vegetation act as a sink by utilizing atmospheric CO_2 through the process of photosynthesis. During the process, energy is harvested from solar radiation and the process exclusively occurs in specialized chloroplast organelle of green leaves, where CO_2 get reduced to glucose. Hence green leaves when exposed to sun can uptake CO_2 from atmosphere and therefore can act as natural sink it has been reported that sun leaves of tropical woody species can up take CO_2 at the rate of 10–15 micromole per m^2 per second (Larcher, 1995). On this basis, quantum of CO_2 uptake by the green vegetation of Hirakud industrial complex was approximated as follows.

Table 9.15: Canopy Coverage in Different Plantation Sites

Sl.No.	Site	Area (ha)	Total Canopy Coverage (m^2)	Canopy Coverage (m^2/ha)	Per cent Contribution to Canopy
1.	Plantation near Main Gate of riverside colony	0.53	31323	62646	1.7
2.	Orchard at Work Manager Bunglow	2.42	108900	45000	6.0
3.	Richshawa para	1.0	17600	17600	1.0
4.	K_2 Plantation Site	18.4	226187	12293	12.6
5.	Ash mound	13.2	575605	43606	32.0
6.	Hillock near rerolling mill	1.0	145600	145600	8.0
7.	Sludge pond area	6.8	117646	17301	7.0
8.	Roadside to Ash mound	5.0	30900	61800	17.0
9.	Colony area	8.0	149600	18700	8.0
10.	Road side Plantation in river side colony	0.8	8081	10101	0.5
11.	Roadside plantation in other areas in Hirakud	4.0	21197	5299	1.2
12.	Executive Bunglow and Guest House	1.0	79300	79300	4.0

The total canopy coverage of the complex was estimated as 1895439 m^2. Assuming a general 40 per cent canopy overlap, expected canopy coverage exposed to sun was calculated to be 1137261 m^2. A square meter of canopy area supports green leaf area of around 10 m^2. This estimated green leaf surface is exposed to solar radiation minimally for a period of 10 hours a day. A square meter of green leaves exposed to sun can uptake 10–15 micromole of CO_2 or 12.5 micromole of CO_2 on average per second. This amounted to a total uptake of $4.5 \times 10_5$ micro mole CO_2 per day or 19.8 g CO_2 per m^2 of sun exposed green leaves per day. Since the Hirakud Industrial Complex has 11372610 m^2 of Sun exposed leaves, the total CO_2 uptake by the green vegetation in a day amounted to 225.18 tons per day or 82189.85 tons per annum. This is a conservative estimate as only green woody vegetation as well as only sun-exposed leaves have been taken in to account. However, the actual sink capacity may be another 25–30 per cent more (total 102737.31 tonnes/annum) if shade leaves are considered as they have also the capability to accept carbon for photo-assimilation with a lower rate.

Dust Collecting Capacity of the Vegetation in and Around Hirakud Industrial Complex

Importance of the plant to reduce the dust in the environment has been emphasized by several workers (Das and Pattanaik, 1978; Das, 1981; Maiti, 1992; Ghosh, 2005). Ghosh (2005) reviewed the dust collecting capability of some Indian tree species and reported that their dust collecting capability varies with species. Tropical plant species show a dust collecting range of 2–5.35 g/m^2 of leaf surface (with an average value of 3.68 g/m^2). Assuming the total green leaf area estimate of Hirakud Industrial complex to be the dust collecting surface (18954390 m^2), dust collecting capacity of the entire vegetation of Hirakud complex was calculated to be 69.75 tonnes. Assuming that the turn over rate of dust collection by leaves is 2–4 times a year because of weather parameters, the actual dust collection may vary from 140–280 tonnes per annum in the complex.

Recommendations

Different site specific recommendation for plantation is suggested on the basis on the reference of various workers (Maiti, 1993; CSIR Publication, 1994; Rawat and Banerjee, 1996; Ghose, 2005).

K2 Plot

During the study it was observed that in the K2 plot (total area of 51 Ac) about 10 per cent of area((5 acres) is lying vacant. The K2 plot contributes 12.6 per cent of the total canopy area is Hirakud. So here there is a scope of creating another 30,000 m^2 canopy area through plantation. The water logged area can be planted with *Eucalyptus maculatus* and *Casunina equisetifolia*. Besides this, on the upland area adjoining river bank the following species are recommended for plantation:

Sl.No.	Scientific Name	Local Name
1.	*Sygigium cumini*	Jamu
2.	*Artocarpus lakoocha*	Panas
3.	*Terminalia arjuna*	Arjun
4.	*Salvodora persika*	Sahada
5.	*Azadirachta indica*	Nimba
6.	*Cassia fistula*	Sunari
7.	*Ficus benghalensis*	Bara
8.	*Ficus religiosa*	Aswastha
9.	*Dalbergia sisso*	Sisu
10.	*Anacandium occidantalis*	Kaju

Road Side Plantation

The density of plants on the road approaching to Hirakud as well as road approaching the ash mund (right side of the road) is very sparse and thin. Presently this plantation is contributing to only 1.2 per cent of the total canopy area of plants in Hirakud. There is ample scope to increase plantation in these areas by another 20,000 sqm. The species recommended for the road side plantation are as follows:

Sl.No.	Scientific Name	Local Name
1.	*Pongamia pinnata*	Karanja
2.	*Alstonia scholaris*	Chhutim
3.	*Azadirachta endica*	Neem
4.	*Ficus benghalensis*	Bara
5.	*Ficus religiosa*	Aswastha
6.	*Cassia fistula*	Sunari
7.	*Spathodea campanulateae*	Panituri
8.	*Lagerostroemia flosreginac*	Patali
9.	*Albizzia labbeck*	Siris
10	*Albizzia procera*	White siris

Avenue/Colony Plantation

The total avenue plantation presently contributes to only 8 per cent of the total canopy area, this can further be maximized by planting the following trees in the open spaces available, keeping in view their canopy structure and phenology.

Sl.No.	Scientific Name	Local Name
1.	*Delonix regia*	Krishnachuda
2.	*Peltophorum pterocarpus*	Radhachuda
3.	*Mimosups elougi*	Baula
4.	*Cassia Fistula*	Sunari
5.	*Lagerostromia flosregineae*	Patali
6.	*Pongamia pinnata*	Karanj
7.	*Ficus benghalensis*	Bara
8.	*Bahunia variegata*	Kanchan
9.	*Spathodea companulatae*	Panituri
10.	*Azadirachta indica*	Neem
11.	*Syzigium cummini*	Jamun

Ash Pond

In the ash pond areas the following species are recommended for their high dust collecting capacity and air pollution tolerance indices.

Species having High Dust Collecting Capacity

1. *Tectona grandis*
2. *Shorea robusta*
3. *Terminalia arjuna*
4. *Bahumia variegata*
5. *Agadirachta indica*
6. *Madhuca indica*

Species having Air Pollution Tolerance Index

1. *Albizzia lebbeck*
2. *Azadirachta indica*
3. *Emblica officinalis*
4. *Tamarindus indica*

Besides this, species like *Delonix regia, Pongamia pinnata. Dalbergia Sisso,* which has successfully been grown in ash mound near the proposed new ash pond may also be selected for plantation mixed plantation type. The plantation should be mixed plantation type.

Acknowledgements

The authors are thankful to their revered teacher Prof. M.C. Dash for his advice to undertake such studies in the interest of Environmental protection. Support of Hindalco in form of research project is gratefully acknowledged.

References

Brahmam, M. and Saxena, H.O., 1995. *The Flora of Orissa*, Vols. 1 to 5. Publ. RRL, OFDC, Bhubaneswar.

CSIR, 1994. *Plants for Reclamation of Wastelands*. Publication and Information Directorate, Council of Scientific and Industrial Research, New Delhi.

Das, T.M. and Pattanayak, P., 1978. Dust filtering property of green Plants and the effect of air borne particles in growth and yield. In: *Proceedings of the Int. Symp. on Env. Agents and their Biological Effects*, Hyderabad, pp. 224–234.

Hanes, H.H., 1961. *Botany of Bihar and Orissa*, Vols. 1 to 3. BSI, Calcutta

Ghosh, A.K., 2005. Green city: Indian perspectives. *Environica*, Vol. 3 No. June, p. 7–12.

Juwarkar, Asha A. and Singh, Sanjeev Kumar, 2006. Carbon sequestration through terrestrial ecosystem: An ecofriendly solution to global warming. In: *Environmental Issues and Options*, (Eds.) C.S.K. Mishra, J.W. Kim and Amita Saxena. Daya Publishing House, New Delhi, p. 1–31.

Larcher, W., 1995. *Physiological Plant Ecology*. Springer, 506 pp.

Maiti, S.K., 1993. Dust collection capacity of plants growing in coal mining areas. *Ind. J. Environ. Protection*, 13(4): 276–280

Misra, R., 1968. *Ecology Workbook*. Oxford & IBH Publishing Co., New Delhi, 238 pp.

Rawat, J.S. and Banerjee, S.P., 1996. *Energy, Environment Monitor*, 12(2): 104.

Chapter 10

Studies on the Liver of Rohu in Relation to Environmental Stress from the Lakes of Bangalore, Karnataka

S.G. Raghu Prasad[1]* and Bela Zutshi[2]**

[1]Research Scholar; [2]Senior Lecturer

Aquatic Biology and Fish Toxicology, Department of Zoology, Jnanabharathi, Bangalore University, Bangalore – 56, India

ABSTRACT

Fishes are indicators of level of water pollution. No lake can be considered non-polluted at the present state of living condition of human beings. In the present investigation industrial effluents and sewage fed lakes have been surveyed and changes in the HSI and histophysiology of liver of *L. rohita* has been recorded.

According to the level of various physico-chemical parameters of water *e.g.*, pH. BOD, COD, DO, phosphates, sulphates, nitrates, salinity, zinc and copper, Chowkalli lake is considered to be more polluted than Hebbal lake of Bangalore. The fishes sampled from the two lakes showed severity of degenerative changes including liver cord disarray, vacoulation of cytoplasm and necrosis of hepatocytes in histology of fiver, which is detoxification organ.

The histopathological changes in liver coupled with significant decrease in hepato-somatic index (HSI) attributed hepatic inactivation in all the various stages of fishes. It was more significant in Chowkalli as compared to Hebbal lake.

Keywords: *Fish, Labeo robita, HSI, Liver, Polluted lake.*

---

* E-mail: bela_zutshi@yahoo.co.in; ** E-mail: sgr_prasad2003@yahoo.com.

Introduction

The quality of water is determined by its physico-chemical parameters. Since all the freshwater bodies are continuously been fed with various non-point source pollutants, no river or lake can be considered to be non-polluted. The level of pollution can be assessed by the health of fish thriving in it. According to Hora (1942) and Verma and Daleia (1975) pollutants in river and lakes affect fish fauna. Changes occurring in the biochemical and histological characteristics of various tissues of fishes provide a comparative measure to know the health of fish fauna, (Saravanan, Aneez and Harikrishan, 2000). Among different organs of fish subjected to pollution, liver is worst affected (Brown, 1954). It is the first organ to be exposed to toxic compounds that have been ingested, but it has the ability to degrade them and can subsequently be damaged. Hence it is a target organ of various xenobiotic substances.

Although work has been done on various tissues of fishes by exposing them to various types of pesticide (Chandra, Ram and Singh, 2004) and heavy metals (Haffor and Abou-Tarboush, 2004), but there is paucity in the work relating to the effect of water quality on HSI and histophysiology of liver of fresh water fishes reared in polluted lakes.

Hence in the present investigation, a study has been conducted on the histological alterations in the liver of *L. rohita* reared in fresh water lakes, *viz.*, Hebbal and Chowkalli lakes of Bangalore.

Materials and Methods

The physico-chemical parameters of the two lakes (Table 10.1), Hebbal and Chowkalli were assessed to analyse the influence of these parameters on the histological alterations of the liver of *L. rohita* reared in the two freshwater lakes.

Fishes of various sizes and weights were collected randomly with the help of gin and drag net from Hebbal and Chowkalli lake and were brought to laboratory in living condition. The live fishes were anaesthetized by MS 222. Weight and length of the fishes was recorded and they were sacrificed for histological studies. Liver was studied and no parasitic infection was observed. It was dissected and weight was recorded. Small pieces of liver were fixed in format-saline and were processed for paraffin embedding. Sections were cut at 5µ and were stained with haematoxylin and counter-stained with eosin for further histological observation.

Hepatosomatic Index (HSI)

The hepatosomatic index was determined by the following formula:

$$HSI = \frac{\text{Weight of liver}}{\text{Weight of Fish}} \times 100$$

Results

Morphology and Histology of Teleost Liver

In teleost, the liver is a bilobed gland and comprises of two tissue compartments, the parenchyma (epithelial cells or hepatocytes that perform the organs major functions) and stroma (hepatopancreas, bile duct, blood vessels and connective tissue) (Figure 10.1).

Histological structure of liver reveals the presence of dual blood supply *i.e.*, the portal vein and the hepatic artery, which enter into the liver for the exchanges between the blood and hepatocytes.

Hepatocytes constitute 80 per cent of cell population of river. Exocrine pancreatic tissue (hepatopancreas) is a pronounced feature in liver and is clearly visible as darkly stained tissue around the hepatic portal veins. It can be differentiated from hepatic tissue by its acinar arrangement and its characteristic stain with H and E. Each exocrine pancreatic cell has a basal portion containing homologous cytoplasm and an apical part containing large number of zymogen granules (Figure 10.1).

Hepatosomatic Index

The lowest HSI was recorded in female *rohu* during spawning season and the highest was observed in the developing phase. Subsequently HSI was found to decrease during maturation phase and an increase during spent and resting phase.

It was observed that in major carps of Hebbal and Chowkalli lake the gonadal differentiation starts at 600–700 g and it starts maturing at 1000–1200 g fish weight. The fish reaches spawning condition at 2000–2300 g of its weight This is evidenced by our investigation which shows an increase in HSI from 600–1000 g and a subsequent decrease from 1200–2300 g. Highest value of HSI is noticed at 1000–1200 g of fish weight which is concomitant with maturation phase of ovarian growth and development which can in turn be correlated to vitellogenesis.

In the present investigation HSI of various stages of *rohu* of Hebballake shows higher values when compared to that of fishes from Chowkalli lake (Table 10.2 and Figure 10.1A).

Table 10.1: Seasonal Changes in Physico-chemical Parameters of Two Lakes Hebbal (A) and Chowkalli (B)

Test	*Winter*		*Summer*		*Rainy*	
	A	*B*	*A*	*B*	*A*	*B*
Temperature (°C)	22	21	26	25	22	23
Colour	7.6	7.8	6.8	8.6	9.5	7.2
Turbidity (NTU)	11	12	6	8	17	58
pH	7.2	7.5	7.2	8.0	7.1	7.5
TDS	577.2	694.2	620	800	380	490
Conductivity (mho s/cm)	660	980	756	1200	799	1037
Total Alkalinity	312	422	354	499	235	377
Sodium	169	191	121	181	112.5	142.7
DO	4.8	3.8	4.2	3.4	6.3	4.9
BOD	5.5	6.6	6.3	7.2	2.8	5.2
COD	27	53	14.4	7.3.04	25	43
Nitrates	13.3	29.2	12.5	14.5	15.5	17.1
Phosphate	1.5	1.05	0.59	0.55	0.39	0.52
Sulphates	25.58	32.82	14.8	24.68	49.36	74.05
Iron	0.39	0.42	0.06	0.13	0.46	0.98
Zinc	0.023	0.033	0.02	0.04	0.2	0.32

Except Temperature, pH, conductivity and Turbidity all values are in mg/l.

Histopathology of Liver in Polluted Lakes

According to the physico-chemical parameters of the two lakes (Table 10.1), Hebbal and Chowkalli, the latter is considered more polluted and the fishes sampled from the latter showed greater magnitude of histopathological changes in the tissue when compared to those of Hebbal lake. Histopathological investigation for 200–300 g fishes sampled from Hebbal lake revealed granulated hepatocytes with distinct centrally located nuclei and nucleoli. Hepatopancreatic cells were smaller in size and were mildly granulated with zymogen granules (Figure 10.2). But in fishes from Chowkalli lake showed mild necrosis and vacuolization in hepatocytes. A few cells showed degranulated condition and had pyknotic nuclei. Release of blood cells from the vein and arteries causing haemorrhage was also observed in the tissue. Large numbers of macrophages were observed among hepatocytes (Figure 10.3). The hepatopancreatic cells show degeneration of variable magnitude, which was marked by darkly stained specks of necrotic nuclei (Figure 10.4).

Table 10.2: Periodic Variation in HSI of *Labeo rohita* According to its Maturation in Hebbal and Chowkalli Lake

Sl.No.	Hebbal Lake				Chowkalli Lake			
	Fish		Liver Weight (g)	HSI	Fish		Liver Weight (g)	HSI
	Total Length (cm)	Weight (g)			Total Length (cm)	Weight (g)		
1.	24.5	200	2.91	1.105	23	200	1.41	0.705
2.	26	300	2.73	0.910	24	300	1.89	0.630
3.	25	350	2.49	0.785	27	350	1.92	0.548
4.	26	400	2.85	0.712	25	400	1.95	0.487
5.	32	500	6.20	1.240	27	500	2.94	0.588
6.	35	600	8.14	1.356	30	600	4.20	0.700
7.	37	700	9.8	1.402	34	700	5.70	0.814
8.	40	800	12.5	1.562	37	800	7.90	0.987
9.	42	900	14.6	1.642	44	900	12.56	1.395
10.	45	1000	17.2	1.727	42	1000	14.30	1.430
11.	47	1200	19.5	1.625	44	1200	15.20	1.265
12.	49	1400	20.2	1.442	47	1400	16.21	1.080
13.	51	1700	23.6	1.388	55	1700	17.90	1.052
14.	52	2000	24.2	1.210	51	2000	19.73	0.986
15.	54	2300	25.0	1.086	54	2300	20.30	0.946

500–700 g fishes from Hebbal lake showed comparatively large and darkly stained hepatocytes. More number of aggregations of granulated hepatocytes was observed in 500 g (Figure 10.5) when compared to that of 200 g of fishes (Figure 10.2). The hepatocytes had a distinct cell wall and prominent rounded nuclei. The granulated cytoplasm attributed induction of its biosynthetic activity. The liver tissue it 500 g fishes from Chowkalli lake, showed various histopathological changes, like vacuolation or hydropic degeneration and aggregation of hepatocytes. To restore the degenerated tissue, regeneration of hepatocytes and accumulation of nucleus without cytoplasm is seen to occur in groups.

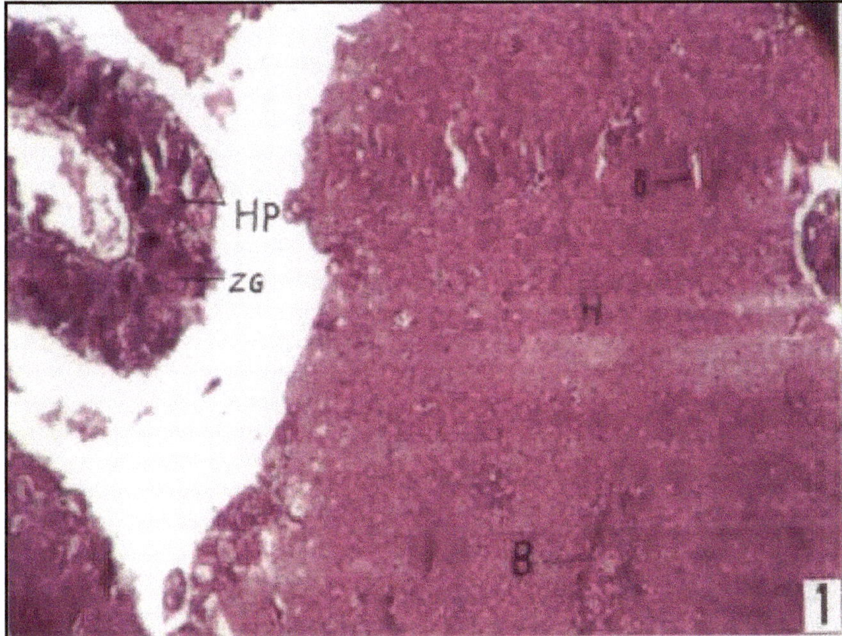

Figure 10.1: Microphotograph of Liver of *L. rohita* Showing Hepatocytes (H), Hepatopancreas (HP), Sinusoids (S) and Blood Cells (B). Note the exocrine pancreatic cell with a basal portion containing homologous cytoplasm and an apical part containing large number of zymogen granules (ZG). H&EX400.

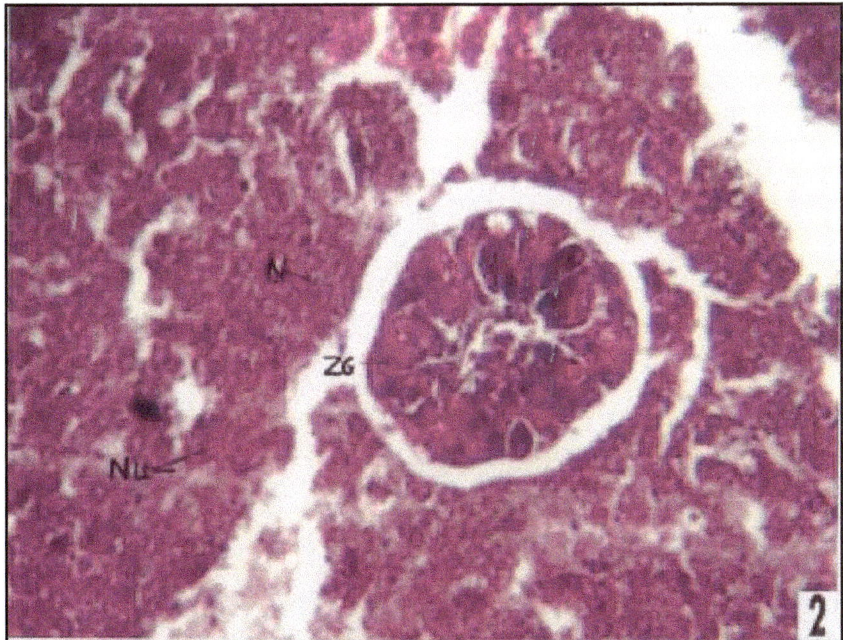

Figure 10.2: Hepatocytes from Hebbal Lake Revealing Scanty Granulation with Centrally Located Nuclei (N) and Nucleoli (Nu). Note small and mildly granulated with zymogen granules (ZG) hepatopancreatic cells. H&E X400.

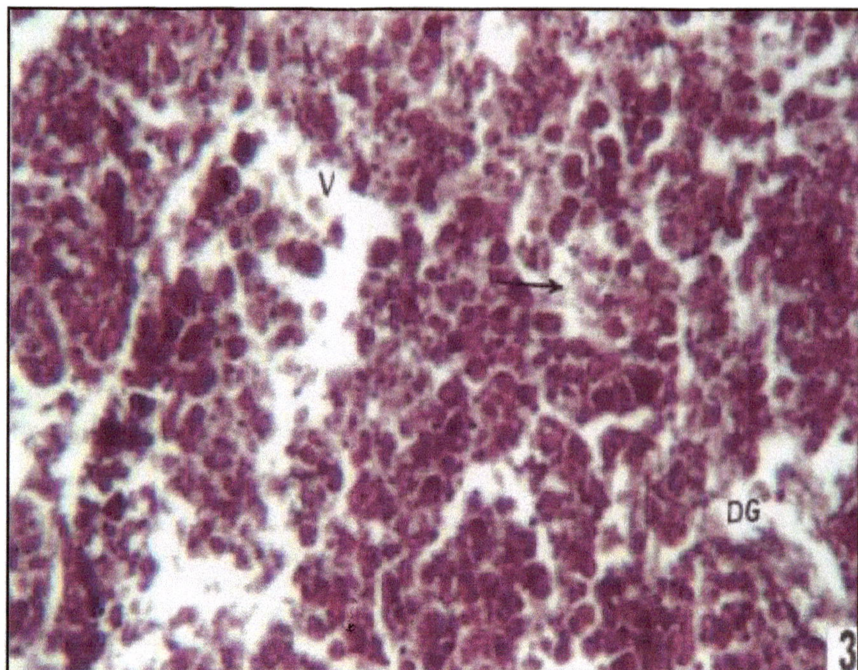

Figure 10.3: Hepatocytes from Chowkalli Lake Showing Mild Necrosis (←), Vacoulation (V) and Degranulation (DG) with Pyknotic Nuclei. H&E X400.

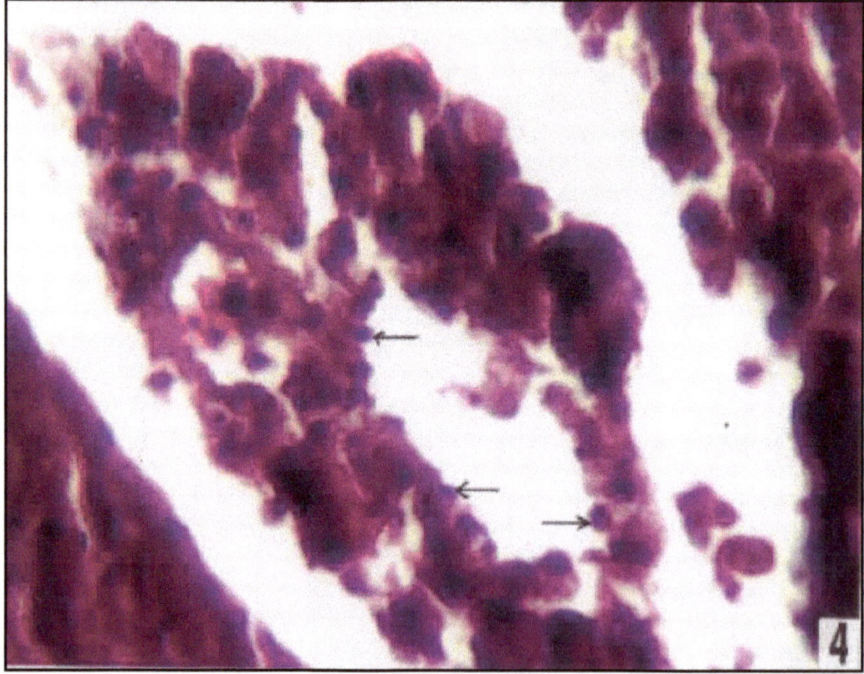

Figure 10.4: Hepatopancreatic Cells from Chowkalli Lake Revealing Degeneration and Darkly Stained Specks of Necrotic Nuclei (←). H&E X400.

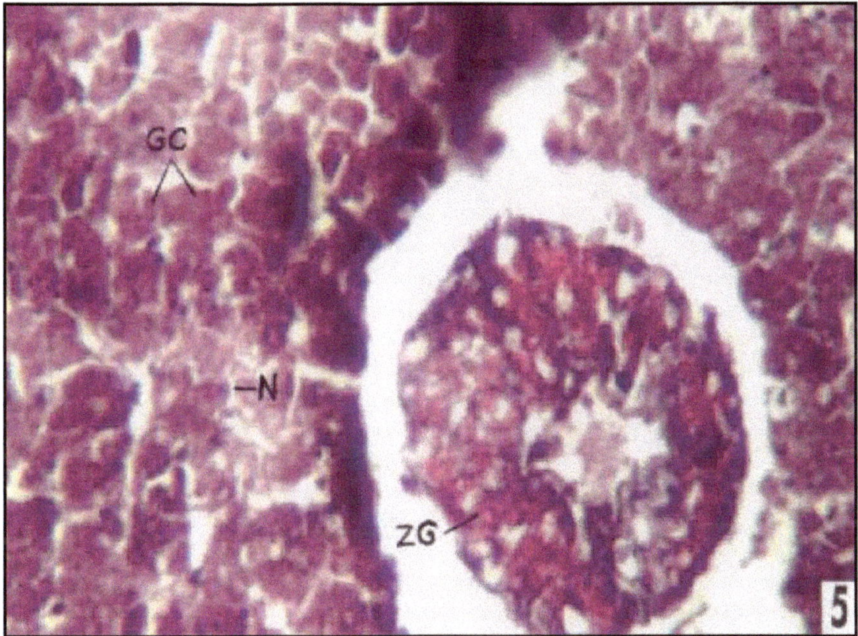

Figure 10.5: Hepatocytes from Hebbal Lake Showing Darkly Stained and Aggregation of Granulated Cells (GC) with Distinct Cell Wall and Prominent Rounded Nuclei (N). Mildly granulated hepatopancreatic cells with zymogen granules (ZG) were seen. H&E X400.

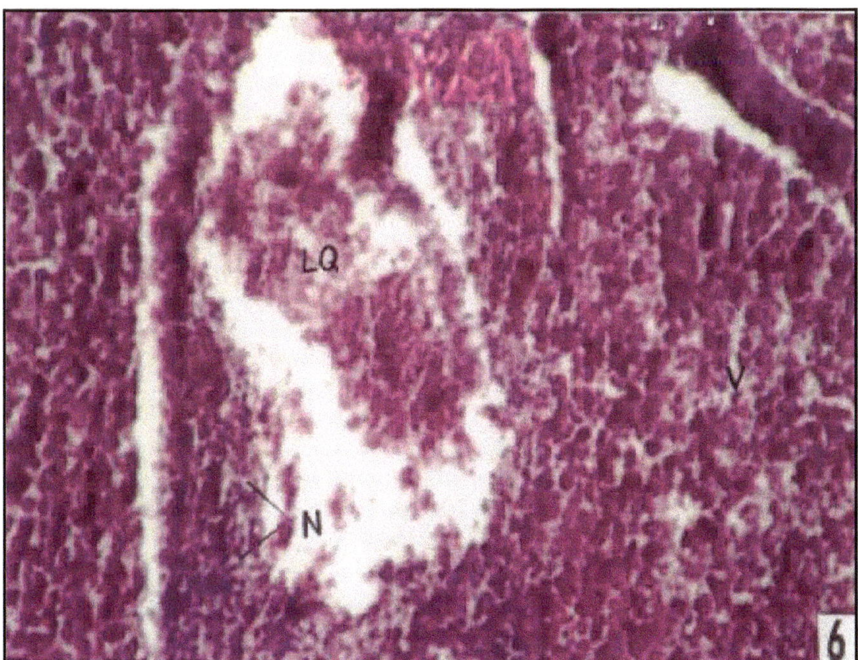

Figure 10.6: Hepatocytes from Chowkalli Lake Showing Vacoulation (V) and Hydropic Degeneration. Cytolytic necrosis, regeneration and accumulation of nucleus (N) without cytoplasm are seen in groups. Note liquefication of hepatocytes (LQ). H&E X400.

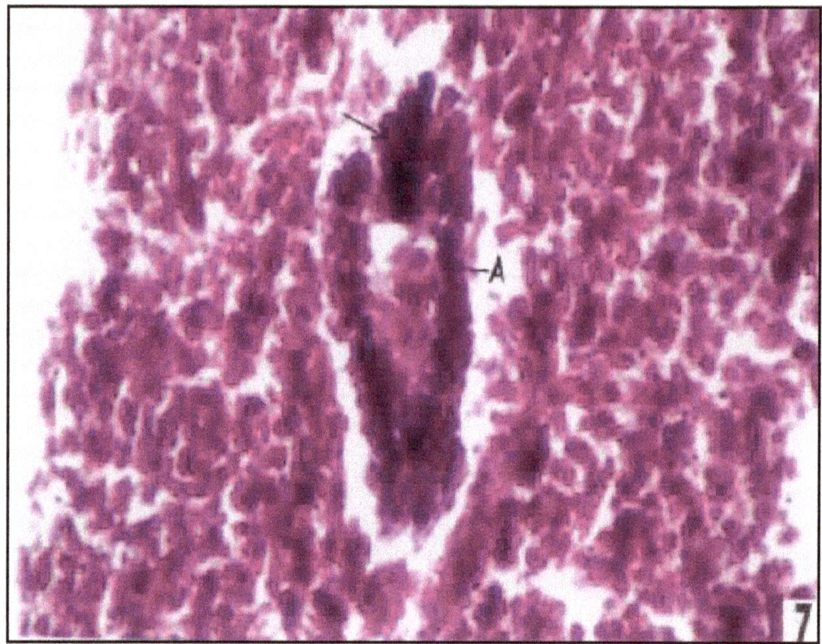

Figure 10.7: Hepatopancreas from Chowkalli Lake Showing Darkly Stained Acinar Cells (A). Few areas of hepatopancreatic cells revealed accumulated coagulum (←) within the focal necrotic zone. H&E X400.

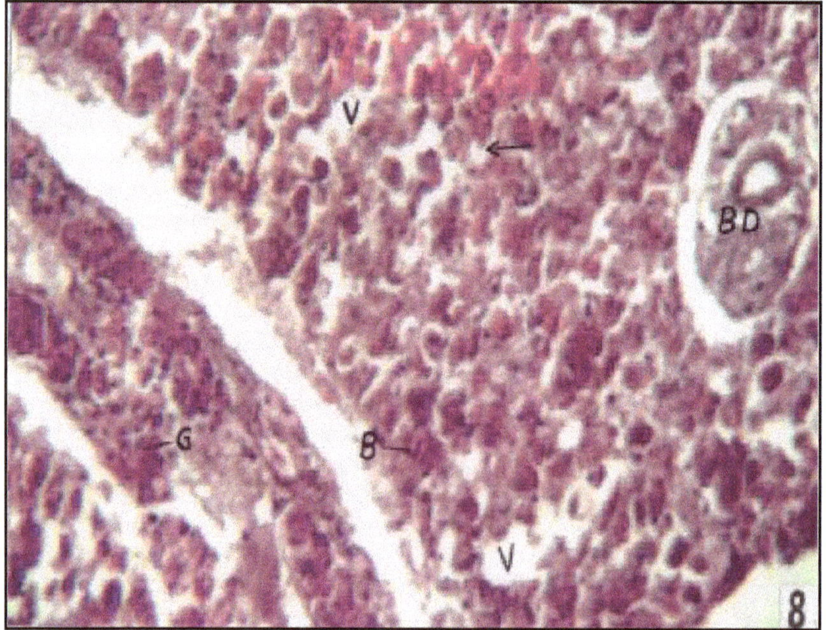

Figure 10.8: Liver Cord Disarray, Collapsed Sinusoids (←), few Blood Cells (B), Large Numbers of Degranulated and Vacuolated (V) Hepatocytes from Hebbal Lake can be Seen. Bile duct (BD) is also present. Active granulated hepatopancreatic cells (G) are noticeable. H&E X400.

Cytolytic necrosis is seen in the parenchyma tissue. It is characterized by liquefication of hepatocytes (Figure 10.6). There is an increase in the size and number of the macrophages. Hepatopancreatic cells of fish sampled from Hebbal lake contained, rounded nuclei and showed vacuolation and lightly stained cytoplasm when compared to that of fishes from Chowkalli lake. Few areas of hepatopancreatic cells revealed accumulated coagulum within the focal necrotic zone in the fishes from Chowkalli lake (Figure 10.7).

Maximum number of hepatocytes are sparsely granulated in 1200–1500 g fishes from Hebbal lake. Liver cord disarray, collapsed sinusoids, large numbers of degranulated and vacuolated hepatocytes are observed. Hepatopancreatic eels are more in number and are granulated it signifies their role in energy mobilization (Figure 10.8). There is an increase in number of blood cells and necrotic hepatocytes in the parenchyma. Severe histopathological changes in the liver tissue are noticed in the fishes collected from Chowkalli lake. There is an increase in cytoplasmic vacuolation, nuclear disintegration and destruction of nuclear membrane leading to complete lysis of hepatocytes. Clumping of nuclear materials in hepatopancreatic cells of Chowkalli lake is evident (Figure 10.9). Haemorrhage of blood vessels and hepatic lesions can be seen due to exposure to xenobiotic substances in the polluted water of Chowkalli. There is an increase in area and number of individual macrophage aggregates, which is a biomarker to indicate stress due to various pollutants and changes in physico-chemical parameters of water (Figure 10.9).

Discussion

Discharge of industrial effluents and other domestic sewage from various sources into the nearby water bodies, poses a threat for human health, survival of fish and other aquatic organisms. Katz and

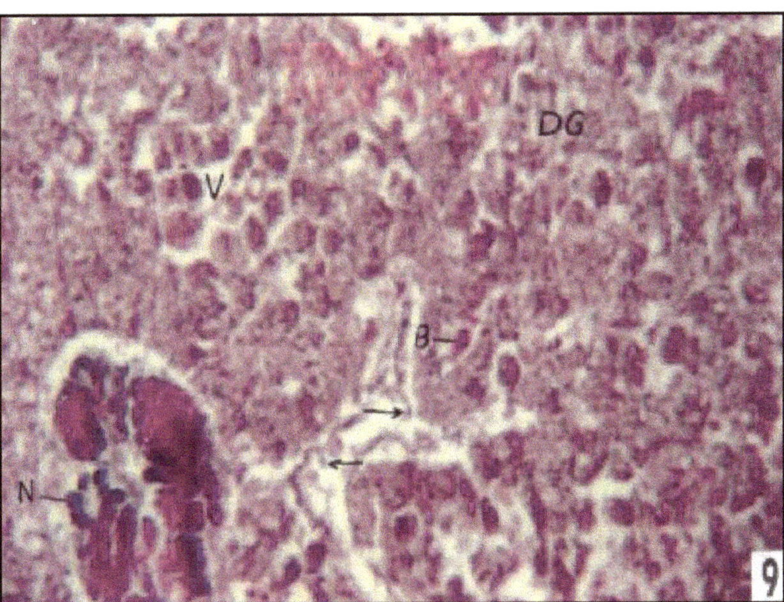

Figure 10.9: Hepatoyytes from Chowkalli Lake Showing an Increased Cytoplasmic Degranulation (DG) and Vacoulation (V), Nuclear Disintegration and Destruction of Nuclear Membrane (←) Leading to its Complete Lysis. Haemorrhages of blood vessels (B) are also seen. Clumping of hepatopancreatic nuclear material (N) can be seen. H&E X400.

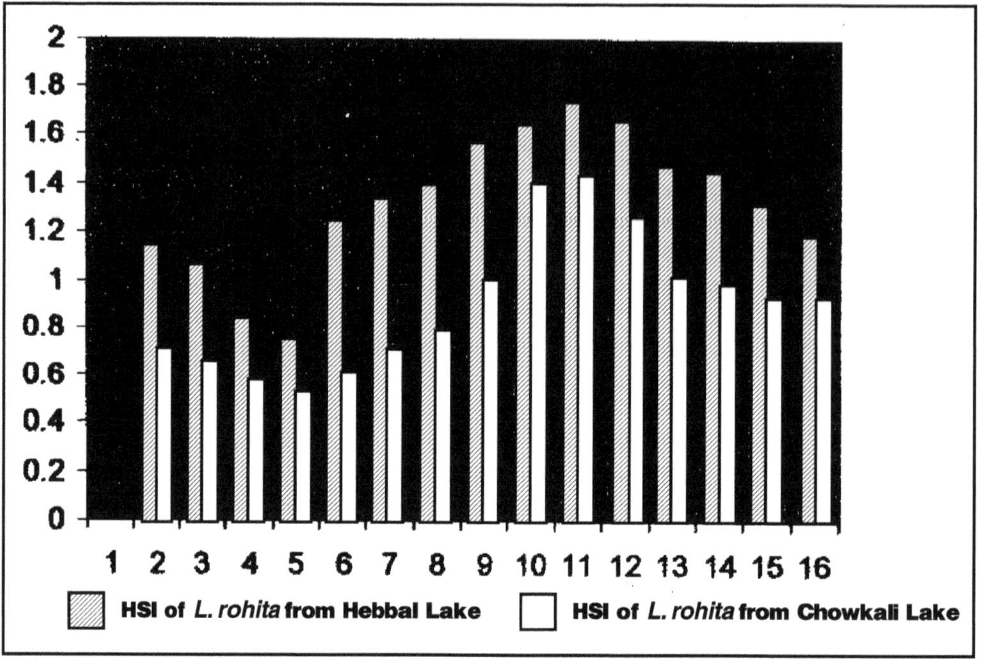

Figure 10.10: Periodic Variation in HSI of Rohu According to its Maturition in Hebbal and Chowkalli Lake

Gaufin (1952) reported that number of species of fish available at a particular place and their relative abundance should be considered as indicators of pollution. Duling the present investigation HSI was noticed to be in increasing pattern in relation to the weight in fishes sampled from Hebbal as well as Chowkalli lake (Table 10.2 and Figure 10.10) suggesting its correlation with phases of ovarian development and induced synthetic activity in liver.

In the present study, HSI value shows a decrease, in the fishes sampled from Chowkalli lake than Hebballake (Table 10.2) and (Figure 10.10). Due to constant exposure to xenobiotic substances it may overwhelm the liver's detoxification capability causing some degree of structural damage within the liver parenchyma. Due to degeneration and necrosis of hepatocytes; the liver tissue loses its weight and thus there is a decrease in HSI of fishes from Chowkalli. Similar observations have been reported in the liver of *C. pundatus* exposed to mercury and ammonia, by Banerjee and Bhattacharya (1997).

Obvious histological changes supportive of activation of hepatopancreatic cells were exemplifying its role in mobilization of energy reserve from liver to gonadal tissue with concurrent advancement with ovarian development and growth, which can be correlated to weight of the fish.

Progressive necrosis and degranulation was seen in hepatic tissue of fishes sampled from Chowkalli lake. Similar observation were reported by Smith and Piper (1975), in trout due to chronic exposure to ammonia. In the present study loosening of hepatocytes and distention of cells, pyknotic hepatocytes, centrolobular degeneration and scattered necrotic cells were observed. Similar changes have been recorded in the hepatocytes of *Anabas testudineus* by Matton and LaHam (1969). This is considered to be the result of tissue hypoxia. Sastry and Sharma (1979), reported degeneration and disintegration of hepatic cells and its nuclei and vacuolation of cytoplasm in *Channa punctatus* due to its exposure to sub-lethal concentration of endrin. Our observations about release of blood cells causing haemorrhage in liver tissue are well corroborated with the insight that pollutants caused initial morphologic alterations and destruction of sinusoidal endothelial cells. With the loss of sinusoidal endothelium, hemorrhage into the space of Disse and interhepatic space was associated with hepatocyte shrinkage and hemorrhagic neaosis (Dray and Hinton, 1988a).

Macrophage aggregates and cytolytic necrosis involving rapid disintegration of hepatocytes are used as biomarker lesions to indicate stress (Roberts, 1975). In the present study similar changes were seen in the hepatic parenchyma of fish sampled from Chowkalli lake due to exposure to industrial effluents and other sewage pollutants.

Thus, the histopathological analysis of liver of fishes from Chowkalli lake reveals the level of cellular alterations and damage caused in the course of detoxifying the fish body so as to acclimatize itself in highly polluted water.

References

Banerjee, S. and Bhattacharya, S., 1997. Histopathological changes induced by chronic non-lethal level of elesan, mercury and ammonia in the liver of *Channa punctatus* (Bloch). *J. Env. Biol.*, 18(2): 141–148.

Brown, V.M., 1954. The prediction of acute toxicity of river waters to fish. In: *Proc. IV British Course Fish Conf.*, Liverpool University.

Chandra, Smita, Ram, R.N. and Singh, I.J., 2004. First ovarian maturity and recovery response in common carp, *Cyprinous carpio* after exposure to carbofuran. *J. Environ. Biol.*, 25(3): 239–249.

Droy, B.E and Hinton, D.E., 1988a. Allyl formate–induced hepatotoxicity in rainbow trout. *Mar. Environ. Res.*, 24: 259–264.

Haffor, A.S. and Abou-Tarboush, E.M., 2004. Testicular cellular toxicity of cadmium: Transmission electron microscopy examination. *J. Environ. Biol.*, 25(3): 251–258.

Hinton, D.E and Lauren, J.L., 1990. Integrative histopathological approaches to detecting effects of environmental stressors on fishes. *Am. Fish. Soc. Symp.*, 8: 51–66.

Hora, S.L., 1942. A short note on the pollution of streams in India and its likely effects on fishes. *Reports of Fish Committee*, I.C.A.R., Delhi.

Katz, N. and Gaufin, A.R., 1952. The effect of sewage pollution on the fish population of mild western stream. *Trans. Amer. Fish Soc.*, 8: 156–162.

Matton, P. and LaHam, A.N., 1969. Effects of the organophospate Dylox on: Rainbow trout larvae. *J. Fish. Res. Board Can.*, 26: 2193–2200.

Roberts, R.J., 1975. Melanin-containing cells of teleost fish and their relation to disease. In: *The Pathology of Fishes*, (Eds.) W. Ribelin and G. Migaki. University of Wisconsin Press, Madison, WI, pp. 399–428.

Roberts, R.J., 1975. The pathophysiology and systematic pathology of teleosts. In: *The Pathology of Fishes*, (Eds.) W. Ribelin and G. Migaki. University of Wisconsin Press, Madison, WI, pp. 399–428.

Saravanan, T.S., Mohamed, M. Aneez and Harikrishan, R., 2000. Studies on the chronic effects of endosulfan on blood and liver of *Oreochromis mossambicus*. *J. Ecol. Res. Biocon.*, 1(2): 24–27.

Sastry, K.V. and Sharma, S.K., 1979. Endrin induced hepatic injury in *Channa punctatus* (Ham.). *Indian J. Fish*, 26: 250–253.

Smith, C.E. and Piper, R.G., 1975. Lesions associated with chronic exposure to ammonia. In: *The Pathology of Fishes*, (Eds.) W. Ribelin and G. Migaki. University of Wisconsin Press, Madison, WI, pp. 497–514.

Verma, S.R. and Dalela, R.C., 1975. Studies on the pollution of the Kalinadi by industrial wastes, Mansurpur. Part 2: Biological characteristics of the river. *Acta. Hydrobiol.*, 3: 259–274.

Chapter 11

Impact of Some Medicinal Plants on the Growth of *Trichosporon begeilii*

S. Goswami, J. Das and R.B. Srivastava
Division of Biotechnology, Defence Research Laboratory, Post Bag No. 2, Tezpur – 784 001, Assam

ABSTRACT

The antifungal activity of methanol extract of eight medicinal plants *viz.*, *Meyna spinosa*, *Duranta repens*, *Sida carpinifolia*, *Tephrosia purpurea*, *Cestrum nocturnum*, *Trachyspermum ammi*, *Lippia nudiflora* and *Ludwigia parvif/ora* were studied against clinically important pathogen *Trichosporon beigelii*. The medicinal characteristics of these plants were compared with commercially used antibiotic clotrimazole. The antimicrobial assay was done by agar well diffusion method. The fruit extract of *Duranta repens* showed the minimum inhibition zone diameter (16 mm). Among various plants used in the present investigation, *Trachyspermum ammi* and *Cestrum nocturnum* were found to be more potent with inhibition zone of diameter 42.3 and 28 mm respectively, hence they can be further investigated as for potent commercial antifungal agents against *Trichosporon beigelii*.

Keywords: *Antifungal activity, Trichosporon beigelii, Medicinal plants.*

Introduction

Northeast region of India is a vast repository of therapeutically potent flora. In the year 2000, Myers *et al.*, discovered a vast area as biodiversity hot spot near Indo-Burma border of N.E. India. People living in this region are the richest in repository of acquired experiences and knowledge. As per an estimate by WHO, about 80 per cent of the population of the developing countries rely on traditional medicines for their health care needs which are mostly the plant based drugs (Prabhuji *et al.*, 2005). The dependence of man on medicinal plants is next to food for life sustenance and as ancient as human civilization. But, this interest becomes vain with the introduction of synthetic and semi synthetic drugs, which give miraculous results. The rapid growth of industrialization, urbanization and ruthless collection of plants leads towards the shrinking of most of the important flora day by day. Therefore, systematic study of medicinal plants, development of drugs through bio-prospecting and systematic conservation of medicinal plants is very important.

The present study is an attempt in this direction to find out potent antifungal plants against *Trichosporon beigelii*, which is the most significant pathogen in the genus Trichosporon. The genus Trichosporon included seventeen species and five varieties. Six of the species were associated with infections in humans (Anonymous, 2006). It causes white piedra characterized by white nodules on hair shaft of axilla, moustache and beard. It is also involved in a variety of opportunistic infections in the immuno suppressed patient. Due the emergence of multi-drug resistant strain and wide spread belief that green medicine is healthier than synthetic product (Pushpangadan, 1996) has revived the interest in the natural drugs. Hence the research focus has now turned towards plant products. The crude extracts of eight medicinal plants *viz., Meyna spinosa, Duranta repens, Sida carpinifolia, Tephrosia purpurea, Cestrum noctumum, Trachispermum ammi, Lippia nudiflora, Ludwigia parviflora* have been screened in the present study against *T. beigelii* under laboratory conditions.

Materials and Methods

Plant Material

The plant materials choosen for this study were collected from the natural habitat in and around Sonitpur district, Assam. The collected plant materials (Leaves, fruits and seeds) were brought to the laboratory in polythene bags. The plant materials so collected were identified by expert of the institute. Fresh plant materials were washed under running tap water, shade dried and homogenized to fine powder and processed accordingly.

Extraction of the Plant Material

50 g plant material of each were extracted with 500 ml methanol using soxhlet extractor at 50°C. The extracts were filtered using Whatman filter paper No. 1 and allowed to concentrate at room temperature. The crude extracts thus obtained were stored in air tight bottles for further studies.

Microorganism

The test organism *T. beigelii*, tested for the biological evaluation was the clinical isolate obtained from School of Tropical Medicine, Kolkata. The fungal culture maintained on agar slant was recovered by inoculating in Saubaraud Dextrose Broth (SDB).

Preparation of Inoculum

Inoculum was prepared by adding one loop full of test pathogen in 50 ml of SDB and then incubated at 28±2°C for 48 hours.

Preparation of Test Extracts

Each test extract was prepared by dissolving 0.2 g of crude extract in 1 ml Dimethyle sulphoxide (DMSO) (w/v), filter sterilized through Millipore filter (0.2 µm pore size) and was used as working stock. Clotrimazole 1000 µg/ml was used as standard for comparison.

Agar Well Diffusion Method

The antifungal activity of the solvent extract was done by agar well diffusion method (Grammar *et al.*, 1976). Sterilized Saubaraud Dextrose Agar medium was poured into sterile petriplates and allowed to solidify. After solidification of the medium the inoculum was vortexed and 100 µl of the inoculum was spread evenly over the surface of the agar plates using a glass spreader. A well of 8 mm diameter was made in each plate with a sterile cup borer. The test extract (100 µl) was introduced into the well. The extracts were allowed to diffuse at room temperature for 2 hours. The plates were then incubated at 28±2°C for 48 hrs. The experiment was replicated thrice. The efficacy was determined by measuring the diameter of zone of inhibition exhibited by the extracts against test pathogen. A negative control containing 100 µ DMSO was also taken.

Activity Index Determination

The activity index was determined by the following formula (Jain and Sharma, 2005):

$$\text{Activity Index} = \frac{\text{Zone of Inhibition of Extract}}{\text{Zone of Inhibition of Standard}}$$

Analysis of Per cent of Inhibition

Percent of inhibition was calculated according to the following formula (Vyas *et al.*, 2006):

$$\text{Per cent of Inhibition} = \frac{\text{Inhibition Zone in mm}}{\text{*Control}} \times 100$$

* Growth zone is equal to plate diameter *i.e.* 80 mm as growth occurs all over the surface of the agar plate.

Results and Discussion

Results of screening of antifungal activity of the crude extracts were summarized in Table 11.1. It was evident from the results that all the plant extracts showed significant growth inhibition ranging from 16 mm to 42.3 mm inhibition zone diameter against the test fungus and was comparable with that produced by the standard drug clotrimazole. The highest zone of inhibition was exhibited by the seed extract of *Trachyspermum ammi* (42.3 mm) followed by extract of *Cestrum nocturnum* (28 mm) and *Lippia nudiflora* (27.3 mm). The negative control was checked and the absence of zone of inhibition indicated susceptibility of the medium.

When we consider the activity index it was clear from the results that all the extracts showed antifungal activity against the tested fungus. The difference in the activity completely depends upon the different secondary metabolites present in different plants. These secondary metabolites act as an antifungal agent, which inhibit the fungal growth. Active compound of most of the medicinal plants have been isolated and introduced in the modern system of medicine (Senthikumar *et al.*, 2005). Medicinal plants having antifungal activity were reported by several authors. (Sauwalak *et al.*, 2005; Nasser *et al.*, 2005). Antifungal agents, currently in market, are limited due to their toxicity, low effectiveness and with cost during prolonged treatment. Therefore, there is a need to develop antifungal agents, which can satisfy the need under present scenario (Nair and Chanda *et al.*, 2006).

Table 11.1: Antifungal Activities of Methanolic Extracts of the Medicinal Plants

Botanical Name	*Part Used*	*Inhibition Zone Diameter (mm)*
Meyna spinosa	Fruit	18
Duranta repens	Fruit	16
Sida carpinifolia	Leaf	24.5
Tephrosia purpurea	Leaf	23
Cestrum nocturnum	Leaf	28
Trachyspermum ammi	Fruit	42.3
Lippia nudiflora	Leaf	27.3
Ludwigia parviflora	Seed	22
Clotrimazole	–	12

The successful search for medicinal property of any plant largely depends on the organic solvent utilized in the process of extraction. There is no precise criteria for selection of organic solvents for extraction as chemical nature of antimicrobial compound present on the plant cannot be forcasted. The activity index as revealed in the Table 11.2, was maximum in seed extract of *Trachyspermum ammi* (3.52) followed by leaf extract of *Lippia nudiflora* (2.89), while it was found least in case of seed extract of *Duranta repens* (1.3). The authors also recorded the antifungal potentiality of these extract against *Candida albicans* (Goswami *et al.*, 2006).

Table 11.2: Activity Index of the Medicinal Plants

Botanical Name	*Activity Index*	*Per cent of Inhibition*
Meyna spinosa	1.5	22.5
Duranta repens	1.3	20
Sida carpinifolia	2.04	30.62
Tephrosia purpurea	1.91	28.75
Cestrum nocturnum	2.33	30.5
Trachyspermum ammi	3.52	52.87
Lippia nudiflora	2.89	27.3
Ludwigia parviflora	2.27	22

The results of the present study supported to a certain degree that the screened plants possess bioactive components which on further investigation can provide a novel source of antifungal agents for new emerging human fungal pathogens resistant to synthetic antifungal drugs. However, further *in vivo* studies are needed to ascertain their clinical applicability.

Acknowledgement

The authors are thankful to Dr. (Mrs) M. Begam, Head of Biotechnology Division, for constant support to carry out this work successfully.

References

Anonymous, 2006.

Goswami, S., Bora, L., Das, J. and Begam, M., 2006. *In vitro* evaluation of some medicinal plants against *Candida albicans*. *J. Cell and Tissue Research*, 6(2): 837–839.

Grammar, A., 1976. Antibiotic sensitivity and assay test. In: *Microbiological Methods* 6th edn., (Eds.) Collins, C.H., Lyne, P.M. and J.M. Grange. Butterworths and Co. Ltd., London, p. 235.

Jain, N. and Sharma, M., 2005. Broad spectrum antimycotic drug. *Curr. Sci.*, 85(1): 30–34.

Myers, N., Russel, A.M., Cristina, G., Gustavo Foneca, A.B. and Kent, J., 2000. Biodiversity hotspots for conservation priorities. *Nature*, 403: 853–858.

Nair, R. and Chanda, S., 2006. Evaluation of *Polyalthia longifolia* (Sonn) THW. Leaf extracts for antifungal activity. *J. Cell and Tissue Res.*, 6(1): 581–584.

Nasser Vahdati Mashhadian and Hassan, Rakhshandeh, 2005. Antibacterial and antifungal effects of *Nigella sativa* extracts against *S. aureus*, *P. aeroginosa* and *C. albicans*. In: *Proceeding International Conference on Botanicals*, Kolkata. India, p. 369–373.

Prabhuji, S.K., Rao, G.P., Singh, S.P. and Singh, S.D., 2005. Emerging medicinal plant of 21st century: Safed Musali, Satawar, Brahmi and Jatamansi. In: *Recent Advances in Medicinal Plant Research: Vision for Twenty First Century*, p. 11–41.

Pushpangadan, P., 1996. Traditional medicine. In: *Supplement to Cultivation of Medicinal Plants*, (Eds.) Handa, S.S. and M.K. Kaul. NISC Publication, p. 689–702.

Senthikumar, M., Gurumoorthi, P. and Janardhanan, K., 2005. Antibacterial potential of some plants used by tribals in Maruthamalai hills, Tamil Nadu. *Nat. Prod. Radiance*, 4(1).

Souwalak, Phongpaichit, Sanan, Subhadhirasakul and Chatchai Wattanapiromsakul, 2005. Antifungal activities of extract from Thai medicinal plants against opportunistic fungal pathogens associated with AIDS patients. *Mycoses*, 48: 333–338.

Vyas, Y.K., Bhatnagar, M. and Sharma, K.J., 2006. Antimicrobial activity of a herb: Herbal based and synthetic dentrifrices, against oral microflora. *J. Cell and Tissue Res.*, 6(1): 639–642.

Chapter 12

Nicotine, Saponin and Purine from Therapeutic *Melothria purpusilla* (Blume) Cogn.: A Well Known Home Remedy Herbal for Humankind

S.R. Singh and M. Neshwari Devi
Post Graduate Studies Centre, HRDRI, Canchipur –795 003, Imphal, Manipur

ABSTRACT

Home remedy herbals, that therapeutically prescribed in various dreadful human diseases, have been practiced since immemorial time in Manipur. The present paper highlight the Nicotine, Saponin and Purine as therapeutic compound of the *Melothria purpusilla,* a well known home remedy herbal by using TLC chromatography, for updating the traditional knowledge by adaptation of science and Technological knowledge to revitalize the aged old believes and wisdom of herbal health care system, a felt need of the present era, have also been described to strengthen the wide and more applicability of therapeutic chemistry of home remedy herbals towards herbal health care system of our precious aged old traditional wealth of indigeneous therapeutic and medical knowledge.

Keywords: *Home remady herbal, Therapeutic chemistry, Melothria purpusilla, Traditional wealth, Thin layer chromatography.*

Introduction

Melothria purpusilla, a wonder crude herbal drug, commonly practiced in Manipur since immomarial time as herbal health care system in jaundice and related liver problems. The plant is bitter and all aerial parts posses a potent of antihepatoxic and miraculous healing capacity from dreadful disease of jaundice, the roots recommends as laxative anti helmintic, diuretic and in constipation (Sinha, 1990, 1996). Other medicinal uses are sudonitic inflatulence by decoction with seed. Seeds in crushed used in sprained backs and when marticated relieves toothache, the tendrils shoots and tender leaves are used as gentle aperient and present bed in vertigo and biliousness.

Ayurveda accord about 2000 plant species having medicinal value, while the Chinese pharmacopoeia lists over 5,700 traditional medicines most of which are of plant origins. About 500 herbs are still employed within conventional medicine, Curative properties of medicinal plants due to the presence of various complex chemical substances of different composition mainly secondary plant metabolites in one or more parts of these plants. These plant metabolites, according to their composition are grouped as alkaloids, glycosides, caticosteroides, essential oil etc. (Purohit and Prajapati, 2003) although whole plants are rarely used.

Further, the therapeutic compounds, the chemical compounds or compounds of drugs, that used in a medicine to treat illness, protect against disease or improve health, being component of pharmaceutical chemistry or Medicinal chemistry its study take into account the determination of structure, synthesis, isolation of compounds involvement to the metabolism, mechanism of action, relationship between the structure (of therapeutic compound) and their biological activity. A review of literature did not reveal any information of therapeutic compounds of Nicotine, Saponin and Purine for *Melothria purpusilla,* henceforth the present paper describes the isolation of these new compounds from the widely used semi processed novel natural products of *Melothria purpusilla.*

Materials and Methods

All the aerial parts of the fresh *Melothria purpusila* were collected from the HRDRI campus Canchipur on the flowering season (June–July, 2006) and authenticated at the Botanical survey of India, (BSI) Shillong. A voucher specimen of the sample has been deposited at the Institute, the plants were washed thoroughly with tap water and then with distilled water. Then they were dried under shade and all the dry aerial parts were cut and ground in a mixture to obtained the plant. The powder mass was divided into three equal portions and each portion was subjected to extraction in soxhlet separately with $CHCl_3$, MeOH and EtOH extraction. 5 ml of concentrated was carried on using TLC on silica Gel G chromatoplates and libermann-barchard reagent (LB), iodine-potassiam iodide-HCL reagent (I/HCL), were used as spraying agent.

Results and Discussion

The TLC photo drug atlas has an immediate clarity of representation that facilitates the learning of TLC drug analysis and photographic reproduction of thin layer separation has a large didactic advantage over mer graphic representation (Wagner, 1996). The present experimentation on the therapeutic compounds scavenge the R_f value 0.75, 0.4, 0.75, in the thin layer chromatographs indicating an organic natural compounds *viz.*, Saponine, Purine and Nicotine respectively. The chromatograms are presented in Figures 12.1–12.3. The colouration and the characteristic of the spot elucidate and put construction on getting the explicatory confirmation of the compounds. The chemical structure of the compounds Saponine, Purine and Nicotine accord in Figures 12.4–12.6 with the formulae $C_{27}H_{39}O_3$, $C_5H_4N_4$ and $C_{10}H_{14}N_2$ respectively.

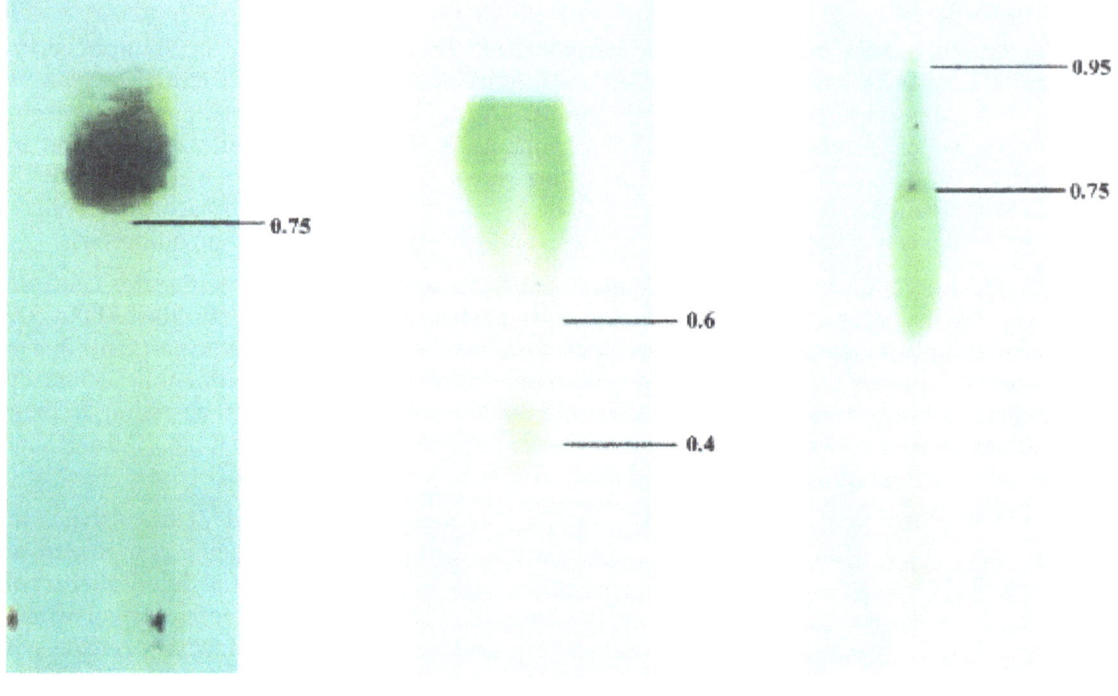

Figure 12.1: Thin Layer Chromatogram of Saponin

Figure 12.2: Thin Layer Chromatogram of Purine

Figure 12.3: Thin Layer Chromatogram of Nictotine

Figure 12.4: Saponin (Sarsaparillae radix)

Figure 12.5: Purine

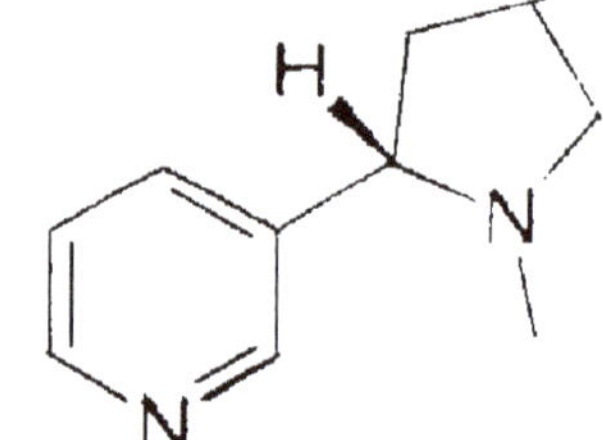

Figure 12.6: Nicotine

Ethanol extract of the aerial parts of *Melothria,* assigned quenching in UV light with hRf of 75 patented the Nicotine. The finding substantiated the therapeutic compound Nicotine as TLC synopsis of important alkaloids under Nicotine folium, of Alkaloid Drug consigned Rf 0.75 to Nicotine (Wagner, Hand Bladt, 1996). Nicotine was first isolated in 1882 by German Chemists Posselt and Reimann from the tobacco plant. Its chemical empirical formula was described by Melsens in 1843 and it was first synthesized by A Pictel and Crepiocex in 1893. Nicotine acts on the central and peripheral nervous systems. Nicotine enters when the body it quickly gets distributed through the blood stream and can cross the blood brain barrier. On average it takes about seven second for the substance to reach the brain (Anonymous, 2006a).

Therapeutically in treating nicotine dependence (smoking) controlled levels of nicotine to a patient through gums, dermal patches, or nasal sprays in an effort to wean them off at their dependence. Nicotine can be used in suffering from outsomal dominant frontal lobe epilepsy and a number of disorder including ADHD and parkinson's Disease. Nicotine acts on the nicotinic acetyl choline receptors. In small concentration it increases the activity of these receptors, among other things leading to an increased flow of adrenaline, a stimulating hormones. The release of adrenaline causes an increase in heart rate, blood pressure and respiration as well as higher glucose levels in the blood. Cotinene being a break-dawn product of nicotine which remain in the blood for upto 48 hours and can be used as an indicators of person's exposure to smoke.

The present investigation detect the chloroform extract of aerial parts of *Metothria* posses a distinct spot with hRf value 75 with yellow brown compound patented to Saponin. The value corroborate to that of the Saponin form other sources (Neher, 1969; Stahl and Jork, 1969). Saponin, a glycosides with aglycon, the sapogenin captivated the rings either steroidal and triterpinoidal (The Merck Index, 1983). The present investigation explored the presents of saponin in the test plant comprising confirmation in all the aerial parts of the plant with Rf value patented the compound from the therapeutic plant. The saponin are glycosidic triterpenoids (Bumouf Radosevich *et al.*, 1985; Mizui *et al.*, 1988, 1990; Ma *et al.*, 1989, Meyer *et al.*, 1990; Ridout *et al.*, 1991). Among the associated properties, saponin are bitter testing compounds, formation of stable forms when affiliated in water, the formation of oil-in-water emulsion, and haemolytic activity however when injected into the blood stream causes dissolving of red corpuscles even in extreme dilution.

Plant saponins also have a wide spectrum of activities including lowering blood cholesterol, inhibiting cancer cell growth, and acting as antifungal and antibacterial agents.

Phytochemists reports that saponins can act by binding with bile acids and cholesterol. It is thought that these chemicals "clean" or purge these fatty compounds from the body (thus lowering blood cholesterol levels).

Glycosidic saponin were antifungal and molluscicidal in vitro and able to kill liverflukes both in vitro and in vivo (Wren 1988). Further saponin trigger responses in gastric mucosa, which in turn activate mucous glands in the bronchi through parasympathetic signalizing to aid in the removal of mucous (Schulz *et al.*, 1998).

The chromatogram of Purine, from the methanol extract of aerial parts of *Metothria* showed the light pale green compound Purine with hRf 40. Authenticated range of value of Purine accord by Mangold (1969) from the nuclic acids and nucleotides bestowed the present finding compound. Purine from the synthetic pharmaceutical products triggered various activities. (Gan Shirt, 1969). Purine, first coined by Emil Fischer, in 1884 and synthesized in 1898, is a heterocyclic aromatic organic compound consisting of a pyramidine ring fused to an imidazole ring. Two of the bases in nucleic acids, adenine and guanine being purines, the metabolites authenticate to the metabolic functional activities of life. Biochemically purines signified as components of DNA and RNA, to other important biomolecules such as ATP, GTP cyclic AMP, NADH and coenzyme A. The present phytochemical investigation on therapeutic plant *Melothria* vividly explored the presence of purine with R_f value 0.4. The finding authenticated to the therapeutic properties of the purine and thereof effect to the diseases and ailments is reinforce.

Purines from food (or from tissue turnover) that metabolized by several enzymes, including xanthine oxidase into uric acid correlate to cause with gout, if the body is lacking an intrinsic uncase enzyme.

It generally synthesized and breakdown by metabolic pathways from the source food as nucleosides (lilainosine monophosphate) to adenine and guanine. Moreover, Purines from turnover of nucleic acids (or from food) sometimes be salvaged and reused in new nucleotides. The enzyme adenine phosphoribosyltransferase salvages adenine, while hypoxanthine guanine phosphoribosyl transferase (HPRT) salvages guanine and hypoxanthine, Genetic deficiency of HPRT causes Lesch-Nyhan syndrome (Anonymous, 2006b). In this connection it is stated that International Conference on Medicinal Plants and Ayurveda (ICMPA) recommend (*a*) standerdization; (*b*) safety; (*c*) quality; (*d*) integrity; (*e*) authenticity of the practices and the products (Anonymous, 2003). The present finding laid a piece and enhance the innovation to modernization of aged old indigenous health care system and Ayurveda, a holistic health science, having diversity flexibility, accessibility, affordability and expedient potential to meet with the new challenges to human life extending upto the molecular level by exploring the chemical compounds contained in the therapeutic plants, *i.e.*, it may supplement a piece among the enormous activities to be done for the vast and unlimited knowledge of Human Healing system.

The present investigation explored the nicotine, saponin and purine from the therapeutic *Melothria purpusilla.* Earlier Singh and Devi (2005, 2006) reported the presence of cucurbitacin glucosides, cucurbitacin aglycone, tocopherol and emetine, an antioxidant from all the aerial parts of the same herbal plant. The finding highlight the therapeutic properties of the test plant and hitherto hi-tech analyses for more information to medicinal chemistry. The present finding also suggest the probability of interactive activities of the precious therapeutic chemical compounds that are present in the same plant. If possible it may act as synergistic effect towards the potentianized action of the healing capabilities of human body functions through the systems of organ and organelles.

The present finding not only boosting the novel approach to drug discovery but also integrating the science of ethno botany, ethno medicine, medicinal chemistry and pharmaceutical chemistry, herbal therapeutic chemistry as reciprocity to the indigenous therapeutic technological knowledge and culture and expedite the revitalization of traditional knowledge of herbal health care system with technological justification.

Acknowledgements

The authors are thankful to the Director BSI, Shillong for information of confirmating the identity of test plant and are grateful to Dr. P.N. Singh Senior Physician for clinical and therapeutic discussion and suggestions.

References

Anonymous, 2003. *International Conference Medicinal plants and Ayurveda,* 16 December, 2002 New Delhi. *AgroBios,* 8: 9–10.

Anonymous, 2006a. *Nicotine: Wikipedia,* The free encyclopedia. File://A:/New%/20%Folder/nicotine%201.htm.

Anonymous, 2006b. http://en.wikepedia.org/wiki/purine.

Bumouf-Radosevich, M., Delfel, N.E. and England, R.E., 1985. Gas chromatography-mass spectrometry of oleanane- and ursane-type triterpenes: Application to *Chenopodium quinoa* triterpenes. *Phytochemistry,* 24: 2063–2066.

Ganshirt, H., 1969. Purine derivatives of various activity in Synthetic pharmaceutical products. In: *Thin-layer Chromatography,* (Ed.) Egon Stahl. Springer-Verlag, Berlin, New York, p. 506–566.

Mangold, H.K., 1969. TLC of nucleic acids and their constituents in nucleic acids and Nucleotides. In: *Thin-layer Chromatography,* (Ed.) Egon Stahl. Springer-Verlag, Berlin, New York, p. 506–566.

Ma, W.-W., Heinstein, P.F. and McLaughlin, J.L., 1989. Additional toxic, bitter saponins from the seeds of *Chenopodium quinoa*. *J. Nat. Prod.,* 52: 1132–1135.

Meyer, B.N., Heinstein, P.F., Bumouf-Radosevich, M., Delfel, N.E. and McLaughlin, J.L., 1990. Bioactivity-directed isolation and characterization of quinoside A: one of the toxic/bitter principles of quinoa seeds (*Chenopodium quinoa* Willd.). *J. Agr. Food Chem.,* 38: 205–208.

Mizui, F., Kasai, R., Ohtani, K. and Tanaka, O., 1988. Saponins from brans of quinoa, *Chenopodium quinoa* Willd. I. *Chem. Pharm. Bull.,* 36: 1415–1418.

Mizui, F., Kasai, R., Ohtani, K. and Tanaka, O., 1990. Saponins from brans of quinoa, *Chenopodium quinoa* Willd. I. *Chem. Pharm. Bull.,* 38: 375–377.

Neher, R., 1969 Saponin and sapogenins in TLC of steroids and related compound. In: *Thin-layer Chromatography,* (Ed.) Egon Stahl. Springer-Verlag, Berlin, New York, p. 311–362.

Purohit S.S. and Prajapati, N.D., 2003 Medicinal plants local heritage with global importance. *Agrobios,* 1(8): 7 –8.

Ridout, C.L., Price, K.R., DuPont, M.S., Parker, M.L. and Fenwick, G.R., 1991. *Quinoa saponins:* Analysis and preliminary investigations into the effects of reduction by processing. *J. Sci. Food Agr.,* 54: 165–176.

Schulz, V., Hansel, R. and Tyler, V.E. 1998. *Rational Phytotherapy: A Physicians' Guide to Herbal Medicine.* Springer, New York.

Sinha, S.C., 1996. *Medicinal Plants of Manipur*. Mass and Sinha, Imphal.

Singh, S.R. and Devi, M.N., 2005. Cucurbitacin Glucosides, Cucurbitacin Aglycone, Tocopherol in therapeutic compounds of *Melothria purpusilla* Research Monograph VII. HRDRI, Imphal.

Singh, S.R. and Devi, M.N., 2006 Emetine an antioxidant from *Melothria purpusilla* (Blume) COGN a well known home remedy herbal for humankind. *J. Curr. Sci.,* 9(1): 421–425.

Sinha, S.C., 1990. Notes on ethnomedicinal plants of Manipur (New report). *Curr. Pam. Letters,* 1: 3–2.

Stahl and Jork, E., 1969. TLC of involatile terpene derivatives, essential oils, balsams, and pesins. In: *Thin-layer Chromatography,* (Ed.) Egon Stahl. Springer-Verlag, Berlin, New York, p. 206–258.

The Merck Index, 10th Edn., 1983. Merck and Co. New Jersey, United States of America, pp. 375/1204.

Wren, R.C., 1988. *Potter's New Cyclopaedia of Botanical Drugs and Preparations.* The C.W. Daniel Company Ltd., Essex.

Wagner, H. and Bladt, S., 1996. *Plant Drug Analysis.* Springer-Verlag, Berlin Heidelberg, pp. 384.

Chapter 13

Effects of Biomass on Environmental Air Pollution

M. Venkateshwarlu, D. Srinivas and B. Mallaiah
Department of Botany, Kakatiya University, Warangal – 506 009, Andhra Pradesh, India

ABSTRACT

The health effects of biomass air pollution by using improved biomass stoves showed that reduction of respiratory infection was 50–60 per cent among women and 30–40 per cent among children over traditional biomass stoves. The smoke irritation to the eye of the cook was expressed by all cooks irrespective of regions or Towns. Eighty percent of them expressed that wood smoke caused respiratory related diseases like cold, cough asthma and wheezing. The symptoms of cold, cough (allergy) and fever with breathing difficulty before and after installation of improved biomass stoves were assessed and are presented in Sharma *et al.* (1998) reported that the respiratory symptom of children varied from 1.6 to 6.3 per cent during biomass air pollution. Smaller respiratory particulates from stoves are more harmful to children, because they get deep in to the lungs Leaderer *et al.* (1990). Significant improvement of breathing difficulty was there due to the reduction of smoke after the introduction of improved biomass stoves. The awareness of smoke emittance in traditional biomass stoves and biomass air pollution related diseases were studied.

Keywords: *Air pollution, Biomass, Environmental effect, Smoke, Pollution, Infection.*

Introduction

The highest exposures are probably experienced by women, infants, and young children. Laryngo tracheobronchitis was defined in fants with brassy cough, respiratory stridor hoar seness and respiratory distress. Locally available fuelwoods, cowdung cake and crop residues are burnt in traditional stoves. Single and double pot conventional biomass stoves with low burning rate are used to prepare food. Exposure to pollution from wood burning stoves for (biomass) heating is associated with severe respiratory symptoms and mortality. Domestic cooking is the main causative factor for biomass air pollution in rural areas.

Materials and Methods

Diagnosis of respiratory infections was made if the infant suffered from more than two episodes of suffocation or wheezing in the year. Wood smoke is the main causative factor for biomass air pollution in rural kitchens. Impact of smoke on health with respect to the symptoms like irritation to the eye respiratory infection and injurious to health was surveyed. Health profile of family members were assessed by a medical practioner before and after installation of improved biomass air pollution. Present of all infants were interviewed by a medical practioner during the visits using a self explanatory structured questionnaire.

Results and Discussion

The awareness in usage of biomass stove biomass air pollution related diseases and health effects when using traditional and improved biomass stoves are briefly discussed in the present study. The survey of awareness of smoke emittance in traditional biomass stove was conducted in all selected Towns and the summary is presented in the moisture content of fuel and incomplete combustion were the major reasons for smoke. The efficiency of the traditional biomass air pollution varied with respect to dimension and design of the stove used in that area. An up ward shift towards a cleaner fuel or improved fuel was postulated to reduce the exposure with less ventilated environment (Saksena *et. al.*, 1996). The intensity of smoke causes irritation to throat which resulted in cough. From there was smoke reduction from improved biomass stove resulting in less biomass air pollution than traditional one the medical practioner addressed prevalence of respiratory infection symptoms among women and children before and after installation of improved biomass stove. Exposure to smoke is directly related to the initial symptom like irritation and watering of eyes. Nearly 40-50 per cent of the symptoms like irritation of eyes and watering in eyes were reduced after the introduction of improved biomass stoves. The clinical assessment survey showed reduction in prevalence of smoke related signs among women. Also cough symptom was found to be 60 per cent when they switched over to improved biomass air pollution. Awareness about smoke related diseases from biomass stoves was gathered and summarized in (Table 13.1).

Table 13.1: Effect of Biomass (Cooking Stoves) Air Pollutants

Sl.No.	*Symptoms*	*Traditional Biomass Stove*	*Improved Biomass Stove*			
			I	*II*	*III*	*Mean*
1.	Irritation of throat or cough during cooking	05	08	04	04	04
2.	Watering of eyes	70	30	36	21	32
3.	Irritation of eyes	80	45	47	32	42
4.	A cute respiratory infection	19	16	15	03	10
5.	Cold	25	19	18	04	16
6.	Cough	20	16	18	14	15

Conclusion

A part from these reasons in sufficient air circulation in the stove for combustion was one among main equsative factors which directly correlates with internal efficiency of the stove. The thermal efficiency of the improved biomass air pollution (40 per cent) was almost double the traditional stove (20 per cent) which directly implied to fuel bum rate. Installation of improved biomass stove resulted in reduction of 33.8 per cent of cough 50 per cent reduction of acute respiratory infection and 30 per cent reduction of cold allergy than the use of traditional biomass air pollution. Although scientific and medical experts seem to realize that biomass air pollution is potentially a significant problems; the people who are really affected by it are the poor women and their children.

References

Leaderer, B., Boone, P.M. and Hammond, S.K., 1990. Total particle and sulphate acid aerosol emission from Kerosene space heaters. *Environment Science Technology,* 24: 908–912.

Saksena, S., Ravishankar, V., Prasad, R.K., Malhotra P., Singh, P.B., Srivastava, K.K. and Saxena, M., 1996. Fuel usage in Delhi slums an exposure to air pollution. In: *Proceedings of the International Conference on Biomass Energy Systems*, TERI, New Delhi, pp. 319–325.

Sharma, S., Sethi, G.R., Rohtagi, A., Chandhary, A., Shankar, R., Bapna, J.S., Joshi, V. and Sapir, D.G., 1998. Biomass air quality and acute respiratory infection in Indian urban slums. *Environmental Health Perspectives,* 106: 291–297.

Suraji, C., Padhmasutra, L. and Wahyunings, D., 1994, *Household Environmental Problems in Jakarta.* Stockholm Environment Institute, Stockholm, pp. 64.

Chapter 14

Studies on Pollen Behaviour at Different Levels of Mulberry Plants

M. Venkateshwarlu[1], A. Komuraiah[1], D. Srinivas[1], K. Sujatha[2] and Ch. Sammaiah[2]
[1]Department of Botany, [2]Department of Zoology (Sericulture Unit), Kakatiya University, Warangal – 506 009, Andhra Pradesh

ABSTRACT

Mulberry is generally a diploid, monongenous, heterozygous tree belonging to the family Moraceae having 28 chromosomes (2n = 28) Roxb., indigenous to India, and *M. tiliaefolia* Makino, originally from Japan and Korea, are known to be hexaploids. *M. nigra* L. is a decosaploid (2n = 22x = 308), with the largest number of chromosomes among phanerogams indicating a lot of variation in chromosome numbers as well as level of ploidy in mulberry (Machii *et al.*, 1999). These variations in ploidy and geographical diversity have resulted in plasticity of mulberry for many characters. Distinguishing ploidy by pollen fertility, pollen size, stigma length and other characters has been well correlated with the cytological confirmation (Koyama, 1978). The present communication deals with floral morphology and pollen behaviour related to ploidy.

Keywords: *Variation, Geographical diversity, Pollen fertility, Stigma, Mulberry plants.*

Introduction

It has 24 species and one sub-species and distributed in wide area of tropical, subtropical, temperate and sub-articzones (Koidzumi, 1917). It is rich in ploidy and a lot of triploid varieties have been found especially among *Morus bombysis* Koidz. In *M. Cathayana* Hemsl. Tetraploid, pentaploid and hexaploid varieties are present. *Alserrata* Roxb. More than 500 pollen grains per plant were scored. Only well filled, stained pollen grains were used for diameter measurement. The greater diameter of each pollen grain was measured at a magnification of 600x with the aid of an ocular micrometer.

For germination study, pollen grains were allowed to germinate in 10 per cent sucrose solution in hanging drop method in the cavity slides. Pollens 1.5 times larger than the normal pollen grain were considered as giant pollen. Data were statistically analysed. ANOVA showed significant difference among the accessions of different ploidy on different characters. Data on catkin length and width of triploids were larger than the diploid and tetraploid accessions ranging from 28.30–30.14 mm in triploids followed by diploid (25.94–29.89 mm) and tetraploid (19.95–21.66 mm). Similar trends were observed in catkin width also. Among the accessions at different levels of ploidy, number of florets in a single catkin was highest in diploids (35.81) which was significantly higher ($P < 0.01$) than the triploids and tetraploids except in Mandalaya (22.65). The flowers in triploids did not show any morphological changes other than superiority in size. The inheritance of bigger size of catkin but less number of florets in the catkins in triploid accessions indicated genetical advance over the diploids and tetraploids where triploids were originated either through crossing of diploid ($2n = 28$) and tetraploid ($2n = 4x = 56$) or a natural triploid ($2n = 3x = 42$). The reduced number of florets per catkin in triploids and tetraploids did not follow the earlier observation (Dwivvedi *et al.*, 1986).

Average size of pollen grain of the accessions significantly varied with the difference in ploidy level. The diameter of pollen ranged from 18.76 to 22.46 µm. triploids showed highest values (20.61–22.46 µm) followed by tetraploid (19.20–20.43 µm) and diploid (18.76–19.03 µm) accessions. The accessions, which were diploid and tetraploid showed uniformity in pollen size with an average size of 18.89 µm in diploid and 19.59 µm in tetraploids but for triploids (21.67 µm) the size varied significantly. The number of germ pores in the pollen also influenced the size of the pollen. Generally, two germpores were recorded in the pollen of diploid, triploid and tetraploid genotypes except three germpores in rare occasions in triploids. The occurrence of different sizes of pollen in the triploids placed them in between the tetraploids and diploids. Stanleyand Linskens reported that in many crop plants, pollen size has positive correlation with the chromosome numbers and ploidy level except few herbaceous plants, where pollen size has negative correlation (Pandey, 1971). The pollen size either increased or decreased with the increase or decrease of ploidy level. Increase in average size of pollen in triploid was due to wide variation with more number of bigger sized pollens. The size variation of pollen in triploid genotypes indicated their hybridity through crossing of diploid and tetraploid genotypes. The tendency of occasional occurrence of giant pollen with good fertility in the diploid genotypes help to explain the natural occurrence of polyploidy cultivars though incidentally no giant pollen were observed in the diploid genotypes.

Result and Discussion

Results on pollen germination showed that pollen viability varied significantly with difference of ploidy. Pollen germination was highest in diploid (89.46 per cent) followed by tetraploid (56.90 per cent) and triploid (46.24 per cent). The reduction of pollen fertility coupled with presence of giant pollen (> 1 per cent) in triploid and tetraploid corroborates the observations of earlier report in mulberry (Table 13.1). The les germination of pollen in triploids may be due to genetical imbalance and also due to variation in chromosomes. Catkin size, number of florets per catkin, pollen germination and giant pollen number showed a wider range. Maximum value of phenotypic co-efficient of variation was observed in occurrence of giant pollen, followed by pollen germination (30–40 per cent).

Table 14.1: Variation of Catkin Size and Pollen Characters at Different Ploidy Levels in Mulberry

Genotypes	*Ploidy*	*Chromosome No.*	*Catkin Length (mm)*	*Catkin Breadth (mm)*	*No. of Florets/ Catkin*	*Pollen Diameter (μm)*	*Pollen Germination (%)*	*Giant Pollen (%)*
T-4	4x	56	19.95	6.86	27.12	19.20	57.50	1.49
			(1.091)	(0.120)	(0.489)	(0.922)	(0.922)	(0.247)
T8	4x	56	21.56	6.16	25.78	19.34	57.80	1.68
			(0.943)	(0.092)	(0.214)	(1.077)	(1.077)	(0.098)
T-11	4x	56	21.66	6.57	22.87	20.43	55.40	1.65
			(0.681)	(0.249)	(0.323)	(0.327)	(1.281)	(0.186)
Mean			21.05	6.55	25.26	19.59	56.70	1.60
Tr-4	3x	42	29.92	9.02	28.07	20.61	45.60	2.63
			(0.631)	(0.117)	(0.870)	(0.466)	(0.917)	(0.450)
Tr-9	3x	42	28.30	8.70	25.99	22.46	46.63	2.96
			(0.856)	(0.126)	(1.057)	(0.543)	(1.114)	(0.835)
Tr-10	3x	42	30.14	8.92	25.52	21.94	46.50	2.63
			(0.594)	(0.140)	(0.802)	(0.934)	(1.688)	(0.390)
Mean			29.45	8.88	19.86	2.1.67	46.24	2.74
Kosen	2x	28	29.89	7.95	50.83	19.03	89.40	–
			(0.608)	(0.287)	(0.902)	(0.469)	(1.020)	
Mandalaya	2x	28	25.94	7.91	22.65	18.89	90.101	–
			(0.626)	(0.104)	(0.951)	(0.587)	(1.044)	
Burma-8	2x	28	26.42	7.55	33.96	18.76	88.90	–
			(0.721)	(0.143)	(1.017)	(0.335)	(1.221)	
Mean			27.41	7.80	35.81	18.89	89.46	–
CD at 5%			0.74**	0.13**	0.73**	0.55**	1.11**	0.35**

Data in parenthesis are Standard Deviation of Mean.

Conclusion

The traits giant pollen, pollen germination and number of florets per catkin indicator high heritability while catkin size as moderate heritability. Therefore, for using the mulberry genetic resources as male parent in breeding, pollen germination, occurrence of.giant pollen, pollen diameter, catkin size and also frequency of florets per catkin may be studied for preliminary screening of ploidy before cytological confirmation.

References

Dwivedi, N.K., Sikdar, A.K., Dandin, S.B., Sastry, C.R. and Jolly, M.S., 1986. Induced tetraploidy in mulberry. I. Morphology, anatomy and cytological investigations in cultivar RFS–135. *Cytologia*, 51: 393–401.

Koidzumi, G., 1917. Taxonomic discussion on *Morus* plants. *Bull. Imp. Sericulture Expt. Stat.*, 3: 1–62.

Koyama, H., 1978. Distribution of diploids, triploids and tetraploids identified by pollen grains in *Lxeris dentate* (Thunb.), Nakai, *Sensu lato,* in Japan. *Proc. Jap. Soc. Pl. Tazonom,* 4: 1–3.

Machii, H., Koyama, A. and Yamaguchi, H., 1999. A list of genetic mulberry resources maintained at National Institute of Sericultural and Entomological Science. Misc. Publ. Nat. Seric. Entomol. Sci., 26: 1–77.

Pandey, K.K., 1971. Pollen size and incompatibility in *Nicotiana.* In: *Pollen: Development and Physiology,* (Ed.) J. heslop–Harrison. Co., Ltd. London, p. 317–322.

Sharma, A., 1976. *The Chromosomes*. Oxford and IBH Publishing Co., New Delhi, pp. 225–236.

Chapter 15

Studies on the Primary Productivity in Two Perennial Tanks from Kolhapur District (Maharashtra) India

Milind S. Hujare[1] and M.B. Mule[2]
[1]Department of Zoology, S.M. Dr. Bapuji Salunkhe College, Miraj, M.S.
[2]Department of Environmental Science, Dr. B.A.M. University, Aurangabad, M.S.

ABSTRACT

The primary productivity of two perennial tanks (Talsande and Vadgaon) Kolhapur district (Maharashtra) India has been studied for two years (May, 1999 to April, 2001) by following light and dark bottle method. The gross primary productivity in Talsande tank ranged from 0.329 g.c/m^3/h to 0.935 g.c/m^3/h during first year and 0.32 g.c/m^3/h to 1.132 g.c/m^3/h during second year. The gross primary productivity in Vadgaon tank during first year ranged from 0.133 g.c/m^3/h to 0.629 g.c/m^3/h and during second year it was 0.187 g.c/m^3/h to 0.601 g.c/m^3/h. The seasonal variations in primary productivity along with some physico-chemical parameters and phytoplankton were studied. In both the tanks productivity showed bimodal peak. The primary peak of productivity was observed in summer and secondary in winter. The productivity was lowest in monsoon. Significant relationships were observed in case of water temperature, phytoplankton, transparency, EC, pH, dissolved oxygen and alkalinity. However relationship with dissolved oxygen was only significant in Talsande tank.

Introduction

The primary production of the plankton is one of the most important sources of energy input in the aquatic ecosystem. The primary productivity studies are essential for estimating the fish production potential of water body. It is well known that the primary productivity is influenced by the interaction of various factors.

Several attempts have been made to assess the primary productivity of ponds, tanks, lakes, and reservoirs of India, notable among those are of Sreenivasan (1965), Ganapati and Phatak (1969), Rao *et al.* (1981), Paulsamy *et al.* (1993) Vijaykumar (1994), Birasal (1996), Gurusamy and Ramadoss (2000).

The Talsande and Vadgaon tanks are being used by local fishermen for pisciculture activities, however there is lack of baseline date pertaining to the primary productivity of these tanks. There fore the present investigation has been carried, out which is the part of hydrobiological studies conducted during May, 1999 to April, 2001.

Material and Methods

Primary productivity studies were conducted for a period of two years (May, 1999 to April, 2001) by employing the light and dark bottle method (Gaarder and Gran, 1927). The operation was carried out *in situ* at two different sites of each tank and results were expressed as average of two sites. Two sets of light and dark bottles were filled with tank water. For estimation of initial dissolved oxygen the water sample was fixed. The sets of light and dark bottles were suspended in subsurface water for the duration of one hour. Meanwhile the initial dissolved oxygen was measured by employing Wrinkler's method. After completion of incubation period of one hour the dissolved oxygen from light and dark bottles was measured. The GPP, NPP and community respiration values were calculated by using following formulae.

Gross Primary Productivity (g.c/m^3/h)

$$= \frac{LB - DB}{T} \times \frac{0.375}{PQ}$$

Net Primary Productivity (g.c/m^3/h)

$$= \frac{LB - IB}{T} \times \frac{0.375}{PQ}$$

Community Respiration (g.c/m^3/h)

$$= \frac{IB - DB}{T} \times \frac{0.375}{PQ} \times RQ$$

where,

IB: Initial oxygen concentration.

LB: Oxygen concentration in light bottle after incubation.

DB: Oxygen concentration in dark bottle after incubation.

PQ: Photosynthetic quotient = 1.0.

RQ: Respiration quotient = 1.0.

0.375: A factor (ratio of molecular weight of carbon and oxygen).

T: Time in hours.

The physico-chemical parameters were determined fortnightly from same sites. The water temperature was measured by using mercury thermometer. The pH was measured by using pH meter (Hanna model champ). The transparency of water to light was measured by using secchi disc. The electrical conductivity was measured by using control dynamics digital conductivity meter, model APX 185-E

The chemical parameters were determined by following standard methods as described by APHA (1985), Trivedy *et al.* (1998). The nitrates (NO_3-N) and phosphate (PO_4-P) were determined by using systronic U.V. visible spectro-photometer 118.

The plankton samples were collected fortnightly by filtering hundred liters water through a plankton net made up of bolting silk No. 125. The concentrated samples were preserved by adding 4 per cent formation and 1 ml of lugols iodine. The quantitative analysis of phytoplankton was done by Lacky's drop count method Lacky (1938). The results were expressed as No/1. Phytoplankton were identified following Fritch (1944) Adoni *et al.* (1985), Cox (1996).

Study Area

Kolhapur is agriculturally well developed district (17°17′ to 15°43′ N and 70°40′ to 70°42′ E) of south western Maharashtra. The Sahyadri ranges on west and river Warana in north forms natural boundaries of district. The climate is tropical monsoon. The Talsande and Vadgaon tanks are minor irrigation tanks of Hatkanangale tahsil of Kolhapur district. The area receives average annual rainfall about 900 to 1000 mm. The Talsande tank is situated in densely populated area. The maximum depth is 5.0 m. The catchment area of tank is above 8.24 km^2. The basin slope is gentle. The water depths around edges are less. The tank shows vast littoral zone which sustains profuse growth of aquatic macrophytes. The tank water is greenish due to permanent bloom of *microcystics*. The anthropogenic activities such as washing, bathing, cattle wading are seen in and around the tank. The Vadgaon tank is located away from human locality. It has maximum length 1362.5 m, maximum width 750.0 m. The maximum depth is about 12.0 m. The water depth around the edges is greater. The tank sustains poor growth of aquatic macrophytes. No human activities except fishing are seen in Vadgaon tank.

Results

The monthly values of primary production (GPP and NPP) and community respiration of Talsande and Vadgaon tank are depicted in Table 15.1.

In Talsande tank during first year of investigation GPP varied from 0.329 g.c./m^3/h (September) to 0.935 g.c./m^3/h (April). The NPP values ranged between 0.23 g.c./m^3/h (September) and 0.753 g.c./m^3/h (April). The community respiration values ranged from 0.0795 (September) to 0.28 g.c./m^3/h (May). The yearly average values of GPP, NPP and community respiration were recorded as 0.687, 0.484 and 0.20 g.c./m^3/h respectively. The GPP and NPP showed bimodal peaks. The primary peak of GPP was recorded in the month of April (0.935 g.c./m^3/h) and secondary peak was observed in the month of January (0.87 g.c./m^3/h). The primary peak of NPP was recorded in April (0.753) and secondary peak was recorded in January (0.658 g.c./m^3/h). During second year of study the productivity followed same seasonal pattern. The GPP varied from 0.32 g.c./m^3/h (September) to 1.132 g.c./m^3/h (May). The NPP values varied from 0.113 g.c./m^3/h (September) to 0.817 g.c./m^3/h (May). The community respiration varied from 0.112 g.c./m^3/h (November) to 0.339 g.c./m^3/h (January). The yearly average values of GPP, NPP and respiration were 0.651, 0.431 and 0.284 g.c./m^3/h respectively.

Table 15.1: Primary Productivity (Gross and Net) and Respiration of Talsande and Vadgaon Tank

	Talsande Tank			*Vadgaon Tank*		
	GPP g.c./m³/h	*NPP g.c./m³/h*	*CR g.c./m³/h*	*GPP g.c./m³/h*	*NPP g.c./m³/h*	*CR g.c./m³/h*
May, 1999	0.903	0.63	0.28	0.401	0.273	0.123
June	0.795	0.58	0.215	0.259	0.183	0.075
July	0.496	0.31	0.185	0.133	0.145	00.385
Aug	0.597	0.41	0.18	0.168	0.127	0.041
Sept	0.329	0.23	0.0795	0.139	00.835	00.56
Oct	0.596	0.347	0.248	0.298	0.202	0.0955
Nov	0.583	0.342	0.240	0.386	0.246	0.140
Dec	0.728	0.574	0.154	0.549	0.383	0.166
Jan	0.87	0.658	0.21	0.358	0.31	0.0485
Feb	0.62	0.378	0.228	0.306	0.240	0.0655
Mar	0.8	0.598	0.2	0.629	0.427	0.202
April	0.935	0.753	0.183	0.476	0.33	0.143
Yearly average	0.687	0.484	0.20	0.346	0.24	0.099
May	1.132	0.817	0.339	0.397	0.294	0.1 03
June	0.976	0.66	0.320	0.380	0.198	0.192
July	0.49	0.305	0.169	0.216	0.172	0.044
Aug	0.41	0.320	0.150	0.187	0.117	0.07
Sept	0.32	0.113	0.206	0.219	0.17	0.049
Oct	0.395	0.202	0.093	0.3 64	0.305	0.058
Nov	0.471	0.358	0.112	0.288	0.223	0.065
Dec	0.74	0.495	0.245	0.563	0.415	0.148
Jan	0.795	0.470	0.327	0.396	0.252	0.144
Feb	0.716	0.547	0.169	0.348	0.196	0.152
March	0.547	0.377	0.169	0.511	0.409	0.103
April	0.829	0.509	0.282	0.601	0.353	0.247
Yearly Average	0.651	0.431	0.284	0.372	0.258	0.114
Two Years Average	0.669	0.457	0.242	0.357	0.252	0.107

The two years average GPP, NPP and respiration was 0.669, 0.457 and 0.242 g.c./m^3/h respectively.

The primary productivity in Talsande tank showed seasonal variation. During both the years of investigation the highest values of GPP, NPP and respiration were recorded during summer seasons and lowest during monsoon season. The seasonal variations in primary productivity and various physico-chemical and biological parameters in Talsande tank are shown in Figure 15.1.

The values of correlation coefficient (r) between gross productivity some physico-chemical parameters, phytoplankton in Talsande and Vadgaon tank are shown in Table 15.2.

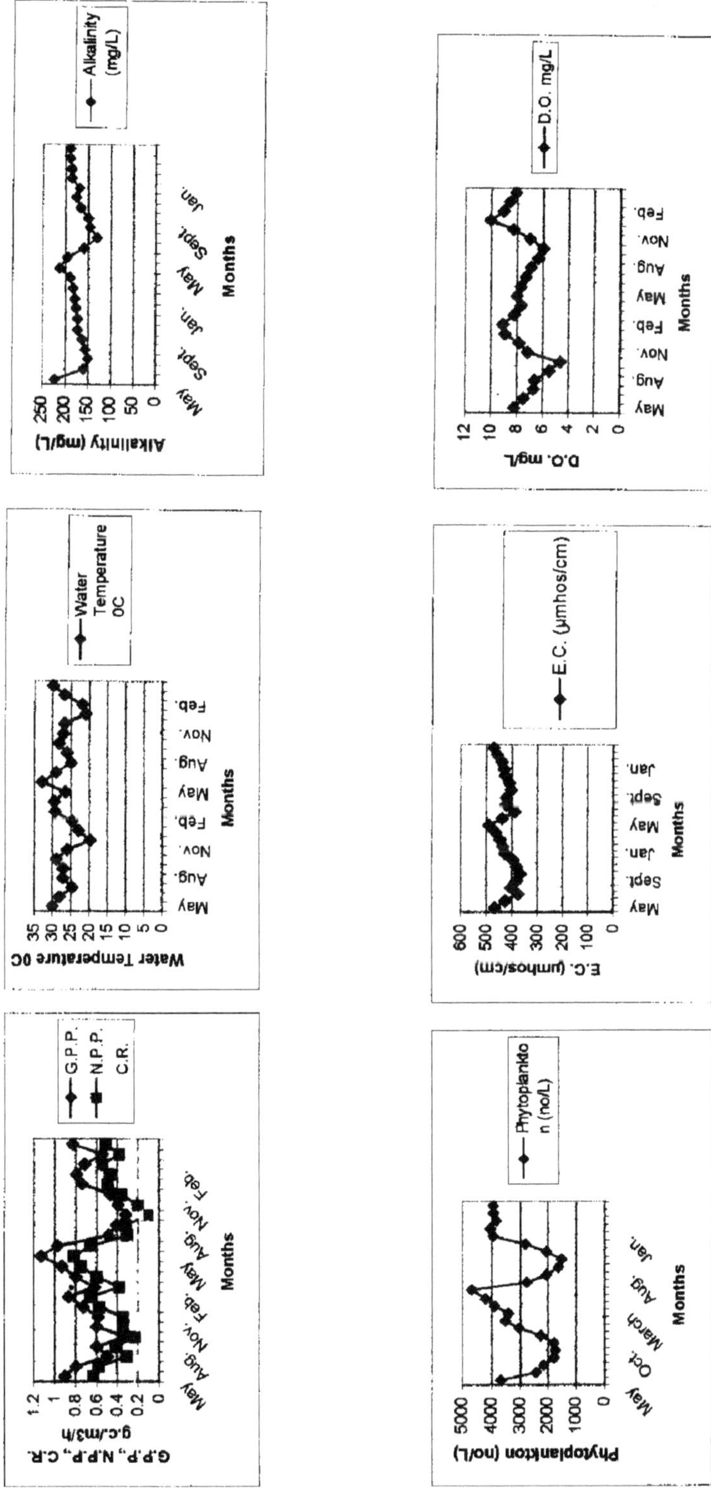

Figure 15.1: Seasonal Variation in NPP, GPP, CR, Some Physico-chemical and Biological Parameters in Talsande Tank

Table 15.2: Values of Correlation Coefficient Between Gross Primary Productivity and Some Physico-chemical and Biological Parameters of Talsande and Vadgaon Tame

Parameters	Gross Primary Productivity	
	Talsade	Vadgaon
Water temperature (°C)	0.36*	0.18
Secchi disc transparency (cm)	–0.65****	0.34*
Electrical conductivity (μ mhos/cm)	0.67*****	0.57****
pH	0.50***	0.69*****
Dissolved oxygen (mg/l)	0.63****	0.16
Alkalinity (mg/l)	0.62****	0.48**
Hardness (mg/l)	0.58****	0.52***
Chlorides (mg/l)	0.32	0.10
Total dissolved solids (mg/l)	0.62****	0.48**
Nitrate NO_3-N (mg/l)	–0.18	–0.26
Posphate PO_4-P (mg/l)	–0.12	–0.30
Phytoplankton (No/l)	0.76*****	0.61****

*****: $p < 0.001$; ****: $P < 0.01$; ***: $P < 0.02$; **: $p < 00.5$; *: $P < 0.10$.

The gross primary productivity showed significant direct correlation with water temperature r = 0.36, E.C. r = 0.67, pH r = 0.50, alkalinity r = 0.62, Dissolved oxygen r = 0.63, total dissolved solids r = 0.62, Hardness r = 0.58 and phytoplantkton r = 0.76.

Significant inverse relationship between GPP and secchidisc transparency (r = –0.65) was observed.

In Vadgaon tank during first year of investigation the GPP varied from 0.133 g.c./m^3/h (July) to 0.629 g.c./m^3/h (March). The NPP ranged form 00.835 g.c./m^3/h (September) to 0.427 g.c./m^3/h (March). The community respiration ranged between 0.0385 g.c./m^3/h (July) and 0.202 g.c./m^3/h (March). The yearly average values of GPP, NPP and CR for first year are recorded to be 0.346, 0.24 and 0.099 g.c./m^3/h respectively.

During second year the GPP values in Vadgaon tank varied from 0.187 g.c./m^3/h (August) to 0.601 g.c./m^3/h (April). The NPP varied from 0.117 g.c./m^3/h (August) to 0.415 g.c./m^3/h (December). The community respiration ranged between 0.044 g.c./m^3/h (July) to 6.247 g.c./m^3/h April. The yearly average values of GPP, NPP and respiration are 0.372, 0.258 and 0.114 g.c./m^3/h.

The two years average values of GPP, NPP and community respiration in Vadgaon tank are 0.357,0.252 and 0.107 g.c./m^3/h.

The productivity in Vadgaon tank showed seasonal pattern being highest in summer and lowest in monsoon. The seasonal variations in primary productivity and physico-chemical and biological parameters of Vadgaon tank are depicted in Figure 15.2.

In the Vadgaon tank significant and direct correlation between gross productivity and transparency r = 0.34, pH r = 0.69, alkalinity r = 0.48, total dissolved solids r = 0.48 and phytoplankton r= 0.61 was observed (Table 15.2).

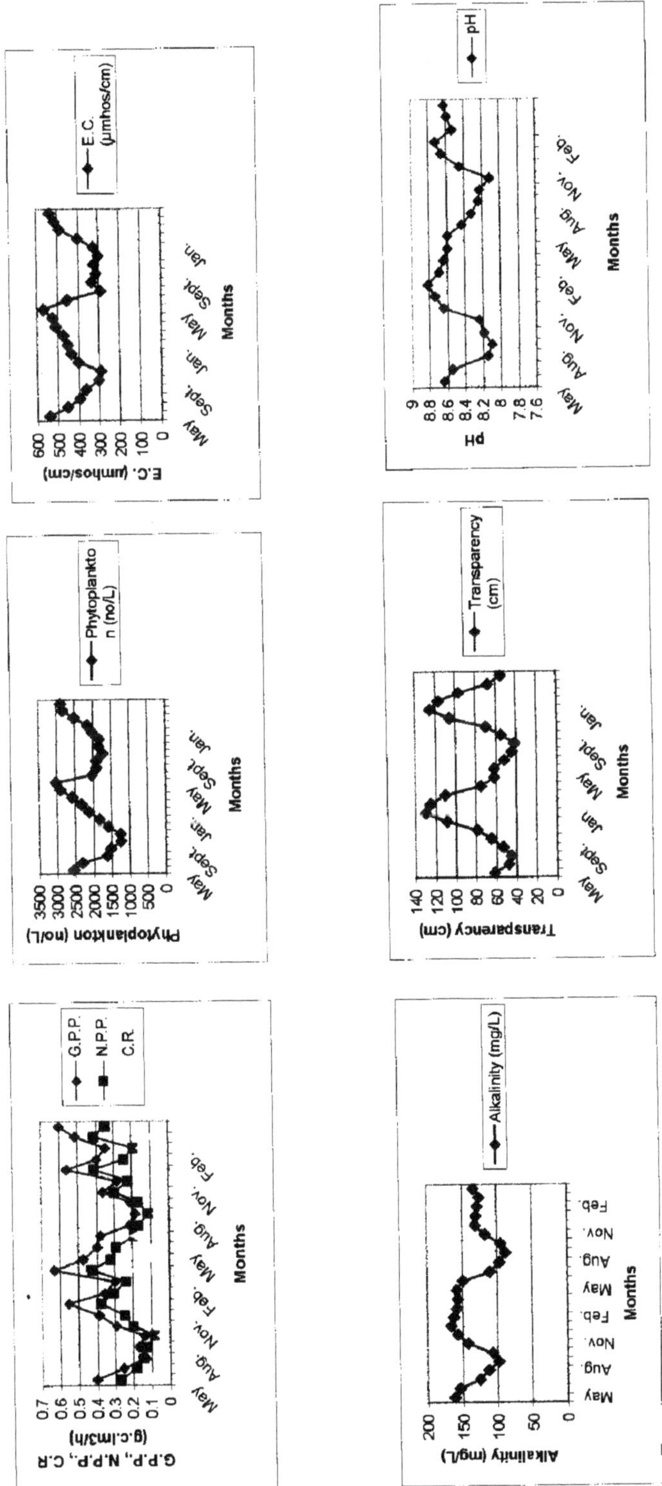

Figure 15.2: Seasonal Variation in NPP, GPP, CR, Some Physico-chemical and Biological Parameters in Vadgaon Tank

Discussion

In the present investigation higher rate of primary productivity was recorded in Talsande tank than Vadgaon tank. This can be attributed to the comparable morphometric features of the two tanks. The Talsande tank is smaller and shallower water body than Vadgaon tank. As the Talsande tank is located in dense human locality, intensive anthropogenic activities are seen in and around Talsande tank. A sizable quantities of nutrients may enter through washing and bathing activities in Talande tank. On the other hand the Vadgaon tank is located away from human locality, the anthropogenic activities are less and the only major source of nutrients may be the run-off. The Talsande tank shows vast littoral zone and some allocanthonous organic matter may also be introduced by aquatic vegetation growing in littoral zone. Ganapati and Sreenivasan (1970) observed higher productivity in smaller water bodies as compared to large lakes.

In the present study definite trend of seasonal variation in primary productivity was observed in both tanks and bimodal pattern of increased productivity was noticed. The primary peak of production was observed during summer season, resulted in to highest gross production. The secondary peak was noticed in winter and the lowest rate of production was recorded during monsoon. The decline in the productivity during rainy seasons may be due to dilution of tank water and subsequent reduction in phytoplankton density. Moreover the cloudy weather in monsoon might have resulted in low production. Clear days generally resulted in relatively greater photosynthesis than the cloudy days. Prasad and Nair (1963), Singh *et al.* (1996) also reported low productivity in ramy season.

In the present study direct and significant correlation between gross productivity and phytoplankton population was observed. The rate of production in both the tanks coincided with peak of phytoplankton population. The fall in phytoplankton density in monsoon reflected in decrease in primary production. Khan (1980) observed that population of phytoplankton is sufficiently related with the rate of production.

A close relationship between temperature and primary productivity was reported by Singh (1990) and Vijaykumar (1994). But in the present study the relatively smaller peak of production was observed in winter season when water temperature was recorded minimum.

Hence although direct relationship between GPP and temperature was observed in Talsande tank, no such relationship was noticed in Vadgaon tank. The bimodal peak of productivity was also reported by Khan and Siddiqui (1971). Rao *et al.* (1981) reported winter peak of productivity in Wagholi pond and observed no direct relationship between temperature and productivity. According to Qasim *et al.* (1969) in tropical waters production remains moderate through out the year with little oscillations. This is clearly evident from the present study and temperature is not seems to be limiting factor under tropical conditions.

The alkalinity can be considered as a significant index for production. According to Jackson (1961) alkalinity below 50 mg/l indicates low rate of photosynthetic rate. In Talsande and Vadgaon tank the alkalinity was much higher and both the tanks can be classified as alkaline tanks. However the alkalinity showed considerable decline during monsoon remained moderate during winter and highest during summer and as such showed significant direct relationship with productivity. Similar relationship was also reported by Sreenivasan (1964) and Sumitra (1971).

Inverse relationship between gross production and transparency as observed in Talsande tank can be attributed to the increase in transparency during monsoon. As the Talsande tank sustains permanent bloom of *Microcystis* the dilution of tank water during monsoon resulted into considerable reduction in *Microcystis* population and subsequently decline in gross productivity during monsoon.

The direct relationship of gross productivity and transparency as noticed in Vadgaon tank may be due to decrease in transparency during monsoon resulting into fall in phytoplankton density which resulted in to low gross production in monsoon. A similar relationship was also reported by Khan (1980) and Singh (1990).

The range of electrical conductivity in both the tanks indicate their high productivity potential. The electrical conductivity is a measure of total dissolved solids. The EC and TDS values of both tanks showed seasonal variation being highest in summer when the rate of primary productivity was maximum. Thus in the present investigation gross productivity was found to be directly and significantly correlated with EC and TDS.

The pH of both the tanks remained alkaline throughout the study period. The annual fluctuations were small. The higher pH is normally associated with high photosynthetic activity in water (Goel *et al.*, 1986). In present study it was observed that pH gradually increased from decreased in monsoon. The seasonal variations in pH coincided with gross productivity. Thus direct and significant correlation between pH and gross productivity was observed.

The direct and significant relationship between gross productivity and dissolved oxygen in Talsande tank can be attributed to increase in dissolved oxygen during summer and winter due to higher photosynthetic activity of phytoplankton.

The nitrate (NO_3-N) and phosphate (PO_4-P) concentration was very poor in both tanks. The nutrients were not found to be limiting the productivity, because lower values of both the nutrients were recorded during summer and winter when the productivity was higher. This can be attributed to the faster recycling of nutrients. A similar observation has been made by Trivedy *et al.* (1993) who recorded high production rates under nutrient poor conditions.

The higher values of community respiration observed during summer months may be due to higher phytoplankton and zooplankton density. Goel and Chavan (1991) also reported higher community respiration in summer. Inspite of clear seasonal trend, in the Talsande and Vadgaon tank, the production was fairly uniform throughout the year except during few monsoon months. The low productivity during monsoon was probably due to heavy rain which resulted in dilution of water. The similar effect have been reported by Ray *et al.* (1966).

As the temperature changes in tropical region are not marked, different patterns of seasonal variation in productivity have been observed in different water bodies depending up on the local conditions. Therefore generalization seems to be difficult. However Ganapati and Sreenivasan (1970) have tried to generalize the seasonal variations in the productivity of some Indian water bodies and observed that the order of magnitude of production, season wise was almost same in different water bodies, maximum during summer and minimum during winter season. This is in contrast to present observations where minimum primary production was recorded during monsoon season.

References

APHA, AWWA, WPCF, 1985. *Standard Methods for the Examination of Water and Wastewater*, 20th edn. American Public Health Association, Washington D.C.

Adoni, A.D., Joshi, G., Ghosh, K., Chourasia, S.K., Vaishya, A.K., Yadav, M. and Verma, 1985. *Workbook on Limnology*. Pratibha Publ., Sagar, M.P., pp. 216.

Birasal, N.R., 1996. The primary productivity of the Supa reservoir during its filling phase. *J. Environ. Biol.*, 17(2): 127–132.

Cox Eileen, J., 1996. *Identification of Freshwater Diatoms from Live Material*. Chapman and Hall, 26 Boundary Row, London.

Fritch, F.E., 1944. The present day classification of algae. *Bot. Rev.*, 10.

Gaarder, T. and Gran, H.H., 1927. *Rappet. Proc. Verb Cons. Expl. Mer.*, 42: 1–48

Ganpati, S.V. and Sreenivasan, A., 1970. Energy flow in natural aquatic ecosystem in India. *Arch. Hydrobiol.*, 66: 458–498.

Ganapati, S.V. and Pathak, C.H., 1969. Primary productivity of Sayaji Sarovar (A man made lake) at Baroda. In: *Proc. Seminar on the Ecology and Fisheries of Freshwater Reservoirs*, Central Inland Fisheries Research Institute, Barrackpore, pp. 37–45.

Goel, P.K., Kulkarni, A.Y. and Trivedy, R.K., 1986. Limnological studies of a few freshwater bodies in South Western Maharashtra with special reference to their chemistry and pollution. *Poll. Res.*, 5M(2): 79–84.

Goel, P.K. and Chavan, V.R., 1991. Studies on the limnology of a polluted fresh water tank. In: *Aquatic Sciences in India*, (Eds.) Gopal B. and V. Asthana. Indian Association for Limnology and Oceanography, p. 51–64.

Gurusamy, K. and Ramdoss, V., 2000. Seasonal variation of chlorophyll and phytoplankton productivity in man made ponds. *J. Ecobiol.*, 12(4): 281–287.

Jackson, D.F., 1961. Comparative studies on phytoplankton photosynthesis in relation to total alkalinity. *Verh. Int. Ver. Limnol.*, 14: 125–133.

Khan, A.A. and Siddiqui, A.Q., 1971. Primary production in a tropical fish pond at Aligarh India. *Ibid*, 37: 447–456.

Khan, Rashid A., 1980. Primary productivity and tropical status of two tropical water bodies of Calcutta, India. *Bulletin of the Zoological Survey of India*, 2 (2&3): 129–138.

Lacky, J.B., 1938. The manipulation and counting of river plankton and changes in same organisms due to formalin preservation. *US Public Health Reports*, 53: 2080–2093.

Paulsamy, Rathinasmy R. and Manian, S., 1993. Studies on primary productivity in a freshwater ecosystem of Bhavanisagar reservoir. *Indian Bot. Contractor*, 10: 29–31.

Prasad, R.R. and Nair, P.V.R., 1963. Studies on aquatic production. I. Gulf of mannar. *J. Mar. Biol. Assoc., India*, 5: 1–26.

Qasim, S.Z., Wellershhaus, S., Bhattathiri, P. and Abidi, S.A.N., 1969. Organic production in a tropical estuary. *Proc. Indian Acadm. Sci.*, p. 51.

Rao, M. Babu, Muley, E.V. and Mukhopadhyay, S.K., 1981. Seasonal fluctuations of primary production in two freshwater ponds at Wagholi, Poona, India. *Geobios*, 8: 270–210.

Ray, P. Singh and Sehgal, K.L., 1966. A study of some aspects of the ecology of river of Ganga and Jamuna at Allahabad (U.P.). *Proc. Nat. Acad. Sci., India*, 37: 235–272.

Singh, Ravindra, 1990. Correlation between certain physico-chemical parameters and primary production of phytoplankton at Jamalpur, Munger. *Geobios*, 17: 229–234.

Sreenivasan, A., 1964. Limnological feature and primary production in a polluted moat at Vellore, Madras State. *Environmental Health*, 6: 237–245.

Sreenivasan, A., 1965. Limnology of tropical impoundments. II. Limnology and productivity of Amaravathy reservoir, (Madras State). *Hydrobiologia,* 26: 501–516.

Sumitra, V., 1971. Seasonal variations in primary production in three tropical ponds. *Hydrobiologia,* 38: 395–408.

Singh, Anil K., Shukla, Arvind N., Saxena, Pankaj and Mendhe, Kavita, 1996. Some observations on primary productivity in river Narmada (Western Zone) M.P. India. *Journal of Environment and Pollution,* 3(384): 203–206.

Trivedy, R.K., Goel, B. and Goel, P.K., 1993. Comparative study of primary production and chlorophyll concentration in a non-polluted and sewage receiving reservoir. *International Journal of Ecology and Environmental Science,* 19: 103–120.

Trivedy, R.K., Goel, P.K. and Trishal, C.L., 1998. *Practical Methods in Ecology and Environmental Science.* Enviromedia Publications, Karad.

Vijaykumar, K., 1994. Seasonal variations in the primary productivity of a tropical pond. *J. Ecobiol.,* 6(3): 207–211.

Chapter 16

Devolatilization Characteristics of Poultry Litter Using Thermogravimetric Analysis

V. Kirubakaran[1], V. Sivaramakrishnan[2], M. Premalatha[3], P. Subramanian[4]

[1]Lecturer in Energy, Rural Technology Centre, Gandhigram Rural Institute, Gandhigram, Dindugal –624 302, Tamil Nadu, India, E-mail: kirbakaran@yahoo.com

[2]Professor, Department of Mechanical Engineering, Saranathan College of Engineering, Trichy, Tamil Nadu, India, E-mail: vsmp1967@yahoo.com

[3]Assistant Professor, CEESAT, National Institute of Technology (NIT), Tiruchirappalli – 620 015, Tamil Nadu, India, E-mail: manic_prem@yahoo.com

[4]Professor and Head (Retired), CEESAT, National Institute of Technology (NIT), Tiruchirappalli – 620 015, Tamil Nadu, India, E-mail: psuubu44@vahoo.com

ABSTRACT

Modelling of combustion of any fuels in furnaces, boiler and industrial process requires adequate knowledge of fuel properties. In past, the burning of biomass fuels has been viewed as an incineration process to remove a waste product, rather than the burning of a fuel. Today, however with the increasing use and shortage of these supplementary fuels, there arises the need to burn the biomass more efficiently. Considering the need for an effective and efficient combustion and differences in the properties, it is aimed to characterize the biomass fuels thoroughly for understanding better behaviour of fuels in the furnace. One of the biomass, Poultry litter instead of being a problem of waste, can and should be a source of energy and nutrients. Future decisions on the disposal of litter should be based on fuel cost, efficiency, capital cost and environmental and regulatory policies. Since combustion route appears to be feasible, litter in large quantities is required for a reasonable unit size. The design of combustors/gasifiers requires a thorough understanding of physio-chemical changes that are occurring in litter, while heated. Thermogravimctric analysis of poultry litter gives the complete picture in the thermal degradation of poultry litter at various temperature zones. This papers deals with complete devolatilization characteristics of poultry litter. This paper also verifies the literature values of the combustible matter to be converted in to chart-fixed carbon and the actual value of fixed carbon available in the poultry litter by proximate analysis of poultry litter.

Keywords: *Poultry litter, Thermogravimetric analysis, Proximate analysis, Ultimate anlysis.*

Introduction

Biomass generally contains carbon, hydrogen, oxygen, nitrogen, sulphur, chlorine and water besides oxides of silicon, sodium, potassium, aluminum, iron etc. These constituents are grouped into three, namely moisture, combustible matter and ash. These are available in varied quantities depending upon the type of biomass. The combustible matter is further divided into volatile matter and fixed carbon. The composition of biomass is analyzed in two ways: Proximate Analysis and Ultimate Analysis.

The proximate analysis provides a quick look into groups that are essential for energy contribution. It comprises of moisture, volatile matter, fixed carbon and ash. The quantities of these constituents are mainly estimated from weight loss obtained at various temperature levels. Even though the availabilities of these groups are known in quantity, their chemical compositions are necessary for arriving at a mechanism of combustion/gasification. An insight into the mechanism would provide operating conditions required for maximizing the specific calorific value of the product. Therefore chemical analysis of each of these groups becomes important.

Ultimate analysis of biomass would yield the compositions of various constituents in the combustible matter.

Raveendran *et al.* (1995) have reported the compositions of various biomasses that are commonly available. A complete analysis of ash has also become important as it is found to catalyze the decomposition. Tables 16.1–16.3 is the reproduction of results reported by Raveendran *et al.* (1995).

Williams and Besler (1993) have reported that the combustible matter consists of hemicellulose, cellulose and lignin. During their decomposition, char known as fixed carbon is formed as end product. Lawrence (1999) reported the cellulose Hemicellulose and Lignin content of various biomasses. Table 16.4 is the reproduction of the results reported by Lawrence (1999).

Any solid fuel combustion can be broadly separated into two main reaction stages:

1. Rapid heating and devolatilization, as a result of pyrolysis, and
2. Combustion/oxidation of char, the residue from the devolatilization. The pyrolysis stage corresponds to reactions related to devolatilization or thermal decomposition that occur during the heating up of the fuel and is completed very fast taking time of the order of 0.1 second (Young and Smith, 1987). The pyrolysis process is represented as follows:

$$\text{Fuel} \rightarrow \text{Oil/Tar} + CO + CO_2 + H_2O + H_2 + \text{Hydrocarbons} + \text{Char}$$

Table 16.1: Properties of Biomass

	Proximate Analysis (wt per cent)		*Ultimate Analysis (Wt. per cent)*				*HHV*[a] *(MJ kg⁻¹)*	*Density (kg m⁻³)*
	VM(daf)	*Ash (db)*	*C*	*H*	*N*	*O*		
Bagasse	84.2	2.9	43.8	5.8	0.4	47.1	16.29	111
Coconut coir	82.8	0.9	47.6	5.7	0.2	45.6	14.67	151
Coconut shell	80.2	0.7	50.2	5.7	0.0	43.4	20.50	661
Coir pith	73.3	7.1	44.0	4.7	0.7	43.4	18.07	94
Com cob	85.4	2.8	47.6	5.0	0.0	44.6	15.65	188
Com stalks	80.1	6.8	41.9	5.3	0.0	46.0	16.54	129
Cotton gin waste	88.0	5.4	42.7	6.0	0.1	49.5	17.48	109
Groundnut shell	83.0	5.9	48.3	5.7	0.8	39.4	18.65	299
Millet husk	80.7	18.1	42.7	6.0	0.1	33.0	17.48	201
Rice husk	81.6	23.5	38.9	5.1	0.6	32.0	15.29	617
Rice straw	80.2	19.8	36.9	5.0	0.4	37.9	16.78	259
Subabul wood	85.6	0.9	48.2	5.9	0.0	45.1	19.78	259
Wheat straw	83.9	11.2	47.5	5.4	0.1	35.8	17.99	222

[a]: Higher heating value.

Source: Raveedran *et al.*, 1995.

Table 16.2: Component Analysis of Biomass (wt per cent db)

	Ash	*Holo-cellulose*	*Cellulose*	*Hemi-cellulose*	*Lignin*	*Extractives*	*Total (holo)*	*Total (hemi)*
Bagasse	2.9	65.0	41.3	22.6	18.3	13.7	99.9	98.8
Coconut coir	0.8	67.0	47.7	25.9	17.8	6.8	111.7	99.0
Coconut shell	0.7	67.0	36.3	25.1	28.7	8.3	98.7	100.1
Coir pith	7.1	40.6	28.6	15.3	31.2	15.8	94.8	98.1
Com cob	2.8	68.2	40.3	28.7	16.6	15.4	102.9	101.8
Com stalks	6.8	63.5	42.7	23.6	17.5	9.8	97.6	100.5
Cotton gin waste	5.4	90.2	77.8	16.0	0.0	1.1	86.7	100.2
Groundnut shell	5.9	55.6	35.7	18.7	30.2	10.3	102.0	100.7
Millet husk	18.1	50.6	33.3	26.9	14.0	10.8	96.5	104.1
Rice husk	23.5	49.4	31.3	24.3	14.3	8.4	96.5	101.8
Rice straw	19.8	52.3	37.0	22.7	13.6	13.1	98.8	106.2
Subabul wood	0.9	65.9	39.8	24.0	24.7	9.7	101.2	99.0
Wheat straw	11.2	55.8	30.5	28.9	16.4	13.4	96.7	100.4

Source: Raveedran *et al.*, 1995.

Table 16.3: Ash Composition of Biomass: Major Elements (ppmw dry biomass)

Al	*Ca*	*Fe*	*Mg*	*Na*	*K*		*P*	*Si*
Bagasse	–	1518	125	6261	93	2682	284	17340
Coconut coir	148	477	187	532	1758	2438	47	2990
Coconut shell	73	1501	115	389	1243	1965	94	256
Coir pith	1653	3126	837	8095	10564	26283	1170	13050
Corn cob	–	182	24	1693	141	9366	445	9857
Corn stalks	1911	4686	518	5924	6463	32	2127	13400
Cotton gin waste	–	3737	746	4924	1298	7094	736	13000
Groundnut shell	3642	12970	1092	3547	467	17690	278	10960
Millet husk	–	6255	1020	11140	1427	3860	1267	150840
Rice husk	–	1793	533	1612	132	9061	337	220690
Rice straw	–	4772	205	6283	5106	5402	752	174510
Subabul wood	–	6025	614	1170	92	614	100	195
Wheat straw	2455	7666	132	4329	7861	28930	214	44440

Table 16.4: Combustible Matter Content of Biomass

Fuel	*Cellulose*	*Hemicellulose*	*Liflnin*
Coir Waste	39.1	24.0	23.2
Groundnut shell	31.2	28.2	25.6
Bagasse	41.5	22.9	18.5
Saw dust	43.6	23.9	21.2
Wheat straw	37.7	23.0	14.8
Rice straw	37.2	22.0	13.7
Rice husk	32.3	22.4	15.3

Source: Lawrence (1999).

The volatile matter's yield is dependent upon pyrolysis temperature and heating rate. Most of the tar is also converted into gas immediately. Devolatilization stage plays vital role in deciding ignitability and flammability, the two important combustion properties of the fuel.

The ignitability refers to the minimum temperature at which ignition starts. The flammability is the temperature at which stable flame can be achieved. Since these two combustion properties are very much related to devolatilization pattern, investigation of pyrolysis or volatile matter release profile by means of Thermogravimetric (TO) analysis will be of great use. Lawrence (1999) reported the following characteristics temperature from volatile release profile of Thermogravimetric analysis for various biomass.

VM_{IT}: The temperature at which release of volatile mater starts

VM_{PT}: The temperature at which rate of volatile release is maximum for the first peak.

$VM_{P''T}$: The temperature at which rate of volatile release is maximum for the second peak

VM_{CT}: The temperature at which release of volatile mater ceases.

Table 16.5 is the reproduction of the results reported by Lawrence (1999).

Table 16.5: Volatile Release Profile Temperature Data

Fuel	VM_{IT}	$VM_{P'T}$	$VM_{P''T}$	VM_{CT}
Coir waste	240	312	358	470
Groundnut shell	192	308	354	520
Bagasse	235	322	375	476
Saw dust	239	310	376	500
Wheat straw	232	–	360	500
Rice straw	220	–	344	520
Rice husk	244	–	320	510

The devolatilization behaviour of palm shell and fibre was investigated using Thermogravimetric analyzer (TOA) by Shamsuddin and Williams (1992) and they concluded that two groups of reactions were involved in this process and where characterized by two overlapping derivate thermogram (DTO) peaks. Based on TOA studies using pure hemicellulose, cellulose and lignin, Williams and Besler (1996) have reported that hemicellulose started decomposing between 200°C and 350°C leaving 20 per cent char approximately, cellulose between 325°C and 400°C leaving 8 per cent char and lignin from 200°C onwards leaving 55 per cent char. Lawrence (1999) studied devolatilization profile of coir waste, groundnut shell, bagasse, sawdust, wheat straw, rice straw and rice husk also confirms the same that coir waste and groundnut shell with more lignin content was found to have more fixed carbon among the biomass which he investigated.

Poultry Litter: A Plausible Bioresidue

More than 90 per cent of world population takes eggs and broiler chicken as food directly or indirectly to meet their protein demand economically. Egg and broiler industries have grown in size considerably during the last decade. Several food processing, preservating and transporting technologies have been developed, demonstrated and adopted successfully.

Each chick in addition to yielding egg or flesh also excretes at least 50 grams of wastes daily (Kirubakaran *et al.*, 2006). The size of poultry litter production in many countries shows a sustained and increasing trend, presenting problems for the management in disposal of litter. The ecofriendly disposal of these wastes poses several problems. Excreta (manure), bedding material or litter (*e.g.* wood shavings or straw), waste feed, dead birds, broken eggs and feathers are the main constituent of waste from the poultry industry. Owing to their high nutritional value, the litter and manure are used as organic fertilizer to recycle the nutrients such as nitrogen, phosphorous and potassium.

The physico-chemical characterization of poultry litter varies widely depending upon the region, method and type of spreading material used and collection. However by judicial analysis of the results, several investigators have reported the analytical results.

Around 200 kg of poultry litter from a nearby farm was collected and dried in the shadow to achieve a constant weight. The dried litter was stored in double polythene bag and sealed to prevent the moisture entering the system. This sample is used for the present investigation.

The poultry litter normally contains moisture, volatile matter, fixed carbon and ash. Fixed carbon is generated from volatile matter during its decomposition. The decomposable matter (Cellulose, Hemicellulose and Lignin) in broiler litter reported by authors (Rasool *et al.*, 1998; and Al-Masari and Zarkawi, 1999) is given in Table 16.6 and is compared with those available in the present sample. Table 16.7 compares the proximate composition of poultry litter reported in the literature as well as that of present investigations, while Table 16.8 those of ultimate analysis and Table 16.9 is the ash characteristics of poultry litter.

Table 16.6: Decomposable Matters in Poultry Litter (wt per cent)

Properties	*Rasool et al. (1998)*	*Al-Masari and Zarkawi (1999)*	*Present Investigation*
Dried mass (per cent) at 103°C	90.16	–	
Organic matter	79.78	–	
Neutral detergent fibre (NDF)	38.97	34.71±0.24	
Acid detergent fibre (ADF)	35.97	15.87±0.06	
Hemicellulose	3.00	18.84±0.28	20.20
Cellulose	18.10	11.28±0.28	11.00
Non–Fibre carbohydrates	16.01	–	
Lignin	2.82	4.59±0.07	5.20
Silica	14.56	–	
Total N	3.84	5.00±0.09	2.5
Protein N	1.28	–	
Ammonia N	0.27	–	
Non-Protein N	2.56	–	

Table 16.7: Proximate Analysis of Poultry Litter

Sl.No	*Source*		*Constituents per cent*			
			Volatile Matter	*Fixed Carbon*	*Moisture*	*Ash*
1.	Report from M/s. Antares Group Incorporated Landover MD 20785 (1999)	As Received	47.30	9.8	27.4	15.70
		Dry	62.73	13.87	0.00	23.40
2.	Henihan *et al.* (2003)		38.91	1.66	43.01	16.42
3.	Patel *et al.*		61.49	13.36	27.01	25.15
4.	John *et al.* (2001)		62.2	14.0	25.5	23.9
5.	Present investigation		50.3 (Combustible) 2.5 (Others)	10.2	8.2	28.8

Table 16.8: Ultimate Analysis of Poultry Litter

Sl.No.	Source	Constituents Per cent							
		C	H	O	N	S	Cl	Ash	Moisture
1.	Report from M/s. Antares Group Incorporated Landover MD 20785 (1999)	27.22	3.72	23.10	2.69	0.33	0.71	15.70	27.4
2.	Henihan *et al.* (2003)	39.57	5.11	48.27	5.31	0.77	0.88	–	–
3.	Davalos *et al.* (2002)								
	(wet)	10.2±0.7	9.1±0.6	79.4	1.3±0.1	–	–	–	–
	(dry)	34.7±0.4	5.2±0.2	54.09	5.6±0.2	–	–	–	–
4.	Report from Maryland Environmental Service (1999)								
	(wet)	21.9	2.7	14.6	2.0	0.50	–	27.5	29.9
	(dry)	31.2	3.85	20.8	2.85	0.71	0.84	39.7	0
5.	John *et al.* (2001)	35.6	4.6	29.8	5.3	0.9	–	23.9	0
6.	Present investigation	24.84	1.9	33.76	2.5	2.5	2.5	28.80	8.2

Table 16.9: Ash Composition of Poultry Litter

Components	Report from M/s. Antares Group Incorporated Landover MD 20785 (1999)	Sharon Falcone Millera, et al.	John et al. (2001)	Present Investigation
SiO_2	8.1	39.4	22.2	18.5
Al_2O_3	1.9	9.14	2.9	3.1
TiO_2	0.2	0.51	0.2	–
Fe_2O_3	1.16	4.04	1.7	2.1
CaO	17.3	12.7	21.3	18.5
MgO	5	4.01	5	4.2
Na_2O	9.2	3.6	3.9	3.6
K_2O	16.3	9.94		
P_2O_5	24.4	14.0	20.2	25.7
SO_3	6.7	2.58	7	5.2
BaO	–	0.05	–	–
SrO	–	0.03	–	–
CO_2/Other	9.74	–	0.3	–
Ash Fusion Characteristics				
Initial Deformation Temperature (°C)	–	–	–	1139
Softening Temperature (°C)	–	–	–	1149
Hemispherical Temperature (°C)	–	–	–	1161
Fusion Temperature (°C)	–	–	–	1163

Characterization of Poultry Litter

Flammability and ignitability, the two combustion properties are very much related to devolatilization pattern, investigation of pyrolysis or volatile matter release profile of poultry litter by means of Thermogravimetric (TG) analysis will be of great use. The detailed specification of TGA Analyzer used for the present investigation is given in Table 16.10. TGA studies were carried out at various heating rate covering the temperature range from ambient to 900°C. All the studies were carried out in an inert atmosphere of flowing nitrogen of 99.99 per cent purity. In addition to the weight loss; rate of weight loss *i.e.* Derivative Thermo Gram (DTG) as a function of temperature was recorded continuously. From the above thermograms of various heating rate, it was observed that there was an initial loss of moisture around 120°C. The thermal decomposition of poultry litter started around 140–165°C and there was a major loss of weight, where the main devolatilization occurred, which was completed approximately 300°C. A further loss of weight occurred until 520°C. After that there was no loss of weight.

Table 16.10: Specification of Thermogravimetric Analyzer

Make	Perkin Elmer, USA
Model	Pyris 7 TGA
Balance Sensitivity	10 μg (10^{-2} mg)
Balance Accuracy	Better than 0.1 per cent
Weighing Precision	Upto 10 ppm
Sample Capacity	Upto 250 microliters
Temperature range	50°C to 1500°C
Heating and cooling rates	0.1 to 200°C/min in 0.1°C increments
Cool down times	1500°C to 100°C in less than 15 min.
Sample Type	Solids, Liquids, Powders, Films for Fibers.
Atmosphere	Static or dynamic including nitrogen, argon, carbon dioxide, air, oxygen or other inert or active gases. Analyses may also be made at normal or reduced pressure.
Reduced Pressure	Operation to 10^{-4} torr optional
Temperature Sensors	Platinum/10 per cent rhodium–platinum thermocouple.
Cooling method	Forced air cooling

Four devolatilization characteristic temperatures as reported by Lawrence (1999) for biomasses, the characteristics temperature for the poultry litter also found out and reported in Table 16.11 for various heating rate. The corresponding TGA curves are shown in Figures 16.1–16.4. For the slow heating rate (10 to 30 °C/min) the initial shoulder-like peaks were observed. Hence, it is possible to measure $VM_{P'T}$ for these heating rates. The initial peak could not be observed distinctly for higher heating rate but it appears that the initial peak might have merged with the main peak and hence $VM_{P'T}$ could not be measured.

Based on Williams and Besler (1993) the char that could have been formed from poultry litter having the composition reported by Al-Masri and Zarkawi (1999) and present investigation reported in Table 16.12 was calculated as follows:

% combustible matter in the poultry litter (Table 16.7)

= V.M (%) + F.C. (%)

= 50.3 + 10.2

= 60.5%

Al-Masri and Zarkawi data:

% char = 60.5 (0.543 × 0.2 + 0.325 × 0.08 + 0.132 × 0.55)

= 12.53%

Present Investigation:

% char = 60.5 (0.555 × 0.2 + 0.302 × 0.08 + 0.143 × 0.55)

= 12.93%

Table 16.11: Devoltilization Characteristics Temperature of Poultry Litter

Heating rate °C/min	*VM_{IT} °C*	*VM_{PT} °C*	*$VM_{P'T}$ °C*	*VM_{CT} °C*
10	165	265	304	504
15	160	272	309	520
20	158	276	312	524
25	158	282	310	537
30	157	290	307	545
35	155	–	300	541
40	156	–	300	540
50	154	–	300	557

Table 16.12: Volatile Matter Composition of Poultry Litter

Sl.No.	*Component*	*Al-Masari and Zarkawi (1999)*		*Present Investigation*	
		%		*%*	
1.	Hemicellulose	18.84	54.3	20.20	55.5
2.	Cellulose	11.28	32.5	11.00	30.2
3.	Lignin	4.59	13.20	5.20	14.3
	Total	34.71	100.00	36.40	100.00

Char in the poultry litter sample is found to be 10.2 per cent as fixed carbon (Table 16.7). The higher percentage of char (12.53 per cent and 12.93 per cent) from the constituent compared that obtained from poultry litter is attributed due to catalytic action of ash on converting char to gas. Thus the results are in line with the investigation made by the Shamsuddin *et al.* (1992) and Lawrence (1999).

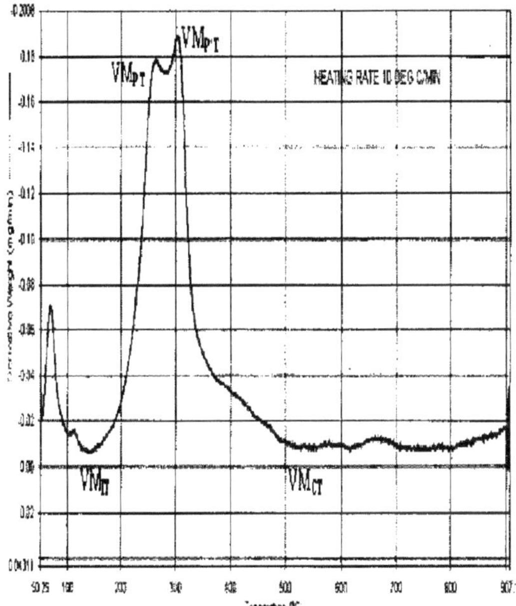

Figure 16.1: Devolatilization Profile of Poultry Litter for Heating Rate of 10°C/min

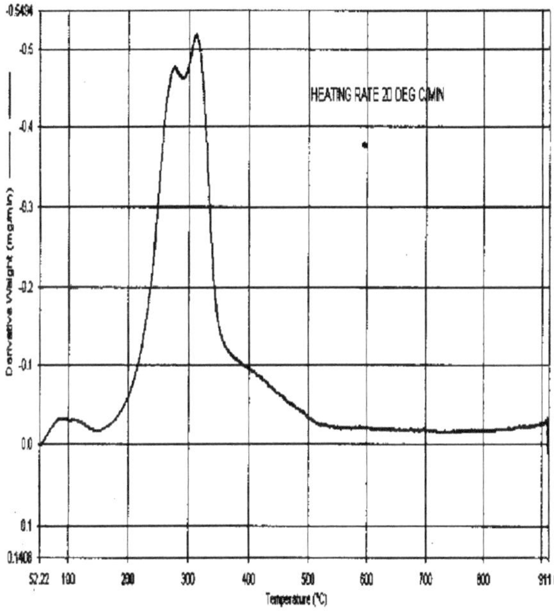

Figure 16.2: Devolatilization Profile of Poultry Litter for Heating Rate of 20°C/min

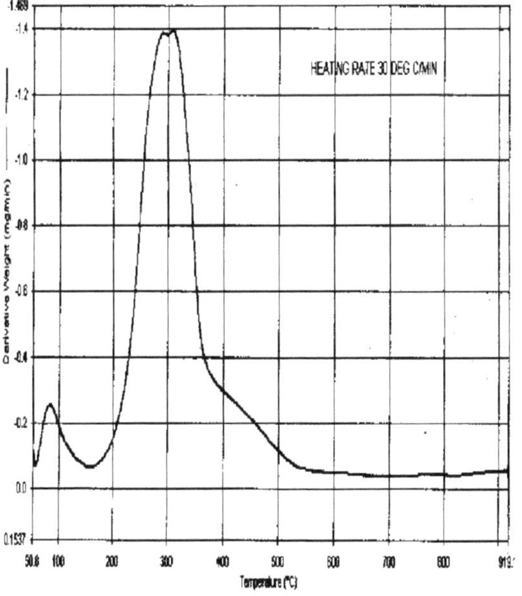

Figure 16.3: Devolatilization Profile of Poultry Litter for Heating Rate of 30°C/min

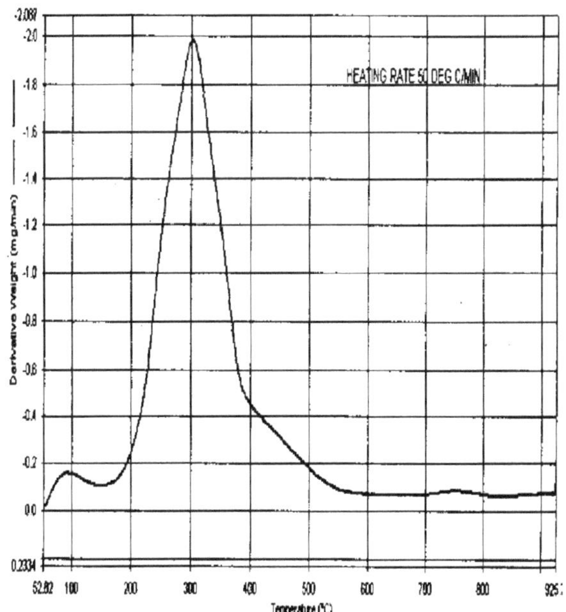

Figure 16.4: Devolatilization Profile of Poultry Litter for Heating Rate of 50°C/min

Conclusion

Proximate analysis of poultry litter reveals that it has less fixed carbon and high ash content. The calorific value of the poultry litter is found to be 2276 kcal/kg. From the proximate and ultimate analysis of poultry litter the composition of litter may be empirically expressed as $C_{2.01}H_{1.99}O_{2.11}$. The composition of combustible matters of poultry litter cellulose, hemicellulose and lignin has been found out and compared with the literature values. Using thermogravimetric analysis the devolatilization characteristics of poultry litter has been carried out and the characteristics temperature for various heating rate is found out. Based on Williams and Besler (1993) the char that could have been formed from poultry litter having the combustible matter composition reported in the literature and the present investigation have been calculated and compared with the fixed carbon values obtained from the proximate analysis. The results are very close and slight deviation in the values is due to catalytic action of ash on converting char to gas.

Acknowledgements

We express our sincere thanks to Science and Engineering Research Council (SERC), Department of Science and Technology (DST), Government of India, New Delhi for sponsoring this research project.

References

Al-Masari, M.R. and Zarkawi, M., 1999. Digestibility and composition of broiler litter, as affected by gamma irradiation. *Bioresource Technology*, 69: 129 –132.

Davalos, J.Z., Maira Victoria, Roux and Jimenez, Pilar, 2002. Evaluation of poultry litter as a feasible fuel. *Thermochimica Acta*, 394: 261–266.

Henihan, A.M., Srinivasa Rao, C.H., Srinivasa Rao, Subba Rao, A. and Singh, Muneshwar, 2003. Emissions modeling of fluidized bed co-combustion of poultry litter and peat. *Bioresource Technology*, 87: 289–294.

Kirubakaran, V., Sivaramakrishnan, V., Premalatha, M. and Subramanian, P., 2001. Environmental impacts of poultry litter disposal and sustainable technology for energy production. In: *Proceedings of International Interdisciplinary Conference on Sustainable Technologies for Environmental Protection* organized by CIT, Coimbatore, Tamil Nadu, India and The University of TOLEDO Ohio, USA, pp. 1–6, http://www.icstep2006.com/new/PDF1/session/session%2001/129.pdf.

Lawrence, A., 1999. Studies on combustion characteristics of biomass fuels. *Ph.D. Thesis*, Bharathidasan University, Tamil Nadu, India, pp. 81–90.

M/s. Antares Group Incorporated, 1999. Landover MD 20785–Report, "Economic and Technical Feasibility of Energy Production from poultry litter and Nutrient Filter Biomass on the Lower Delmarva Peninsula" pp. 25–28. http://www.mbp.org/pdfs/abstr20.pdf.

Maryland Environmental Service, 2000. Report, Eastern correctional institution cogeneration facility full-scale poultry litter test bum", pp. 18–22, http://www.esm.versar.com/pprp/eci/2-Test%20Burn% 20Report%20PDF.pdf.

Patel, B., Mc-Quigg, K. and Tome, R. Integration of poultry litter gasification with conventional pulverized coal fired power plant. http://bioproductsbioenergy.gov/pdfs/bcota/abstracts/11/z176.pdf, pp 1–2.

Rasool, S., Sial, M.A., Ahsan-Ul-Haq and Ameth Jamil, 1998. Chemical changes during ensiling of sudax fodder with broiler litter. *Animal Feed Science and Technology*, 72: 347–354.

Raveendran, K., Anuradda Ganesh and Khilar, K.C., 1995. Influence of mineral matter on biomass pyrolysis characteristics. *Fuel*, 74: 1812–1822.

Reardon, John P., Lilley, Art, Browne, Kingsbury and Beard, Kelly, 2001. Demonstration of a Small Modular BioPower System Using Poultry Litter, DOE SBIR Phase-I, Final Report, Contract: DE–FG03–01ER83214, Community Power Corporation, pp 22–26. http://www.nrel.gov/docs/fy00osti/27984.pdf.

Shamsuddin, A.H. and Williams, P.T., 1992. *J.I. Energy*, 65: 31.

Sharon Falcone Millera, Millera, Burce G. and Jawady, Cutis M. The occurrence of inorganic elements in various biofuels and its effect on combustion behavior. pp. 1–2. http://bioproducts-bioenergy.gov/pdfs/bcota/abstracts/18/z178.pdf.

Williams, P.T. and Besler, S., 1993. The pyrolysis of rice husks in a thermogravimetric analyzer and static batch reactor. *Fuel*, 72: 151–159.

Williams, P.T. and Besler, S., 1996. The influence of temperature and heating rate on the slow pyrolysis of biomass. *Renewable Energy*, 7: 233–250.

Young, B.C. and Smith, I.W., 1987. In: *International Symposium on Coal Combustion*, Tsinghua University.

Chapter 17

Study of Groundwater in Ludhiana District

Ritesh Jain[1], Jaspal Singh[2] and S.K. Singh[3]
[1]Assistant Professor, [2]Associate Professor,
Department of Civil Engineering, College of Agricultural Engineering,
P.A.U. Ludhiana – 141 004
[3]Professor, Department of Civil Engineering,
Delhi College of Engineering, Delhi – 110 042

ABSTRACT

In order to evaluate the problems of pollution hazards of groundwater and to ascertain its suitability for drinking and agricultural purposes in Ludhiana district of Punjab, the data relating to groundwater chemical quality are analyzed from 23 places covering the whole district. The available data is analyzed and compared with various standards laid down by various agencies. The analysis shows in general, the quality of the groundwater is fresh except in and around Ludhiana city, where the groundwater is polluted due to industrial effluents.

Introduction

The quality of groundwater of any area is of great importance for human beings and irrigation. All the groundwaters irrespective of their source of origin contain mineral salts and their chemical properties. The kind and concentration of these constituents depend upon various geological and physical factors. Since most of these factors are varying from place to place, the groundwater of any region are characterized by marked difference in their chemical properties. Since the quality of the groundwater is directly or indirectly dependent related to its intended use, there is always a need to classify the groundwaters of an area on a regional basis.

Introduction of Ludhiana District

The district covers area of 3760 km^2 and forms a part of Indo-Gangetic alluvial plain. Sutlej river demarcates the western boundary of the district. Budha nallah, probably an old course of river Sutlej passes through district. Low-lying alluvial tract along river Sutlej and south of it constitutes 'BET' area. The district experiences continental type of climate.

The district is occupied by Indo-Gangetic alluvium of Quarternary age. The subsurface geological formation comprise sand, silt, clay and kankar in various proportions. The depth to water level in area ranges from 4–16 m below ground level.

In this study, the analysis of various data on groundwater collected from Pollution Control Board, Patiala (Punjab), Public Health Engineering Department (Punjab) and Irrigation Department, Punjab has been presented which gives an idea about the quality of groundwater of the Ludhiana district (punjab).

Criteria of Water for Drinking Purposes

The evolution of standards for the quality control of public water supplies depends upon the limitations imposed by the local factors. The water should be free from any risk and suitable for drinking purposes. Table 17.1 shows the standards for drinking purposes.

Table 17.1: Standards for Potable Water

Sl.No.	*Characteristics*	*World Health Organization 1971*		*Govt. of India, Ministry of Works and Housing 1975*	
		Highest	*Maximum Desirable*	*Acceptable Permissible*	*Cause of Rejection*
1.	Turbidity (JTU Scale)	5.0	25.0	2.5	10.0
2.	Colour (Hazen Unit)	5.0	50.0	5.0	25.0
3.	Taste and Odour	Nothing	Disagreeable	Nothing	Disagreeable
4.	pH	7.5–8.0	6.0–9.2	7.0–8.5	8.5–9.2
5.	Total dissolved (mg/l) solids	500	1500	500	1500
6.	Total hardness (mg/l)	100	500	200	600
7.	Chloride as Cl (mg/l)	200	600	200	100
8.	Sulphate as SO_4 (mg/l)	200	400	200	400
9.	Fluoride as F (mg/l)	1.0	1.5	1.0	1.5
10.	Nitrate as NO_3 (mg/l)	45	45	45	45
11.	Calcium as Ca (mg/l)	75	200	75	200
12.	Magnesium as Mg (mg/l)	30	150	30	150
13.	Iron as Fe (mg/l)	0.1	1.0	0.1	1.0
14.	Manganese as Mn (mg/l)	0.05	0.5	0.05	0.5
15.	Copper as Cu (mg/l)	0.05	1.0	0.05	1.5
16.	Zinc as Zn (mg/l)	5.0	15.0	5.0	15.0
17.	Phenolic compounds as phenol (mg/l)	0.001	0.002	0.001	0.002

Contd...

Table 17.1–Contd...

Sl.No.	Characteristics	World Health Organization 1971		Govt. of India, Ministry of Works and Housing 1975	
		Highest	Maximum Desirable	Acceptable Permissible	Cause of Rejection
18.	Mineral oil (mg/l)	0.01	0.30	0.01	0.30
19.	Arsenic as As (mg/l)	–	0.5	0.05	0.05
20.	Cadmium as Cd (mg/l)	–	0.01	0.01	0.01
21.	Chromium as Cr (mg/l)	–	0.01	0.05	0.05
22.	Cyanides as CN (mg/l)	–	0.05	0.05	0.05
23.	Lead as Pb (mg/l)	–	0.10	0.1	0.01
24.	Selenium as Se (mg/l)	–	0.01	0.01	0.01
25.	Mercury as Hg (mg/l)	–	0.001	0.001	0.001

Generally the total concentration of the dissolved solids (Electrical Conductivity) and Sodium Content reported as Sodium Absorption Ration (SAR) values more or less indicate the suitability of the water for irrigation purpose.

The criteria for classification of irrigation water as recommended by the U.S. Salinity Laboratory Department of Agriculture are based on the electrical conductivity (Total Soluble Solids or TSS) and SAR value, limits of which have been indicated in Table 17.2 and Table 17.3 below.

Table 17.2

Sl.No.	Water Class	Electrical Conductivity (MICRO MHOS)
1.	Excellent	Less than 250
2.	Good	250–750
3.	Permissible	750–2000
4.	Doubtful	2000–3000
5.	Unsuitable	More than 3000

Table 17.3

Sl.No.	Water Class	SAR Value
1.	Excellent	Less than 10
2.	Good	10–18
3.	Fair	18–26
4.	Poor	More than 26

SAR Value reported in the Table 17.3 is expressed as follow:

$$SAR = \frac{\text{Sodium}^{+}}{\{(\text{Calium}^{++} + \text{Magnesium}^{++})/2\}^{1/2}}$$

where, the concentrations of the constituents are expressed in milli-equivalent per litre.

Data Analysis

The data considered here has been taken from the various agencies involved in the work of underground water. Mainly these data are collected from the Pollution Control Board, Patiala (Punjab), Public Health Engineering Department (Punjab) and Irrigation Department, Punjab.

Analysis of Data for Drinking Purposes

If we analyze the data for drinking purposes then we have to compare the limits of various constituents of groundwater found in the samples with the standards for the potable water given in the Table 17.1.

Here, we find that except Calcium (Ca) all other characteristics of water in all the 23 samples are within the acceptable limits. Calcium less than 75 ppm was found in 14 samples out of 23 samples *i.e.* in Sarduala, Niloni, Gujarwal, Raikot, Mullanpur, Dehlok, Badowal, Marsura, Khanka, Seah, Chaunta, Hadon, Mandema and Kohara. Thus we can say that the risk of lesser calcium present in the underground water (which is the drinking water for population) is present in about 61 per cent area of the district. Hence, we can conclude that in general except the small belt of the district, the groundwater is fit for drinking purposes.

Analysis of Data for Irrigation Purposes

In order to check the suitability of the groundwater for agricultural purposes the chemical data were presented on the sodium (alkali) hazard versus salinity hazard diagram by the US Salinity Laboratory Staff. On the basis of rating or irrigation water falling in various groups is given below:

Salinity Groups	Quality for Irrigation Purposes
C_1S_1, C_2S_1	Good
C_1S_2, C_2S_2, C_3S_1, C_3S_2	Moderate
All other groups	Bad

Samples of 8 locations fall in C_2S_1 group and hence the groundwater of these locations is good for irrigation purposes. Samples of 10 locations fall in C_1S_2, C_2S_2, C_3S_1 and C_3S_1 group so the water of these locations is suitable for irrigation purposes.

About 9 per cent of the area are not suitable for irrigation purposes, the samples of which fall in C_4S_1, C_4S_2, C_1S_3, C_2S_3, C_3S_3, C_4S_3, C_4S_4, C_3S_4, C_2S_4 and C_1S_4 groups.

Conclusions

From the above study we conclude that:

1. There is no major pollution hazard in the groundwater of Ludhiana District except in and around Ludhiana city.
2. In the smaller belt of the district, the problems of lesser calcium were found.
3. The quality of groundwater in most of the areas of the district is suitable for drinking purposes.

References

Brown, R.H. *Groundwater studies*. An International Guide for Research and Practice.

Gautam, Mahajan. *Groundwater Surveys and Investigation*. Ashish Publishing House.

Kalra, C., Pal. S., Kataria, V. and Verma, G., 1997. *The Status of Surface and Groundwater in Punjab State*.

Singh, S.K. and Anand, Vivek, 1991. *Study of Groundwater Quality in Jhunjhunu District (Raj.).*

Singh, S.K., Chopra, Pankaj, Goyal, Vivek Krishan, Singh, Parminder and Vij, Sachin, 1998. Status of quality of groundwater in Punjab state. *Project Report*, GZSCET, Bathinda.

Chapter 18

Effect of Pesticides and FYM on Microbial Activity and CO_2 Evolution in Acid Alfisol of Ranchi

D.K. Shahi, Kumari Nisha and P.K. Singh*
Department of Soil Science and Agricultural Chemistry, Birsa Agricultural University, Ranchi – 834 006, Jharkhand

ABSTRACT

The field experiments were carried out to find out the effect of endosulfan, monocrotophos and FYM on soil micro flora (bacteria, fungi, actinomycetes) microbial biomass carbon, nitrogen and CO_2 evolution. Adwrse effect of 3 sprays each of endosulfan and monocrotophos was noticed on bacteria (5×10^6 cells), actinomycetes (1×10^6 cells) and fungi (2×10^4 propagules) g^{-1} soil in absence of FYM. The minimum microbial biomass C (15.10 mg/100 g soil) and biomass N (1 mg/100 g soil) were found in the treatment receiving three sprays of monocrotophos. However, CO_2 evolution reduced to 17.60 mg/100 g soil in case of 3 sprays of endosulfan from the initial 47.74 mg/100 g soil. Addition of FYM @ 20.0 t ha^{-1} increased the microbial population and related properties than that of untreated plots

Keywords: *Endosulfan, Monocrotophos, FYM, Soil microflora, CO_2 evolution, Biomass–C and N.*

Introduction

Endosulfan [a,b,1,2,3,4,7,7-Hexachloro bicycle (2,2,1)-heptene-(2) bishydroxy methylene-(5-6)-sulphate trade name Endocel–35 EC and LD_{50} for rat is 80–110 mg kg^{-1} of body weight is an

---

* Present Address: Associate Professor, Department of Entomology, BAU., Ranchi.

organochlorine compound. Monocrotophos (3-Dimethoxy-Phosphinyloxy-N-methyl isocrotonamide) is an organophosphate compound. Its trade name is Monosil–36 SL and LD_{50} for rat is 8.0 mg kg^{-1} of body weight. Both the insecticides are used for control of insect pest in vegetable crops like okra, cauliflower, brinjal etc. Both the pesticide are extensively used in Indian agriculture specially in Jharkhand region. Soil microorganism also playa vital role in degrading some of the insecticides thereby affecting its persistency. Organochlorine groups insecticides have been found to be more resistant to microbial degradation than organophosphates groups (IIRC, 1979).

The relationship between soil microorganisms and insecticides can be view from to angles (*i*) suppression or stimulation of microbial population and their activity and (*ii*) microorganism mediated modification in the potency and persistence of the applied chemicals. Several, workers have reported adverse effect of higher dose of insecticide on soil microbial population their activity Pandey and Rai (1995) and Shukla and Das (2001). Ahmad Sahid (2000) reported that nuvacrone. endosulfan and pendimethylene completely inactivated the fungal population up to 45 days of treatments. No work has so far been done on the effect of endosulfan and monocrotophos on soil microbial community (particularly bacteria, actinomycetes and fungi). Hence the present investigation, was planned to study the effect of different concentration of insecticides and FYM on total microbial population, biomass C and CO_2 evolution in acid soil (Alfisol) of Jharkhand.

Materials and Methods

A field experiment was conducted in randomized block design with three replications at Horticulture Research Farm, Birsa Agricultural University, Ranchi during *Kharif* 2003. The soil was pH 5.4, organic carbon 0.30 per cent available nitrogen 214.87 kg ha^{-1}, available P_2O_5 10.94 kg ha^{-1} and available K_2O 160.0 kg ha^{-1} Farm yard manure was mixed well with soil in the respective treatments prior to sowing of the crop @ 2 l ha^{-1} and different number of sprays of endosulfan and monocrotophos was given after 20, 30 and 50 days of sowing.

The soil sample were analysed for total bacterial actinomycetes and fungal population by standard pour plate technique using soil extract agar (Subba Rao, 1977) and Ken Knight and Munaiers media (Subba Rao, 1977) for bacteria and actinomycetes, respectively, and Rose Bengal agar medium (Martin, 1950) for fungi. Triplicate plates (for each soil microbial group) were incubated at 30°C±1°C for a week.

Microbial biomass C in the soil was determined by chloroform fumigation and incubation teclullque outlined by Jenkinson and Powlson (1976). Four, 100 g fresh weighed portions of soil were taken from each soil sample and each portion was placed in a petridish. Half of the portions were fumigated with chloroform and the rest half portions were left unfumigated.

After completion of microbial biomass C analysis, all fumigated and unfumigated samples were analysed for ammonical and nitrate nitrogen.

Microbial Biomass N was computed as:

$$\text{Microbial biomass N} = \frac{\begin{array}{c}(NH_4\text{-N} + NO_3\text{-N mineralized by fumigated soil in 0–10 day}) - \\ (NH_4\text{-N} + NO_3\text{-N mineralized by unfumigated soil in 0–10 day})\end{array}}{K_N}$$

$K_N = 0.57$

The CO_2 evolution was determined by the method of Bremner and Schmidt (1964). 100 g soils maintained at 1/3rd moisture holding capacity were kept in one litre capacity incubation bottles along with a test tube containing 20.0 ml of N/10 NaOH solution. The entire assembly was made air tight and was incubated at 30±1°C. After 7 days interval excess NaOH was titrated against N/10 HCl using phenolphthalein as indicator. The amount of CO_2 was calculated by the factor:

1 ml of N/10 HCl = 2.2 ml of CO_2/100 g soil.

Results and Discussion

Application of different concentrations of pesticide endosulfan showed more suppressive effect on bacteria, actinomycetes and fungi than monocrotophos. The decline in mean population was more rapid with endosulfan either with increasing number of sprays or in absence of FYM. The initial population (2.5 × 10^6 cells g^{-1} soil) of actinomycetes was maintained even with up to two sprays. However, it declined to 1.5 × 10^6 cells g^{-1} soil and was lower than initial value with both the insecticides, The use of organic manure @ 20 t ha^{-1} reduced adverse impact as mean population were 3.4 and 2.7 × 10^6 cells g^{-1} soil in monocrotophos and endosulfan treated plots containing FYM over corresponding mean population of 1.7 × 10^6 cell g^{-1} soil in plots having no FYM. Recommended dose of single spray of both insecticide endosulfan and monocrotophos showed no adverse impact on fungal population (Table 18.1), But with increasing number of sprays *i.e.* second and third sprays, the fungal count decreased from 34.0 to 9.5 × 10^4 propagules g^{-1} soil with endosulfan. It was further observed that addition of FYM helped in antagonizing the inhibitory effect of insecticides on fungal population. Between two insecticides, endosulfan was noted with relatively more inhibitory effect on fungal population than that of monocrotophos.

Table 18.1. Effect of Insecticides and Manure on Soil Microbial Population in Okra Rhizosphere

Soil Microbes	*Organic Manure*	*Pesticides*							
		I_1 (Monocrotophos)				*I_2 (Endosulfan)*			
		S_1	*S_2*	*S_3*	*Mean*	*S_1*	*S_2*	*S_3*	*Mean*
Bacteria	M_0	18	10	5	11	7	5	5	9
(x10^6 g^{-1} soil)	M_1	25	15	5	15	22	9	8	13
	(20 t/ha)								
	Mean	215	12.5	5.0		19.5	7.0	6.5	
	Initial	17.0							
Actinomycetes	M_0	2	2	1	1.7	2	2	1	17
(x10^{6}' cells g^{-1}	M_1	5	3	2	3.4	3	3	2	2.7
soil)	(20 t/ha)								
		Mean	3.5	2.5	1.5		2.5	2.5	1.5
		Initial	2.5						
Fungi	M_0	30	27	7	21.3	25	14	2	11.5
(x 10^4	M_1	38	22	12	24.0	35	16	3	21.0
propagules	(20 t/ha)								
g^{-1} soil)	Mean	34.0	24.5	9.5		30.0	15.0	2.5	
	Initial	28.0							

The values of biomass carbon and nitrogen as affected by insecticide sprays and organic manure have been depicted in Tables 18.2 and 8.3. The highest mean value (28.84 and 4.0 mg/100 g soil) were recorded in the treatment with one spray of monocrotophos in conjunction with farm yard manure. Decreased biomass C and N were found with increasing number of insecticidal (endosulfan/monocrotophos) sprays and the lowest value of biomass carbon was found 18.44 mg/100 g soil which was even lower than the initial value (20.4) when three sprays of endosulfan were applied. However addition of FYM @ 20 t ha^{-1} increased its value to some extent (19.33 mg/100 g soil).

Table 18.2: Effect of Pesticides and Manure on Microbial Biomass C (mg/100 g Soil) in Okra Rhizosphare

Organic Manure	Pesticides							
	I_1 (Monocrotophos)				I_2 (Endosulfan)			
	S_1	S_2	S_3	Mean	S_1	S_2	S_3	Mean
M_0 (No FYM)	27.33	18.24	15.10	20.22	26.49	20.03	18.44	21.66
M_1 (FYM 20 t/ha)	28.84	24.69	20.82	24.79	27.73	21.56	19.33	22.87
Mean	28.09	21.47	17.96		27.11	20.79	18.88	
Initial	26.40							

Table 18.3: Effect of Pesticides and Manure on Microbial Biomass N (mg/100 g soil) in Okra Rhizosphare

Organic Manure	Pesticides							
	I_1 (Monocrotophos)				I_2 (Endosulfan)			
	S_1	S_2	S_3	Mean	S_1	S_2	S_3	Mean
M_0 (No FYM)	2.0	1.0	10	1.3	2.0	1.0	1.0	1.3
M_1 (FYM 20 t/ha)	4.0	4.0	2.0	3.3	4.0	3.0	1.0	2.7
Mean	3.0	2.5	1.5		3.0	2.0	1.0	
Initial	2.8							

Decrease in biomass C and N with increasing number of sprays may be attributed to cumulative effect caused by increased concentration of pesticide. Addition of organic manure reduced the adverse effect of insecticides because of its adsorption on organic colloids. Munshi and Raghu (1991) also reported les total biomass in carbendazim treated soils than untreated soils.

Soil respiration measured in terms of evolution of carbon dioxide within specified time per unit of soil as influenced by application of insecticides in presence and absence of FYM has been presented in Table 18.4. No marked change in microbial respiration was noticed over initial value of 47.74 mg CO_2/100 g soil with use of recommended one spray of either of the two insecticides which remained 521.4 and 50.60 mg CO_2 evolved/100 g soil in presence of monocrotophos and endosulfan, respectively. However, it reduced to the minimum 31.24 and 25.63 mg CO_2/100 g soil at the end of third sprays of monocrotophos and endosulfan respectively. The higher mean CO_2 evolved (41.07 mg/100 g soil) was found in manured plots than that in unmanured plots (33.51 mg CO_2 evolved/100 g soil). Sai Lakshmi *et al.* (1999) also reported decreased CO_2 evolution with increased rate of pesticides.

Table 18.4: CO_2 Evolution (mg/100 g soil) as Affected by Pesticide and Manure in Okra Rhizosphere

Organic Manure	*Pesticides*							
	I_1 (Monocrotophos)				*I_2 (Endosulfan)*			
	S_1	*S_2*	*S_3*	*Mean*	*S_1*	*S_2*	*S_3*	*Mean*
M_0 (No FYM)	47.96	36.08	30.14	38.06	49.06	33.88	17.60	33.51
M_1 (FYM 20 t/ha)	56.32	40.7	32.34	43.12	52.14	46.92	33.66	44.07
Mean	52.14	38.39	31.24	50.60	40.15	25.63		
Initial	47.74							

References

I.T.R.C., 1999. *Indo-US Workshop on Biodegradation and Pesticides.*

Jenkinson and Powlson, 1976. *Soil Biology and Biochemistry,* 8: 205–213.

Martin, J.P., 1950. *Soil Sci.,* 69: 215–232.

Munshi, Meenakshi and Raghu, K., 1991. *Adv. Pl. Sci.,* 4: 61–67

Pandey, R.K. and Rai, S.N., 1995. *Journal of Research* (BAU), 7(1): 71–73.

Sahid, Ahmad, 2000. *Am. Agric. Research,* 21(4): 527–529.

Sai Lakshmi, R., Sankara Rao, V. and Seshagiri Rao, M., 1999. *Andhra Agric. J.,* 46(3 and 4): 161–164.

Shukla, L. and Das, T.K., 2001. *Pestology,* 25(11): 33–38.

Subba Rao, N.S., 1977. *Soil Microorganisms and Plant Growth.* Oxford and IBH Publishing Co., New Delhi.

Chapter 19

Rural Urban Interface in Relation to Industrial Development and Environment Security

V.K. Sharma[1] and Shiv Raj Singh[2]

[1]STPC, UHF, Solan – 173 230, H.P.

[2]Department of Public Administration, H.P. University, Shimla – 171 005, H.P.

ABSTRACT

The mountain areas are considered to be fragile eco-systems whereby livelihood avenues of the rural masses are strategically integrated with the resource endowments of the area. The recent mountain development policies adopted by national policy planners and the respective state governments have laid emphasis on the generation of increased employment avenues through rapid industrialization. The state of Himachal Pradesh is also actively marching towards this pursuit. It is in this context that the dynamism of rural-urban interface in relation to industrial development and environment security must be considered for sustainable development. The most prominent impact of industrialization is visualised in this research paper through changes in rural-urban interface which if not desirable leads to rupturing of the socio-eco-cultural and ecological settings of the area. The findings have recommended involvement of local community in project planning and implementation so that their ownership rights and environment security can be sustained in the process of industrial development.

Keywords: *Rural-urban interface, Industry, Mountain area, Environment security, Sustainability, Policy formulation, Social set up.*

Introduction

Historically, rural-urban interface has received significant and vibrant changes during era of industrial revolution. The liberal and free minded approach to promote small and large scale industries has not only resulted in economic development but also changed the rural-urban interface. The migration of rural people to urban areas for employment and livelihood avenues is emerging as a regular feature. It is leading to unplanned growth of urban areas resulting in the deteriorating quality of urban atmosphere and adjoining countryside to a significant extent.

This trend of interfacial change in rural-urban can also be seen in the hill state of Himachal Pradesh where the state government is continuously encouraging entrepreneurs for establishing increased number of industries. Presently there are 327 medium and large-scale industries and 32,830 small-scale units in the state. Medium and large-scale industries include Cement plants, pharmaceuticals, paper mills, fertilizer plants, electronic units and fruit and vegetable processing units. These industries are directly or indirectly causing the environmental problems. Rural community suffers heavily because of their close affinity and interaction with local biotic and abiotic components of nature. Urban concentrations intensify the deficiencies of housing, transport capacity, food storage and distribution which in turn magnify crowding, noise, insect and rodent populations, air pollutants and street filth.

It is in this context that the study of Gujarat Ambuja Cement Industry at Darlaghat (Solan) and Himachal Futuristic Communication Limited, Chambaghat (Solan) was undertaken. This study is likely to contribute towards the development of effective industrial policy. The changes in social structure, social cohesiveness, changes in development necessities and their impact on rural-urban linkages in Darlaghat and Chambaghat areas are discussed.

Research Methodology

The study area of Solan district with the population of 4,99,380 is one of the major industrial centres of Himachal Pradesh having 3194 small and large industrial units. This is highest in any district of the state. The rapid industrial and population growth has disturbed the ecosystem and also shown visible impact on rural-urban interface. 2001 Census data of Himachal Pradesh has also showed an increase of 1.10 per cent in migration of rural people to urban during the last decade.

Hence, Solan district of Himachal Pradesh was purposively selected as it is being recognised as industrial district of the state. Kunihar and Solan development blocks representing transitional phase of industrial development were selected as the final sampling unit. Two industrial units M/S Gujarat Ambuja Cement limited (GACL), Darlaghat and M/S Himachal Futuristic Communication Limited (HFCL), Chambaghat were selected for their likely impact. In Kunihar Development Block, the study was carried out in the vicinity of the GACL, Darlaghat, covering Nagar Panchayat, Arki from urban side and the Gram Panchayat, Daria (including some areas of Kashlog, Parnu, Giana and Mango Panchayats) from rural side. While, in Solan Development Block the study was concentrated in the vicinity of the HFCL, Chambaghat, covering all wards of Municipal Committee area of Solan from urban side and the Gram Panchayat, Basal (including some areas of Dangri, Padag and Saproon Panchayats) from rural side. Gender perspectives were carefully observed while understanding the complexity of the research problem. Applying the technique of random sampling, three hundred forty four respondents comprising of farmers, educated, businessmen and professionals were interviewed from rural (168) and urban (176) areas (Table 19.1).

Table 19.1: Number of Samples Taken from Different Categories in Rural-urban Areas

Category	Locations				Total
	Rural		Urban		
	Daria	Basal	Arki	Solan	
Farmers	32	32	18	18	100
Educated	22	22	18	18	80
Businessmen	16	16	34	34	100
Professionals	14	14	18	18	64
Total	84	84	88	88	344

Eighteen administrative heads of various government development and welfare departments from each Kunihar and Solan Development Blocks were interviewed and data were collected on different aspects (a total of 36 respondents).

Data was collected through personal interview method, Triangulation was also done to check the reliability of data. The collected data was tabulated and analysed on the basis of differently designated broad categories of farmers, educated, businessmen and professionals.

Results and Discussion

The results of the present investigations are as follows:

Occupational Changes

The rise in population in the study district as depicted in the Table 19.2 clearly support the fact that the industrialization is one the factors for significant change in population graph. With the industrial development, rural skills came in demand resulting thereby rural migration to the urban centers. This pace of migration earlier was slow but the rapid industrial development increased the rate of rural migration to urban areas, thus changing the rural-urban interface.

Table 19.2: Comparative Changes in the Rural and Urban Population

Area	Rural		Per cent Change	Rural		Per cent Change
	1991	2001		1991	2001	
Solan district	3,34,989	4,08,205	21.85	40,279	91,175	126.36
Himachal Pradesh	47,21,681	54,82,367	16.11	4,49,196	5,94,881	32.43

Source: Census of India, 1991, Himachal Pradesh Primary Census Abstract Census of India, 2001, Himachal Pradesh, Paper 2 of 2001.

The comparative temporal and spatial change in occupation pattern were considered to be helpful in analysis of the rural-urban interfaces. The results of the study are presented in Tables 19.3 and 19.4. The occupation pattern in rural and urban areas was found to be distinctive as the agriculture component continues to serve as dominant occupation for the rural area.

Table 19.3: Change in Occupational Pattern of the Respondents in the Rural Areas

Category	Location	Agriculture		Govt./Private Sector Employment		Business		Transporters		No. of Respondents
		Before	After	Before	After	Before	After	Before	After	
Farmers	Daria	32 (100.00)	24 (75.00)	14 (43.75)	10 (31.25)	2 (6.25)	10 (31.25)	2 (6.25)	14 (43.75)	32
	Basal	32 (100.00)	30 (93.75)	13 (40.63)	9 (28.13)	2 (6.25)	10 (31.25)	2 (6.25)	8 (25.00)	32
Educated	Darla	22 (100.00)	22 (100.00)	14 (63.64)	14 (63.64)	4 (18.18)	12 (54.55)	0 (0.00)	8 (36.36)	22
	Basal	22 (100.00)	22 (100.00)	12 (54.55)	13 (59.10)	4 (18.18)	12 (54.55)	0 (0.00)	0 (0.00)	22
Business-men	Daria	16 (100.00)	14 (87.50)	8 (50.00)	8 (50.00)	12 (75.00)	16 (100.00)	0 (0.00)	12 (75.00)	16
	Basal	16 (100.00)	16 (100.00)	7 (43.75)	7 (43.75)	10 (62.50)	16 (100.00)	0 (0.00)	2 (12.50)	16
Professionals	Daria	14 (100.00)	14 (100.00)	0 (0.00)	1 (7.14)	1 (7.14)	2 (14.29)	0 (0.00)	2 (14.29)	14
	Basal	14 (100.00)	14 (100.00)	0 (0.00)	0 (0.00)	1 (7.14)	2 (14.29)	0 (0.00)	0 (0.00)	14
Total	Daria	84 (100.00)	74 (88.10)	36 (42.85)	33 (39.29)	19 (22.62)	40 (47.62)	2 (2.38)	36 (42.85)	84
	Basal	84 (100.00)	82 (97.62)	32 (38.10)	29 (34.52)	17 (20.24)	40 (47.62)	2 (23.81)	10 (11.90)	84
Overall		168 (100.00)	156 (92.86)	68 (40.48)	62 (36.90)	36 (21.43)	80 (47.62)	4 (2.38)	46 (27.38)	168

Figures in parentheses are per cent of the total.

From the perusal of Table 19.3, it is revealed that 92.86 per cent of sample respondents in rural area were associated with agricultural occupation, 36.90 per cent in employment sector, 47.62 per cent in business and 27.38 per cent in transportation work. An overall decline in agricultural and service sector was noticed in comparison to other occupations. Maximum decline in agriculture was observed under the farmers category in Daria area (25.00 per cent) followed by businessmen category (12.5 per cent) in the same region. The employment occupation in Daria and Basal areas showed a declining trend. Business was found to be the dominant activity under all the categories in both Daria and Basal areas. In Daria area, it has been observed that the sample respondents got involved with transportation work showing,thereby the demand of such work in the GACL. But this was not tlie case in respect of Basal area. The increase in transporters in Basal area was due to expansion of vegetable production.

In case of urban areas, the occupational trend among the sample respondents was found in increasing order except in agriculture side. The overall figure shows the increasing per cent in employment occupation (42.05 to 46.02 per cent), business (51.41 to 64.20 per cent) and transportation (5.68 to 13.64 per cent) occupations, whereas the decline from 62.50 to 56.25 per cent was observed in agricultural occupation. The study of Tables 19.3 and 19.4 show that with the establishment of industries, the transportation has emerged as major occupation particularly in Daria area. It can be analysed from the tables that business dominated over all other occupations in all the categories of sample respondents.

Table 19.4: Change in Occupational Pattern of the Respondents in the Urban Areas

Category	Location	Agriculture		Govt./Private Sector Employment		Business		Transporters		No. of Respondents
		Before	After	Before	After	Before	After	Before	After	
Farmers	Arki	18 (100.00)	18 (100.00)	8 (44.44)	8 (44.44)	4 (22.22)	4 (22.22)	0 (0.00)	0 (0.00)	18
	Solan	18 (100.00)	18 (100.00)	9 (50.00)	9 (50.00)	4 (22.22)	5 (27.78)	0 (0.00)	1 (5.56)	18
Educated	Arki	16 (88.89)	15 (83.33)	16 (88.89)	18 (100.00)	0 (0.00)	2 (11.11)	0 (0.00)	0 (0.00)	18
	Solan	0 (0.00)	0 (0.00)	18 (100.00)	18 (100.00)	0 (0.00)	2 (11.11)	0 (0.00)	2 (11.11)	18
Business-men	Arki	22 (64.71)	18 (52.94)	17 (50.00)	14 (41.18)	28 (82.35)	34 (100.00)	4 (11.76)	8 (23.53)	34
	Solan	18 (52.94)	14 (41.18)	6 (17.65)	10 (29.41)	32 (94.12)	34 (100.00)	6 (17.65)	12 (35.29)	34
Professionals	Arki	18 (100.00)	16 (88.89)	0 (0.00)	2 (11.11)	10 (55.56)	16 (88.89)	0 (0.00)	0 (0.00)	18
	Solan	0 (0.00)	0 (0.00)	0 (0.00)	2 (11.11)	12 (66.67)	16 (88.89)	0 (0.00)	1 (5.56)	18
Total	Arki	74 (84.09)	67 (76.14)	41 (46.59)	42 (47.73)	42 (47.73)	56 (63.64)	4 (4.55)	8 (9.09)	88
	Solan	36 (40.91)	32 (36.36)	33 (37.50)	39 (44.32)	48 (54.55)	57 (64.77)	6 (6.82)	16 (18.18)	88
Overall		110 (62.50)	99 (56.25)	74 (42.05)	81 (46.02)	90 (51.14)	113 (64.20)	10 (5.68)	24 (13.64)	176

Figures in parentheses are per cent of the total.

Ecological Impact

The respondents from rural and urban areas were asked about the change in temperature, water and wildlife due to the industries (GACL and HFCL) selected for the present study and the results are shown in Table 19.5. The local edaphic factors were affected due to industrial expansion. The resultant air and noise pollution deteriorate the quality of life. The biotic components of the area showed stunted growth as dust particles settles on them. Many other human ailments are also the direct consequences. About 50.00 per cent respondents opined increase in temperature due to industries. In case of water sources, 28.49 per cent responded decrease. Data shows interesting picture in terms of increase and decrease of wildlife in the study area. About 14.00 per cent respondents showed a decrease, 25.00 per cent viewed increase in wildlife. However, majority respondents (61.05 per cent) considered no change.

All the respondents from rural and urban areas, who were interviewed with regard to the GACL impact, reported increase in temperature, whereas no change was reported due to the HFCL. About

57.00 per cent opined decrease in water sources/levels due to the GACL whereas no change was reported due to the HFCL. It was observed that due to mining and decrease in forest cover, the wild animal and monkeys were shifted to human localities for their food etc.

Table 19.5: Impact of Industries on Environmental Status in the Study Areas (Rural/Urban)

Name of Industry	Area	Temperature			Water			Wildlife			No. of Respondents
		D	C	I	D	C	I	D	C	I	
GACL	Rural	0 (0.00)	0 (0.00)	84 (100.00)	66 (78.57)	18 (21.43)	0 (0.00)	18 (21.43)	12 (14.29)	54 (64.28)	84
	Urban	0 (0.00)	0 (0.00)	88 (100.00)	32 (36.36)	56 (63.64)	0 (0.00)	26 (29.55)	30 (34.09)	32 (36.36)	88
	Total	0 (0.00)	0 (0.00)	172 (100.00)	98 (56.8)	74 (43.02)	0 (0.00)	44 (25.58)	42 (24.42)	86 (50.00)	172
HFCL	Rural	0 (0.00)	84 (100.00)	0 (0.00)	0 (0.00)	84 (100.00)	0 (0.00)	4 (4.76)	80 (95.24)	0 (0.00)	84
	Urban	0 (0.00)	88 (100.00)	0 (0.00)	0 (0.00)	88 (100.00)	0 (0.00)	0 (0.00)	88 (100.00)	0 (0.00)	88
	Total	0 (0.00)	172 (100.00)	0 (0.00)	0 (0.00)	172 (100.00)	0 (0.00)	4 (2.33)	168 (97.67)	0 (0.00)	172
Total	Rural	0 (0.00)	84 (50.00)	84 (50.00)	66 (39.29)	102 (60.71)	0 (0.00)	22 (13.10)	92 (54.76)	54 (32.14)	168
	Urban	0 (0.00)	88 (50.00)	88 (50.00)	32 (18.18)	144 (81.82)	0 (0.00)	26 (14.77)	118 (67.05)	32 (18.18)	176
Overall		0 (0.00)	172 (50.00)	172 (50.00)	98 (28.49)	246 (71.51)	0 (0.00)	48 (13.95)	210 (61.05)	86 (25.00)	344

Figures in parentheses indicate per cent of the total number of respondents in each case.

D: Decrease; C: Constant; I: Increase.

Policy Implications

It can thus be seen that with the industrial expansion the rural-urban balance in terms of human population is disturbed. There is always a large scale migration of rural people to urban areas for employment. The study has also clearly brought into focus the changes in the occupation pattern due to advent of industries in the studied area. The change is more towards ecologically sensitive employment avenues where by decline in agriculture occupation was observed. The different pollution hazards have also been reported during the study. The policy reformulations are, therefore, needed for sustaining the rural-urban linkages. The following policy suggestions can help in this direction.

Rural-urban linkages need specific research input in the projects like industrial expansion for ensuring economic sustainability. The changed scenario towards increasing urban migration is thwarting the existence of sustainable society. The rise of urban population with the scarcity of basic amenities like health, drinking water, education, sanitation, etc. are the direct consequences of this unplanned industrial development. It is, therefore, suggested that any future policy on industry be formulated and designed in such a way that not only the rural-urban linkages are strengthened but also the rural eco-systems are made self-sustained.

Many community based organisations and local bodies are effectively working in rural areas for upliftment of socio-economic conditions. These grass root agencies are fully conversant with the local impulse and existing challenges. Therefore, their rich experience in terms of people involvement should be ensured in present administrative setup. Action is further required for administrative initiatives to fill the existing gaps and challenges.

The study clearly brought into focus the importance of administrative and policy reforms in the light of changes in rural-urban interface for achieving the goal of sustainable development. The developmental strategy needs to be formulated as per the perspectives of the hill areas and should have complementary with inherited socio-economic and ecological settings. The administrative role can further be strengthened if there is local input in the form of people participation through community based organisations. Finally, it is concluded that for achieving the sustainability goal, the issues of local capacity building, local employment generation, control of environmental degradation and deeper understanding of rural-urban interface must be ensured. Further research can be initiated for comprehensive assessment of the model of community based interventions as suggested by the present study.

References

Chakarborty, S.S., 1999. Rural-urban interface for sustainable development. *Environment and People*, Hyderabad, 5(11): 15.

Economic Survey of Himachal Pradesh, 2002. Department of Economics and Statistics, Shimla.

Statistical Summary of Solan District, 2001. District Statistical Office, Solan, Himachal Pradesh.

Census of India, 2001, Himachal Pradesh, Provisional Totals, Paper-2 of 2001, Directorate of Operations, Shimla.

www.himachal.nic.in (official website of Himachal Pradesh Government).

Chapter 20

Evaluation of New Insecticides Against Sorghum Shoot Fly, *Atherigona soccata* Rondani

S.T. Aghav[1], A.D. Jagdhani[1], M.A. Sushir[1], H.S. Baheti[2] and S.D. Rajput[1]

[1]Oilseeds Research Station, M.P.K.V., Jalgaon – 425 001, Maharashtra

[2]Department of Entomology, M.P.K.V., Rahuri – 413 722, Maharashtra

ABSTRACT

Six insecticides *viz.*, thiamethoxam, imidacloprid, acetamiprid, profenofos 40 per cent + cypermethrin 4 per cent, endosulfan @ 0.07 per cent and carbofuran with various concentrations were evaluated against shoot fly of Sorghum (*Sorghum bicolor* L. Monech). The shoot fly oviposition, plants with eggs and dead hearts were ranged from 0.95 to 2.88 eggs/plant, 32.57 to 59.12 per cent and 6.34 to 29.14 per cent, respectively in various treatments as against to 1.78 eggs/plant, 35.76 per cent and 48.96 per cent, respectively in untreated control. Maximum grain and fodder yields were obtained from imidacloprid @ 1.2 per cent ST treated plots *i.e.* 27.59 and 91.67 q/ha, respectively and closely followed by thiamethoxam @ 0.75 per cent ST *i.e.* 27.50 and 81.92 q/ha, respectively.

Keywords: *Shoot fly, Atherigona soccata, Dead hearts.*

Introduction

Sorghum (*Sorghum biecolor* L. Monech) is the fourth most important cereal crop in India after rice, wheat and maize. Green revolution attempts in other crops including sorghum have failed due to a number of major insect-pests as reported by Jotwani and Young (1971). Shoot fly, (*Atherigona soccata* Rondani) is serious pest in India. Shoot fly causes damage at seedling stage by killing central shoot called dead hearts. This pest is internal feeder and due to this its control has become difficult. Earlier, the number of insecticides were tested by the scientist against this pest (Jotwani and Prem Kishor, 1982; Prem Kishor, 1996). To evolve cost effective control of such pest, various commonly used and easily available insecticides were evaluated.

Materials and Methods

A field experiment was conducted at Sorghum Research Station, MAU, Parbhani during Rabi season of 2001–2002. The trial was laid out in randomised block design with three replications. Each treatment plot was sown at a spacing of 45 × 15 cm with gross plot size of 6.5 × 3.25 m^2 and net plot size of 5.00 × 2.25 m^2. The variety used for this experiment was Maldandi (M-35-1). There were 16 treatments which includes seed treatment, foliar sprays and soil application of various insecticides by different concentrations along with untreated plot (Table 20.1). The sprayings were under taken with hand operated sprayer on 7th and 17th day after emergence. Observations were recorded at 6, 11, 16, 21, and 28 days after emergence on number of eggs, percentage eggs laid per plant, dead hearts due to shoot fly and grain and fodder yield in quintal per ha. The statistical analysis was carried out as per standard procedure.

Results and Discussion

The shoot fly oviposition ranged from 0.95 to 2.88 eggs/plant in different treatments as compared to 1.78 eggs/plant in untreated control. Maximum egg laying per seedling was observed in the plots of imidacloprid 70 WS @ 1. 2 per cent ST (2.88) followed by thiamethoxam 70 WS @ 0.75 per cent ST (2.81), 0.5 per cent ST (2.75) and carbofuran 30 @ 0.9 kg a.i./ha soil application (2.74). Untreated control recorded (1.78) more eggs per seedling than profenofos 40 per cent + cypennethrin 4 per cent @ 0.12 per cent (0.95), 0.08 per cent (1.12) and 0.04 per cent (1.20), respectively. The percentage of plants with eggs ranged from 32.57 to 59.12 per cent in different treatments as compared to 35.76 per cent in untreated control. Significantly highest per cent of plants with eggs were observed in the plots with seed treatment of thiamethoxam 70 WS @ 0.75 per cent ST (59.12 per cent) and 0.5 per cent ST (58.73 per cent) and carbofuran 30 @ 0.9 kg a.i./ha (58.70 per cent). Untreated control (35.76 per cent) recorded maximum per cent of plants with eggs than imidacloprid 70 WS @ 0.4 per cent ST (32.57 per cent), profenophos 40 per cent + cypermetluin 4 per cent @ 0.12 per cent (33.23 per cent), acetamiprid 20 SP @ 0.009 per cent (34.34 per cent) and endosulfan @ 0.007 per cent (34.51 per cent). The average eggs per plant and percentage of plants with eggs showed that shoot fly preferred the plants of treated seeds for egg laying. However, less number of eggs were recorded in untreated control plots. Similar results were reported by Srivastava and Jotwani (1976), Chandurwar and Karanjkar (1979), Sukhani and Jotwani (1980) and Naitam and Sukhani (1985). The dead hearts caused by shoot fly ranged from 6.34 to 29.14 per cent in different treatments as compared to 48.96 per cent in untreated control. Significantly less dead hearts were observed in the treatment of imidacloprid 70 WS @ 1.2 per cent ST (6.34 per cent), 0.75 per cent ST (14.28 per cent) and the other treatments *viz.,* with acetamiprid 20SP @ 0.003 per cent (29.14 per cent), 0.006 per cent (26.66 per cent), 0.009 per cent (26.14 per cent) and profenophos 40 per cent + cypetmetluin 4 per cent @ 0.04 per cent (26.13 per cent) were effective as less number of dead hearts were recorded as compared to control. Similar results were reported by Men *et al.* (1986); Patil *et al.* (1994), Mote *et al.* (1995), Shanna *et al.* (1996) and Satpute (1999). Maximum grain and fodder yields were obtained from imidacloprid 70 WS @ 1.2 per cent ST treated plots (27.59 and 91.67 q/ha) followed by thiamethoxam 70 WS @ 0.75 per cent ST (27.50 and 81.92 q/ha) and imidacloprid 70 WS @ 0.8 per cent ST (27.46 and 80.55 q/ha). Similar results were obtained by Hiremath *et al.* (1995) and Borad *et al.* (1995).

Table 20.1: Effect of Insecticidal Treatments on Oviposition of Sorghum Shoot Fly, Percentage of Plants with Shoot Fly Eggs, Dead Hearts Caused by Shoot Fly and Yield

Sl.No.	Treatments	Shoot Fly			Yield (q/ha)	
		Average No. of Shoot Fly Eggs/Plant	Mean Percentage of Plants with Eggs	Mean Percentage of Dead Hearts	Grain	Fodder
1.	Thiamethoxam 70 WS @ 0.25 per cent ST	2.33 (1.68)	46.36 (42.90)	21.87 (27.88)	23.00	71.80
2.	Thiamethoxam 70 WS @ 0.5 per cent ST	2.75 (1.81)	58.73 (50.02)	17.90 (25.02)	26.66	80.60
3.	Thiamethoxam 70 WS @ 0.75 per cent ST	2.81 (1.82)	59.12 (50.30)	14.28 (23.22)	27.50	81.92
4.	Imidacloprid 70 WS @ 0.4 per cent ST	2.45 (1.71)	32.57 (34.79)	19.71 (26.35)	24.87	72.24
5.	Imidacloprid 70 WS @ 0.8 per cent ST	2.70 (1.79)	40.56 (39.75)	14.07 (22.02)	27.46	80.55
6.	Imidacloprid 70 WS @ 1.2 per cent ST	2.88 (1.85)	45.32 (42.31)	16.34 (24.58)	27.59	91.67
7.	Acetamiprid 20 SP @ 0.003 per cent	1.59 (1.46)	47.76 (43.71)	29.14 (32.67)	17.33	51.64
8.	Acetamiprid 20 SP @ 0.006 per cent	1.54 (1.44)	45.79 (42.57)	26.66 (31.08)	18.09	54.00
9.	Acetamiprid 20 SP @ 0.009 per cent	1.48 (1.40)	34.34 (35.83)	26. 14 (30.75)	18.52	55.37
10.	Profenofos 40 per cent + cypermethrin 4 per cent @ 0.04 per cent	1.20 (1.30)	34.64 (36.05)	26.13 (30.74)	17.40	51.83
11	Profenofos 40 per cent + cypermethrin 4 per cent @ 0.08 per cent	1.12 (1.27)	34.57 (36.00)	23.43 (28.94)	18.39	56.80
12.	Profenofos 40 per cent + cypermethrin 4 per cent @ 0.12 per cent	0.95 (1.20)	33.23 (35.19)	18.56 (25.51)	20.80	60.65
13.	Endosulfan @ 0.07 per cent	1.68(1.47)	34.51 (35.97)	22.17 (28.08)	21.97	65.17
14.	Carhofuran 3G soil application @0.9 kg a.i./ha	2.74 (1.80)	58.70 (50.00)	17.85 (24.98)	26.50	80.83
15.	Carbofuran 3G @ 0.03 kg a.i./ha 3T	2.37 (1.69)	46.50 (43.00)	21.90 (27.89)	23.50	73.10
16.	Untreated control	1. 78 (1.50)	35.76 (36.67)	48.96 (44.59)	I 1.20	34.47
	SE±	0.06	0.62	0.39	0.47	0.17
	CD	0.18	1.79	1.13	1.36	2.23

Standard statistical procedure was followed for the analysis.

References

Akashe, V.B., 1983. Efficacy of some newer insecticides for control *Atherigona soccata* Rondani. *M.Sc. Agri. Thesis,* M.A.U., Parbhani, pp. 51–53.

Borad, P.K., Patel, S.N., Patel, R.H. and Parikh, R.K., 1995. Bio-efficacy of various insecticidal formulations against sorghum shoot fly, *Atherigona soccata. Gujarat Agric. Univ. Res. J.,* 21(1): 75–77.

Chandurwar, R.D. and Karanjkar, R.R., 1979. Effect of shoot fly infestation levels on grain yield of sorghum hybrid CSH–8R. *Sorghum Newsl.,* 20: 70.

Hiremath, L.G., Bhuti, S.G. and Lingappa, S., 1995. Imidacloprid a new promising seed dress for the management of sorghum shoot fly. *Karnataka J. Agric. Sci.,* 8(2): 163–167.

Jotwani, M.G. and Prem Kishore, 1982. Insect pest problems on millets: Progress strategies for eighties seed farms, 8(1–2): 15–19.

Kishore, Prem, 1999. Evolving management strategies for pest of millets in India. *J. Ent. Res.,* 20(4): 287–297.

Men, H.B., Mundiwale, S.K. and Borie, M.N., 1986. Efficacy of certain insecticides against sorghum shoot fly. *Pesticides,* 20(8): 59–60.

Mote, U.N., Mohite, A.P. and Lolage, G.R., 1995. Effect of imidacloprid as seed dresser and foliar sprays against sorghum shoot fly. *Pestology,* 19(5): 24–28.

Naitam, N.R. and Sukhani, T.R., 1985. Ovipositional behaviour of sorghum shoot fly, *Atherigona soccata* Rondall, under different soil, plant, weather parameters. *Indian J. Ent.,* 47(2): 195–200.

Patil, B.V., Bheemanna, M., Thulsi Ram, K. and Hugur, P.S., 1994. Bioefficacy of polytrin–C–44 EC on cotton insect pests. *Pestology,* 17(5): 16–19.

Satpute, N.S., 1999. Effect of seed treatment of some insecticides in the management of sucking pests of cotton. *M.Sc. (Agri.) Thesis,* Dr. PDKV, Akola.

Shanna, M., Kapoor, K.N. and Bharaj, G.S., 1996. Effect of seed treatment of sorghum with some new insecticides for the control of shoot fly. *Crop Res.,* 11(1): 90–92.

Srivastava, K.P. and Jotwani, M.G., 1976. Efficacy of granular insecticides for the control of shoot fly, *Atherigona soccata* Rondani. *Entomologists Newsl.,* 6(10): 58–59.

Sukhani, T.R. and Jotwani, M.G., 1980. Efficacy of mixture of carbofuran treated and untreated sorghum seed for the control of sorghum shoot fly, *Atherigona soccata* Rondani. *J. Ent. Res.,* 4(2): 186–189.

Chapter 21

A Case Study of Lake Jaisamand with Special Reference to Algal Flora, India

Premlata Vikal* and Sandhya Tyagi
Department of Botany, Mohanlal Sukhadia University, Udaipur – 313 002, Rajasthan, India

ABSTRACT

The present article deals with the study of algal flora in Lake Jaisamand during the year, 2005. A total of 29 genera and 43 species belonging to Cyanophyceae, Chlorophyceae, Euglenophyceae, Bacillariophyceae and Xanthophyceae have been recorded. During the study period blooming conditions developed twice. The bloom forming algae were *Gloeothece palaea* and *Microcystis aeruginosa*, respectively. Various physico-chemical parameters, *e.g.*, surface water temperature, pH, dissolved oxygen, bio-chemical oxygen demand, alkalinity, hardness, sulphate, nitrate, phosphate, total dissolved solids and conductivity were also conducted as it affects aquatic flora and fauna.

Keywords: *Gloeothece palaea, Microcystis aeruginosa, Aquatic ecosystem, Physico-chemical parameters.*

Introduction

Lake Jaisamand, the chief source of water supplies to fulfil various requirements like domestic and irrigational, etc. It covers an area of 36 square kilometers and its capacity is 27.5 ft. It is 51 kilometers far from Udaipur City. It was constructed by Maharaja Jai Singh during 1685–1691 in the form of dam at river Gomati and later converted to lake. It is also known for fish cultivation. The man-

* E-mail: vikal16@yahoo.co.in

made aquatic ecosystems play a vital role in inland fishery development in India due to their vast area and huge production potential (Fasihuddin and Puttaiah, 2004).

Algae form the base of food chain in an aquatic ecosystem. Therefore, it plays a major role in maintaining aquatic life and water quality. As phytoplanktons are the basis of aquatic production, any serious alteration in their constitution may have detrimental consequences on the food chain and concurrently entire community structure (Adesalu and Nwankwo, 2005). Today algal assays are very commonly used in assessing water quality (Wong, 1989 and Whitton *et al.*, 1991). Micro- as well as, macroalgae are known to be most sensitive organisms to pollutants (Kapoor and Solanki, 1992). Type of algae growing in a water body and its abundance or any changes in their population may helps in assessing the quality of water and its nutrient status, respectively as they are sensitive indicators of water quality. Therefore, present study was designed to study the algal flora of the lake in relation to its water characteristics.

Materials and Methods

At Lake Jaisamand two sampling stations were selected, *i.e.*, Boating site (Station I) and Shiva temple (Station II). Algal and water samples were collected monthly in the morning between 9.00 to 10.00 A.M. About 1.5–2.0 litres of water sample was collected in polythene cans and brought to laboratory for analysis as per standards (Trivedi and Goel, 1986; Manivasakam, 1996 and APHA, 1998).

Physico-chemical Analysis

Water was analyzed for various physico-chemical parameters, *e.g.*, air and water temperature, pH, dissolved oxygen (DO), biochemical oxygen demand (BOD), total alkalinity, hardness, sulphate, nitrate, phosphate, total dissolved solids (TDS) and conductivity. Air and water temperature were measured in the field using a mercury thermometer of 110°C and DO was also fixed on the spot with Winkler A and Winkler B. TDS and conductivity were determined using Systronics TDS-Conductivity meter-308 and pH with Systronics Digital pH meter-335.

Identification of Algae

Algal samples were identified using Desikachary (1959), Palmer (1980) and Bold and Wynne (1978).

Results and Discussion

Algae represent an essential and basic part of the cycle of living organisms as, small aquatic organisms feed on algae. Algae may prove to be nuisance, if it starts forming bloom or mats. Blooms may cause problems of tastes and odours, filter and screen clogging and slime accumulation in pipes. In the year 2005, a variety of algal species have been recorded which belong to Cyanophyceae, Chlorophyceae, Euglenophyceae, Xanthophyceae and Bacillariophyceae. Water temperature, pH and available nutrients play an important role in the growth of particular alga: The capacity of synthesizing organic matter by any aqualic ecosystem depends largely on physico-chemical parameters like temperature, nutrients and solar radiations besides producer populations (Fasihuddin and Puttaiah, 2004).

In January, 2005 different members of cyanophyceae and diatoms were recorded. *Gloeothece palaea* were abundant in January, 2005 which formed blooming conditions in February, 2005 at Station I. During that period, the water temperature was between 19.0–19.5°C, which favoured its growth. It vanished completely in March, 05 may be due to increased temperature. Liken (1975) observed a close relationship between temperature and productivity. The pH range 8.5 to 9.0 favours the growth of blue-green algae. Blooming has a direct correlation with pH, temperature, DO and available phosphates contributed by sewage and sedimentary cycle (Kodarkar *et al.*, 1991). The vigorous growth of algae has the ability to reduce hardness, as much as, one-third (Palmer, 1980) and it can be observed very clearly after February, 2005 (Table 21.2).

In February and March 05, abundance of diatoms was recorded. *Fragilaria capucina* were found attached to *Cladophora aegagropila*. Later, these diatoms were found free floating. The presence of *Synedra ulna*, *Cymbella ventricosa* and *Achnanthes* sp. indicated polluted water condition (Table 21.1). During this period TDS were high ranging between 517 to 766 ppm. High dissolved solids may be attributed for higher electrical conductivity values. The growth of algae depends on the physico-chemical characteristics of water. In turn, the abundance of algae is also capable of altering these water characteristics (Palmer, 1980). Total alkalinity was highest in April, 05 at both these stations I and II *i.e.*, 210 and 204 ppm, respectively. Alkalinity below 500 ppm indicates low photosynthetic rate (Das, 2002). Decreased photosynthetic activity lowers dissolved oxygen.

Table 21.1: Algal Flora in Lake Jaisamand during the Year 2005

Algal Species	*Jan.*	*Feb.*	*Mar.*	*Apr.*	*May*	*June*	*July*	*Aug.*	*Sept.*	*Oct.*	*Nov.*	*Dec.*
Cyanophyceae												
Gloeothece palaea	++	++	–	–	–	–	–	–	–	–	–	–
Microcystis aeruginosa	–	–	–	–	–	–	–	–	+	++	+	+
M. scripta	–	–	–	–	–	–	–	–	–	++	+	+
Oscillatoria anguina	+	+	–	–	–	–	–	–	–	–	–	–
O. agardhii	–	–	–	–	+	–	–	–	–	–	–	+
O. chlorina	–	–	–	–	+	–	–	–	–	–	–	–
O. lauterborni	–	–	–	–	+	–	–	–	–	–	–	–
O. limosa	–	–	–	–	–	–	–	–	–	–	–	+
Phormidium anomala	–	–	+	–	–	–	–	–	–	–	–	
P. cevennense	+	–	–	–	–	–	–	–	–	–	–	–
Lyngbya digueti	–	–	–	–	+	–	–	–	–	–	–	–
L. gracilis	–	–	+	+	–	–	–	–	–	–	–	–
L. polysiphonia	–	–	+	–	–	–	–	–	–	–	–	–
L. corbierei	–	–	–	+	–	–	–	–	–	–	–	–
Anabaena planktonica	–	–	–	–	–	–	–	–	–	–	–	+
Stigeoclonium lubricum	+	+	+	+	+	–	–	–	–	–	–	–
Chlorophyceae												
Chlorococcum multinucleatum	–	+	–	–	–	–	–	–	–	–	–	–
Pediastrum tetras	–	+	+	–	–	–	–	–	–	–	–	–
P. simplex	–	–	+	+	–	–	–	–	–	–	–	–
Chlorela vulgaris	–	–	–	–	–	–	–	–	–	–	–	+

Contd...

Table 21.1–Contd...

Algal Species	*Jan.*	*Feb.*	*Mar.*	*Apr.*	*May*	*June*	*July*	*Aug.*	*Sept.*	*Oct.*	*Nov.*	*Dec.*
Tetraedron muticum	–	–	–	–	+	–	–	–	–	–	–	–
Chlamydomonas reinhardi	–	–	–	–	–	–	–	–	–	–	–	+
Tetraspora gelatinosa	–	–	–	–	–	–	–	–	+	+	+	–
Planktosphaeria gelatinosa	–	–	–	–	–	–	–	–	–	–	+	–
Scenedesmus quadricauda	–	–	+	+	–	–	–	–	–	–	–	–
Cladophora aegagropila	–	+	+	+	+	–	–	–	–	–	–	–
C. glomerata	–	–	–	–	–	–	–	–	–	–	–	+
C. crispata	–	–	–	–	–	–	–	–	–	–	+	–
Oedogonium suecium	–	–	–	+	–	–	–	–	–	–	–	–
O. boscii	–	–	–	+	–	–	–	–	–	–	–	–
Spirogyra fluviatilis	–	–	–	–	+	+	–	–	–	–	+	+
Microspora wittrockii	–	–	–	–	–	–	–	–	–	–	+	–
Euglenophyceae												
Euglena viridis	–	–	–	–	–	–	–	–	–	–	–	+
E. sangvlnea	–	–	–	–	–	–	–	–	–	–	–	+
E. stellata	–	–	–	–	–	–	–	–	–	–	–	+
Bacillariophyceae												
Nitzschia palaea	+	+	+	–	–	–	–	–	–	–	+	+
Navicula gracilis	+	+	–	+	–	–	–	–	–	–	+	+
Fragilaria capucina	–	+	+	+	+	+	–	–	–	–	–	–
Gomphonema parvulum	–	+	+	+	–	–	–	–	–	–	–	–
Cymbella ventricosa	–	+	–	–	–	–	–	–	–	–	–	–
Achnanthes sp.	–	–	+	–	–	–	–	–	–	–	–	–
Synedra ulna	+	+	–	–	–	–	–	–	–	–	–	–
Xanthophyceae												
Tribonema bombycinum	–	–	–	–	+	–	–	–	–	–	–	–
Desmidaceae												
Cosmarium sp.	–	–	+	–	–	–	–	–	–	–	–	–

++: Bloom forming algae.

In May, 2005 among cyanophyceae different species of *Oscillatoria,* a species of *Lyngbya* and *Stigeoclonium,* among chlorophyceae *Tetraedron muticum, Spirogyra fluviatilis* and one species of xanthophyceae, *i.e., Tribonema bomycinum* were recorded.

After rains in August, 2005 species of *Microcystis* (*M. aeruginosa* and *M. scripta*) developed which formed bloom in October, 2005. According to Pathak (1979), free CO_2 will be limiting where there is a permanent bloom of BGA, especially *Microcystis* sp. Sabu and Azis (1995) emphasized on strong positive relationship between phosphorus and productivity. The foremost cause attributed to the triggering and spread of blooms is increasing pollution through anthropogenic activities (Bhat and

Matondkar, 2004). The abundance or *Microcystis* spp. might be responsible for the increased concentration of alkalinity and hardness in November–December, 2005. Throughout the year, the alkalinity was found to be more than 100 ppm, which indicated highly productive water (Alikunhi, 1957). The blooms of *Microcystis* spp. are harmful as it manufactures microcystin, which yield 7 (or 14) amino acids on hydrolysis. It causes enlargement and congestion of the liver followed by haemorrhage and may also exhibit neurotoxic activity. In recent years, there has been considerable scientific attention towards harmful algal blooms (Blaxter and Southward, 1997; Chattopadhyay *et al.*, 2002b; Sarkar and Malchow, 2005). Harmful algal blooms refer to blooms, which cause discomforts or toxic effects on human health directly or indirectly. Algal blooms hurt the system in two ways. First, they cloud the water and block sunlight causing underwater grasses to die, which provide food and shelter for aquatic creatures. Second, when these algae die and decompose, oxygen is used up which is essential to most organisms living in the water.

Table 21.2: Selected Physico-chemical Characteristics of Lake Jaisamand in the Year, 2005

Sl.No.	Parameters	Jan.	Feb.	Mar.	Apr.	May	Jun.	Jul..	Aug.	Sep.	Oct.	Nov.	Dec.
1.	Air temperature	I	18.0	24.0	25.0	24.5	34.0	31.0	–	27.25	24.5	21.0	14.0
		II	19.0	24.0	25.0	24.0	36.0	31.0	–	27.5	27.5	21.0	15.0
2.	Water temperature	I	19.0	19.5	26.5	29.5	28.5	30.0	–	30.0	25.0	22.5	19.5
		II	20.0	19.0	26.5	29.0	29.5	30.0	–	29.59	26.5	23.0	19.5
3.	pH	I	8.39	8.69	8.71	8.64	8.22	8.59	–	9.15	8.38	8.04	8.06
		II	8.35	8.65	8.75	8.63	8.52	8.66	–	9.12	8.49	8.14	7.99
4.	D.O.	I	9.20	7.95	7.65	7.04	6.734	6.53	–	8.23	4.59	4.08	5.714
		II	8.90	8.05	7.96	7.35	6.53	6.99	–	8.299	5.71	4.49	5.714
5.	B.O.D.	I	1.56	1.43	5.002	2.55	1.40	6.326	–	4.558	1.531	1.041	1.63
		II	2.01	1.96	1.735	2.45	3.469	6.734	–	4.524	3.572	0.918	2.04
6.	Total alkalinity	I	148.0	166.0	168.0	210.0	208.0	188.0	–	131.0	138.0	160.0	163.0
		II	146.0	164.0	174.0	204.0	200.0	192.0	–	130.0	142.0	167.0	160.0
7.	Hardness	I	157.0	162.5	160.0	163.0	162.0	155.0	–	117.0	130.0	142.0	145.0
		II	155.0	202.0	164.0	162.0	162.0	155.0	–	115.0	132.0	141.0	141.0
8.	Sulphate	I	70.0	150.0	150.0	190.0	140.0	55.5	–	125.0	80.0	355.0	400.0
		II	90.0	101.0	101.0	220.0	115.0	115.0	–	230.0	195.0	350.0	340.0
9.	Nitrate	I	3.21	1.33	–	0.77	3.987	3.323	–	3.101	0.775	4.21	2.879
		II	4.02	6.31	–	2.75	3.101	6.756	–	7.42	0.332	12.105	2.215
10.	Phosphate	I	0.035	0.049	0.032	0.006	0.052	0.005	–	0.018	0.061	0.021	0.042
		II	0.038	0.043	0.028	0.008	0.049	0.011	–	0.024	0.098	0.021	0.042
11.	T.D.S.	I	320.0	517.0	749.0	317.0	287.0	313.0	–	2210	246.0	225.0	240.0
		II	326.0	527.0	766.0	320.0	289.0	311.0	–	223.0	248.0	2270	244.0
12.	Conductivity	I	0.640	1.03	1.49	0.633	0.579	0.627	–	0.443	0.493	0.450	0.992
		II	0.652	1.05	1.53	0.641	0.583	0.623	–	0.448	0.498	0.452	0.489

–: No sampling done, I and II are sampling stations.

All parameters are in ppm except for temperature (°C), pH and conductivity (mS/cm).

The highest number of species was recorded in December, 2005 (Table 21.2). At the end it may be concluded that as the Lake Jaisamand is an important source of water supply therefore, care should be taken so that blooms of harmful algae may not develop. 4 per cent copper sulphate solution is useful to eradicate nuisance planktonic forms. Use of specific algicides will prove to be useful from the standpoint of fisheries management, as algae constitute the base of food pyramid. Thus, a monitoring of a water body from time to time is essential to maintain the water quality and its aesthetic value.

Acknowledgements

The author (P. Vikal) is grateful to UGC, New Delhi for giving research fellowship and to Prof. N.C. Aery, Head, Department of Botany, M.L.S. University, Udaipur for providing laboratory facilities. Author also acknowledges the encouragement of Prof. K. Kapoor. I am very much thankful to my father Shri P.D. Vikal and brothers Ghanshyam Singh and Rokesh Kumar Singh for their continuous help and support throughout.

References

Adesalu, T.A. and Nwankwo, D.I., 2005. Studies on the phytoplankton of Olero creek and parts of Benin river, Nigeria. *The Ekol.*, 3(2): 21–30.

Alikunhi, K.H., 1957. Fish culture in India. *Fram. Bull.*, 20, ICAR, New Delhi, pp. 144.

APHA, AWWA and WPCF, 1998. *Standard Methods for the Examination of Water and Wastewater*, 20th edn. American Public Health Association, New York.

Bhat, S.R. and Matondkar, S.G.P., 2004. Algal blooms in the seas around India: Networking for research and outreach. *Curr. Sci.*, 87(8): 1079–1083.

Blaxter, J. and Southward, A., 1997. *Advances in Marine Biology*. Academic Press, London.

Bold Harold, C. and Wynne Michael, J., 1978. *Introduction to Algae*. Prentice Hall of India Pvt. Ltd., New Delhi, pp. 706.

Chattopadhyay, J., Sarkar, R. and Mandal, S., 2002b. Toxin producing plankton may act as a biological control for planktonic blooms-field study and mathematical modeling. *J. Theor. Biol.*, 215: 333–344.

Das, A.K., 2002. Phytoplankton primary production in some selected reservoirs of A.P. *Geobios*, 29: 52–57.

Desikachary, T.V., 1959. *Cyanophyta*. Indian Council Agricultural Research, New Delhi, pp. 686.

Fasihuddin, S. and Puttaiah, E.T., 2004. Spatio-temporal trends in phytoplankton primary production in marikanave reservoir. *Indian J. Environ. and Ecoplan.*, 8(1): 179–184.

Kapoor, K. and Solanki, P.S., 1992. In water management problems related to water resources in Udaipur. *Nat. Sem. Strat. Env. Prot. Industrialization*, M.L.S. University, Udaipur, pp. 119–126.

Kodarkar, M.S., Muley, E.V. and Rao, V., 1991. Toxic algal blooms in the Lake Hussainsagar, Hyderabad. *J. Aquatic Biol.*, 6(1&2): 13–18.

Likens, G.E., 1975. Primary production of inland aquatic ecosystems. In: *The Primary Productivity of the Biospheres*, (Eds.) Lieth, H. and R.H. Whittaker. Springer-Verlag, New York.

Manivasakam, N., 1996. *Physico-chemical Examination of Water, Sewage and Industrial Effluents*. Pragati Prakashan, Meerut (India), pp. 257.

Palmer, C.M., 1980. *Algae and Water Pollution.* Castle House Publication Ltd., England, pp. 123.

Pathak, V., 1979. Evaluation of productivity in Nagarjunasagar reservoir (Andhra Pradesh) as a function of hydrological and limnological parameters. *J. Inland Fish. Soc.*, India, 11: 49–68.

Sabu, T. and Abdul Azis, P.K., 1995. A study on the primary production in Peppara reservoir. In: *The Proc. of Seventh Kerela Science Congress,* Palakkad, pp. 76–77.

Sarkar, R.R. and Malchow, H., 2005. Nutrients and toxin producing plankton control algal blooms: A spatio-temporal study in noisy environment. *J. Biosci.*, 30(5): 101–113.

Trivedi, R.K. and Goel, P.K., 1986. *Chemical and Biological Methods for Water Pollution Studies.* Environmental Publications, Karad, pp. 248.

Whitton, B.A., Rott, E. and Friedrick, G., 1991. *Use of Algae for Monitoring Rivers.* Institute of Botanik, Universitat Innsbruck, Germany.

Wong, S.L., 1989. Algal assays to interpret toxicity guidelines for natural waters. *J. Env. Sci. Health,* 24: 1001–1010.

Chapter 22

Limnobiotic Study of Kanhan River at Mouda, Near Nagpur (Maharashtra)

R.R. Khapekar*, P.R. Chaudhari and S.R. Wate
National Environmental Engineering Research Institute (NEERI), Nehru Marg, Nagpur – 440 020 (India)

ABSTRACT

Limnobiotic investigation was conducted for qualitative assessment of Kanhan River at Mouda Town near Nagpur city of Maharashtra state during March 2004 to February 2005. The study was concentrated on three sampling sites of Kanhan River namely Site 1: Kesalpur, Site 2: Kiranapur and Site 3: Sawangi (confluence point of polluted Nag River and Kanhan River). The present investigation aimed to study the physico-chemical and biological features of 3 sites of Kanhan River. For the investigation of water quality, total seventeen physico-chemical parameters were analyzed. It was found that there is gradual increase in pollution level in river water quality from Site-1 to Site-3 as evident from increase in levels of C.O.D., T.D.S., chlorides and sulphate and decrease in D.O. concentration. The highest degrees of pollution were found at Site-3. The results revealed that water quality at Site-3 was more polluted due to addition of sewage and various industrial effluents of Nagpur city into Kanhan River through polluted Nag River.

Biological studies revealed an increase in species diversity and population density of members belonging to Chlorophyceae, Bacillariophyceae (phytoplanktons) and insects (macro-invertebrates) from Site-1 to Site-3. During the investigation period it is observed that few clear water species recorded at Site-1 were disappeared and certain pollution tolerant species of phytoplanktons and zooplanktons were appeared at Site-3. The present investigation indicate progressive pollution levels in the river present near urban environment due to unscrupulous discharge of wastewater.

***Keywords**: Kanhan River, Water pollution, Physico-chemical parameters, Biological parameters, Urban area.*

* E-mail: rkhapekar@rediffmail.com dr.khapekar@gmail.com

Introduction

Among all the other natural resources, water is the most precious one, and life is not possible without water. Although the global freshwater in our planet is about 1500 million cubic kms (Penman, 1970) but most of it is useless for us as it contains too much of salt. Total amount of estimated available freshwater on our planet is only 84.4 million cubic kms. In recent years, rapid population growth increasing living standard, wide spheres of human activities and industrialization have resulted in greater demand of good quality water, while on other hand pollution of water resources is also increasing steadily (Murugesan and Sukumaran, 1999).

India is a country having various landforms and rivers. There are fourteen major rivers in India, which share about 83 per cent of the total drainage basin and contribute 85 per cent of total surface flow Nilay (Chaudhary, 1983). The total annual flow in all the rivers is estimated to be 18,18,100 million m^3 according to the Central Water and Power Commission (Kayastha, 1989). Enormous anthropogenic activities make all this water resources unfit for any purpose. Eutrophication of water bodies is rapidly increasing due to rapid increase in quality and quantity of sewage discharge, which in turn enhance algal proliferation. A vast literature exists on the limnological studies of algae. Some of the important contributors are Phillipose (1960), Zafar (1964), Venkateswarlu (1969), Munawar (1970), Kant and Kachroo (1971), Kumar (1977), Khan and Zutschi (1980).

The present investigation was carried out by exploring the various sites of Kanhan River of Mouda Town near Nagpur City in Maharashtra. The present investigation includes the study of physico-chemical and biological characteristics of different sites of Kanhan River.

Materials and Methods

Three sampling sites, namely S-1 (Keslapur); S-2 (Kiranapur) and S-3 (Sawangi) were selected for present study. The area near Kanhan River is predominantly agricultural interspersed by various kinds of grasses, herbs, shrubs and trees. The 3 sites of Kanhan River were selected with purpose to collect water samples for physico-chemical analysis and biological parameters at seasonal intervals during March 2004 to February 2005.

Analysis of physico-chemical and biological parameters were made after following the work of Edmondson (1959), Needham and Needham (1962), APHA (1992). For physico-chemical analysis water samples were collected in 1 lit polyethylene bottle during morning hours. For biological analysis, samples were collected in 100 ml polyethylene bottles. Fifty liter of water was passed through zooplankton net to collect the concentrated water sample for zooplankton analysis. Water samples directly collected for phytoplankton was preserved in Lugol's Iodine solution and zooplankton samples were preserved in 5 per cent formaldehyde solution at the sampling sites. Samples were brought to the laboratory for microscopic examination. Identification of organisms was done by following the work of Ward and Whipple (1959), Mellanby (1963), Tonapi (1980), Gaur *et al.* (2000).

Results and Discussion

Physico-chemical River Water Quality

Seasonal physico-chemical river water quality at Site-1, 2 and 3 are given in Tables 16.1–16.3 respectively. The river water temperature at Site-3 remained comparatively higher (31.50°C–34.0°C) than Site-1 (30.50°C–33.50°C) and Site-2 (30.5°C–33.0°C) through out the period of investigation. Site-3 is a confluence point of Nag and Kanhan River and Nag River contains effluents from various areas of Nagpur City including domestic and industrial effluents. Klien (1956), Jolly and Chapman (1966),

Venkateswarlu and Jayanti (1968) and Munawar (1970 a) also made similar observations and explained heat is generated when chemicals react with each other and discharged into water and thus bringing about an increase in water temperature.

Table 22.1: Seasonal Physico-Chemical River Water Quality at Site-1

Sl.No.	Parameters	Seasons		
		Summer	Monsoon	Winter
1.	Temperature	33.00	32	30.5
2.	pH	7.22	7.20	7.50
3.	TS	290	410	350
4.	TDS	165	320	255
5.	TSS	125	90	95
6.	DO	3.1	3.8	3.4
7.	Free CO_2	–	–	–
8.	Total alkalinity	134	131	152
9.	Carbonate	7	8	10
10.	Bicarbonate	45	31	30
11.	Total hardness	32	42	37
12.	Calcium hardness	3.727	4.008	2.926
13.	Magnesium hardness	33.116	37.992	29.236
14.	Chloride	25.72	22.92	19.72
15.	Sulphate	70.10	81.81	80.70
16.	Phosphate	0.132	0.147	0.283
17.	COD	138.36	153.53	115.312

All values in mg/l except temperature and pH.

The recorded pH values were comparatively lower at Site-1 (7.20–7.50) and Site-2 (7.00–7.20) and higher at Site-3 (7.00–8.50). Chandra *et al.* (1981) also reported an increase in pH values with addition of effluents from various industries. Similarly, Site-2 showed higher values of total alkalinity (143 mg/l–167 mg/l) than at Site-1 (131 mg/l–152 mg/l) and Site-2 (115 mg/l–127 mg/l). Chandra *et al.* (1981) and Badrinath *et al.* (1987) also recorded higher total alkalinity where a different factory effluent was discharged in water.

Total solid content at different river sites were estimated in the range of 290–350 mg/l, 205–255 mg/l and 430–560 mg/l respectively at Site-1, Site-2 and Site-3. During present investigation it was observed that the highest values of TS, TDS and TSS were recorded at Site-3. It might be due to higher percentage of suspended as well as other impurities present at this particular site. It imparts slight turbidity to the water. The similar observations were also recorded by Bhosle *et al.* (1994).

DO contents at Site-1 remained high (3.1 mg/l–3.8 mg/l) as compared to Site-2 (3.0 mg/l-3.2 mg/l). Whereas, at Site-3, the DO values 3 (1.5 mg/l–2.0 mg/l) were comparatively low during the period of investigation. Addition of sewage and industrial effluent might be responsible for low value of DO. A similar impact of effluents from fertilizer factory on the DO contents in water has been observed by Woodward (1984) and Mathuthu *et al.* (1993).

Table 22.2: Seasonal Physico-Chemical River Water Quality at Site-2

Sl.No.	*Parameters*	*Seasons*		
		Summer	*Monsoon*	*Winter*
1.	Temperature	33.0	32	30.5
2.	pH	7.20	7.10	7.00
3.	TS	205	320	255
4.	TDS	95	245	170
5.	TSS	80	70	68
6.	DO	3.0	3.2	3.1
7.	Free CO_2	–	–	–
8.	Total alkalinity	127	115	121
9.	Carbonate	19	24	27
10.	Bicarbonate	109	91	90
11.	Total hardness	36	44	40
12.	Calcium hardness	6.012	6.412	5.112
13.	Magnesium hardness	35.432	37.588	31.182
14.	Chloride	35.62	33.99	31.19
15.	Sulphate	78.62	87.22	86.87
16.	Phosphate	0.148	0.188	0.211
17.	COD	157.36	163.42	161.53

All values in mg/l except temperature and pH.

Table 22.3: Seasonal Physico-Chemical River Water Quality at Site-3

Sl.No.	*Parameters*	*Seasons*		
		Summer	*Monsoon*	*Winter*
1.	Temperature	34.0	32	31.5
2.	pH	8.50	7.00	7.20
3.	TS	430	560	490
4.	TDS	290	450	360
5.	TSS	140	110	130
6.	DO	1.8	2.0	1.5
7.	Free CO_2	–	–	–
8.	Total alkalinity	167	143	156
9.	Carbonate	26	28	29
10.	Bicarbonate	112	115	110
11.	Total hardness	49	64	58
12.	Calcium hardness	8.126	9.619	6.828
13.	Magnesium hardness	49.861	54.381	42.738
14.	Chloride	60.10	56.98	51.10
15.	Sulphate	99.13	90.48	83.48
16.	Phosphate	2.160	3.166	3.837
17.	COD	182.10	194.16	142.61

All values in mg/l except temperature and pH.

During the whole investigation period, CO_2, was absent at all the three sites. The absence of free CO_2 might be due to cause of pollution and photosynthetic activities of algae (Sreenivasan, 1971). Similarly nil free CO_2 values in most of the month were also recorded by Nandkar, (1979) and Shankhadarwar (2002), which resembles with the present study.

During the entire study period, the bicarbonate values were higher than carbonate values. It was also found that the Site-3 has always higher values of carbonate (19 mg/l–27 mg/l) and bicarbonate (90 mg/l–109 mg/l). The higher values of carbonate and bicarbonate might be due to high pollutional load at Site-3. The present observation is coinciding with the results of Shankhadarwar, (2002) and Khapekar (2005).

Total hardness, calcium hardness, magnesium hardness, chloride and sulphate were found to be increased at Site-3 because Site-3 is a confluence point of Nag and Kanhan. River and it contains effluents from various industries indicating presence of salts of these ions. Similar observations were made by Wetzel (1975), Chandra *et al.* (1991) and Chapman and Kimstach (1992).

Concentration of chloride were found to vary within 19.72 mg/l to 25.72 mg/l, 31.19 mg/l–35.62 mg/l and 51.10–60.10 respectively at Site-1, Site-2 and at Site-3. During the whole study period, concentrations of chloride were high at Site-3. The higher concentration of chloride at Site-3 is because of addition of sewage and industrial effluents. Chloride increases degree of eutrophication (Sinha, 1986). Klein (1957) found a direct correlation between chloride concentration and pollution load. Similar observations were also recorded by Chatterjee *et al.* (2002).

The values of sulphate ranged within 70.10 mg/l to 81.81 mg/l, 78.62 mg/l to 87.22 mg/l and 83.48 mg/l to 99.13 mg/l respectively at Site-1, Site-2 and Site-3. Higher values were always recorded at Site-3. Higher values can be attributed to addition of sewage and industrial effluents into the river. The present observation finds support with the work of Kannan (2000) and Chatterjee *et al.* (2002).

Phosphate content of river water were observed comparatively higher at Site-3 (2.160 mg/l–3.8373 mg/l) than Site-1 (0.132–0.285 mg/l) and Site-2 (0.148–0.211 mg/l). Similar observations were recorded by De Smet and Evans (1972), Michael (1984), Das (1972), Dogra and Kaw (1989), Ouzounis *et al.* (1989), Gopal (1990), Chapman and Kimstach (1992), Meybeck *et al.* (1992) and Mathuthu *et al.* (1993).

The COD concentrations of Site-1 (115.31 mg/l–153.55 mg/l) and Site-2 (157.30 mg/l–163.42 mg/l) were comparatively lower as compared to the Site-3 (142.61 mg/l–194.16 mg/l). The COD content of water increased near pollution input point. Hence, during the whole investigation period, highest values of physico-chemical parameters were found at Site-3. The results of present investigation were coincides with the work of Vijay Kumar and Sagwal (2000), Asif Khan *et al.* (2002) and Shankadarwar (2002).

Biological River Water Quality

The observations on distribution of phytoplanktons and zooplanktons population at three sampling sites of Kanhan River are given in Tables 16.4 and 16.5. In all, 20 phytoplankton species belonging to Chlorophyceae, Cyanophyceae and Bacillariophyceae were observed in Kanhan River. In terms of population density, Bacillariophyceae was found to be the most dominant group at all 3 sampling sites. Species diversity and population density of phytoplankton were found to be higher at Site-3 where effluents from Nag river is poured in Kanhan river. The effluents of Nag River contains various industrial sewage and domestic wastes which enrich water with a wide variety of nutrients required for growth of various phytoplankton.

Table 22.4: Distribution of Phytoplanktons at Three Sites

Sl.No.	Species/Genera	Sampling Sites		
		Site-1	Site-2	Site-3
I	**Chlorophyceae**			
1.	*Spirogyra* sp.	+	+	+
2.	*Pithophora* sp.	+	+	+
3.	*Stigeochlonium* sp.	+	+	+
4.	*Pediastrum* sp.	+	+	+
5.	*Hydrodictyon* sp.	+	+	+
6.	*Closterium* sp.	–	–	+
7.	*Scenedesmus* sp.	–	–	+
8.	*Cosmarium* sp.	+	–	+
9.	*Oedogonium* sp.	–	–	+
10.	*Mougeotia* sp.	+	+	–
11.	*Vaucheria* sp.	+	+	–
II	**Cyaophyceae**			
1.	*Oscillatoria* sp.	–	+	+
2.	*Lyngbya* sp.	+	+	+
3.	*Phormidium* sp.	–	–	+
III	**Bacillariophyceae**			
1.	*Melosira* sp.	+	+	+
2.	*Synedra* sp.	+	+	+
3.	*Navicula* sp.	+	+	+
4.	*Pinnularia* sp.	+	+	+
5.	*Fragilaria* sp.	+	+	+
6.	*Gomphonema* sp.	–	–	+

Table 22.5: Distribution of Zooplanktons at Three Sites

Sl.No.	Species/Genera	Sampling Sites		
		Site-1	Site-2	Site-3
I	**Protozoa**			
1.	*Euglena* sp.	+	+	+
2.	*Difflugia* sp.	+	+	+
3.	*Ceratium* sp.	+	+	+
4.	*Chilomonas* sp.	+	+	–
5.	*Euglypha* sp.	+	+	+
6.	*Paramecium* sp.	+	+	+

Contd...

Table 22.5–Contd...

Sl.No.	*Species/Genera*	*Sampling Sites*		
		Site-1	*Site-2*	*Site-3*
II	**Rotifera**			
1.	*Monostyla* sp.	+	+	–
2.	*Philodina* sp.	+	+	–
3.	*Trichotoria* sp.	+	+	+
4.	*Hexartha* sp.	+	+	–
5.	*Notholca* sp.	+	+	+
6.	*Neptunis* sp.	+	+	–
7.	*Gastropus* sp.	–	+	+
8.	*Asplanchna* sp.	+	+	–
III	**Crustacea**			
(*a*)	**Copepoda**			
1.	*Cyclops* sp.	+	+	+
2.	*Mesocyclops* sp.	+	+	+
3.	*Diaptomus* sp.	+	+	+
(*b*)	**Cladocera**			
1.	*Daphnia* sp.	+	+	+
2.	*Chydorus* sp.	+	+	+
3.	*Bythotrephes* sp.	–	–	+
4.	*Moina* sp.	+	+	–
5.	*Simocephalus* sp.	+	+	+
6.	*Eurycercus* sp.	–	–	+
IV	**Insecta**			
1.	*Chironomus* sp.	+	+	+
2.	*Culex* sp.	+	+	+
3.	*Hydrophilis* sp.	+	+	+
4.	*Eristalis* sp.	–	–	+
5.	*Baetis* sp.	–	–	+
6.	*Corixa* sp.	+	+	+
7.	*Tipula* sp.	–	–	+
8.	*Tabanus* sp.	–	–	+
9.	*Perla* sp.	–	–	+
10.	*Tanypus* sp.	–	–	+
11.	*Ephemera* sp.	+	+	–

In the present investigation among Chlorophyceans, *Mougoetia* sp. and *Vaucheria* sp. were found only in clean water, similar observations were also recorded by Sladeck (1961), David and Ray (1966),

Bulusu *et al.* (1967) and Gunale and Balkrishnan (1981), *Closterium* sp., *Scenedesmus* sp. and *Oedogonium* sp. were observed only at Site-3 where water is contaminated with different types of wastes. Whereas, *Spirogyra* sp., *Pithophora* sp., *Stigeoclonium* and *Pediasterum* sp. were observed at all the three sites. All these species have been reported in a wide variety of polluted waters by various authors (David and Ray, 1966; Jolly and Chapman, 1966; Venkateswarlu and Jayanti,1968; Singh *et al.*, 1969; Munawar, 1970b; Verma *et al.*, 1978; Govindan and Sudaresan, 1979; Zutschi *et al.*, 1980 and Gunale and Balkrishnan, 1981).

Among Cyanophyceans *Oscillatoria* sp. was observed at Site-2 and Site-3, but profuse growth of this species was found at site-3 indicating the higher degree of pollution. The same species have been observed in clean as well as in polluted waters by Kolkwitz and Marsson (1908), Sladeck (1961), David and Ray (1966), Jolly and Chapman (1966), Govindan and Sundareean (1979), Zutshi *et al.* (1980) and Paramasivam and Sreenivasan (1981). *Lyngbya* was found at all the three sites. *Lyngbya* was found in clean water by Verma and Shukla (1970) and Paramsivam and Sreenivasan (1981). In the present investigation, *Lyngbya* sp. was also reported in polluted water. This might be due of mixing of clean water with polluted water. *Phormidium* sp. was observed only at Site-3 and was conspicuous by its absence at other sites. Thus this species as well as *Oscillatoria* can be served as the indicator of polluted water.

Among Bacillariophyceans, *Melosira* sp., *Synedra* sp., *Navicula* sp. *Fragilaria* sp. and *Pinnularia* sp. were reported at all the three sites where as *Gomphonema* sp. were reported only at Site-3. Among all observed species of Bacillariophyceae, *Gomphonema* may be labeled as a pollution indicator species because it was observed only at Site-3.

In all, 34 zooplanktons species belonging to Protozoa, Rotifera, Crustacea, Cladocera and Insecta were recorded from Kanhan River. Amongst all the reported protozoans *i.e. Euglena* sp., *Difflugia* sp. *Ceratium* sp., *Chilomonas* sp., *Euglypha* sp., and *Paramecium* sp. only *Chilomonas* sp. was not reported at Site-3. Species diversity and population density of zooplanktons were more or less similar at all 3 sites. It was found that growth of aquatic vegetation and amount of nutrients were favorable for growth of photosynthetic protozoans. These findings were in accordance with some other workers *viz.* Zafar (1959a). Phillipose (1960), Venkateswarlu (1969) and Munawar (1970b). Amongst the rotifers, the presence of *Trichotoria* sp. and *Notholca* sp. was common at all the three sites. *Monostyla* sp., *Philodina* sp., *Hexartha* sp., *Asplanchna* sp. and *Neptunis* sp. were observed in clean water at Site-1 and Site-2. Among this only *Neptunis* sp. were earlier reported to be tolerant species (Algarsamy *et al.*, 1967, Bilgrami and Datta Munshi, 1979; Vasisht and Sra, 1979; Sladecek, 1983; Goel and Chavan, 1991). *Trichotoria* sp., *Notholca* sp. and *Gastropus* sp. were found at Site-3. These species were reported to be pollution tollerant species. (Arora, 1966b; Krishnamoorthi and Visvesvara, 1966; De Smet and Evans, 1972; Sandhu *et al.*, 1948).

In the present investigation, in all 9 species of crustaceans were found at Site-1, Site-2 and Site-3. Out of these, 6 species were common to all the three sites. *Cyclops* sp., *Mesocyclops* sp., and *Diaptomus* sp. were the representative of Copepoda. Among cladocerans, *Daphnia* sp., *Chydorus* sp. and *Simocephalus* sp. were present at all the sites, *Moina* sp. were found only at Site-1 which has no direct source of pollution and *Bythotrephes* sp. and *Eurycercus* sp. were observed only at Site-3. Krishnamoorthi and Visvesvara (1966), Sitaramaih (1966), Liberaman (1970), Verma *et al.* (1978, 1984) and Batish and Kumari (1986) also reported *Moina* to be a clean water species. Rest of the crustaceans are reported to be tolerant species (Dextar, 1959; Mallanby, 1963; Jolly and Chapman, 1966; Das, 1989). Hence, in this investigation Site-3 is characterized by disappearance of a clean water species and appearance of two pollution tolerant species.

At Site-1 and Site-2 only 5 insects species were found whereas at Site-3, 10 species were found. The insects discovered at Site-1 and Site-2 were *Chironomus* sp. (larvae), *Culex* sp. (larvae), *Hydrophilis* sp., *Corixa* sp. and *Ephemera* sp. In present study, *Ephemera* sp. was found only in clean water at Site-1 and Site-2. Similar type of observation was also reported by Verma and Shukla (1969), Gaufin (1974), Badala and Singh (1981) and Das (1989). Site-3 is characterized by the presence of *Eristalis* sp., *Baetis* sp., *Tabanus* sp., *Perula* sp. and *Tanypus* sp. All these species of insects are pollution tolerant species as also reported by Daviet and Ray (1966), Gaufin (1974) and Das (1989).

Conclusion

During the investigation period, it was observed that the water quality at confluence point of Nag and Kanhan river *i.e.* Site-3 (Sawangi) was grossly polluted because of the addition of all domestic wastewater and different industrial effluents from Nagpur city into the Kanhan river. Physico-chemical investigation showed that Site-3 is highly polluted as compared to Site-1 and Site-2. Biological evaluation of water quality also supports this observation. The pollution tolerants groups of phytoplanktons (Cyanophyceae) and zooplanktons (Insecta) were more dominant at Site-3. The clean water indicator species *viz. Mougetia, Vaucheria* (green algae), *Lyngbya* (blue-green algae), *Chilomonas* (protozoa), *Monosyla, Phyilodina, Hexartha, Alsplanchna* (rotifera), *Moina* (crustacea), *Culex, Corixa, Ephemera, Hydrophilas* (insecta) were reported from Site-1 and Site-2 which are relatively unpolluted region of Kanhan River. While pollution indicators species *viz. Closterium, Scenedesmus, Oedogonium* (green algae), *Phormidium* (blue-green algae), *Gomphonema* (diatom), *Euglena, Parameacium* (protozoa), *Trichotoria, Notholca, Gastropus* (rotifera), *Bythtrephes, Eurycercus* (crustaca), *Eristalis, Baetis, Tabanus, Perula, Tanypus* (insecta), were recorded from Site-3 of Kanhan River which is highly polluted due to confluence of polluted Nag River.

Acknowledgement

The authors wish to thank Director, NEERI, Nagpur for providing facilities to carryout this work.

References

Algarsamy, S.R. Abdulappa, M.K. and Bopardikar, M.V., 1967. *High Rate Deep Stabilization Pond, Part II: Studies on the Faunal Assemblage of the System.*

Arora, H.C., 1966 b. Rotifers as indicators of trophic nature of environments. *Hydrobiologia*, 27(1–2): 146–159.

APHA, 1992. *Standard Methods for the Examination of Water and Wastewater*, 18th Ed. American Public Health Association, AWWA, WPCF, Washington, D.C.

Asif Khan, Untoo, S.A. and Parveen, S., 2002. Limnology of a reservoir receiving effluents from a thermal power plant. In: *Ecology of Polluted Water,* (Ed.) Arvind Kumar. A.P.H. Publishing Corp., New Delhi.

Badrinath, S.K., Kaul, S.N. and Gadkari, S.K., 1987. Fertilizer industries, an environmental appraisal. *T.J.E.P.*, 7(3): 201–208.

Batish, S.K. and Kumari, P., 1986. Effect of physico-chemical factors on the seasonal abundance of cladocera in typical pond at village of Raaba, Ludhiana. *Indian. J. Ecol.*, 13(1): 146–151.

Bhosle, A.B. Agarkar, S.V. and Ambore, N.E., 1994. A brief pollution survey of godavari river water at Nanded. In: *Nat. Sympo. on Eco-Environmental Impact and Organism Response.*

Bilgrami, K.S. and Datta Munshi, J.S., 1979. Limnological survey and impact of human activities on river Ganges (Barauni to Farakka range). *Tech. Rep. Man and Bios. Prog. Project–5,* UNESCO, Bhagalpur University, Bhagalpur.

Bulusu, K.R., Arora, H.L. and Aboo, K.M., 1967. Certain observations on self-purification of Khan river and its effect on Kshipra river. *Indian. Environ. Hlth.*, 9: 275–295.

Chandra, K., Singh, B., Shrivastava, G.N. and Meharotra, S.N., 1981. Pollution from wastes of industries manufacturing nitrogenous fertilizer: A case study from river Ganga near Allahabad. *Int. Symp. Water Res. Cons. Poll. and Abatement,* (Ed.) Chituranashi, V.R., Roorkee.

Chapman, D. and Kimstach, V., 1992. The selection of water quality variables. In: *Water Quality Assessment,* (Ed.) D. Chapman. Chapman and Hall, London, New York, Tokyo.

Chatterjee, Chinmoy and Raziuddin, M., 2002. Determination of water quality index (WQI) of a degraded river in Asansol industrial area (W.B.). *Ecology of Polluted Water,* Vol. II, (Ed.) Arvind Kumar. APH Publishing Corporation, New Delhi.

Chaudhari, N., 1982. *Water and Air Quality Control: The Indian Context.* Central Board for the Prevention and Control of Water Pollution, New Delhi, India.

Das, S.M., 1989. *Handbook of Limnology and Water Pollution*. South Asian Publishers Pvt. Ltd., New Delhi.

De Smet, W.H.O. and Evans, F.M.J.C., 1972. A hydrobiological study of the polluted river Lieve (Ghent, Belgium). *Hydrobiologia,* 31(1): 91–154.

Dexter, R.W., 1959. *Anosteraca*: 558–571. Quoted from Ward. H. B. and Whipple, G.C., 1959.

Dogra, S. and Kaw, J.L., 1989. Industrial effluents and aquatic ecosystem: A toxicological over view. In: *Management of Aquatic Ecosystem,* (Eds.) Agrawal, V.P., Desai, B.N. and Bidi, S.A.M. Soc. of Biosci., Muzaffarnagar, India, p. 271–279.

Edmondson, W.T., 1959. *Freshwater Biology*. John Wiley and Sons, Inc., New York, 1248 pp.

Gaur, R.K., Asif, A. Khan, Sultanat Parveen and Sayeed, A. Untoo, 2001. Sediment quality characteristics of a leach ate reservoir receiving effluents from a thermal power plant. *J. Ecophysial. Occup. Hlth.,* 1: 16–178.

Goel, P.K. and Chavan, V.R., 1991. Limnological study of a polluted fishery tank at Kolhapur. In: *Proc. Nat. Symp. Planning for Environ. Sus.*, New Delhi.

Gopal, B., 1990. Indian subcontinent and the aquatic habitats. In: *Ecology and Management of Aquatic Vegetation in the Indian Subcontinent*, Kluwer Academic Publ., Netherlands, p. 7–28.

Govindan, V.S. and Sudaresan, B.B., 1979. Seasonal succession in algal flora in polluted region at Adyar river. *Indian J. Environ. Hlth.,* 21(2): 131–142.

Gunale, V.R. and Balkrishnan, M.S., 1981. Biomonitoring of eutrophication in the Pavana, Mula and Mutha rivers flowing through Poona. *Indian. Environ. Hlth.,* 23(4): 316–321, *Hydrobiologia,* 27: 501–514.

Jolly, V.H. and Chapman, V.A., 1966. A preliminary biological study of the effects of pollution on farmers Creek and Cox's river, New South Wales. *Hydrobiologia*, 27: 160–187.

Kannan, K., 1991. *Fundamentals of Environmental Pollution.* S. Chand and Co. Ltd., New Delhi.

Kant, S. and Kachroo, P., 1974. Limnology studies in Kashmir lakes IV seasonal dynamics of phytoplankton in the Dal and Nagin. In: *Proc. Indian Nat. So. Aca.*, 40: 77–99.

Khapekar, R.R., 2005. Ecophysiological studies on algae of thermal power station. *Ph.D. Thesis*, R.T.M., Nagpur University, Nagpur.

Khatavkar, S.D., Kulkarni, A.Y. and Goel, P.K., 1989. Limnological study on two lentic freshwater bodies at Kolhapur with special reference to pollution. *J.E.P.*, 9(3): 198–203.

Klein, L., 1956. The problems of river pollution in industrial area. *Ray. Sec. Prov. Hlth. F.*, 76: 348–357.

Klein, L., 1957. *Aspects of River Pollution*. Butterworths Scientific Publication, London.

Kolkwitz, R. and Marsson, M., 1908. Oekologider Pflanzlichen Sprolien. *Ber. Bot. Ges.*, 26: 505–519.

Krishnamoorthi, K.P. and Visvesvara, G., 1996. Hydrobiological studies in Gandhisagar (Jammu Tank). *Seasonal Variations in Plankton* (1961–62).

Lieberman, M.E., 1970. The response of *Maina brachiata* (Jurine) 1820 to biological oxygen demand and light. *Hydrobiologia*, 36(1): 9–16.

Mathuthu, A.S., Zaranyika, F.M. and Jannalagadde, S.B., 1993. Monitoring of water quality in upper Mukuvisi river in Harare, Zimbabwe. *Environ. Int.*, 19(1): 51–61.

Mellanby, H., 1963. *Animal Life in Freshwater*, 6th ed. Chapman and Hall Ltd., London.

Meybeck, M., Friedrich, G., Thomas, R. and Chapman, D., 1992. Rivers. In: *Water Quality Assessments*, (Ed.) D. Chapman. Chapman and Hall, London, New York, Tokyo, p. 239–316.

Michael, P., 1984. *Ecological Methods for Field and Laboratory Investigations*. Tata McGraw Hill. Corp. Ltd., New Delhi.

Munawar, M., 1970. Limnological studies on freshwater ponds of Hyderabad, I: Biotope. *Hydrobiologia*, 35: 127–162.

Munawar, M., 1970b. Limnological studies on freshwater ponds of Hyderabad, II: Biocenose. *Hydrobiologia*, 36(1): 105–128.

Munawar, M., 1970. Limnology on freshwater ponds of Hyderabad, India: The biotops. *Hydrobiologia*, 35C 17, IR 7–162.

Muragesan, A.G. and Sukumaran, N., 1999. Impact of urbanization and industrialization on the river Tamirabarani, the life line of Tirunelveli and Thoothukudi districts. *Proc. Sem. Env. Prob.*, p. 1–6.

Nandkar, P.B. and Marathe, K.V., 1983. Estimation of the oxygen produced by dominant algae at six polluted streams. *J. Indian Bot. Soc.*, 62: 245–252.

Needham, J.G. and Needham, P.R., 1962. *A Guide to the Study of Freshwater Biology*. Holden Day Inc., Francisco, 108 pp.

Ouzounis, K.G., Zachariadis, G.A. and Straits, J.A., 1989. Preliminary environmental studies of New Karvali Bay (Kaval Gulf. North Greece). *Toxicol. Environ. Chem.*, 20(21): 227–232.

Paramasivam, M. and Sreenivasan, A., 1981. Changes in algal flora due to pollution in Cauvery river. *Indian J. Environ Hlth.*, 23(3): 222–238.

Penman, H.L., 1970. *The Water Cycle Scientific American*, 233(3): 99.

Philipose, M.R., 1960. Freshwater phytoplankton of Inland fisheries. *Proc. Sym. Algology*, ICAR, p. 272–291.

Sandhu, J.S., Sehgal, N.S. and Toor, N.S., 1984. Rotifer species diversity in manured experimental fish tanks. *Indian J. Ecol.*, 3(2): 303–308.

Shankhadarwar, S.D., 2002. Eco-physiological studies on toxic wastewater algae. *Ph.D Thesis,* R.T.M., Nagpur University, Nagpur.

Singh, V.P., Saxena, P.N., Tiwari, A., Lonsane, B.K., Khan, M.A. and Arora, I., 1969. Algal flora of sewage Uttar Pradesh in relation to physico-chemical variables. *Environment. Hlth.,* 11: 208–219.

Sinha, A.K., Shrivastava, R.K., Diwivedi, A. and Pande, D.P., 1991. Sediment characteristics of river Ganga at Shuklaganj (Unnao), Gagasso (Rae Bareli) and Phakhamau (Allahabad). *Proc. Nat. Conf. on Ag. Science in India,* New Delhi.

Sinha, M.P., 1986. Limnobiotic study on trophic status of a polluted freshwater reservoir of coal field area. *Poll. Res.*, 5(1): 13–17.

Sitarmaiah, P., 1960. Studies on the ecology of a freshwater pond community. *Hydrobiologia,* 27: 529–547.

Sladeck, J., 1961. A guide to limno-saprobial organisms. *Sc. Proc. Inst. Tech. Progr. Technol. Water,* 7: 543–562.

Tonapi, G.R., 1980. *Freshwater Animals of India: An Ecological Approach*. Oxford and IBH Publication Co., New Delhi.

Trivedy, R.K. and Goel, R.K., 1986. *Chemical and Biological Methods for Water Pollution Studies*. Environ. Publis., Karad, India.

Vasisht, N.S. and Sra, G.S., 1979. *Proc. Symp. Environ. Biol.*, p. 429–440. Quoted from Saksena, D.N. and Mishra, S.R., 1990.

Venkateshwarlu, V., 1969. An ecological study of river Moosi, Hyderabad (India) with special reference to water pollution, I: Physico-chemical complexes. *Hydrobiologia,* 32: 352–363.

Venkateshwarlu, Y., 1969. An ecological study of the algae of the river Moosi. Hyderabad (India) with special reference to water pollution, I: Physico-chemical complexes. *Hydrobiologia,* 33: 117–143.

Ventakeswarlu, V. and Jayanti, R.V., 1968. Hydrobiological studies of river Sabarmati to evaluate quality. *Hydroblologia,* 31: 442–449.

Verma, S.R. and Shukla, G.R., 1969. Pollution in a perennial stream, Khala by the sugar factory effluent near Laksar, Shaharanpur, *India. Environ. Hlth.,* 11: 145–162.

Verma, S.R. and Shukla, G.R., 1970. The physico-chemical conditions of Kamala Nehru tank, Muzaffarnagar in relation to the biological productivity. *Environ. Hlth.,* 12: 110–128.

Verma, S.R. Tyagi, A.K. and Dalela, R.C., 1978. Physico-chemical and biological characteristics of Kadarabad drain in U.P. *Indian J. Environment. Hlth.*, 20(1): 1–13.

Verma, S.R., Sharma, P., Tyagi, A., Rani, S., Gupta, A.K. and Dalela, R.C., 1984. *Limnologica* (Berlin), 15: 69–133. Quoted from Saksena, D.N. and Mishra, S.R., 1990.

Vijay Kumar and Sagwal, O.P., 2000. Recent studies on soil and water pollution in some parts of India: A Review. *Agric. Rey.*, 21(3): 186–192.

Ward, H.B. and Whipple, G.C., 1959. *Freshwater Biology,* 2nd ed. John Wiley and Sons. Inc., New York, London.

Wetzel, R.G., 1975. *Limnology*. W.B. Saunders, Co., Philadelphia, London and Toranto.

Woodward, G.M., 1984. Pollution control in Hamber estuary. *Water Pollution Control*, 83(1): 82–90.

Zafar, R., 1964. On the ecology of algae in certain fish ponds of Hyderabad. India, Physico-chemical complex. *Hydrobiologia*, 23: 179–195.

Zutshi, D.P., Subla, B.A., Khan, M.A. and Waganeo, A., 1980. Comparative limnology of nine lakes of J&K Himalayas. *Hydrobiologia*, 72: 101–112.

Chapter 23

Densities and Excess Volumes of Binary Liquid Systems of PEG 200 and PEG 400 with Isobutanol and Iso-Amyl Alcohol at 303K, 308K and 318K

D.N. Vora and F.J. Jani
Chemistry Department, Mithibai College, Vile Parle (West), Mumbai – 400 056

ABSTRACT

The densities of binary liquid systems of PEG 200 and PEG 400 with isobutanol and iso-amyl alcohol have been studied. The excess volume (V^E) has been calculated at 303K, 308K and 313K by using the standard relation over the entire range of composition for the systems. The data has been fitted to an empirical equation:

$$V^E/X_1X_2 = [a + b\,(2X_1 - 1) + c\,(2X_1 - 1)^2]$$

where, constants a, b and c have been evaluated by the least square method, along with standard deviation (σ). The negative excess volumes (V^E) have been observed for all the systems. The temperature effect has been studied in the light of specific interaction phenomena.

Introduction

A number of researchers (Rastogi *et al.*, 1967; Jain *et al.*, 1977; Oswal and Rathnam, 1984; Rathnam 1986; Sharma and Maken, 1992; Vora and Dodwad, 1995) have studied the thermodynamic properties of binary liquid systems to understand the nature of molecular interaction in systems. The binary mixtures with alcohols (Campbell *et al.*, 1975; Campbell *et al.*, 1976) have been earlier studied. The study of densities and excess volumes of binary systems with polar and non-polar solvents like cyclohexane and carbon disulphide have been reported (Rastogi *et al.*, 1967; Jain *et al.*, 1977). The densities arid excess volumes of PEG 200 and PEG400 with isobu1anol and iso-amyl alcohol have been reported in this article.

Experimental

PEG200, PEG 00, isobumnol and iso-amyl alcohol (E. Merck Grade) have been purified by conventional methods (Riddick and Bunger, 1970) and used. The purities of liquids have been further checked by measuring the densities and refractive indices at 303K and are in close agreement with the literature values (Timmemans, 1965; Weissberger, 1959). The density data has been correlated upto ±0.0001 units. The liquids have been weighed in ground stoppered flasks on a precision balance accurate upto 0.001 units to prepare binary liquid systems by taking due precautions to minimize evaporations. The temperature of the water thermostat bath has been controlled upto ±0.1K.

Results and Discussion

The densily measurements have been made al 303K, 308K and 313K for binary liquid systems of PEG 200 and PEG 400 with isobutanol and iso-amyl alcohol over the entire range of composition. The excess volume (V^E) has been calculated by using the relation:

$$V^E = [(M_1X_1 + M_2X_2)/d_{12}] - [M_1X_1/d_1 + M_2X_2/d_2] \quad (1)$$

where, M, X and d indicate the molecular weight, mole fraction and density respectively. The subscripts 1, 2 and 12 represent the pure liquid and binary liquid system respectively. The density and excess volume values have been reported in Tables 23.1–23.4.

Table 23.1: Densities (d) (gcm^{-3}) and Excess Volumes (V^E) (cm^3mo^{-1}) for PEG 200 with Isobutanol at 303K, 308K and 313K

X1	*303K*			*308K*			*313K*		
	d	$V^E_{expt.}$	$V^E_{cal.}$	*d*	$V^E_{expt.}$	$V^E_{cal.}$	*d*	$V^E_{expt.}$	$V^E_{cal.}$
0.0000	0.7838	–	–	0.7712	–	–	0.7599	–	–
0.1963	0.8869	–0.1577	–0.1713	0.8769	–0.2506	–0.2722	0.8681	–0.3586	–0.3724
0.26S1	0.9170	–0.2518	–0.2580	0.9078	–0.3503	–0.3568	0.8996	–0.4558	–0.4634
0.3025	0.9328	–0.3502	–0.3341	0.9240	–0.4530	–0.4492	0.9160	–0.5505	–0.5507
0.4890	0.9982	–0.4500	–0.4682	0.9913	–0.5555	–0.5522	0.9850	–0.6507	–0.6353
0.5353	1.0122	–0.4586	–0.4734	1.0058	–0.5572	–0.5706	0.9009	–0.6582	–0.6556
0.6579	1.0458	–0.4521	–0.4620	1.0406	–0.5505	–0.5085	1.0358	–0.6584	–0.6709
0.7910	1.0772	–0.3597	–0.3402	1.0731	–0.4558	–0.4620	1.0693	–0.5516	–0.5714
0.8611	1.0917	–0.2510	–0.2403	1.0882	–0.3525	–0.3519	1.0850	–0.4508	–0.4024
0.9101	1.1010	–0.1504	–0.1600	1.0980	–0.2500	–0.2495	1.0952	–0.3564	–0.3387
1.0000	1.1174	–	–	1.1144	–	–	1.1116	–	–

Table 23.2: Densities (d) (gcm^{-3}) and Excess Volumes (V^E) (cm^3mo^{-1}) for PEG 400 with Isobutanol at 303K, 308K and 313K

X1	303K			308K			313K		
	d	$V^E_{expt.}$	$V^E_{cal.}$	d	$V^E_{expt.}$	$V^E_{cal.}$	d	$V^E_{expt.}$	$V^E_{cal.}$
0.0000	0.7838	–	–	0.7712	–	–	0.7599	–	–
0.1161	0.8919	–0.1095	–0.0925	0.8820	–0.2063	–0.1997	0.7350	–0.3003	–0.2966
0.2413	0.9634	–0.2016	–0.2297	0.9555	–0.3094	–0.3272	0.9482	–0.4004	–0.4237
0.3417	1.0032	–0.3001	–0.3257	0.9964	–0.4054	–0.3988	0.9903	–0.5057	–0.5444
0.4005	1.0381	–0.4029	–0.3907	1.0324	–0.5088	–0.4967	1.0273	–0.0010	–0.0231
0.5600	1.0602	–0.4082	–0.4078	1.0552	–0.5004	–0.4820	1.0508	–0.6012	–0.6126
0.6898	1.0828	–0.4046	–0.3909	1.0787	–0.5024	–0.4970	1.0751	–0.6030	–0.6035
0.7597	1.0924	–0.3086	–0.2900	1.0888	–0.4029	–0.4136	1.0855	–0.5047	–0.5152
0.6038	1.0978	–0.2045	–0.2231	1.0944	–0.3044	–0.3146	1.0613	–0.4055	–0.4147
0.9192	1.1107	–0.1005	–0.1011	1.1078	–0.2011	–0.2100	1.1052	–0.3055	–0.2906
1.0000	1.1183	–	–	1.1154	–	1.1128	–	–	

Table 23.3: Densities (d) (gcm^{-3}) and Excess Volumes (V^E) (cm^3mo^{-1}) for PEG 200 with Iso-amyl Alcohol at 303K, 308K and 313K

X1	303K			308K			313K		
	d	$V^E_{expt.}$	$V^E_{cal.}$	d	$V^E_{expt.}$	$V^E_{cal.}$	d	$V^E_{expt.}$	$V^E_{cal.}$
0.0000	0.8022	–	–	0.7990	–	–	0.7952	–	–
0.1906	0.8874	–0.1325	–0.1950	0.8849	–0.2315	–0.2400	0.8820	–0.3324	–0.3521
0.2097	0.8956	–0.2316	–0.2299	0.8931	–0.3314	–0.3205	0.8902	–0.4340	–0.4188
0.3060	0.9324	–0.3326	–0.3285	0.9299	–0.4326	–0.4452	0.9271	–0.5359	–0.5362
0.4854	0.9924	–0.4337	–0.4382	0.9899	–0.5337	–0.5382	0.9872	–0.6369	–0.6382
0.5476	1.0108	–0.435.4	–0.4414	1.0083	–0.5347	–0.5343	1.0057	–0.6375	–0.6343
0.6199	1.0130	–0.4376	–0.426.3	1.0285	–0.5356	–0.5334	1.0260	–0.6396	–0.6334
0.7414	1.0617	–0.3334	–0.3168	1.0593	–0.4336	–0.4168	1.0569	–0.5369	–0.5188
0.8431	1.081	–0.2328	–0.2414	1.0827	–0.3328	–0.3214	1.0804	–0.4363	–0.4214
0.9581	1.1096	–0.1325	–0.1117	1.1073	–0.2324	–0.2590	1.1051	–0.3357	–0.3590
1.0000	1.1174	–	–	1.1144	–	–	1.1116	–	–

The excess volume (V^E) have been fitted to an empirical equation:

$$V^E/X_1X_2 = [a + b\,(2X_1 - 1) + c\,(2X_1 - 1)^2] \quad (2)$$

where, constants a, b and c have been evaluated by the least square method at three different temperatures and have been reported in Table 23.5, along with standard deviation (σ) defined by equation:

$$\sigma = \Sigma\,(V^E_{expt.} - V^E_{cal.})/(n - 3)^{1/2} \quad (3)$$

where, n is the total number of measurements.

Table 23.4: Densities (d) (gcm^{-3}) and Excess Volumes (V^E) (cm^3mo^{-1}) for PEG 400 with Iso-amyl Alcohol at 303K, 308K and 313K

X1	303K			308K			313K		
	d	$V^E_{expt.}$	$V^E_{cal.}$	d	$V^E_{expt.}$	$V^E_{cal.}$	d	$V^E_{expt.}$	$V^E_{cal.}$
0.0000	0.8022	–	–	0.7990	–	–	0.7952	–	–
0.1826	0.9319	–0.0813	–0.1085	0.9293	–0.1823	–0.2023	0.9264	–0.2812	–0.3066
0.2178	0.9495	–0.1834	–0.1978	0.9470	–0.2833	–0.2628	0.9441	–0.3625	–0.3769
0.3510	1.0015	–0.2867	–0.2939	0.9989	–0.3867	–0.3869	0.9962	–0.4837	–0.4837
0.4597	1.0331	–0.3856	–0.3741	1.0305	–0.4876	–0.4601	1.0279	–0.5848	–0.5679
0.5372	1.0512	–0.3879	–0.3951	1.0486	–0.4887	–0.4849	1.0460	–0.6869	–0.5611
0.6375	1.0708	–0.3866	–0.3720	1.0682	–0.4875	–0.4769	1.0657	–0.5868	–0.5771
0.7228	1.0845	–0.2848	–0.2904	1.0820	–0.3896	–0.4000	1.0795	–0.4855	–0.5090
0.8399	1.1006	–0.1848	–0.1816	1.0960	–0.2878	–0.3014	1.0957	–0.3875	–0.3954
0.9181	1.1098	–0.0877	–0.0857	1.1072	–0.1720	–0.1720	1.1049	–0.2878	–0.2629
1.0000	1.1183	–	–	1.1154	–	–	1.1128	–	–

Table 23.5: Parameters of Eq. (2) and Standard Deviation σ (V^E) (cm^3mol^{-1}) of Eq. (3) of Binary Liquid Systems

Temperatures	a	b	c	σV^E
	Binary liquid system of PEG 200 with isobutanol			
303K	–1.8916	–0.7950	–0.0446	0.0252
306K	–2.2295	–0.9017	–0.1172	0.0259
313K	–2.5628	–0.9489	–1.1860	0.0284
	Binary liquid system of PEG 400 with isobutanol			
303K	–1.6211	–3239	–1.8603	0.0146
308K	–1.9760	–0.3601	–2.4324	0.0380
313K	–2.1012	–4894	–2.5416	0.0388
	Binary liquid system of PEG 200 with iso-amyl alcohol			
303K	–1.5224	–0.7642	–0.8024	0.0118
308K	–1.5598	–0.9210	–0.8132	0.0225
313K	–1.7962	–1.0198	–0.8416	0.0399
	Binary liquid system of PEG 400 with iso-amyl alcohol			
303K	–1.6567	–0.7987	–1.2257	0.0288
300K	–1.9044	–0.8571	–1.7130	0.0319
313K	–2.2200	–1.4421	–1.8200	0.0324

The graphs of excess volumes (V^E) against mole fractions (X_1) have been plotted at 303K, 308K and 313K for all the systems and have been represented in Figures 23.1–23.4.

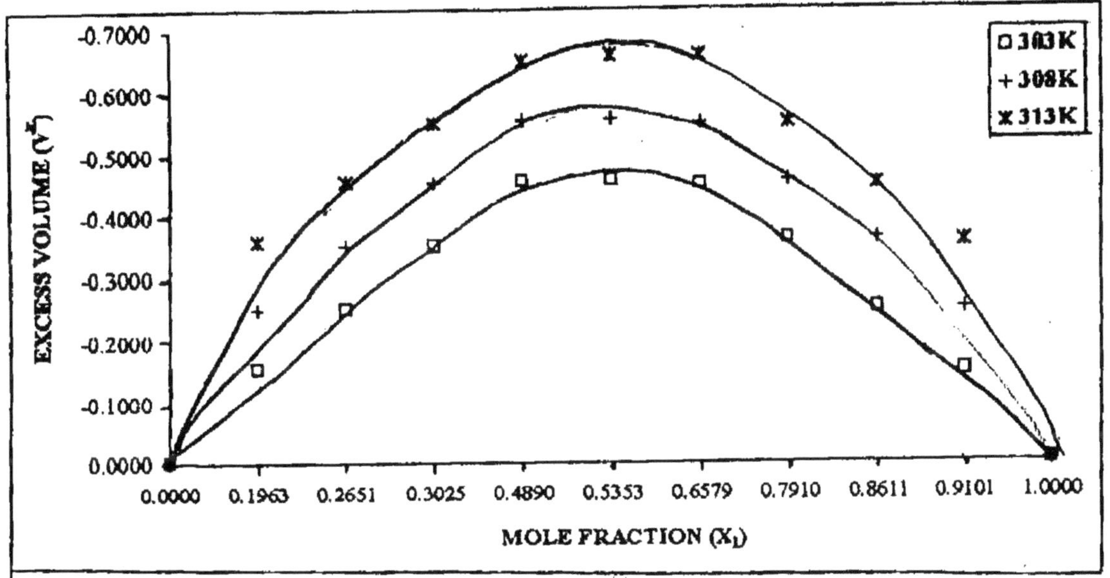

Figure 23.1: Binary Liquid System of PEG 200 with Isobutanol at 303K, 308K, 313K

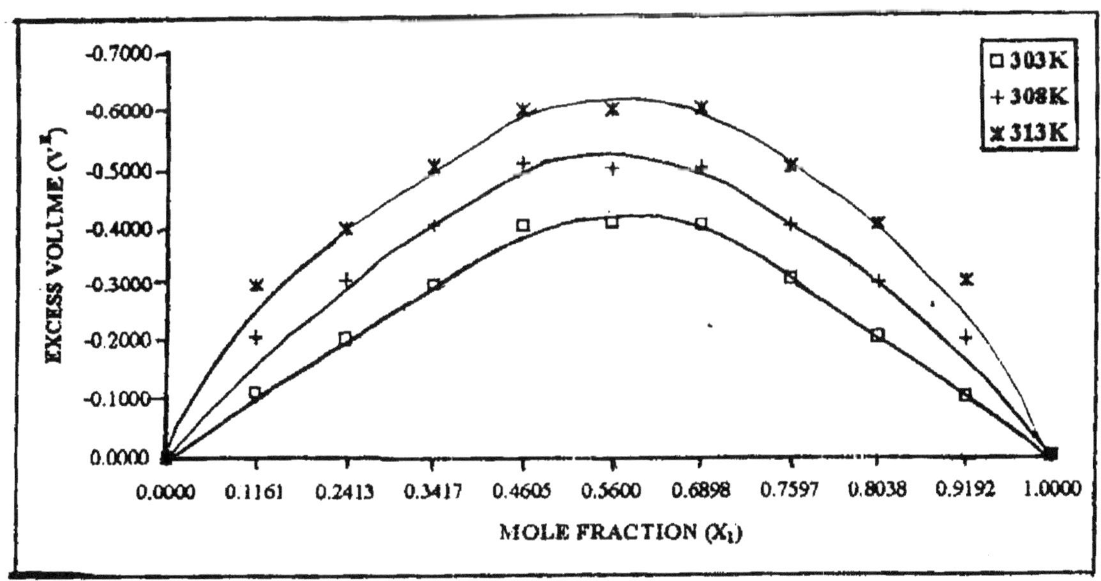

Figure 23.2: Binary Liquid System of PEG 400 with Isobutanol at 303K, 308K, 313K

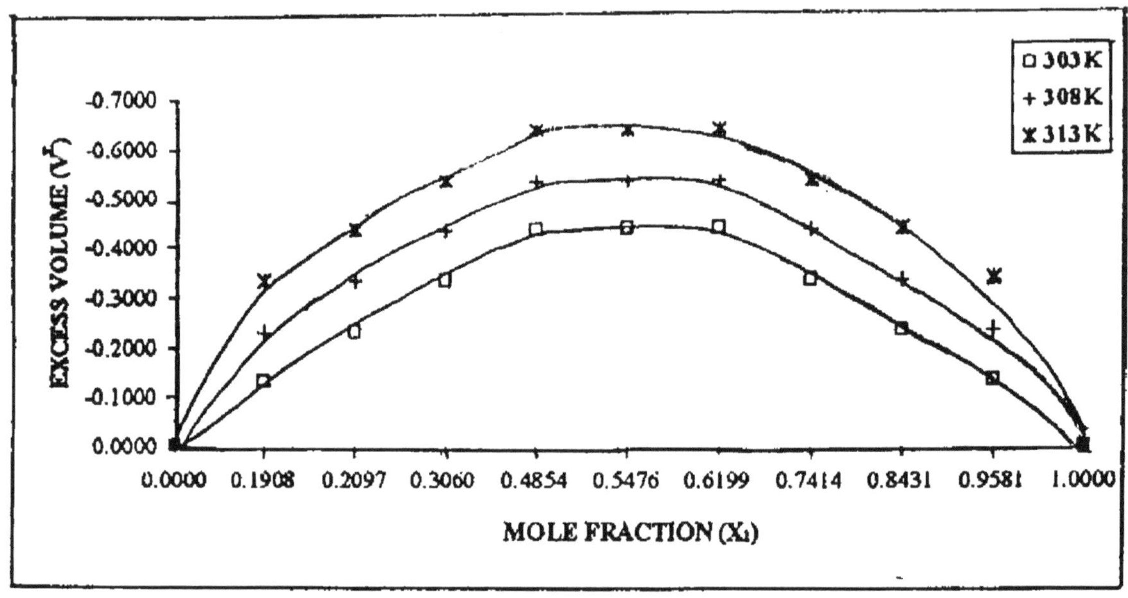

Figure 23.3: Binary Liquid System of PEG 200 with Iso-amyl Alcohol at 303K, 308K, 313K

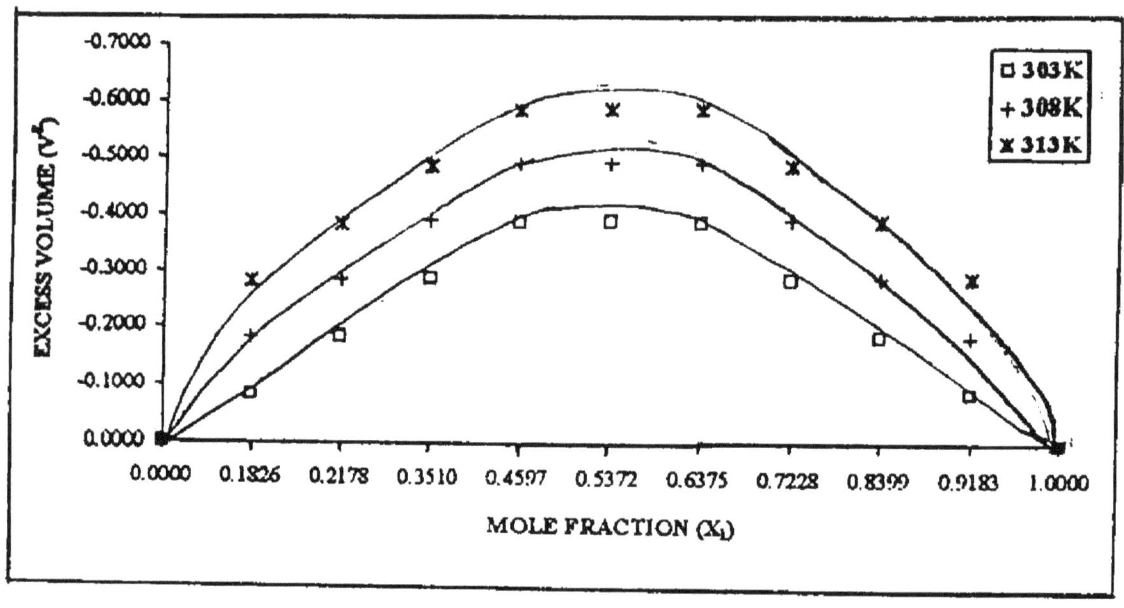

Figure 23.4: Binary Liquid System of PEG 400 with Iso-amyl Alcohol at 303K, 308K, 313K

The excess volumes (V^E) for all the systems at mole fraction ($X_1 = X_2 = 0.5$) at 303K, 308K and 313K have been reported in Table 23.6.

Table 23.6: Excess Volumes (VE) for Binary Liquid Systems at Mole Fraction ($X_1 = X_2 = 0.5$)

Binary liquid systems	*303K*	*308K*	*313K*
PEG 200 with isobutanol	–0.4409	–0.1	–0.6414
PEG 400 with isobutanol	–0.3924	–0.4913	–0.5904
PEG 200 with iso-amyl alcohol	–0.4224	–0.5224	–0.6249
PEG 400 with iso-amyl alcohol	–0.3711	1.4726	–0.5719

The excess volume (V^E) for all the systems over the entire of composition at 303K, 308K and 313K has negative values. The observed negative values are due to the depolymerization of pure glycol aggregates, specific interaction between the components and size of the molecules of the two components of the system.

References

Campbell, C., Brink, G. and Glasser, L., 1976. *J. Phy. Chem.*, 80: 686.

Campbell, C., Drink, G. and Glasser, L., 1975. *J. Phy. Chem.*, 79: 660.

Jain, D.V., Wadi, R.K. and Sharma, S.K., 1977. *J. Chem. Thermodynamics*, 9: 743.

Oswal, S.L. and Rathnam, M.V., 1984. *J. Ind. Chem. Soc.*, 61: 269.

Rastogi, R.P., Nath, J.A. and Mishra, J., 1967. *J. Phy. Chem.*, 71: 2524

Rathnam, M.V., 1986. *Ind. J. Chem.*, 25A: 1145.

Riddick, J.A. and Blulger, W.B., 1970. *Organic Solvents Physical Properties and Methods of Purification*. Wiley Interscience, N.Y.

Sharma, V.K. and Maken, S.K., 1992. *Ind. J. Chem.*, 31A: 721.

Timmennans, J., 1950, 1965. *Physico-chemical Constants of Pure Organic Compounds*, Vol. 1 and 2. Elsevier, London.

Vora, D.N. and Dodwad, S.S., 1995. *Asian. J. Chem.*, 7(1): 213.

Weissberger, A., 1959. *Techniques of Organic Chemistry*. Interscience, New York.

Chapter 24

Diversity and Composition of Insecta in Rice Agroecosystem in Barak Valley of Assam (N.E. India)

D.C. Ray and Partha P. Bhattacharjee
Department of Ecology and Environmental Science, Assam University, Silchar – 788 011

ABSTRACT

Diversity and density of insects were studied in 5 (five) sites of Cachar district of Assam during 2003–04. Order-wise study of insect showed maximum abundance by Coleoptera (35–53 per cent) followed by Hemiptera (9–37 per cent). Site-wise and season-wise study of Shannon-diversity index (H) proved moderate to high diversity. Sorensen Co-efficient of similarity (Cs) showed more similarity of insect population distribution. Diurnal activity of insect revealed maximum (159.0) activity at 17–18 hrs during Boro season followed by Amon season (115.0) at 15–16 hrs 't'–test proved insignificant regarding the activities of insects at peak hrs in all the three seasons. Density was found to be more by *Coccinelid sp.* (659/m^2) in Boro season followed by *Oxya hyla hyla* (136/m^2) in Amon season.

Keywords: *Diversity index, Sorensen Co-efficient, Insects, Agroecosystem, Diurnal activity, Density.*

Introduction

Insects are the dominant components of most ecosystems and agroecosystems are no exception. They play a key role in pollination and natural pest control, but it is a fact that most of the agricultural pests are insects (LaSalle, 1999). Thus, insect biodiversity form a vital component of a variety of ecosystem functioning.

According to the status report of India' s faunal wealth, large proportion of Indian fauna constitute Insecta (more than 40 per cent of the total fauna) (Anonymous, 1991). But this faunal diversity is under estimated for improper surveys and lack of systematic identifications, particularly insects whose only larval and pupal stages occur in aquatic habitats (Brij Gopal, 1997). A number of workers worked on insect diversity on various ecosystem levels (Engels and Wood, 1999; Wood and Lenne, 1999; Chesalmah *et al.*, 2000; Ogbeibu, 2001).

Barak valley-southern part of Assam belongs to North-east hotspot area and Himalayan foothills where the study is being carried out. A great insect diversity was noticed from rice agroecosystem. Considering the importance of insects in agroecosystem an attempt has been made to study the diversity and composition of insects during different seasons.

Description of the Study Sites

The present investigation was undertaken in Cachar district southern part of Assam. The geographical location of study sites are situated at longitudes of 24°41′29.9″ N and latitude of 92°45′25.9″ E. Altitude of the study areas are about 36 meter above msl. The entire study areas are low plain land and wetland with small hillocks. Rainfall is moderate to high. Paddy is cultivated in medium extent and mostly depends upon climatic conditions. The region has a plenty of faunal biodiversity. Vegetation is evergreen semi-evergreen and subtropical. The overlying soil cover is mostly sandy laterite. In most of the areas land has poor nutrient status except the riverine plains of Barak River. Most of the hillocks are covered by tea plantation. Paddy is cultivated in the low plains between the hillocks.

Materials and Methods

Five sites *viz.*, Irongmara, Machghat, Bariknagar, Dudpatil and Binnakandi were selected for the study purpose during 2003–04. A quadrat. (1 m × 1 m) was used to collect the insect sample as a standard device (Ghosh, 1985). The diversity of insects were determined by counting the numbers and kept in 70 per cent alcohol for preservation. The known insects were released to the field in living conditions and unknown specimens were kept in preservative for identification and further study. The sample was drawn at an interval of 10 (ten) days from all the sites. From each site 5 (five) quadrats were drawn randomly from 0.5 hectare are cropland during the Boro season.

Diurnal studies were conducted in Irongmara paddy field during Amon, Aus and Boro seasons on Sataki, Mayamoti and Basful varieties, respectively. The period of study were confined to 7–17 hrs at Amon season, from 6 to 18 hrs at Aus and Boro seasons at an interval of 1 (one) hour. Here also 5 (five) quadrats of same size were randomly drawn from 0.5 hectare of cropland.

Shannon-diversity index (H) was employed with the following formula (Shannon and Weaver, 1949)

$$H = -\Sigma\, Pi\,.\, Log\, Pi$$

Sorensen Co-efficient (Similarity) (Cs) was employed for the determination of similarity and dissimilarity of insects between communities of different seasons by the following formula (Sorensen, 1948)

$$Cs = \frac{2a}{(2a + b + c)}$$

't'–test was employed to assess the significance of diurnal activities of insects at various peak hours.

Results and Discussion

Group-wise abundance of insects was incorporated in Table 24.1. As regards the abundance it was found that more dominant group was Coleoptera (35–53 per cent) followed by Hemiptera (9–37 per cent) in all the sites out of seven available orders. Lowest group was considered as Odonata (4–14 per cent) (Table 24.1). Shannon diversity index indicated moderate diversity of insects in all the study sites during Boro season only (Table 24.2). Whereas season-wise diversity index (H) indicated also moderate diversity of insects between the community (Table 24.3).

Table 24.1: Group-wise Diversity of Insects on Rice Ecosystem During Boro Season (2003)

Orders	*Sites*				
	Irongmara	*Machghat*	*Bariknagar*	*Dudpatil*	*Binnakandi*
Coleoptera	35.55%	52.14%	47.37%	53.40%	36.49%
Lepidoptera	19.05%	3.11%	1.75%	4.85%	0
Hymenoptera	3.81%	3.11%	3.51%	0	5.40%
Diptera	3.49%	19.07%	0	0	2.70%
Hemiptera	12.38%	9.34%	0	13.01%	37.84%
Orthoptera	21.59%	5.06%	33.33%	3.88%	12.16%
Odonata	4.13%	8.17%	14.03%	4.85%	5.40%

Note: '0' means nil.

Table 24.2: Showing the Shannon-diversity Index (H) in Five Sites of Barak Valley (2003)

Study Sites	*Boro Season*	
	Diversity Index	*Paddy Varieties Studied*
Irongmara	H = 1.72547	Basful
Machghat	H = 1.9050	Basful
Barik nagar	H = 1.5441	Tupa Sali
Dudpatil	H = 1.6339	Asatti
Binnakandi	H = 1.6757	Jaya

Table 24.3: Shannon-diversity Index (H) Showing Insect Communities During Three Seasons (2003–04)

Diversity Index	*Seasons*		
	Aus	*Amon*	*Boro*
Shannon diversity index (H)	1.8694	1.8765	1.8581

As regards the Sorensen Co-efficient (Cs) of similarity between the two seasons it was found more similarity (0.75-0.92) of insect population distribution (Table 24.4). Diurnal activities of insect population was studied during three season (Tables 26.5–26.7 and Figure 24.1). During Aus season study was undertaken at hourly interval from 6–18 hrs. Maximum activities were recorded at 6–7 hrs (67.0) and 17–18 hrs (66.0). Most part of the day the activities of insects were moderate (Table 24.5). Diurnal activities were studied during Amon season, which indicated maximum activity (115.0) at 15–16 hrs followed by 14–15 hrs (100.0). Least activity 57.0 was noticed at 7–8 hrs (Table 24.6). In Boro season the peak activity was recorded 159.0 at 17–18 hrs followed by 16–17 hrs where the insect population was recorded 132.0. The lowest activity (78.0) was recorded at 7–8 hrs (Table 24.7). However, a 't'–test was employed between the peak hour activity of insects between two seasons. Although the peak hour activities varied in different seasons it did not significantly different from Aus vs. Amon, Aus vs. Boro and Amon vs. Boro ($P < 0.05$) (Tables 26.5–26.7).

Table 24.4: Sorensen Co-efficient of Insect Communities During Three Seasons (2003–04)

	Aus/Amon	*Amon/Boro*	*Boro/Aus*
Sorensen Co-efficient (Cs)			
Similarity	0.758	0.846	0.928
Dissimilarity	0.242	0.154	0.072

Table 24.5: Diurnal Variation of Insects from Paddy Field of Irongmara During Aus Season (2003)

No. of Obs.	*Time of Obs.*	*A*	*B*	*C*	*D*	*E*	*F*	*G*	*H*	*I*	*J*	*K*	*L*	*M*	*N*	*O*	*Total*
1.	6–7	18	5	2	2	5	3	7	6	1	1	0	0	0	0	0	67
2.	7–8	13	2	3	2	5	0	3	4	5	2	2	0	0	0	0	54
3.	8–9	17	1	2	2	8	0	2	7	3	0	0	1	1	1	0	57
4.	9–10	17	1	4	2	4	0	4	4	3	0	0	2	1	0	0	47
5.	10–11	10	1	3	7	11	1	7	6	3	0	0	1	1	1	2	64
6.	11–12	13	1	2	4	10	0	6	5	3	0	0	0	0	1	2	54
7.	12–13	12	0	1	1	8	1	4	7	0	1	0	1	0	2	0	43
8.	13–14	8	0	2	0	10	0	5	5	2	0	0	0	0	1	1	38
9.	14–15	13	0	2	0	7	0	3	3	2	0	0	0	0	1	1	37
10.	15–16	15	2	1	2	6	0	3	6	0	1	0	0	0	0	0	45
11.	16–17	15	0	2	2	10	0	3	6	2	0	0	0	0	1	2	52
12.	17–18	16	1	4	0	7	0	7	7	2	3	0	0	0	1	2	66
Total		**167**	**14**	**28**	**24**	**91**	**5**	**54**	**66**	**26**	**8**	**2**	**5**	**3**	**9**	**10**	**624**

Note: '0' means nil.

A: *Coccinelid* sp., B: *Bagrada crusifararum;* C: Demsel fly; D: *Neurothemis* sp.; E: *Oxya hyla hyla;* F: *Tryporyza incertula;* G: *Nephotettix virescens* (Dist); H: *Musca domestica;* I: *Alstonae* sp.; J: *Attractomorpha crenulata;* K: *Dicladispa armigera;* L: *Apis* sp.; M: *Vespa orientalis;* N: *Demoleus* sp.; O: *Leptocoryza acuta.*

Aus vs. Boro at 15–16 hrs: $t = 1.24$ (insignificant: $P < 0.05$).

16 17 hrs: $t = 1.72$ (insignificant: $P < 0.05$).

17 18 hrs: $t = 0.91$ (insignificant: $P < 0.05$).

Table 24.6: Diurnal Variation of Insects from Paddy Field of Irongmara During Amon Season (2003)

No. of Obs.	Time of Obs.	A	B	C	D	E	F	G	H	I	J	K	L	M	N	Total
1.	6–7	12	22	6	0	2	4	0	8	2	10	2	0	0	0	68
2.	7–8	5	21	7	0	0	0	4	2	0	13	5	0	0	0	57
3.	8–9	16	39	0	0	0	4	2	0	5	13	2	0	0	0	81
4.	9–10	20	44	2	1	0	3	4	0	0	10	3	0	0	1	88
5.	10–11	13	51	2	1	0	0	0	7	3	12	4	0	0	0	93
6.	11–12	12	31	1	1	0	0	5	2	5	10	3	0	1	0	71
7.	12–13	10	23	3	2	0	3	2	8	0	4	3	1	0	0	59
8.	13–14	14	39	3	4	0	0	3	1	8	22	5	0	0	0	99
9.	14–15	20	39	1	14	1	3	0	3	3	16	0	0	0	0	100
10.	15–16	14	51	8	19	1	6	1	5	3	7	0	0	0	0	115
Total		**136**	**360**	**33**	**42**	**4**	**23**	**21**	**36**	**29**	**117**	**27**	**1**	**1**	**1**	**831**

Note: '0' means nil.

A: *Oxya hyla hyla*; B: *Coccinelid* sp.; C: *Neurothemis* sp; D: *Attractomorpha* sp.; E: Demsel fly; F: *Leptocoryza acuta*; G: *Sitophilus oryzae*; H: *Musca domestica*; I: *Bagrada; crusiferarum*; J: *Tryporyza incertula*; K: *Galerucella* sp.; L: *Demoleus* sp; M: *Dicladispa armigera*; N: *Apis* sp.

Aus vs. Amon at 15–16 hrs: t = 1.19 (insignificant: P < 0.05).

Table 24.7: Diurnal Variation of Insects from Paddy Field of Irongmara During Boro Season (2003–04)

No. of Obs.	Time of Obs.	A	B	C	D	E	F	G	H	I	J	K	L	M	Total
1.	6–7	49	11	4	1	3	9	12	2	2	0	0	3	0	96
2.	7–8	43	9	3	2	5	4	7	3	0	1	0	1	0	78
3.	8–9	60	8	6	4	5	11	14	2	3	0	0	3	1	117
4.	9–10	55	9	3	5	12	6	9	4	3	1	2	2	0	111
5.	10–11	63	5	2	3	12	4	17	1	1	1	2	4	0	115
6.	11–12	47	5	2	9	6	4	8	1	2	1	3	2	0	90
7.	12–13	42	8	5	3	8	3	10	1	5	0	0	1	1	87
8.	13–14	41	8	6	3	7	3	6	2	2	1	2	1	0	82
9.	14–15	51	8	5	7	6	6	10	0	4	3	0	5	1	106
10.	15–16	59	8	5	10	13	9	9	3	5	1	2	6	0	130
11.	17–18	61	8	9	14	12	11	6	7	0	0	0	4	0	132
12.	18–19	88	7	11	7	11	4	16	2	2	1	4	6	0	159
Total		**659**	**94**	**61**	**68**	**100**	**74**	**124**	**28**	**29**	**10**	**15**	**38**	**3**	**1303**

Note: '0' means nil.

A: *Coccinelid* sp.; B: *Leptocoryza acuta*; C: *Neurothemis* sp; D: *Oxya hyla hyla*; E: *Tryporyza incertula*; F: *Musca domestica*; G: *Dicladispa armigera*; H: *Bagrada crusiferarum*; I: Demsel fly; J: *Attractomorpha crenulata*; K: *Nephotettix virescens* (Dist); L: *Alstonae* sp.; M: *Demoleus sp.*

Amon vs. Boro at 15–16 hrs: t = 0.0043 (insignificant: P < 0.05).

Table 24.8: List of Insects with Their Density (m^{-2}) in Paddy Field During Three Crop Seasons (2003–04)

Taxa Scientific Name	Common Name	Aus	Amon	Boro
Coccinelid sp.	Lady bird beetle	167	360	659
Alstonae sp.	Blue beetle	26	–	38
Dicladispa armigera Ol.	Rice hispa	2	1	124
Sitophilus oryzae Linn.	Rice weevil	–	21	–
Galerucella sp.	–	–	27	–
Tryporyza incertula	Yellow rice borer	5	117	100
Demoleus sp.	Butter fly	9	1	3
–	Demsel fly	28	4	29
Neurothemis sp.	Dragon fly	24	33	61
Oxya hyla hyla	Grasshopper	91	136	68
Attractomorpha crenulata	Grasshopper	8	42	10
Bagrada crusiferarum	–	14	29	28
Leptocoryza acuta	Mahua	10	23	94
Nephotettix virescens (Dist)	Green leafhopper	54	–	15
Musca domestica	Fly	66	36	74
Apis sp.	Bee	5	1	–
Vespa orientalis	Wasp	3	–	–

–: Means nil.

During three crop seasons the insect density was studied m^2 area basis (Table 24.8). Study indicated that maximum numbers were *Coccinelid* sp. in all the seasons. Peak population was recorded as 659/m^2 during Boro season followed by 360/m^2 during Amon season. Least numbers (167/m^2) were recorded in Aus season. The second dominant insect was *Tryporyza incertulus* was found in all the study seasons. Some insects *viz., Sitophilus oryzae, Gallerucella* sp. and *Vespa orientalis* were not found in all the seasons.

Among the crops, rice cultivation is predominant in Barak Valley of Assam. Insect diversity and abundance are very rich in rice agroecosystem. It is perhaps tropical area, moderate to high humidity and rainfall prevails during the rainy season which helps in multiplication of insect population. Group-wise study reveals that out of seven orders Coleoptera occupies the dominant: group. Ogbeibu (2001) studied group-wise insect on aquatic ecosystem in four sites in Nigeria. Site-wise study of Shannon diversity index (H) reveals moderate to high diversity in all the sites but in season-wise it also proves moderate to high diversity. Diversity index also reported by Ogbeibu (2001) and Nautiyal *et al.* (2000) on aquatic system and found higher diversity index.

Sorensen Co-efficient of similarity reveals high similarity of insect population among the three seasons. Ogbeibu (2001) reported significant dissimilarity among the four study sites. This may be because of aquatic ecosystem where the work is performed. As regards the diurnal study it reveals out of three seasons maximum activities (159.0) reported during Boro season (2003–04) at 17–18 hrs

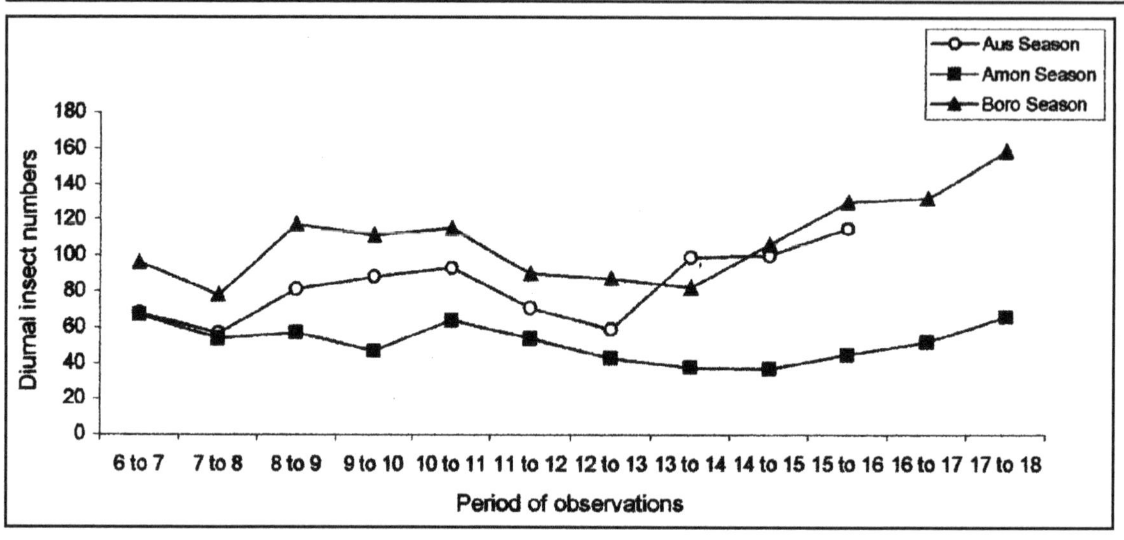

Figure 24.1: Showing Diurnal Insect Abundance During Three Seasons (2003–04)

followed by Amon season (115.0) at 15–16 hrs (Figure 24.1). This may be because of less temperature preferred by insects which prevails at that time. 't'–test was employed to assess the activities of insect among the peak hour period between Aus vs. Boro, Aus vs. Amon and Amon vs. Boro seasons. It proves the difference of activities insignificant. ($P < 0.05$) among all the seasons although the numbers of insect abundance is varying at different hour intervals. Density of insect taxa is found to be maximum in case of *Coccinelid* sp. (659/m^2) in Boro season followed by *Oxya hyla hyla* (136/m^2) in Amon season. It may be noted that in all the three seasons *Coccinelid* sp. is found to be dominating. Dominating numbers recorded may be because of species which can sustains in all types of environmental conditions. *Galerucella* sp., *Vespa orientalis* species may not available because of either seasonal insect or non-availability of food in off-season.

References

Anonymous, 1991. *Animal Resources of India: Protozoa to Mammalia*. State of the Art, Zoological Survey of India, Calcutta, 694 pp.

Chesalmah, M.R., Hassan, S.T.S. and Abu Hassan, A., 2000. Local movement and feeding pattern of adult *Neurothemis tullia* (Drury) (Odonata : Libellulidae) in a rainfed rice field. *Tropical Ecology*, 41(2): 233–241.

Engels, J.M.M. and Wood, D., 1999. *Conservation of Agro-biodiversity*. CAB International Agro-biodiversity, pp. 355–385.

Ghosh, A.K., 1985. *Sampling Techniques in Rice Research*. ICAR Publication, New Delhi, 623 pp.

Gopal, B., 1997. Biodiversity in Inland Aquatic Ecosystems in India: An overview. *International Journal of Ecology and Environmental Sciences*, 23(40): 305–313, International Scientific Publications, New Delhi.

LaSalle, J., 1999. *Insect Biodiversity in Agroecosystems: Function, Value and Optimization*. CAB International Agro-biodiversity, pp. 155–182.

Nautiyal, C., Nautiyal, P. and Singh, H.R., 2000. Species richness and diversity of epilithic diatom communities on different natural substrates in the cold water river Alaknanda. *Tropical Ecology*, 41(2): 255–258.

Ogbeibu, A.E., 2001. Composition and diversity of Diptera in temporary pond in southern Nigeria. *Tropical Ecology*, 42(2): 259–268.

Shannon, C.E. and Weaver, W., 1949. *The Mathematical Theory of Communication*. Urbana, IL, University of Illinois Press.

Sorensen, T., 1948. A method of establishing groups of equal amplitude in plant sociology based on similarity of species count and its application to analyses of the vegetation on Danish commons. *Biol. Skr*. (K. Danske Vidensk. Selsk. NS), 5: 1–34.

Wood, D. and Lenne, J.M., 1999. *Agro-biodiversity and Natural Biodiversity: Some Parallels*. CAB International Agro-biodiversity, pp. 425–445.

Chapter 25

Use of Biogas Slurry as a Fish Pond Manure

Sarbjit Singh Sooch[1*], Asha Dhawan[2] and Davinder Pal Singh[3]

[1]Assistant Professor, Department of Civil Engineering, College of Agricultural Engineering, PAU, Ludhiana

[2]Zoologist (Fisheries), Department of Zoology and Fisheries, PAU, Ludhiana

[3]B. Tech. Student (Ex.), PAU, Ludhiana

ABSTRACT

Effects of animal manures (Poultry droppings, biogas slurry and cow dung) on water quality parameters plankton production and survival and growth of fish were studied at PAU, Ludhiana. The water quality parameter were remained in optimum range. Plankton production (Phytoplankton and Zooplankton) was observed maximum in ponds manured with biogas slurry. However the growth of *Cyprinus carpio* in terms of percentage increase in length and weight gain was observed maximum in the pond manure with poultry droppings, followed by biogas slurry and cow dung.

Keywords: *Organic manures, Water quality, Fish production and Fish growth.*

Introduction

India has made impressive and rapid strides in the development of fisheries, particularly in the area of aquaculture. Increasing aquaculture production is no longer the main target but its sustainability is more important. The sustainable aquaculture depends upon the provision of eco-friendly, economically and socially viable culture system.

E-mail: sarbjit_sooch@sify.com.

The supplementary diet contribute to 40–60 per cent of the total production cost in intensive fish culture system. Increasing cost of feed stuffs (grains, fish meal, oil, cakes etc.) used in the preparation of supplementary diet has stressed the scientists to look for some cheaper substitutes. Many wastes and by-product of agriculture, animal husbandry and agro-industries have food value and can be used as pond manures and fish feed ingredients.

Moreover fish is an excellent converter of poor quality ingredients into high level animal protein *i.e.* the nutrients for natural food of fish (Plankton) which had a significant and positive influence on fish growth. Some fish species also feed directly upon the undigested fractions of these wastes (Schroedar, 1980) which may be low in nutritional value, but microorganisms adhere to them are of high protein value. Further these wastes can also be incorporated into the production of cheap low protein pelleted feed (Devaraj and Keshavappa, 1985).

Review of Literature

Many organic manures like cow dung, poultry manure, duck manure, pig dug and sewage etc., have been tried. Much work has been done on the productivity potential of cow dung, which is now universally used as the organic manure (Banerjee *et al.*, 1969) in pisiculture practice.

Schoonbee *et al.* (1979) observed marked fluctuations in dissolved oxygen level in manured ponds (cattle manure) but anerobic conditions did not prevail at any time. Burn and Stickney (1980) reported increase in ammonia with increase in level of poultry waste in the ponds. Sandhu (1982) found that phosphate contents of water increased by manuring (farm yard manure and poultry droppings of 5–17 t/ha/yr).

Yadav and Garg (1992) observed that addition of poultry droppings @ 30–428 kg dry matter/ha and cow dung @ 24 t/ha/yr respectively did not affect the water quality and hydrobiological factors remained within the optimum level.

The animal manure increase the biological productivity of the pond through various pathways, which is judged through the qualitative and quantitative estimation of plankton production (Schroeder, 1980).

The type of manure also influence the plankton production. Biogas slurry has been found to be one of the potential manurial input in enhancing the growth of plankton than cow dung (Rai, 1981, Kalayani and Shetty, 1987; Sehgal *et al.*, 1991).

Animal manures have direct or indirect effect on the survival and growth of fish. Above 90 per cent survival of *C. carpio* was reported in ponds fertilized with poultry manure (Kaur and Kaur, 1987) whereas 100 per cent survival of *C. carpio* was reported from ponds receiving cowdung and biogas slurry (Kaur *et al.*, 1987). Dhawan and Toor (1989) recorded maximum growth of *C. mrigala* in ponds treated with poultry droppings followed by cow dung (both @ 20 g/ha/yr) than unmanured pond. Growth rate of 71.15, 48.33, 38.15 and 29.67 g of *C. carpio* was reported from ponds receiving poultry droppings (18 t/ha/yr), cow dung (10 t/ha/yr), biogas slurry (19 t/ha/yr) and control (without manure) respectively (Kaur, *et al.*, 1989).

Materials and Methods

Experiments were conducted from 19th April 2002 to 25th June, 2002 at the Fisheries Research Complex of Punjab Agricultural University, Ludhiana to study the effect of biogas slurry on the growth of *Cyprinus carpio* (common carp). Physico-chemical parameters of water, plankton production were also analysed to assess whether biogas slurry had any effect on them.

Six cemented tanks (1–6) each with an area of 0.002 hectare (5.85 × 2.77 m) and a depth of 1.22 m were selected and manuring was done with cow dung, poultry waste and biogas slurry. Cow dung, poultry waste and biogas slurry were added at desired levels *i.e.*, @ 18 t/ha/yr in a tank (700 g/ha/week). The tanks were stocked with fingerlings of *Cyprinus carpio* (common carp). The different water quality parameters like water temperature, secchi disc transparency, hydrogen ion concentration (pH), dissolved oxygen (DO), alkalinity and free carbon dioxide were analysed. Pond productivity was assessed in terms of plankton production which include both phytoplankton and zooplankton by taking. Plankton samples simultaneously at the time of sampling water for its physico-chemical analysis.

Results and Discussion

The effects of different manures on water quality parameters, pond productivity and fish growth are shown in Tables 25.1–25.3 respectively.

Table 25.1: Effect of Different Manures on Water Quality Parameters

Parameters	Days	Manures			Optimum Range of Water Quality Parameters
		Biogas Slurry	Cow Dung	Poultry Droppings	
Water temperature (°C)	0	28.75	28.75	27.75	
	20	31.50	31.00	31.50	15–37.5
	40	27.50	27.50	26.50	
	60	33.00	32.00	31.50	
Secchi Disc	0	23.25	19.87	24.62	–
Transparency (cm)	20	20.95	18.60	21.75	
	40	17.87	17.05	19.60	
	60	17.45	16.60	19.10	
Hydrogen Ion	0	8.87	8.42	8.19	
Concentration (pH)	20	8.95	8.91	8.63	6.5–9.6
	40	9.07	8.22	8.17	
	60	9.41	8.18	8.59	
Dissolved Oxygen (mg/l)	0	7.40	6.60	5.50	
	20	15.30	12.00	7.90	7.8–9.0
	40	12.40	5.30	5.60	
	60	13.40	10.40	12.90	
Alkalinity (mg/l)	0	275	247	365	
	20	249	231	223	137–230
	40	130	152	158	
	60	166	162	206	
Free Carbon dioxide (mg/l)	0	0	0	0	
	20	0	0	0	15.00
	40	0	8	16	
	60	0	8	0	

Table 25.2: Effect of Different Manures on Pond Productivity

Pond Productivity	*Days*	*Manures*		
		Biogas Slurry	*Cow Dung*	*Poultry Droppings*
Phytoplankton (ml/l)	0	34	5	7
	20	10	7	18
	40	27	20	28
	60	22	5	13
Mean		23.25	9.25	16.5
Zooplankton (ml/l)	0	30	10	15
	20	44	22	18
	40	48.52	47	67
	60	35	42	50
Mean		39.38	30.25	37.5

Table 25.3: Effect of Different Manures on Fish Growth

Days	*Fish Growth*	*Manures*		
		Biogas Slurry	*Cow Dung*	*Poultry Droppings*
Initial (0 day)	Length (cm)	12.50	14.08	12.30
Final (60 days)		13.76	14.58	14.60
Length gain (%)		10.08	3.55	18.70
Initial (0 day)	Weight (gm)	35	39	38
Final (60 days)		42	42	50
Weight gain (%)		20	7.69	31.57

In the present study, phytoplankton was observed maximum in pond manured with biogas slurry (23.25 ml/l). Similarly zooplankton was observed to be maximum in the pond manured with biogas slurry (39.38 ml/l) followed by poultry droppings (37.5 ml/l) and cow dung (30.25 ml/l). Biogas slurry had been observed to be one of the potential manurial input enhancing the growth of plankton (Phytoplankton and Zooplankton) than cow dung (Rai, 1981; Kalayani and Shetty, 1987). This high value of plankton in pond manured with biogas slurry was because of high nutrients contents of organic manures which enhance the plankton production.

Fish growth (in terms of percentage increase in length and weight gain) was observed to be maximum in the pond manured with poultry droppings (length gain = 18.7 per cent, weight gain = 31.57 per cent) followed by biogas slurry (length gain=10.08 per cent), weight gain=20 per cent) and cow dung (length gain = 3.55 per cent, weight gain=17.6 per cent). Maximum growth of fish was observed in case of poultry droppings, this has also been observed by Dhawan and Toor (1989).

Conclusion

Water quality parameters *viz.*, water temperature, pH, dissolved oxygen, alkalinity, secchi disc transparency and free carbondioxide were analysed, which remained in optimum range. The most critical parameter *i.e.*, dissolved oxygen in all the treatments was in optimum range and maximum dissolved oxygen values were in case of biogas slurry. Plankton production (Phytoplankton and Zooplankton) was observed maximum in ponds manured with biogas slurry followed by poultry droppings and cow dung. The growth of *Cyprinus carpio* in terms of percentage increase in length and weight gain was observed maximum in case of pond manured with poultry droppings (length gain = 18.7 per cent, weight gain = 31.57 per cent) followed by biogas slurry (length gain = 10.08 per cent, weight gain = 20 per cent) and cow dung (length gain = 3.55 per cent, weigh gain = 17.6 per cent).

The organic wastes are available in large quantities all over the world can be used for supplementary diet for fish production. Moreover recycling of organic wastes for fish culture served the dual purpose of cleaning the environment and reaping economic benefits.

References

Banerjee, R.K., G.S. Singit and P. Ray. 1969. Some observations on the use of poultry manure as fertilizer in rearing major carp fry. *Indian J. Fish.*, 16: 29–34.

Burn, R.P. and Stickney, R.P. 1980. Growth of *Tilapia auria* in ponds receiving poultry wastes. *Aquaculture*, 20: 117–121.

Devaraj, K.V. and Keshavappa, G.Y. 1985. Preliminary studies on poultry-cum-fish culture. *Mysore J. Agric. Sci.*, 19(3): 189–193.

Dhawan, A. and Toor, H.S. 1989. Impact of organic manure and supplementary diet on plankton production and fish growth and feeding of an Indian major carp, *Cirrhina mrigala* (Ham.) in fish ponds. *Biol. Waste*, 29: 289–298.

Kalayani, R. and Shetty, H.P.C. 1987. Use of biogas slurry as a fertilizer for production of carps. *The First Indian Fisheries Forum*, December 4–8.

Kaur, K. and Kaur, S. 1987. Efficacy of poultry droppings in common carp *Cyprinus carpio* (Linn) culture: Effect on survival and growth. *Zoologica Orientalis*, 4(1&2): 13–18.

Kaur, K., Kaur, K. and Dhawan, A. 1989. Effects on organic wastes on the growth and ovarian maturation of *Cyprinus carpio. Proc. Nat. Sem. Freshwat. Aqua.*, p. 81–83.

Kaur, K., Sehgal, G.K. and Sehgal, H.S. 1987. Efficiency of biogas slurry in carp. *Cyprinus carpio* var. *communis* (Linn.) culture. Effects on survival and growth. *Wastes*, 22(2): 139–146.

Rai, S.P. 1981. Role of biogas plant in integrated aquaculture for rural development. In: *Summer Institute on Farming System Integrating Agriculture Livestock and Fish Culture*, 6th July to 4th August, CIFRI, Barrackpore, West Bengal.

Sandhu, J.S. 1982. Role of organic wastes in composite fish culture. *Ph.D. Thesis*, P.A.U., Ludhiana.

Schoonbee, H.J., Nakani, V.S. and Prinsloo, J. 1979. The use of cattle manure and supplementary feeding in growth studies of the Chinese Silver Carp in Transkri, South Africa. *S. Afr. J Sci.*, 75(2): 489–495.

Schroeder, G.L. 1980. Fish farming in manure loaded ponds. ICLARM–SEARCH *Conf. Agri. Aquaculture Fmg. Systems*, Manila, Phillipines, p. 73–86.

Sehgal, H.S., Kaur, K. and Sehgal, G.S. 1992. Zooplankton response to biogas slurry in carp ponds. *Bioresource Technology*, 41: 111–116.

Yadav, N.K. and Garg, S.K. 1992. Relative efficacy of different doses of organic fertilizer and supplement feed utilization under intensive fish farming. *Bioresource Technology*, 42: 61–65.

Chapter 26

Soil and Groundwater Pollution by Agrochemicals: A Review

D.S. Kler, Navneet Kaur and R.S. Uppal
Department of Agronomy and Agrometeorology, Punjab Agricultural University, Ludhiana - 141 004

ABSTRACT

Increased consumption of agrochemicals for attaining self-sufficiency in the developing countries, resulted in soil and groundnut pollution. Agrochemicals *i.e.* pesticides and herbicides disrupts the activity of soil microorganisms like bacteria, fungi, algae and nematode which affect the nutritional quality of soils. These agrochemicals affect the important biological processes (Ammonification, Nitrification, Denitrification and Nitrogen fixation) and enzymatic activities of soil. The leaching of nitrogenous fertilizers to the groundwater also has many biological consequences. These agrochemicals affected the health of human beings. To achieve the desired goal. We will have to take special measure to solve the problems of the future. Use less persistent pesticides, run off, volatilization and leaching may be reduced to the maximum extent possible. To achieve the desired goal we will have to take special measures to solve the problems of the future considering future ecology more important than present economy.

Introduction

An increased scale of man's agricultural, industrial and social activities has brought with it an undesirable change in the chemical, physical and biological properties of the environment. According to Miller (1991), "Any undesirable changes in the characteristics of the air, water, soil and food that can adversely affect the health, survival or activities of humans or other living organisms is called pollution". The developing countries for attaining self-sufficiency made massive investments in the development of new varieties of crops, fertilizers, pesticides and agricultural machinery, storage techniques and marketing facilities. In intensive cropping system, the increased consumption of fertilizers and herbicides has created many environmental problems among which groundwater and soil pollution are main.

Soil Pollution by Pesticide and Fertilizers

Soil development and soil fertility results from a combination of abiotic and biotic factors. The abiotic factors include the composition of parent rock, temperature regime, hydrology etc. Biotic factors are cyclic processes involving the synthesis of complex organic molecules by plants, death of plant tissue, utilizing its energy content by decomposer organisms, circulation of nutrients etc. The process of decomposition of dead plant tissue and release of the nutrients of plant growth is performed by microorganisms and animals in soil. The pesticides and fertilizers reaching the soil environment may persist and alter the structure and function of these beneficial non-target organisms. Nutrients and pesticides are the primary contaminants of groundwater. The persistence of any chemical depends upon the factors like nature of the chemical, temperature, humidity, light and other soil factors including the activities of microorganisms which breakdown the chemical and render it less harmful.

Fertilizer Application in Relation to Environment

Some major concerns of fertilizer application in relation to pollution problems are pollution of drinking water, eutrophication of lakes and rivers, volatilization of NH_3 and nitrogen oxides, imbalanced quality of plant products and agricultural land as sinks for methane. The nutrient of primary concern in agriculture is N but P is also a concern. Nitrate affects the quality of drinking water. The presence of other plant nutrients such as phosphate, Potassium and magnesium can improve the quality of drinking water because these ions are essential for human and animal nutrition. Nitrate itself is not toxic but nitrite originating from the reduction of nitrate induces methaeno-globinemia in infants by inhibiting oxygen transport in the blood. Nitrate and amines can react to form nitrosamines which are cancinongenic. Acceptable values for NO_3 concentrations in drinking water have been set at 50–100 mg nitrate per litre (WHO), a maximum of 50 mg nitrate per litre (EL) and a maximum of 45 mg nitrate/litre (US Public Health Service). The recovery of fertilizer N is incomplete. Nitrogen balance in terms of per cent of fertilizer N showed that 58 per cent was taken up by the crop while 5 per cent was leached out of the rooting profile and 21 per cent was found as residual soil mineral N and 16 per cent was considered as unaccounted (violatile) losses (Mengel *et al.*, 2001). For arable crops 20–60 per cent of applied N is taken up while for grass is 40–80 per cent depending soil, season and rates. Nitrate leaching in arable soils can be reduced by catch crops. Most of the excessive N applied is lost either by leaching or by volatilization in the form of NH_3 or after denitrification in the form of N_2 and N_2O. Dinitrogen oxide (N_2O) is involved in the destruction of the ozone layer in the stratosphere. Ammonia released from the intense livestock farms in to the atmosphere is brought back in to the soils by precipitation when it may be oxidized to nitrate and thus also contribute to the nitrate in lakes, rivers and aquifers. Nitrate leaching, N_2O release in to the atmosphere and the loss of phosphate because of too high rates application are a hazard to the environment. Phosphorus may also become a limiting factor along O_2 due to eutrophication and the eutrophic threshold level below which algal growth is limited is considered to be in the region of 10 microgram per litre of P. Soil phosphates contribute only to about 4 to 5 per cent to the total phosphate level leached in to the surface water and other comes from detergents and urban wastes.

Pesticides Residue in Soil Organism

Edwards (1973) gave following four effects of pesticides on living organisms (soil microorganisms, soil invertebrates and plants) in soil:

1. They may be directly toxic to animal in soil.
2. They may affect the soil microorganisms genetically to produce population resistant to the pesticides.
3. They may have sublethal effects that result in alterations in behaviour changes in metabolic or reproductive activity.
4. They may be taken into bodies of soil flora or fauna and passed on to the other organisms.

The soil microorganisms like bacteria, fungi, algae and nematodes play important role in soil nutrient through their role in decay of plant and other organic matter in soil as nitrifiers. Anything that disrupts their activity could be expected to affect the nutritional quality of soils. Most of the fungicides, nematicides and insecticides applied in the soil also suppress the microbial activity in soil. Microorganisms play a major role in the break down or degradation of pesticides in soil. Soil treatment with pesticides at high dosage has been shown to persist at levels toxic not only to harmful pests but also to beneficial soil arthropods. Populations of predaceous and parasitic insects in soil have been found to reduce by the repeated application of pesticides, especially by using organochlorine insecticide. High residues of pesticides in invertebrates are found only when these are applied at higher dosages.

Microorganisms that live in soil can be killed not only by chemicals applied directly to the soil, but also by those that reach the soil in drift from sprays off washed of from foliage. It is of great problem when pesticides applied in the soil kill some soil organisms which are essential in the breakdown of some kinds of dead leaf material into its organic and inorganic constituents for the incorporation of these materials into the soil. Earthworm feeding is very important in the breakdown of deciduous tree litter. *Enchytraeid* worms, *Collembola,* some *Acarina* and Dipterous larvae help to disintegrate plant material. Hence, any chemical, which affects the soil organisms, will certainly influence the soil fertility.

Persistence of Herbicides in Soil

Herbicides help in solving weed problems in crop production. Herbicides differ with regard to their period of persistence in the soil (Table 26.1). In intensive cropping system where cropping intensity is 200 to 300 per cent, the prolonged period of persistence of herbicides in the soil, can adversely affect the sensitive crop in the rotations (Brar and Randhawa, 1992).

Table 26.1: Relative Persistence of Some Herbicides in the Soil (Brar and Randhawa, 1992)

Persistence Period (Months)		
< 3	*2–6*	*> 6*
Aminotriazol	Chlorbromuron	Atrazine
Aziprotryn	Chlorotoluron	Bromacil
Carbetamide	Diallate	Dichlorbenil
Chlorprophan	Dinitriamine	Diuron
Cynazine	Isoproturon	Metribzin
Dalapon	Linuron	Simazine
Metaxuron	Methabenzthiauron	Terbacil
Prometryn	Triallate	Trifluralin
Propachlor		
Prophan		
Terbutryn		

Fate of Pesticide in Soil Environment

The factors like adsorption to clay and organic matter, leaching with the downward percolation of water, volatilization to the atmosphere, uptake by soil organisms, microbial degradation, chemical and photolysis affect the fate of pesticide in soil (Mishra *et al.*, 1989). The chemical nature of the pesticides and soil type are the important factors, which decide the behaviour of pesticides.

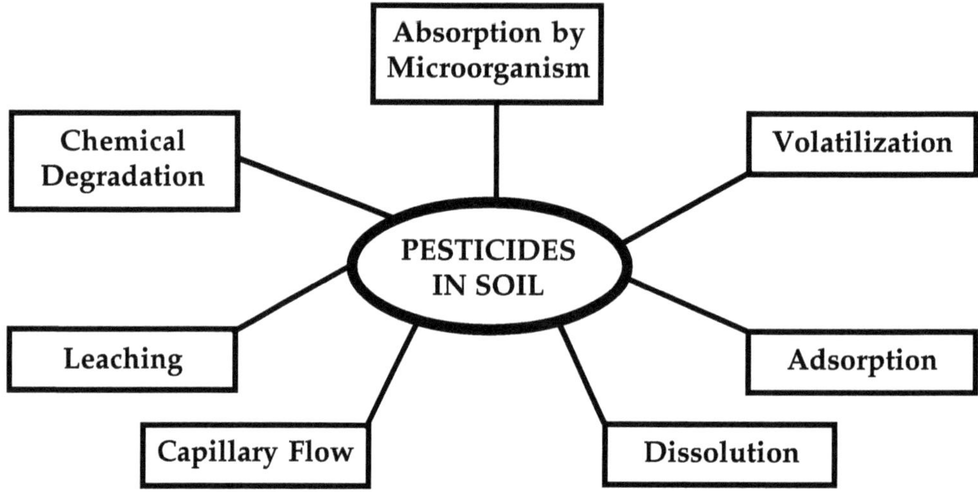

Effect of Herbicides on Soil Biological Processes

Herbicides affect the important soil biological processes like ammonification, nitrification, denitrification, nitrogen fixation and soil enzymatic activities (Rao *et al.*, 1993 and Kulshrestha, 1993).

Ammonification

Ammonification is the important process for supplying readily available nitrogen to plants and microorganisms. The process of ammonification is simulated by some of the herbicides. Phenoxy acids, Triazines and urea type compounds have positive affect on these microorganisms. 2,4-D or its ester Bladex has neither an adverse nor a favourable effect on ammonification in soils at field application rates but at 10 times the field rate ammonification was inhibited. Simazine and Atrazine had no significant effect on ammonification in cropped or uncropped plots.

Nitrification

At normal field application, 2,4-D does not affect nitrification. Soil application of Simazine @ 6 kg/ha stimulates both *Nitrosomonas* and *Nitrosococcus*. Debona and Andus (1970) established a decreasing order of inhibiting effectiveness for ten herbicides against nitrification as–Propanil > Ioxynil > Chlorflurazole > Bromoxynil > Chlorthiamid = 2,3,6–TBA > Picloram > Paraquat > Dchlobenil.

Denitrification

Dissimilating nitrate—a nitrite reduction is a common and important activity in soil. Dalapon at field rates is not toxic to denitrification in culture. Atrazine at concentrations of 3.4 and 2.6 µg/kg causes an increase in the NO_3^- reduction to NO_2 compared to N_2. Diallate, Cycloate and

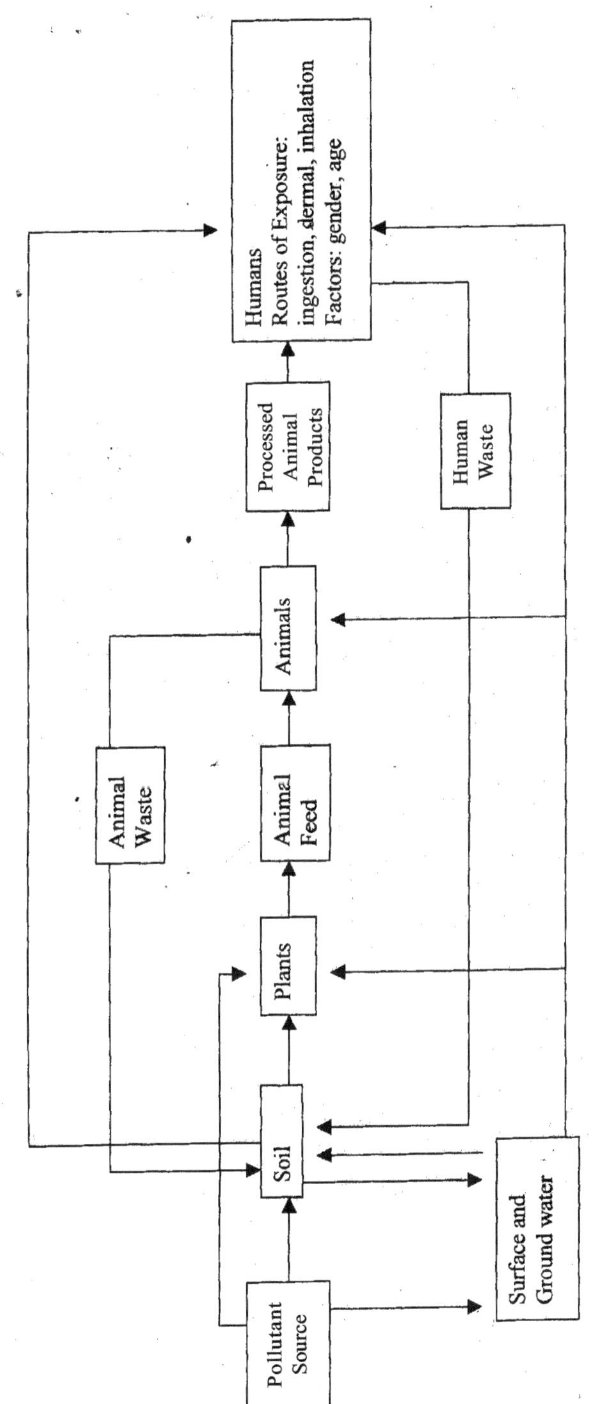

Figure 26.1: Pathways for Human Exposure to Pollutants Applied to Soils (Adapted from Brams *et al.*, 1989)

phenomedaphon had no influence on denitrifying bacteria and their activity in paddy soil at field rates. Halogen substitution in phenyl urea and aniline herbicide affected differential inhibiting reaction on the activity of denitrifying bacteria.

Nitrogen Fixation

Free living bacteria such as blue-green algae, *Azotobacter* sp. and *Clostridium pasteurianum* and those having symbiosis with higher plants like *Rhizobium* sp. fix atmospheric nitrogen for plant growth. The blue-green algae play important role in soil fertility in low land rice cultivation. Propanil and EPTC at low concentrations stimulated algal growth and N_2 fixation. Butachlor and Alachlor have detrimental effect on growth of *Nosto muscorum* and *Anabena doliolum*. Nitrogen and MSMA at excessive rate inhibited growth and activity of *Azotobacter* sp in cotton but at normal rate they were higher (100 μg/ml) than blue green algae (10 or 20 μg/ml). DNOC, Dinosoeb and Linuron are most inhibiting herbicides to *Rhizobium* under the influence of Dinoseb, *Rhizobium trifoli* showed uncoupling of oxidative phosphorylation. Similarly Dinitroanilines and Carbamates showed negative effects on *Rhizobia* and nodulation.

Effect on Soil Enzymatic Activities

The biochemical activity in soil is the resultant of various reactions which are catalyzed by enzymes. These enzymes may be intracellular component of the microbial community or extracellular (cell-free) enzymes existing in soil. Soil enzymes are often referred as an index of soil fertility. There are many enzymes which serve as useful and reliable indicators of soil fertility and soil health, because of their sensitivity, accuracy and ease of assay. Certain herbicides inhibit the enzymatic activities while some stimulate their activities (Table 26.2).

Table 26.2: Effect of Herbicide on Soil Enzymatic Activity (Hicks *et al.*, 1990)

Compound Class	*Enzyme*							
	Dehydrogenase	*Urease*	*Catalase*	*Phosphatase*	*Protease*	*Invertase*	*Asparaginase*	*Proteinase*
Substituted urea	I	I	I	S	S	I	I, S	U
Triazines	I	I	I, S	S	I	S	S	I
Nitrophenols	I, S, N	I, S	I	U	U	N	U	U
Carbamates	N	I	U	S	I	I	U	I
Hetero Cyclic Compounds	I	I	U	S	I	I	U	I
Benzoic acid phenyl acetates	I	I	S	S	U	S	U	U

I: Inhibition; S: Stimulation; U: Effects unknown; N: No effect.

Effect on Soil Microorganism Population

Anthropogenic activities such as indiscriminate use of fertilizers and herbicides, disposal of industrial and domestic effluents, municipal sewage sludge and solid waste, cause undesirable changes in the environment. The pollutants in the form of chemicals and waste products affect all forms of soil organisms ranging from protozoans to small mammals. Kuhnelt (1963) was among the first to state that the soil pollution by industrial waste adversely affects the soil fauna. These pollutants affect the qualitative and quantitative composition of soil organisms. Many of these pollutants can increase in concentration while passing through the food web of ecosystem and therefore, become serious threat to organisms occupying higher trophic levels.

Effect of Agrochemicals on Population Structure

Organic and inorganic compounds of agrochemicals used as pesticides and fertilizers in agroecosystem to boost the production of food materials have created many environmental problems. When used indiscriminately these are left as residues in the soil and crop system.

Herbicides

The herbicides applied to control unwanted vegetation, turn into potential pollutants in the environment. Most of the effects of herbicides on soil arthropods appear to be indirect. The Paraquat and Atrazine have toxic effects on the secondary decomposers like micro arthropods. The herbicide combination of Praquat, Glyphosate, Alachlor, Linuron, Fluanzifopbutyl, Acifluorfen and Bentazon decreased the decomposition rate of wheat straw residue and lowered the total number of microarthropods. At field conditions, Dichlobeini (7.5 kg/ha), Dalapon (1.5–7.5 kg/ha). Flumeturon (1 kg/ha) inhibited the growth of bacteria (El-Shahawy *et al.*, 1982). While Gupta and Kahlon (1987) reported that *Rhizobium phaseoli, Azotobacter chroococcum, Pseudomonas convexa* and *Bacillus sp* were unaffected by Isoproturon and Metoxuron at normal rates in pure culture. While the *Actinomycetes* showed relatively resistant behaviour to herbicides and only high concentrations cause change in growth rates. Simazine @ 8 kg/ha caused stimulation of growth of *Actinomycetes.* Five years of annual application of Diuron in sugarcane was inhibitory to *Actinomycetes* in soil (Heydrich and Fernandez, 1984). Some species of fungi are sensitive and susceptible to herbicides while some show resistant behaviour for herbicides. Some of the substituted Ureas, Carabamates, Bipyridyls, Diphenyl ethers and PCP reduce the number of propagules in soil and inhibit vegetative growth and spore formation in pure cultures (Pathak *et al.*, 1988). Certain pathogenic species such as *Sclerotium erolfsii* and *Pythium* sp. are susceptible to most herbicides, while *Trichoderrna* and *Penicillium* are resistant. Most pathogenic *Fusarium* sp. are susceptible while *Foxysporum* is not (Pope and Hort, 1981). Algae are also affected by various herbicides. Atrazine and Metauron inhibited the photosynthesis. 2,4-D significantly decreased the heterotrophic algae while *Perometrayne* decreased the population of *Anabaena* sp. (Yee *et al.*, 1985).

The herbicides affect the population function. The functioning of soil arthropods population is an important process in ecosystem dynamics and is affected when the soil substratum faces any population stress. Litter breakdown is one of the major functions of soil arthropods. Since these populations influence litter decomposition process and nutrient cycling in the ecosystem, any change in the structure of the soil arthropods population would influence this function. Herbicides such as Paraquat and Atrazine showed toxic effects on the secondary decomposers like microarthropods. The herbicide combinations of Paraquat, Glyphosate, Alachlor, Linuron, Fluanzifopbutyl, Acifluorfen and Bentazon decreased the decomposition rate of wheat straw residues and lowered the total number of micro arthropods.

Fertilizers

Considerable amounts of different types of fertilizers such as N, P, lime etc. are used annually to improve productivity as a common practice in various agroecosystems, grasslands and forests. The fertilizers change the physicochemical properties of the soil. These changes affect soil arthropods both qualitatively and quantitatively and the fertilizers tend to increase the population of *Collembola* (Miller, 1991).

Nitrogen

Application of NPK increased the number of *Collembola* two-fold. However, a decrease in the population densities of *Aptergola* insects and other soil arthropods was recorded due to the application of N. Such changes are attributable to the changes in pH and toxicity caused by high concentration of ammonia and other salts. The abundance of microarthropods increased considerably due to the application of NPK incorporations under maize and wheat, as well as under the monoculture of wheat, paddy and jute cultivation. Such increase in total population was mainly due to the increase in *Tarsonemina, Acaridal, Oribataidae* after 4 and 8 years of fertilizer application. However, *Collembola* and *Proshgmata* and *Endoshgmata* changed in significantly in abundance compared to the control plots (Miller, 1991).

Lime

Lime application decreased the abundance of cryphostigmated mites and increased that of *Gamasina, Uropodina* and *Trembidoidea* in pure stand which was probably due to changes in soil pH. The oribated mite *Tectocepheus clatus* was significantly reduced due to the effect of lime. Lime treatment increased the population of *Diptera* larvae more than two-fold. However, it did not show any significant effects upon *Coleoptera* and spiders. Similarly, the density of *Chilopda* was slightly reduced in the fertilized plots.

Besides some other products sold as fertilizers may be enriched in certain trace elements. The primary routes of exposure of humans to trace elements in soil are through food chain transfer and of microorganisms by direct ingestion of soil particle. Cadmium, Arsenic, Lead etc. and other non-essential poisoning elements being impurities in the ore may exist in the materials that were used to make the fertilizers (Table 26.3). The soil is the primary recipient of global emissions of trace elements. The Pb and Cd have profound health effect on humans, but their means of exposure are quite different due to their chemical behaviour in the soil environment and concentration ranges at which they occur in contaminated soils. The Cd is readily taken up by plants (a high bioavailability for plant uptake) and food chain transfer is the primary route exposure (Table 26.4). The Cd toxicity causes renal dysfunction and *itai-itai* disease. The disease is characterized by Cd induced bone loss that produces localised, severe pain in the joints of victims.

Table 26.3: Average and Range of Arsenic, Cadmium and Lead Concentrations (mg/kg) in Selected Inorganic Commercial Fertilizers and Soil Amendments, Liming Materials Sold in California (Department of Food and Agriculture, 1998)

Material	*Grade*	*As*	*Cd*	*Pb*
Limestone	–	2.2 (0.3–5.1)	6.2 (4.0–8.1)	33.9 (3.2–52.0)
Gypsum	–	1.3 (0.2–3.4)	0.9 (0–2.5)	4.4 (0.8–9.2)
Triple super phosphate	0–20–0	4.8 (0.5–14.0)	110 (0–163)	2.0 (0–4.0)
Monoammonium phosphate	11–23–0	8.1 (1.5–16.0)	74.1 (0–166)	2.6 (0–9.0)
Diammonium phosphate	18–20–0	4.9 (0.3–8.5)	38.1 (5.0–125)	4.0 (0–15.6)
Soft/rock phosphate	–	13.0	130	13.0
Muriate of potash	0–0–50	0.4 (0.1–1.6)	0.8 (0–1.6)	3.3 (2.5–4.0)
Urea	49–0–0	0	0	0
Blended	11–5–0	11.2	95.0	3600
Zinc sulfate	–	2.6 (0–29.5)	32.3 (0–233)	203 (0–1430)

Table 26.4: Estimated Magnitude of the Extent of Trace Element Poisonings (Nriagu, 1988)

Elements	Global Emissions (1000 mg/year)			People Affected	Comments
	Air	Water	Soil		
Pb	332	138	796	>1 billion	Blood Pb > 20 μg/dL
Cd	7.6	9.4	22	500,000	Producing renal dysfunction
Hg	3.6	4.6	8.3	80,000	Certified Hg poisonings
As	18.8	41	82	> 100,000	Skin disorder and H_2O As > 2 μg/L

For trace element contaminated soils, Pb concentrations are generally much higher than Cd concentrations. Despite the higher Pb concentrations IN soil, plant uptake of Pb is very limited (a low bioavailability for plant uptake). Lead has a number of physiological effects on the human body and infant children are at much greater risk than adults. One such effect is impairment of mental development that causes children with chronic exposure to Pb to score lower. Hg acids also cause serious diseases. The Hg caused *Minamata* disease manifested itself in mental retardation, cerebral palsy and fetus mortality in the children of the exposed population. Similarly human health also affected from As. General symptoms include weakness, muscular aches, hearing loss in children and possibly skin cancer. Exposure most often occurs from drinking water, although soils that have received arsenical pesticides, such as orchards and area where vegetables were produced can be highly contaminated and on consumption become primary route of exposure. In plants trace elements cause phytotoxicity *i.e.* reduced yields or death of plants by substances in soil. Symptoms for trace elements induced phytotoxicities can include stunting, chlorosis, necrosis and death of plants. Trace elements most often associated with phytotoxicity are B, Cu, Ni and Zn leading to reduced both the quantity and quality of the food produced from the soil (Nriagu, 1988).

Groundwater Pollution

The water contained in the saturated layer below the water table is called groundwater. Soil acts as filter but its capacity is not unlimited. Moreover contaminants can easily move through a saturated flow to the groundwater thus adversely affecting its quality. Groundwater seeps from upland to lowland areas, eventually discharging in streams, lakes, ponds and wetlands. The rates of movement of groundwater vary depending upon size of cracks in rocks and size of the pores between soil and rock particles. Groundwater seldom moves hundreds of miles. Most precipitation which recharges groundwater moves only a few miles from the point of recharge to the point of discharge. Groundwater mostly stays within same runoff watershed surface. That means agricultural activities could contaminate groundwater. Untreated industrial waste effluents when discharged in unlined drains or low lying area can percolate down thereby affecting the quality of groundwater. Although a considerable portion of metallic cations (Cr, Cu, Pb, Cd, Ni etc.) are likely to be retained in the top soil, however, if the soil is light textured and rate of down movement of water is high, then migration of these pollutants to the saturated zone may occur. A study of groundwater of industrial area of Ludhiana revealed pH was as low as 2.0 (characteristic of strong acids). *Totawat* (1987) reported in Rajasthan, the well water near a Zn smelter plant had very high concentration of flouride and zinc.

Nitrate Pollution in Groundwater

Depending on the rate and method of application of fertilizer nitrogen, the kind of crop, soil and climatic variables, 35–60 per cent of the applied N is usually recovered by crop plants, 10–20 per cent may be converted to N_2 gas, nitrous oxide or ammonia and volatilize into the atmosphere and if there is excess water percolating through the soil, part of the remaining N may move below the root for leaching to groundwater. So the nitrogen fertilizer is applied carefully and properly. Certain circumstances could affect nitrogen leaching to the groundwater:

1. *Nitrogen rate*–at one pound of nitrogen per 1000 square feet, no leaching occurred.
2. *Nitrogen source*–slow release nitrogen source fertilizers can reduce the chance of leaching.
3. *Application timing*–in late fall, plants take up less nitrogen for a greater chance for leaching to occur.
4. *Irrigation practices*–the more irrigation, the greater the chances for nitrate leaching.
5. *Soil texture*–the sandier the soil the more chances for nitrate leaching.
6. *Age of site*–younger sites usually have less organic matter and need to be fertilized more therefore increasing the chance of leaching.

Problems Associated with High Nitrate Levels

When nitrate nitrogen concentrations reach excessive levels there can be harmful biological consequences for the organisms, which depend on groundwater. The United States Environmental Protection Agency established the current drinking water standards (Table 26.5) health advisory level of 10 mg/l nitrate nitrogen (equivalent to 10 ppm nitrate-nitrogen or 45 ppm nitrate) based on the human health risks due to nitrate consumption (Kross, 1993). The disease methemoglobinemia (also infant cyanosis or blue baby syndrome) is the resultant of high nitrate (above 10 ppm) concentrations (Kross, 1998).

Table 26.5: Maximum Permissible Limits of Nitrate in Potable Water Prescribed by Different Countries and Organizations (Lunkad, 1994)

Country/Organization	*Concentration as NO_3–N (mg/l)*	*Concentration as NO_3 (mg/l)*
WHO	10	45
US Environmental Protection Agency	10	45
ICMR (India)	10	45
Canada	10	45
Poland	10	45
EEC	11.30	50
Bulgaria	6.7	30
Belgium	11.30	50
Denmark	11.30	50
Finland	6.8	30
Hungary	9.0	40
UK	11.3	50
USA	10	45

Blue Baby Syndrome

Methemoglobinemia is the condition in the blood, which causes infant cyanosis or blue baby syndrome. Methemoglobin is probably formed in the intestinal tract of an infant when bacteria converts the nitrate ion to nitrite ion (Comly, 1987). One nitrite molecule then reacts with two molecules of hemoglobin to form methemoglobin. In acid mediums, such as the stomach, the reaction occurs quite rapidly (Comly, 1987). This altered form of blood protein prevents the blood cells from absorbing oxygen which leads to slow suffocation of the infant which may lead to death (Gustafsen, 1993; Finely, 1990). Because of oxygen deprivation, the infant will often take on a blue purple tinge in the lips and extremities, such as vomiting and diarrhoea, relative absence of dietress when severely cyanetic but irritable when mildly cyanotic and chocolate-brown coloured blood (Johnsen *et al.*, 1987; Comly 1987). The US Environment Protection Agency (EPA) has established a maximum contaminant level of 10 mg NO_3–N/l (45 mg NO_3/l) to protect the safety of US drinking water supplies. Animals can also be susceptible to methemoglobinemia, although the health advisory level for livestock is much higher, ~ 140 mg NO_3 –N/l (180 mg NO_3^-/l), the other major human health concern with N is the potential carcinogenic effect of nitrosamine compounds with the general chemical structure R_2–N–N = O, where R represents any carbon group.

Gastric Cancer

Nitrate, once produced form different derivatives, which can react with availability of organic compounds in the human body to form nitrosamines and nitrosamides through nitrosation reactions in the stomach. Nitrosamines and nitrosamides are carcinogenic as they can be metabolised to protect electrophilic alkylating agents (Archer, 1989). It is thought carcinogenic nitros compounds are involved in the etiology of some types of cancers in humans. Human oral cancer caused by tobacco chewing and snuff dipping and cancers related to smoking are thought to be results of nitrosamines in tobacco and tobacco products (Moncada, 1988; Hecht, 1989). Nitrosamines are formed by the reaction, under highly acidic conditions (pH < 4) of secondary and tertiary amines (R_2NH, R_3N) with nitrous acid anhydride (N_2O_3). Nitrosamines have produced tumors in laboratory animals, but conclusive evidence for a causative role of nitrosamines in human gastric cancer does not exist.

Other Health Effects

There are reports relating high nitrate intake and increasing incidence of goitre, because high nitrate reduces assimilation of iodine by human body causing goitre but is yet to be proved (ECETOC, 1988). It is also suggested that high nitrate intake with drinking water leads to the birth of a malformed child (Dorsche, 1984). But it is also not proved yet. There are reports of other health disorders namely non-Hodkin's lymphoma (Weisnberger, 1991), increased infant mortality (Super *et al.*, 1981) and hypertension (Malberg *et al.*, 1978).

Groundwater Contamination by Pesticides

Pesticides are classified according to the target organisms to be controlled. Weeds cause the greatest economic loss due to their interference with crop production and without the use of these chemicals, it is often very difficult for the agriculturist to produce significant yields of crop. Once the pesticide has been applied to a field, its movement is controlled by several processes like retention, transformation transport and plant uptake (Burner *et al.*, 1977).

Retention

Retention, commonly known as adsorption is the soil's ability to hold or adsorb, the pesticide on its surface. The soil characteristics like pH, moisture content, clay content, oxide content, cation exchange capacity, specific surface area and organic matter content etc affect the adsorption rate of a pesticide. Retention affects the characteristics of the pesticides for several reasons. First, adsorption provides the pesticide with a place that it can connect and force chemical transformations. These chemical changes can create very complex problems that may make remediation very difficult. Adsorption can also affect the pesticide's ability to be transported. The more pesticide adsorbed by the soil, the less that will be transported.

Transformation

When the pesticides enter the soil/groundwater, are degraded, transformed or stored by microorganism, plants and animals. The major transformations are photochemical, chemical and microbial processes. When pesticides are absorbed into the soil these affect the metabolic processes in the soil by affecting the metabolic processes in the soil microorganisms. Microbes are major controlling factor in the subsurface movement of pesticides because they breakdown the pesticide into H_2O and CO_2 molecules. The five basic processes involved in the microbial transformation of pesticides include biodegradation, cometabolism, polymerization or conjugation, accumulation and secondary effects of microbial activity. Chemical transformation takes place on pesticide molecules as soon as the pesticide enters the water and is affected by pH, buffering, general acid and base catalysts, temperature, dissolved organics and suspended solids, metal ions and the redox state of the water column. The type of soil to which the pesticide is applied, hydrolysis and redox reactions in sediments are the most important modes of transformations. Photochemical transformation is an important degradation pathway for many pesticides, especially the surface applied.

Transport

A pesticide can be transported in several different ways. The most common being mass flow and diffusion. Mass flow is when a pesticide is transported by a flow of water in which the pesticide has been retained and can transport the pesticide molecules very quickly over large distances. Mass flow in soil is a function of the rate of water movement and the soil's characteristics. Diffusion occurs in a much more random manner than mass flow. It occurs when a pesticide travels from an area of high concentration to an area of low concentration by way of random molecular movement. Diffusion is a much slower process than mass flow because of its randomness. It is dependent upon the soil characteristics and molecular structure of the pesticides.

Plant Uptake

Plant uptake is a process by which pesticides are transported into and within the plant structure. This process involves sorption by the roots of the plant and adsorption with subsequent movement to the plant's super surface structure. The solubility of the pesticide is the most important factor which governs the sorption and movement within the plants. The content of the surrounding soil is also important to the plant uptake. For non-polar pesticides the volume of organic factors such as pH, clay and microbial activity are more important as the polarity of the pesticide increases. The accumulation of pesticide through plant uptake can have severe consequences through the food chain if the pesticide is translocated into the section of the plant that will be subsequently harvested. Besides these pesticides, the soil characteristics and water volume also affect the risk of groundwater contamination (Byers and Fall, 1996). Contaminated groundwater by herbicides poses threat to water supplies used by public. Various studies showed that drinking water was found contaminated with several herbicides (Table 26.6). Most herbicide concentrations are observed in the range of 0.1–100 μg/l. Atrazine was the most commonly detected herbicide in groundwater in nearly all maize belt in US (Erickson and Lee, 1989).

	Low Risk	*High Risk*
Pesticide Characteristics		
Water solubility	Low	High
Soil adsorption	High	Low
Persistence	Low	High
Soil Characteristics		
Texture	Fine clay	Coarse sand
Organic matters	High	Low
Macropores	Few, small	Many, large
Depth of groundwater	Deep (100 ft or more)	Shallow (20 ft or less)
Water Volume		
Rain/Irrigation	Small volumes at infrequent intervals	Large volumes at frequent intervals

Table 26.6: Direct Pollution of Wells and Well Bearings with Herbicides

Herbicide	*No. of Fractions*	*Concentrations (μg/l)*
2,4-D	2	< 0.2–800
Dichlorprop	5	< 0.5–3800
2,4-Dichlorophenol	2	0.2–0.4
Diquat	2	0.2–0.4
Paraquat	1	< 3–50
Atrazine	2	< 0.5–5.0

Management Practices to Control Groundwater Pollution

Nutrient Management

Soil testing is basic foundation for fertilizer recommendations. Testing manures for nutrient content allows accurate crediting for fertilizer replacement. Split applications of N, fertigation, slow release fertilizers, injections and deep placement of volatile N fertilizers are directed toward improving synchrony between N availability and crop N uptake pattern. Use of a nitrification inhibitor on fine-textured soils where nitrogen is full applied may reduce leaching of nitrate-nitrogen. Adding nitrapyrin (N-serve) to applied nitrogen reduced nitrate leaching on an average by 10–15 per cent.

Integrated Pest Management

It is generally assumed that reduced pesticide use results in a reduced probability of groundwater contamination. Integrated pest management reduced unnecessary use of pesticides. The use of crop rotation than continuous corn is advised to reduce the need for corn rootworm insecticides and use of crop rotation with tolerant varieties could help to control plant diseases.

Conservation Tillage

Reducing tillage and retaining crop residues on the soil surface limit the runoff and overland flows that carry pesticides and nutrients out of the field. Reduction of runoff and erosion is accomplished by increasing infiltration of water. Increased infiltration particularly through earthworms formed macropores, offers a transport system to the subsoil for soil applied pesticides to follow. Conservation tillage methods are most important in controlling soil erosion on sloping land. Ridge tillage system on sandy soil also reduces groundwater pollution. Placing chemicals on and in the ridges and timing of application to the crop's growth cycle significantly reduced the amount that moves to groundwater. Reduced volume of run off water that leaves the field or farm will help. Thus keeping the chemicals in the field or trapping them in biologically active areas provide opportunity for microorganism to degrade the pesticides, eventually rendering them harmless. The practices that reduce surface losses of pesticides by increasing infiltration may actually increase the amount of pesticide that reach groundwater supplies.

Cover Crops

A cover crop such as small grain or legume may provide water quality benefits from several standpoints. The effectiveness of cover crops in controlling erosion is well documented and controlling erosion is an important component of surface water quality protection. Small grain cover crops have shown some efficiency at retrieving fertilized corn or vegetable crops. Cover crops may recover significant quantity of soil profile NO_3 to prevent N transport to groundwater.

Chemical Properties and Selection of Pesticide

The agricultural chemicals are most critical for producers on vulnerable complex task that must take into account the crop, the tillage system, target species and a host of other variables. Useless persistent chemicals. There is little rise of groundwater contamination by those chemicals that absorb strongly to soil particles except in very sandy areas. Those chemicals may be held near the soil surface by erosion control practices until they are broken down by natural processes. Chemical properties of the herbicides are important to consider when evaluating their potential to leach to the groundwater. The three most important characteristics of a pesticide that influence leaching potential are solubility in water, ability to bind with the soil (adsorption) and the rate at which it breaks down in the soil. High solubility (dissolves readily), low binding ability and slow breakdown help increase a pesticide ability to move to the groundwater. Among the frequently used herbicides that have a greater potential to leach and are labelled with groundwater advisories are those that contain acetochlor, alachlor, atrazine, sulfetrazone, disulfuron, acifluorfen, dimethenamid, chloransulam, carbofuran, flumetsulam, simazine, metribuzin, metalachlor and clopyralid.

Precautions for Irrigation

Chemigation refers to the application of fertilizers and pesticides through an irrigation system and is a management tool that has benefits and potential drawbacks for groundwater protection. Similarly fertigation a way of application of fertilizers with irrigation water particularly nitrogen application can be more carefully spread out in the vegetative growth period of grain crops, thereby minimizing the susceptibility of leaching. Chemigation system should be equipped with backflow–prevention devices as they reduce the threat of back siphoning undiluted chemicals into irrigation well. Backflow prevention devices are mandatory on irrigation systems that injects fertilizers and pesticides.

References

Brams, E., Anthony, W. and Witherspoon, L. 1989. Biological monitoring of an agricultural food chain: soil cadmium and lead in ruminant tissues. *J. Environ. Qual.*, 18: 317.

Byres Dan and Fall Dob Harris. 1996. *Organic Pesticide Contaminants in Groundwater.*

California Department of Food and Agriculture 1998. Development of risk based concentrations for arsenic, cadmium, and lead in inorganic commercial fertilizers, prepared for the California Department of Food and Agriculture by Foster Wheeler Environmental Corporation, Scramento, CA.

Chopra, P. and Magu, S.P. 1985. Effect of selected herbicides and city compost on the rhizospheric microflora of wheat and maize. *Indian J. Agron.*, 30(1): 5–9

Comly, H.H. 1987. Cyanosis in infants caused by nitrates in well water. *J. American Med. Assoc.*, 257: 2788–2792.

Debona, A.C. and Andus. L.J. 1970. Studies on the effects of herbicides on soil nitrification. *Weed Res.*, 10: 250–263.

Edward, C.A. 1971. *Pesticides and the Soil Fauna.* Report Rothamated Experiment Station, 1970: 194.

El-Shahawy, R.M., Mostafa, M.T., Retad, M.S. and Saleh, S.A. 1982. Response of cotton plant and soil microflora to cotoran Multi. *Ann. Agric. Sci., Mashtohar*, 18: 352.

Gustafson, D.I. 1993. *Pesticides in Drinking Water.* Van Nostrand Reinhold, New York, pp. 241.

Heydrich, M. and Fernandez, C. 1984. Charges in the composition of the actinomycete community in a soil treated with diuron. In. *Soil Biology and Conservation of Biosphere*, (Ed.) J. Szegi. Akademiai-Kiado, Budapest, Hungary, p. 209–215.

Hicks, R.J., Stotzky, G. and Van Voris, P. 1990. Review and evaluation of the effects of xenobiotic chemicals on micro-organisms in soil. *Adv. Appl. Microbiol.*, 35: 195–253.

Kross, B.C., Hallberg, G.R., Bruner, R., Cherryholmes, K. and Johnson, K.J. 1993. The nitrate contamination of private well water in Iowa. *American J. Pub. Hlth.*, 83: 270–272.

Kuhnelt, W. 1963. Soil inhating arthropoda. *Ann. Rev. Entomol.*, 8: 115–136.

Kulshrestha, G. 1993. Ecological implications of herbicide use in Agriculture. In: *Pesticides: Their Ecological Impact in Developing Countries*, (Eds.) G.S. Dhaliwal and Balwinder Singh. Commonwealth Publishers, p. 190–216.

Lunkad, S.K. 1994. Rising nitrate levels in groundwater and increasing N-fertilizer consumption. *Bhu Jal News*, p. 4–10.

Mengel, K., Kirkby, E.A., Kosegarten, H. and Appel, T. 2001. *Principles of Plant Nutrition.* Kluwer Academic Publishers, London, p. 383–390.

Miller, G.T. 1991. *Environmental Science: Sustaining the Earth.* Wadsworth Publishing, Belmont, CA.

Mishra, S.D., Prasad, D. and Dwivedi, B.K. 1989. Pesticide residue in soil and soil organisms. In: *Soil Pollution and Soil Organisms.* p. 17–29.

Nriagu, J.O. 1988. A silent epidemic of environmental metal poisoning. *Environ. Pollut.*, 50: 139–161.

Pope, P.E. and Holt, H.A. 1981. Paraquat influences development and efficacy of the mycorrhizal fungus. If Glomus fasciculatus. *Can. J. Bot.*, 59(4): 518–521.

Rao, V.R., Adhya, T.K. and Sethunathan, N. 1993. Effect of pesticides on soil health. In: *Pesticides: Their Ecological Impact in Developing Countries,* (Eds.) G.S. Dhaliwal and Balwinder Singh. Commonwealth Publishers, p. 112–130.

Totawat, K.L. 1987. Effect of treated effluent discharged from zinc smelter on field crops and well waters adjoining to effluent stream. In: *Tech. Report,* Sukhadia University Udaipur, p. 4–12.

Yee, D., Weinberger, P., Johnson, D.A and Dechain, C. 1985. *In vitro* effects of s-triazine herbicide prometryne, on growth of terrestrial and aquatic microflora. *Arch. Envir. Contam. Toxicol.,* 14 (1): 25–31.

Chapter 27

Sustainability through Environmental Conservation on Agricultural Land Resources

S.R. Singh and Th. Manimala Devi
Post Graduate Studies Centre, HRDRI, Canchipur, Imphal – 795 003

ABSTRACT

Sustainability of Thoubal district based on agricultural land explored about the planning development of the area spontaneously. Agricultural land resource, a unit of sustainable development accorded with 21496.50 hectares in 1991 and 33215.0 hectares in 2001 highlighted the provision view. In depth analysis, an ecologically productive land per capita in hectares (EPLC) compute 0.073127 in 1991 and 0.09066 in 2001; ecologically deficit land per capita (EDLC) in hectare as 1.926872 and 1.90933 in 1991 and 2001 respectively; ecologically deficit land per capita (EDLC) in per cent available in decades and in annum compute 2634.96 and 263.496 in 1991 and 2106.03 and 210.603 in 2001 respectively scrutinize the ways of strengthening the upshot planning and programmes in substantiation with full conservation of nature and environment.

Keywords: *Sustainability, Agricultural land, Ecologically productive, Ecologically deficit land per capita, Conservation of nature and environment.*

Introduction

The World Commission on Environment and Development, the Brundtland Commission in 1987 recommended sustainable development is the way to achieve global environmental security. More empathetically the term itself had been used before the Brundtland commission endorsed it, notably in the "World Conservation Strategy', jointly prepared by the World Conservation Union (IUCN), the United Nations Environment Programme (UNEP), the World Wide Fund for Nature (WWF) in 1980. (Ruth and William, 2000).

The World Commission on Environment and Development in 1993 was created, published "Our Common Future Report" in which it stated that sustainable development is a process of change in which it defines sustainable development as "Development that meet the needs of the present without compromising the ability of future generations to meet their own needs."

Moreover, for practical approach "Sustainable development is a process of change in which the exploitation of resource, the direction of investments, the orientation of technological and institutional change are all in harmony and enhance both current and future potentials to meet human needs and aspirations".

Agenda 21 emphasizes that sustainable development will reverse poverty and environment degradation thus provides a plans for combating degradation of land, air, water and for conserving forests and biological diversity (Tewari, 1994).

The Earth Summit held in Rio de Janeiro in 1992 also targeted towards the sustainable development. However, no such investigation have been worked out in Manipur. The present study implies on the multifacet meaning of the sustainable development based on the following objective to determine the sustainable development on the capabilities of agricultural area based on EPLC in hectares, EDLC in hectares, EDLC in percentage available in decades and annum etc.

Materials and Methods

Surveys being to produce information to desirable, compare and predict attitudes, opinions, values and behaviour based on what people say or see and what is contained in records about them and their activities the present survey's aim is fixed to the goal (Fink, 1995; Yadava, 1995). During the present work samples in surveys was adopted according to convenient for the data mainly to probability sampling like simple random sampling, stratified random sampling, systematic sampling and cluster sampling whereas non-probability sampling like convenience sampling, snowball sampling, quote sampling and focus groups. Then measure survey and their reliability and validity was checked adopting reliability *i.e.* test-retest, alternate-form, internal consistency, interobserver and validity *i.e.* face, content, criterion and construct. All the criterion include concurrent and predictive. The construct also include convergent and divergent.

According to P.V. Young, data has categorised as documentary sources and field sources. Documentary sources include material already collected whether published or unpublished. Among documentary sources the first place may be given to the books written by different experts. And reports of surveys, memories, accounts of travels, historical accounts, official published data and other unpublished record are also belongs to documentary sources. Field sources including living persons, scholars, scientists, research workers, leaders who have worked with the social group or have studied the problem.

For the determination of sustainable development on the capabilities of agricultural land of the Thoubal district under mention formulae have adopted following Rees, 1992.

1. Ecologically productive land per capita, EPLC (in Ha) = $\frac{\text{TA (in Ha)}}{\text{TP (in nos.)}}$

where,

EPLC: Ecologically productive land per capita in hectares.

TA: Total area in hectares.

TP: Population in numbers.

2. Ecologically deficit land per capita (EDLC) in hectares

   EDLC = FP – EPLC in hectares

   FP: Foot print = 2 Ha since India belongs to (2–3) Ha foot print.

   EPLC: Ecologically productive land per capita in hectares.

3. Ecologically deficit land per capita (EDLC) in per cent available in decades.

$$\text{EDLC in \% available in decades} = \frac{\text{EDLC (in Ha)}}{\text{EPLC (in Ha)}} \times 100$$

4. Ecologically deficit land per capita (EDLC) in per cent available in annum.

$$\text{EDLC in \% available in annum} = \frac{\text{EDLC (in Ha)}}{\text{EPLC (in Ha)}} \times 10$$

Results and Discussion

The population of Thoubal district in 1991 recorded 293958 persons and 21496.50 hectares in agricultural area. The calculated ecologically productive land per capita accord 0.073127 and ecologically deficit per capita as 1.926872. The percentage available of ecologically deficit per capita in decades and annum was 2634.96 and 263.496 respectively. The district recorded a population of 366341 persons and 33215 hectares in agricultural area during 2001. The calculated value of ecologically productive land per capita accord 0.09066 hectares and ecologically deficit land per capita in hectares compute as 1.90933. The ecologically deficit land per capita in percentage available in decades and annual was computed as 2106.03 and 210.603 respectively (Table 27.1). The illustrated data were plotted in Figure 27.1.

Table 27.1: Determination of Sustainability on Agricultural Area for Developmental Planning of Thoubal District and Manipur State

District/ State	*Year*	*Total Population (TA) in Nos.*	*Total Agricultural Area (TAA) in Ha*	*Ecologically Productive Land per Capita (EPLC) in Ha*	*Ecologically Deficit Land per Capita (EDLC)*			*Remarks*
					in Hectares	*in % Available in Decades*	*in% Available in Annum*	
Thoubal	1991	293958	21496.50	0.073127	1.926872	2634.96	263.496	
	2001*	366341	33215.0	0.09066	1.90933	2106.03	210.603	Assuming 2 Ha as footprint
Manipur	1991	1837149	148572.62	0.080871	1.919128	2373.06	237.306	
	2001	2388634	155232.0	0.06498	1.93502	2977.87	297.787	

Sources: 1. Statistical abstract of Manipur, 2001; 2. Directorate of Remote Sensing, Manipur; 3. Department of Agriculture, Manipur; 4. Provisional Population.

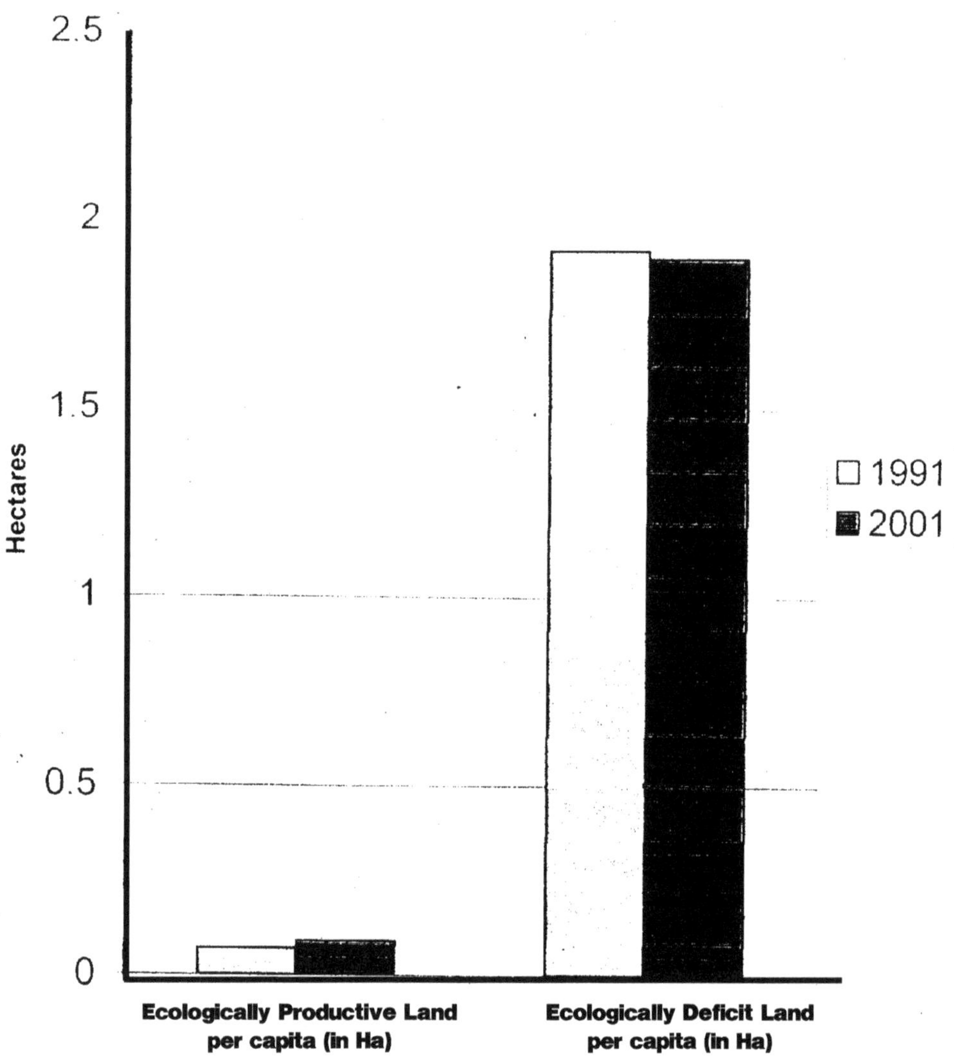

Figure 27.1: Sustainability Development on Agricultural Area of Thoubal District

Manipur state recorded 1837149 persons of population and 148572.62 hectares of agricultural area during 1991. The calculated value of ecologically productive land per capita in hectare accord 0.080871 and 1.919128 hectares as ecologically deficit land per capita. The population of Manipur state in 2001 recorded 2388634 persons and 155232 hectares in agricultural land. The calculated value of ecologically productive land per capita in hectares accord 0.06498 and ecologically deficit land per capita in hectare as 1.93502. The percentage available of ecologically deficit land per capita in decades and in annum accord 2977.87 and 297.787 respectively (Table 27.1). The exemplified data were plotted in Figure 27.2.

Table 27.1 revealed the increased area of cultivable land in the district with an area of 11718.5 hectares in year 1991 year than that of the year 2001 and increased area of agricultural land in the

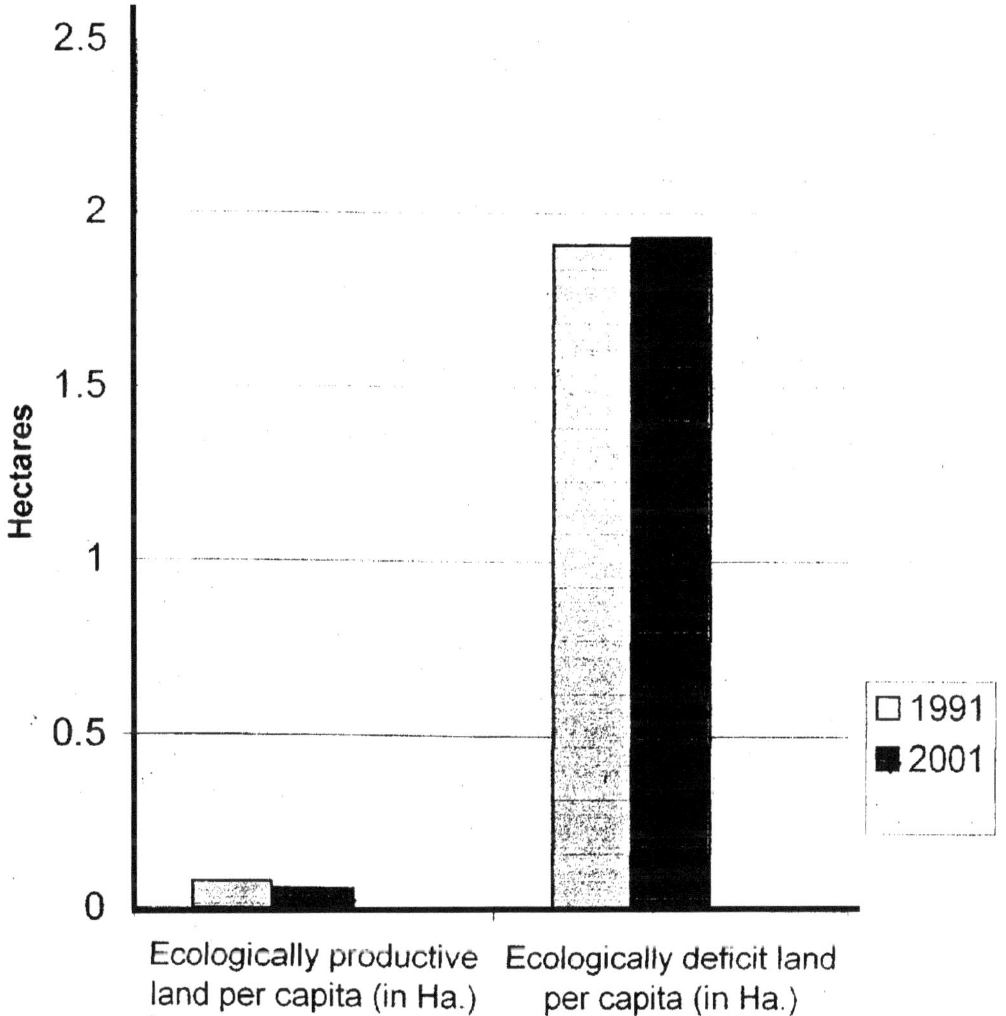

Figure 27.2: Sustainability Development on Agricultural Area of Manipur State

Manipur state. The district accord the area of cultivable land 21496.50 hectares, the ecologically productive land per capita in hectares 0.073127, ecologically deficit land per capita in hectare 1.926872, ecologically deficit land per capita in percentage available in decade 2634.96, and ecologically deficit land per capita in percentage available in annum as 263.496 during 1991. During 2001, the district observed the total area of cultivable land of 33215.0 hectares, ecologically productive land per capita 0.09066 hectares, ecologically deficit land per capita in hectare 1.90933, ecologically deficit land per capita in percentage available in decade 2106.03 and ecologically deficit land per capita in percentage available in annum 210.603.

Manipur observed 148572.62 hectares of cultivable land in 1991 with a calculated value of ecologically productive land per capita in hectare as 0.080871, ecologically deficit land per capita in hectare as 1.919128, ecologically deficit land per capita in percentage available in decade as 2373.06 and ecologically deficit land per capita in percentage available in annum as 237.306.

During 2001, the state record the population of 2388634 persons and agricultural land 155232.0 hectares. The computed value of ecologically productive land per capita in hectares as 0.06498, ecologically deficit land per capita in hectares as 1.93502, ecologically deficit land per capita in percentage available in decades as 2977.87 and ecologically deficit land per capita in percentage availab1e in annum as 297.787.

The increase area in agricultural land naturally deduct from the natural land area of the district mainly from forest, scrub, wasteland etc. The situation is compelled by human being for producing more food production for feeding the ever increasing population. The present analysis clearly showed that the population pressure have great impact on the environment as nature keep a vast area of natural vegetation or resources to maintain the stability of ecological balance in an area. It maintains the system of all ecological components in a very well manage living system in that particular area for restoration of lifelines of any disturbance comes to that system nature suffers a lot of problems to maintain them and took many years in restoring the system sometimes it may not be in such a situation that enable to maintain them in that state the ecological balancing system fails and caused the eco-disturbances and consequence to unwanted environment. Thus, invite the special needs of human kind to check the consistent increase of disturbing ecological balance and destroying the natural resources to a state of newer restoring back. The investigating area highlighted the consciousness of mankind towards the right use of natural resources and immediate action of maintenance. The finding was in collaboration with the results of other workers in different places on earth.

Kircher *et al.* (1984) suggested land tenure problems on the lowland plains and technical difficulties/prejudices in the extensive low hilly areas have directed the in-migrants on the steep, often mountainous land to practice their "Kaingin" or shifting cultivation, resulting in accelerating destruction of the forests and an inevitable degradation of natural resources in all parts of the island and its adjoining marine waters (the 'present trends' scenario).

Sustainable development is underlined by an ethical norm focussing on a new relationship of co-operative and co-existence of humans with nature as compared to the prevailing conquest and exploitation of nature by humans (Batie, 1989). For the current practitioners of agricultural science, including agricultural economists, many of the concepts and tools employed are rooted in technically efficient resource development (or utilization). But this quality of agricultural systems, although definable and amenable to measurement, has been incriminated as inadequate or inappropriate for a sustainable perspectives that embodies a sense of ecological connections between action of humans and the consequences for them.

Acknowledgements

The authors thank to the Head, Department of Economics and Statistics, Remote Sensing and Agriculture, Government of Manipur, for supplying the required data.

References

Arlene, F., 1995. *How to Ask Survey Questions*? Sage Publications India Pvt. Ltd.

Batie, S., 1989. Sustainable development: Challenges to the profession of agricultural economics. *Amer. J. Agric. Econ.*, 71: 1083–1101.

Kirchner, J., Leduc, G., Goodland, R. and Drake, J., 1984. *Carrying Capacity Population Growth and Sustainable Development*. Population and Development Series. Washington, D.C., The World Bank.

Rees, W., 1992. Ecological footprints and appropriated carrying capacity: What urban economics leaves out. *Environment and Urbanization*, 4(2): 121–130.

Ruth, A.E. and William, R.E., 2000. *The Environment Encyclopedia,* Vol. 9. Salmonella Tropical Dry Forest.

Tewari, D.N., 1994. ICFRE and sustainable development. *Indian Forester*, November.

Yadava, S.S. and Yadava, K.N.S., 1995. *Statistical Analysis for Social Sciences*. Manak Publications.

Chapter 28

Antibacterial Activity of Plant Extracts on Different Human Pathogens

Mamta Rawat
Department of Zoology, J.N.V. University, Jodhpur – 342 005, Rajasthan

ABSTRACT

Because of the appearance of bacterial resistant to antimicrobial agents, more effort is being made to find alternative antimicrobial component, so, the natural products are more preferred to the synthetic ones. In this series, the antibacterial activity of plant extract was evaluated against gram-negative organisms, *Escherichia coli, Klebsiella pneumoniae, Pseudomonas aeuroginosa, Proteus* spp. and Gram-positive Cocci, *Staphylococcus aureus.* The extracts from following plants were used *Nerium indicum, Acacia nilotica, Zizipus glabrata, Ocimum americanum, Ficus bengalensis,* and *Mimosa hamata.* The extracts were prepared ether and distill water. The test was performed using nutrient agar. The results showed that methanol and acetone extracts were most effective and aqueous extract were the least.

To sum up, these extract showed different inhibitory effect against all tested strains.

Keywords: *Antibacterial activity, Pathogens, Plant extracts.*

Introduction

Inspite of production of a number of new antibodies during last three decades by pharmacological industries, which are resident to the drugs by microorganisms to drugs are being utilized as therapeutic agents. (Cohen, 1992 and Nascimento *et al.*, 2000).

Several microorganism-derived antibiotics are currently in use to treat a variety of infectious human disease. Most of them have a limited antibacterial species of the pathogen; some even lead to serious side effects. Therefore actions must be taken to reduce this problem, for *e.g.*, to control the use of antibiotic, develop research to better understand the genetic mechanism of research and its continue studies to develop new drugs, either synthetic or natural. The ultimate goal is to offer appropriate and efficient antimicrobial drugs to the patient. Efforts are thus being directed to identify antibiotic sources other than traditional microorganisms.

For a long period of time, plants have been a valuable source of natural products for maintaining human health, especially in the last decade, with more intensive studies for natural therapies. India has a rich tradition in use of medicinal plants to develop drugs from plants. The use of plant compound for pharmaceutical purposes has gradually increased here. Says WHO, medicinal plants would be the best source to obtain a variety of drugs. Medicinal and aromatic plants constitute a major source of natural organic compound and serve as the immediate source of drugs (Dobhal, 1994). About 80 per cent of individuals from developed countries use traditional medicine, which has compound derived from medicinal plants. The ayurvedic system of medicine has descriptions of thousands of plants for the management of curable diseases (Tripathy, 1996). Therefore, such plants should be investigated to better understand their properties, safety and efficiency.

In this present investigation various solvent extracts from various plants were screened for their antibacterial activity against selected human pathogens.

Materials and Methods

Plant Material

Leaves of different plants were collected from the different localities; Rajasthan, India and plant were identified.

Preparation of Extract

For the evaluation of the antibacterial effect of plants against selected human pathogens, acetone, methanol, petroleum ether and aqueous of leaf of each plant were extracted. For extraction, fresh leaves were collected, washed, air-dried individually in shade, powdered separately and stored at temperature range of 15–20°C. The materials were extracted with above solvents using soxhlet apparatus. The extracts were concentrated in a rotary vacuum evaporator.

Application of Extract

The test organisms were obtained from the institute of microbial technology (IMTECH), Chandigarh, India and maintained on nutrient agar. The Kirby-Bauer disk diffusion method (Bauer *et al.*, 1966) was used for testing antibacterial activity. The media (25 ml) inoculated with suspension of each test organisms was poured into sterilized petty dishes and left to gel at room temperature. Whatman's no. 1 filters paper discs (5 mm diameter) were impregnated with 25 mg of each solvent extract individually. The filter paper discs were placed equidistantly on inoculated media and diffusion of solution was allowed to occur for 30 min at room temperature. After incubation at 37°C for 24 hrs the diameter of zones of inhibition in the bacterial growth were measured to detect and quantify the antibacterial activity. Microorganisms, which showed a zone of inhibition > 8 mm were sensitive.

Results and Discussion

The data pertaining to the antimicrobial potential of plant extracts in acetone, methanol, petroleum, ether and distil water are presented in Table 28.1–284.4 respectively.

Table 28.1: Antibacterial Activity Caused by the Plant Extract (in Acetone) through Agar Diffusion Method

Microorganism	*Escherichia coli*	*Klebsiella pneumoniae*	*Pseudomonas aeuroginosa*	*Proteus Species*	*Staphylococcus aureus*
Nerium indicum (Kaner)	–	+	–	–	+
Acacia nilotica (Babul)	–	–	+	–	+
Zizipus nummularia (Ber)	–	+	–	–	+
Ocimum Americana (Jungli tulsi)	+	+	+	+	–
Ficus bengalensis (Bergad)	–	–	–	–	–
Mimosa hamata (Alaili)	+	–	+	+	–

(+) Susceptibility (inhibition zone ≥ 7 mm)

(–) Absence of susceptibility.

Table 28.2: Antibacterial Activity Caused by the Plant Extract (in Methanol) through Agar Diffusion Method

Microorganism	*Escherichia coli*	*Klebsiella pneumoniae*	*Pseudomonas aeuroginosa*	*Proteus Species*	*Staphylococcus aureus*
Nerium indicum (Kaner)	–	–	–	–	–
Acacia nilotica (Babul)	–	+	+	+	–
Zizipus nummularia (Ber)	–	–	+	–	–
Ocimum Americana (Jungli tulsi)	+	+	–	–	+
Ficus bengalensis (Bergad)	–	–	–	–	–
Mimosa hamata (Alaili)	+	+	–	–	–

(+) Susceptibility (inhibition zone ≥ 7 mm)

(–) Absence of susceptibility.

Table 28.3: Antibacterial Activity Caused by the Plant Extract (in Petroleum Ether) through Agar Diffusion Method

Microorganism	*Escherichia coli*	*Klebsiella pneumoniae*	*Pseudomonas aeuroginosa*	*Proteus Species*	*Staphylococcus aureus*
Nerium indicum (Kaner)	+	+	+	–	–
Acacia nilotica (Babul)	–	–	–	–	–
Zizipus nummularia (Ber)	–	–	–	–	–
Ocimum Americana (Jungli tulsi)	–	+	+	–	–
Ficus bengalensis (Bergad)	–	–	–	–	–
Mimosa hamata (Alaili)	–	–	–	+	–

(+) Susceptibility (inhibition zone ≥ 7 mm)

(–) Absence of susceptibility.

Table 28.4: Antibacterial Activity Caused by the Plant Extract (in Petroleum Ether) through Agar Diffusion Method

Microorganism	*Escherichia coli*	*Klebsiella pneumoniae*	*Pseudomonas aeuroginosa*	*Proteus Species*	*Staphylococcus aureus*
Nerium indicum (Kaner)	–	+	+	–	–
Acacia nilotica (Babul)	–	–	–	–	–
Zizipus nummularia (Ber)	–	–	–	–	–
Ocimum Americana (Jungli tulsi)	+	+	–	–	–
Ficus bengalensis (Bergad)	–	–	–	–	–
Mimosa hamata (Alaili)	–	+	+	–	–

(+) Susceptibility (inhibition zone ≥ 7 mm)

(–) Absence of susceptibility.

The extracts from various plants species (*Nerium indicum, Acacia nilotica, Zizipus rummularia, Ocimum americana, Ficus bengalensis* and *Mimosa hamata*) presented; antimicrobial activity to at least one of the tested microorganisms. The extracts from *Ocimum americana* (Jungli tulsi) presented highest activities *i.e.* they were able to inhibit highest percentage of microorganism.

Whereas the extract from *Ficus bengalensis* did not show any antimicrobial activity. Rajendhran *et al.* (1998) reported that *E. coli* was resistant to acetone extract of *Ocimum* which is similar to present findings. The acetone fraction of plant extract was found to be most effective of all and distill water extract to be the least.

K. pneumoniae, the common nosocomial pathogens is resistant to many antibiotics such as ceftazidine and gentamicin, the infections caused by them not only require expensive antibiotic treatment but also increase the morbidity and mortality in patients. This bacterial strain was observed most sensitive against all solvents of plant extract.

E. coli, resistant to ampicillin, sulphonamides, tetracyclines and trimethoprim, was observed most effective to *Ocimum americanum* followed by *Mimosa hamata* and *Nerium indicum* respectively.

The variable susceptibility of microorganism to plant extracts was reported by several authors. Ramasamy *et al.* (2001) studies the effects of plants namely *Anisometes indica, A. malabreca, Blumea lacera* and *Melio azadirachta* on different human pathogens. Apart from this, different work on various indigenous plants of various regions of India have been recently reported by Darokar *et al.* (1998), Rajendaran *et al.* (1998), Dushyent and Bohra (2000).

From this, it is obvious that all plant extracts that were selected are having antibacterial activity with difference in their magnitudes. But the effective biomolecules which act as antibacterial substance have to be identified, isolated and subjected to extensive scientific and pharmacological screening that can be used as sources of new drugs.

Finally the study expresses that plant extracts have great potential as antimicrobial compounds against microorganisms. Thus, they can be used in the treatment of infectious disease caused by resistant microbes. Also the present study support the use of alcoholic and acetone extract of different plant extracts.

We are trying to extend the study of inhibitory effect of these plants extract via their chemical compositions. Also the antibiotic resistant species of bacteria, will be practiced against synergistic effect from the association of antibiotic with plant extract.

This effect will enable use of respective antibiotic when it is no longer effective by itself during therapeutic treatment.

References

Bauer, A.W., Sherris, T.M. and Kirly, W.H.M., 1966. Antibiotic susceptibility testing by a standardized single disk method. *American Journal of Clinical Pathology*, 45: 493.

Cohen, M.L., 1992. Epidemiology of drug resistance: Implications for a post-antimicrobial era. *Science*, 257: 1050–1055.

Darokar, M.P., Mathur, A., Dwivedi, S., Bhalla, R., Khaniya, S.P.S. and Kumar, S., 1993. Detection of antibacterial activity in the floral petals of some higher plants. *Current Science*, 76: 187–189.

Dobhal, R., 1994. Medicinal and aromatic plants. *Current Science*, 71: 397–398.

Dushyent, G. and Bohra, A., 2000. Toxic effect of various plant part extracts on the causal organism of typhoid fever. *Current Science*, 78: 780–781.

Nasciment G.G.F., Locatelli, J., Freitas, P.C., Silva, G.L., 2000. Antibacterial activity of plant extracts and phytochemicals on antibiotic-resistant bacteria. *Brazilian Journal of Microbiology*, 31(4): 285–299.

Rajendhran, J., Mani, M.A. and Navaneetha Kannan, K., 1998. Antibacterial activity of some selected medicinal plants. *Geobios*, 25: 280–282.

Ramasamy, S., Chales Manoharan, A. and Rajendran, J., 2001. Antibacterial effect of various plants extracts on the human pathogens. *Indian Journal of Environment and Ecoplanning*, 5(2): 317–319.

Tripathy, Y.B., 1996. Research strategy and ayurveda. *Current Science*, 73: 418.

Chapter 29

Recovery of Membrane Bound Enzymes in Soleus and EDL Muscles of Palm-Squirrel after Experimentally Induced Sciatic Neuropathy

K. Pratap Reddy and K. Praveen Kumar*
Neurobiology Lab, Department of Zoology, University College of Science, Osmania University, Hyderabad – 500 007, India

ABSTRACT

This study reports the time course changes of weight and activities of membrane bound enzymes, acetyl cholinesterase (AChE), $Na^{+}K^{+}$ ATPase, Mg^{++} ATPase and Ca^{++} ATPase in tonic soleus and phasic Extensor digitorum longus (EDL) muscles of palm squirrel at 1,4,8,15,30,60 and 120 days after experimentally induced peripheral (sciatic) mono neuropathy. The muscles showed distinct and characteristic progressive atrophy from day 1 to 60, the rate of decrease is being more in soleus (63 per cent) than EDL (44 per cent), followed by recovery of weights to control levels at day 120. The AChE activity of soleus increased progressively up to day 60 (156 per cent) with reversal to control on day 120, in contrast to decrease of AChE in EDL up to day 60 (71 per cent) with recovery to control on day 120. The Na^{+} ATPase activity shows a contrasting pattern in both muscles to that of AChE. The Ca^{++} ATPase was not altered up to day 60 in soleus and EDL with the decreased activity in EDL on day 120. Mg^{++} ATPase was not significantly altered in soleus while in EDL it gradually decreased (35 per cent). The results show that the time dependent and muscle specific differential reversal of atrophy, transport ATPases and AChE activities in soleus and EDL over 120 days after mononeuropathy.

Keywords: *Mononeuropathy, ATPases, AChE, EDL, Soleus, Sciatic nerve.*

* Corresponding Author: E-mail: pratapkreddy@rediffmail.com.

Introduction

Neuropathy represents a pathological state of the nervous system with the potential to interfere with normal end organ function and metabolism. The neuropathy is initiated by primary lesion of nerve due to trauma and neuroma or dysfunction of the nervous system or chronic diseases like diabetes, post therapeutic (viral) neuralgia or rheumatoid inflammation (Bridges *et al.*, 2001). An animal model of peripheral mono neuropathy in the rat has been described which has pain sensation reminiscent of those seen in man (Attal *et al.*, 1990; Bennet and Xie, 1983; Jazat *et al.*, 1998; Xie and Xiao, 1990). The end organs like muscles exhibit corresponding responses in neuropathies, such alterations in nerve physiology may have impact on muscle metabolism and manifest as quantitative enzymatic changes (Korr, 1981). The different type of muscles exhibit characteristically varied enzymatic responses during disease or experimental denervation and reinnervation (Gutamam, 1976). There are no discernable literature on the responses of fast phasic and slow-tonic type of muscles after neuropathic pain in experimentally produced mono neuropathy with sciatic nerve ligature, which mimics the human neuropathies. Therefore, this study is designed to report the time course alterations in atrophy, membrane transport ATPases, and acetyl cholinesterase activities in slow tonic soleus and fast phasic EDL of Indian palm squirrel with experimentally induced sciatic mononeuropathy over 120 days.

Materials and Methods

Adult male Indian palm-squirrels, *Funambulus palmaram* were used in all experiments. The animals were housed in groups of 6 in standard laboratory cages. The animals were anesthetized with sodium pentobarbital (40 mg/kg) and mono neuropathy was induced as described by Bennet and Xie (1983). The common sciatic nerve was exposed at the level of middle-thigh region and about 1 cm nerve was freed of adhering tissue and 4 ligatures (4.0 chromic gut) were tied loosely around it with about 1.5 mm spacing. An identical dissection was performed in each animal on the contra lateral side except that the sciatic nerve was not ligated. The animals were returned to cages and maintained in aseptic conditions for selective time intervals up to 120 days.

The animals were sacrificed at 1, 4, 8, 15, 30, 60 and 120 days and the soleus as well as EDL were dissected out for experimental analysis. The ligated limb was not in proper use up to 30 days, after which it had shown sign of usage and its usage was almost normal at 120 days. The rate of atrophy of muscles (dry-weight) was obtained by allowing the muscles in a dessecator containing hygroscopic P_2O_5 at 60°C for 24 h. The acetyl cholinesterase activity was assayed according to Ellmann's method (1961). The muscles were homogenized in phosphate buffer containing EDTA, Triton X100 and centrifuged at 16,000 rpm to obtain crude membrane fraction which was used for enzyme assay. The Na^+K^+ ATPase and Mg^{++} ATPase activities were estimated as described by Kaplay (1979). The muscles were homogenized in sucrose buffer containing Histidine and centrifuged at 16,000 rpm to obtain crude membrane preparation. The Na^+K^+ ATPase and M^{++} ATPase activities were assayed by using ouabain, which inhibits the formers activity. The Ca^{++} ATPase activity was assayed as described by Samaha and Yunis (1973). The phosphorous content in ATPases activity was estimated as described by Tasuaky and Shorr (1953) and protein content in enzyme preparations was estimated as described by Lowry method (1951).

Results and Discussion

As shown in Figure 29.1, the weights of soleus and EDL were progressively decreased up to day 60 but the rate of the decrease was more in former. The weights of both muscles were however recovered to that of control muscle on day 120. The AChE activity (Figure 29.2) in soleus was not significantly

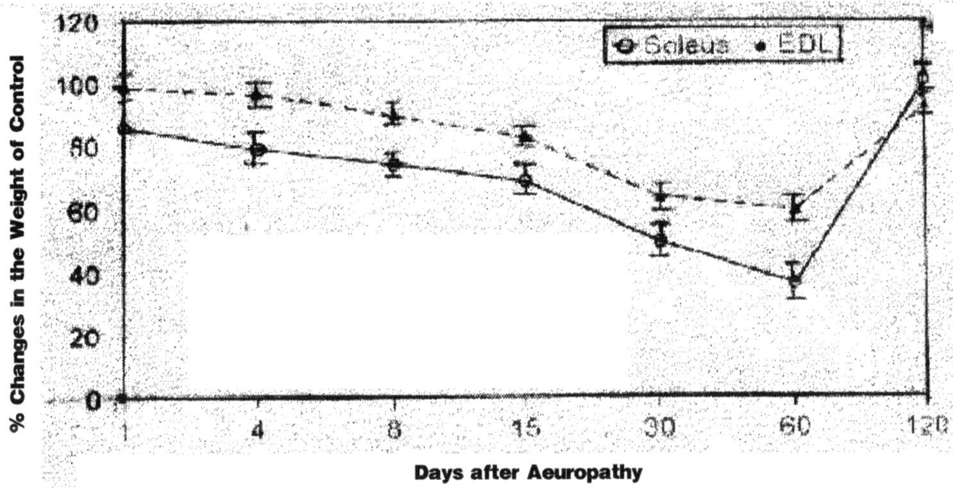

Figure 29.1: The Change in Dry Weights of Soleus and EDL Muscles after Experimentally Induced Sciatic Mononeuropathy

The dry weights of muscle calculated as mg of muscle/animal. The numbers of the ordinate represent per cent of control value and on abscissa represent time in days after neuropathy. Each point represent mean ÷ S.D of 6 animals. The dry weights of soleus and EDL are 12.01÷2.0 and 19.5÷2.05 mg of muscle/animal respectively.

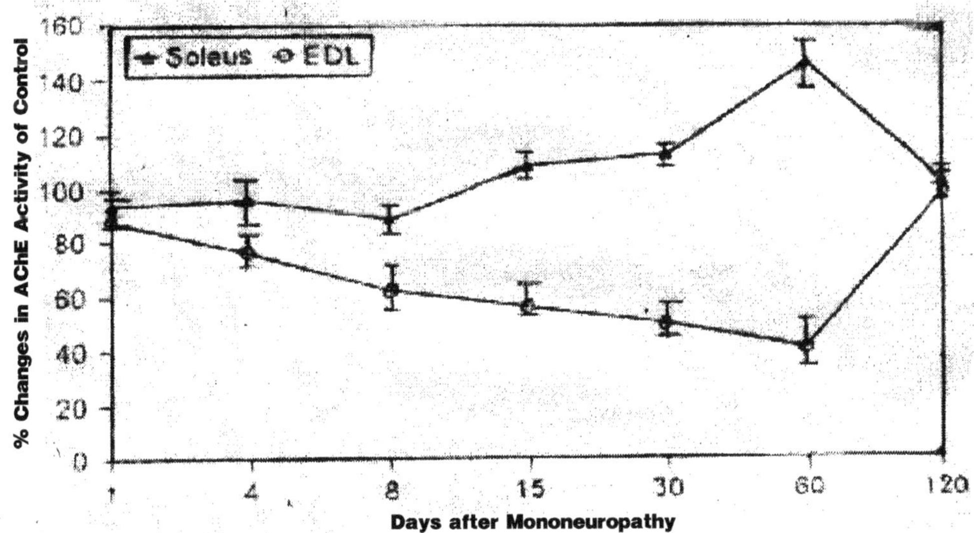

Figure 29.2: The AChE Activity of Soleus and EDL Muscle of Mice During Experimentally Induced Sciatic Mononeuropathy

The Acetylcholinesterase activity is calculated as nano moles of AChE hydrolyzed mg of muscle/ min. Numbers of the ordinate represent per cent of control value and on abscissa represent time in days after neuropathy. Each point represent mean÷S.D of 6 animals. The AChE activity in control soleus and EDL is 0.4642÷0.032 and 0.7972÷0.024 n moles of AChE/mg of muscle/min respectively.

altered from control up to day 8 followed by gradual increase upto day 60 with reversal to control value on day 120. The EDL exhibited progressive decrease of AChE activity upto day 60 with reversal to control value on day 120. Na^+K^+ ATPase (Figure 29.3) activity was increased on day 1 in soleus, with reversal to control value upto 15 with progressive decrease up to day 60 followed by reversal to control on day 120. In contrast, the enzyme activity increased (22 per cent) in EDL on day 60 followed by reversal to control on day 120. Mg^{++} ATPase activity (Figure 29.4) showed insignificant decrease from control up to day 60 with reversal to control value. The EDL showed elevated level on day 4 followed by progressive decrease. The Ca^{++} ATPase activity (Figure 29.5) was not altered upto day 15 followed by progressive decrease in both muscle up to day 60. However, soleus showed recovery on day 120 in contrast to decreased activity in EDL (35 per cent).

The time dependent and muscle specific alterations and reversal of weight and membrane bound enzyme activities of soleus and EDL of palm squirrel to normal levels over 120 days were not reported earlier are in line with behavioral and neuroanatomical evidence of peripheral mono neuropathy found in rat (Attal *et al.*, 1990; Xie and Xiao, 1990; Jazat *et al.*, 1998; Gautron *et al.*, 1990; Sugimoto *et al.*, 1989). The recovery of weight and AChE in soleus and EDL in long term reinnervation after single crush in rat (Misulis and Dettbarn, 1985) and palm-squirrel (Reddy, 1994) were reported. The observed elevated AChE activity in soleus and sudden decrement in EDL up to 60 days may be due to conversion of slow fibers to fast type of fibers but not vice versa (Guth Brown, 1965), conversion of slow muscle to neonatal state which dominated by fast fibers (Rubinstein *et al.*, 1978) or preferential faster reinnervation of slow muscle than fast muscle (Guth *et al.*, 1965) and reversal of activities to control values during further period (120 days) may be related to reestablishment of neural connection with simultaneous elimination of hyper neuralization (Misulis and Dettbarn, 1985). In addition the differential pattern of AChE in soleus and EDL may be due to characteristic neural regulation of muscle specific AChE molecular forms in soleus and EDL (Groswald *et al.*, 1985).

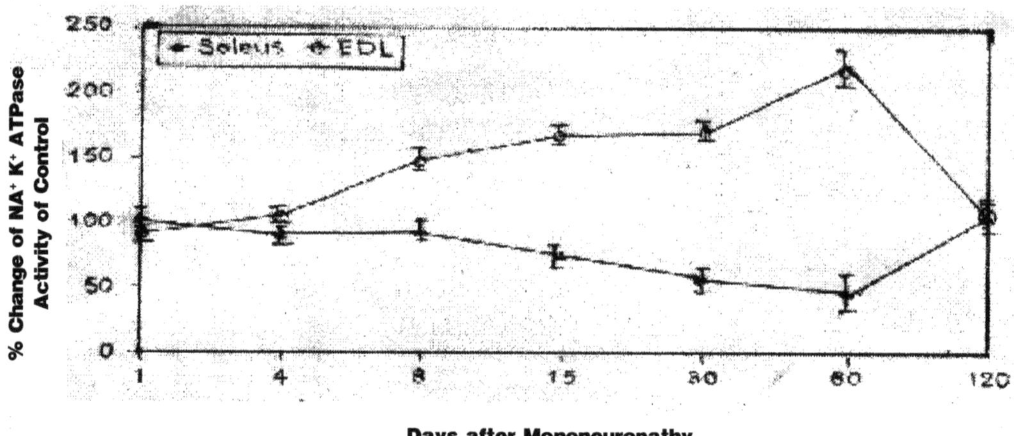

Figure 29.3: The Na^+K^+ ATPase Activity of Soleus and EDL Muscle of Mice During Experimentally Induced Sciatic Mononeuropathy

The Na^+K^+ ATPase activity is calculated as n moles of Pi/mg protein/hr. Numbers of the ordinate represent per cent of control value and on abscissa represent time in days after neuropathy. Each point represent mean÷S.D of 6 animals. The Na^+K^+ ATPase activity in control soleus and EDL is 116÷4.30 and 92÷5.70 n moles of Pi/mg of muscle/min respectively.

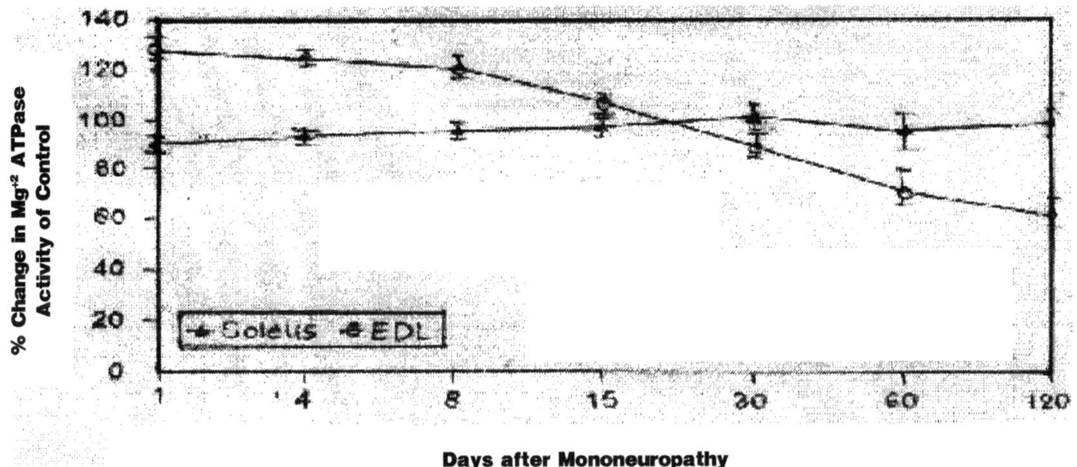

Figure 29.4: The Mg^{2+} ATPase Activity of Soleus and EDL Muscle of Mice During Experimentaly Induced Sciatic Mononeuropathy

The Mg^{2+} ATPase activity is calculated as n moles of Pi/mg protein/hr. Numbers of the ordinate represent per cent of control value and on abscissa represent time in days after neuropathy. Each point represent mean÷S.D of 6 animals, The Mg^{2+} ATPase activity in control soleus and EDL is 828÷24.0 and 1415÷23.0 n moles of Pi/mg of muscle/min respectively.

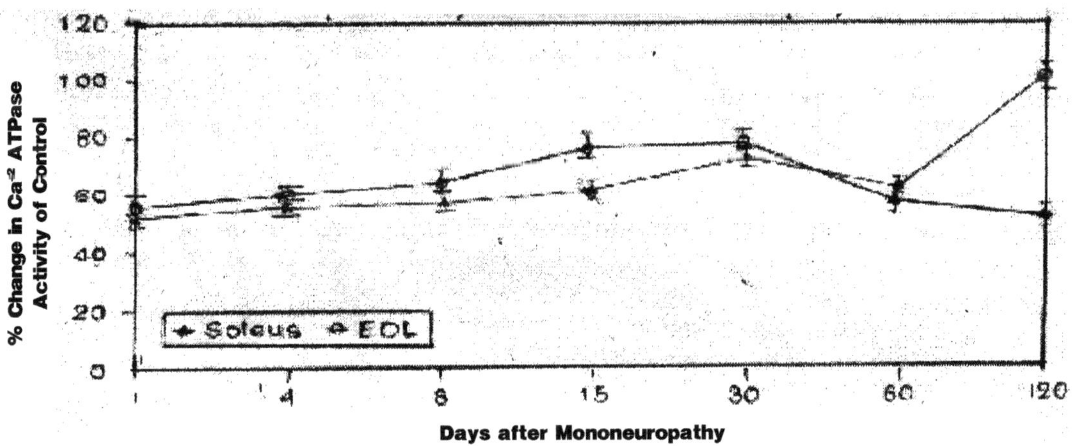

Figure 29.5: The Ca^{2+} ATPase Activity of Soleus and EDL Muscle of Mice During Experimentally Induced Sciatic Mononeuropathy

The Ca^{2+} ATPase activity is calculated as nano moles of Pi/mg protein/hr. Numbers of the ordinate represent per cent of control value and on abscissa represent time in days after neuropathy. Each point represent mean÷S.D of 6 animals. The Ca^{2+} ATPase activity in control soleus and EDL is 828÷24.0 and 1415÷23.0 n moles of Pi/mg of muscle/min respectively.

The disturbance in neural connection found to alter the membrane permeability and its active capacity to pump cations (Broadie and Sampson 1985). The conduction and contractile properties of slow and fast muscles observed to show varied altered pattern after neural disturbance and these are due to alterations in ion specific transport, ATPases (Brandel *et al.*, 1987). The observed activity pattern of $Na^{+}K^{+}$ ATPase, Mg^{++} ATPase and Ca^{++} ATPase in soleus and EDL of palm squirrel, which serves as an important tool for exploring the potential pathogenic role of metabolic factors, confirms the involvement of altered membrane function in time dependent progressive degeneration and reversal of muscle structure and function during the long-term mononeuropathic pain. The differential responses observed in membrane bound enzymes of soleus and EDL may be due to characteristic membrane, structural and neural properties of the muscles. In this animal mode the loosely constrictive ligatures around sciatic nerve may result in loss or damage of nerve fibers consequently showing the symptoms of pain like hyperalgesia, allodynia, dysasthesia which may appear in about 2 days and last for 2–3 months (Attal *et al.*, 1990; Xie and Xiao, 1990; Gautron *et al.*, 1990). The dramatic loss of myelinated fibres in the lesioned nerve and reduction in their conduction velocities could result in observed responses (Xie and Xiao, 1990; Gautron *et al.*, 1990). Abnormal ectopic discharges in primary myelinated afferent fibres of ligated nerve and consequential occurrence of more dark neurons in spinal cord with trans synaptic degeneration after experimental painful neuropathy may account for some aspects of observed pathological significance (Sugimoto *et al.*, 1980). The tendency of normalization of altered properties could be due to the involvement of neural factors particularly regeneration and remyelination of sciatic nerve in later period after possible demyelination and axonal degeneration as a result of ligation (Gautron *et al.*, 1990). These altered probable mechanisms are in confirmation to observed time course changes reported in this animal model.

Acknowledgement

Authors express thanks to UGC [No. F: 30–191/2004(SR)] for financial assistance.

References

Attal, N., Jazat, F., Kayser, V. and Guilbaud, G., 1990. Further evidence for pain related behaviours in a model of unilateral peripheral mononeuropathy. *Pain,* 41: 235.

Bennett, G.J. and Xie, Y K., 1983. A peripheral mononeuropathy in rat that produces disorders of painsensation like those seen in man. *Pain,* 33: 87.

Brandel, C.J., de Legn, S., Martin, S. and MacLehnan, B.H., 1987. Adult forms of Ca^{++} ATPase of *Sarcoplasmic reticulum* expression in developing skeletal muscle. *J. Biol. Chem.,* 262(8): 3767.

Bridges, D., S.W.N. Thompson and A.S.C. Rice, 2001. Mechanisms of neuropathic pain. *Br. J. Anaesth.,* 87(1): 12–26.

Broadie, C. and Sampson, S.R., 1985. Contribution of electrogenic $Na^{+}K^{+}$ ATPase to resting membrane potential of cultured rat skeletal myotubes. *Brain Res.,* 28: 35.

Ellmann, G.L., Courtney, D.K., Amders Jr. V. and Eestherstone, Rm., 1961. A new and rapid calorimetric determination of acetylcholineesterase activity. *Biachem. Pharmacal.,* 7: 88

Gautron, M., Jazart, F., Ratinahirana, H., Hanco, J.J. and Gilbard, G., 1990. Are there some natomical correlates of hyperalgesia and anodynia exhibited by rats with a unilateral mononeuropathy: A preliminary light microscopic study of the sciatic nerve. *Neurosci. Lett.,* 111: 28.

Groswald, D.E. and Dettbarn, W.D., 1985. Nerve induced changes in AChE molecular forms of soleus and EDL muscles. *Exp. Neural,* 89: 198.

Gutamam, E., 1976. Neurotrophic relation. *Ann. Rev. Physial.,* 38: 177.

Guth, L., and Brown, W.C., 1965. The sequence of changes in cholinesterase activity during reinnervation of muscle. *Exp. Neurol,* 29: 198.

Jazat, F., Attal, N., Gautron, H. and Guilbaud. G., 1998. Behavioural evidence that a peripheral experimental neuropathy in rat induces abnormal pain sensation. *Eur. J. Neurasci. Supp.,* pp. 179.

Kaplay, S.S., 1979. Erythrocyte membrane $Na^{+}K^{+}$ activated ATPase in protein calorie malnutrition. *Am. J. din. Nutri.,* p. 579

Korr, I.M., 1981. The spinal cord as organizer of disease processes, 1 V Axonal transport and neurotrophic function in relation to somatic dysfunction. *Jam Ostepath Assoc.,* 84: 415.

Lowry, O.H., Rosebrough, N.J., Farr, A.L. and Pandel, R.J., 1951. Protein measurement with Folin Phenol reagent. *J. Biol. Chem.,* 193: 265.

Misulis, K.E. and Dettbam, E.D., 1985. Is fast fibre innervation responsible for increased acetylcholinesterase activity in reinnvervating soleus mucle. *Exp. Neural,* 89: 204.

Reddy, K.P., 1994. Transient elevations of AChE in soleus, plantaris and EDL muscles of palm squirrel during prolonged reinnervation. *Indian J. Exp. Biol.,* 32: 364.

Rubinstein, N.A. and Kelly, A.M., 1978. Myogenic and neurogenic contribution to the development of fast and slow twitch muscles in rat. *Develop. Biol.,* 62: 473.

Samba, E.J. and Yunia, F.J., 1973. Quantitative and histochemical demonstration of calcium activated ATPase in skeletal muscle. *Exp. Neural.,* 41: 431.

Sugimoto, T., Bennett, G.J. and Kajander, K.C., 1989. Strychnine enhanced transynaptic degeneration of dorsal horn neurons in rats with and experimental painful peripheral neuropathy. *Neuroscience Lett.,* 98: 139.

Tussaky, E.M. and Shorr, E., 1953. A micro-calorimetric method for the determination of inorganic phosphorus. *J. Biol. Chem.,* 220: 675.

Xie, Y.K., and Xiao, W.H., 1990. Electrophysiological evidence for hyperalgesia in the peripheral neuropathy. *Sci. China B.,* 33: 663.

Chapter 30

Preliminary Screening of Endophytic Fungi from Medicinal Plants in India for Antimicrobial and Antitumor Activity

M. Rajasekar Pandian[1], G. Sharmila Banu[1], and G. Kumar[2]
[1]Centre for Biotechnology, Muthayammal College of Arts and Science, Rasipuram, Namakkal – 637 408, Tamil Nadu
[2]Department of Biochemistry, Selvamm Arts and Science College, Namakkal – 637 030, Tamil Nadu

ABSTRACT

The screening of antimicrobial activity against Gram-positive bacteria, Gram negative bacteria, yeast and fungi was carried out on isopropanol extracts prepared from 121 isolates of endophytic fungi isolated from medicinal plants in India. Sensitivity was found to vary among the microorganisms. *Bacillus subtilis, Saccharomyces cerevisiae* and *Alternaria* sp. were susceptible to extracts from three, two and two isolates of endophytic fungi, respectively. None were found effective against *Salmonella typhimurium.* Sixteen endophytic fungal isolates tested were also found to exhibit antitumor activity in the yeast cell-based assay.

Keywords: *Medicinal plants, Endophytic fungi, Antimicrobial, Antitumor.*

Introduction

The nature and biological role of endophytic fungi with their plant host is variable. Endophytic fungi are known to have mutualistic relations to their isolation of endophytic fungi hosts, often protecting plants against herbivore, insect attack or tissue invading pathogens (Siegel *et al.,* 1985;

Clay, 1986; Yang *et al.*, 1994); and in some instances the endophyte may survive as a latent pathogen, causing or quiescent infections for a long period and symptoms only when physiological or ecological conditions favors virulence (Carroll, 1986; Bettucci and Saravay, 1993). In India, extract from many types of local plants are used in traditional manner for treatments of various ailments (Beulah Jerlin *et al.*, 2004; Rajasekara Pandian *et al.*, 2006). The question is whether they are produce by the plant itself or as a consequence of a mutualistic relationship with beneficial organisms in their tissue. Many reports showed that in a microbe-plant relationship, endophytes contribute substances that possess various types of bioactivity, such as antibacterial and antifungal. Thus in this study, we focus on the isolation of endophytic fungi and screening them for bioactivity.

Materials and Methdos

Isolation of Endophytic Fungi

A random sample from each plant consisting of asymptomatic leaves and branches was taken. Leaves and branches portion were thoroughly washed in running tap water, after which they were surface sterilized by submerging them in 75 per cent ethanol for 2 min. The branch portions were further sterilized sequentially in 5.3 per cent sodium hypochlorite solution for 5 min, and 75 per cent ethanol for 0.5 min. After drying, each leaf was divided into three segments and placed on potato dextrose agar (PDA) supplemented with 50 mg/l chloramphenicol to suppress bacterial growth. Branch portions were cut to expose their inner tissue and placed on the same medium. All the plates were incubated at 27°C for up to 3 weeks. Emerging fungi were transferred to fresh PDA plates, incubated for 1 week and periodically checked for purity.

Antimicrobial and Antitumor Activity Tests

The endophytic fungi were grown at 27°C with shaking in 5 ml F-4 medium (Glycerol 40 g/l, Soy bean meal 25 g/l, Yeast extract 5 g/l, Com steep liquor 1 g/l, NaCl 0.5 g/l) and PD-Y medium (potato dextrose broth 24 g/l, Yeast extract 2 g/l) for 5 days. For extraction, an equal volume of isopropanol was added to the culture broth and vortexed vigorously for 1 min followed by a centrifugation at 3000 rpm for 10 min.

About 80 ml of supernatant was applied per sterile paper disc (5 mm diameter). After drying, the extract impregnated discs were used in a disc diffusion assays using *Alternaria* sp. in potato dextrose agar (PDA), *Bacillus substilis, Salmonella typhimurium* in PMg agar, and *Saccharomyces cerevisiae* in YPG agar as test microorganisms for antimicrobial activity. Five millilitre of spore suspension of *Alternaria* sp. grown in vegetable juice (tomato juice, 200 ml $CaCO_3$, 4.5 g agar, 3.0 g in a total vol. of 300 ml) were used. *S. cerevisiae* was grown in YPG broth (yeast extract, 20 g/l; peptone 20 g/l; glucose, 20 g/l), and the *B. subtilis* and *S. typhimurium* were grown in PMg broth (peptone 10 g/l; $MgSO_4.7H_2O$, 2 g/l). Chloramphenicol (50 mg/ml) and nystatin (100 mg/ml) were used as positive controls.

To prepare assay plate in a 21.5 cm × 21.5 cm square plate, Yeast Nitrogen Base (YNB) broth (1.4 g) and Bacto agar (2.4 g) were dissolved in 150 ml of sterile distilled water. The pH was adjusted to 6.5 prior to autoclaving at 121°C for 15 min. When the agar is about 42°C, the following components were added: 20 ml of 50 per cent galactose, 2 ml of 20 per cent sucrose, 10 ml of 20X concentrated adenine (0.5 mg/ml), 2 ml of 100X concentrated histidine (2 mg/ml), 2 ml of 100X concentrated tryptophan (2 mg/ml), 2 ml of 100X concentrated uracil (2 mg/ml) or 2 ml of 100X concentrated leucine (2 mg/ml) when using the test strain DCK or DCS, respectively; 10x concentrated 4 dropout amino acid (containing each at lOx concentrated of arginine at 240 mg, methionine at 240 mg, tyrosine at 360 mg, isoleucine at 360 mg, lysine at 360 mg, phenylalanine at 600 mg, aspartic acid at 1000 mg, valine at 1500 mg and thymine at 2000 mg per 1000 ml sterile distilled water) and yeast glycerol (30–40 per cent) stock of the yeast test strain W303–1AY18 containing either the plasmids pMR438–CyclinA D 24–62 (UCK) or Yep51–SRX5 src (UCS) (Sikder *et al.*, 1997), and poured into the square plate on a horizontal place. The 5 mm paper disc impregnated with the supernatant described above were placed on the agar plate and incubated at 30°C for 3–4 days. The growth circle around the disk that indicates positive results for anti-tumor activity was measured.

Glucose (50 per cent) was used as a positive control. All the screenings procedures were performed twice in duplicates. Inhibition zone (for antimicrobial test) or growth zone (for antitumor test) around the disk of 6 mm or more were defined as positive for biological activity.

Results and Discussion

Plants have long provided mankind with a source of medicinal agents, with natural products once serving as source of all drugs (Balandrin *et al.*, 1993). Though synthetic chemical also have long been used as active agents in reducing the incidence of plants, animals and humans diseases, they are costly, have potentially harmful effect on the environment and may induce pathogen resistance. Thus, biological controls or the use of microorganisms or their secretions to prevent diseases offer an attractive alternative or supplement to disease management without the negative impact of chemical control.

In natural product discovery programmes, typical procedures included isolating microorganisms from samples, growing at various temperatures in a variety of selective or nonselective media and testing the extracts in a spectrum of targeted screens for activity for potential industrial or pharmaceutical applications. For a successful fungal screening, a varied and novel repertoire of either well-known or unexplored fungi is desirable. The most promising trend in isolating new fungi is the move towards investigating novel endophytes, with the idea that unusual endophytes may produce untapped natural products.

A total of 121 endophyte isolates were obtained from 62 of 72 (86.1 per cent) different types of medicinal plants use by the local population in India (Table 30.1). The results of this study showed that endophyte fungi were more prevalent in the leaves (110/121 or 90.9 per cent) than the branches. Further and more intensive samplings are necessary to clarify the fungal assemblages of the leaves and branches, as in traditional practice; the local population used mostly the extract from the leaves of the plants (Ong and Nordiana, 1998, 1999). Though there are still a lot of subjects to be explained in the mutualistic association of endophytic fungi and their plant host, more reports indicated the occurrence of endophytes in plants especially in relation to the possible origin of the plant metabolites detected (Strobel *et al.*, 1996; Cacabuono and Pomilio, 1997; Rizzo *et al.*, 1997). The culture residue of the isopropanol extract of the endophytes cultures 12L and 22L in F-4 medium and IB, 19L, 21L2 and 27Ll in PD-Y medium yielded impressive anti-fungal, anti-bacterial or anti-yeast activities. However, only extracts from endophytic fungal cultures of 5L, 24L2, 37L, 41Ll, 50ML, b34Ll, b53L, b20L, b69L2, b30L, b91, b70L2, b14L, b49L and b29L in PD-Y broth showed positive activity for anti-tumor in the UCK/UCS yeast cell-based assay (Table 30.2). The basis of the anti-tumor screening using a yeast cell-based assay was that the hyper activation of cylcin-dependent-kinase (CDK) resulted in growth arrest of the yeast harboring the genetically engineered recombinant plasmids, and a compound from the extract that can rescue the cyclinA1-induced growth arrest is viewed as a potential antitumor candidate. In this study, we demonstrated that crude extracts from the culture broth of endophytic fungi grown aerobically in PD-Y or F4 medium displayed antibacterial, antifungal, anti-yeast or antitumor activity. These results suggest the presence of either good antimicrobial potency of the extract or of a high concentration of an active principle in the extracts of strains showing positive biological activities. Other endophytic fungal extracts, which showed low antimicrobial or antitumor activity in the bioassay, may have active compounds but probably in smaller amounts and/or the screened crude extracts could yield more potent compounds once they had undergone some purification (Fabry *et al.*, 1998). Also extracts which showed no antimicrobial or antitumor activity in the disc diffusion bioassay might be active against other microbes, which were not tested. Looking at the differing activity of test results obtained, additional modes of action should be explored for those isolates that do not have antimicrobial activity, as it is possible that some of these endophytes may produce substances that may ward off microbial infections by stimulating the host immune system rather than by antimicrobial activity.

Table 30.1: Endophytic Fungi Isolated from Medicinal Plants

Sl.No.	Local Name	Scientific Name	First Isolation	Second Isolation
1.	Kuppai meni	*Acavpha indica*	–	b1L
2.	Vasambu	*Acarus calcmrus*	2L, 2ML	b2L
3.	Nayumvi	*Acyranthes aspera*	–	b3L
4.	Sirupulai	*Aerva lanala*	4L	b4L
5.	Vellai poondu	*Alium sativum*	–	–
6.	Kumari	*Aloe vera*	–	–
7.	Malai vembu	*Andrographis paniculata*	7L	–
8.	Karuvel	*Areca catechu*	–	–
9.	Kaduku	*Brassica juncea*	9L, 98	b9L
10.	Usihthakarai	*Cassia tora*	10L, 10ML	b10L
11.	Nitbya kalyani	*Catharantus roseus*	–	b11L
12.	Valuluvai	*Celassrus paniculatus*	12L	b12L1, b12L2
13.	Ilavangam	*Cinnamomum zeylamicum*	13L	b13L
14.	Athondai	Cap *paris zeylanica*	–	b14L1, b14L2
15.	Nallavelai	*Cleome gynandra*	–	b15L1, b15L2
16.	Kattukodi	*Cocculus hirsutus*	–	b16L
17.	Omavalli	*Coleus aromaticus*	17L	b17L
18.	Kanavalai	*Commelina bengalensis*	18L/B	–
19.	Agasa karudan	*Corallocarpus epigaeus*	19L	–
20.	Naruvili	*Cordia oblique*	–	–
21.	Kothamalli.	*Coriandrum sativum*	–	b21L, b21B
22.	Kottam	*Costus speciosus*	–	b22L
23.	Nilappanaik-kilangu	*Curcuigo orchioides*	23L	b23L1, b23L2
24.	Manjai inji	*Curcuma amada*	–	b24L
25.	Kua	*Curcuma angustifolia*	–	b25L
26.	Kasthuri manjal	*Curcuma aromatica*	26L, 26B	b26L1, b26L2
27.	Karuumathai	*Datura metal*	27L	–
28.	Karisilankanni	*Eclipta prostrata*	28L, 28ML	–
29.	Anashovadi	*Elephanlopus scaber*	–	b29L
30.	Kirambu	*Eugenia caryophyllata*	–	b30L1, b30L2
31.	Palperukki	*Euphorbia heterophylla*	31L1, 31L2, 31ML	b3IL1, b31L2
32.	Amman pacharisi	*Euphorbia hirta*	–	–
33.	Athi	*Ficus racemosa*	–	b33L

Contd...

Table 30.1–Contd...

Sl.No.	Local Name	Scientific Name	First Isolation	Second Isolation
34.	Kanmochi	*Gendarussa vulgaris*	34L1, 34L2, b34L	34BI, 34B2
35.	athumathuram	*Glycyrrhiza glabra*	–	b35L1, b35L2
36.	Nannari	*Hemidesmus indicus*	36L	b36L
37.	Kasarai	*Hibiscus cannabinus*	–	b37L
38.	Nanmthali	*Ipomoea obscura*	–	b38L
39.	Mallai	*Jasminum sambac*	39L, 39B	
40.	Adalai	*Jatropha glandulifera*	40L	b40L
4L	Kattamanakku	*Jatropha gossypifolia*	41L	MIL
42.	Kacholum	*Kaempferia angustifolia*	–	–
43.	Tindiyam	*Lager.vtroemia indica*	43L1, 43L2	b43L1, b43L2
44,	Eluthani poondu	*Launaea sarmentosa*	44L1, 44L2, 44B	b44L, b44b
45.	Mamthontri	*Lawsonia inermis*	–	b45B
46.	Arai	*Alarsilea quadrifolia*	–	b46L
47.	Elikathilai	*Merremia emarginata*	47L1, 47L2	b47L1, b47L2
48.	Sathikkay	*Myristica fragrans*	–	b48L
49.	Karuveppilai	*Murraya koenigii*	–	b49L, b49B
50.	Sadamaniil	*Nardostachys jatamansi*	–	b50L
51.	ThuJasi	*Dcimum basilicum*	51B	–
52.	Inbura	*Oldenlania umbellata*	–	b52L
53.	Anai	*Padalium murex*	–	b53L
54.	Uppilankodi	*Pentatropis capensis*	–	b54L, b54B
55.	Sathakuppai	*Peucedanum grande*	b55L	
56.	Thippili	*Piper longum*	56L, 56B1, 56B2	b56L
57.	Milagu	*Piper nigrum*	–	–
58.	llanthalari	*Plume ria ruhra*	58L1, 58L2	b58L
59.	Kodi veli	*Plumbago zeylanica*	59L1, 59L2	b59L1, b59L2
60.	Narumunnai	*Premna corymbosa*	60L1, 60L2, 60ML	b60L1, b60L2
61.	Nagamalli	*Rhinacanthus nasutus*	–	b61L1, b61L2
62.	Alingi	*Rhododendron arboreum*	62LI, 62L2	b62L1, b62L2
63.	Amanakku	*Ricinus communis*	63ML	–
64.	Rosa	*Rosa centifolia*	–	b64LI, b64L2
65.	Asogam	*Saraca indica*	65L	b65L
66.	Nathai suri	*Spermacoce hispida*	66L	b66L1, b66L2
67.	Fuji	*Tamarindus indica*	67B	b67L
68.	Kadukkay	*Termanalia chebula*	–	b68L
69.	Manakkarat	*Vangueria spinosa*	b69L	
70.	Amukkiran	*Withania somniftra*	–	b70L
71.	Iratchai	*Zanthoxylum budrunga*	71L, 72ML	–
72.	Inji	*Zingziber officinale*	–	b72L
73.	llanthai	*Zizyphus jujuba*	–	b73L
74.	Korgodi	*Zizyphus nummularia*	74L	b74L
75.	Suraimullu	*Zizyphusoenoplia*	75L, 75ML	b75L

–: No endophytic fungi isolated.

Table 30.2: Endophytic Fungal Isolates Showing Biological-Activity Against Test Organisms

	Antimicrobial Activity (mm)[a]				*Antitumor Activity (mm)*[b]	
Endophytes	*Bs*	*St*	*Sc*	*Al*	*UCS*	*UCK*
12L	15	–	–	–	–	–
19L	19.5	–	–	20	–	–
22L	19.2	–	–	–	–	–
21L2	–	–	13.2	–	–	–
27LI	–	–	13.5	–	–	–
IB	–	–	–	21	–	–
5L	–	–	–	–	8	8.5
24L2	–	–	–	–	8	8
37L	–	–	–	–	9.2	9.1
41LI	–	–	–	–	8.3	8
50ML	–	–	–	–	10	10.3
b34L	–	–	–	–	7.2	7.1
b53L	–	–	–	–	8	8.1
b20L	–	–	–	–	7.5	7.5
b69L2	–	–	–	–	8.1	8
b9L	–	–	–	–	10	10.3
b30L	–	–	□	–	9	9
b7L	–	–	–	–	9.3	9.2
b70L2	–	–	–	–	8.6	8.3
b14L	–	–	–	–	8.2	8.2
b49L	–	–	–	–	8.7	8.5
b29L	–	–	–	–	8.3	8.1

[a]: Test microorganisms: Bs: *Bacillus subtilis;* St: *Salmonella typhimurium;* Sc: *Sacchoromyces cerevisiae;* and Al: *Alternaria* sp.

[b]: Yeast test strain W303-1AYI8 containing: plasmid pMR438-Cyc1inAD 24-62 (UCK) or Yep51-SRXS see (UCS).

–: None detected.

In addition, there is also the possibility that substances present in the extract can stimulate the growth of the microorganisms, as was evident by several isolates showing good bacterial growth forming wide zone of inhibition around the disk, thus counteracting the effect of inhibitory substances. The observation that antibacterial and antifungal, although in crude extract, were detectable in several isolates may indicate, but not prove, that these isolates produce bioactive substances. In traditional natural products screening programs extracts that are 'hits' in a screen of interest require follow-up analysis, typically involving analytical chemists. This aspect will be further investigated as in any natural product screening, the "referm" problem (rare cultures that produce an activity of interest the first time they are grown often cannot be made to produce that activity again when they are refermented) need to be addressed to enhance production of the secondary metabolites of interest. Therefore, any information and/or research on endophyte-plant symbiosis, such as in this study is of value, especially taking into account the positive biological activity as antimicrobial and antitumor agents. Effective extracts could provide potential leads towards the development of novel and environmental friendly biologically active agents.

References

Balandrin, M.F., Kinghorn, A.D. and Farnsworth, N.R., 1993. Plant-derived natural products in drug discovery and development. In: *Human Medicinal Agents from Plants*, (Eds.) Kinghorn, A.D. and M.F. Balandrin. American Chemical Society, Washington, D.C.

Bettucci, L. and Saravay, M., 1993. Endophytic fungi in *Eucalyptus globulus*: A preliminary study. *Myco. Res.*, 97: 679–682.

Beulah Jerlin, S., John De Britto, A., Balasingh, J., Swamidoss, D.P. and Prakash, A.A., 2004. Medicinal flora of Koonthakulam bird Sanctuary. *J. Ecobiol.*, 16: 7–16.

Cacabuono, A.C. and Pomilio, A.B., 1997. Alkaloids from endophyte-infected *Festua argentina*. *J. Ethnopharmacol.*, 57: 1–9.

Carroll, G.C., 1986. The biology of endophytism in plants with particular reference to woody perennials. In: *Microbiology of the Phyllosphere*, (Eds.) N. Fokkema and J. van den Heuval. Cambridge University Press, Cambridge, pp. 392.

Clay, K., 1986. Grass endophytes. In: *Microbiology of the Phyllosphere*, (Eds.) N. Fokkema and J. van den Heuval. Cambridge University Press, Cambridge, pp. 392.

Fabry, W., Okemo, P.O. and Ansorg, R., 1998. Antibacterial activity of East African medicinal plants. *J. Ethnopharmacol.*, 60: 79–84.

Ong, H.C. and Noralina, J., 1998. Malay herbal medicine in Gemencheh, Negeri Sembilan, Malaysia. *Fitoterapia*, 70: 10–14.

Ong, H.C. and Nordiana, M., 1999. Malay ethno-medico botany in Machang, Kelantan, Malaysia. *Fitoterapia*, 70: 502–513.

Rajasekara Pandian, M., Sharmila Banu, G., Kumar, G. and Smila, K.H., 2006. Medicinal plants of ethnobotanical importance curing diabetes from Namakkal district (Tamil Nadu), India. *Indian J. Environ. and Ecoplan.*, 12: 201–205.

Rizzo, I., Varsavky, E., Haidukososki, M. and Frade, H., 1997. Macrocyclic trichothecence in *Baccharis coridifolia* plants and endophytes and *Baccharis artemisioides* plants. *Toxicon.*, 35: 753–757.

Siegel, M.R., Latch, G.C.M. and Johnson, M.C., 1985. Acremonium fungal endophytes of tall fescue and perennial ryegrass: Significance and control. *Plant Dis.*, 69: 179–183.

Sikder, H., Fukakoshi, M., Nishimoto, T. and Kobayashi, H., 1997. An altered nuclear migration into the daughter bud is induced by the cyclin Al-mediated Cdc28 kinase through an aberrant spindle movement in *Saccharomyces cerevisiae*. *Cell Struct. Funct.*, 22: 465–476.

Strobel, G.A., Hess, W.M., Ford, E.J., Sidhu, R.S. and Yang, X., 1996. Taxol from fungal endophytes and the issue of biodiversity. *J. Industr. Microbiol.*, 17: 417–423.

Yang, X., Strobel, G., Stierle, A., Hess, W.M., Lee, J. and Clardy, J., 1994. A fungal endophyte-tree relationship: *Phoma* sp. in *Taxus wallachiana*. *Plant Sci.*, 102: 1–9.

Chapter 31

Changes in GPC Level of Human Seminal Plasma in Alcoholic Drug Abusers and Smokers

M. Singh[1], P. Parashar[2], K.N. Mishra[2], P.K. Sinha[2], A.K. Dubey[3] and G.N. Trivedi[2]

[1]Department of Zoology, M. P. College, Mohania, Rohtas

[2]Reproduction Physiological and Biochemical Laboratory, P.G. Department of Zoology, H.D. Jain College Campus, V.K.S. University, Ara – 802 301, Bihar

[3]Principal, R.K.S. Degree, Baruin, Zamania, U.P.

ABSTRACT

When 335 human seminal plasma samples of normal as well as alcoholic, drug abusers and smokers of age group 25–35 yrs. Who were addicted since last 6–7 years were biochemically analysed for Olyceryl Phosphoryl Choline, (OPC) a highly significant ($P < 0.001$) decrease was observed in ganja smokers, alcoholic and drug abusers, in comparison to control/normal subjects. Comparatively this decrease was at the extent of ($P < 0.001$) from alcoholic to ganja smokers and drug abusers to alcoholic subjects. Such significant decreasing pattern in seminal OPC level of drugs abusers, alcoholic and smokers is an indicative of toxicological effect of drug of abuse, alcoholism and ganja smoking on epididymal metabolism and sperm function and might be a causative factor in making a men infertile.

Keywords: *Glyceryl phosphoryl choline, Seminal plasma, Drugs, Infertile.*

Introduction

Seminal plasma is the non-gametic portion of human ejaculate having strong capacity to maintain a relatively natural and protective environment for sperm function in comparison to the intracellular compartments of (Kurpisz *et al.*, 1996). Various biochemical constituents, Olyceryl phosphoryl Choline is identified as a secretory product of the epididymis more than two decades ago (Dawson *et al.*, 1957) and is predominantly formed by the metabolism of phospholid in the vesicular gland of epididymis of mammals and partly split very readily by an enzyme from the prostat gland in the liberation of choline in men. Epididymis represents the major source of seminal OPC (Brook *et al.*, 1974 (a) Brown-woodman *et al.*, 1976; Hinton and Setchell, 1980; Riar *et al.*, 1978 and Wallace *et al.*, 1966). Chilitham Ashman *et al*, (1956) and Dawson *et al.* (1959) and reported that OPC level can be effectively suppressed by castration and strongly stimulated by exogenous administration of Teststome. But to know whether the level of OPC was affected by alcoholism, smoking and in drug abusers, the present proposed invertisation has been planed.

Materials and Methods

For this proposed investigation, 335 human volunteers of age group 25–35 years were associated. All the volunteers were divided into 4 groups–Control, Ganja smokers, drug abusers and alcohol users. Control group comprised of those volunteers (110) in number, who were healthy, normal free from any veneral diseases and neither have taken nor taking any intoxicants. Ganja smokers comprised of those 80 in number who were taking ganja since last 6–7 years. They took Ganja 5 times a day by way of smoking. Alcohol abusers comprise of those (75 in no.) who were taking alcohol since last 7–8 years. They took alcohol two times a day. Drug abusers comprised of those (70 in no.) who were taking heroin since last 5–6 years. They took drug by way of smoke inhalation 4 to 5 times a day.

The semen of all categories of volunteers were collected by masturbation with an abstinence of minimum 5 days. The semen was collected in dry clean and sterilized graduated centrifuge tube and left for an hour to Liquify at room temperature. The GPC in semen was estimated by the method of White (1959).

Results and Discussion

As indicated in the Table 31.1 a highly significant decrease in glyceryl. Phosphoryl Cholin concentration in human seminal plasma were observed in ganja smokers, alcoholic and drug abusers in comparison to control subjects. This highly significant decrease was comparatively low in alcoholic to ganja smokers and drug abusers to alcoholic as well as in drug abusers to ganja smokers. Earlier report of Mann and Lutwak-Maun (1987) suggested that sperm phospholipids in general and plasnlalogens in particular play an important role in the metabolism of mammalian semen, mainly as substrate for endogenous respirating during epididymal transit. Human semen contains a high amount of acid phosphatese that arises from prostatie secretion and directly influenced by and rogen titer. The bounded cholin in the GPC has physiological function in semen involve lipotrophic activity, stimulation of phospholipid turnover, transmetthylation. It plays some role in sperm transit in female genital tract and affect mortality. A highly significant reduction in the GPC level of ganja smokers, alcoholic and drug abusers might be due to its abnormal effect on testis, epididymis and accessory reproductive glands which ultimately affect the fertility of human beings. Earlier findings of Lloyd and Wilhams (1948) Schrag and Dixon, (1985). Stillman *et al.* (1986) also indicated that the smoking, alchohlism and drug of abuse affects the structure as well as marking of the Testis and accessory glands of reproduction. Roy *et al.* (2001) also reported that in vasectomiged subjects as well as in obstructive azoospermia OPC level showed its decreased level.

Table 31.1: Level of GPC in the Seminal Plasma of Alcoholic, Drug Abusers and Smokers

Sl.No.	Categories of Volunteers	Glyceryl Phosphoryl Chain in Semen (in mg/l 00 ml)	P-Value
1.	Control[a]	1379.50 + 4.55 (110)	++[ab]
2.	Ganja Smokers[b]	1116.86+3.48 (80)	++[ac], ++[bc]
3.	Alcoholic[c]	1064.00 + 3.60 (70)	
4.	Drug abusers[d]	959.46 + 3.15 (75)	++[ad], ++[bd], ++[cd]

Number in Parenthesis indicates number of samples.

++: $P < 0.001$; ab: Between control and ganja smokers; ac between control and alcoholic; ad: Between control and drug abusers; bc: Between ganja smokers and alcoholic; bd: Between ganja smokers and drug abusers; cd: Between alcoholic and drug abusers.

References

Brawn, Woodman, P.D.C., Sale, D. and Calhite, I.G., 1976. The GPC content of the rat edpididymis after injection of chlorohydrine and liqating the *vasa efferretia*. *Acta Eur. Fertil.*, 7: 155.

Brooks, D.E., Hamiltox, D.W. and Mallek, Att., 1974. Camitine and glyceryl phosphoryl choline in the reproductive tract of the male rat. *J. Reprod. Fertil.*, 36: 141.

Dawson, R.M.C. and Ronilands, I.W., 1959. *Quart. J. Exp. Physiol.*, 44: 26.

Dawson, R.M.C., Mann, T. and Calhite, I.G., 1957. GPC and phosphoryl choline in semen and their relation to chlorine. *Biochem. J.*, 65: 627.

Hinton, B.T. and Setchell, B.P., 1980. Concentration of GPC, phosphocholine and free inorganic phosphate in the leumiual fluid of the rate testis and epididymis. *J. Reprod. Fertil.*, 58: 401.

Kurprisz, M., Miesel, R., Sancoka, D. and Zedrzejegak, P., 1996. Seminal plasma can be a predictive factor for male infertility. *Hum. Reprod.*, 11(6): 1222–1226.

Lloyd, C.W. and Williams, R.H., 1948. Endocrine changes associated with Lanccnel's Ciorhosis of the liver. *Amer. J. Med.*, 4: 315–330.

Mann, T. and Lutwak-Mann, C., 1981. Biochemistry of seminal plasma and male accessory fluids: Application to andrological problems. In: *Male Reproductive Function and Semen.* Spring-Verlag, Berlin, Heidelberg, New York, p. 269–326.

Riar, S.S. Setty, B.S. and Kar, A.B., 1973. Studies on the physiology and biochemical of mammalian epididymis: Biochemical composition of epididymis. A comperative study. *Fertil. Steril.*, 24: 355.

Roy, S., Banerjee, A., Pandey, H.C. Singh, G. and Kumari, G.L., 2001. Application of seminal germ morphology and semen biochemistry in the diagnosis ad management of azoospermic subjects. *Asian J. Andrology*, 3(1): 55–62.

Schrag, S.D. and Dixon, R.L., 1985. Reproductive effect of chemical agents. In: *Reproductive Toxicology*, (Ed.) R.L. Dixon. Raven Press, New York, pp. 301.

Stillman, R.J., 1986. *Fertil. and Steril.*, 46(1): 554.

Wallace, J.C., Wales, R.G. and White, I.G., 1986. The respiration of the rabbit expididymis and synthesis of GPC. *Aust. J. Bioch. See.*, 19: 411

White, I.G., 1959. Studies on the estimation of glyceroly, fructose Lactic acid with particular reference to semen. *Aust. J. Expte. Biochem.*, 37: 441.

William-Ashman, H.G. and Banks, J., 1956. *Bioc. Chem.*, 223: 509.

Chapter 32

A Preliminary Review of *Pseudosecodes* Girault (Hymenoptera : Chalcidoidea : Eulophidae)

T.C. Narendran, M. Sheeba, S. Santhosh, M.C. Jiley and Abhilash Peter
Systematic Entomology Laboratory, Department of Zoology, University of Calicut, Kerala – 673 635, India

ABSTRACT

Two new species of *Pseudosecodes* Girault *viz., P. malabaricus* Narendran sp. nov.and *P. calicuticus* Narendran sp. novo are described from India. The only known species *P. splendidus* Girault is commented upon. A key to species of *Pseudosecodes* is provided.

Keywords: *Pseudosecodes, New species, Review, Eulophidae.*

Abbreviations

CC: Costal cell; F1–F4: Funicular segments 1 to 4; MY: Marginal vein; OOL: Ocellocular line; POL: Postocellar line; SMY: Submarginal vein; STY: Stigmal vein; T1–T8: Gastral tergites 1–8; PMY: Postmarginal vein; DZUC: Department of Zoology, University of Calicut; ZSIK: Western Ghats Regional Station, Zoological survey of India, Kozhikode; QM: Queensland Museum, Brisbane, Australia.

Introduction

The genus *Pseudosecodes* was raised by Girault (1915) from Australia. Boucek (1988) redefined the genus and stated that besides the type species two undescribed species are present in Australia and one in Brunei. The author is also of the opinion that the genus *Beornia* Hedqvist (Hedqvist, 1975) is very close to *Pseudosecodes*, but in *Beornia* transverse frontal groove is absent. In this article two new species from India are described and the known species reviewed. A key to separate the species is also provided. The holotypes of the new species described are deposited at DZUC but eventually will be transferred to ZSIK at a later stage.

Materials and Methods

The specimens were collected using specially made collecting net and curated by methods described by Narendran (2001). The figures were drawn by using the drawing tube of Leica Stereozoom Microscope (Switzerland made) and enlarged using enlarger of B2M model (Indian made).

Results and Discussion

Pseudosecodes mlalabaricus Narendran sp. nov.

Holotype

Female: Length 3.5 mm, with dark metallic blue reflection except following parts: eyes and ocelli pale reflecting yellow with dark tinge on eyes and brown tinge on ocelli; antennal scape pale brownish yellow; tegula pale yellow; legs concolorous with body except first three tarsal segments pale whitish yellow; ovipositor brown.

Head

Width 1.49x its length in anterior view; width in dorsal view 6x its median length; clypeus well marked; scrobe moderately deep delimited in front of front ocellus with a cross groove; frons and vertex reticulate, with short white sparse hairs; malar space about 0.5x height of eye in profile; POL 17x OOL; OOL 3x OD. Antenna inserted a little above lower ocular line; antennal formula 11252; scape not quite reaching front ocellus; relative measurements of length: width of antennal segments: scape = 24 : 6; pedicel = 6 : 4; Fl = 14 : 7; F2 = 13 : 7; F3 = 13 : 8; F4 = 12 : 9; FS = 10 : 8.5; clava + apical spine (14+4) = 18 : 8.

Mesosoma

Pronotum subvertical; meso scutum and scutellum well reticulate; mesoscutum as long as scutellum; propodeum with a short and broad median carina sub equal in length to dorsellum; propodeal spiracle almost touching posterior margin of metanotum; rim of spiracle exposed, partially covered by anterolateral Cold. Hindcoxa distinctly reticulate on dorsal side; hind femur about 3.6x as long as wide; hind tibia without ventral carina; forewing 2.2x as long as wide, hyalinc, pilosity very sparse, mostly restricted to apical half; marginal fringe shorter than half the STV; SMV with S dorsal setae; wing disc with a row of 3 characteristically long setae a little distance below MV; relative length of CC and veins: CC = 27; SMV = 17; MV = 27; PMV = 7; STY = 2; apex of hind wing rounded.

Gaster

Sessile, 3.1x as long as wide in dorsal view; longer than head + mesosoma (84 : 60); relative lengths of tergites in dorsal view: T1 = 13; T2 = 6; T3 = 7; T4 = 6; T5 = 11; T6 = 13; T7 = 10; T8 = 9; ovipositor sheath as long as T1 or T6 in side view; hypopygium not reaching middle level of gaster.

Male

Unknown.

Host

Unknown.

Holotype

Female, INDIA, Kerala, Nanminda (11°26′ N 75°50′ E), Near Calicut, 30. viii. 2004, coil. T.C. Narendran and party.

Discussion

The genus *Pseudosecodes* contains only a single known species from Australia *viz.*, the type species *P. splendidus* Girault [Girault, 1915 (230)]. It differs from this new species in having: (1) F2 distinctly longer than F1 (in *P. malabaricus* F1 longer than F2); (2) F5 subequal in length to F1 (distinctly shorter than F1 in *P. malabaricus*) and 3) first claval joint as long as F5 (distinctly shorter than F5 in *P. malabaricus*).

Pseudosecodes calicuticus Narendran sp. nov.

Holotype

Female: Length 3.2 mm. Dark metallic green reflection with following parts as follows: eye pale yellow with a dark tinge; ocelli pale brownish reflecting yellow; tegula pale brownish yellow; legs with coxae concolorous with body; femora concolorous with body with base and apex paler; trochanter yellow; tibiae mostly yellow with weak brownish band in middle; first three tarsal segments yellow, fourth brown. Wings hyaline with veins pale brown.

Head

Width 1.21x its length in anterior view; width in dorsal view 5.25x as long as wide; clypeus not well delimited; scrobe relatively narrower than *P. malabaricus;* frons and vertex well reticulate, with sparse white hairs; malar space 0.29x height of eye in profile; POL 5.33x OOL; OOL 0.75x OD. Antenna inserted middle; antennal formula 11253; scape hardly reaching front ocellus; relative lengths of antennal segments: scape = 30 : 6; pedicel = 8 : 4; Fl = 14 : 6; F2 = 14 : 7; F3 = 17 : 7; F4 = 14 : 7; F5 = 10 : 7; clava + apical specula (18 + 3.5) = 21.5.

Mesosoma

Pronotum very short, vertical, mostly hidden in dorsal view; pronotum, mesoscutum and scutellum with raised reticulations; mesoscutum as long as scutellum in dorsal view; propodeum with a short and broad median carina, a little longer than dorsellum; propodeal spiracle very near posterior margin of metanotum, less than half its diameter away from metanotum; rim partly covered anterolateral fold. Hind coxa well reticulate on dorsal and part of ventral sides; hind femur 3.13x as long as wide; hind tibia without ventral carina. Forewing 2.57x as long as wide; pilosity very sparse mostly restricted to apical half; marginal fringe shorter than half length of STY; SMY with 6 dorsal setae; wing disc with a ventral row of 5 characteristic longer setae a little distance below MV; relative lengths of CC and veins: CC = 26; SMV = 20; MV = 33; PMV = 6; STY = 2.1; apex of hind wing rounded.

Gaster

Sessile; 3.5x as long as wide in dorsal view; longer than combined length of head + mesosoma (77 : 54); relative lengths of tergites: T1 = 12; T2 = 5; T3 = 7; T4 = 6; T5 = 11; T6 = 13; T7 = 11; T8 = 7; ovipositor sheath in side view as long as T6; hypopygium not reaching middle of gaster.

Male

Unknown.

Host

Unknown.

Holotype

Female, INDIA, Kerala, Calicut (11°18′ N 75°48′ E), 28. viii. 2003, coll. T.C. Narendran and party.

Discussion

This new species can be separated from the related species by using the key given below:

Key to Species of *Pseudosecodes* Girault

1. Head and body with dark metallic blue or bluish green reflections; tibiae concolorous with body; CC equal or sub equal to MV; PMV 2.5x as long as STY; clava two segmented

  2

   Head and body dark metallic green (without blue reflections), tibiae mostly yellow with weak brownish tinge in middle; CC distinctly shorter than MV (26 : 33); PMV 2.9x as long as STY; clava 3 segmented; (India)

   .... *calicuticus* Narendran sp. nov.

2. F2 distinctly longer than F1; F5 subequal in length to F1; first claval joint as long as F5; (Australia)

   .... *splendidus* Girault

   F2 distinctly shorter than F1; F5 distinctly shorter than F1; first claval joint distinctly shorter than F5; (India)

   .... *malabaricus* Narendraq sp.nov

Pseudosecodes splendidus Girault, 1915: 209–210, Queensland, Australia (QM)

Female

Length 3.14 mm dark metallic blue-green; legs concolorous with body except first 3 tarsal segments of all legs pale strow yellow; antenna concolorous with head with scape reddish; antennal formula 11252, pedicel distinctly shorter than F1; F2 distinctly longer than F1; F3 subequal to F2; F4 shorter than F3; F5 shorter than F4; subequal to F1 which is about 2x as long as wide; clava some what longer than F2, the first club joint as long as F5; longer than the second claval joint.

Male

Unknown.

Host

Unknown.

Type Locality

Australia: Queensland (13°32′ S 131°26′ E) Harvey's Creek, Type No. Hy 2234, Queensland Museum Brisbane, the specimen on tag, the head and hind leg on a slide.

Remarks

The above account is based on the original description of Girault (1915).

Acknowledgements

We are grateful to the Authorities of University of Calicut for the facilities to work.

References

Boucek, Z., 1988. *Australasian Chalcidoidea*. CAB International, Wallingfora. Oxon, pp. 8.

Girault, A.A., 1915. Australian Hymenoptera Chalcidoidea-IV. Supplement. *Mem. Qd. Mus.*, 3: 180–299.

Hedqvist, K.J., 1975. Notes on Chalcidoidea. VIII. The Swedish Chrysolampini with descriptions of a new genus and species (Hymenoptera : Chalcidoidea : Pteromalidae). *Ent. Tidsks*, 96: 133–136.

Chapter 33

Medicinally Important Orchids of North-East India

C.M. Sarma, R.K. Bora* and N. Basumatary
Department of Botany, Pragjyotish College, Guwahati – 781 009, Assam

ABSTRACT

Besides ornamental value, the use of orchids as medicine reported from various parts of the world can be linked with the remote past. In North-East India, the region of many ethnic tribes and races, the orchids have been used in many cultural and religious practices and to cure ailments as well. The present study was undertaken to find out the role and use of native orchids in ethno-medicine among different tribes and races of the region. Forty-six species of orchids have been enumerated in the investigation from the region, which are commonly used for their medicinal properties. Majority of plants enumerated from the region are used as simple preparation from a single plant source. In a few preparations more that one type of plants are used where more than one active principles may be involved. Out of the enumerated species *Geodorum densiflorum* is used in veterinary medicine and also have insecticidal property commonly used to kill flies of the cattle.

Introduction

North East India, the biological paradise unexplored ranging from Sikkim and North Bengal in the West to Arunachal Pradesh, Manipur and Nagaland in the East including Assam, Meghalaya, Mizoram and Tripura is the repository of an unravelled wealth of biodiversity. It is considered as one of the richest mega-biodiversity hot spot area of the existing mega-biodiversity hot spots in the world.

---

* E-mail: ranjankumarbora@yahoo.com.

NE India and Eastern Himalayas along with Myanmar are positioned in such a way that the convergence of flora and fauna from East and South East Asia has taken place before entering mainland India leading to an assemblage of both native and migratory species. The region is located along with foothills and ranges of the Eastern Himalayas, splendid valleys, turbulent rivers, sub tropical climate with high rainfall and humidity, which makes the region congenial for the luxuriant growth of different types of orchids.

Orchids, extremely diversified, are found in virtually all regions around the world except perhaps Antarctica, but their greatest diversity occurs in the tropical and subtropical climates where positive factors for growth and reproduction like thick vegetation, high humidity prevail. Orchids by habit are mostly perennial herbs and occur on variety of substratum as epiphytes, lithophytes, terrestrial and saprophytes except aquatic.

Orchids belong to the family Orchidaceae, are among most advanced flowering plants, showing an incredible range of diversity in shape, size) colour and beauty of their flowers. Orchidaceae is the second largest family of Angiosperms in the world comprising of about 788 genera and 18500 species (Mabberley, 1988). Karthikeyan (2000) reported the occurrence of about 1229 species of orchids wider 184 genera in India, considered as second largest family of angiosperms in India after Poaceae. Orchids in India showing maximum diversity in the Eastern Himalayas and NE India alone contributed to about 870 species under 189 genera. Arunachal Pradesh is the richest state of orchids comprising of about 544 species in 126 genera (Singh, 2001) followed by Sikkim with 525 species in 137 genera (Kumar, 2001). Besides, other states of NE India are also rich in orchid diversities, *e.g.* Meghalaya 332 species (Kataki, 1986), Manipur 250 species (Ghatak and Devi, 1986), Mizoram 253 species (Singh *et al.*, 1990), Nagaland 241 species (Hynniewta and Kataki, 2000). Orchid flora of Assam is yet to explored fully. From the available published reports, the number of orchid species in Assam appears to be 216.

Orchids, the most wondrous plants of the world, are extensively studied plants among the angiosperms. Orchids are highly valued for the bewitching beauty and long vase life of their flowers that extends up to six to ten weeks. Orchids have been attracting the people particularly naturalists, taxonomists and floriculturists since time immemorial to their fads, fancies and fashions leading to 'Orchid Mania' throughout the world. It provides a source of aesthetic pleasure to both owners and visitors and constitutes an order of royalty in the world of ornamental plants.

Besides ornamental value, orchids are rich in alkaloids, flavanoids, glycosides, carbohydrates and other phyto-chemical contents. From the dawn of cultivation, mankind has been in search for plant that could be used to take care of pains, deformities, ailments and diseases that affect some of the unfortunate members of the society. The use of orchids as medicine reported from various parts of the world can be linked with the remote past. Various ancient literatures, folklores, religious practices of the world clearly indicated the use of orchids as medicine for curing of various diseases. Ancient literature on medicine from China reported that dried stems of *Dendrobium* were used as tonic, astringent, analgesic, anti-inflammatory etc. during 200 BC to 200 AD. In India, the use of orchids as medicine dates back from Vedic period. A group of eight drugs, known as 'Ashtawarga' is used in preparation of various rejuvenating formulations and tonics such as 'Chyavanprash' in the Ayurvedic system of medicine. Among the eight drugs of 'ashtawarga' group 'Jeevak' [*Malaxis muscifera* (Lindl.) Kuntze], 'Rishbhak' (*M. acuminata* D Don), 'Ridhi' (*Habenaria intermedia* D. Don) and 'Vridhi' (*H. edgeworthii* Hook. f) are derived from the orchids (Singh, 2001). In 'Rasna', and 'Jeevaniyo Dashko Mahakshay', two other Ayurvedic drugs, orchid is the main constituent. The leaves of Vandika (*Vanda tessellata*) were used to cure rheumatism in Ayurveda.

In NE India, the region of many ethnic tribes and races, the orchids have been used in many cultural and religious practices and to cure various ailments as well. Different tribes of the region are still following their own traditional medicines. Accordingly, the composition, method of preparation, administered doses of the drugs etc. are different among various tribes. However, a good number of orchid species have been used as a remedy of number of ailments by various tribes according to their own traditional knowledge.

A retrospection of the healing power of plants and a return to natural remedies are now emphasized, as they do not have serious side effects. Synthetic drugs produced by pharmaceutical industries have created side affects, which appear to be more dangerous than the disease they claim to cure. Herbal medicine is based upon the premise that plants contain natural substances that can promote health and alleviate illness. Recently, therefore, 'indigenous system of medicines' or 'ethno-medicines' has attracted modem scientists in finding out solutions for many challenging diseases. The possibility of invention of new drugs from the plants used in ethno-medicines cannot be ignored.

Though the NE India is the richest region of orchid in the country, thorough and systematic study of the role of orchids in ethno-medicine is not yet fully explored. Published available literatures provide a few reports throwing light on a handful of orchids of this region and their uses in ethnomedicine. Keeping this view in mind, the present investigation was undertaken to find out the role and use of native orchids in ethno-medicine among different tribes of the region.

Methodology

Data on medicinal use of orchids were mainly collected from direct interactions with the local people including tribal and from folk medicine practitioners of the region. Ancient literature, scripts, folklore etc. were also consulted as the source of information.

Name of the orchid species of NE India having medicinal values, either used in ethno-medicine of NE India or used in other indigenous system of medicine are enumerated here. Their therapeutic uses and parts used are also listed in Table 33.1.

Table 33.1

Sl.No.	*Botanical Name*	*Parts Used*	*Therapeutic Use*
1.	*Acampe papillosa* (Lindl) Lindl.	Roots	Rheumatism
2.	*Aerides odorata* Lour.	Whole plant	Anti tuberculosis
3.	*Anthogonium gracile* Lindl.	Roots	Adjubant
4.	*Arundina graminifolia* (D. Don.) Hochr.	Roots	Analgesic
5.	*Bulbophyllum umbellatum* Wall.	Whole plant	Tonic, vitalizer
6.	*Calanthe triplicata* (Willem.) Ames.	Roots	Anti diarrhoeal, tooth decay
7.	*Cleisostoma williamsonii* (Reichb. F.) Garay.	Leaves and stems	Fracture of bones
8.	*Coelogyne punctulata* Lindl.	Pseudobulbs	Burn injury and healing of wounds
9.	*Coelogyne corymbosa* Lindl.	Pseudobulbs	Burn injury
10.	*Coelogyne ovalis* Lindtl.	Whole plant	Aphrodisiac
11.	*Cymbidium aloifolium* (L) Sw.	Leaves	Blood coagulant, Madness
12.	*Cymbidium ensifolium* (L) Sw.	Rhizome flower	Anti gonorrheal, Eye sore
13.	*Cymbidium iridioides* D. Don.	Leaves	Blood coagulant

Contd...

Table 33.1–Contd...

Sl.No.	Botanical Name	Parts Used	Therapeutic Use
14.	*Cymbidium longifolium* D. Don.	Whole plant	Salep as emetic
15.	*Cypripedium elegans* Reichb. F.	Whole plant	Neural disorder
16.	*Dactylorhiza hatagirea* (D. Don.) Soo	Bulbs	Astringent, expectorant
17.	*Dendrobium fimbriatum* Hook.	Whole plant	Liver complain, nervous debility
18.	*Dendrobium monticola* PF Hunt. et. Summerh.	Pseudobulbs	Emollient, treating boils, skin eruption, pimple etc.
19.	*Dendrobium nobile* Lindl.	Seeds	Astringent, antiseptic
20.	*Dendrobium ovatum* (Willd) Krnzl.	Whole plant	Stomachic, purgative, laxative
21.	*Epipactis helleborine* (L) Craintz.	Whole plant	Nervous disorder
22.	*Eria museicola* (Lindl.) Lindl.	Whole plant	Cardiac
23.	*Eria pannea* Lindl.	Whole plant	Ague
24.	*Eria spicata* (D. Don) Hand.	Stem	Analgesic, stomach ailment
25.	*Euiophia debta* Roehr.	Tuber	Stomach ailment
26.	*Euluphiu nuda* Lindl.	Tuber	Appetiser, aphrodisiac, blood purifier
27.	*Flickingeria macraei* (Lindl.) Seidenf.	Pseudobulbs	In Xerophthalmia, Aphrodisiac, expectotant and astringent
28.	*Goodvera renens* (L.) R.Br.	Whole plant	Eye disorder, Stomach injury
29.	*Geodorum densiflorum* (Lam.) Schltr.	Roots	Flies kilter and anti diarrhoeal in goat
30.	*Gymnadaenia orchidis* Lindl.	Tuber	Anti dysenteric, Anti diarrhoeal, antipyretic
31.	*Habenaria ensifolia* Lindl.	Whole plant	Rejuvenating
32.	*Malaxis acuminata*	Pseudobulbs	Enhance sperm production, Tonic and anti tuberculosis
33.	*Nervilia aragoana* Gaud.	Leaves	Early recovery from post natal sickness, abate thirst
34.	*Papileonanthe teres* (Roxb.) Schltr.	Flower	General debility, antipyretic
35.	*Pecteliis susannae* (L.) Rafin.	Tuber	Boils, skin eruption
36.	*Pholidota ariculata* Lindl.	Pseudobulb	Tonic
37.	*Pholidota imbricata* Lindl.	Pseudobulb	Analgesic: in abdominal pain and rheumatism
38.	*Pleione maculata* Lindl. et. Paxton	Rhizome	In liver complain and stomach ache
39.	*Rshynchostylis retusa* (L.) Bl.	Whole plant	In otorrhoea, ear pain, madness, stomach disorder and as emollient
40.	*Satyrium nepalenses* D. Don.	Tuber	Anti malarial
41.	*Spiranthes sinensis* (Pers.) Ames.	Shoot and bulb	Sores
42.	*Tropidia curculigoides* Lindl.	Whole plant	Anti diarrhoeal and anti malarial
43.	*Vanda cristata* Lindl.	Whle plant	Expectorant
44.	*Vanda tessellata* Hook. f.ex. Don.	Whole plant	Rheumatism, sedatives, neurological disease, anti syphilis, coolant in high fever
45.	*Vanda testacea* (Lindl.) Reichb. f.	Whole plant	Rheumatism, antiviral, anti cancerous
46.	*Zeuxine strateumatico* (Lindl.) Schltr.	Roots	Salep as tonic

In most cases the local people know only the uses of the orchids butare unable to relate them with the local names. They identified the proper plants from their long observations. Again, the same species are known by various names in various languages and tribes of the region. So, in the following enumeration, only the scientific na.1i1es are used and the species are arranged in alphabetical order.

Enumerations

Acampe papillosa **(Lindl.) Lindl.**

The crushed roots are applied externally to cure rheumatism. In ayurveda it is called as 'Rasna'.

Aerides odorata **Lour.**

Plant extract is used as anti tuberculosis.

Anthogonium gracile **Lindl.**

Paste of the roots is used as binding material or gummy substances in medicine.

Arundina graminifolia **(D. Don.) Hochr.**

Root decoction is used in body ache.

Bulbophylium umbellatum **Wail.**

Plant extract is used as tonic to enhance longevity of life.

Calanthe triplicata **(Willem.) Ames.**

Extract of the roots is used in diarrhea and tooth decay.

Cleisostoma williamsonii **(Reichb. f.) Garay.**

Leaves and stem are used for treating bone fracture. Pulp is applied on the fractured bone area and plastered, and decoction is taken orally.

Coelogyne punctulata **Lindl.**

The powder obtained from dried pseudobulbs is used in burn injuries and healing of wounds.

Coelogyne corymbosa **Lindl.**

Juice of fresh pseudobulbs is applied on bum wounds as painkiller.

Coelogyne ovalis **Lindl.**

The plant extract is used as aphrodisiac.

Cymbidium aloifolium **(L.) Sw.**

Fresh extract of the leaves are thrown over the face of the patient and to take scent of it to cure madness. Fresh juice of leaves used as quick blood coagulant for large bleeding wounds. 'Salep' used as nutrient and demulcent.

Cymbidium ensifolium **(L.) Sw.**

Decoction of the rhizome is used in gonorrhoea. Extract of flowers is used in eyesores.

Cymbidium iridioidtts **D. Don.**

Fresh extract of leaves is utilized for clotting of blood in wounds.

Cymbidium longifolium **D. Don.**

'Salep' prepared from this plant used as demulcent aquatic solution of dried bulb powder is taken orally in empty stomach condition as emetic.

Cypripedium elegans **Reichb.f.**

Plant extract is used to cure nervous disorder.

Dactyiorhiza hatagirea **(D. Don.) Soo.**

Bulb paste applied externally in bone fracture; severe cuts and wounds to act as an astringent. Lepchas take it orally in paste form for curing cough.

Dendrobium fimbriatum **Hook.**

Plant extract is used for liver complains and nervous debility.

Dendrobium monticola **PF Hunt. and Summerh.**

The poultice of the pseudobulbs is used for treating boils, pimples and other skin eruptions. In ayurveda the plant is known as 'jiwanti'.

Dendrobium nobile **Lind.**

Seeds are applied to freshly cut wounds for quick healing.

Dendrobium ovalum **(WiHd) Kranzl.**

Entire plant is used for curing stomache, bile secretion and laxative.

Epipactis helleborine **(L.) Craintz.**

Plant used to cure nervous disorder.

Eria muscicola **(Lindl.) Lindl.**

Plant is used as a remedy for chest and heart disorder.

Eria pannea **Lindl.**

Paste of whole plant is mixed in bathing water for ague.

Eria spicata **(D. Don) Hand**

Stem paste applied externally on forehead to cure headache. The same is taken orally for stomach ailments.

Eulophia debia **Hochr.**

Tubers are used for stomach ailments.

Eulphia nuda **Lindl.**

Tubers are used as tonic to cure aphrodisiac, blood purification and as appetiser.

Flickingeria macraei **(Lindt) Seidenf.**

In ayurveda it is known as 'Jeevanti'. Pseudobulbs are used as aphrodisiac, astringent and expectorant. Plant is also used in asthma, bronchitis, and night blindness.

***Geodorum densiflorum* (Lamk.) Schltr.**

Crushed root stock rubbed on cattle skin for killing flies. The same is given orally to goats in diarrhoeal symptoms.

***Goodyera repens* (L.) R Br.**

Plants use in eye disorder and stomach injury.

***Gymnadaenia orchidis* Lindl.**

Tubers used in diarrhoea, dysentery, fever etc.

***Habenaria ensifolia* Lindl.**

The plant is used as the constituent of *'Ridhi'* and *'Vridhi'* in ayurveda.

Malaxis acuminata

Pseudobulbs used as nutritive tonic. In ayurveda it is known as *'Rishbhak'*. It is an ingredient of ashtawarga group of drugs, used as tonic and cure tuberculosis. It also enhances sperm production.

***Nervillia aragoana* Gaud.**

Decoction of leaves mixed with the leaves of *Tenagocharis latifolia* (Butomaceae) is fed to women after childbirth to get physical strength and early recovery from postnatal sickness. Tubers chewed to abate thirst.

***Papileonanthe teres* (Roxb.) Schltr.**

The decoction of its flower and stem alongwith flowers of *Heteropanax fragans* is applied in general debility. It is also applied on forehead to keep it cool during high fever. Pieces of its stem tied on loin of the body to protect from colds and cough.

***Pectelis susannae* (L.) Rafin.**

The tubers are used in bolls.

***Pholldota articulata* Lindl.**

Pseudobulbs are used as tonic.

***Pholidota imbricata* Lindl.**

Extract of pseudobulbs is used in abdominal pain and body pain. Crushed pseudobulbs are applied as remedy of rheumatism.

***Pleione maculata* Lind. and Paxton.**

Rhizome of this plant used in liver complain and stomach ache.

***Rshynchostylis retusa* (L.) Bl.**

Decoction of aerial roots is effectively used against Otorrhoea. Basal portion of young leaves is slightly heated; crushed and aqueous extract is prepared, filtered and poured drop by drop into the ear. Fresh extract of the leaves are thrown over the face of the patient and to take scent of it to cure madness. Fresh extracts of young stems use to cure stomach disorder. Paste of its flower is applied as emollient on face.

***Satyrium nepalenses* D. Don.**

Dried tuber used in malaria.

***Spiranthes sinensis* (Pers.) Ames.**

Paste made *of* shoot and bulb is applied as poultice in sores.

***Tropidia curculigoides* Lindl.**

Plant extract is used to cure diarrhoea and malaria.

***Vanda cristata* Lindl.**

Plant extract is used as expectorant.

***Vanda tessellata* (Roxb.) Hook. f. ex. Don**

Roots are bitter in taste and used in rheumatic and other pains. Also used in neurological diseases and syphilis. Plant extract used in ear infection. Paste of leaves is applied on forehead during high fever as coolant.

Powdered roots are used as antidote for poisoning. Also uses in abdominal complaints. In ayurveda it is known as 'Rasna'.

***Vanda testacea* (Lindt) Reichb.f.**

Plant extract is used in rheumatism. The plant is said to be utilized as antiviral and anti-cancerous.

***Zeuxine strateumatica* (Lindl.) Schltr.**

Fleshy roots are used as salep.

Conclusions

Both epiphytic and terrestrial orchids have the medicinal properties which are used in folk medicine or indigenous system of medicine. Forty-six species of orchids have been enumerated in the study from the region, which are commonly used for their medicinal properties.

Out of 46 species of medicinally important orchids of this region, majority of plants enumerated here are used as simple preparations from a single common plant source. In a few preparations more than one type of plant are used. where more than one active principle may be involved. Their uses are: 6 numbers in stomach disorder, 4 numbers each in rheumatism and neural disorders, 3 numbers as expectorant, anti-diarrhoeal and general tonic, analgesic, 2 numbers as anti-tuberculosis, antimalarial, antipyretic, astringent, coolant and emollient, 1 number as anti-virus, anti-gonorrhoea, anti-cancerous and anti-syphilis. Besides, they are also used in bum injuries xerophthalmia, madness, eyesore and as aphrodisiac and vitilizer. One species is used in veterinary medicine and one have insecticidal property commonly used to kill flies of the cattle.

Decreasing efficacy along with toxic side effects of the allopathic drugs becomes headache to the pharmaceutical scientists and they are in constant search for new formulations and active principles with great efficacy and little or no side effects. In this situation studies of ethno-medicine play the key role and thousands of new formulations and active principles have come out which are more effective to a number of prominent diseases (Nagasampagi, 2006). There is often a direct relationship between the plants used by a folk and indication from scientific analysis of its active constituents data which supports such claims (Farnsworth *et al.*, 1985). In the present study also, a few species of orchids have been reported to be used in some prominent diseases like tuberculosis, malaria, cancer etc. which indicates the possibility of development and extraction of new formulations with more effective active principles against these serious diseases. For the purpose, a comprehensive study to evaluate the active principles is required.

Some significant indications of therapeutic utility emerged from the present study *viz., Dendrobium monticola* and *Rhynchostylis retusa* are used as emollient, *Malaxis acuminata* is used against azoospermia, *Flickingeria macraei* is aginst xerophthalmia, *Cleisostoma williamsonii* against bone fracture etc. These reports remain to be tested scientifically for human use.

Most of the sedative, analgesics and antipyretics increase gastric juice in the stomach causing gastritis. There are 6 species of orchids of NE India which are used as sedatives and antipyretics that further study may lead to discovery of safest sedatives, analgesics and antipyretics from these species cannot be ruled out.

Ethno-medicine is the system of medicine, which gives the initial medicinal information about a particular plant clinically. These therapeutic values of the said medicinal properties are justified by screening the plant with different pharmacological aspects (Khare, 1999).

Moreover, at present, the international market for medicinal plants is valued at over US$60 billion per year with an annual growth of 7 per cent. The turn over of herbal medicines in India is about US$1 billion with a meager export of about US$80 million (Kamboj, 2000). India at present exports herbal medicines and materials to the tune of Rs. 446 crores which can be raised to Rs. 3000 crores annually (Kumar and Nagiyan, 2006). But, unfortunately, till to date, there is hardly any authentic information on quantitative availability of the plants their current and future consumption demands. This study reports the medicinally used orchid species available in NE India and their uses, which can be commercially exploited in a sustainable way. The production of herbal medicine may be developed as a cottage industry in a scientific way in the developing countries like India. These sorts of attempts will surely enhance the economy of India.

The practice of folk medicine generally prevails in the area where the modern amenities of health care are not available. NE India is a remote area where tough terrain and natural barriers cause some diseases to appear as epidemic. At the beginning of the 21st century hundreds of human lives are lost due to malaria. That is why the documentation of this traditional use of herbal resources in various ailments is important. This primary information may lead to serious pharmacological research and if even one such plant could be identified which would lead to the design of a new drug then all the efforts of traditional medicine man in preservingthe1r traditional knowledge will be rightly rewarded (Singh, 2006). In this connection, it is necessary to conserve our natural heritage and wealth from destruction. Most of the orchid species are either endangered or threatened or vulnerable. Reckless destruction of forests and loss of habitats push the orchids to the verge of extinction. Unscientific collection for trading also enhances the loss of orchid population. So, there is an urgent need to conserve this natural wealth from extinction. Proper management, ex *situ* and *in situ* conservation, awareness of the local people may help to a great extent achieving the success.

References

Farnsworth, N.R., Akerele, O., Bingel, A.S., Soejarto, D.D. and Guo, Z., 1985. Medicinal plants on theory. *Bull. World Health Organization*, 63: 965–981.

Ghatak, J. and Devi, R.K., 1986. Orchids of Manipur. In: *Biology Conservation and Culture of Orchids*, (Ed.) S.P. Vij. East-West Press, New Delhi, pp. 357–362.

Hynniewta, T.M. and Kataki, S.K., 2000. *Orchids of Nagaland*. BSI, Kolkata.

Kamboj, V.P., 2002. Herbal medicine. *Curr. Sci.*, 78: 35–39.

Karthikeyan, S., 2000. A statistical analysis of flowering plants of India. In: *Flora of India, Introductory Volume II*, (Eds.) N.P. Singh, D.K. Singh, P.K. Hajra and B.D. Sharma. New Delhi, India, pp. 201–217.

Kataki, S.K., 1986. *Orchids of Meghalaya.* Forest Department, Government of Meghalaya, Shillong.

Khare, P.K., 1999. Ethnomedicine: History and future scope. *Adv. Plant Sci.*, 12(2): 611–614.

Kumar, C.S., 2001. Orchids of Sikkim: A historical perspective. In: *Orchids: Science and Commerce*, (Eds.) P. Pathak, R.N. Sehgal, N. Shekbhr, M. Sharma and A. Sood. BSMPS, Dehra Dun, pp. 100–143.

Kumar, S. and Nagiyan Paradhi, 2006. Assessment and conservation of medicinal plant wealth of Haryana. In: *Medicinal Plants: Ethnobotanical Approach*, (Ed.) P.C. Trivedi. Agrobios, Jodhpur, India, pp. 147–199.

Mabberley, D.J., 1998. *The Plant Book*, 2nd edn. Cambridge University Press, Bath, UK.

Nagasampagi, B.A., 2006. Utilization and development of drugs from medicinal plants. In: *Medicinal Plants: Ethnobotanical Approach*, (Ed.) P.C. Trivedi. Agrobios, Jodhpur, India, pp. 1–30.

Singh, D.K., 2001. Orchid diversity of India. In: *Orchids: Science and Commerce*, (Eds.) P. Pathak, R.N. Sehgal, N. Shekbhr, M. Sharma and A. Sood. BSMPS, Dehra Dun, pp. 35–65.

Singh, D.K., Wadhwa, B.M. and Singh, K.P., 1990. A conspectus of orchids of Mizoram: Their status and conservation. *J. Orch. Soc. Ind.*, 4(1–2): 51–64.

Chapter 34

Herbal Medicine in India with Special Emphasis on Commonly Used Herbs

M. Rajasekara Pandian[1], G. Sharmiia Banu[1], G. Kumar[2]
[1]Centre for Biotechnology, Muthayammal College of Arts and Science, Rasipuram, Namakkal – 637 408, Tamil Nadu, India
[2]Department of Biochemistry, Selvamm Arts and Science College, Namakkal – 637 003, Tamil Nadu, India

ABSTRACT

To assess the current situation of sales and uses of herbal medicines in India, more than 100 herbalists throughout the country were interviewed. The collected data included the types of herbs present in the market, the recommendations made by the herbalists in the treatment of ailments, the level of education and training of the herbalists, and miscellaneous observations. One hundred and fifty medicinal plant species were present in the local market. Based on their availability in the market and on the herbalists' recommendations, 9 plant species were considered very common and 17 were considered as common. The survey indicated that most of the herbalists were not educated or trained in the field of herbal medicine except for their expertise gained from their predecessor, none were licensed for this particular purpose; several odd or unprecedented recommendations were passed to the customers. This survey emphasizes the necessity of proper handling of herbal medicines that requires proper regulations and licensing.

***Keywords**: Herbal medicine, Herbalists, India, Licensing.*

Introduction

Herbs have been used for medical treatment since the earliest days of mankind. Various Chinese,

Egyptian, Babylonian, Greek, Arab and Moslem-scientists enriched our today's knowledge with their expertise in herbal medicine. However, the astounding advancements of synthetic medicine overshadowed traditional herbal medicine for over 50 years. Thus, herbal medicine became less common in the last decades, but existed in remote areas or for small and poor sections of the population. However, the recent call for "back to the nature" has affected all public sectors, either because some synthetic drugs failed to prove effective with serious side effects, or could not cure recently discovered illnesses including AIDS. The majority of the world's population in developing countries still relies on herbal medicines to meet their health needs (Abu Irmaileh and Afifi, 2000). Furthermore, in cases where synthetic medicine could not relieve patients who suffer from hard-to-cure illnesses, they have found peace of mind and psychological comfort with folk medicine practitioners.

Following this trend, herbal medicine sprouted at all levels in the society. Unfortunately, the numbers of the scientifically well-oriented and experienced herbalists are few, but many who have found a prosperous trade, started to deal with herbal medicine without proper background. In this research we surveyed the current situation in selling herbal medicines in India in order to salvage the existing knowledge about plants and their properties which exist in the hands of rapidly diminishing number of experienced people but are more and more practiced by mainly incompetent dealers. Additionally, the improper handling of herbal medicine in India was pointed out. The approximate number of medicinal plants sold in the herbalists' shops was recorded based on the survey carried out during a two-year period. Hence, this paper, which is intended to be a basic study of herbal medicine in India, emphasizes the most commonly and commonly used medicinal herbs to complement the previously published paper on less commonly used herbs in India (Afifi and Abu-Irmaileh, 2000).

Materials and Methods

Over a two-year survey, almost all herbalists in India were interviewed. The collected data from the questionnaire included the types of herbs present in the market, recommendations made by the herbalists in the treatment of ailments,' method of preparation, level of education and training of the herbalists, and miscellaneous observations. List of names of plants sold as single medicines are recorded. The macroscopical identification of the plants was based on taxonomic studies of the collected samples using recognized descriptive references (Bruneton, 1993; Boulos, 1983; Evans, 2002; Feinbum-Dothan, 1978; Post and Dinsmore, 1932; Robbers *et al.*, 1996). Documented uses were based on several accepted references in herbal medicine (Blumenthal *et al.*, 1998, 2000; Bruneton, 1993; Duke, 1989; Evans, 2002; Newall *et al.*, 1996; Robbers *et al.*, 1996; Tyler, 1993; WHO, 1999).

All plant materials were authenticated by comparison with the herbarium specimens of the Rapinet Herbarium, St. Joseph's College, Tiruchirapalli, Tamil Nadu, India. Plant mixtures or ready-prepared decoctions, honey, and incorrectly identified plants, are excluded from the list. Medicinal plant species recommended by all herbalists were considered as most commonly used, and those recommended by more than 40 per cent of the herbalists were considered as commonly used medicinal plants. Furthermore, the recommended uses of each plant species were ranked as high (H), medium (M), and low (L). When a particular plant species was recommended for the same use by more than 50 per cent of the herbalists was considered "H," between 25 and 50 per cent as "M," and less than 25 per cent as "L".

Results and Discussion

More than 100 herbalists throughout the country were interviewed. The number of handled medicinal plant materials in these herbalist shops exceeds 150 plants (Abu Irmaileh and Afifi, 2000).

Twenty-seven plant species were found as the commonly recommended and sold as medicinal plants in most herbalist shops in India. These plants belonged to 14 botanical families. Nine plant species can be categorized as most common (Table 34.1), since they are very well known in the local folk medicine and found in all herbalist shops. They are considered safe without documented adverse effects, except when misused, or not properly handled. The remaining 17 plant species which were sold, to a certain extent, less than those considered as commonly used plants. This group includes some plant species which contain toxic constituents.

Table 34.1: The Uses of the Most Commonly and the Commonly Used Herbs Prescribed by the Indian Herbalists

Plant Name	*Parts Used*	*Recommended Uses by the Local Herbalists*[a]
		Most Commonly Used Herbs[b]
Salvia triloba (Labiatae)	Leaves	**Headache*, flatulence**, toothache*,** abdominal pain***, common cold*
Matricaria aurea (Compositae)	Flowers and shoot	**Constipation*, allergy*, kidney stones*, hypertension*, diabetes*, gynecological disorders*,** inflammations***, urinary tract infections*, abdominal pain***, flatulence*, diarrhoea*, common cold***, general weakness*, nervosity*, hyperactivity*
Teucrium polium (Labiatae)	Shoot	**Constipation*, kidney stones*,** diabetes***, abdominal pain***, urinary tract inflammations*, flatulence**, indigestion*, hypertension*, obesity*
Cassia senna (Leguminosae)	Leaves and pods	**Headache*, edema**, anurea*,** constipation***, flatulence*, abdominal pain***, obesity***
Pimpinella anisum (Umbelliferae)	Fruits	**Constipation**, arthritis*, male hyper-sexuality*,** common cold***, abdominal pain**, indigestion*, nervosity**, hyperactivity**, flatulence**, urinary tract infections*, general weakness*
Nigella sativa (Ranunculaceae)	Seeds	**Arthritis*,** general weakness***, gynecological disorders*, lactation deficiency*, hypercholestrolemia*, common cold*, inflammations*
Artemisia herba-alba (Compositae)	Shoot	**Constipation*, hypercholesterolemia*, Jaundice*,** abdominal pains***, diabetes***, parasitic worms**, flatulence*, inflammations*, common cold*, kidney sand and stones*
Achillea fragrantissima (Compositae)	Shoot	**Arthritis*, kidney stones*,** abdominal pains***, diabetes**, flatulence*, common cold*, parasitic worms*, urinary tract infections*
Origanum syriacum (Labiatae)	Shoot	Abdominal pains***, common cold**, blood coagulation*, constipation*, loss of memory*, kidney sand and stones*, diabetes*
		Commonly Used Herbs
Laurus nobilis (Lauraceae)	Leaves	**Hypertension*, diarrhoea*,** general weakness***, arthritis*, hair loss and Dandruff*
Zingiber officinalis (Zingiberaceae)	Rhizomes	**Anemia*,** common cold***, abdominal pains*, indigestion*, gynecological disorders*, impotence**, general weakness*

Contd...

Table 34.1–Contd...

Plant Name	*Parts Used*	*Recommended Uses by the Local Herbalists*[a]
Cinnamomum ceylanicum (Lauraceae)	Stern bark	**Ulcer*, blood coagulation*, constipation*, kidney stones*,** diabetes*, common cold*, flatulence*, nervosity*, general weaknees*, postpartum conditions***
Foeniculum vulgare (Umbelliferae)	Fruits	**Alcoholism*,** common cold**, abdominal pains***, constipation**, anurea*, nervosity***
Rosmarinus officinalis (Labiatae)	Shoot	**Obesity*, constipation*, kidney stones*,** hypertension*, common cold***, abdominal pains**, ulcer**, flatulence*, toothache*, edema*, gynecological disorders*, nervosity*
Hibiscus sabdariffa (Malvaceae)	Calyx	Hypertension***, blood purifier***, abdominal pains*, diabetes*, arthritis*
Cuminum cyminum (Umbellifetae)	Fruits	Flatulence***, general weakness*
Paronychia argentea (Caryophyllaceae)	Shoot	Kidney sand and stones***, urinary tract infections*
Artemisia vulgaris (Compositae)	Shoot	Headache*, abdominal pains***, indigestion*, diabetes**, dysentery*
Trigonella foenum-graecum (Leguminosae)	Seeds	Blood purifier*, common cold*, indigestion*, ulcer*, diabetes*, kidney sand and stones**, lactation deficiency***, general weakness***, inflammations*, poisoning*
Peganum harmala (Zygophyllaceae)	Seeds	Hypertension*, blood purifier*, common cold**, back pain*, ulcer*, impotence**, nervosity*, arthritis**
Melilotus officinalis (Leguminosae)	Shoot	Jaundice*, impotence*, general weakness***, arthritis*
Thymus vulgaris (Labiatae)	Shoot	Hypertension*, blood purifier*, common cold***, abdominal pain*
Zea mays (Gramineae)	Corn silk	Common cold*, constipation*, kidney sand and stones**, edema**, Obesity**
Ruta chalepensis (Rutaceae)	Shoot	Indigestion*, nervosity*, general weakness*, arthritis***
Ricinus communis (Euphorbiaceae)	Oil	**(Seeds are used as abortive and contraceptive)*,** constipation*, skin diseases and eczema***, psoriasis*
Rheum ribes (Polygonaceae)	Rhizomes	Hypertension*, diabetes***, kidney sand and stones*, obesity***

[a]: Uses in bold letter are not documented.

[b]: Ratings of uses: *: Low; **: Medium; ***: High.

Among them are *Peganum harmala* and *Ricinus communis.* One of the conditions for which *Peganum harmala* is recommended, is the treatment of male impotence. For this purpose *Peganum harmala* is used in the form of a decoction. Overdose of the prepared decoction can be harmful as the plant contains many potent alkaloids such as harman, harmalan, harmalin which are extremely toxic and may accumulate to dangerous levels in the body of the patient, since in most cases of folk medicine the activity. The polypeptide ricin, the toxic compound of the seeds, is a heat labile compound, and is absent in the medicinally used fixed oil.

In addition to this common use of the oil by the Indian herbalists, oral intake of the seeds are recommended by some herbalists as a female contraceptive despite the fact that five to eight seeds may kill a mature person (Knight, 1979; Salhab *et al.*, 1996). Although a literature survey of well known references indicated identical uses for these groups of plants, yet our data obtained from Indian herbalists showed some undocumented uses for some of these commonly used plants (Table 34.1) (Bruneton, 1993; Boulos, 1983; Duke, 1989; Evans, 2002; Newall *et al.*, 1996; Robbers *et al.*, 1996; Tyler, 1993; WHO, 1999). In general, in herbalists' shops in Indian, medicinal herbs are sold to customers either upon the request of the patient or as recommended by the herbalist (personal observations).

Our observations indicated that many herbalists promote the herbs they have in their shops regardless of the appropriateness of the illness. In some cases the herb is recommended to the patient by another patient who had a previous experience with it. The plant materials are sold as single drugs, pre-packed mixtures or as freshly mixed preparations. An additional observation was that unsuitable storage conditions for the plant materials exist in some herbalist shops. In a few cases, plant materials were stored in inadequately ventilated stores, with poor sanitary conditions where they are liable to rodent attack and some plant material could be noticed as rotted or attacked by insects. Other herbalists apply insecticides mainly pyrethroids on the stored plants in order to prevent the attack of the insects.

Although many herbalists claimed their expertise in folk medicine, none of them was licensed as herbalist, and many are not well-educated in this field. An interesting observation is that over 70 per cent of the interviewed herbalists do not have a high school degree, and a considerable percentage had no schooling at all. Hence, the scientific background of the herbalists is surely not sufficient for the diagnosis or understanding the etiology of the diseases, and prescription of the herbal medication. Our findings indicate a level of illiteracy in handling of herbal medicine. In addition to herbal medicine, few herbalists also practice cunning, spirituality, and magic. It was noticed that some practitioners are highly successful in treating conditions such as hysteria, neurotic cases, madness, and other psychic disorders using the above mentioned practices.

There are many books in the market written in English language on herbal medicine by non-specialized authors, who largely depend on old references in this subject for describing the plants and recommending the treatment which in turn used by the herbalists. The possible mistakes in the non-specialized books include the misnaming of the plant species in question. Furthermore in the old references, a medicinal plant were only known by their local common names and, based on the fact that one plant can be known by more than one local name in different locations within the country or even in the same location by different people, brings more confusion to the recognition of the correct species. Additionally, different plants may share the same common name. Neither the consumer nor the herbalists are aware of the fact that these different plants have different uses. Presumably misnomination/misidentification of the intended plant could lead to mistreatment, and could subject the patient to harmful effects.

Even though traditional herbal medicine has many virtues, it seems that it has many shortcomings, especially in view of the recent rise in the trade of herbal medicine which is managed by the non-experienced herbalists. This survey showed that herbal medicine is prescribed by the herbalists symptomatically based on signs and symptoms alone rather than as a result of a full understanding of the underlying disease. Proper diagnosis is totally absent.

As any plant, medicinal herbs contain many chemicals that are subjected to change with changing conditions of the environment, especially storage. If not properly stored, the valuable medicinal chemicals in plants could vanish by hydrolysis, oxidation or other means, or could be changed into rather toxic compounds. Storing conditions should be inspected and controlled by the medical symptoms, hence the suggestion for the treatment should base on proper diagnosis. In certain mild conditions, herbal medicine could be useful, independent of diagnosis, but this is unlikely in the majority of cases which require history of the patient, thorough physical examination and proper laboratory data for correct diagnosis and adequate treatment.

For example *Matricaria aurea*, or *Matricaria chamomilla* can be beneficial in the treatment of mild chest pain, sore throat and common cold symptoms provided that these are not symptoms of a serious underlying disease, in which these plants are useless in curing them. Moreover, symptoms may be misleading, and the herbalists are unable to judge the actual cause of such. Local common names are misleading as they are not exclusive authorities in the country. Also, many cultivated medicinal herbs went through different breeding procedures throughout the years with the aim of increasing the yield. Presumably, the new cultivated and hybridized strains of the same species may not correspond to the plants described originally in the old books.

Despite these shortcomings many herbalists in Indian rely on these books which could be the source in the differences among the documented and recommended uses of the commonly used medicinal plants. In addition, vague description of the preparation methods accompanied the herbalists' recommendations. Such findings support the necessity of proper handling of herbal medicine which requires proper regulation and licensing. WHO regulations (WHO, 1989, 1991, 1993) can be considered a good basis for such regulations.

References

Abu-Irmaileh, B.E. and Afifi, F., 2000. Treatment with medicinal plants in India. *Dirasat*, 27: 53–74.

Afifi, F.U. and Abu-Irmaileh, R., 2000. Herbal medicine in India with special emphasis on less commonly used medicinal herbs. *J. Ethnopharmacol.*, 72: 101–110.

Blumenthal, M., Busse, W.R., Goldberg, A., Gruenwald, J., Hall, T., Riggins, W. and Rister, R.S., 1998. *The Complete German Commission E Monographs.* American Botanical Council, Austen, Texas.

Blumenthal, M., Goldberg, A and Brinckmann, J., 2000. *Herbal Medicine Expanded Commission E Monographs.* American Botanical Council, Austen, Texas.

Boulos, L., 1983. *Medicinal Plants of North Africa*. Reference Publications Inc., Algonac, Michigan.

Bruneton, J., 1993. *Pharmacognosy, Phytochemistry, Medicinal Plants,* 2nd edn. Intercept Ltd., Andover, England, UK.

Duke, J.A., 1989. *CRC Handbook of Medicinal Herbs,* 7th edn. CRC Press Inc., Boca Raton, Florida.

ElBahri, L. and Chemli, R., 1991. *Peganum haramala* L.: A poisonous plant of North Africa. *Veterinary and Human Toxicol.,* 33: 276–277.

Evans, W.E., 2002. *Trease and Evans Pharmacognosy*. W.B. Saunders, London.

Feinburn-Dothan, N., 1978. *Flora of Palaestina.* The Israel Academy of Sciences and Humanities, Jerusalem.

Knight, B., 1979. Ricin: A potent homicidal poison. *British Med J.,* 1: 350–351.

Newall, A.C., Anderson, L.A. and Philipson, J.D., 1996. *Herbal Medicines: A Guide for Health Care Professionals*. The Pharmaceutical Press, London.

Post, G.E. and Dinsmore, J.E., 1932. *Flora of Palestine and Sinai*, 2nd edn. AUB, Beirut.

Robbers, J.E., Speedie, M.K. and Tyler, Y.E., 1996. *Pharmacognosy and Pharmaco-biotechnology*. Williams and Wilkins, Baltimore.

Salhab, A.S., Issa, A.A. and Alhougog, I., 1996. On the contraceptive effect of castor beans. *Int. J. Pharmacog.*, 34: 1–3.

Tyler, Y.E., 1993. *The Honest Herbal: A Sensible Guide to the Use of Herbs and Related Remedies*, 3rd edn. Pharmaceutical Products Press, Binghamton, N.Y.

WHO, 1989. Regional Office for the Eastern Mediterranean. An Act Aimed at Ensuring the Safety and Quality of Herbal Remedies (The Herbal Remedies Act) and Notes for Guidance.

WHO, 1991. Report on the Inter country Expert Meeting of Traditional Medicine and Primary Health Care, 30th November–3rd December. Cairo, Egypt.

WHO, 1993. Report on the Inter country Meeting on the Development of Guidelines for National Policy on Traditional Medicine, 21–23 December. EMRO, Alexandria, Egypt.

WHO, 1999. *Monographs on Selected Medicinal Plants*, Vol. 1. WHO, Geneva.

Chapter 35

Combined Effect of Environmental Factors and Nutritional Status on the Weight of Testis and Testicular Total Protein in Albino-rats

S.B. Gupta[1], P. Parashar[1], A.K. Dubey[2], K.N. Mishra[1], P.K. Sinha[1] and G.N. Trivedi[1]

[1]Reproduction Physiological and Biochemical Laboratory, P.G. Department of Zoology, H.D. Jain College Campus, V.K.S. University, Ara – 802 301, Bihar

[2]Principal, R.K.S. Degree, Baruin, Zamania, U.P.

ABSTRACT

When effect of some aspects of environmental factors like temperature, photoperiodic length and nutritional status (in their low and high level quantitatively) were studied on testicular mass and total protein level, a highly significant ($P < 0.001$) increase in testicular mass and a decrease in total testicular protein was observed in experimental rat groups kept on high temperature, long photoperiodic length and high nutritional status in comparison to their control group. But in reverse experimental condition *i.e.* on low temp., short photoperiodic length and low nutritional status, the decrease in the level of testicular protein was also highly significant ($P < 0.001$) in comparison to their control group. All such finding of the above investigation might be an indication of responsiveness of male gonads and its activity 10 certain aspects of physical environment and nutritional status which ultimately affect the male reproductive unction, adversely.

Keywords: *Testicular, Protein, Nutritional, Environmental factors.*

Introduction

Natural selection has provided the mammals with a rich variety of signaling systems; each of which couple environmental variational of some kind with appropriate neuro-endocrine response. In recent years the study of low nutrients and their metabolism as well as various environmental factors affect reproductive processes has gained a great momentum.

Saldier (1969) reported that environmental factors modulate reproduction in mammals. Bronson (1985) was the opinion that the evolutionary forces that have shaped the breeding success of males usually are fundamentally different form those acting on females and by implicating the environmental controls governing reproduction probably, also often differ either qualitatively or quantitatively in the two sexes. In general, the male reproductive systems show much more resistant to malnutrition and photoperiodic length of environmental factors alone influences the seasonal changes in testicular volume in sheep (Martion, 1984). According to Williams (1999) in mature male sheep and goats a nutrition-gonadotropin connection is affected by factors of environment. Therefore in this proposed investigation it has been tried to know the combined effects of some aspects of environmental factors (temperature) and photoperiodic length) and nutritional status on the testicular weight and protein level in albino rats.

Materials and Methods

For the present investigation healthy male albino rats of Chari's Foster strain of 120–130 gm body weight were employed. Rats were divided into four groups each consisting of six rats. Two groups of six rats were regarded as control group and other two groups of rats were considered as experimental groups. The experimental rat groups were maintained at different environmental and dietary conditions. One group of experimental rats were provided with low nutritional status (two times in 24 hours) with long duration dark (18hD : 6hL) and low environmental temperature (experimentation) were set up in month of November to February month). Second group of experimental rats were provided with long duration light (18hL : 6hD) and high temperature (Experimentation were set up during April to June month) and high nutritional status (3 times in 24 hours).

After one month of experimental set up the control as well as experimental rat groups were autopsied, their testis were weighted and testicular protein were measured by the method of Lowary *et al.* (1951) and results were computed by student's t-test.

Results and Discussion

Testicular Protein

As indicated in Table 35.2 the total protein level in testicular tissue of rats kept on high temperature and long photoperiodic condition showed a highly significant decrease in comparison to control, 'indicating low testicular activity while rats were of high nutritional status. This might be due to inhibitory effect on reproduction and circardian photogonado sensitivity.

Recent findings of Bensaad and Maurel (2002) also support our findings that long day experimental photoperiodic condition in mammals (wild rabbit) effects testicular activity by inhiting reproduction.

In case of low temperature and short day experimental photoperiodic condition the level also showed a highly significant decrease in comparison to control clue to the low nutritional status and due to the maximum utilization of protein degradation for energy utilization for maintain body temperature constant. As earlier reports of Bennett (1987) indicates that body temperature regulation in mammals expensive in time, and energy and temperature conformity entails variability in all physiological processes in general and reproduction in particular sense.

Table 35.1: Effects of Environmental Factors and Nutritional Status on the Weight of Male Reproductive Organ (Testis) in Rats

Sl.No..	*Status of Rat*	*Weight of Rat (in gm)*	*Environmental Condition*	*Nutritional Status*	*Weight of Testis (in gm/100 gm Body Weight) Mean + S.E.*
1.	Control	(120–120)	Normal	Adlibitum	1.00±0.02[a] (5)
2.	Experimental	120–120)	Low temperature (winter) long duration dark 18hD : 6hL)	Low food (2 times in 24 h)	0.92±0.05[b] (5)
3.	Control	(120–120)	Normal	Adlibitum	0.51±0.05[c] (5)
4.	Experimental	(120–120)	High temp. (Summer) long duration light 18hL : 6hD)	High food. (3 times in 24 h.)	0.92±0.05[d] (5)

Number of parenthesis indicates the number of Sampls

P Value: a to b: Non-significant; c to d : (P > 0.001) Highly significant; b to d: Non-significant.

Table 35.2: Effects of Environmental Factors and Nutritional Status on the Level of Testicular Total Protein in Alpine Rats

Sl.No	*Status of Rat*	*Weight of Rat (in gm)*	*Environmental Condition*	*Nutritional Status*	*Level of Total Protein Level in (μg/100 mg) Mean± S.E.*
1.	Control	(110–140)	Normal	Adlibitum	57.056±1.74[a] (5)
2.	Experimental	(120–120)	Low temp. (winter) long duration dark 16hD : 6hL)	Low food (2 times in 24 h.)	41.369±1.368[b] (5)
3.	Control	(120–120)	Normal	Adlibitum	73.524±1.491[C] (5)
4.	Experimental	(120–120)	High temp. (Summer) long duration light 18hL : 6hD)	High food (3 times in 24 h.)	61.172±1.540[d] (5)

Number of parenthesis indicates the number of Samples.

P Value: a to b: (P < 0.001) Highly significant; c to d: (P < 0.001) Highly significant; b to d: (P < 0.001) Highly significant.

Mammals have to face several environmental variables such as day length, temperature rainfall, food availability etc. In order to survival of the species a synchronization of their reproducive activity in the period of year when conditions are most conducive for the off spring is a necessary demand.

Weight of Testis

Classical studies on the role of environmental factors that affects functioning of reproductive system have focused mainly on photoperiodic and temperature separately with respect to endogenous regulation, the existence of reproductive hormones in many groups of mammals in well documented.

In the overwhelming majority of mammalian species inhabiting the earth their reproductive functions are restricted to a well defined and after quite short breeding season. The animal cycles of transitions between reproductive activity and quiescence are driven by the environmental signals that influences male reproductive activity in mammals, inhabiting the temperate zone, the role of photoperiod is best understand and probably most important.

As indicating in Table 35.1 as in significant increasing trend of testicular weight in experimental rat groups which were subjected to low temperature and short photoperiodic condition of environment and survive on low nutritional status of food supply were observed. This decrease testicular weight might be due to reduced testicular activity. As earlier finding of Bartke and Steger (1992) in case of Golden hamster indicated that short photoperiodic length of light inhibits prolactin and gonadotropin which ultimately suppress spermatogenesis and inhibit testicular activity. But a highly significant increase in the testicular weight in experimental rat groups subjected to high temperature and long photoperiodic condition of environment and live on high nutrition status might be due to high food availability. The finding of present investigation may be correlated with the finding of Reiter *et al.* (1971) and Wallen and Turck (1981), that albino rats is non-photoperiodic species and high nutritional status renders progressive effects on testicular activity.

Acknowledgement

Authors are thankful to Prof. U.S. Sinha, Head, P.G. Department of Zoology, V.K.S. University, Ara for providing Laboratory facilities for the present work.

References

Bartke, A. and Steger, R.W., 1992. Seasonal changes in the function of the hypothalamic-pituitory testicullar Axis in the Syrian hamster: Mini review. *Proc. Soc. Expr. Biol. Med.*, 91: 139–148.

Bennett, A.F., 1987. The accomplishment of ecological physiology In: *New Directions in M.E. Feeder, Ecological Physiology*, (Eds.) A.F. Bennett, W.W. Burggren and R.B. Huey. Cambridge University Press, Cambridge, pp. 79–86.

Bensaad, M.M. and Maurel, D.L., 2002. Long day inhibition of reproduction and circardian photogondao sensitivity in Zebra Island wild rabbit (*Oryctologus canicules*). *Biological Abs.*, 109(5): 72–89.

Bronson, F.H., 1985. Mammalian reproduction: An ecological perspective. *Biol. Reprod.*, 32: 1–26.

Lowary, O.H., Rosebrough, N.R., Farr, A.L. and Landall, R.T., 1951. Protein measurement with Folin-Phinol reagent. *J. Biol. Chem.*, 193: 265–275.

Martion, G.B., 1984. Factors affecting the secretion of L.H. in the *ewe. Biol. Review*, 59: 1–87.

Reiter, R.J., Sorrentino, Jrs., Rulph, L.L., Lynch, H.J., Mull, D. and Jarrow, E., 1971. Some endocrine effects of blinding and anosmia in adult male rats with observation an pineal melatonin *Endocrinology*, 88: 895–900.

Saldlier, R.M.F.S., 1969. *The Ecology of Reproduction in Wild and Domestic Mammals*. Methuen, London, pp. 1–88.

Wallen, E.P. and Turck, F.W., 1981. Photoperiodicity in the male albino laboratory rat. *Nature,* 289: 402–404.

Williams, G.L., 1999. Nutritional factors and reproduction. In: *Encylopaedia of Reproduction,* (Eds.) by Kniel and Knobip. Academic Press, New York, 3: 412–422.

Chapter 36

Varietal Screening of *Catharanthus roseus* (L.) G. Don. for Root Alkaloid Ajmalicine Content

C. Abdul Jaleel, R. Gopi, R. Somasundaram,
R. Sridharan, B. Sankar and R. Panneerselvam
Division of Plant Physiology, Department of Botany,
Annamalai University, Annamalainagar – 608 002, Tamil Nadu, India

ABSTRACT

Two varieties (rosea and alba) of *Catharanthus rose us* (L.) G. Don. an important medicinal plant, were selected for the present investigation. The alkaloid ajmalicine was isolated and quantified from the roots of both the verities. Out of the two varieties, the rosea variety was found to contain more amount of ajmalicine when compared to alba. The increase in root alkaloid content paralleled the increase in production of more foliage and roots in rosea variety.

Keywords: *Catharanthus roseus, Rosea, Alba, Alkaloid, Ajmalicine.*

Introduction

It has been estimated that just 10–15 per cent of the world's higher order plants have been investigated for bioactive compounds (Balandrin *et al.*, 1993). The production of secondary metabolites by plants has become an active field of study because of its potential as a source of valuable pharmaceutical compounds (Zarate *et al.*, 2001). *C. roseus* (Apocynaceae) is an important medicinal and a popular ornamental plant-with pink or white flowers (Filippini *et al.*, 2003). It is one of the most extensively investigated medicinal plants. It was found to contain a very large number of alkaloids, about 100 of which have been isolated so far (Verpoorte *et al.*, 1997; Samuelsson, 1999). This plant produces widely used alkaloids such as the anticancer drugs vinblastine and vincristine as well as the antihypertensive compounds ajmalicine and serpentine (Tikhomiroff and Jolicoeur, 2001; El-Sayed and Verpoorte, 2005). The present investigation was carried out with an objective of evaluating two varieties of *C. roseus* plants for the content of root alkaloid ajmalicine.

Materials and Methods

Plant Culture

The field experiment was conducted in the botanical garden during the months of February–May, 2005. The seeds of both rosea and alba varieties were sown separately in prepared fields. The plants were uprooted carefully on 90 DAS, keeping the root system intact and washed thoroughly with tap water and surface dried. The roots were separated from the plants of both the varieties, shade dried and finely powdered in an electronic blender and kept in separate containers for ajmalicine extraction.

Ajrnalicine Extraction and Quantification

Ajmalicine extraction was carried out by following standard extraction method (Zhao *et al.*, 2000). Identification and quantification of ajmalicine was done by preparative Thin Layer Chromatography using silica gel G (Merck) in chloroform:methanol (98 : 2 v/v) (Renaudin, 1984) by comparison of R_f values with authentic ajmalicine standard (Himedia, Mumbai). Ajmalicine was spotted by Dragendorffs reagent (Stahl, 1969) (Figure 36.1). The spot corresponding to the R_f value of standard ajmalicine was scraped and eluted with 2 ml methanol. The silica gel was removed from sample by centrifugation at 1000 rpm, and the final volume of the supernatant was made to 5.0 ml the concentration of ajmalicine separated from the crude extract was determined at 254 nm using a spectrophotometer. A standard graph was prepared using authentic ajmalicine standard at various

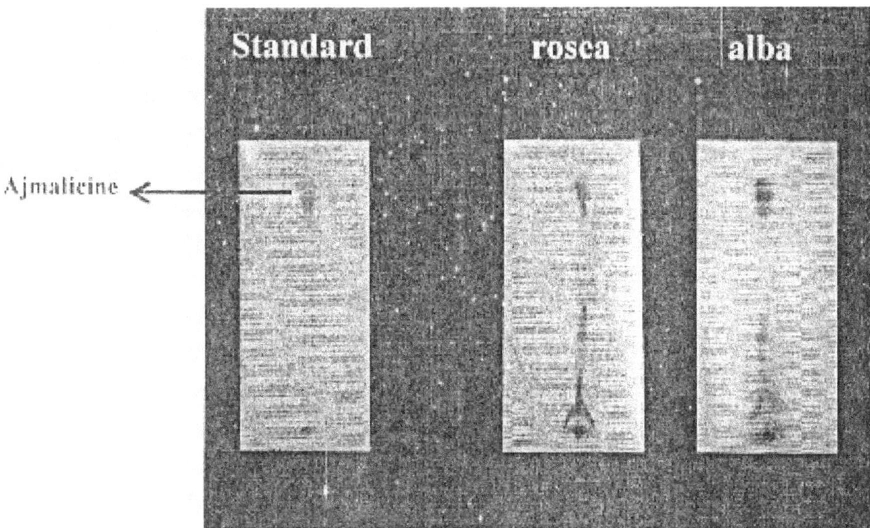

Figure 36.1: TLC Profile of Ajmalicine Standard and Ajmalicine Samples from Two Varieties of *Catharanthus roseus* (L.) G. Don

dilutions and the amount of ajmalicine in each sample was determined. Three experiments were performed independently for each sample and ajmalicine content was expressed as mean of three replicates in mg/g dry weight (DW).

Results and Discussion

The results of this study revealed that the root alkaloid ajmalicine content significantly varied in different varieties. The content of ajmalicine was 0.467 mg g^{-1} DW in rosea variety. But in alba variety, there found a significant reduction in ajmalicine content and it was 0.266 mg g^{-1} DW (Figure 36.2).

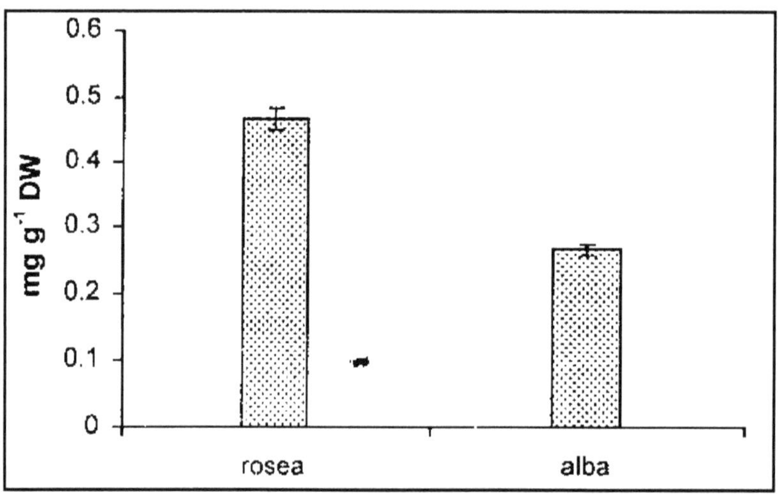

Figure 36.2: Ajmalicine Content of Two Varieties of *Catharanthus roseus* (L.) G. Don. (Values are mean±SD of 3 samples)

Many literatures are there, regarding the varietal variation in the total alkaloid content. In *C. roseus,* the pink flowered (rosea) variety gives higher yield of foliage and roots and total alkaloids (CSIR, 1992). The alkaloid content in plants varies according to species and also in varieties (Kong *et al.*, 2003). According to Gillian *et al.* (1998), the differences in plants in terms of morphology, metabolite content etc. may be due to genetic variation and not by geographical location. Here from this study, it can be concluded that, the pink flowered rosea variety of *C. roseus* is more suitable for cultivation for root alkaloids, due to the increased production of secondary metabolites in the plant level.

References

Balandrin, M.F., Kinghorn, A.D. and Farnsworth, N.R., 1993. *Human Medicinal Agents from Plants,* (Eds.) Kinghorn, A.D. and M.F. Balandrin. American Chemical Society, Washington D.C.

CSIR, 1992. *The Wealth of India: A Dictionary of Indian Raw Materials and Industrial Products, Raw Materials,* Volume 3: Ca–Ci, p. 389–396.

El-Sayed, M. and Verpoorte, R., 2005. Methyl jasmonate accelerates catabolism of monoterpenoid indole alkaloids in *Catharanthus rose us* during leaf processing. *Fitoterapia,* 76: 83–90.

Filippini, R., Vamiato, R., Piovan, A. and Cappelletti, E.M., 2003. Production of anthocynasis by *Catharanthus roseus. Fitoterpia,* 74: 62–67.

Gillian, A., Coopper, D. and Madhumita, B., 1998. Role of phenolic in plant evolution. *Phytochemistry*, 49: 1169–1174.

Kong, J.M., Goh, N.K., Chia, L.S. and Chia, T.F., 2003 Recent advances in traditional plant drugs and orchids. *Acta Pharcol. Sin.*, 24: 7–21.

Renaudin, J., 1984. Reversed-phase high-performance liquid chromatographic characteristics of indole alkaloids from, cell suspension cultures of *C. roseus*. *J. Chromatogr.*, 291: 165–174.

Samuelsson, G., 1999. *Drugs of Natural Origin*, 4th edn. Swedish Pharmaceutical Press, p. 487.

Stahl, E. (Ed.), 1969. *Thin Layer Chromatography: A Laboratory Handbook.* Springer-Verlag, Berlin.

Tikhomiroff, C. and Jolicoeur, M., 2002. Screening of *Catharanthus rose* us secondary metabolites by high-performance liquid chromatography. *J. Chromatography*, 955A: 87–93.

Verpoorte, R., Heijden, R. and Moreno, P.R.H., 1997. Biosynthesis of terpenoid indole alkaloids in *Catharanthus roseus* cells. In: *Alkaloids*, (Ed.) G.A. Cordel. Academic Press; 49: 221.

Zarate, R., Driks, C., Heijden, R. and Verpoorte, R., 2001. Terpenoid indole alkaloid profile changes in *Catharnathus pusillus* during development. *Plant Science*, 160: 971–977.

Zhao, J., Zhu, W. and Hu, Q., 2000. Enhanced Ajmalicine production in *Catharanthus roseus* cell cultures by combined elicitor treatment from Shake-flask to 20–1 airlift bioreactor. *Biotechnol. Lett.*, 22: 509–551.

Chapter 37

Puberty Delay Effects of Grouped Female Urine in Female Lesser Bandicoot Rat *Bandicota bengalensis* (Gray and Hardwicke)

Gurpreet Kaur and V. R. Parshad
Department of Zoology and Fisheries,
Punjab Agricultural University, Ludhiana – 141 004, India

ABSTRACT

Juvenile females (30 days old) of the lesser bandicoot rat, *Bandicota bengalensis,* exposed individually daily to grouped adult female urine soaked in filter paper pieces for nasal contact during bedding, attained the onset of puberty 50 days latter than females juvenile exposed to water soaked filter paper pieces. Females treated with grouped adult female urine showed vaginal opening (VO) at age of 175.0±23.8 days and first oestrous (FO) at 188.75±25.7 days with corresponding body weights of 179.50±26.9 g and 191.75± 25.4 g compared to VO at 131.0±18.9 days and FO 138.50±19.7 days of age with corresponding body weights of 132.0±8.98 g and 135.75±5.45 g in untreated females. Individual variations in these parameters occurred irrespective of the origin of females from the colonies and uniformity in initial body weights and experimental conditions of treated and untreated females.

Keywords*: Puberty, Delay, Urine, Pheromone, Bandicota bengalensis.*

Introduction

Though the timings for the onset of puberty in mammals occurs at a species typical age and thus is thought to be under genetic control while heredity plays an important role in establishing the "approximate time of puberty"(Stone and Barker, 1940) other factors such as nutrition and social stimuli like urine odour also play an important role in scheduling the time of puberty. In contrast to the ability of the adult male to hasten sexual maturation in juvenile female mice, females are able to delay sexual maturation in other females. This war discovered when the onset of puberty was found to be later among female mice living in a group than those living in social isolation (Vandenbergh *et al.*, 1972). Female stimulus delays puberty in females, either by the presence of grouped females as in house mouse *Mus musculus* (Drickamer, 1979), hopping mouse *Notomys alexis* (Breed, 1976), Mongolian gerbil *Meriones unguiculatus* (Payman and Swanson, 1980) or by its urinary pheromones as in house mouse *Mus musculus* (Drickamer, 1979), prairie deer mouse *Peromyscus ochrogaster* (Haigh *et al.*, 1985). The lesser bandicoot rat, *Bandicota bengalensis,* which is the most predominant agricultural pest in south Asia (Parshad, 1999) breeds throughout the year with large litters and maintains its large population (Parshad *et al.*, 1989). However, little is known about the factors regulating its puberty, reproduction and population dynamics. This paper describes the puberty delay effects of urine of grouped adult females in female *Bandicota bengalensis.*

Materials and Methods

Adult *B. bengalensis* caught from the fields were paired in laboratory breeding cages each of (76 × 38 × 40 cm size) with food and water *ad libitum.* Food consisted of a mixture of cracked wheat, skimmed milk powder, powdered sugar and groundnut oil (85 : 5 : 5 : 5). Green leafy vegetables and vitamin B complex syrup (Glaxo India Limited) in water were also provided 2 to 3 times in a week. Breeding cages for the experiment were kept in separate rooms so as to avoid the effect of odours of strange males or females. The males were separated from females at about 12 days of pregnancy. Young ones from three colonies were used for the experiment (Table 37.1).

Table 37.1: Composition of Three Colonies and Number of Juvenile Females Used in Control and Treatment Groups

Colony Number	Litter Size	Number of Males	Number of Females	Female Used	
				Control	Treated
C_1	9	5	4	2	2
C_2	6	4	2	1	1
C_3	8	6	2	1	1
Total	23	15	8	4	4

On day 30, the young ones were weaned from their mothers. Female young ones were separated, weighed and divided into 2 groups of 4 each. Only 4 rats each in 2 groups with similar body weights could be derived from 3 different colonies. They were transferred in individual cages. Before it, the cages were washed and deodourized with detergent. Mean body weight (gms) of 30 day old rats in each group was almost similar in treatment and control groups (that was 56 gm and 57 gm) Cages of both groups were placed in separate rooms in which no other animal was housed so as to avoid the effect of strange odours on puberty.

Urine Donors

Urine of grouped rats was used because the previous studies in mice indicated that urine from isolated animals do not initiate puberty delay (McIntosh and Drickamer, 1977). Therefore, urine of grouped bandicoot rats was used in the present study for which 4 cages each with 2, 6–12 month old females were placed closely to each other in isolated room. Grouping of 3–4 adult females in a cages caused aggression, injury and killing of individuals and therefore to avoid this problem 2 female rats each will be in separate cages were grouped. The bottom of each cage was lined with filter paper which absorbed excreted urine. Urine soaked parts of filter paper were cut into small pieces and placed inside the cage of each recipient female rat.

Exposure of Young Females to Urine Odour

Group of 4 individually caged 30 day old female rats derived from the colonies, as described above, were presented to urine soaked filter paper pieces daily till they attained puberty and during this process they got exposed urine odour. This method was earlier adopted by (Massey and Vandenbergh, 1981). Quantity of urine soaked in filter paper pieces could not be measured. However, filter paper pieces which had soaked urine of 8 donor rats were pooled and divided into 4 equal amounts which were placed in the cage of each female rat. The filter paper pieces were changed daily so that the rats of the treatment group has a fresh source of urinary pheromone.

Observations

In both groups (control and treated) daily feed intake and body weights at weekly intervals were recorded. All females were examined daily for vaginal opening and vaginal smears were prepared from each female daily after introitus. Day of first oestrous was recorded from the presence of fully formed cornified cells in the vaginal smear of rats.

Statistical Analysis

The data of body weights, daily food consumption, day of vaginal opening and day of first oestrous of treated and control groups were compared by student's T test (Gupta, 1991).

Results and Discussion

Food Consumption

Mean daily food consumption in g/day/100 g body weight of female rats exposed to the urine of grouped female rats was 9.79±0.13 which was significantly more than of the untreated rats 9.01±0.12 as calculated on the basis of daily food consumption from day 1 of the experiment (30 day old rats) to the day of first oestrous which ranged from day 100 to day 234 in different rats (Table 37.2). Though the difference in mean consumption was just 0.78g/day/100 g body weight but were statistically significant with 433 and 470 number of observations on treated and control rats. On daily basis each treated rat consumed 8.6 percent more food daily than the control rat. Mechanism by which the urinary pheromones from grouped adult females induced more food consumption by weaning and subadult females is not clear. Possibly the odour of adult females urine provides a familiar environment to the young females similar to their maternal environment or the pheromone by suppressing physiological events relating to maturity affecting the daily food requirement.

Body Weight

Mean body weight of rats on day 1 (30 day old rats) of the experiment was 57.5 g in control and 61.2 g in treated groups and the difference was non significant. Treatment with urine induced significantly more increase in body weight compared to rats of the control group (Table 37.3). The body weight at first oestrous of treated rats was 191.75±25.4 g which was significantly more than of the untreated rats (135.75±5.45 g).

Previous studies by (Lombardi and Whitsett, 1980) in prairie deer mice showed no effect of urine on increase in body weight. But the previous studies clearly show increased growth rate of treated rats whose sexual maturity was delayed. Apparently most of energy, which is derived from food, is used in increasing the body weight but not in sexual maturity in urine treated rats as compared to water treated rats.

Vaginal Opening

Urinary pheromones from the grouped adult female rats were found to delay the sexual maturity in young female rats as in treated rats vaginal opening (VO) and first oestrous were delayed by about 45 days compared to that in control rats. The mean day of vaginal opening in treated rats was 1 75±23.8 which was significantly more than untreated rats in which it was at 131±18.9 day. Individual variations in the age of VO were due to their inheritance from 3 different colonies. Age of rats at the time of VO was 113 and 119 days in treated rats from colony C_1, 223 From C_2 and 165 from colony C_3 compared with 94 and 105 days of their sisters from colony C_1, 174 from colony C_2 and 151 from colony C_3 in control group. However, minor differences occurred in the age at VO in rats from the same colony in both treated and control groups. The present results can be compared with those of (Haigh *et al.*, 1985) in *Peromyscus leucopus* and (Lombardi and Whitsett, 1980) in female prairie deer mice. According to them the VO were significantly delayed in females treated with female urine.

These observations clearly showed that delay in puberty is due to hormonally mediated social factors, but the actual age of VO is genetically dependent which may vary in different colonies.

First Oestrous

All rats in the control group showed first oestrous (FO) 7 days after VO and 33 days in treated group. The mean age at first oestrous of treated rats was 188.75±25.7 days which was significantly more than that of untreated rats (138.5±19.7) (Table 37.2). Treated female rats attained first oestrous on an average of 50 days latter than untreated female rats (Kaur, 2005). The results are comparable to those of (McIntosh and Drickamer, 1977) in which they observed 5 day delay in FO after VO in house mice. Similar delay in FO was reported by (Drickamer, 1982), (Cowley and Wise, 1972) and (Vandenbergh *et al.*, 1972) in mice and (Lombardi and Whitsett, 1980) in prairie deer mice.

According to (Lepri *et al.*, 1985) chemical cues of unknown identity or origin are produced by female mice when densely crowded and are released in the urine. These cues are detected by other females in the population via the vomeronasal organ and the signal is transmitted to central nervous system via the accessory olfactory bulb. This signal results in the release of ACTH and possibly prolactin from the pituitary. These changes activate the adrenal cortex to secrete adrenocortical hormones which act in an unknown manner on a target organ that produces a compound found in the bladder urine of the female. Recently it was found that 2,5-dimethylpyrazine excreted from grouped caged female mice urine functions as puberty delaying pheromone and it was adrenal mediated (Ma *et al.*, 1998 and Novotny, 2003).

From the present study it is concluded that heredity plays an important role in establishing the approximate time of puberty in *B. bengalensis*, but other factors such as social stimuli (urine odour) also play role in scheduling the time of puberty within the range of ages limited by heredity.

Table 37.2: Effect of Application of Urine of Grouped Female Rats on Growth and Maturity of Young Females

	Control (Water)	*Treated (Grouped Female Urine)*	*t-value*	*Level of Significance*
Body weights of rats (g) on different days (n=4)				
Day 30	57.5±4.21	61.2±4.83	1.57	NS
Day 60	95.2±8.95	109.5±10.0	3.14*	P < 0.10
Day 90	113.75± 15.4	147.5±17.8	3.18*	P < 0.05
Day 120	122.5±9.09	153.7±14.6	2.70*	P < 0.10
Body weight (g) on days of vaginal opening (n=4)	132.0±8.98	179.5±26.9	1.87*	P < 0.20
Body weight on day of first oestrous (n=4)	135.75±5.45	191. 75±25.4	2.24*	P < 0.20
Mean daily food consumption (g/100 g) body weight)	9.01±0.12 (n=433)	9.79±0.13 (n=4 70)	7.47*	P < 0.05
Day of vaginal opening (n=4)	131.0± 18.9	175.0±23.8	3.03*	P < 0.10
Day of first oestrous (n=4)	138.50±19.7	188.75±25.7	3.03*	P < 0.10

*: Based on the actual number of rats till the day of first oestrous in rats from different colonies.

Table 37.3: Body Weight (g) Differences in Individual Rats of Different Colonies on the Day (Age) of Vaginal Opening and First Oestrous

			Colony C_1			*Colony C_2*			*Colony C_3*	
			Body Weight	*Day*		*Body Weight*	*Day*		*Body Weight*	*Day*
Control	Vaginal	UR_1	150	94	UR_3	130	174	UR_4	140	151
	opening	UR_2	108	105						
	First	UR_1	138	100	UR_3	140	106	UR_4	145	155
	oestrous	UR_2	120	113						
	Vaginal	TR_1	118	113	TR_3	240	223	TR_4	205	165
	opemng	TR_2	155	119						
Treated	First	TR_1	124	124	TR_3	243	226	TR_4	215	171
	oestrous	TR_2	185	234						

R_1 to R_4 are the number assigned to individual rats and U and T represented untreated and treated rats.

Acknowledgement

Thanks are due to Professor Dr. J.G. Vandenbergh, Department of Zoology, N.C. State University, Raleigh, NC 27650, USA, Dr. L.C. Drickamer, Department of Biological Sciences, Northern Arizona University, Flagstaff, AZ 86011-5640, USA for providing literature and to Indian Council of Agricultural Research, New Delhi for providing support to centre of All India Network Project on rodent control under which this work was carried out.

References

Breed, W.G., 1976. Effect of environment on ovarian activity of wild hopping mice *Notomys alexis*. *J. Reprod. Fertil.*, 47: 395–397.

Cowley, J.J. and Wise, O.R., 1972. Some effects of mouse urine on neonatal growth and reproduction. *Anim. Behav.*, 20: 49–506.

Drickamer, L.C., 1979. Acceleration and delay of first oestrous in wild *Mus musculus*. *J. Mammal*, p. 215–216.

Drickamer, L.C., 1982. Acceleration and delay of first vaginal oestrous in female mice by urinary chemosignals: Dose levels and mixing urine treatment sources. *Anim. Behav.*, 30: 456–460.

Gupta, S.P., 1991. *Statistical Methods*. Sultan Chand and Sons, New Delhi.

Haigh, G.H., Lounsburg, D.M. and Gordon, T.A., 1985. Pheromone induced reproductive inhibition in young female *Peromyscus leucopus*. *Biol. Reprod.*, 33: 271–276.

Kaur, G., 2005. Urinary signals in the lesser bandicoot rat *Bandicota bengalensis* (Gray) and their application for management of its population. *Ph.D. Dissertation*, Punjab Agricultural University, Ludhiana, India, pp. 1–132.

Lepri, J.J., Wysocki, C.J. and Vandenbergh, J.G., 1985. Mouse vomeronasal organ: Effects on chemosignal production and maternal behaviour. *Physiol. and Behav.*, 35: 809–814.

Lombardi, J.R. and Whitsett, J.M., 1980. Effects of urine from conspecifics on sexual maturation in female prairie deermice *Peromyscus maniculatus bairdii*. *J. Mammal*, 61: 766–768.

Ma, W., Miao, Z. and Novotny, M.V., 1998. Role of the adrenal gland and adrenal mediated chemosignals in suppression of oestrous in the house mouse: The lee boot effect revisited. *Biol. Reprod.*, 59: 1317–1320.

Massey, A. and Vandenbergh, J.G., 1981. Puberty acceleration by a urinary cue from male mice in feral population. *Biol. Reprod.*, 24: 523–527.

McIntosh, T.K. and Drickamer, L.C., 1977. Excreted urine, bladder urine and the delay of sexual maturation in female house mice. *Anim. Behav.*, 25: 999–1004.

Novotny, M.V., 2003. Pheromones, binding proteins and receptor responses in rodents. *Biochem. Socie. Transec.*, 31: 117–123.

Parshad, V.R., 1999. Rodent control in India. In: *Pest. Mgmt. Rev.*, 4: 97–126.

Parshad, V.R., Kaur, P. and Guraya, S.S., 1989. Reproductive cycles of mammals: *Rodentia*. In: *Reproductive Cycles of Indian Vertebrates*, (Eds.) S. K. Saidapur. Allied Publishers Limited, pp. 347–408.

Payman, B.C. and Swanson, H.H., 1980. Social influence on sexual maturation and breeding in the female Mongolian gerbil *Meriones unguiculatus*. *Anim. Behav.*, 28: 528–535.

Stone, C.P. and Barker, R.G., 1940. Change of the age or puberty in albino rats by selective mating. *Proc. Soc. Exp. Biol. Med.*, 44: 48–50.

Vandenbergh, J.G., Drickamer, L.C. and Colby, D.R., 1972. Social and dietary factors in the sexual maturation of female mice. *J. Reprod. Fertil.*, 28: 397–405.

Chapter 38

Trophic Status and Fishery Potentials of Irumbi River, Manipur

Laishram Kosygin[1], Haobijam Dhamendra[1] and Nongthombam Premananda[2]

[1]Lake Monitoring and Research Laboratory, Loktak Development Authority, Ningthoukhong – 795 126, Manipur, India

[2]Zoology Department, Presidency College, Motbung – 795 107, Manipur, India

ABSTRACT

This article deals with the trophic status and fishery potentials of Irumbi river. The river is considered eutrophic as it has high concentration of nutrients (NPK) and fecal coliform bacteria. The collection included 17 species of fishes (belonging to 15 genera and 11 families) and 4 species of shell-fishes (belonging to 3 genera and 3 families) which can thrive well in eutrophic water bodies. Fishery potentials of the river and prospects of paddy cum fish and shell-fish culture in the adjacent agricultural fields of the river are discussed.

Keywords: Irumbi river, Water quality, Eutrophic, Fishery potential.

Introduction

Water quality assessment is one of the first steps required in the rational development and management of water resources. There are a wide variety of beneficial uses of water and their suitability depends on the type and amounts of impurities present in it. In India, rivers are most important resources of water and are considered to be the lifeline as they served the purpose of drinking water supply, irrigation and fisheries (Khatavkar and Trivedy, 1993; Prakash *et al.*, 2005; Devi *et al.*, 2005).

The river ecosystems are usually rich in fish biodiversity and may be considered one of the most productive systems, if used wisely. However, due to reckless exploitation and misuses, river water getting polluted and affects its physical, chemical, biological and aesthetic values beyond the desirable limits (Reginaa and Nabi, 2003).

In Manipur, there are numerous rivers and streams, which have high potential of fisheries. However, information about the water quality and fish fauna of these rivers and streams are scanty. In the present study an attempt has been taken up to study the water quality and fish fauna of Irumbi river. The river originates from the western catchment of Loktak Lake and finally discharged into the Loktak Lake near Moirang in Bishenpur District of the state. Before entering the lake the river flows through paddy fields and serve as a source of water for irrigation to local communities. Fishery potentials of the river and prospects of paddy cum fish and shell-fish culture in the adjacent fields of the river are discussed.

Materials and Methods

Sub-surface water samples were collected from the down stream of Irumbi river at Naransena on quarterly basis (September/post-monsoon, December/winter, March/pre-monsoon and June/monsoon) during 2000 and 2002. For each season, five replicates of water samples were collected and their physico-chemical and microbiological parameters were analyzed following standard methods (APHA, 1989; Trivedy and Goyal, 1986). The average of five samples for each parameter studied was considered as one reading. The water temperature, pH, dissolved oxygen (DO) and free CO_2 were determined in the field and other parameters were analysed in the laboratory within 48 hours. Water temperature was measured using a mercury thermometer and pH by digital pH meter. DO was estimated by the azide modification of Winkler's method. Total dissolved solid (TDS) was determined as the residue left after evaporation of filtered sample. Free CO_2, chloride, calcium and magnesium were determined by titration methods. Sodium and potassium were estimated by flame photometer. Nitrate was estimated calorimetrically using Phenol disulfonic acid method and total nitrogen by micro-kjeldahl distillation method. Phosphorus was determined calorimetrically by stannous chloride method. Organic phosphorus was calculated as the difference between the total phosphorus and inorganic phosphorus. Coliform counts were made using membrane filter (MF) technique. Fishes were identified following Jayaram (1981) and Talwar and Jhingran (1991) and shell-fish following Rao (1989).

Results and Discussion

Physico-chemical Characteristics

The results of physico-chemical characteristics of the Irumbi river are summarized in Table 38.1. Surface water temperature ranged from 16.8°C to 29.1°C with an average value of 24.46°C. The pH, dissolved oxygen (DO) and free CO_2 concentrations showed marked seasonal fluctuations (Figure 38.1). The lower value of pH was observed during monsoon and post-monsoon periods. The average value of pH was slightly alkaline. The pH value is varied from 6.2 to 8.3. Highest value of DO was observed during December and lowest during June. The free CO_2 concentration was usually high during the present study with an average value of 8.325 mg/L. Its concentration was highest during June and lowest during December.

The total dissolved solid ranged from 95 to 140 mg/l with an average value of 116.25 mg/l. The concentration of Potassium,was highest during the pre-monsoon period and declines during the monsoon and post-monsoon periods (Figure 38.2). The calcium content was usually low and may be

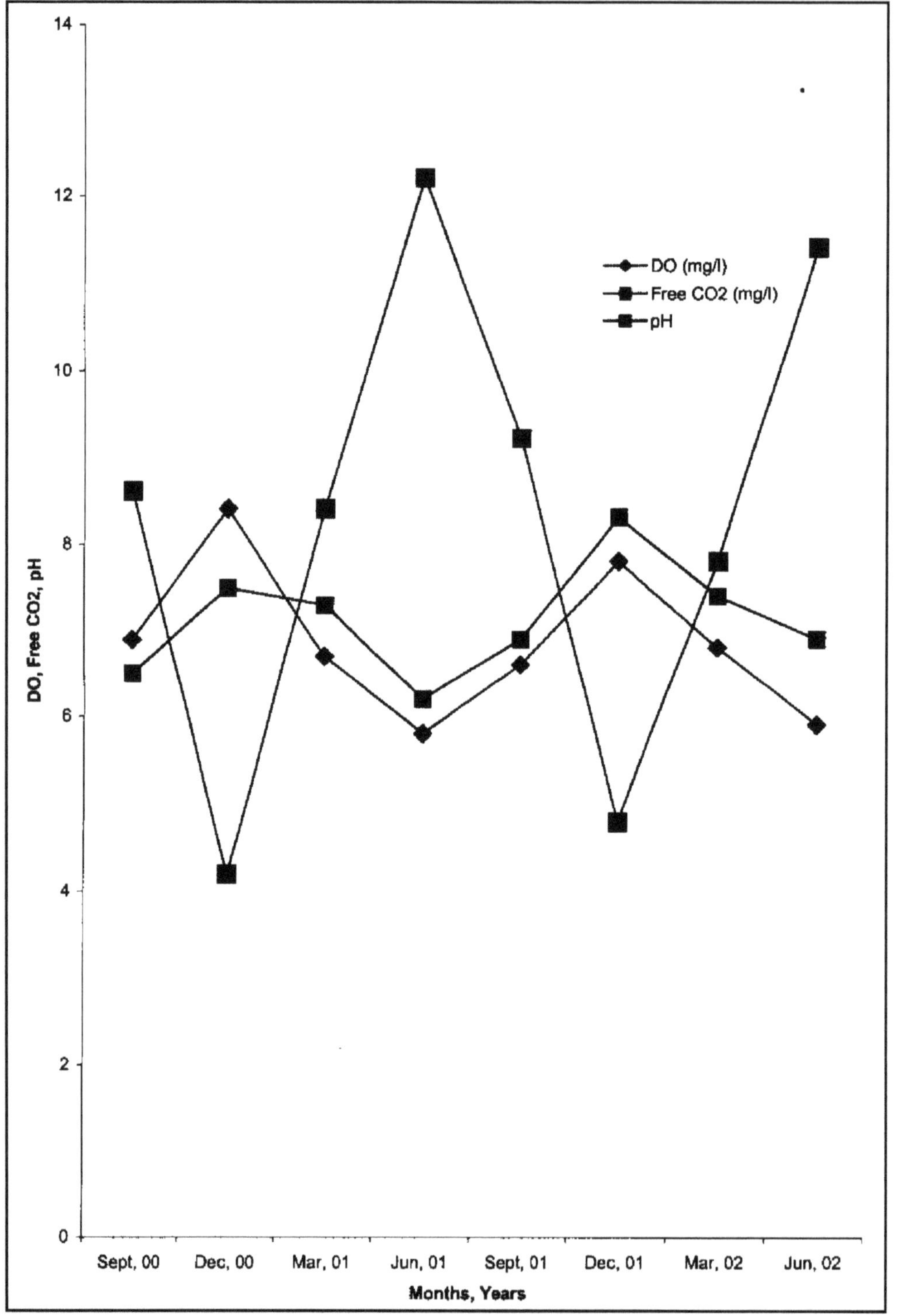

Figure 38.1: Seagional Variations in Dissolved Oxygen (DO), Free CO_2 and pH of Irumbi River (September, 2000 to June, 2002)

categorized as calcium poor water. Concentrations of sodium, chloride, and magnesium occur within the tolerance limit of ISI (1982).

Table 38.1: Data on Physico-chemical Characteristics of Irumbi Rivers (September, 2000–June, 2002)

Sl.No.	*Parameters*	*Range*		*Mean*	*SD*
		Min.	*Max.*		
1.	Water temperature (°C)	16.8	29.1	24.46	4.594
2.	pH	6.2	8.3	7.125	0.651
3.	Dissolved Oxygen	5.8	8.4	6.862	0.878
4.	Free CO_2	4.2	12.2	8.325	2.803
5.	Total Dissolved solids	95	140	116.25	23.358
6.	Chloride	6.8	9.5	8	1.052
7.	Calcium	4.0	5.5	4.65	0.534
8.	Magnesium	3.8	6.1	5.1	0.853
9.	Sodium	13	19	16	2.203
10.	Potassium	1	3	1.625	0.744
11.	Nitrate-nitrogen	0.191	0.585	0.344	0.138
12.	Total Nitrogen	1.034	2.098	1.541	0.370
13.	Inorganic phosphorus	0.035	0.108	0.064	0.026
14.	Organic phosphorus	0.119	0.361	0.222	0.084
15.	Total phosphorus	0.181	0.469	0.287	0.095

Note: All the values are in mg/l, except temperature and pH.

The nitrate-nitrogen varied from 0.191 to 0.585 mg/l with an average value of 0.344 mg/l. The nitrate-nitrogen concentration was highest during the June and lowest during the September. The phosphorus content was comparatively higher during the pre-monsoon period (Figure 38.2). The average concentration of inorganic phosphorus in the present investigation was found to be higher than the standard permissible limit recommended by USEPA (1976) for any tributary discharging to a lake. The concentration of total nitrogen varied from 1.034 mg/l to 2.098 mg/l with an average value of 1.541 mg/l. The concentrations of nutrients were higher in winter (December) and pre-monsoon (March), decreasing in the monsoon season and lowest in post-monsoon (September). This may be attributed to the fact that during winter and pre-monsoon period the river water is more or less stagnant with less water volume, thereby increasing the nutrient concentration. One of the main sources of nutrients in the river is agricultural residues which discharged into the river through surface runoff and leaching from the surrounding agricultural fields. Sand-Jensen (2001) remarked that the nutrient input was highest from the agricultural fields and it has increased dramatically during the last few decades. The river also receives nutrients from the fertile soils from the catchment hills, domestic and animal wastes from the surrounding villages.

Microbiological Characteristics

Microbial analysis in terms of Most Probable Number (MPN) of total coliform bacteria and fecal coliform/100 ml ranges from 32 to 84 and 16 to 46 respectively. The concentration was highest during

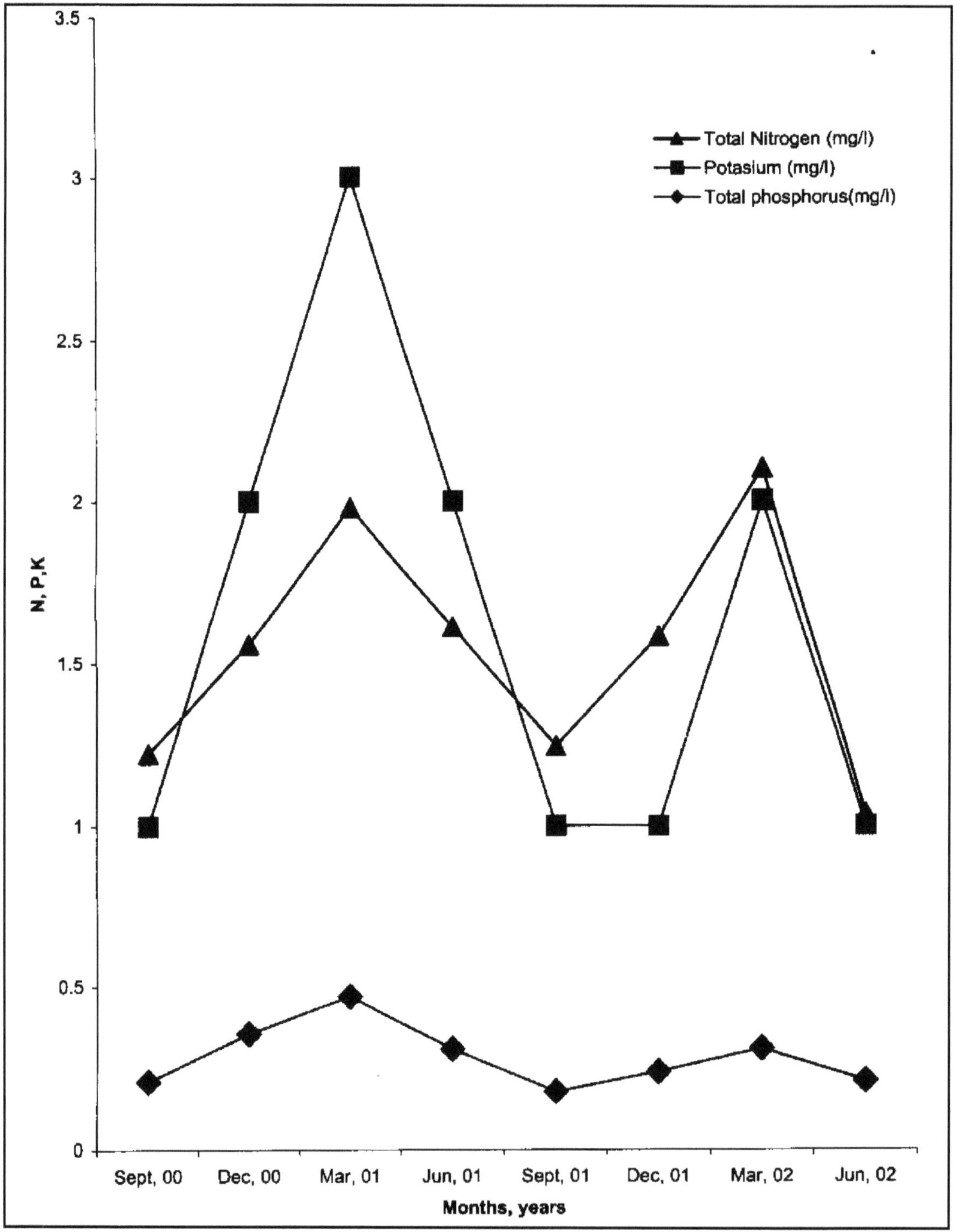

Figure 38.2: Seagional Variations in NPK of Irumbi River (September, 2000 to June, 2002)

the monsoon and lowest during post monsoon. The present findings indicate contamination of microorganisms in the river water and may be considered unsuitable for drinking purposes without treatment (BIS, 1991).

Fish and Shell Fish Fauna

A list of fishes collected and identified during the present study is shown in Table 38.2. The collection included 17 species of fishes (belonging to 15 genera and 11 families). Most of the species are widely distributed forms, which can thrive well in the eutrophic waters. The malaco fauna was represented by 4 species of shell-fishes belonging to 3 genera and 3 families. They are *Pila globosa* Swainson, *Pila* sp., *Brotia costula* (Rafinesque) and *Lamellidens* sp. These species of shell-fish are edible and sold in the markets of Manipur. There is a high potential of edible shell-fish production in Manipur through aquaculture and this may provide additional food resources in the region.

Table 38.2: Fish Fauna of Irumbi River and their Fishery Potential

Sl.No.	Fishes	Local Name	Fishery Potential
Family: Cyprinidae			
1.	*Cyprinus carpio* (Linnaeus)	Puklaobi	High
2.	*Ctenopharyngodon idellus* (Val.)	Napichabi	Medium
3.	*Puntius manipurensis* Menon, Rema and Vishwanath	Ngakha meingangbi	Low
4.	*Puntius sophore* (Hamilton)	Phabounga	Low
5.	*Ampblypharyngodon mola* (Hamilton)	Mukanga	Low
6.	*Esomus danricus* (Hamilton)	Ngasang	Low
Family: Cobitidae			
7.	*Lepidocephalus gun tea* (Ham)	Ngakijou	Medium
Family: Bagridae			
8.	*Mystus bleekeri* (Day)	Ngasep	Medium
Family: Clariidae			
9.	*Clarias batrachus* (Linnaeus)	Ngakra	Medium
Family: Heteropneustidae			
10.	*Heteropneustes fossilis* (Bloch)	Ngachik	Low
Family: Symbranchidae			
11.	*Monopterus albus* (Zuiew)	Ngaprum	Medium
Family: Chandidae			
12.	*Chanda nama* (Hamilton)	Ngamhai	Low
Family: Gobiidae			
13.	*Glossogobius giuris* (Ham)	Nailon ngamu	Low
Family: Anabantidae			
14.	*Anabas testudineus* (Bloch)	Ukabi	Medium
Family: Belontiidae			
15.	*Colisa* sp.	Ngabemma	Low
Family: Channidae			
16.	*Channa orientalis* (Hamilton)	Meitei ngamu	Medium
17.	*Channa striatus* (Bloch)	Ngamu bogra	Medium

Fishery Potential

The Irumbi river flows through the paddy fields of Bishenpur district of Manipur before entering the Loktak lake. The fishes and shell-fishes of the river may be cultured *in situ* or *ex situ* in the river ecosystem in small scale using scientific techniques. Biswas, (1990) remarked that culture of fish in paddy fields is of grate significance in the economy of rural areas. It can provide a supply of cheap and wholesome protein food, besides affording an additional income. The practice of paddy cum fish culture is common in Italy, Japan, Malaysia, several African countries and to some extent in India. In certain areas, paddy fields adjacent to the Irumbi river remain flooded with river water for a period of 3-4 months in a year. During this period fishes like common carp and live fishes which can withstand fairly turbid and shallow water may be cultured. These fishes can grow to a marketable size in a few months. Further, a continuous cultivation process of suitable fishes and shell-fishes may be made on trial basis in the surrounding paddy fields by transferring the fishes to specially prepared channels of the river or directly to the river course using case culture techniques during the paddy harvest or at other times where there is less or no water in the surrounding fields. From the present findings the river water may be considered suitable for fish culture and wild life propagation as the pH and DO values are within the limit of class 'D' water (ISI, 1982). High concentration of nutrients in the river water is also favourable for the growth of planktons and other aquatic weeds which are important foods of some fishes. However, concentration of free CO_2 is slightly higher than the permissible limit (6 mg/l) of fish culture. Therefore, river water still needs to be improved, if it is to be used for fishery purposes.

From the above discussion, it is clear that down stream of the lrumbi river is eutrophic as it has high concentration of nutrients, free CO_2 and coliform organisms. Occurrence of fish fauna which can thrive well in the polluted environment further confirmed eutrophic nature of the river. In order to save the river from further process of eutrophication and also to enhance the field yield, proper scientific conservation measures may be taken up.

Acknowledgement

We are grateful to Shri Th. Manihar, former Project Director, Loktak Development Authority, Manipur for providing laboratory facilities. We are also thankful to India Canada Environment Facility (ICEF) for extending financial assistance to carry out the work.

References

APHA, 1989. *Standard Methods for the Examination of Waters and Wastewaters,* 17th edn. American Publication Health Association, Inc., New York.

BIS, 1991. *Indian Standard: Drinking Water–Specification* (*First Revision*). IS 10500: Bureau of Indian Standards publication, pp. 8.

Biswas, K.P., 1990. *A Textbook of Fish, Fisheries and Technology.* Narendra Publishing House, Delhi, pp. 531.

Devi, R.K.R., Dhamendra, H., Gyaneshwari, R.K. and Kosygin, L., 2005. Limnological studies of Nambol river, Manipur with a note on its aquatic bio-resources. *Indian J. Environ. and Ecoplan.,* 10: 831–834.

ISI, 1982. *Indian Standard, Tolerance Limit for Inland Surface Waters Subject to Pollution* (Second revision). Indian Standards Institution publication, pp. 18.

Jayaram, K.C., 1981. *Freshwater Fishes of India, Pakistan, Bangladesh, Burma, and Sri Lanka.* Handbook, Zoological Survey of India, Calcutta, pp. 475.

Khatavkar, S.D. and Trivedy, R.K., 1993. Ecology of freshwater reservoirs in Maharashtra with reference to pollution. I. The water bodies and pollution load. In: *Ecology and Pollution of Indian Lakes and Reservoirs,* (Eds.) Mishra P.C. and R.K. Trivedy. Ashish Publishing House, New Delhi, pp. 27–55.

Prakash, K.L., Nataraj, A.G., Somashekar, R.K. and Rao, N.M., 2005. A model approach for the water quality: A case study of river Cauvery. *Indidn J. Environ. and Ecoplan.,* 10: 557–564.

Sand-Jensen, K., 2001. Freshwater ecosystems, human impact on. In: *Encyclopedia of Biodiversity,* (Ed.) S.A. Levin. Academic Press, New York, 3: 89–108.

Rao. N.V.S., 1989. *Handbook: Freshwater Molluscs of India.* Zoological Survey of India, Calcutta, pp. 289.

Reginaa, B. and Nabi, B., 2003. Physico-chemical spectrum of the Bhavani river water collected from the Kalingarayan Dam, Tamil Nadu. *Indian J. Envion. and Ecoplan.,* 7: 633–636.

Talwar, P.K. and Jhingran, A.G., 1991. *Inland Fishes of India and Adjacent Countries.* Oxford and IBH Publ. Co. Pvt. Ltd., New Delhi, pp. 541.

Trivedy, R.K. and Goyal, P.K., 1986. *Chemical and Biological Methods for Water Pollution Studies.* Environmental Publication, Karad, pp. 248.

USEPA, 1976. *Quality Criteria for Water.* EPA–440/9–76–023, Washington, D.C.

Chapter 39

Growth and Characterization of Rare Earth Metal Ion (La^{3+}, Pr^{3+}, Sm^{3+}) Doped Potassium Acid Phthalate Single Crystals

K. Uthayarani[1], R. Sankar[2], C.K. Shashidharan Nair[1]***

[1]Department of Physics, P.S.G. College of Technology, Coimbatore – 641 004, India

[2]Crystal Growth Centre, Anna University, Chennai – 600 025, India

ABSTRACT

Single crystals of potassium acid phthalate (KAP) and rare earth metal ion (La, Pr, Sm) doped KAP were grown from aqueous solutions by slow cooling method. The grown crystals were characterized using powder XRD and FTIR analysis. EDAX analysis confirmed the incorporation of the dopants in the parent KAP lattice. Thermal stability of KAP in the presence of dopants was analysed using TG/DTA studies and the maximum temperature for non-linear optical application of this compound in the presence of dopants was found out. The SHG measurements were carried out for all the crystals.

Keywords: Single crystal growth, KAP, Non-linear optical material.

Pacs: 81.10 Dn, 42,70 Mp, 78.20.-C.

---

* Corresponding Author: E-mail: ck_shashidharan@gmail.com; **uthayakalarikkal@rediffmail.com.

Introduction

In the field of plasma physics and X-ray fluorescence spectroscopy, the single crystals of potassium acid phthalate (KAP) are of very great interest (Bearden and Huffman, 1963). This crystal has a non-centrosymmetric C^5_{25}- P_{21ab} space group with lattice parameters a = 9.605 A°, b = 13.331 A° and C = 6.473 A° (Okaya, 1965). They have excellent physical properties and a good record for long term stability in devices (Beck *et al.*, 1995; Jason *et al.*, 2003; Kejalakshmy and Srinivasan, 2003). Its electro-optical and nonlinear optical properties have been reported (Miniewicz, 1993). The influence of difference impurities on KDP showed that the trivalent cations greatly influence on the process of step growth (Hottenbhis and Lucasius, 1986). It was pointed out that there is a close relation between impurity effectiveness of an arbitrary cation on the KAP (010) face and its dehydration frequency. The lower this frequency, the stronger is its inhibiting effects. The influence of La^{3+} on the growth and properties of KDP single crystals have been carried out (Kannan *et al.*, 2006) and similarly in TGS crystals rare earth ion dopants like, La, Ce and Nd have been incorporated and their influence on various properties have been investigated'). But no report is available on the influence of rare-earth ion doping on KAP crystals.

Hence in this present work the influence of rare earth metal ions like lanthanum (La^{3+}), praseodymium (Pr^{3+}) and samarium (Sm^{3+}) on the various properties of KAP crystals were studied and reported.

Experimental Details

Analar grade KAP was used as solute. Milli-pore water of resistivity 18.2 MQ/cm was used as solvent. The dopant materials were taken in the form of their nitrates. Mother solution was prepared according to the solubility condition. Equal weight percentage (5 per cent by weight) of all the dopants were taken and added to the mother solution and seeds were obtained. Each solution was saturated at 45°C. Slow cooling method was adopted for the growth ot crystals using an optically heated constant temperature bath of control accuracy ±0.01°C. The growth parameters were kept constant for all the dopants. After the growth nm, the pure and doped crystals were harvested. The grown crystals were characterized with powder XRD for structural properties. The incorporation of dopants in parent KAP crystals is confirmed using EDAX analysis. Further FTTR analysis was carried out for pore as well as doped crystals. The second harmonic generation was observed using Nd: YAG laser for all the crystals.

Results and Discussions

XRD Analysis

The structure of the crystals KAP, (La, Pr, Sm) doped KAP was examined by X-ray powder diffraction analysis using Rich Seifert (Model 2002) diffractometer with a CuK_a radiation ($\lambda = 1.5418$ A°). The peaks of KAP powder XRD pattern were indexed for pure KAP crystal using the single crystal XRD data recorded for pure KAP crystal using ENRAF nonius CAD4 diffractometer. The indexed peaks matches with powder XRD data of file number 15-805 of KAP (PDIS). The lattice parameters were in good agreement with the literature. The recorded spectra are shown in the Figures 39.1a–d. In order to establish the entry of the rare earth metal ions La^{3+}, Pr^{+} and Sm^{3+} into the KAP lattice, the 28 values of the characteristics peaks were compared. There is a characteristic shift indicated in the Table 39.1.

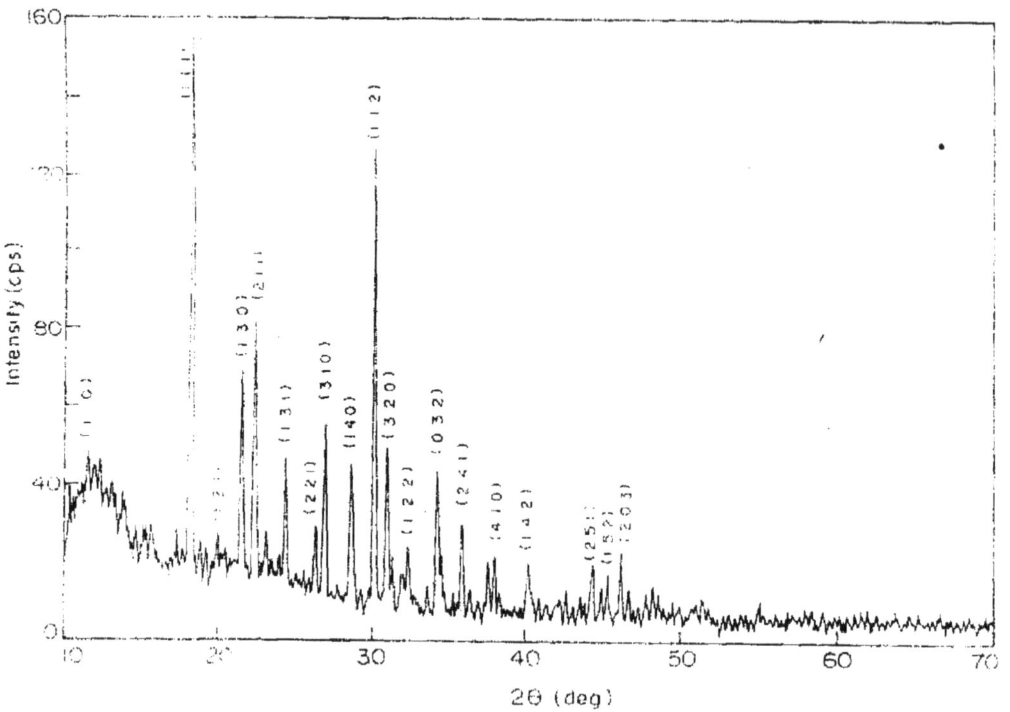

Figures 39.1(a): XRD spectra of pure KAP

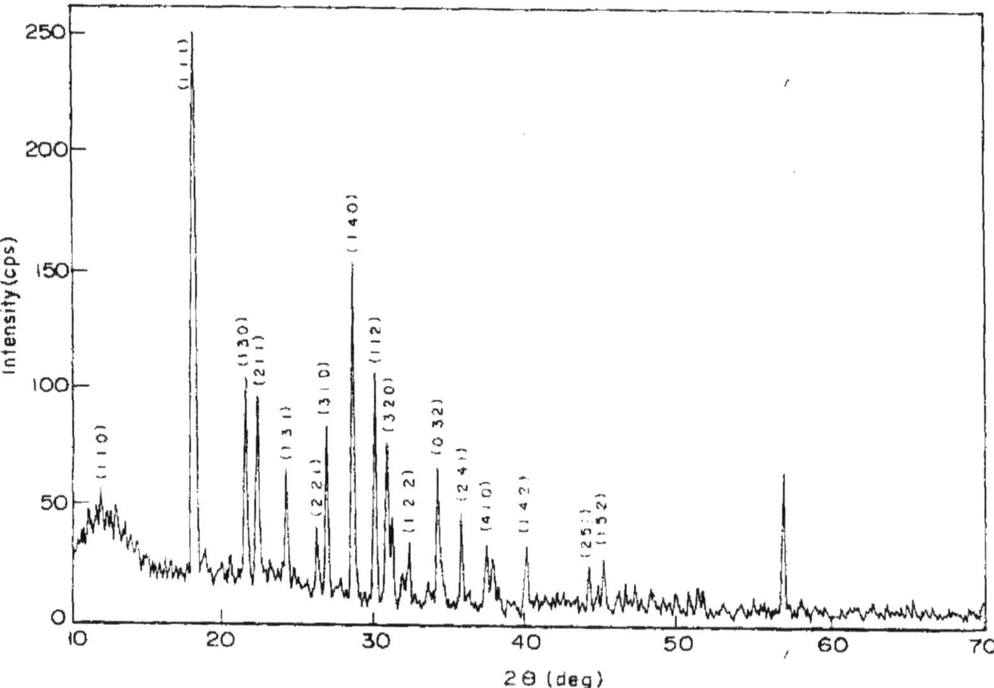

Figures 39.1(b): XRD spectra of La : KAP

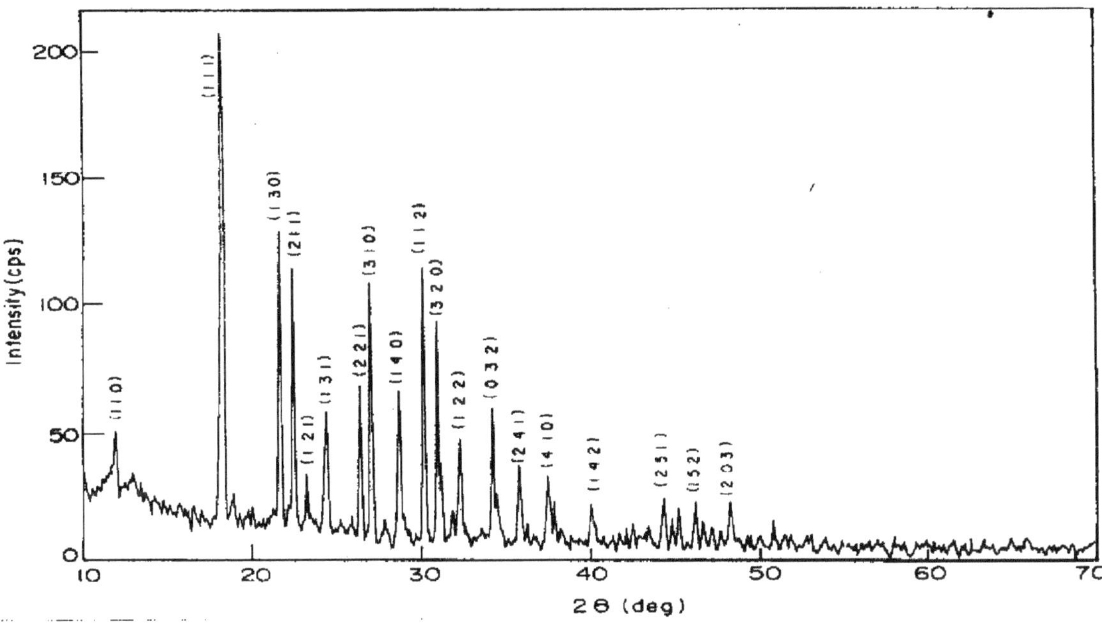

Figures 39.1(c): XRD spectra of Pr : KAP

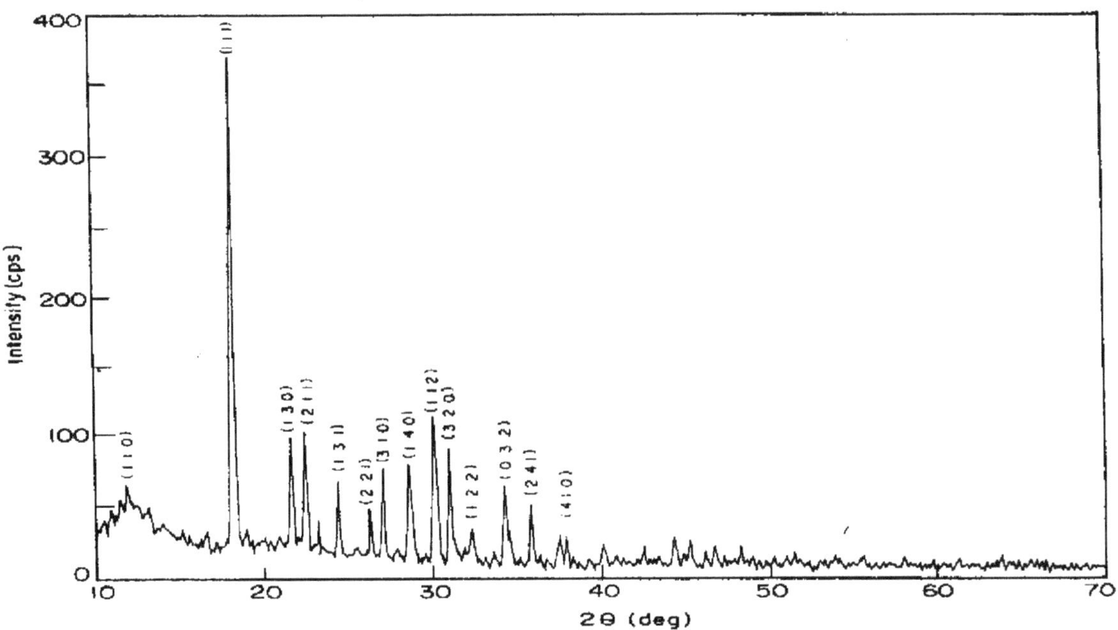

Figures 39.1(d): XRD spectra of Sm : KAP

Table 39.1: 2θ Values of Pure and Doped KAP

KAP	La : KAP	Pr : KAP	Sm : KAP
11.4518	11.8371	11.8440	11.8588
18.2229	18.1274	18.1772	18.1545
22.4272	22.4044	22.3713	22.3865
44.2907	44.3127	44.2584	44.2867

This shift in '2θ' values clearly indicate the entry of lanthanum, praseodymium and samarium into the parent crystal lattice.

Energy Dispersive Analysis by X-ray Diffraction (EDAX)

The energy dispersive X-ray diffraction analysis was employed to find the stoichiometric composition using KWA 200 system connected to a LEO-stage scan 440 scanning electron microscope. EDAX spectra of La : KAP, Pr : KAP and Sm : KAP are shown in the Figures 39.2a–c. The atomic percentage of metals in each crystal is indicated in the Table 39.2. The level of La, Pr and Sm doping in KAP is observed to be nearly equal.

Table 39.2: Atomic Per cent of Dopants

Element	Atomic Per cent of Metals	Weight (Per cent) of Metals
Pure KAP	100	100
KK		
La : KAP	1.16	4.00
KK	98.84	96.00
Pr : KAP	1.10	4.07
KK	98.90	95.93
Sm : KAP	1.61	5.91
KK	98.39	94.09

FTIR Analysis

The FTIR spectra of La^{3+}, Pr^{3+} and Sm^{3+} doped KAP are shown in Figures 39.3b–d respectively. The spectrum of acid phthalate is also given (39.3a). Comparison of the spectra of KAP and that of La^{3+} doped KAP, although shows similar features, the C = O stretching vibration of La^{3+} doped KAP is shifted to lower wave number (1670 cm^{-1}) than KAP with La^{3+}. As the COO^- vibration (1283 cm^{-1}) appears broad, the shift in peak position as a result of La^{3+} interaction is not clearly evident. Even then, there is a slight shift of about 2 cm^{-1} to lower wave number in La : KAP compared to parent KAP.

Similarly the spectrum of Pr^{3+} doped KAP also shows such shift in vibrations, thus proving entry of Pr in the lattice of KAP. The spectrum of Sm doped KAP, although presence of Sm is established by EDAX analysis, does not show significant shift in C = O stretching vibration, but the asymmetric COO^- vibration of Sm doped KAP is shifted by 5 cm^{-1} compared to KAP. The asymmetric COO^- stretch of Sm : KAP occurs at 1565.4 cm^{-1} whereas the same vibration in KAP occurs at 1560 cm^{-1}. It appears to be characteristic evidence of presence of Sm in the lattice of KAP.

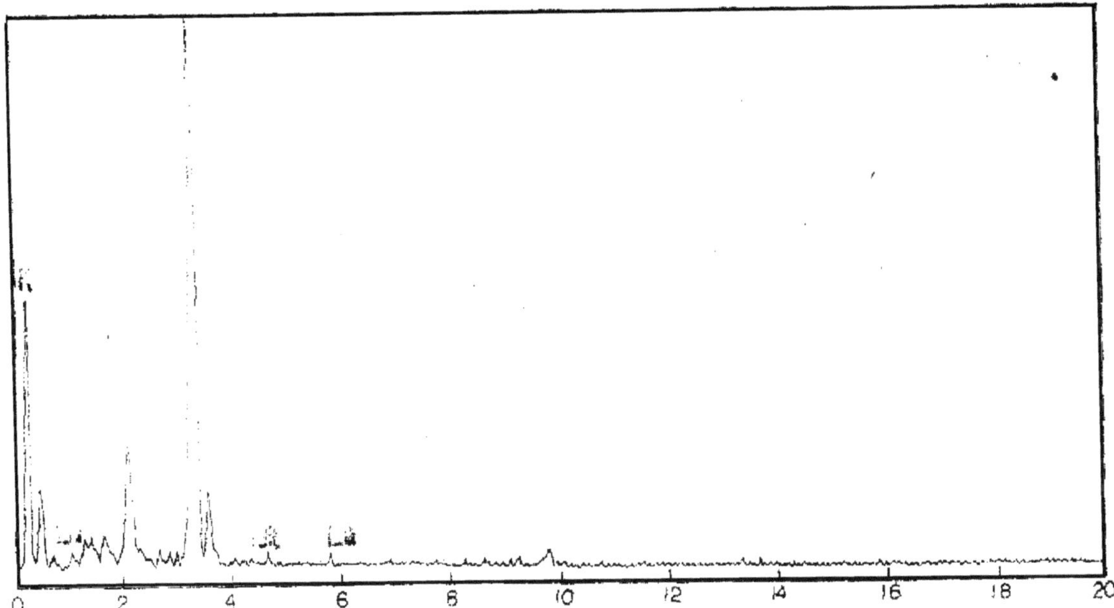

Figure 39.2(a): EDAX spectra of La : KAP

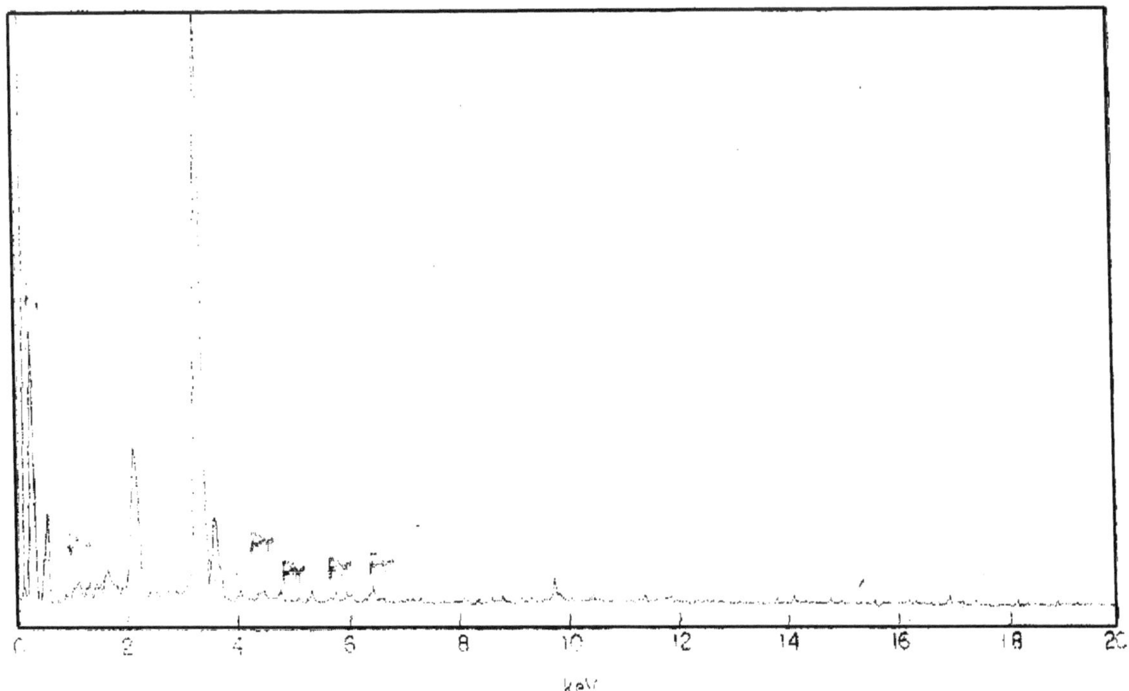

Figure 39.2(b): EDAX spectra of Pr : KAP

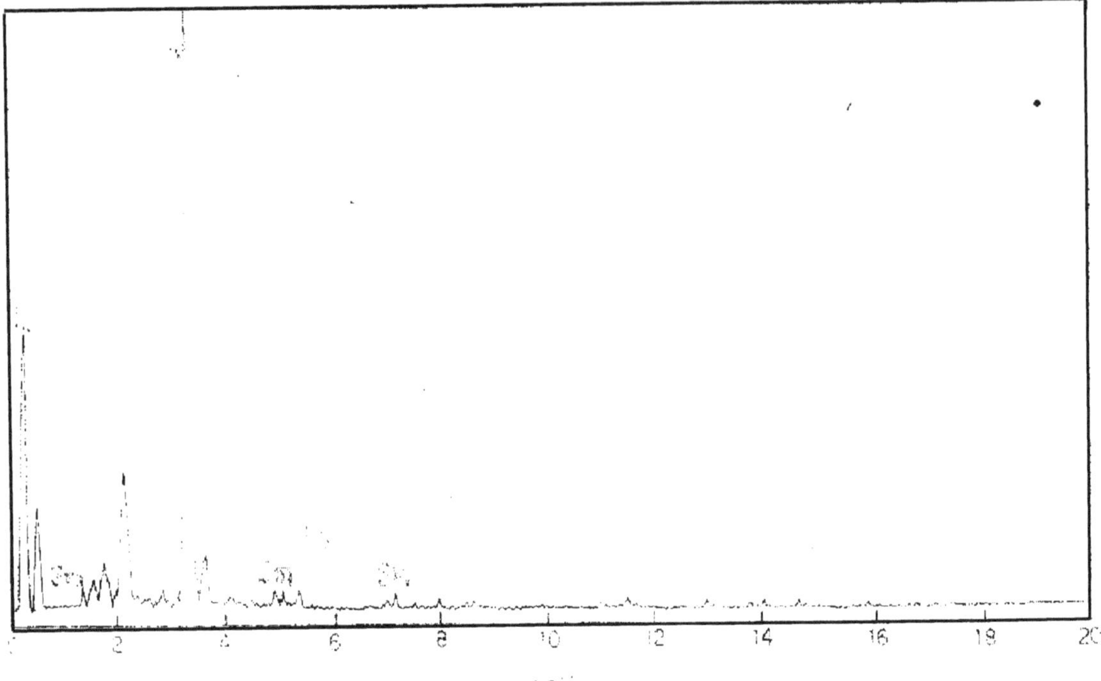

Figure 39.2(c): EDAX spectra of Sm : KAP

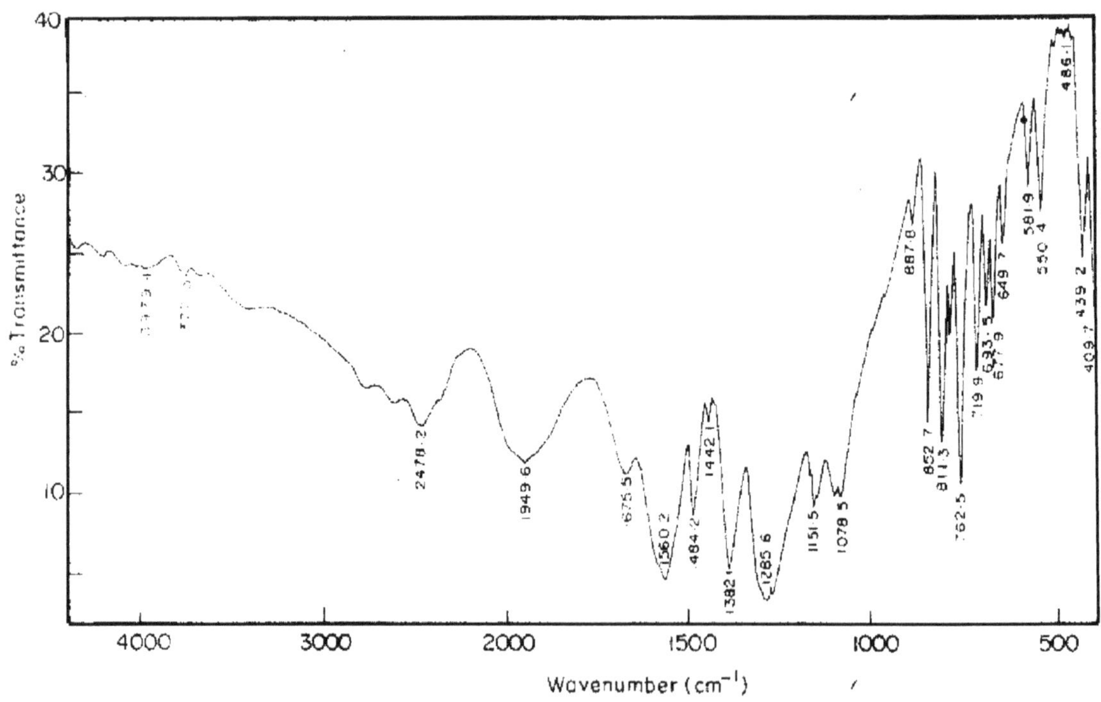

Figure 39.3(a): FTIR spectra of pure KAP

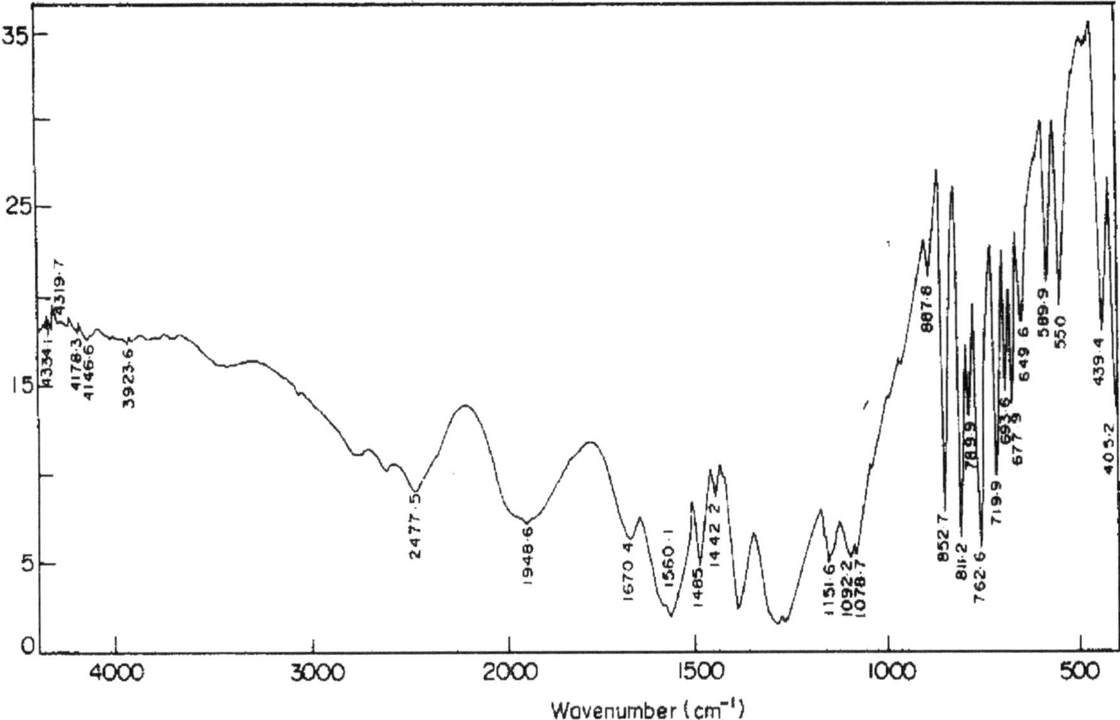

Figure 39.3(b): FTIR spectra of La : KAP

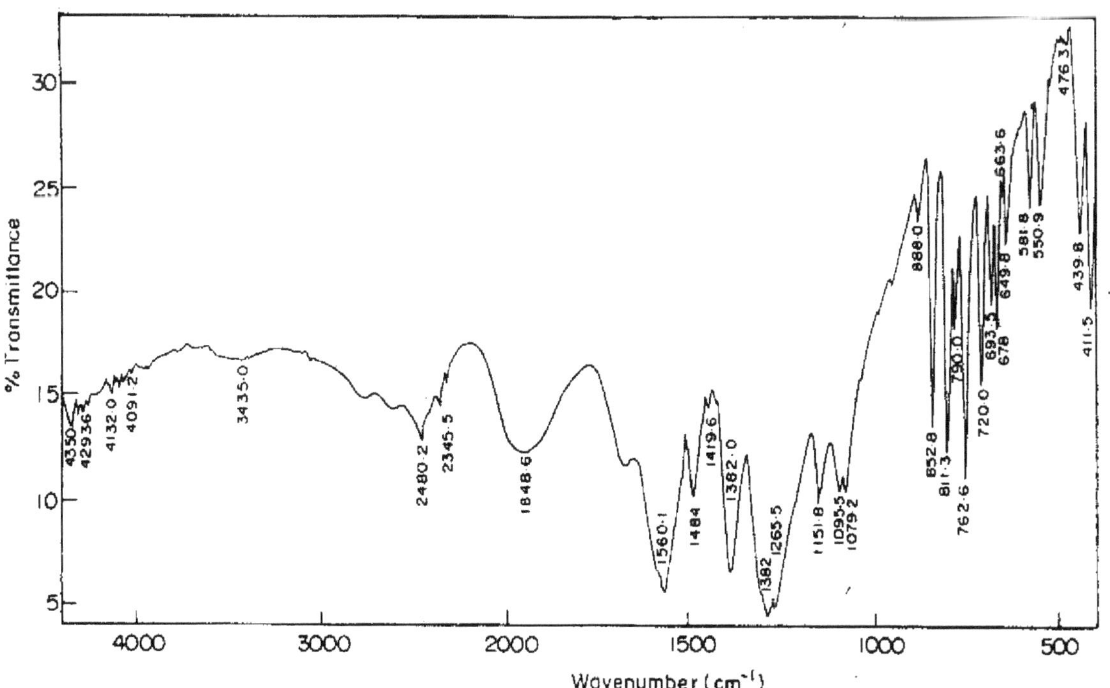

Figure 39.3(c): FTIR spectra of Pr : KAP

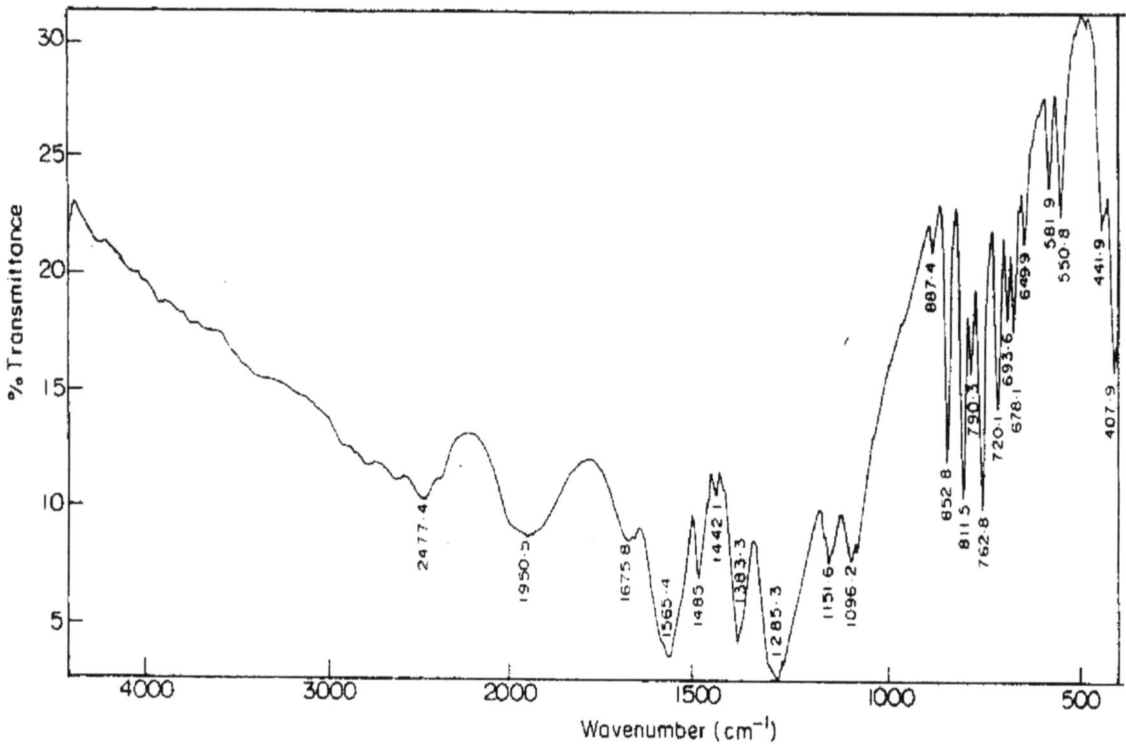

Figure 39.3(d): FTIR spectra of Sm : KAP

From the foregoing analysis, it is therefore established, KAP appears to have affinity to accommodate rare earth metal ions in its lattice.

Conclusion

Good quality single crystals of pure and raxre-earth metal ion doped KAP single crystals were grown by slow cooling method. The slight shift in the 28 values observed using powder X-ray diffraction analysis reveals the entry of the dopants in the crystal lattice of KAP. The incorporation of the dopants was further confirmed by EDAX and FTIR analysis. The SHG emission was confirmed for all the doped KAP crystals.

References

Bearden, A.J. and Huffman, F.N., 1963. *Rev. Sci. Inst.*, 34: 1233.

Okaya, Y., 1965. *Acta. Crystallog.*, 19: 879.

Beck, L., Stemmler, P. and Legrand, F., 1995. *Rev. Sci. Instru.*, 66: 1601.

Jason, B., Benedict, Paul M. Wallace, Philip T. Reid, Sci-Hum Tang and Kahr, B., 2003. *Adv. Mater.*, 15: 1068.

Kejalakshmy, N. and Srinivasan, K., 2003. *J. Phys. D: Appl. Phys.*, 36: 1778.

Miniewicz, A., 1993. *Mol. Cryst. Liq. Crys.*, 229: 13.

Hottenbhuis, M.H.J. and Lucasius, C.B., 1986. *J. Cryst. Growth,* 78: 379.

Kannan, V., Bairava Ganesh, R., Satyalakshmi, R., Rejesh, N.P. and Ramasamy, P., 2006. *Crys. Res. Technol.,* 41(7): 678–682.

Muralidharan, R., Mohan Kumar, R., Ushashree, P.M., Jayavel, R. and Ramasamy, P., 2002. *J. Cryst. Growth,* 234: 545–550.

PDIS, Power diffraction file, Organic Volume no. PDIS–15iRB, file No. 15–805, JCPDS, Pennsylvania, pp. 340.

Chapter 40

Metal Induced Augmentation, Stabilization and Transistion of a Thermophillic Xylose Isomerase Activity in *Opuntia vulgaris*

S. Ravikumar, S. Shyamala and K. Srikumar*
Department of Biochemistry and Molecular Biology, School of Life Sciences, Pondicherry University, Pondicherry – 605 014

ABSTRACT

Xylose isomerase (OVXI) enzyme activity investigated in Xerophytic *Opuntia vulgaris* cladode homogenate (10 per cent w/v) 10,000 xg supernatant comprised of three temperature isoforms (I, II and III) and three pH isoforms when assayed in the temperature range 30°C–100°C and in the pH range 3.5–10.5 respectively. Isoform I was active at 40°C, II at 70°C and III at 90°C when assayed at pH 7.5. The OVXI activity at pH 10 was found to be 50 per cent of its activity at pH 7.5. Determinatior of isomerase activity was also carried out in the presence of metal ions. (Mg^{2+}, Mn^{2+}, Co^{2+} and Ca^{2+}), each at 1 mM, 5 mM, and 10 mM concentration and in the temperature range cited earlier. Metal ion induced shift in optimal activity of the OVXI at 80°C was observed. Mg^{2+} and Co^{2+} augmented this isomerase activity and yielded greater temperature stability for isoforms II and III, where as Ca^{2+} was found inhibitory to all isoforms studied. Metal

---

* Corresponding Author: Phone: +91-413-2655991 Ext. 422; E-mail: kotteazeth_srikumar@excite.com.

stabilization of the OVXI activity was in the order $Mg^{2+} > Co^{2+} > Mn^{2+} > Ca^{2+}$. We report that the OVXI activity comprises of thermostable isoforms exhibiting broad pH specificity.

Keywords: *Xylose Isomerase,* Mg^{2+}, Mn^{2+}, Co^{2+}, Ca^{2+}, *Opuntia vulgaris.*

Introduction

D-Xylose/D-Glucose isomerase (EC 5.3.1.5) in microbes and in plants (Viellie *et al.*, 2001) catalyzed the isomerization of either D-Xylose to D-Xylulose in vivo or D-glucose to D-fructose *in vitro* (Takasaki *et al.*, 1969). Xylose isomerase is a catalytic activity that finds application in the industry for the production of high fructose corn syrup, and for the production of ethanol and Xylitol from Xylan (Antrim *et al.*, 1979; Crabb and Mitchinson, 1997). This enzyme reportedly requires divalent cations for its catalytic activity (Meng *et al.*, 1993) while Xylose isomerase from different organisms utilized different cations for their function (Marg and Clark, 1990). Metal ions reportedly stabilized the enzyme's native structure (Chen, 1980; Whitlow *et al.*, 1991). The identification of thermostable isoforms of Xylose isomerase in *Opzmlia ntlgaris* that exhibited metal ion dependency for its' activity is reported here.

Materials and Methods

All chemicals used were of analytical grade purchased from suppliers in India. Freshly collected cladodes of *Opuntia vulgaris* were used for the preparation of a 10,000 x g supernatant of the cladode homogenate (20 per cent w/v) used as the enzyme source in our studies.

Enzyme Assay

The OVXI activity in the enzyme source was measured in the temperature range 30°C–100°C by the method of Lee *et al.*, (1990). The specific activity (units mg^{-1}) was calculated using the total protein concentration in the enzyme source as estimated by the method of Lowry *et al.* (1951).

Effect of Metal Ions on Enzyme Activity

The effect of metal ions on the OVXI activity of the enzyme source was determined in independent experiments employing fixed 1 mM, 5 mM, and 10 mM concentration of Mg^{2+}, Mn^{2+}, Co^{2+} and Ca^{2+}, each in their chloride form and in the temperature range 30°C–100°C (Lee *et al.*, 1990) as before.

Results and Discussion

Results of our studies indicated the existence of three temperature isoforms (Figure 40.1) for the OVXI enzyme activity, namely at 40°C, 70°C and 90°C respectively. The pH profile yielded broad activities (Figure 40.2), at pH 4.0, pH 7.5 and pH 10.0 respectively. In the presence of a given concentration (1, 5, 10 mM) of Mg^{2+}, Mn^{2+}, Co^{2+} and Ca^{2+} variable OVXI activity varied compared to that of the respective control (Table 40.1). The optimum concentration for use was found to be 1 mM for Mg^{2+} and Co^{2+} and 5 mM for Mn^{2+}. Mg^{2+} augmented the OVXI activity at 70°C as well as at 90°C and stabilized the enzyme activity at 80°C. The metal ion treated samples XI yielded reduced activity at 100°C. The presence of Mn^{2+} generally reduced the XI activity at 70°C and 80°C, whereas a moderate increase in the enzyme activity was observed at 90°C for the sample containing 5 mM Mn^{2+}, Ca^{2+} reduced the XI activity by 30 per cent at 90°C. The metal ion activation effect was found to be in the order $Mg^{2+} > Co^{2+} > Mn^{2+} > Ca^{2+}$. Although the enzyme activity was observed for control and the metal ion containing samples at each temperature studied. XI activity at 30°C, 50°C and 60°C fell on the declining slopes of the peak activity at 40°C (isoform peak).

Table 40.1: Effect of Divalent Cation Concentration on OVXI Activity

Metal Ions Added in the Form of Chloride	*Temperature (°C)*							
	30	*40*	*50*	*60*	*70*	*80*	*90*	*100*
				Mgcl$_2$				
Control Enzyme	14.6	42.05	32.617	21.459	42 839	31.326	37.474	24.32
Enz+1mM	23.604	43.204	34.327	34.048	47.066	45.207	44.631	22.603
Enz+5mM.	22.603	42.918	31.178	22.603	44.777	36.48	43.487	21.602
Enz+10mM	30.61	39.627	30.896	24.892	45.493	40.486	42.918	18.597
				CoCl$_2$				
Control Enzyme	10.872	38.912	17.593	22.603	28.039	21.315	25.752	11.444
E nz+1mM	17.167	43.776	21.04	24.606	32.211	34.363	27.178	10.583
E nz+5mM	23.175	38.912	20.468	24.034	30.042	28.492	25.601	9.249
E1Z+10mM	22.174	26.609	19.174	18.311	29.62	21.689	13.733	10.3
				MnCl$_2$				
Control Enzyme	11.955	42.059	11.444	10.447	28.894	18.025	21.315	13.876
Enz+1mM	12.073	46.63	20.886	12.16	31.473	22.174	25.464	15.307
Enz+5mM	12.589	52.646	26.17	16.451	38.197	27.038	29.176	19.885
Enz+10mM	11.437	44.634	21.609	14.559	30.05	24.892	27.609	17.596
				CaCl$_2$				
Control Enzyme	15.164	37.767	20.028	13.733	35.622	18.311	38.34	11.874
Enz+1mM	13.304	42.345	14.162	12.017	30.328	13.733	29.899	12.011
Enz+5mM	10.872	34.62	15.164	7.153	29.184	12.589	23.891	8.011
Enz+10mM	9.585	30.042	11.874	5.722	26.609	11.157	26.037	9.012

Values are expressed in enzyme activity (U/ml).

Thermostable enzymes find use in industrial processes that utilized biocatalytic activities for processing raw materials (or) reaction intermediates at high temperature (Bently and Williams, 1996; Nichaus *et al.*, 1999). Thus far, thermophillic enzymes from thermophillic microbial sources have found great use for the purpose (Vieille *et al.*, 1996). Identification of thermophillic enzymes from alternate natural sources had also remained elusive thus far. Current investigations indicate the existence of a high temperature stable XI enzyme activity in the xerophytic species, *Opuntia vulgaris*. In addition to being thermostable up to 90°C, OVXI exhibited alkali stability at pH 10.0, retaining greater than 50 per cent of its' catalytic activity, at pH 7.5 (Callens *et al.*, 1985). Even though the free enzyme is active at 40°C, 70°C and 90°C (temperature isoforms), the addition of divalent metal ions to the assay cocktail yielded enhanced activity for temperature isoformes. Although both Mg^{2+} and Co^{2+} were essential for activity, Mg^{2+} was found to be an activator, while Co^{2+} responsible for stablised the enzyme activity at different temperatures. In contrast Ca:' was found to be an inhibitor. Further the temperature optima of XI isoforms I and II, varied significantly with the nature of metal ion added to

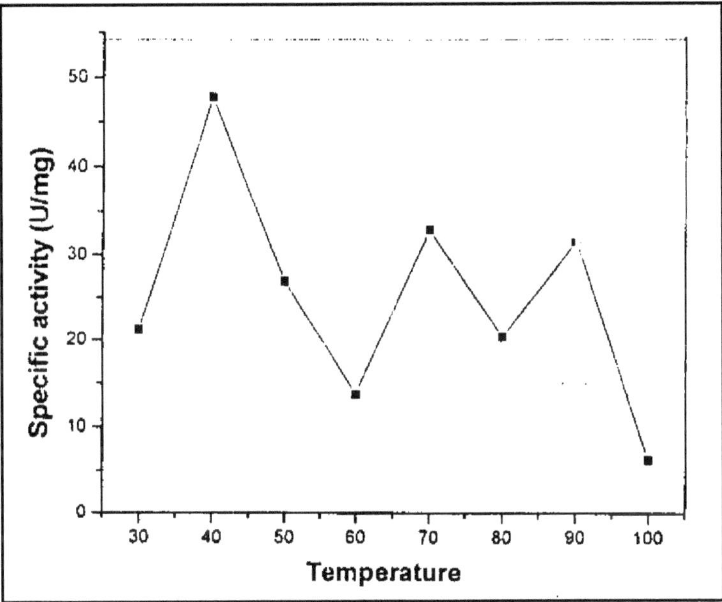

Figure 40.1: Effect of Temperature on the Activity of OVXI

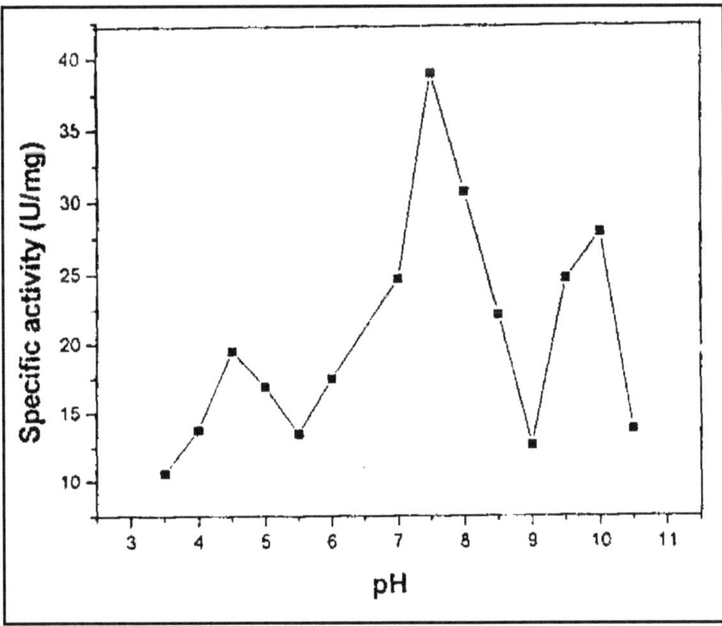

Figure 40.2: Effect of pH on the Activity of OVXI

the reaction. While isoform I, remained without appreciable loss of enzyme activity. It was interesting to note that the addition of Mg^{2+} and Co^{2+} to the assay cocktail at 100°C enhanced XI activity by 20 per cent. Thus, three peaks of XI activity were discerned from the temperature study. Interestingly repeated determinations of the temperature profile with the presence of Mg^{2+} and Co^{2+} indicated a shin (transition) in the catalytic optimum or this enzyme from 90°C to 80°C. The XI activity with the temperature optimum at 40°C remained as a distinct isoform.

References

Antrim, R.L., Colilla, W. and Schnyder, B.J., 1979. Glucose isomerase production of high-fructose syrup. *Appl. Biochem. Bioeng.*, 2: 97–155.

Bentley, I.S. and Williams, E.C., 1996. Starch conversion. In: *Industrial Enzymology*, 2nd edn. (Eds.) T. Godfrey, and S.I. West. Stockton Press, New York, p. 339–357.

Callens, M., Kersters-Hilderson, H., Vangrysperre, W., Van Opstal, O. and De Bruyne, C.K., 1985. Metal ion binding to D-xylose isomerase from Streptomyces violaceoruber. *Arch. Int. Physiol. Biochim.*, 93: 129–136.

Chen, W.P., 1980. Glucose isomerase (a review). *Process Biochem.*, 15: 36–41.

Crabb, W.D. and Mitchinson, C., 1997. Enzymes involved in the processing of starch to sugars. *Trends Biotechnol.*, 15: 349–352.

Lee, C., Bhatnagar, L., Saha, B.C., Lee, Y.E., Takagi, M., Imanaka, T., Bagdsarian, M. and Zeikus, J.G., 1990. Cloning in expression of the Clostridium-Thermosulfurogenes Glucose-isomerase gene in *Escherichia coli* and *Bacillus subtilis*. *Applied and Environmental Microbiology*, 56: 2638–2643.

Lowry, O.H., Rosebrough, N.J., Farr, A.L. and Ranuall, R.J., 1951. Protein measurement with the Folin phenol reagent. *J. Biol. Chem.*, 193: 265–275.

Marg, G.A. and Clark, D.S., 1990. Activation of the glucose isomerase by divalent cations: Evidence for two distinct metal-binding sites. *Enzyme Microb. Technol.*, 12: 367–373.

Meng, M.H., Bagdasarian, M. and Zeikus, J.G., 1993. Thermal stabilization of xylose Isomerase from Thermoanaerobacterium-Thermosulfurigenes. *Biotechnology*, 11: 1157–1161.

Niehaus, F., Bertoldo, C., Kahler, M. and Antranikian, G., 1999. Extremophiles as a source of novel enzymes for industrial application. *Appl. Microbiol. Biotechnol.*, 51: 711–729.

Takasaki, Y., Kosugi, Y. and Kanbayashi, A., 1969. Studies on sugar isomerizing enzyme. *Agric. Biol. Chem.*, 33: 1527–1534.

Vieille, C., Burdette, D.S. and Zeikus, J.G., 1996. Thermozymes. *Biotechnol. Annu. Rev.*, 2: 1–83.

Vieille, C., Sriprapundh, D., Kelly, R.M. and Zeikus, J.G., 2001. In: *Methods in Enzymology*, Vol 330, (Eds.) Adams, M.W.W. and R.M. Kelly. Academic Press, New York, pp. 215–224.

Whitlow, Moo Howard, A.J., Finzel, B.C., Poulos, T.L., Winborne, E. and Gilliland, G.L., 1991. *A* metal-mediated hydride shin mechanism for xylose isomerase based on the 1.6 Å Streptomyces rubiginosus structures with xylitol and D-xylose. *Proteins*, 9: 153–173.

Chapter 41

Effect of Weed Control Treatments on Dry Matter Production and Crop Growth Rate in Direct Sown Wet Seeded Rice

B. Rajendra Kumar* and A. Christophrer Lourduraj**
Department of Agronomy, Tamil Nadu Agricultural University, Coimbatore

ABSTRACT

A Field experiment was conducted at Tamil Nadu Agricultural University, Coimbatore during rabi season 2002–2003. The performance of ASD 19 was superior than CO 43 and ADT 38. The crop growth rate and dry matter production were higher in ASD 19 at all the stages of crop growth. Combination of ASD 19 and weed control treatment Pretilachlor + Safener 0.45 kg.a.i. ha^{-1} at 4DAS + 1 hand weeding at 35 DAS resulted in low weed density, lesser total weed dry matter production and higher NPK uptake by crop resulting in higher' grain yield and straw yield. Initial suppression of weeds with Pretilachlor + Safener led to an increase in dry matter production of rice and higher values of Crop growth rate at all the stages of crop growth.

***Keywords**: Dry matter Production, Crop growth rate, Rice.*

Introduction

It is very difficult to produce rice economically without good crop management including weed control in direct seeded rice even from varieties with high yield potential. Thus, crop establishment

---

* Corresponding Author and Agricultural Officer, Agricultural Research Station, Amadalavalasa, ANGRAU, A.P.

* * Associate Professor, Department of Agronomy, Tamil Nadu Agricultural University, Coimbatore.

and weed control techniques are critical in rice farming especially in direct seeded rice. It is not uncommon to see wet seeded rice fields smothered by weeds, mainly grasses and broad leaf weeds. In transplanted rice, young seedlings establish early and compete well with emerging weeds. Standing water also reduced the density of many of the weed species. However, weed competition is more severe in wet seeded rice as weed seeds are in advantageous position than rice seeds. Weed competition reduces plant height, dry matter accumulation, productive tillers, grain number and ultimately the grain yield. Several workers have reported the critical period of weed competition is the first quarter of the crop growth. Morphological characteristics of rice and some important rice weeds have been studied to examine how morphological traits affect their competitiveness for light in direct seeded rice Caton *et al.* (1997), Yamasue *et al.* (1997).

Materials and Methods

A Field experiment was conducted at Tamil Nadu Agricultural University, Coimbatore during rabi season 2002–2003. The experiment was laid out in split plot design with three varieties ASD 19, ADT 38 and CO 43 in the main plot and weed management practices (Pretilachlor + safener 0.45 kg a.i. ha^{-1} at 4 DAS + 1 hand weeding at 35 DAS; Butachlor 1 kg a.i. ha^{-1} at 8 DAS + 1 hand weeding at 35 DAS; Butachlor + Propanil 0.84 + 0.84 kg a.i. ha^{-1} at 10 DAS (tank mix) + 1 hand weeding at 35 DAS; two hand weedings at 20 and 35 DAS; unweeded check) in the subplots with three replications.

In each plot, outside the net plot, but within the border rows, plants from 0.25 m^2 area were cut close to ground level, chopped and dried. The air dried samples were oven dried at 80°C for 72 hours. Dry weight of the samples were recorded at active tillering, flowering and at maturity. The dry matter production was expressed in kg ha^{-1}. Crop growth rate was worked out at tillering to flowering and from flowering to maturity and expressed as gram of dry matter produced per day.

Results and Discussion

There was significant difference in dry matter production (DMP) among the varieties at tillering, flowering and harvest stages. At tillering stage, highest DMP was recorded by CO 43 (1596 kg ha^{-1}), followed by ASD 19 and did not differ significantly with each other. Lowest DMP (1498 kg ha^{-1}) was recorded by ADT 38. At flowering stage, highest DMP was recorded by ASD 19 (M_1) (9800 kg ha^{-1}), followed by CO 43 and did not differ significantly with each other. Lowest DMP (8697 kg ha^{-1}) was recorded by ADT 38. At harvest stage CO 43 recorded significantly highest DMP of 10794 kg ha^{-1}. The lowest DMP (11304 kg ha^{-1}) was recorded by ASD 19.

The weed control treatments significantly influenced the DMP at all the stages of crop growth, *i.e.* tillering, flowering and harvest. At tillering stage, hand weeding twice at 20 and 35 DAS (S_4) recorded significantly higher DMP of 1719, which was not significantly different from Pretilachlor + safener 0.45 kg a.i. ha^{-1} + 1 hand weeding at 35 DAS (S_1) and Butachlor +Propanil 0.84 + 0.84 kg a.i. ha^{-1} +1 hand weeding at 35 DAS (S_3). The lowest DMP (1072 kg ha^{-1}) was recorded in unweeded check (S_5). At flowering stage, S_1 recorded significantly higher DMP of 11581 kg ha^{-1}. Lowest DMP of 4685 kg ha^{-1} was recorded by S_5. At harvesting also same trend was observed with S_1 recording a DMP of 13450 kg ha^{-1} and S_5, 5617 kg ha^{-1}. Interaction effect was significant only in the early stages *i.e.* at tillering and flowering stages. At tillering stage, highest DMP was recorded by CO 43 with two hand weedings (M_3S_4) (1832 kg ha^{-1}) and the lowest was recorded by ADT 38 when left unweeded (M_2S_5) (1005 kg ha^{-1}). At flowering stage, highest DMP was recorded by ASD 19 with Pretilachlor + safener (M_1S_1) (12253 kg ha^{-1}) and the lowest was recorded by CO 43 when left unweeded (M_3S_5) (4479 kg ha^{-1}). Interaction effect was not significant at harvest stage.

Table 41.1: Effect of Weed Control Treatments on Dry Matter Production (kg ha^{-1}) in Direct Sown Wet Seeded Rice

Treatments	*Tillering*				*Flowering*				*Harvest*			
	M_1	M_2	M_3	*Mean*	M_1	M_2	M_3	*Mean*	M_1	M_2	M_3	*Mean*
S_1	1717	1643	1772	1710	12253	10843	11649	11581	14066	12711	13573	13450
S_2	1610	1595	1632	1612	10509	8171	9355	9345	12039	10016	10917	10990
S_3	1629	1621	1624	1625	10608	8804	9514	9642	12355	10764	11208	11442
S_4	1698	1629	1832	1719	10968	10775	10602	10775	12506	12646	12769	12640
S_5	1091	1005	1120	1072	4661	4915	4479	4685	5557	5787	5507	5617
Mean	1549	1498	1596		9800	8697	9120		11304	10384		10794

	SEd	CD (P=0.05)	SEd	CD (P=0.05)	SEd	CD (P=0.05)
M	28.80	79.96	113.09	314.01	106.74	296.36
S	49.88	102.96	183.35	378.43	191.54	395.32
M at S	82.48	177.39	305.74	660.84	315.35	NS
S at M	86.41	178.34	317.58	655.46	331.76	NS

M_1: ASD 19; M_2: ADT38; M_3: CO 43.

S_1: Pretilachlor + safener 0.45 kg a.i. ha^{-1} + 1 hand weeding at 35 DAS; S_2: Butachlor 1 kg a.i. ha^{-1} + 1 hand weeding at 35 DAS; S_3: Butachlor + Propanil 0.84 + 0.84 kg a.i. ha^{-1} + 1 hand weeding at 35 DAS; S_4: Two hand weedings at 20 and 35 DAS; S_5: Unweeded check.

Table 41.2: Effect of Weed Control Treatments on Crop Growth Rate (g m^{-2} d^{-1}) in Direct Sown Wet Seeded Rice

Treatments	*Tillering to Flowering*				*Flowering to Harvest*			
	M_1	M_2	M_3	*Mean*	M_1	M_2	M_3	*Mean*
S_1	210.71	167.53	200.49	192.91	60.44	62.26	65.13	62.61
S_2	178.66	131.52	154.46	154.88	50.94	61.42	52.05	54.80
S_3	180.24	143.67	155.50	169.80	58.22	65.29	56.46	59.99
S_4	185.40	182.52	175.41	181.11	51.27	63.04	72.20	62.17
S_5	71.40	78.20	67.18	72.26	29.87	29.07	34.20	31.07
Mean	170.09	140.69	150.61		50.148	56.21	56.00	

	SEd	CD (P=0.05)	SEd	CD (P=0.05)
M	4.71	13.09	5.50	NS
S	5.27	10.88	6.07	12.54
M at S	9.43	21.14	10.95	24.63
S at M	9.13	18.84	10.53	21.73

M_1: ASD 19; M_2: ADT38; M_3: CO 43.

S_1: Pretilachlor + safener 0.45 kg a.i. ha^{-1} + 1 hand weeding at 35 DAS; S_2: Butachlor 1 kg a.i. ha^{-1} + 1 hand weeding at 35 DAS; S_3: Butachlor + Propanil 0.84 + 0.84 kg a.i. ha^{-1} + 1 hand weeding at 35 DAS; S_4: Two hand weedings at 20 and 35 DAS; S_5: Unweeded check.

Significant difference in crop growth rate was observed in the varieties at tillering to flowering stage only. The highest crop growth rate (170.09) was recorded in ASD 19 (M_1). The lowest CGR (140.69) was recorded in ADT 38 (M_2), which did not differ significantly with CO 43 (M) (150.61). At flowering to harvest stage there was no significant difference among the varieties. Crop growth rate (CGR) was significantly influenced by the weed control treatments in both the stages of observation *i.e.* tillering to flowering and flowering to harvest stage. At tillering to flowering stage, CGR differed significantly in each of treatments. Highest CGR of 192.91 was recorded in Pretilachlor + safener 0.45 kg a.i. ha^{-1} + 1 hand weeding at 35 DAS (S_1), followed by hand weeding twice at 20 and 35 DAS (S_4). The lowest CGR of 72.26 was recorded in unweeded check (S_5). At flowering to harvest stage the CGR was significantly higher in all the treatments excepting S_5. Highest CGR of 62.61 was recorded in S_1 and the lowest CGR (31.07) was recorded in S_5. The interaction effect between varieties and weed control treatments was significant. At tillering to flowering stage, highest CGR was recorded in ASD 19 with Pretilachlor + safener (M_1S_1) (210.71), followed by CO 43 with Pretilachlor + safener (M_3S) (200.49). Lowest CGR (67.18) was recorded in the treatment CO 43 when left unweeded (M_3S_5) (67.18). At flowering to harvest stage, highest CGR was recorded in CO 43 with two hand weedings (M_3S_4) (72.2). Lowest CGR was recorded by ADT 38 when left unweeded (M_2S_5) (29.07).

Increased dry matter production was observed in ASD 19 at flowering and harvest stages when compared with other varieties. The comparatively weed free condition provided by this treatment had enhanced the tiller production and LAI and resulted in higher DMP of rice. This is consistent with the findings of Karuppiah. (1995). Initial suppression of weeds with Pretilachlor + safener led to an increase in dry matter production of rice. This might be due to control of grasses, sedges and broad leaved weeds upto the critical periods of crop weed competition (50 DAS) by Pretilachlor + safener, which led to better plant growth and resulted in increased dry matter production. Increased dry weight of weeds in the unweeded plot exerted a negative effect and reduced plant dry matter production. This is in line with findings of Janardhan *et al.* (1999). Amarjith S. Bali *et al.* (1994) reported that the dry matter accumulation of crop was inversely proportional to the weed dry matter. The dry matter production at tillering stage was reduced in the herbicide treatments, when compared to hand weeding treatments, which could be attributed to the stand reduction and phytotoxicity of the rice seedlings due to the herbicides. But at later stages, crop dry matter increased due to better control of weeds by herbicides than unweeded check. Pretilachlor + safener resulted in higher values of CGR at all the stages of crop growth. At tillering to flowering stage, the increase in crop growth rate was 6.5 per cent in Pretilachlor + safener over hand weeding. At flowering to harvest the values were almost similar in hand weeding and Pretilachlor + safener treatment.

References

Bali, Amarjit S., Ganai, B.A., Singh, K.N. and Kotro, R., 1994. Weed control in transplanted rice (*Oryza sativa*). *Indian J. Agron.*, 39(1): 16–18.

Caton, B.P., Folin, T.C. and Hill, J.E., 1997. Phenotypic plasticity of red stem (*Ammania* spp.) in competition with rice. *Weed Res.*, 37: 33–38.

Janardhan, G., Muniyappa, T.V., Reddy, V.C., Ramachandra, C. and Bhaskar, S., 1999. Studies on weed control and crop toxicity ratings of pre-emergence herbicides in transplanted rice. *Mysore J. Agric. Sci.*, 33: 333–337.

Karuppiah, P., 1995. Integrated nitrogen and weed management practices for wet seeded rice. *M.Sc. (Ag.) Thesis,* Tamil Nadu Agricultural University, Coimbatore.

Yamasue, Y., Murayama, H., Inove, H., Matsui, T. and Kusanagi, T., 1997. Productive structures of rice and *Echinochloa oryzicola* using mixed stands. *J. Weed Sci. and Tech.*, 42: 351–365.

Chapter 42

Evaluation of the Efficacy of Garlic Bulb Extract and Selected Animal Excrements on the Mycelial Growth, Sporulation and Conidial Germination of *Helminthosporium oryzae* (Breda de Haan) Subram and Jain (*in vitro*)

M. Thamarai Selvi, P. Balabaskar and V. Kurucheve
Department of Plant Pathology, Faculty of Agriculture, Annamalai University, Annamalainagar 608 002, Chidambaram, Tamil Nadu

ABSTRACT

Efficacy of certain natural products like garlic bulb extract, sheep urine, buffalo urine, goat urine, cow urine, pig dung, goat dung, cow dung, buffalo dung and hen litter were tested against *Helminthosporium oryzae* under *in vitro* conditions. The results revealed that sheep urine (5 per cent), buffalo urine (20 per cent), goat urine (20 per cent), cow urine (40 per cent), hen litter (80 per cent), goat dung (100 per cent) and garlic bulb extract (20 per cent) were found to completely inhibit the mycelial growth of *Helminthosporium oryzae*. The natural products at their respective MIC's showed complete inhibitory effect on the sporulation and conidial germination of the test fungus.

Keywords: *Helminthosporium oryzae, Percent inhibition, Animal excreta, Inverted Petri plate method, Reinoculation, MIC's (minimum inhibitory concentration).*

Introduction

Rice brown spot disease caused by *Helminthosporium oryzae* Cav. is generally considered as the principal disease of rice because of its wide occurrence and destructiveness under favourable conditions. Eventhough the chemical fungicides contributed greatly for the management of the disease, (Fawcett, and Spencer, 1970) the possible pollutions and residual effects due to chemicals necessitated the search for non-chemical methods of managing the disease. Among the non-chemical methods, use of plant products and animal excrements has gained importance in the recent years. With this background the present experiment was conducted to test the efficacy of garlic bulb extract, sheep urine, buffalo urine, hen litter and goat dung extracts against *Helminthosporium oryzae* under *in vitro* conditions. The products were tested for the presence of non-volatile and volatile antifungal principles using poisoned food technique and inverted Petri plate techniques and the reduction of sporulation and conidial germination by cavity slide method.

Materials and Methods

Preparation of Garlic Bulb Extract

For the preparation of garlic bulb extract, the method suggested by Gerard Ezhilan *et al*. (1994) was followed. Fresh bulbs were collected, washed with tap water followed by sterile distilled water, processed @ 1 ml per g of tissue (1 : 1 v/w) with pestle and mortar and filtered through a double layered cheese cloth. This formed the standard bulb extract solution (100 per cent).

Preparation of Animal Dung Extracts

For the preparation of animal dung extract, the method of Sundarraj *et al*. (1996) was followed. Animal dung were collected fresh, shade dried for one week and made into powder. The powdered animal dung was soaked in sterile distilled water @ 5 ml g^{-1} (5: 1 v/w) and kept ever night; blended and filtered through cheese cloth. This formed the standard animal excreta solution (100 per cent). After extraction, they were subjected to low speed centrifugation (5000 rpm for 20 minutes) to avoid any contamination (Jagarinathan and Narasimhan, 1998).

Urine

Freshly collected urine was used as such forming the standard extract (100 per cent) (Raja and Kurucheve, 1998).

Poisoned Food Technique

PDA medium was prepared in 250 mo Erlenmeyer flasks and sterilized. Aqueous extracts of 2.5 and 10 ml and were added to 47.5 and 40 ml of broth respectively in the. flask so as to get the final concentration of 5 and 20 per cent of the extracts in the medium. For getting 100 per cent concentration of extract, the standard extract as such was used. The fungicide Minosan 50 per cent EC at 0.2 per cent concentration was used for comparison. PDA medium without any extract served as control. Each Petriplate was poured with 15 ml of natural product amended medium. The culture disc (9 mm) of *Helminthosporium oryzae* obtained from the periphery of the actively growing seven days old culture was inoculated in the centre of the Petri plates aseptically and incubated for seven days. Five replications were maintained for each treatment. The diameter of the colony was measured when the mycelium fully covered the Petri plates on anyone of the treatments. The percent inhibition of growth was calculated as per Vincent (1947) for each treatment and expressed in mm.

$$\text{Percent inhibition} = \left[\frac{C-T}{C}\right] \times 100$$

where,

C: Diameter of growth in control

T: Diameter of growth in treatment.

Inoculum discs showing no growth due to treatment with garlic bulb extract and animal excrements were tested again for their viability by re inoculating them in to fresh medium.

Inverted Petri Plate Method

Fifteen ml of PDA medium was seeded with three ml of spore suspension (1 × 10″ spores/ml) and allowed to solidify in Petri plates. These plates were then inverted upside down. Extracts of garlic bulb and animal excrements (20 ml) at different concentrations *viz.*, 10, 20, 30, 40, 60, 80 and 100 were poured into the lower lid of the inverted Petri plates. The Petri plates were then incubated at room temperature (25±3°C) for seven days. Sterile distilled water was poured in control plates. Fungicide Hinosan 50 per cent EC (0.2 per cent) was used for comparison. After the incubation period, mycelial growth was recorded. Inoculum discs showing no growth due to treatment with garlic bulb extract and animal excrements were tested again for their viability by re inoculating them in to fresh medium.

Sporulation and Conidial Germination of *H. oryzae* (Cavity Slide Method)

Sporulation

The effect of garlic bulb extract and animal excrements on the sporulation of *H. oryzae* was determined by the method of Bera and Saba (1983) with slight modification.

Medium containing natural products at their respective MIC were poured into Petri plates aseptically at 15 ml/plate. Fungal disc of nine mm size was inoculated in the centre of Petri and plate and incubated at 25±3°C for seven days. Five replications were maintained for each treatment and a suitable control was also maintained. A single mycelial disc from the periphery of Petri plate was removeq with the aid of a sterile cork borer after incubation and transferred into one ml of sterile distilled water and shaken vigorously. A drop of spore suspension was placed on a glass slide and the spore produced was estimated using a haemocytometer.

Spore Germination

Double the required strength aguous solution of the test fungicides were prepared, one drop of spore suspension (2.5 × 10^4 Conidia ml^{-1}) was placed in the well of each cavity slide together with a drop of individual fungicidal solutions and incubated at 22°C in moist Petri dishes for 12 h. Hinosan 50 per cent EC at 02 per cent concentration was used for comparison. Five replications were kept for each treatment. Observations were taken from ten random fields under microscope and per cent inhibition of conidial germination was calculated and the inhibition of germ tube length was expressed as per cent reduction over control (Mustaq Ahmad, 1992).

Results and Discussion

Poisoned Food Technique

The results on the effect of water extracts of natural products at various concentrations on the mycelial growth of *H. oryzae* are presented in Table 42.1. Among the various natural products tested,

Table 42.1: Effect of Some Natural Products Against Mycelial Growth of *P. oryzae*

Sl.No.	Sources	Diameter Mycelial Growth (mm)									% Decrease (–) Over control							
		2.5%	5%	10%	20%	40%	60%	80%	100%	Mean	2.5%	5%	10%	20%	40%	60%	80%	100%
1.	Garlic bulb extract	81.67	71.31	20.00	0.00	0.00	0.00	0.00	0.00	21.62	9.25	20.77	77.78	100	100	100	100	100
2.	Sheep urine	20.00	0.00	0.00	0.00	0.00	0.00	0.00	0.00	3.90	65.43	100	100	100	100	100	100	100
3.	Goat urine	88.00	35.56	28.00	0.00	0.00	0.00	0.00	0.00	18.95	2.22	60.48	68.89	100	100	100	100	100
4.	Cow urine	86.00	45.00	33.00	15.00	0.00	0.00	0.00	0.00	22.38	4.44	50.00	63.33	83.33	100	100	100	100
5.	Buffalo urine	82.00	40.00	24.00	0.00	0.00	0.00	0.00	0.00	18.25	8.89	55.56	73.33	100	100	100	100	100
6.	Hen letter	89.00	60.00	46.00	36.00	33.00	30.00	0.00	0.00	39.75	1.11	33.33	48.89	60.00	63.00	66.67	100	1001
7.	Pig dung	90.00	61.00	45.00	37.00	31.00	27.00	24.00	13.00	41.00	0.00	32.22	50.00	58.89	65.56	70.00	73.33	100
8.	Goat dung	90.00	60.00	30.00	28.00	26.00	18.00	13.00	0.00	33.13	0.00	33.33	66.67	68.89	71.11	80.00	85.53	100
9.	Cow dung	85.53	62.00	39.00	37.00	30.00	26.00	21.00	11.00	38.94	4.97	31.11	56.67	58.89	66.67	71.11	76.67	87.78
10.	Buffalo dung	87.00	60.00	41.00	39.00	38.00	36.00	15.00	13.00	41.13	3.33	33.33	54.44	56.67	57.78	60.00	83.33	85.56 I
11.	Hinosan 50% EC (0.2%)	0.00	–	–	–	–	–	–	–	0.00	–	–	–	–	–	–	–	–
12.	Control	90.00	–	–	–	–	–	–	–	90.00	–	–	–	–	–	–	–	–
	Mean	75.02	41.24	25.50	16.00	13.171	11.42	8.08	3.08									

Factors	S.E	CD (p=0.05)
Main treatment	0.2480	0.7260
Sub-treatment	0.1864	0.5166
ST × MT	0:6456	1.7895
MT × ST	0.6528	1.8242

* Mean of five replications

Garlic bulb extract at 20 per cent concentration completely inhibited the mycelial growth of *H. oryzae* and it was on par with Hinosan (0.2 per cent). Among the animal excrements tested, sheep urine at 5 per cent, buffalo and goat urine at 20 per cent concentrations completely inhibited the mycelial growth of *H. oryzae* and all these were on par with Hinosan (0.2 per cent). In case of dung, hen litter and goat dung extracts gave 100 per cent inhibition of mycelial growth at '100 per cent concentration. The fungicide Hinosan at 0.2 per cent recorded complete inhibition of the growth of the test fungus and the untreated control recorded the maximum mycelial growth (90 mm). All the inhibited discs were inoculated in fresh medium and observed for mycelial growth for 7 days as given Table 42.2. No growth was observed indicating the fungitoxic nature of all the tested natural products.

Table 42.2: Evaluation of Fungitoxicity of Selected Natural Products by Reinoculating the Inhibited Discs of *H. oryzae* into Fresh Medium

Sl.No.	*Sources*	*Conc. (per cent)*	*Mycelial Growth**				*Nature of Toxicity*
			1	*2*	*3*	*4*	
1.	Garlic bulb	20	0	0	0	0	Fungicidal
2.	Sheep urine	5	0	0	0	0	Fungicidal
3.	Buffalo urine	20	0	0	0	0	Fungicidal
4.	Hen litter	100	0	0	0	0	Fungicidal
5.	Goat dung	100	0	0	0	0	Fungicidal
6.	Hinosan (0.2%)	0.2	0	0	0	0	Fungicidal
7.	Control	–	90				

*: Mean of five replications.

Raja and Kurucheve (1997) found that buffalo urine even at the lowest concentration tested (2.5 per cent) recorded 82 per cent inhibition, of *M. phaseolina*. The effect of cow urine on the in *vitro* inhibition of *F. subglutinans* was reported by Alonoso *et al.* (1994). Hen litter extract (10 per cent) recorded the maximum inhibition of the mycelial growth of *R. solani* (Sundarraj *et al.*, 1996). Buffalo urine (10 per cent) was found to have fungicidal activity against *M. phasolina* (Raja and Kurucheve, 1997). *F. oxysporum* f. sp. Lycopersici (Raja and Kurucheve, 1998). Pig dung extract (40 per cent) exhibited inhibitory action against *P. aphanidernafum* (Sivaprakash and Kurucheve, 1999), inhibited disc of *S. oryzae* replanted into fresh medium there was no growth indicating the fungicidal nature of cattle urine (Rajendraprasad and Kurucheve, 2002). The difference in the inhibitory effect of natural products at various concentrations may be due to the quantitative differences in the antifungal principles present in them.

Inverted Petri Plate Method

All the treatments *viz.*, sheep urine and buffalo urine (60 per cent), hen litter and goat dung (100 per cent) and garlic (10 per cent) were found to inhibit the growth of *H. oryzae* when compared to control. Garlic bulb extract completely inhibited the growth of *H. oryzae* at 10 per cent concentration while all the cattle urines needed 40 per cent concentration for complete inhibition of mycelial growth of the test fungus. In case of goat dung and hen litter, the mycelial growth was completely inhibited at

80 and 100 per cent concentrations. Animal urines needed higher concentrations than their MICs for complete inhibition except in garlic bulb extract, which needed only 100 per cent concentration for complete inhibition of pathogen on the upper lid of the Petri plate (Table 42.3). The results of the present study the efficacy of natural products to inhibit the pathogen through volatile compounds similarly Wani (2001) reported that cow urine, buffalo urine, sheep urine and garlic bulb extract completely inhibited the mycelial growth of *A. niger* and *A. flavus* by inverted petri plate method. This might be due to the emission of volatile substances from these natural products.

Table 42.3: Effect of Garlic Bulb Extract and Animal Excrements on the Growth of *H. oryzae* (Inverted Petri plate method)

Sl.No.	Sources*	Mycelial Growth*					
		Conc. 10%	20%	40%	60%	80%	100%
1.	Garlic	–	–	–	–	–	–
2	Sheep urine	++	+	+	–	–	–
3.	Buffalo urine	++	++	+	–	–	–
4.	Hen litter	+++	+++	++	++	+	–
5.	Goat dung	+++	– +++	++	++	+	–
6.	Hinosan (0.2%)	–					
7.	Control		+++				

+++: Profuse growth; ++: Moderate growth; +: Scare growth; –: No growth.

*: Mean of five replications.

Effect of Garlic Bulb Extract and Animal Excrements on Sporulation and Conidial Germination of *H. oryzae* (Cavity Slide Method)

Garlic bulb extract and buffalo urine at 20 per cent concentration, sheep urine at 5 per cent concentration and hen litter and goat dung at 100 per cent concentration were found to completely inhibit the sporulation of *H. oryzae* and they were on par with Hinosan 0.2 per cent. Among the natural products garlic bulb extract at 10 per cent, sheep urine at 5 per cent, buffalo urine at 20 per cent, hen litter 60 per cent and goat dung at 80 per cent concentrations recorded complete inhibition of the conidial germination of *H. oryzae*. Several earlier workers have reported about the efficacy of natural products to inhibit spore germination and sporulation (Table 42.4).

Santhosh Kumar (2000) reported that bulb extract of garlic leaf extract of lawsonia and buffalo urine completely inhibited the conidial germination and sporulation of *C. falcatum*. Resmy Jayaraj (2002) reported that garlic bulb extract, cow urine, buffalo urine, goat urine and hen litter completely inhibited the conidial germination. sporulation of *H. oryzae*. The same products inhibited the conidial germination and sporulation of *C. capsid* (Krishna Kumar, 2002). These reports are in line with our findings. Reduction in spore germination and. growth rate might be due to the presence of some inhibitory substances in aqueous extracts, which affected the physiological metabolism of the fungus as reported by Bera and Saba (1983).

Table 42.4: Effect of Natural Products on the Conidial Germination and Sporulation of *P. oryzae*

Sl.No.	*Sources Concentration (%)*	*Conidial Germination (per cent) **								**Sporulation at MICs (after 7 days)*
		2.5%	*5%*	*10%*	*20%*	*40%*	*60%*	*80%*	*100%*	
1.	Garlic	24.54 (29.69)	22.14 (28.06)	0	0	0	0	0	0	–
2.	Sheep urine	22.52 (28.32)	0	0	0	0	0	0	0	–
3.	Buffalo urine	67.72 (55.37)	51.58 (45.92)	34.61 (36.03)	0	0	0	0	0	–
4.	Hen litter	98.34 (82.58)	82.26 (65.09)	75.51 (60.36)	62.82 (54.42)	38.68 (38.47)	0	0	0	–
5.	Goat dung (82.08)	98.12 (70.73)	89.12 (62.65)	78.92 (56.04)	68.82 (45.71)	51.24 (34.94)	32.78	0	0	–
6.	Hinosan (0.2%)	0.00								–
7.	Control	98.34 (82.51)								+++

Factors	S.E.	CD (p=0.05)
Main treatment	0.0102	0.0309
Sub treatment	0.0095	0.0265
ST × MT interaction	0.0267	0.0749
MT × ST interaction	0.0268	0.0759

Figures in parenthesis are transformed values.

*: Mean of five replications; –: No sporulation; +++– Heavy inoculum

MIC's Garlic (bulb extract)	20%
Sheep urine	5%
Buffalo urine	20%
Hen litter	80%
Goat dung	100%

References

Alonso, S.K. de, Baretto, M., Costa, A.F., Silva Acuna, R., Zambolim, L., Ventura, J.A. and De Alonoso, S.K., 1994. Effect of cow urine on the growth of *Fusarium subglutinans in vitro*. *Fitopatologia brasileria,* 19: 235–238.

Bera, A.K. and Saha, A., 1983. Antifungal activity of an aqeous extract of *Catharanthus roseus*. *J. Pl. Dis. Prot.,* 90: 363–365.

Fawcett, C.H. and Spencer, O.M., 1970. Plant chemotherapy with natural products. *Ann. Rev. Phytopathol.,* 8: 403–416.

Gerard Ezhilan, J., Chandrasekhar, V. and Kuruchcve, V., 1994. Effect of six selected plant products and oil cakes on the sclerotial production and germination of *Rhizoctonia solani*. *Indian Phytopath.,* 47: 183–185.

Jagannathan, R. and Narasimban, V., 1998. Effect of plant extracts/products on two fungal pathogens of finger millet. *Indian J. Mycol. Pl. Pathol.,* 18: 250–254.

Krishnakumar, B., 2002. *In vitro* evaluation of efficacy of some natural products against. *Colletotrichum capsici* (Syd.) Batter and Bisby, the incitant of fruit rot disease of chilli. *M.Sc. (Ag.) Thesis,* Annamalai University, India.

Ahmad, Mustaq, 1992. Control of rice blast by foliar fungicidal sprays. *Plant. Dis. Res.,* 7(1): 24–32.

Raja, J. and Kuruchevc, V., 1997. Antifungal properties some animal products against *Macrophomina phaseolina* causing dry not of cotton. *Plant Dis. Res.,* 12: 11–14.

Raja, J. and Kurucheve, V., 1998. Influence of plant extracts and buffalo urine on the growth and sclerotial germination of *Maqrophomina phaseolina*. *Indian Phytopathology,* 51: 102–103.

Rajendraprasad, K. and Kurucheve, L., 2002. Effect of human and animal excreta on sheath rot of rice. In: *54th Annual Meeting and National Symposium on Crop Protection and WTO: An Indian Perspective,* January 22–25. Organized by Central Plantation Crops Research Institute, Kasaragod, Kerala, India.

Resmy, J., 2002. *In vitro* evaluation of antifungal activity of some natural products against *Helminthosporium oryzae*. *M.Sc. (Ag.) Thesis,* Annamalai University, Annamalai Nagar, India.

Santhoshkumar, K., 2000. *Mycotoxic* activity of natural products against *Collelotrichum falcatum* (went): The incitant of sugarcane red rot. *M.Sc. (Ag.) Thesis,* Annamalai University, Annamalai Nagar, India.

Sivaprakash, K.E. and Kurucheve, V., 1999. There any possibility to manage sheath blight of rice without synthetic fungicides? In: *Agro Annual Meeting China 99, Plant Protection and Plant Nutrition,* April 13–19. Organized by Beijing International Convention Centre, Beijing, China.

Sundarraj, T., Kurucheve, V. and Jayaraj, J., 1996. Screening of higher plants and animals faeces for the fungi toxicity against *Rhizoctonia solani*. *Indian Phytopath.,* 49: 398–403.

Vincent, J.M., 1947. Distortion of fungal hyphae in the presence of certain inhibitors. *Nature,* 159: 850.

Wani, M.A., 2001. *In vitro* evaluation of some natural products against storage fungi of paddy *Aspergillus niger* Van and Teigh and *Aspergillus flavus* Link. *M.Sc. (Ag.) Thesis,* Annamalai University, Annamalainagar, India.

Chapter 43

Manifestation of Physiological and Biochemical Variations in Rice Cultivars Under Effluent Irrigation During Early Emergence

G. Panduranga Murthy[1], G. Chidananda Murthy[2] and Shivalingaiah[3]

[1]P.G. Department of Studies and Research in Biotechnology, J.S.S. College, (Autonomous) (Affiliated to University of Mysore), Mysore – 570 025, Karnataka, India

[2]Lecturer, Department of Botany, Siddaganga College for Women, Karnataka, India

[3]Lecturer, P.F.S. College, Mandya, Karnataka, India

ABSTRACT

A study was conducted to assess the variations in different concentrations of (0 per cent, 5 per cent, 10 per cent, 15 per cent, 25 per cent, 50 per cent, 75 per cent and 100 per cent) paper mill effluent on physiological parameters, relative water turgidity, pigment contents and biochemical constituents like total contents of proteins, amino acids, soluble sugars, phenol and enzyme, amylase activity in the seedlings of two paddy cultivars, Jaya and IR64. All the parameters were declined with an increase in effluent concentrations. In both cultivars, at the highest concentrations (50 per cent, 75 per cent and 100 per cent), the deleterious effects on relative water turgidity and pigment (Chlorophyll 'a', 'b' and total and carotenoid) contents in leaves, and total content of protein, amino acid, soluble sugar, phenol and on the amylase activity were noticed. All these parameters were drastically decreased along with increased concentrations. The effluent in the lower concentrations enhanced all the above parameters of rice seedlings and are significantly superior over other treatments. So, it can be concluded that, the paper mill effluent can be used for irrigating the fields only after the proper dilution *i.e.,* upto 15 per cent, and finally reduce the problem of scarcity of water during cultivation of the crop plants.

Keywords: *Effluent, Paddy cultivars, Germination, Physiology, Biochemistry.*

Introduction

Among major industries, in India; pulp and paper is one of those that contribute to water pollution to a significant level. The paper industry is one among the 20 highly polluting industries in our country as notified by the department of Environment and forest within the purview of Industrial licensing. The effluent mixed water not only contains Nutrients that enhance the growth of Various Crop Plants, but due to presence of heavy metals and other toxic substances which disturb the Physiological process thus, ultimately affects the biochemical constituents of Crop Plants irrigated with effluents (Bhatnagar, 1986; Agarwal, 1991; Hari *et al.,* 1994; Dutta, 1999; Krishnan, 2001 and Panduranga Murthy *et al.,* 2004). Besides, the undiluted paper mill effluent not only effects the growth of crop plants but can also cause severe problems to environment as well as public health (Ramana *et al.,* 2002 and Asamudo *et al.,* 2005). Hence, keeping the above fact in mind, the study was initiated to investigate the effects of paper mill effluents on some physiological and biochemical constituents of two rice cultivars during early growth respectively.

Materials and Methods

Two cultivars of Paddy Crop Plant was selected for the present study *i.e.* Jaya and IR64, from local farmers, Nanjangud, Mysore, Karnataka.

The effluent was subjected to various physico-chemical analyses according to standard procedures. (APHA, 1995)

The experiments were conducted in laboratory condition, nearly 100 seeds of paddy cultivars were arranged at equal distance in sterilised petri plates lined with filter paper (Top of the paper method). A known and equal volume of various concentrations of paper mill effluent [5 per cent (T1), 10 per cent (T2), 15 per cent (T3), 25 per cent (T4), 50 per cent (T5), 75 per cent (T6) and 100 per cent (17)] were poured into respective petriplates. The seedling raised in the tap water treated as control (T0). For effluent irrigation all the sets were subjected to germination in germinator at 28±2°C.

On termination day (ISTA, 1999), the physiological parameters like Relative water turgidity in leaves, chlorophyll and carotenoids and stability index of chlorophyll were estimated.

The biochemical parameters like total contents of protein, amino acids, soluble sugars, phenol coment were estimated according to the standard procedures. Besides, the enzyme, amylase activity. was also recorded.

Statistical Analysis of Data

All the experiments were repeated four times at different time of intervals and the data were collected from all the experiments and were subjected to two way analysis of variance (*Anova*).

Results and Discussion

Hence in the present investigation, it was aimed to find out the effects of Paper mill effluent on physiological, biochemical and enzymatic changes induced by the afore said effluent treatments during early seedling growth of Paddy (*Oryza sativa* L.) seed varieties under different concentrations.

Physico-chemical Analysis

The results of various physico-chemical parameters of paper mill effluents was represented in the Table 43.1.

Table 43.1: Physico-chemical Characterstics of the Paper Mill Effluent Used in the Present Investigation

Sl.No.	*Parameters*	*Effluent*
	I. Physical Characters	
1.	Color	Dark Brown
2.	pH	8.2
3.	Temperature (°C)	36
4.	Suspended solids	
	Total (mg/l)	940.00
	Fixed (mg/l)	501.00
	Volatile (mg/l)	419.00
5.	Dissolved solids	
	Total (mg/l)	1665.00
	Fixed (mg/l)	601.00
	Volatile (mg/l)	89.00
6.	Total volatile solids (mg/l)	96,88
7.	Dissolved Oxygen (DO)	0.81
8.	BOD 5 pays 24°C (mg/l)	191.56
9.	COD (mg/l)	540.23
10.	Total alkalinity (mg/l as Na_2O)	626
11.	Oil and grease (mg/l)	09
	II. Chemical Characters	
1.	Ammonical Nitrogen (mg/l as N)	26.01
2.	Nitrates (mg/l as N)	Nil
3.	Chlorides (mg/l as Cl)	1471
4.	Phosphates (mg/l as P)	Nil
5.	Phenolic Compounds (mg/l as Phenol)	0.07
6.	Cyanides (mg/l as Cn)	Nil
7.	Sulphides (mg/l as S)	32
8.	Sodium (mg/l as Sd)	119
8.	Aourides (mg/l as F)	1.03
9.	Percent Sodium	4.6
	III. Heavy Metals	
1.	Copper (mg/l as Cu)	0.7
2.	Zinc (mg/l as Zn)	1.06
3.	Other heavy metals	Nil

Physiological Study

Relative Water Turgidity in Leaves

A noticeable reduction was recorded at higher dosages (T6 and T7) of effluent. The reduction of 40.65 per cent in Jaya and 39.60 per cent in IR64 were observed over the lower dosage (T3) of 86.68 per cent and 83.14 per cent respectively (Table 43.2).

Table 43.2: Relative Water Turgidity in Leaves of Paddy Cultivars Exposed to Different Concentrations of Paper Mill Effluent

Treatments		Paddy Cultivars		Mean
		Jaya	*IR-64*	
Control	T0	61.00	62.55	61.78[c]
5 per cent	T1	65.54	63.00	64.27[c]
10 per cent	T2	68.51	70.15	69.33[b]
15 per cent	T3	86.68	83.14	84.91[a]
25 per cent	T4	66.00	71.08	68.54[b]
50 per cent	T5	55.00	58.60	56.80[d]
75 per cent	T6	52.46	52.64	52.55[e]
100 per cent	T7	47.80	43.20	45.50[f]
	Mean	62.07[b]	63.045[a]	62.55

F Value: *: Varieties; **: Treatments; **: Interaction.

*: Significant at 5 per cent level; **: Significant at 1 per cent level.

The observed reduction in relative water turgidity in leaves may be due to loss of membrane integrity in response to higher concentrations (T4–T7) of effluent, causing partial or complete desiccation in the plant system leading to complete or partial death of the plant. Similar observations were also made by Eayers and Weatherley (1968), Patil (1990) and Murthy *et al.*

Chlorophyll Content and Stability Index of Chlorophyll (CSI)

When compared to the other parameters, there was an anomalous contrariety in the pigment content. A non-significant decrease in chlorophyll 'a' chlorophyll 'h' and total chlorophyll was noticed at increased effluent dosages (T4–T7). There was an influence of Chlorophyll 'b' in both the varieties was noticed respectively (Tables 43.3–43.5).

In both the cultivars of Paddy, the chlorophyll stability index was decreased marginally at increased concentrations of paper mill effluent. The reduction of 48.21 per cent in Jaya and 47.10 per cent was noticed in IR 64. The lower dosages (T1–T3) are significantly higher over other treatments (Table 43.6).

Increasing concentrations of the effluent correspondingly decreased the chlorophyll pigments in the treated seedlings of both varieties of paddy. Increasing salinity levels are known to gradually reduce the pigment content and chloroplast activity. The other possible reasons for the decrease in chlorophyll content in response to the: higher concentrations may be due to the formation of *'chlorophyllase'* enzyme that is responsible for chlorophyll degradation. Similar facts were reported by Murthy and Majumdar (1962), Sabater and Rodrique (1978), Rodriquez *et al.* (1987), Majumdar *et al.* (1991), Gadallah (1986), Kumar and Singh (1996) and Murthy *et al.* (2004).

Table 43.3: Chlorophyll 'a' Content (mg.g^{-1}) of Paddy Cultivars Exposed to Different Concentrations of Paper Mill Effluent

Treatments		*Paddy Cultivars*		*Mean*
		Jaya	*IR-64*	
Control	T0	0.0656	0.0541	0.0599[c]
5 per cent	T1	0.0711	0.0613	0.0662[b]
10 per cent	T2	0.0726	0.0629	0.0677[b]
15 per cent	T3	0.0841	0.0791	0.0816[a]
25 per cent	T4	0.0691	0.0541	0.0616[c]
50 per cent	T5	0.0512	0.0490	0.0501[d]
75 per cent	T6	0.0421	0.0444	0.0432[e]
100 per cent	T7	0.0337	0.0365	0.0351[f]
	Mean	0.0611[a]	0.0551[b]	0.0581

F Value: *: Varieties; **: Treatments; **: Interaction.

*: Significant at 5 per cent level; **: Significant at 1 per cent level.

Table 43.4: Chlorophyll 'b' Content (mg.g^{-1}) of Paddy Cultivars Exposed to Different Concentrations of Paper Mill Effluent

Treatments		*Paddy Cultivars*		*Mean*
		Jaya	*IR-64*	
Control	T0	0.0751	0.0639	0.0697[e]
5 per cent	T1	0.0806	0.0824	0.0815[c]
10 per cent	T2	0.0819	0.0910	0.0865[b]
15 per cent	T3	0.0966	0.1041	0.1003[a]
25 per cent	T4	0.0714	0.0806	0.0761[d]
50 per cent	T5	0.0656	0.0529	0.0592[f]
75 per cent	T6	0.0490	0.0501	0.0496[g]
100 per cent	T7	0.0424	0.0479	0.0451[g]
	Mean	0.0703[b]	0.0716[a]	0.0709

F Value: *: Varieties; **: Treatments; **: Interaction.

*: Significant at 5 per cent level; **: Significant at 1 per cent level.

Carotenoid

Both the cultivars of Paddy were displayed highest carotenoid content at increased (T4–T7) dosages of the effluent and with the least at lower dosages (T1–T3) (Table 43.7).

The increased carotenoid content was due to influence of increased nitrogen content in response to various effluent treatments (Ninnaladevi and Janardhanan, 1988 and Murthy *et al.*, 2004).

Table 43.5: Total Chlorophyll ($mg.g^{-1}$) of Paddy Cultivars Exposed to Different Concentrations of Paper Mill Effluent

Treatments		Paddy Cultivars		Mean
		Jaya	IR-64	
Control	T0	0.1411	0.1180	0.1295[d]
5 per cent	T1	0.1517	0.1433	0.1475[c]
10 per cent	T2	0.1545	0.1541	0.1543[b]
15 per cent	T3	0.1803	0.1832	0.1817[a]
25 per cent	T4	0.1407	0.1348	0.1377[d]
50 per cent	T5	0.1163	0.1019	0.1091[c]
75 per cent	T6	0.0904	0.951	0.0927[f]
100 per cent	T7	0.0764	0.0845	0.0804[f]
	Mean	0.131[a]	0.1261[b]	0.128

F Value: *: Varieties; **: Treatments; **: Interaction.

*: Significant at 5 per cent level; **: Significant at 1 per cent level.

Table 43.6: Chlorophyll Stability Index of Paddy Cultivars Exposed to Different Concentrations of Paper Mill Effluent

Treatments		Paddy Cultivars		Mean
		Jaya	IR-64	
Control	T0	81.81	84.64	83.22[b]
5 per cent	T1	79.65	80.00	79.83[c]
10 per cent	T2	82.90	83.33	83.11[b]
15 per cent	T3	92.00	90.73	91.37[a]
25 per cent	T4	73.11	77.70	75.40[c]
50 per cent	T5	66.01	61.54	63.77[d]
75 per cent	T6	56.44	60.00	58.22[c]
100 per cent	T7	51.66	48.64	50.15[f]
	Mean	72.94[b]	73.32[a]	73.13

F Value: *: Varieties; **: Treatments; **: Interaction.

*: Significant at 5 per cent level; **: Significant at 1 per cent level.

Biochemical Study

The biochemical analyses were done on 8th day old seedlings of both Jaya and IR64 at all the effluent treatments *viz.*, 5 per cent, 10 per cent, 15 per cent, 25 per cent, 50 per cent, 75 per cent and 100 per cent respectively.

Table 43.7: Carotenoides ($mg.g^{-1}$) of Paddy Cultivars Exposed to Different Concentrations of Paper Mill Effluent

Treatments		*Paddy Cultivars*		*Mean*
		Jaya	*IR-64*	
Control	T0	0.865	0.765	0.815[a]
5 per cent	T1	0.749	0.619	0.684[b]
10 per cent	T2	0.811	0.520	0.666[b]
15 per cent	T3	0.605	0.465	0.535[c]
25 per cent	T4	0.629	0.545	0.587[c]
50 per cent	T5	0.515	0.702	0.608[d]
75 per cent	T6	0.544	0.660	0.602[d]
100 per cent	T7	0.660	0.606	0.633[c]
	Mean	0.672[a]	0.610[b]	0.641

F Value: *: Varieties; **: Treatments; **: Interaction.

*: Significant at 5 per cent level; **: Significant at 1 per cent level.

Total Content of Protein and Free Amino Acid

The data of total protein content (mg/gm of fresh weight) as affected by different concentrations of effluent on 8th day after treatment are furnished in the Table 43.8. It was observed that, the total protein content of seedlings of both the cultivars decreased with an increased concentrations (T4–T7) of effluents and a linear increase in IR64 than Jaya at lower dosage (T3) was observed.

Table 43.8: Total Protein Content of Paddy Cultivars Exposed to Different Concentrations of Paper Mill Effluent (mg/gm of fresh weight)

Treatments		*Paddy Cultivars*		*Mean*
		Jaya	*IR-64*	
Control	T0	6.5160	7.638	7.077[c]
51ft	T1	7.041	8.091	7.566[c]
10 per cent	T2	7.161	8.556	7.858[b]
15 per cent	T3	8.55'4	9.164	8.859[a]
25 per cent	T4	6.721	6.564	6.642[d]
50 per cent	T5	4.109	4.654	4.381[e]
75 per cent	T6	3.296	3.344	3.321[f]
100 per cent	T7	2.996	3.014	3.005[f]
	Mean	5.799[b]	6.378[a]	6.088

F Value: *: Varieties; **: Treatments; **: Interaction.

*: Significant at 5 per cent level; **: Significant at 1 per cent level.

The highest protein content of 8.554 mg in Jaya and 9.164 mg in IR64 per gram fresh weight of seedlings at 15 per cent effluent treatment (T3) was recorded than any other concentrations (T1, T2 and T4–T6) and untreated set (T0) on 8th day of treatment respectively (Table 43.8).

The total free amino acid was marginally decreased along with the increase in concentrations (T4– T1) in the seedlings of Paddy cultivars. In control (T0), there was a slight increase in free amino acid content on 8th day after treatment (1.481 mg in Jaya and 1.503 mg in IR64 per gram fresh weight). The highest content of amino acid at lower dosage (T3) of effluent was noticed *i.e.*, 1.685 mg in Jaya and 1.702 mg in IR64 respectively (Table 43.9).

Table 43.9: Total Free Amino Acid of Paddy Cultivars Exposed to Different Concentrations of Paper Mill Effluent (mg/gm of fresh weight)

Treatments		*Paddy Cultivars*		*Mean*
		Jaya	*IR-64*	
Control	T0	1.481	1.503	1.492[b]
5 per cent	T1	1.304	1.561	1.432[b]
10 per cent	T2	1.369	1.499	1.434[b]
15 per cent	T3	1.685	1.702	1.693[a]
25 per cent	T4	1.299	1.412	1.355[c]
50 per cent	T5	0.910	1.014	0.962[d]
75 per cent	T6	0.721	0.888	0.804[e]
100 per cent	T7	0.289	0.679	0.484[f]
	Mean	1.132[b]	1.282[a]	1.207

F Value: *: Varieties; **: Treatments; **: Interaction.

*: Significant at 5 per cent level; **: Significant at 1 per cent level.

The reduction in total protein content may be due to destruction in the protease activity at increased concentrations of effluent during germination. The stored food materials in the cotyledons are degraded due to imbibition of the toxic effluent water thus inhibits the protein synthesis. Further, the inhibition of proteolytic enzyme leads to destruct the translocation of reserve food materials to the growing embryonic axis also causing reduction in the total free amino acid needed for normal protein synthesis and thus growth was reduced. Similar results are reported by Somashekar *et al.* (1984), Behera and Misra (1996), Vasantha and Perumal Samy (1998) and Panduranga Murthy *et al.* (2004).

Total Soluble Sugar

The data on total soluble sugar content of seedlings indicated that the effluent with different concentrations has differential response in both the cultivars of *Paddy i.e.*, Jaya and IR64.

The total soluble sugar content was increased at lower dosage (T3) and was decreased at increased dosages (T4–T7) on 8th day after effluent treatment. But the lowest content was recorded in Jaya (9.6/ mg) and IR64 (12.03 mg) at 100 per cent (17) effluent treatment (Table 43.10).

The reduction of total soluble sugar may be perhaps because of both phytotoxic and phototoxic effects on many metabolic and physiological process like reduced Chlorophyll content. In addition, the decrease in total soluble sugar are mainly related to decreased starch degradation. The toxic elements present in the effluents inhibits the hydrolytic enzymes, which are responsible for the breakdown of reserve food materials. These results are in close conformity with the findings of Patil (1990), Behera and Misra (1996) and Vasamha and Perumalsami (1998).

Table 43.10: Total Soluble Sugars of Paddy Cultivars Exposed to Different Concentrations of Paper Mill Effluent (mg/gm of fresh weight)

Treatments		*Paddy Cultivars*		*Mean*
		Jaya	*IR-64*	
Control	T0	27.65	33.84	30.74[c]
5 per cent	T1	30.97	37.51	34.24 [c]
10 per cent	T2	32.82	40.09	72.91[a]
15 per cent	T3	42.69	49.50	46.24[b]
25 per cent	T4	24.13	32.04	28.08[d]
50 per cent	T5	19.60	26.68	23.14[e]
75 per cent	T6	14.34	19.11	16.72[f]
100 per cent	T7	09.61	12.03	10.82[g]
	Mean	25.22[b]	31.38[a]	28.30

F Value: *: Varieties; **: Treatments; **: Interaction.

*: Significant at 5 per cent level; **: Significant at 1 per cent level.

Total Phenol Content

The data indicated that, the total phenol content in seedlings is differed significantly. Among the treatments. It was observed that the content was increased at lower dosages. (T1–T3) and untreated set (T0) on 8th day after treatment. The decreased trend was noticed at higher dosages of the effluent. (T4–T7) *i.e.,* 0.540 mg in Jaya and 0.477 in IR64 as compare to control (T0) (Table 43.11).

Table 43.11: Total Phenol Content of Paddy Cultivars Exposed to Different Concentrations of Paper Mill Effluent (mg/gm of fresh weight)

Treatments		*Paddy Cultivars*		*Mean*
		Jaya	*IR-64*	
Control	T0	1.246	1.971	1.308[c]
5 per cent	T1	1.304	1.416	1.360[b]
10 per cent	T2	1.369	1.384	1.376[b]
15 per cent	T3	1.426	1.614	1.520[a]
25 per cent	T4	1.299	1.306	1.302[c]
50 per cent	T5	0.981	1.016	0.998[d]
75 per cent	T6	0.871	0.801	0.836[e]
100 per cent	T7	0.540	0.477	0.509[f]
	Mean	1. 129[b]	1.173[a]	1.151

Phenols are aromatic compounds with an hydroxyl group substituent. They are common constituents of many plants and are known to participate in a number of physiological process which are essential for the growth and development. Phenols play an important role in determining the resistance or susceptibility of plants to any type of toxicity. The resistant varieties have been found to contain more phenols than susceptible ones. The injurious action of toxic materials present in the effluent may inhibited the poly-phenoloxidase, which is responsible for the conversion of phenols into quinones. Thus, this reduces the total amount of phenol in the plant system. Similar findings are reported by Vanja *et al.* (2000) and Panduranga Murthy *et al.* (2004).

Enzyme Study

Amylase Activity

Amylase is of prime importance in the initial stage of starch degradation in germinating seeds. Any deviation like effluent toxicity may disturb the activity of this enzyme, this intern affects the rate of germination. The data on amylase activity in seedlings of paddy cultivars showed non-significant reduction at increased dosages (T4–T7) of effluent. After that, there was a increase in the activity at lower dosages (T1–T3) was observed. The reduction of 47.69 mg in Jaya and 36.51 mg of glucose realesed in IR64 per gram fresh weight of seedlings on 8th day after treatment at higher dosage (T7) of effluent. Further marginal reduction was observed along with increased (T4–T7) concentrations of the effluents (Table 43.12).

Table 43.12: Amylase Activity in Two Paddy Cultivars Exposed to Different Concentrations of Paper Mill Effluent (mg of glucose/30 min/gram fresh weight)

Treatments		Paddy Cultivars		Mean
		Jaya	IR-64	
Control	T0	63.996	74.210	69.100[f]
5 per cent	T1	81.614	90.601	86.107[d]
10 per cent	T2	93.065	109.59	101.62[b]
15 per cent	T3	114.15	127.50	120.83[a]
25 per cent	T4	89.06	91.54	90.31[c]
50 per cent	T5	71.53	84.65	78.og[e]
75 per cent	T6	55.82	49.71	52.76[g]
100 per cent	T7	47.69	36.51	42.10[h]
	Mean	77.11[b]	83.16[a]	80.13

F Value: *: Varieties; **: Treatments; **: Interaction.

*: Significant at 5 per cent level; **: Significant at 1 per cent level.

The major enzyme involved in the starch degradation is amylase, this may be the predominant enzyme involved during germination. Mainly, the inhibition of protein synthesis may result in the decreased amylase production in response to increased doses of the effluent (T4–T7). Behera and Misra (1996).

Conclusion

From the discussion, it is observed that the paper mill effluent had an evident toxic effect on physiological and biochemical parameters in both varieties of paddy. Among both cultivars, the relative susceptibility of cultivar Jaya than IR64 at higher dosages (T4–T7) of effluent with respect to all the parameters have been noticed.

The study reveals that, the effluent contains various toxic chemicals, suspended solids, dissolved solids, heavy metals like copper, zinc etc., and other unwanted materials. Besides, an excess of various forms of *cations and anions* which are injurious to germination and subsequent growth of the plant. The concentrations of these constituents are not in favour on both physiological and biochemical parameters. It should be reduced to beneficial level by diluting the effluents upto 15 per cent which can be used for irrigation purposes as substitutes for chemical fertilizers and can minimise the application of chemical pesticides also.

Acknowledgement

The Authors thank the Research supervisor for his valuable guidance, encouragement and support in completing the research paper.

References

Anon., 1999. *International Rules for Seed Testing, Seed Science and Technology*, 27: 421–463.

APHA, AWWA and WPCF, 1995. *Standard Methods for the Analysis of Water and Wastewater*. APHA Inc., New York.

Arnon, D.I., 1949. Copper enzymes in isolated chloroplast, poly-phenoloxidase in *Beta vulgaris. Plant Physiol.*, 24: 1–15.

Asamudo N.V., Daba, A.S. and Ezeronye, O.U., 2005. Bioremediation of textile effluent using *Phanerochaete chrysosporium., African Journal of Biotechnology*, 4(13): 1548–1553.

Barrs, H.D. and Weatherley, P.E., 1962. A re-examination of the relative turgidity technique for estimating water deficits in leaves. *Austr. J. Biol. Sci.*, 15(3): 413–428.

Behera, B.K. and Misra, B.N., 1985. The effect of sugar mill effluent on enzyme activities of rice seedlings. *Industrial Research*, 37: 390–398.

Clegg, K.M., 1956. The application of the anthrone reagent to the estimation of starch in cereals. *J. Sci. Food Agric.*, 7: 40–44.

Gadallah, M.A.A., 1996. Phytotoxic effects of industrial and sewage wastewater on growth, chlorophyll content, transpiration rate and relative water content of potted sunflower plants. *Water, Air and Soil Pollution*, 89: 33–47.

ISI, 1974. Tolerance limits for industrial effluents discharged into Inland, surface waters. *Indian Standard Institutions*, New Delhi, I.S.–2490.

Kannabiran, B. and Prakasam, A., 1993. Effect of distillery effluent on seed germination, seedling growth and pigment content of *Vigna mungo* (L.) Hepper (CV19). *Geobios*, 20: 108–112.

Kauskik, A., Bala, R. Kadyan and Kaushik, C.P., 1994. Sugar mill effluent effects on growth, photosynthetic pigments and nutrient uptake in wheat seedlings in aqueous Vs soil medium. *Water, Air and Soil Pollution*, 87(1–4): 39–46.

Kirk, J.T.O. and Allen, R.I., 1965. Dependence of chloroplast pigment synthesis on protein synthesis, effect of acitidione. *Biochemical and Biophysical Research Communications*, 21(6): 523–530.

Lowry, O.H., Rosebrough, N.J., Farr, A. and Rendall, R.J., 1951. Protein measurement with the folin phenol reagent. *J. Biol. Chem.*, 193: 265–275.

Madappan, K., 1993. Impact of tannery effluent on seed germination, morphological characters and pigment concentrations of *Phaseolus mungo* L. and *Phaseolus aureus* L. *Poll. Res.*, 12(3): 159–163.

Majumdar, S., Ghosh, S., Glick, B.R. and Dumbroff, E.B., 1991. Activities of chlorophyllase, phosphoenol, pyruvate, carboxylase and ribulase, 1,5 bi-phosphatecarboxylase in the primary leaves of soybean during scenesence and drought. *Physiol. Plant*, 81: 473–480.

Malick, C.P. and Singh, M.B., 1980. In: *Plant Enzymology and Histo-Enzymology*. Kalyani Publishers, New Delhi, pp. 286.

Miller, G., 1972. *Annal. Chem.*, 31: 456.

Moore and Stein, W.H., 1948. In: *Methods in Enzymology*, (Eds.) S.P. Olowkk and N.D. Kalpan. Academic Press, New York, 3: 468.

Murthy, K.S. and Majumder, S.K., 1962. Modification of the technique for determination of chlorophyll stability index in relation to studies of drought resistance in rice. *Curr. Sci.*, 13(11): 470–471.

Neelam, S. and Sahai, R., 1988. Effect of fertilizer factory effluent on seed germination, seedling growth, pigment content and biomass of *Sesamum indicum* Linn. *J. Environ. Biol.*, 9: 45–50.

Nirmala Devi, J. and Janardhanan, K., 1988. Effect of South India viscose factory effluent on seed germination, seedling growth and chloroplast pigments content in five varieties of Maize (*Zea mays* L). *Madras Agric. J.*, 75(1–2): 41–47.

Pandit, B.R., Prasanna Kumar, P.G. and Mahesh Kumar, R., 1996. Effect of diary effluent on seed germination, seedling growth and pigment of *Pennisetum typhoides* Barm. and *Sorghum bicolor* L. *Poll. Res.*, 15(2): 103–106.

Panduranga Murthy, G., Sudarshana, M.S., Prakasha and Manjunatha, R.A., 2004. Studies on the physico-chemical properties of sugar mill effluent and their effects on germination, seedling morphology and biochemical constituents of some commercial crops. *Biosciences, Biotech. Research Asia*, 2(1): 59–63.

Rodriquez, M.T., Gozatez, M.P. and Linares, J.M., 1987. Degradation of chlorophyll and cholorophyllase in senescing barley leaves. *J. Plant Physiol.*, 127: 369–374.

Sahai, R., Jabeen, S. and Saxena, P.K., 1983. Effect of distillery waste on seed germination, seedling growth and pigment contents of rice. *Indian J. Ecol.*, 10: 7–10.

Thamizhiniyan, P., Sundar, Moorthy P. and Lakshamanachary, A.S., 2000. Effect of Sugar mill effluent on germination, growth and pigment contents of ground nut (*Arachis hypogea* L.) and paddy (*Oryza sativa* L.). *Indian J. Applied and Pure Biol.*, 15(2): 151–155.

Vasanthy, M. and Lakshamana Perumal Samy, 1998. Effect of untreated and carbon treated dyeing industry effluent on seed germination and biochemical parameters of Maize. *Poll. Res.*, 17(2): 173–176.

Velo, V., Arumugam, A. and Arunachalam, G., 1999. Impact of paper mill effluent irrigation on the yield, nutrient content and their uptake in rice. *Madras Agric. J.*, 86(1–3): 118–124.

Wuncheng Wang and Keturi, Paul H., 1990. Comparative seed germination tests using ten plant species for toxicity assessment of a metal engraving effluent sample. *Water, Air and Soil Pollution*, 52: 369–376.

Chapter 44

Hypoglycaemic Effect of *Trichosanthes dioica* Roxb. in Normal and Streptozotocin Induced Diabetic Rats

G. Sharmila Banu[1], M. Rajasekara Pandian[1] and G. Kumar[2]
[1]Centre for Biotechnology, Muthayammal College of Arts and Sciences, Kakkaveri, Rasipuram, Namakkal – 637 408, Tamil Nadu, India
[2]Department of Biochemistry, Selvamm Arts and Science College, Namakkal – 637 003, Tamil Nadu, India

ABSTRACT

The hypoglycaemic effect of fruit extract of *Trichosanthes dioica* (Cucuribitaceae) was investigated in normal glucose load conditions and streptozotocin (STZ)-induced diabetic rats. In normal rats the fruit extracts of *T. dioica* (25, 50 ml/kg p.o) significantly (P < 0.001) reduced the blood sugar levels from 64.5 to 48.5 and 67 to 47 mg per cent 2h after administration and also significantly lowered the blood glucose in STZ diabetic rats from 68 to 105 and 66 to 85.5 mg% 21 days after daily oral administration of the fruit extract (P < 0.001). The results suggested that the fruit extract of *T. dioica* possesses a potential hypoglycaemic effect in diabetic rats.

Keywords: *Trichosanthes dioica (Cucurbitaceae), Streptozotocin, Blood glucose levels, Hypoglycaemic effect.*

Introduction

A large number of plants in India have been traditionally used for the treatment of diabetes as household remedies administered in many ways as per the knowledge and experience of the experts. The impressive and classical works were done on this line (Chopra *et al.*, 1956; Nadkarni, 1976;

Kirtikar and Basu, 1980; Chakraberty, 1983; Umadevi *et al.*, 1989), who worked on medicinal properties of the plants possessing hypoglycaemic activities. Nevertheless, many medicinal plants claimed effective by folk medicine require scientific investigation to ascertain their effectiveness, toxicity and then provide alternative drugs and therapeutic strategies (Xavier, 1997).

Trichosanthes dioica Roxb. (Cucurbitaceae) grows as a vegetable in all over India. It is given to improve appetite and digestion (Chora *et al.*, 1956). The decoction of *T. dioica* is useful as a valuable alternative tonic and febrifuge given in boils and other skin diseases (Kirtikar and Basu, 1980). The leaf juice is applied to patches of *alopecia areata* (Narayan Das Prajapati *et al.*, 2003). The root is used as hydragogue cathartic tonic and febrifuge (Kirtikar and Basu, 1995). The fruits used as a remedy for spermatorrhoea and the juice of unripe fruits, tender shoot used as a cooling and laxative (Kirtikar and Basu, 1975). Fruits and seeds show some prospects in the control of some cancer like conditions and hemagglutinating activity (Asolkar *et al.*, 1992).

To our knowledge, no detailed investigations had been carried out to shed light on the hypoglycaemic effect of *T. dioica* fruit extract. Thus, the present study has been planned to investigate the fruit of *T. dioica* on blood glucose levels of glucose fed hyperglycemic, STZ induced diabetic and normal rats and to compare it with tolbutamide as a reference standard.

Materials and Methods

Plant Material

The fruit of *T. dioica* was collected from Palayapalayam, Namakkal District, Tamil Nadu, India and authenticated by Fr. K.M. Matthew, Director, Rapinat Herbarium, St. Joseph's College, Tiruchirapalli. Voucher Herbarium specimens have been deposited in the (collection number 23644) Rapinat Herbarium for future references.

Experimental Animals

Wistar albino rats (200–250 g) of either sex bred in Animal house, Selvamm Arts and Science College, Namakkal were used. Animals were housed in an air-conditioned animal room at 23 ← 2°C with 12 h light/dark photoperiod, and maintained with free access to water and ad libitum feeding. All studies were conducted in accordance with the National Institute of Health guide (1985).

Chemicals

Streptozotocin was obtained from Sigma Chemical Co., St. Louis. MO, USA. All other chemicals were of analytical grade.

Preparation of the Fruit Extract

Fresh raw deseed fruits of *T. dioica* (1 kg) were peeled, washed, cut into small pieces and homogenized in a warring blender with 2 litres of distilled water. The extraction was carried out in a cold room (20°±1°C) with constant stirring overnight. The homogenate was then squeezed through cheese cloth and centrifuged at 2000 rpm for 10min at 0–4°C. The supernatant being the *T. dioica* fruit extract (yield 210 per cent w/w) was decanted and kept at 4°C until used.

Experimental Design

Assessment of Hypoglycaemic Response of Extract in Normal Rats

The hypoglycaemic effect was evaluated orally by force feeding of fruit extract in 24 normal rats, previously fasted for 12 h and randomly divided into four groups: (n = 6 per group).

Group I: Received distilled water and served as a control group

Group II: Treated with fruit extract of *T. dioica* at a dose of 25 ml/kg.

Group III: Treated with fruit extract of *T. dioica* at a dose of 50 ml/kg

Group IV: Treated with Tolbutamide at a dose of 250 mg/kg, and served as a reference standard drug.

Blood samples were collected by tail nipping before the administration of the glucose and at 30 min, 60 min, and 120 min later.

Assessment of Fruit Extract on Glucose Tolerance

After overnight fasting, the rats were given the reference drug and fruit extracts orally and 30min later, glucose (10 g/kg) was administered orally. Blood samples were collected before the administration of the glucose and at 30 min, 60 min, and 120 min after.

Assessment of Fruit Extract on STZ Diabetic Rats

The diabetic rats were divided into 5 groups of 6 animals each. Group I received vehicle alone and served as control. Group II received STZ (60 mg/kg/i.p) dissolved in 0.1 M-citrate buffer. Group III and Group IV received the fruit extracts of *T. dioica* (25 ml, 50 ml/kg/p.o) suspended in vehicle followed by single intra-peritonial administration of STZ. Group V received Tolbutamide (250 mg/kg/p.o) followed by single intra-peritonial administration of STZ. Blood samples were drawn at weekly intervals till the end of the study (*i.e.* 3 weeks)

Biochemical Analysis

After 21 days treatment, Blood samples were collected by cervical dislocation. The plasma was separated and used to determine the biochemical parameters. Blood glucose level was measured by glucose oxidase method (Triender, 1969) and oral glucose tolerance test (OU Vigneaud and Karr. 1985) was determined.

Statistical Analysis

All experimental data were expressed as Mean ← S.D. and statistically assessed by one-way analysis of variance (ANOVA). The difference between test animals and controls were evaluated by Student's t-test (Scheffe, 1953).

Results

Acute Hypoglycaemic Effect

The effect of fruit extract on fasting blood sugar was assessed in normal rats shown in Table 44.1. Percentage reduction in glucose levels after 2h in the *T. dioica* (25, 50 ml/kg/p.o) treated groups were 24.8, 29.5 per cent respectively. Tolbutamide caused a significant ($P < 0.001$) reduction of 25.38 per cent in glucose levels 2h after its administration while control rats did not exhibit any significant alteration in their glucose levels through the duration of the experiment. The hypoglycaemic effect of fruit extract of *T. dioica* (50 ml/kg/p.o) was higher than that seen in the tolbutamide treated rats.

Effect on Glucose Tolerance

Results of the glucose tolerance test conducted on normal rats fed with fruit extracts of *T. dioica* (25, 50 ml/kg) are shown in Table 44.1. Thirty minutes after feeding glucose, the blood sugar rose by 138.5 per cent in normal controls while the same rise was only 120.6 per cent in rats receiving tolbutamide. Administration of fruit extract of *T. dioica* produced a maximum dose dependent reduction in blood sugar at 60, 120m in comparisons to normal controls (15.68, 19.64 per cent and 20.74, 28.05 per cent). The effect was less pronounced at 30 m indicating a late onset of effect. The percentage fall in tolbutamide treated rats as compared to normal controls at 60 and 120 m was 15.2, 27.54 per cent respectively.

Table 44.1: Effect of Fruit Extracts of *T. dioica* on Blood Glucose Level in Fasted Normal and Oral Glucose (10 g/kg) Tolerance Test in Rats

Group	Treatment (/kg/p.o.)	Blood Glucose (mg/dl)			
		Fasting	30 m	60 m	120 m
		Fasted normal			
I	Control	64.3±3.7	64.1±2.1	63.5±32	62.2±2.9
II	*T. dioica*–25 ml	64.5±5.7	59.5±24 (7.75)*	544±2.6) (15.65)**	48.5±4.1 (24.8)**
III	*T. dioica* –50 ml	67.1±2.9	58.4±2.6 (12.96)**	51.5±3.4 (23.24)**	47.3±5.8 (29.5)**
IV	Tolbutamide 250	65.4±3.6	60.6±1.4 (7.33)*	55.2±2.3 (15.59)**	48.8±2.2 (25.38)**
		Oral glucose tolerance test			
I	Control + glucose–10.0	67.3±1.8	142.5±6.9	131.1±6.4 (8.0)	122.5±7.7 (6.55)
II	*T. dioica* 25 ml+ glucose–10.0	67.8 * 1.6	126.2±5.2*	106.4±4.6 (15.68)**	95.5±5.1 (19.64)**
III	*T. dioica* 50 ml +glucose–10.0	65.1±1.9	126.8±2.8*	100.5±5.3 (20.74)**	72.3±4.8 (28.05)**
IV	Tolbutamide 0.25 + glucose–10.0	65.4±1.6	127.6±1.4*	108.2±4.3 (15.2)**	78.4±4.2 (27.54)**

*: Values are statistically significant at $P < 0.01$; **: Values are statistically significant at $P < 0.001$. Values are given in Mean ← SEM for groups of six animals each. In fasted normal, values in parenthesis indicate the percentage lowering of blood sugar in comparison to the fasting reading. In oral-glucose tolerance test, values in parenthesis indicate the percentage lowering of blood sugar in comparison to the previous reading. Experimental groups were statistically compared with the corresponding values of the controls. All groups were compared with their fasting values.

Effect on STZ Induced Diabetic Rats

The effect of fruit extracts of *T. dioica* in STZ induced diabetes in rats is given in Table 44.2. Administration of STZ (60 mg/kg/i.p) led to over three-fold elevation of blood glucose levels ($P < 0.001$), which were maintained over a period of 3 weeks. After 3 weeks of daily treatment with fruit extract (25, 50 ml/kg) led to a dose dependent fall in blood sugars by 56.3, 64.9 per cent respectively. Tolbutamide caused a significant ($P < 0.001$) reduction of 61.4 per cent in plasma glucose levels 3 weeks after its oral administration.

Discussion

A clinically used tolbutamide (a sulphonylurea drug) is known to lower the blood glucose level by stimulating ← cells to release insulin (Jafari *et al.*, 2000). Since STZ can act as diabetogenic agent (Gunnarsson, 1975) and can be used to induce experimental diabetes in rodents (Agarwal, 1980), when animals are injected with streptozotocin (65 or 75 mg/kg), it produces a significant hyperglycemia (Kumar *et al.*, 2006). In the present study tolbutamide exhibited mild hypoglycaemic activity in the STZ diabetic rats. However, in these rats the fruit extracts of *T. dioica* showed marked hypoglycaemic effect. The blood glucose lowering effect in the absence of a significant change in plasma insulin concentration suggests that the *T. dioica* treatment may involve an insulin independent mechanism.

Table 44.2: Effect of 21 Days Treatment with Fruit Extract of *T. dioica* on Blood Sugar in Normal and STZ Diabetic Rats

Group	*Treatment*	*Blood Glucose (mg/dl)*				
		Basal	*2 Days*	*7 Days*	*14 Days*	*21 Days*
1	Control	67.3±1.8	65.8←2.1	66.4←2.4	64.8←1.9	66.5←1.8
II	STZ treated	678±16	2428←6.5	238.4←58**	240.5←6.2**	244.5←6.5**
II	*T. dioica* (25 ml/kg/p.o)	67.8±1.6	240.6←7.8	166.7←8.2** (30.7)	128.5←6.8** (46.6)	105.2←9.2** (56.3)
III	*T. dioica* (50 ml/kg/p.o)	66.1±1.9	243.5←5.8	153.7←7.5** (36.9)	102.3←6.9** (58)	85.5←8.8** (64.9)
IV	Tolbutamide (250 mg/kg/p.o)	65.4±1.6	248.5←8.5	165.8←7.8** (33.3)	126.8←5.4** (49)	95.87.5** (61.4)

*: Values are statistically significant at P < 0.01; **: Values are statistically significant at P < 0.001. Values are given in Mean ← SEM for groups of six animals each. Values in parenthesis indicate the percentage lowering of blood sugar in comparison to the reading at 48h. STZ treated was compared with the normal control and experimental groups were compared with the corresponding values at 48h.

The present investigation showed that treatment with *T. dioica* reduces the lowering of blood sugar level, may be due to stimulating effect on insulin release from regenerated beta cells of the pancreas or may be due to increased cellularity of the islet tissues and regeneration of the beta cells The aqueous extract might be producing its hypoglycaemic effect by an extra-pancreatic action (Dabis *et al.*, 1984) *e.g.* possibly by stimulating glucose utilization in peripheral tissues (Naik *et al.*, 1991; Obatomi *et al.*, 1994). Also, it could be the result of an increase in glycolytic (Steiner and Williams, 1959) and/or glycogenic enzymes activity in peripheral tissues (Naik *et al.*, 1991). Although it might be possible that the aqueous extract may decrease the secretion of the counter regulatory hormones (glucagons, cortisols and growth hormones) (Roman-Ramos *et al.*, 1995), further studies are required to confirm this.

Since the blood glucose lowering effect of the fruit extract of *T. dioica* was observed in STZ diabetic rats as well as in fasted normal rats, this effect could, possibly be due to inhibition of the proximal tubular reabsorption mechanism for glucose in the kidney, if any, which can also contribute towards blood lowering effect (Sharma *et al.*, 1983).

From the above results, we observed that the fruit extract of *T. dioica* produced a significant decrease of blood glucose levels, and this effect was more potent after repeated oral administration than after single oral administration. These results suggest that the effectiveness of the drugs depend, probably, on the accumulative effect of active principles (Obatomi *et al.* 1994; Peungvicha *et al.*, 1998).

The fruit of *T. dioica* contains an amorphous saponin, a non-nitrogenous bitter amorphous glucoside called Trichosanthin (Shastri, 1952) and Terpenes, fixed oil, starch, reducing sugars (Chopra *et al.*, 1956) and bitter natural principles, cucurbitacins (Narayan Das Prajapati *et al.*, 2003). Further more, comprehensive chemical and pharmacological research is required to reveal the mechanism of the hypoglycaemic effect and to identify the active constituent(s) responsible for this effect.

References

Agarwal, M., 1980. Streptozotocin: Mechanism of action. *FEBS Lett.*, 120: 1–3.

Asolkar, L.V., Kakkar, K.K. and Chakre, O.J., 1992. *Second Supplement to Glossary of Indian Medicinal Plants with Active Principles*, Part I (A–K) (1965–1981): 134–135.

Chakraberty, C., 1983. *A Comparative Hindu Materia Medica*. Neeraj Publishing House, New Delhi.

Chopra, R.N., Nayar, S.L. and Chopra, I.C., 1956. *Glossary of Indian Medicinal Plants*. CSIR, New Delhi.

Dabis, G., Michon, D., Gazenav, J. and Ruffie, A., 1984. Interet clinique d'une strategie biologique adaptee au diabete sucre. *La vie medicale*, 8: 277–290.

Du Vigneaud, V. and Karr, S., 1985. Carbohydrate utilization and disappearance. *J. Biol. Chem.*, 66: 281–300.

Gunnarsson, R., 1975. Inhibition of insulin biosynthesis by alloxan, streptoztocin and N nitrosomethylurea. *Mol. Pharmacol.*, 11: 759–765.

Jafari, M.A., Aslam, M., Kalim Javad and Singh, Surender, 2000. Effect of *Punica granatum* L. (flowers) on blood glucose level in normal and alloxan induced diabetic rats. *J. Ethnopharmacol.*, 70: 309–314.

Kirtikar, K.R. and Basu, B.D., 1975. *Indian Medicinal Plants*, 2nd edn., Vol. 2. Jayyed Press, Allahabad, p. 1110–1111.

Kirtikar, K.R. and Basu, B.D., 1980. *Indian Medicinal Plants*. Lalit Mohan Basu Publications, Allahabad.

Kirtikar, K.R. and Basu, B.D., 1995. *Indian Medicinal Plants*, Vol. 1. International Book Distributors, Dehradun, India.

Kumar, G., Sharmila Banu, G., Murugesan, A.G. and Rajasekara Pandian, M., 2006. Hypoglycaemic effect of *Helicteres isora* bark extracts in rats. *J. Ethnopharmacol.*, 107: 304–307.

Nadkarni, K.M., 1976. *Indian Materia Medica*. Popular Prakashan, Mumbai.

Naik, S.R., Filho, J.M.B., Dhuley, J.N. and Deshmukh, A., 1991. Possible mechanism of hypoglycaemic activity of basic acid, a natural product isolated from *Bumelia sartorum*. *Journal Ethnopharmacology*, 33: 37–44.

Prajapati, Narayan Das, Purohit, S.S., Sharma, Arun K. and Kumar, Tamn. 2003. *A Handbook of Medicinal Plants*. Agrobios (India), Jodhpur, pp. 521.

NIH (National Institute of Health), 1985. *Guide for the Care and Use of Laboratory Animals*. DREW Publication (NIH), Revised, Office of Science and Health Reports: DRR/NIH; Bethesda; USA.

Obatomi, D.K., Bikomo, E.O. and Temple, V.J., 1994. Anti-diabetic properties of the African Mistletoe in streptozotocin induced diabetic rats. *J. Ethnopharmacol.*, 43: 13–17.

Peungvicha, P., Thirawarapan, S.S., Temsiririrkkul, R., Watanabe, H., Prasain J.K. and Kadota, S., 1998. Hypoglycaemic effect of the water extract of *Piper sarmentosum* in tars. *J. Ethnopharmacol.*, 60: 27–32.

Roman-Ramos, R., Flores-saenz, J.L. and Alarcon-Agilar, F.J., 1995. Antihypertensive effect of some edible plants. *J. Ethnopharmacol.*, 48: 25–32.

Scheff'e, H.A., 1953. A method for judging all contrasts in the analysis of variance. *Biometrika*, 40: 87–104.

Sharma, M.K., Khare, A.K. and Feroz, H., 1983. Effect of neem oil on blood glucose levels on normal, hyperglycemic and diabetic animals. *Ind. Med. Gazette*, 117: 380–383.

Shastri, B.N., 1952. *Wealth of India: Raw materials*, Vol. 10. CSIR Publication, New Delhi, pp. 289–290.

Stainer, D.F. and Williams, R.B., 1959. Action of phenetyl biguanide and related compounds. *Diabetes*, 8: 154–157.

Triender, P., 1969. Determination of glucose using glucose oxidase with an alternative oxygen acceptor. *Ann. Clin. Biochem.*, 6: 24–27.

Umadevi, A.J., Parabia, M.H. and Reddy, M.N., 1989. Medicinal plants of Gujarat: A survey. In: *Proceedings of All India Symposium on the Biology and Utility of Wild Plants*, Professor G.L. Shah Commemoration Volume, Department of Biosciences, South Gujarat University, Surat.

Xavier, L., 1997. Les medicaments. Pour la SCIENCE. Edit franchise DE scientific American, p. 341–349.

Chapter 45

Yield, Quality and Nutrient Uptake of Soybean (*Glycine max* L.) as Influenced by Different Planting Methods and Irrigation Schedules

S.S. Mahal, C.S. Maan, A.S. Brar and A.K. Atwal
Department of Agronomy, Agrometeorology and Forestry, Punjab Agricultural University, Ludhiana – 141 004, India

ABSTRACT

A field experiment was conducted during the *kharif* season of 2005 on loamy sand soil, to assess the effect of different planting methods [*i.e.* bed planted with roller with mulch (M_1), bed planted with roller without mulch (M_2), bed planted without roller with mulch (M_3), bed planted without roller without mulch (M_4), conventional (flat) with mulch (M_5), and conventional (flat) without mulch (M_6)] and three irrigation schedules *viz.*, irrigation applied at 50 mm CPE (I_1), 65 mm CPE (I_2) and 80 mm CPE (I_3) on yield, quality and nutrient uptake of soybean. The maximum seed, stover and oil yield of soybean were obtained with M_1 method of planting, which was at par with M_2, but significantly superior to all other. Protein yield was numerically higher in M. planting method as compare to all other planting methods. The N, P, K uptake in seed and stover were maximum under M_1, which was at par with M_2 and significantly higher than other planting methods. The highest seed, stover, oil and protein yields were obtained under I. irrigation schedules, but results were non-significant with other irrigation schedules. The nutrient uptake was maximum under I_1 irrigation schedules, which was at par with I_2 and significantly higher than I_3.

***Keywords**: Soybean, Planting method, Irrigation schedule, Yield, Quality, Nutrient uptake.*

Introduction

Soybean has been termed as miracle bean because it contains about 40 per cent protein with essential amino acids in balanced proportions, 20 per cent oil rich in polyunsaturated fatty acids, 6–7 per cent total minerals, 5–6 per cent crude fibre and 17–19 per cent carbohydrates (Chauhan *et al.*, 1988). Since planting methods playa major role in maximizing the infiltration, minimizing soil erosion and improving water use efficiency, which leads to influence the nutrient absorption and quality of the produce from field crops. Thus present investigation was carried out to assess the yield, quality a!ld nutrient uptake of soybean.

Materials and Methods

The present investigations were carried out during *Kharif* season of 2005 at Department of Agronomy, Agrometeorology and Forestry, Punjab Agricultural University, Ludhiana on loamy sand soil, to study the effect of planting methods and irrigation schedules on yield, quality and nutrient uptake. The available nutrient status of the soil was low in organic carbon and available nitrogen, medium in available phosphorus and potassium. The field capacity and wilting point the soil were 27.40 and 12.01 per cent respectively. The bulk density was 1.56 g cc^{-1}; the soil pH and EC were 7.9 and 0.23 dSm^{-1}, respectively. Soybean (SL525) with duration of 144 days was used, as the test crop. Irrigation scheduling was based on the cumulative pan evaporation (CPE) recorded for USWB open pan evaporimeter. The total rainfall received during the crop season was 589.9 mm. The experiment was laid in split plot design with four replications comprising six planting methods *i.e.* bed planted with roller with mulch (M_1), bed planted with roller without mulch (M_2), bed planted without roller with mulch (M_3), bed planted without roller without mulch (M_4), conventional (flat) with mulch (M_5), and conventional (flat) without mulch (M_6) in main plots while three irrigation schedules *viz.*, irrigation applied at 50 mm CPE (I_1), 65 mm CPE (I_2) and 80 mm CPE (I_3) were kept in sub plots. Recommended package of practices was adopted. The seed yield was recorded at harvest and seed samples were analysed for oil and protein content with Nuclear Magnetic Resonance spectroscope Newport-Analyser (Model MK-IIIA) and Microkjeldhal methods, respectively. The total oil and protein yield per hectare were computed based on seed yield (kg/ha) and percentage of oil and protein content. The samples were analysed for N (Microkjeldhal method), P (Vanadomolybdo phosphate yellow colour by Colorimeter) and K (Flame photometer) in both soybean seed and stover. The total N, P and K uptake per hectare were computed based on seed and stover yield of soybean.

Results and Discussion

Seed and Stover Yield

The treatment M_1 being at par with M_2 both treatments produced significantly (Table 45.1) higher seed yield of soybean as compare to all other planting treatments. The treatment M_2 differed non-significantly with other planting treatments except M_6, which produced lowest seed yield. Soybean sown with method of planting M_1 produced 10.8, 14.1, 14.9, 20.7 and 26.7 per cent higher seed yield as compare to M_2, M_3, M_4, M_5, and M_6, respectively. Raut *et al.* (2000) also reported similar results. Different scheduling of irrigation at I" hand 13 resulted non-significant differences as in seed yield of soybean. Jagathambae and Karivaratharaju (1998) also observed similar results with respect to different irrigation schedules. The treatment M_1 produced significantly higher stover yield as compare to M_4 and M_6 but at par with M_2, M_5 and M_3. Kaur (2003) also observed lower stover yield under flat method of planting. In case of irrigation treatments different levels of irrigation schedule had non-significant effect of stover yield.

Table 45.1: Effect of Different Planting Methods and Irrigation Schedules on Seed, Stover, Protein, Oil Yield and N, P, and K Uptake in Soybean

Treatments	Seed Yield (q ha^{-1})	Stover Yield (q ha^{-1})	Protein Yield (q ha^{-1})	Oil Yield (q ha^{-1})	N Uptake (kg ha^{-1})	P Uptake (kg ha^{-1})	K Uptake (kg ha^{-1})
			Planting Methods				
M_1	25.28	55.81	9.86	5.75	193.14	32.76	139.04
M_2	22.81	53.72	8.67	5.16	171.90	28.92	124.49
M_3	22.15	49.58	8.38	5.26	165.36	29.85	125.88
M_4	22.00	47.04	8.26	5.07	160.90	27.04	119.60
M_5	20.94	50.53	7.88	4.75	157.36	28.37	126.47
M_6	19.96	47.63	7.45	4.50	147.71	25.17	110.86
CD (p = 0.05)	2.55	5.93	NS	0.70	25.93	3.68	
			Irrigation schedules				
I_1	22.76	52.71	8.72	5.09	172.93	31.06	132.54
I_2	22.03	51.16	8.42	5.03	166.48	28.84	124.27
I_3	21.78	48.29	8.10	5.11	158.77	26.16	116.35
CD (p = 0.05)	NS	NS	NS	NS	9.07	1.63	8.10
Interaction	NS	NS	NS	NS	NS	NS	NS

Oil and Crude Protein Yield

There was significant effect of planting methods (Table 45.1) on oil yield of soybean. The maximum oil yield was recorded in M_1 planting method followed by M_3, M_2, M_4, M_5, and M_6 methods of planting. All bed planted treatments with roller and mulch, without roller and mulch were at par with each other, but significantly differed as compare to conventional flat with and without mulch with respect to treatment MI and M3, respectively. Irrigation schedules had non-significant effect on oil yield this might be due to higher [July, August, September, the crop received 183.5 mm (8 days), 200.1 mm (9 days), 166.5 mm (7 days)] rainfall during the crop season. However, the crude protein yield was non-significantly influenced by different planting methods and irrigation schedules.

N Uptake

N uptake was affected significantly by (Table 45.1) different planting methods. The treatment M_1 improved N uptake by 14.4, 16.7, 18.5 and 23.5 per cent over M_3, M_4, M_5 and M_6 methods of planting, respectively. The treatments M_1 and M_2 were at par with each other. Similar results were reported by Kaur (2003). Nitrogen uptake was affected significantly under I_1 over I_3 and at par with I_2. The irrigation schedule I_1 increased the nitrogen uptake by 3.7 and 8.2 per cent as compare to I_1 and I_3 irrigation schedules, respectively. Kumawat *et al.* (2000) also reported similar results.

P Uptake

The treatment M_1 improved (Table 45.1) phosphorus uptake significantly by 11.7, 13.4, 17.5 and 23.2 per cent over M_2, M_5, M_4 and M_6 methods of planting, respectively. Raji and Oyekan (2002) also reported similar results. Phosphorus uptake affected significantly under different irrigation schedules.

The irrigation schedule I_1 increased the phosphorus uptake by 7.1 and 15.8 per cent over I_2 and I_3 irrigation schedules, respectively. The significant increase in phosphorus uptake by I_1 level of irrigation might be due to more availability of water, which favours more uptake and translocation of nutrients within plant. These results are inconformity with those reported by Kumawat *et al.* (2000).

K Uptake

K uptake was affected significantly by (Table 45.1) different planting methods. The treatment M_1 improved potassium uptake significantly by 9.0, 9.5, 10.5, 14.0 and 20.3 per cent over M_5, M_3, M_2, M_4 and M_6 methods of planting, respectively. Raji and Oyekan (2002) also reported similar results. The irrigation schedule II increased the potassium uptake by 6.2 and 12.2 per cent than I_1 and I_3 irrigation schedules, respectively. This might be due to more availability of water, which leads to more absorption and translocation of nutrient with plant system.

References

Chauhan, G.S., Verma, N.S. and Bains, G.S., 1988. Effect of extrusion processing on the nutritional quality of protein in rice-legume blends. *Nahrung*, 32(1): 43.

Jegathambal, R. and Karivaratharaju, T.V., 1998. Influence of irrigation and phosphorus levels on seed quality characteristics in soybean. *Madras Agric. J.*, 85: 177–178.

Kaur, M., 2003. Studies on seed rate, irrigation, weed control and their interactive effects in bed planted soybean (*Glycine max* L.). *Ph.D. Dissertation*, Punjab agricultural University, Ludhiana, India.

Kumawat, S.M., Dhakar, L.L., and Maliwal, P.L., 2000. Effect of irrigation regimes and nitrogen on yield, oil content and nutrient uptake of soybean (*Glycine max*). *Indian J. Agron.*, 45: 361–366.

Raji, J.A. and Oyekan, P.O., 2002. The agronomic assessment of the nutrient production potentials of soybean/sorghum intercrops in peasant farming system. *Crop Res.*, 24: 242–246.

Raut, V.M., Taware, S.P., Halvankar, G.B. and Varghere, P., 2000. Comparison of different sowing methods in soybean. *J. Maharashtra Agric. Univ.*, 25: 218–219.

Chapter 46

Molecular Characterization of Isabgol (*Plantago ovata* Forsk) by RAPD Markers

Ramesh Solanki and Harshal E. Patil*
Department of Agricultural Biotechnology and Biochemistry, B.A. College of Agriculture, Anand Agricultural University, Anand – 388 110, Gujarat

ABSTRACT

The RAPD profile generated by seven random primers showed that the existing molecular variability within the species *Plantago ovata* is low, reflecting the narrow genetic base of the species. Out of eight genotypes studied JI-2 was found to be lying alone in a single cluster, whereas, the other seven genotypes formed the second cluster. Two state released varieties of Isabgol *viz.*, GI-1 and GI-2 were found nearly similar to JI-5, JI-6, JI-7 and JI-8, whereas, genotype JI-1, JI-3 and JI-4 formed different sub clusters showing presence of little molecular diversity. It seems that the result of RAPD analysis may also be due to parentage similarity of the genotypes studied as they may have been selected from a common population on the basis of phenotypic expression but the parent population might be having low genetic advance of the trait.

Keywords: *Isabgol, RAPD, DNA polymorphism and Molecular diversity.*

Introduction

---

* Corresponding Author: E-mail-harry_biotech@yahoo.com.

Plantago ovata locally called, as "Isabgol" is an important medicinal crop. The genus *Plantago* L. comprises of 265 species and has cosmopolitan distribution. Out of these, 10 are found in India (Wealth of India, 1969). Most of the species of the genus *Plantago* belongs to series with basic chromosome number x = 6 (Heitz, 1927). The forms with x = 5 are very rare, *Plantago ovata* Forsk (Blond Psyllium) is the only species in the genus having x = 4 chromosome number. Previous studies have reported that very less amount of phenotypic variability exists in the population may be due to presence of narrow genetic base creating a problem in performing efficient selection. As genetic diversity evaluation is very much important to design plans for improvement of a population structure, their are various levels *viz.*, phenotypic, cytogenetical, biochemical and molecular at which a genetic diversity can be analysed.

To overcome the phenotypic barriers molecular tools such as RAPD (Random Amplified Polymorphic DNA), Inter-SSR (Inter Simple sequence Repeats) *etc.* can be used for diversity analysis. Using molecular tools some attempts were made in other species of Plantago (Wolff and Morgan, 1998) but compared to other species studies on *Plantago ovata* are very limited (Bharati, 2002). Knowing the importance of the crop, considering the present need the present experiment was laid to study the molecular diversity present in the species *Plantago ovata*

Materials and Methods

Ten accessions were used for the experiment. The seeds were obtained from Medicinal and Aromatic Plant Research Unit, Anand Agricultural University, Anand, Gujarat. Out of the ten accessions two were state released varieties *viz.*, Gujarat Isabgol-1 (GI-1), Gujarat Isabgol-2 (GI-2) and six were germplasm lines *viz.*, Jagudhan Isabgol-1 (JI-1, JI-2 JI-3 JI-4 JI-5 JI-6 JI-7 and JI-8).

The DNA extraction buffer (DEB) consisting of 3 per cent CTAB (w/v), Tris HCl pH 8.0 (1.0 M); EDTA pH 8.0 (0.4 M); NaCl (1.5 M) was prepared. 3 M Sodium acetate (NaOAc) solution (pH 5.2), Ribonuclease A (10 mg/ml), Chloroform: Isoamylalcohol (24 : 1), Phenol : Chloroform : Isoamylalcohol (25 : 24 : 1 v/v/v), Ethanol (70 per cent, 100 per cent) and TE buffer [Tris HCI (10mM) and EDTA (1 mM) pH 8.0] were the additional solutions used.

DNA Extraction

The Doyle and Doyle (1990) method with modifications was used for DNA extraction. DNA was extracted from one gram of young Plantago leaves harvested on ice. The leaf tissue was ground in liquid nitrogen using autoclaved pre chilled mortar and pestle. The fine powder was transferred into 10 ml polypropylene tube. Two extractions with Chloroform : Isoamyalcohol (24 : 1 v/v) were performed, second extraction with Chloroform : Isoamyalcohol (24 : 1 v/v) contained 10 per cent CTAB and 1.5 M NaCl. DNA was precipitated with equal volume of pre chilled isopropanol. The pellets were dissolved in excess of TE (pH 8), and RNase treatment was given for 1 hour at 37°C in a water bath. The re-precipitation of DNA was done with double volume of EtOH (Absolute Ethanol) and 1/10th vol. of 3M NaOAc. The pellets were washed twice with 70 per cent ethanol and allowed to dry at room temperature and then dissolved in 100 µl of TE overnight and then preserved at 4°C for analysis.

Statistical Analysis

Band position for each of the *Plantago ovata* accessions and primer combination were scored as either present (1) or absent (0). The scores were entered into a database programme and complied in a binary matrix for phonetic analysis using NTSYSpc version 2.2 (Multivariate Analysis System version 2.2) by Exeter Software. The SIMQUALK programme was used to calculate Jaccard's similarity coefficient and graphic phenogram (Oendogram) of the genetic relatedness among eight accessions was produced by means of the unweighted pair groups method with arithmetic average (UPOMA) analysis (Sneath and Sokal, 1973).

Results and Discussion

The PCR analysis was carried out to detect polymorphism for the marker-binding site among the genotypes. The set of genotypes were screened for presence and absence of bands. Out of 40 primers used for detecting polymorphism, seven gave amplified products. Out of the seven, two primers *viz.*, OPC-1, OPC-4 showed monomorphic bands, whereas the remaining five *viz.*, OPC-2, OPC-9, OPC-10, OPO-13 and OPO-15 gave polymorphic bands.

The RAPD profile generated by seven random primers (Figure 46.1) showed that the existing molecular variability within the species *Plantago ovata* is low, reflecting the narrow genetic base of the species (Bhagat, 1980; Bhagat and Hardas, 1980). These seven primers amplified a total 250 bands, of which only 8 (3 per cent) were polymorphic. Using Jaccard's similarity coefficient data a dendogram was drawn (Figure 46.2). Out of eight genotypes studied JI-2 was found to be lying alone in a single cluster, whereas, the other seven genotypes formed the second cluster. The second cluster containing the nine genotypes were sub clustered in three groups. Two state released varieties of Isabgol *viz.*, 01-1 and 01-2 were found nearly similar to JI-5, JI-6, JI-7 and JI-8. Whereas, genotype JI-1, JI-3 and JI-4 formed different sub clusters showing presence of little molecular diversity for the allele amplified.

The primer specific amplification profile was observed with minimum of 1 amplicons using OPC-02 primer across seven hybrids and a maximum of 4 amplification with OPC-09 primer. The polymorphism shown by five primers varied from 33.00 to 60.0 per cent (Table 46.1). While, the genotype specific amplification profile was observed with minimum of 5 amplicons by JI-4 genotype across five primers and a maximum of 12 amplification with JI-1 genotype. The polymorphism shown by ten genotypes varied from 22.72 to 50.00 per cent (Table 46.2).

Table 46.1: Primer Wise Scorable Bands, Total Loci, Polymorphic Loci and Percentage Polymorphism

Sl.No.	*Primers Used*	*Sequence (5' to 3')*	*Percentage Polymorphism [P/T x 100]*	*Scorable Bands*	*Total Loci*	*Polymorphic Loci*
1.	OPC–02	ACCTGAACGA	33.00	2–3	3	1
2.	OPC–09	TTGGCACGGG	40.00	9–10	10	4
3.	OPC–10	AGGCGGTAAG	33.00	5–6	6	2
4.	OPD–13	CTGGGGACTT	60.00	4–5	5	3
5.	OPD–15	ACCCGGTCAC	50.00	5–7	6	3

T: Total Number of Bands; P: Number of Polymorphic Bands.

There is lot of work going on diversity analysis in the population of *Plantago* species (Wolff and Morgan, 1998; Ronsted *et al.*, 2002) but out of the various *Plantago* sp. work on *Plantago ovata* is limited. As the crop is having high commercial importance in the world market attempts have been made for its improvement to develop high yielding genotypes through selection on the basis of phenotypic expression (Punia *et al.*, 1985). Due to presence of very low amount of variability existing in the population chances for selection becomes less. The present study was done to study the molecular

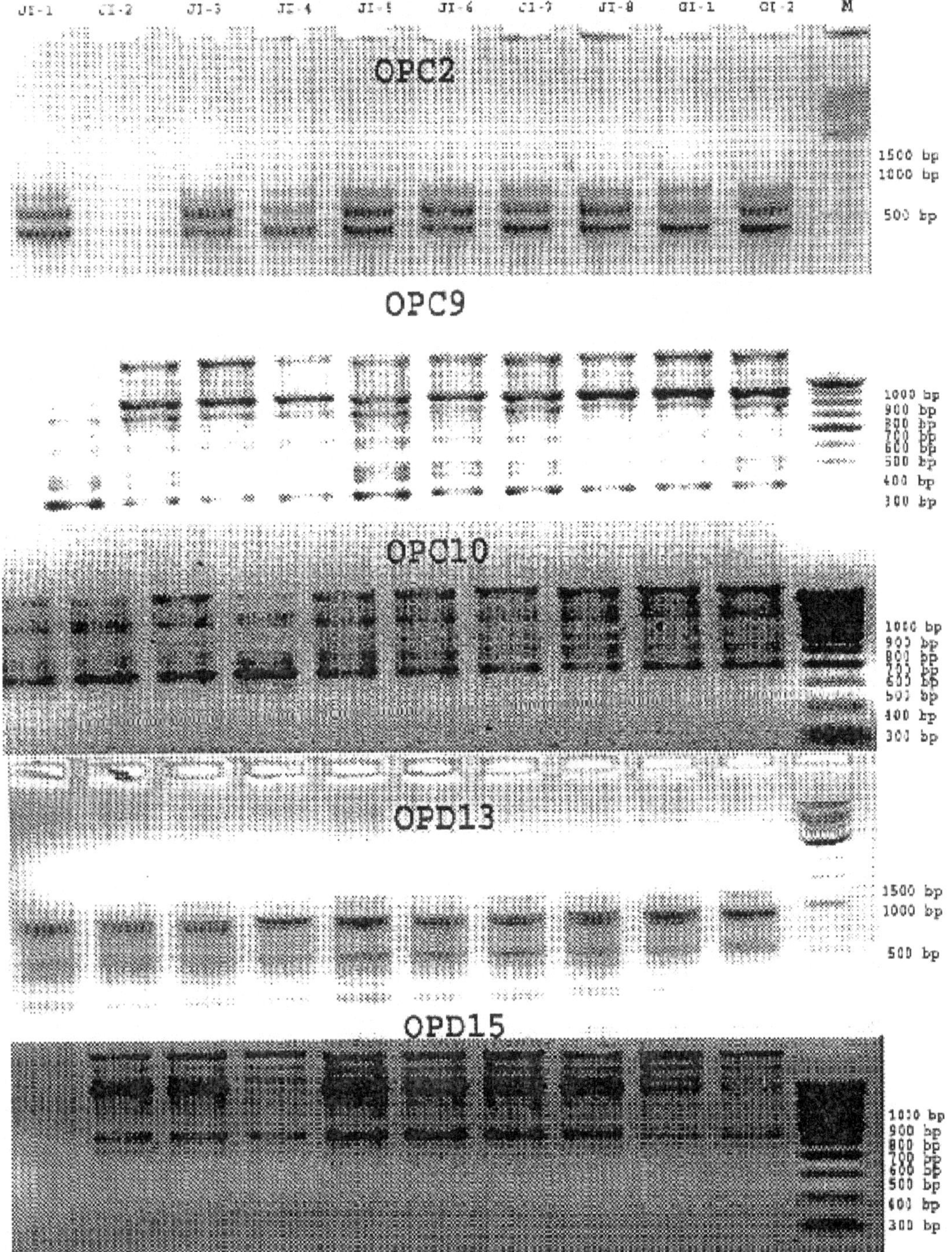

Figure 46.1: RAPD Amplification Pattern of Different *Plantago ovata* Genotypes

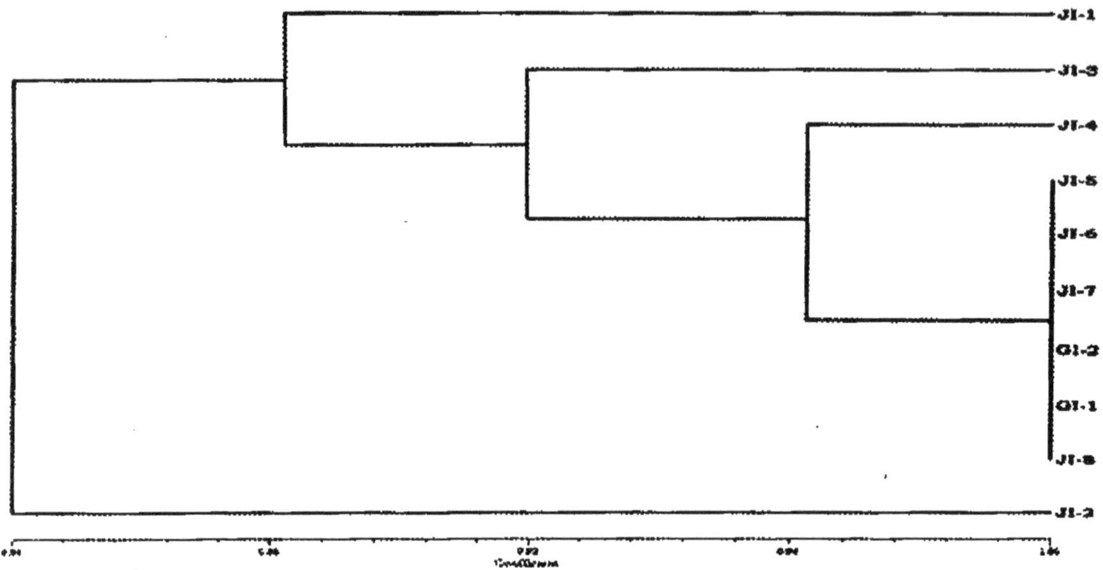

Figure 46.2: Dendogram Based on Jaccard's Similarity Coefficient of Different *Plantago ovata* Lines

diversity existing in the genome, but its seems that the genome of *Plantago ovata* is very conserved for the alleles amplified by using seven different decamer primers. It seems that the result of RAPD analysis may also be due to parentage similarity of the genotypes studied as they may have been selected from a common population on the basis of phenotypic expression but the parent population might be having low genetic advance of the trait.

Table 46.2: Genotype Wise Polymorphic Loci and Percentage Polymorphism

Sl.No.	Genotype	No. of Polymorphic Loci	Total No. of Loci	Percentage Polymorphism (P/T x 100)
1.	GI-1	12	24	50.00
2.	GI-2	10	22	45.45
3.	JI-1	9	20	45.00
4.	JI-2	7	19	36.84
5.	JI-3	6	24	25.00
6.	JI-4	5	22	22.72
7.	JI-5	8	22	36.36
8.	JI-6	6	24	25.00
9.	JI-7	10	21	47.61
10.	JI-8	8	20	40.00

Based on these results it was concluded that, no significant enhancement in genetic diversity among the Isabgol genotypes was developed during different periods.

References

Bhagat, N.R., 1980. Studies on variation and association-among yield and some component traits in *Plantago ovata* Forsk. *Crop Improvement*, 7: 60–63.

Bhagat, N.R. and Hardas, M.W., 1980. Studies on induced and natural variation in *Plantago ovata* Forsk. *Indian Drugs*, 17: 376–380.

Bharati Mittal, 2002. Biochemical and molecular characterization of isabgol. *Unpublished Ph.D. Thesis*, Department of Biochemistry, Anand Agricultural University, Anand, Gujarat.

Doyle, J.J and Doyle, J.L., 1990. Isolation of plant DNA from fresh tissue. *Focus*, 12: 13–15.

Heitz, E., 1927. Das-heterochromatic der Moose. Jabrb. Wiss. *Botany*, 69: 762–818.

Punia, M.S., Sharma, G.D. and Verma, P.K., 1985. Genetics and breeding of *Plantago ovata* Forsk.: A review. *International Journal of Tropical Agriculture*, 8(4); 255–264.

Ronsted Nina, Mark, W. Chase, Dirk C. Albach and Maria Angelica Bello, 2002. Phylogenetic relationship within *Plantago* (Plantaginaceae): Evidence from nuclear ribosomal ITS and plastid *tmL-F* sequence data. *Botanical Journal of the Linnean Society*, 139: 323–338.

Sneath, P.H. and Sokal, R.R., 1973. *Numerical Taxonomy: The Principles and Practice of Numerical Classification*. W.H. Freeman and Company, San Francisco, USA.

Wealth of India, 1969. *Raw Material*. Publication and Documentation Directorate, CSIR, New Delhi, 8: 148–153.

Wolff, K. and Morgan-Richards, M., 1998. PCR markers distinguish *Plantago major* subspecies. *Theoretical and Applied Genetics*, 96(2): 282–286.

Chapter 47

Studies on the Impact of Rubber Plantations (*Hevea brasiliensis* Muell Arg.) on the Soil Quality and its Role in the Rehabilitation of Degraded Lands in North-East India

Samiran Roy* and B.K. Dutta
Microbial and Agricultural Ecology and Biodiversity Conservation Laboratory, Department of Ecology, Assam University, Silchar – 788 011, Assam
***Corresponding Author: Present address: Department of Forestry, NERIST, Nirjuli, Arunachal Pradesh – 791 109**

ABSTRACT

The present study was undertaken in three vegetation type *viz.*, rubber plantation, tea plantation and forest in Barak valley of south Assam, North-East India. The study on the soil, under the three vegetation types in the three depths, revealed that the soil physical properties like texture, bulk density, water holding capacity, soil organic matter content and chemical properties like soil pH, Organic C, available P and K content of soil under rubber plantation is comparable with the natural forest and better than the tea plantation soil. Thus the rubber plantation can be suggested for rehabilitation of denuded/degraded land as it improves soil quality and socio-economy of marginal farmers in gel1eral and tribal farmers in particular.

Keywords: *Hevea brasiliensis, Degraded land, Nutrient status, Soil organic mater, Soil moisture.*

* Corresponding Author: E-mail: samir_oy@yahoo.co.in.

Introduction

Mature rubber plantation provides a simple forest like ecosystem with closed canopy and understory vegetation. It has been reported by Krishnakumar *et al.* (1991) that rubber plantation improves the soil quality. At the same time the natural rubber is a highly economically important commodity. There are various plant species that contains rubber as a constituent in their latex but commercially the *Hevea brasiliensis* Muell Arg. is the main source of rubber (Radhakrishnapillay, 1980).

Natural rubber is nature's most versatile plant product. There are so many uses of rubber in our day-to-day life. It has a good prospect as it is derived from the natural plant product as compared to synthetic rubber, which is derived from non-renewable petroleum product (Krishna Kumar, 1992). Rubber is extensively grown in the southern India and recently it is getting importance in the north-east India as well.

The ecological problems of north eastern states have been intensified since the last few decades, due to indiscriminate felling of forest trees and traditional shifting cultivation system, which is preceded by burning of the felled trees and organic derbies. The burning leads to the destruction of soil structure by burning away the humus and with no major agricultural inputs. This cultivation practice ends up in the depleted soil, once the frequency of jhum cultivation (Shifting cultivation) at the same site increases. The frequency of jhum cultivation is increasing because of the decreased forest areas and increase in tribal population. Thus, this primitive cultivation practices now cannot sustain the tribal families. Therefore, they are now searching for alternative means to sustain their livelihood. In this situation of socio-economic problems, rubber plantations in North Eastern states and in the tribal areas may help to solve the problem considerably. At the same time encouraging rubber plantations, environmental problems can also be taken care of. The process has already been started in some of the areas of South Assam (*i.e.* Cachar, Karimgang and hailakandi Districts) by Rubber Board. Therefore, scientific investigations about the potentiality of rubber plantation in relation to the existing environmental problem, *i.e.* soil erosion, deforestation etc. and the socio-economic conditions of the valley deserve special attention. In South Assam, land is available for planting rubber, as the barren tillahs (hillocks) are scattered throughout the Valley and are left fallow. These lands can be regenerated through rubber plantation. Besides these, even some hilly areas of South Assam, where forest does not exist or depleted due to over exploitation (*i.e.*, felling of trees and jhum cultivation) can be brought under rubber plantation.

There is a belief that large-scale rubber plantation has negative impact on the environment and human health (Paul and Dey, 2000). Therefore, it is necessary to investigate whether this belief has any scientific basis? It is necessary to evaluate the impact of rubber plantation on the environment and also to establish the fact of superiority, if any, of rubber plantation over others, in relation to environmental conditions and degraded land of the North East in general and of South Assam in particular. Therefore, the present work has been taken up to assess the impact of rubber plantation on the environment and its potential to rehabilitate the degraded land.

Materials and Methods

Soil samples were collected from ten different sites of Barak Valley, South Assam India. The samples were colleted after removal of the debris and litter at 0–15, 15–30 and 30–45 cm depths. They were air dried, sieved through 2 mm mesh and analysed for organic carbon, available phosphorous, potassium and physico-chemical parameters etc.

Soil pH

The pH value of the soil samples was determined by the 'Electrometric method'. The pH was studied in 1 : 2.5 (soil : water,) solution.

Soil Texture

Soil texture was determined by the Bouyoucos soil Hydrometer method (Allen, 1989). The different tractions of soil particle (sand, silt and clay) percentage were calculated based on the hydrometer reading at 40 second and at 2 hour. The soil texture was calculated by the following formula:

(Silt + Clay) per cent = (Hydrometer reading at 40 second/weight of soil) × 100

Clay per cent = (Hydrometer reading at 2 hrs/weight of soil) × 100

Silt per cent = (Silt + Clay) per cent – Clay per cent

Sand per cent = 100 – (Silt + Clay) per cent

To determine the soil class, triangle for percentage of sand, silt and clay in major soil texture class, method described by Brady (1969) was followed.

Bulk Density

The bulk density was determined from the core samples as per the method described by Black (1965). Core samples were collected from different depths of different sites by using soil corer. The soil was sieved through 2 mm mesh and oven dried at 105°C. The bulk density was determined by using the following formula:

$$\text{Bulk density (g cm}^{-3}) = \frac{\text{Weight of oven dried soil}}{\text{Volume of soil}}$$

Moisture Content

To determine the moisture content, soil of different sites and depths were collected in polythene bags. The soil samples were then sieved through 2 mm mesh and 20 gm of fresh soil samples were taken in the moisture boxes of known weight. The moisture boxes with soil samples were then oven dried to a constant mass. The moisture content was determined by the following formula:

$$\text{Moisture content (per cent)} = \frac{\text{Weight of fresh soil} - \text{weight of oven dried soil}}{\text{Weight of fresh soil}} \times 100$$

Water Holding Capacity

To determine the water holding capacity the Keen's box method was followed. The water holding capacity was determined by the formula as follows:

$$\text{Water holding capacity (per cent)} = \frac{B - (C + D)}{C - A} \times 100$$

where,

A: Weight of empty Keen's box

B: Weight of saturated soil + Keen's box

C: Weight of oven dried soil + Keen's box

D: Weight of wet filter paper

Organic Carbon and Organic Matter

The percentage of organic carbon of the soil samples was determined by wet oxidation with sulphuric acid and potassium dichromate solution. The percentage of organic matter present in the soil sample was determined by the following formula:

$$\text{Organic matter (per cent)} = \text{Organic carbon} \times 1.724$$

The factor 1.724 is based on the assumption that the carbon is only 58 per cent of the organic matter.

Potassium

Potassium content of soil samples was determined by the flame photometer method. The characteristic radiation for potassium is 768 mm. Therefore, the K20 content of the soil was found out using filter of this wavelength.

Phosphorous

The available soil phosphorous of the soil samples was estimated by the Molybdenum blue method with a photoelectric colorimeter using 660 mm red filter.

Results and Discussion

Physico-chemical Parameters of Soil

The texture of soil under rubber, tea and natural forest was found to be sandy loam, for all the three layers of soil analysed. The data on the particle size distribution of soil is summarized in the Table 47.1. It suggests a more or less homogeneous soil type in all the three sites and in all the respective depth of soil layers (*i.e.*, 0–15, 15–30, 30–45 cm respectively). The soil under rubber, tea and natural forest in different sites of Barak Valley was found to be moderately coarse. Coarse soil has good drainage and aeration, thus the soil is suitable for rubber plantation (Radhakrishnapillay, 1980; Krishnakumar, 1992, 1993). Most soils of the agricultural importance are a type of loam (Brady, 1969). The dominance of sand particles in the soils of hillocks may be the consequences of erosion of topsoil during the period, when no plantation cover existed in the said areas. Due to the degradation of forest the soil had been exposed to rainfall and rainwater may have washed down the fine textured soil and humic substances. Thus, the forest degradation in this area may have affected the soil texture.

Bulk density, was found to have increased with depth in all the three types of plantation/forest soils under study. The surface soil was found to be most loose and the compactness of soil increased with depth. Bulk density and porosity were significantly negatively correlated ($p < 0.05$). The bulk density was found to be lowest in the surface soil, under rubber plantation, followed by the soil under natural forest and tea plantation, respectively (Table 47.2). The lower compaction of soil may be due to high surface feeder root density of rubber plants and good organic matter content, which moderates the soil structure. The lower compactness of soil under rubber has also been reported by Krishnakumar *et al.* (1991). Decrease of organic matter content with the depth of soil was recorded in all the three cases (Table 47.2). It was found to be the highest in the surface layer of the soil under natural forest followed by the soil under rubber plantation. The difference among the percentage of organic matter content was found to be insignificant for the surface layer (0–15 cm) and in the subsequent two layers.

Krishnakumar *et al.* (1991) also reported that the organic matter content is highest in the rubber plantation compared to the natural forest and teak plantation. Present observation conforms to the above findings. The result indicates that the organic matter content of the soil under rubber plantation is comparable with that of natural forest. This may be attributed to the large amount of leaf litter deposition from the rubber plants. As the rubber tree is a deciduous one, it shades all its leaves annually, which contributes to the organic matter content of the soil in the plantation area. High density of the surface feeder roots also contributes to the organic matter content. Organic matter moderates the physical properties of the soil. Soils with high organic matter content generally have low bulk density and high porosity with good water holding capacity (Brady, 1969). Organic matter was significantly positively correlated with moisture content, porosity, organic C, available P and K, while negatively correlated with bulk density ($P < 0.05$) (Table 47.3).

Table 47.1: Particle Size Distribution and Textural Class of the Soil Under Rubber, Tea and Natural Forest

Vegetation	Depth (cm)	Particle Size Distribution			Textural Classification
		Sand (per cent)	Silt (per cent)	Clay (per cent)	
Rubber	0–15	66.90 (±3.45)	22.60 (±2.85)	10.40 (±1.05)	Sandy loam
	15–30	66.30 (±2.54)	21.40 (±3.48)	11.90 (±1.54)	Sandy loam
	30–45	67.80 (±4.56)	20.10 (±1.59)	12.10 (±2.04)	Sandy loam
Tea	0–15	62.00 (±5.03)	28.33 (±4.01)	9.75 (±1.67)	Sandy loam
	15–30	63.75 (±2.56)	23.50 (±3.93)	12.75 (±2.63)	Sandy loam
	30–45	63.00 (±3.32)	22.23 (±5.45)	17.00 (±1.87)	Sandy loam
Natural forest	0–15	67.75 (±3.51)	23.75 (±2.61)	8.50 (±1.03)	Sandy loam
	15–30	69.25 (H.22)	21.00 (±2.19)	9.25 (±2.02)	Sandy loam
	30–45	72.00 (±5.67)	18.00 (±2.76)	1.25 (±1.04)	Sandy loam

n: 10, ±: SE.

Table 47.2: Physical Properties of the Soil Under Rubber, Tea and Natural Forest

Vegetation	Depth (cm)	Organic Matter (per cent)	Bulk Density (g/cc)	Porosity (per cent)	Moisture Content	Water Holding Capacity (per cent)
Rubber	0–15	2.72 (±0.16)	0.90 (±0.09)	66.09	21.40 (±1.13)	54.50 (±2.42)
	15–30	1.90 (±0.11)	1.08 (±0.03)	59.25	19.15 (±1.41)	–
	30–45	1.38 (±0.11)	1.14 (±0.05)	56.98	17.86 (±1.52)	–
Tea	0–15	2.69 (±0.3)	1.03 (±0.04)	61.13	20.38 (±0.35)	54.05 (±2.74)
	15–30	1.93 (±0.15)	1.11 (±0.05)	58.11	19.11 (±1.83)	–
	30–45	1.69 (±0.14)	1.13 (±0.04)	57.36	18.34 (±1.94)	–
Natural forest	0–15	2.86 (±0.18)	0.98 (±0.02)	63.02	19.63 (±2.9)	57.00 (±1.52)
	15–30	1.90 (±0.17)	1.18 (±0.06)	55.47	18.19 (±2.41)	–
	30–45	1.53 (±0.14)	1.33 (±0.03)	49.81	16.50 (±1.25)	–

n: 10; ±: SE.

Table 47.3: pH and Nutrient Status of the Soil Under Rubber, Tea and Natural Forest

Vegetation	*Depth (cm)*	*pH (1 : 2.5)*	*Organic Carbon (per cent)*	*Organic Matter (per cent)*	*Available Phosphorous (ppm)*	*Available Potassium (ppm)*
Rubber	0–15	4.80 (±0.08)	1.58 (±0.16)	2.72	0.36 (±0.06)	9.16 (±1.5)
	15– 30	4.76 (±0.09)	1.10 (±0.11)	1.90	0.21(±0.04)	5.40 (±1.04)
	30– 45	4.84 (±0.08)	0.80 (±0.11)	1.38	0.18 (±0.03)	5.38 (±1.02)
Tea	0–15	4.49 (±0.10)	1.56 (±0.3)	2.69	1.05 (±0.21)	9.05 (±1.70)
	15– 30	4.39 (±0.07)	1.12 (±0.15)	1.93	0.41 (±0.08)	5.90 (±1.20)
	30– 45	4.55 (±0.03)	0.98 (±0.14)	1.69	0.21 (±0.05)	4.40 (±1.00)
Natural forest	0–15	4.90 (±0.18)	1.66 (±0.18)	2.86	0.46 (±0.05)	10.90
	(±2.10)					
	15– 30	4.81 (±0.15)	1.1 0 (±0.17)	1.90	0.26 (±0.05)	5.96 (±1.20)
	30– 45	4.95 (±0.1 0)	0.89 (±0.14)	1.53	0.19 (±0.04)	4.42 (±1.02)

n: 10; ±: SE.

Table 47.4: Correlation Between Different Physico-chemical Parameters of Soil

	Bulk Density	*Porosity*	*Moisture content*	*Sand*	*Silt*	*Clay*	*pH*	*Organic Carbon*	*Available Phos-phorous*	*Available Potas-sium*
Organic matter	–0.80**	0.82**	0.85**	–0.13	0.63*	–0.06	–0.09	1.00**	0.69**	0.95**
Bulk density		–1.00**	–0.95**	0.40	–0.72**	–0.40	–0.80**	–0.82**	–0.43	–0.80**
Porosity			0.95**	–0.40	0.72**	0.40	–0.18	0.82**	0.42	0.80**
Moisture content				–0.38	0.71**	0.32	–0.32	0.85**	0.57*	0.77**
Sand					–0.82**	–0.66*	0.69**	–0.13	–0.24	0.77**
Silt						0.42	–0.57*	0.63	0.61 *	0.58*
Clay							–0.61 *	–0.06	–0.04	–0.11
PH								–0.08	–0.48	0.07
Organic carbon									0.67*	0.95* *
Available phosphorous										0.95**

*: $p < 0.01$; **: $p < 0.05$.

Moisture content was found to be decreasing with the depth *i.e.*, top layer (0–15 cm) was found to have the highest moisture content and with increased depth (30–45 cm), the least moisture content has been observed in both the cases of soil under rubber and tea plantations. In the natural forest the middle layer (15–30 cm) of the soil was found to have the highest moisture content followed by the top layer (0–15 cm) and the third layer (30–45 cm) respectively (Table 47.2). When the top layer of soil (0–15 cm) under rubber tea and natural forest is compared, it is found that the soil under rubber retains the highest amount of moisture followed by the tea soil and natural forest soil respectively. The good moisture content in the topsoil under rubber plantations may be due to the thick canopy cover, which prevents sunlight to reach the ground directly. High litter fall during the winter season also covers the ground and acts as mulch, which helps to conserve moisture. The organic matter content and the low compaction of soil also contribute to the good moisture retention capacity of the soil under rubber. Higher root density under rubber plantation allows maximum absorption of the precipitation and prevents excess run off. Good water retention characteristics of rubber soil have also been reported by Krishnakumar *et al.* (1991) and Mathew (1999).

The water holding capacity was found to be the highest in the surface soil under forest, closely followed by the soil under rubber plantation (Table 47.2). This may be correlated with the organic matter content pattern of the soil under different vegetation. Organic matter improves the soil structure, which in turn improves the water holding capacity of the soil.

The better moisture content of the soil under rubber encourages the growth of understorey vegetation. This was confirmed during the field visit, when it was observed that a good number of understorey vegetation was growing in the rubber plantation area. During the winter season the rubber plants sheds all its leaves. Thus, the evaporation of water through leaves during winter is also reduced considerably. Further, the leaf litter covers the ground and thus arrests the soil moisture from evaporation loss. This may also reduce the drought effect during the winter months (*i.e.*, December–March).

The above results revealed that the surface soil under rubber plantation could retain rainwater comparatively better due to the moderate soil structure. Thus, the soil under rubber plantation has better effect on the reduction of run off water, and as a result the chance of soil erosion is reduced, considerably. The texture of the soil in the study area was found to be moderately coarse. This type of soil is prone to erosion. Therefore, it is very important to take care about the soil erosion aspect during the planting activities in the hillocks of Barak Valley. The Rubber Board recommendations should be followed carefully to check soil erosion (*i.e.*, contouring digging silt pits, cover crops etc.).

pH and Nutrient Status of the Soil

In the present study the soil under natural forest was found to have the highest pH value (pH values were above 4.8 in all the three layers) and the soil under tea was found to be having the lowest pH value *i.e.*, the soil under tea plantation was slightly more acidic (pH values were below 4.55 in all the three layers) (Table 47.3).

In all the three cases the pH values slightly decreased in the middle layer (15–30 cm) and then increased in the 3rd layer. In the surface soil the pH value was highest in the natural forest (4.9) and lowest in the soil under tea plantation (4.49). The low pH of soil under tea plantation of Barak Valley was also reported by Chaudhury (1999) and Dey *et al.* (1969). In Barak Valley moderate to high intensity rainfall is one of the factors that contributes to the acidification of soil. Due to the excess precipitation calcium, magnesium and other bases are leached out leaving a nitrogen saturated acidic soil (Fuchs, 1989), which is relevant to the Barak Valley agro-climatic conditions.

In the natural forest and rubber plantation areas, soil pH was observed to be comparatively higher than tea soil. This can be attributed to the good ground cover in the rubber plantation and natural forest areas, which reduced the run off. The acidification of tea soil may also be due to the intense agricultural practices including the use of inorganic fertilisers for a prolonged period (at the rate of 60–120 kg/ha/yr) (Dutta, 1995).

Organic carbon and available potassium was found to be the highest in the surface soil under natural forest. Available potassium status in the surface soil under rubber plantation follows the surface soil under natural forest. Surface enrichment of nutrients has been found in the soils under all the three vegetation type. Available phosphorous was highest in the surface soil (0–15 cm) under tea plantation (Table 47.3). The reason could be due to the application of phosphatic fertiliser on a regular basis to the soil under tea plantation (at the rate of 20–60 kg P_2O_5/ha/yr) (Dutta, 1995). Fertiliser requirement of rubber plants is comparatively less. In the rubber plantation area of Barak Valley, manuring is generally not done on a regular basis, especially in the mature plantation areas as compared to tea plantations. From the above findings it can be said that there is not much of a difference in the nutrient status of the soils except the available phosphorous status under the three different vegetation types (*i.e.*, rubber, tea and natural forest) under study.

The percentages of variation of nutrient status between the different layers of soil under different vegetations are presented graphically (Figure 47.1). Available phosphorous content in soil under tea plantation was found to be the highest. The surface enrichment of phosphorous may be due to the phosphatic fertiliser used in the tea plantation on regular basis. Except phosphorous all other nutrients show almost similar pattern of percentage decline in the different layer of soil under different vegetation type. The decline of nutrients in soil from 0–15 cm layer to 15–30 cm was found to be the highest in case of natural forest closely followed by the soil under rubber plantation. The percentage decline of nutrients between the next two layers (15–30 cm to 30–45 cm) does not show any particular trend.

From the above study of physico-chemical parameter and nutrient status of the soil it can be suggested that the rubber plantations have a positive impact on the soil. Therefore, planting of rubber in the degraded and/or deforested jhum lands of the North Eastern Region in general and in Barak Valley in particular can be recommended. Krishnakumar and Potty (1992) reported that under the

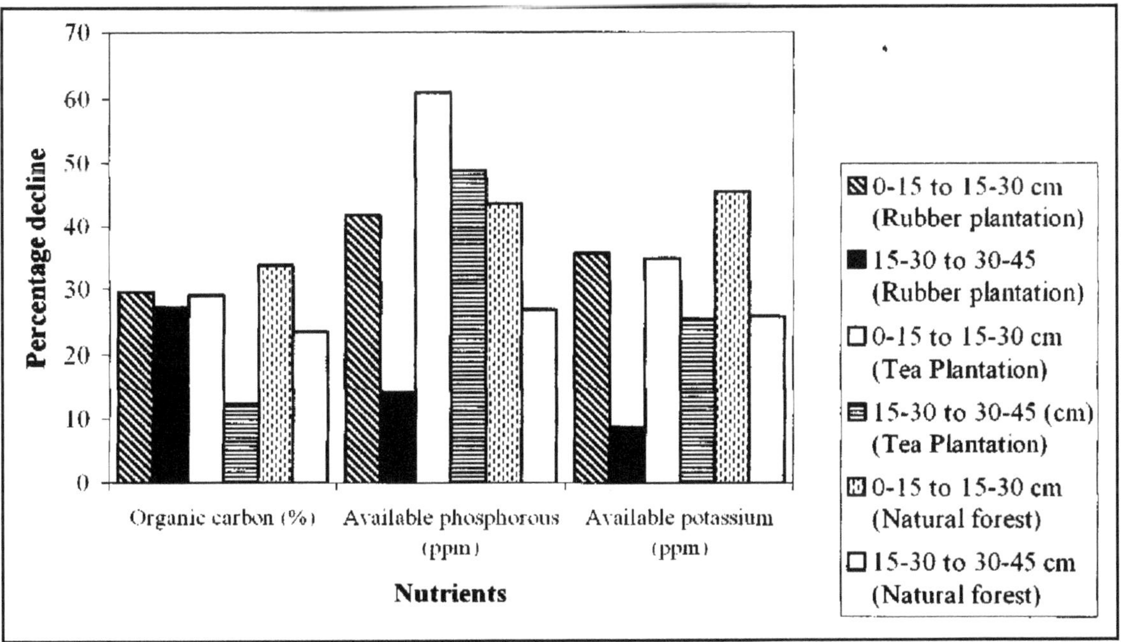

Figure 47.1: Percentage Decline in Available Nutrient with Soil Depth

Indian conditions, soil physical properties in the denuded forests as well as in the areas subjected to continuous shifting cultivation are considerably improved, once a rubber plantation is established. They also observed that the rubber plantations present almost a closed ecosystem, in a near steady state, during their life span. The findings of the present work corroborates with the above observations.

Acknowledgements

We thankfully acknowledge the help provided in soil analysis by Dr. Sunil Kumar Dey, Director, Regional Rubber Research Institute, Kunjaboan, Tripura. We also thankful to Mr. Mridul Bhattacharjee, Executive of Rubber Board Silchar branch for his generous help in taking me to remotely located fields with his vehicle.

References

Alderson, J.M. and Ingram, J.S.I., 1993. *Tropical Soil Biology and Fertility: A Handbook of Methods*, 2nd edn. CAB International, Wallingford, UK.

Allen, S.E., 1989. *Chemical Analysis of Ecological Materials*. Blackwell-Scientific Publications, London.

Black, C.A., 1965. *Methods of Soil Analysis*, Part I. American Society of Agronomy. U.S.A.

Brady, Nyle C., 1969. *Nature and Properties of Soil*, 10th edn. Prentice-Hall, New Delhi.

Chaudhury, P., 1999. Studies on the pests of *Camelia sinensis* L(O) Kuntze and their control in the tea agro-ecosystems in Barak valley, Assam. *Ph.D. Thesis*.

Dey, S.K., Bhattacharyya, N.G., Sengupta, A.K., Barua, D.N. and Ghose, P., 1969. Soil Survey 1967/68, Results for the Cachar District Tocklai Occasional Scientific Papers, No. 4, TRA Publications.

Dutta, B.K., 1995. *Problems and Prospects of Cachar Tea Plantations*. JASC Meeting, Cachar, Assam.

Fuchs, H.J., 1989. Tea environment and yield in Sri Lanka. In: *Tropical Agriculture*, (Ed.) D. Knuth.

Krishnakumar, A.K., 1992. Ecological and socio-economic desirability of rubber plantation in North-east. *Technical Newsletter* 3/1992, Forest Research Division, Tripura.

Krishnakumar, A.K. and Potty, S.N., 1992. Nutrition of hevea. In: *Natural Rubber Biology, Cultivation and Technology*, (Eds.) M.R. Setturaj and N.M. Mathew. Elsevier, Amsterdam.

Krishnakumar, A.K., Gupta, Chandra, Sinha, R.R., Sethuraj, M.R., Potty, S.N., Thomas, Eappen and Das, Krishna 1991. Ecological impacts of rubber plantations in Northeast India: Soil properties and biomass recycling. *Indian Journal of National Rubber Research*.

Krushnakumar, A.K., 1993. Environmental factors that influence rubber plantation. *Rubber Chem. Review*, Rubber Research Institute of India.

Mathew, K.J., 1999. Promise of small holdings. *The Hindu Survey of Indian Agriculture*, pp. 85–89.

Radhakrishnapillay, P.N. (Ed.) 1980. *A Handbook of Rubber Plantations*. Rubber Research Institute of India, Kottoyam.

Chapter 48

Effect of Weed Control Treatments on Crop Stand and Phytotoxicity Rating in Direct Sown Wet Seeded Rice

B. Rajendra Kumar[1] and A. Christophrer Lourduraj[1]
Department of Agronomy, Tamil Nadu Agricultural University, Coimbatore

ABSTRACT

A Field experiment was conducted during the rabi season 2002–2003 to study the weed management of anaerobic tolerant cultivars raised under wet seeding in low land direct sown rice. Early application of Pretilachlor + safener 0.45 kg a.i. ha^{-1} at 4 DAS did not cause any phytotoxicity and thereby crop stand was not affected. ASD 19 is the best suited rice variety for anaerobic seeding under low land direct wet seeded condition. Application of Pretilachlor + safener 0.45 kg a.i. ha^{-1} at 4 DAS + 1 hand weeding at 35 DAS can be advocated to realize crop safety, superior weed control and growth and productivity of wet seeded rice.

Keywords: Crop stand, Phytotoxicity rating, Direct sown wet seeded Rice.

Introduction

Direct seeding of rice is gaining importance as an alternative method to rice cultivation for transplanting. Many Asian rice farmers are now switching from transplanting to direct seeding because

---

[1] Corresponding Author and Agricultural Officer, Agricultural Research Station, Amadalavalasa, ANGRAU, A.P.

[2] Associate Professor, Department of Agronomy, Tamil Nadu Agricultural University, Coimbatore.

of increased labour cost coupled with introduction of short duration cultivars and availability of suitable herbicides. A large number of herbicides are available for controlling weeds in rice, the mode of action of each herbicide may vary with different varieties of rice and a single herbicide may not be suitable to all varieties. Difference in performance of rice cultivars to herbicides was reviewed by Moody and Drost (1983).

Materials and Methods

A Field experiment was conducted at Tamil Nadu Agricultural University, Coimbatore during rabi season 2002, 2003. The experiment was laid out in split plot design with three varieties ASD 19, ADT 38 and CO 43 in the main plot and weed management practices (Pretilachlor + safener 0.45 kg a.i. ha^{-1} at 4 DAS + 1 hand weeding at 35 DAS; Butachlor 1 kg a.i. ha^{-1} at 8 DAS + 1 hand weeding at 35 DAS; Butachlor + Propanil 0.84 + 0.84 kg a.i. ha^{-1} at 10 DAS (tank mix) + 1 hand weeding at 35 DAS; two hand wee dings at 20 and 35 DAS; unweeded check) in the subplots with three replications.

Table 48.1: Weed Control and Phytotoxicity Rating Chart

Treatments	7 Days After Herbicide Spray	
	Weed Control	Crop Injury
S_1: Pretilachlor + safener 0.45 kg a.i. ha^{-1} + 1 hand weeding at 35 DAS	8	0
S_2: Butachlor 1 kg a.i. ha^{-1} + 1 hand weeding at 35 DAS	7	1
S_3: Butachlor + Propanil 0.84 + 0.84 kg a.i. ha^{-1} + 1 hand weeding at 35 DAS	7	1
S_4: Two hand wee dings at 20 and 35 DAS	–	–
S_5: Unweeded control	–	–

Crop: 0: No injury; 1: Slight stunting, injury or discolouration

Weed: 7: Satisfactory control; 8: Good control.

Plant population was counted in a marked area of one meter square using a quadrat at 15 days after sowing. After one week of application of each herbicide, population counts were also taken in order to know the phytotoxicity rating. Weed control and phytotoxicity rating was worked out 7–10 days after herbicidal spray. The phytotoxicity rating was estimated with the help of a weed control and phytotoxicity rating chart.

Results and Discussion

There was no significant difference in plant population among the varieties at 7 days and 15 days-after herbicide application. Significant difference was observed-among the weed control treatments. Unweeded check (S_5) and hand weeding twice at 20 and 35 DAS (S_4) recorded significantly higher plant population m^{-2} compared to other treavnents and did not significantly differ with each other. Among the chemical weed control treatments, Pretilachlor + safener 0.45 kg a.i. ha^{-1} + 1 hand weeding at 35 DAS (S_1) recorded highest plant population of 188.88 and 182.00 m^{-2} at 7 and 15 days after herbicidal application.

The population reduction in Slover control (S_5) was 2.96 per cent at 7 days and 4.59 at 15 days after herbicide application. Lowest plant population was observed in Butachlor + Propanil 0.84 + 0.84 kg a.i. ha^{-1} + 1 hand weeding at 35 DAS (S_3) with 184 and 174.22 m^{-2}, at 7 and 15 days after herbicide application. The population reduction in S_3, over control was 5.47 and 8.67 per cent at 7 and 15 days after herbicide application, respectively. Interaction effect was not significant between varieties and weed control treatments.

Table 48.2: Effect of Weed Control Treatments on Rice Population (m^{-2}) at 7 Days and 15 Days After Herbicide Application in Direct Sown Wet Seeded Rice

Treatments	*7 days*				*15 days*				*Percentage Population Reduction Over Control*	
	M_1	M_2	M_3	*Mean*	M_1	M_2	M_3	*Mean*	*7 Days*	*15 Days*
S_1	189.66	187.33	189.66	188.88	183.00	181.66	181.33	182.00	2.96	4.59
S_2	183.33	186.33	185.66	186.77	180.00	177.66	177.00	178.22	4.05	6.57
S_3	184.33	183.33	184.33	184.00	174.00	173.33	175.33	174.22	5.47	8.67
S_4	194.66	194.33	191.66	193.55	191.00	190.66	191.00	190.77	0.57	0.05
S_5	194.00	194.66	195.33	194.66	190.66	190.33	191.33	190.88	–	–
Mean	190.2	189.20	189.33		183.73	182.73	183.20			
		SEd	*CD (P=0.05)*			*SEd*	*CD (P=0.05)*			
M		1.16	NS			0.52	NS			
S		1.15	2.36			0.81	1.69			
M at S		2.14	NS			1.37	NS			
S at M		2.00	NS			1.41	NS			

M_1: ASD 19; M_2: ADT 38; M_3: CO 43

S_1: Pretilachlor + safener 0.45 kg a.i. ha^{-1} + 1 hand weeding at 35 DAS; S_2: Butachlor 1 kg a.i. ha^{-1} + 1 hand weeding at 35 DAS; S_3: Butachlor + Propanil 0.84 + 0.84 kg a.i. ha^{-1} + 1 hand weeding at 35 DAS; S_4: Two hand weedings at 20 and 35 DAS; S_5: Unweeded check.

Weed control rating was recorded at 7 days after herbicide spray. Good weed control with no injury to crop was observed in Pretihichlor + safener 0.45 kg a.i. ha^{-1} + 1 hand weeding at 35 DAS (S_1). Whereas, the other two herbicides (S_2 and S_3) showed satisfactory control of weeds with slight crop injury.

Herbicide application at different doses and combinations influenced the population of rice crop. Moody (1984) and Mathew and Jagadesh Kumar (1999) stated that the selectivity of herbicides depends upon the rate and times of application. Application of Pretilachlor + safener 0.45 kg a.i. ha^{-1} on 4 DAS did not affect germination and stand of wet seeded rice. Good control of weeds without any reduction in plant stand of rice due to Pretilachlor + safener on 3 DAS was also earlier reported by IRRI (1986). Application of Butachlor alone or and Butachlor mixed with Propanil (Tank mix) showed mortality of rice seedlings.

From the stand point of phytotoxicity rating, Pretilachlor + safener treated plot the effect was rated as "none". The selectivity of Pretilachlor + safener is attributed to the fenclorim (CGA 123407) mediated reduction in root uptake of Pretilachlor and an enhancement in the metabolism of Pretilachlor via enzymatic or non-enzymatic conjugation with glutathione which offered protection to rice seedlings Deng *et al.* (1997).

References

Deng, F., Nagao, A., Shim, I.S. and Usui, K., 1997. Introduction of glutation S-tranferase isozymes in rice shoot treated with a combination of pretilachlor and fenchlorim. *J. Weed Sci. Technol.*, 42(3): 277–283.

IRRI, 1986. *Annul Report*, 1985, IRRI, Los Banos, Philippines.

Mathew, J. and Kumar, T.N. Jagadesh, 1999. Chemical weeding strategies with improved crop safety in direct seeded paddled rice. *Oryza*, 36(4): 355–357.

Moody, K. and Drost, D.C., 1983. Rice cultivar tolerance to herbicide. Technical Bulletin. AS PAC Food and Fertilizer Technology Centre, 76: 1–14.

Moody, K., 1984. Possible ways of reducing herbicide toxicity in wet seeded rice. In: *Proc. First. Trop. Weed Sci. Conf.*, pp. 15–25.

Chapter 49

Studies on the Phytosociology of the Macrophytes in the Oksoipat Lake, Bishnupur (Manipur)

S. Umeshwari Devi and B. Manihar Sharma
Ecology Laboratory, Department of life Sciences, Manipur University, Canchipur – 795 003

ABSTRACT

The investigation analyses the phytosociological present characteristics of the freshwater macrophytes of the Oksoipat Lake Oksoipat lake is a shallow and very old semi-terrestrial freshwater lake situated at Oinam village in the Bishnupur district of Manipur. It is located at about 21 km on the south-western side of Imphal city. The various phytosociological characteristics include Frequency, Density, Abundance and IVI of the macrophytes. A total of 35 macrophytic species were recorded during the present study. Some of the dominant during the present macrophytic plants species recorded are *Alternanthera philoxeriodes, Echinochloa stagnina, Ludwidgia adscendens, Marsilea quadrifoliata, Nymphoides cristatum* and *Salvinia cucullata* etc.

Keywords: *Oksoipat lake, Phytosociology, Macrophytes, Biodiversity, Morphometry.*

Introduction

Freshwater environment is a treasure indispensable to all human activities and it has attracted mankind since time immemorial by their essential elements to the living organisms. Macrophytes in freshwater play major ecological role and help in the regulation and stabilization of trophic state and mineral cycling in the aquatic ecosystem (Melzer, 1981, Weiegleb, 1984). They serve as the bioindicators for the possible degree of damage in the aquatic ecosystem (Pieczynska and Ozimek, 1976). Some of the aquatic weeds can effectively be used for wastewater treatments as they can scavenge some organic and inorganic compounds including heavy metals from wastewater (Gopal 1967). Excessive growth of macrophytes again deteriorate water quality as in the case of Kashmir (Kaul *et al.*, 1978). Almost all the freshwater bodies of Manipur are subjected to a number of environmental stress and are much degraded and most of them have already reached the status of wetlands (Sharma 2000). So, understanding of the structure, function and dynamics of the wetland require relevant and interrelated ecological approaches and environmental relationships, succession and the nature of the community (Knight, 1975). Phytosociology is the study of the vegetation occurring in a particular ecosystem (Ambast, 1969). The interaction of the species, their phenological behaviour, climatic conditions and edaphic factors have a profound effect on the composition as well as overall performances of the species in the aquatic communities. Thus, phytosociological data (quantitative) serve as a prerequisite for understanding the detailed knowledge about the vegetational structure and dynamics of an ecosystem.

Earlier studies on the phytosociology of the freshwater ecosystems in India were undertaken by Vyas (1965), Jha (1965, 1968) and Abbas (1979) which were followed successively by a number of workers like Kaul and Handoo, 1989. However in north-east India, studies on the ecology of freshwater ecosystems is found to be scanty currently. Likewise in Manipur also only a few intensive synecological and phytosociological investigations on such lines were undertaken by Devi, N.B. (1993) in Loktak lake (A Ramsar site), Devi, O.I. (1993) in Waithou Lake; Devi, K.I. (1998) in Utrapat lake etc.

Materials and Methods

The present study site is located in the Bishnupur district of Manipur at about 21 km on the south-western side of the Imphal city. It is situated in the intersection of 24°25′ N–24°40′ N latitude and 93°45′ E–93°55′ E longitude. The morphometry of the lake has been furnished in Table 49.1.

Table 49.1: Morphometry of Oksoipat Lake

Altitude (m)	783
Longitude	93°45′–93°55′ E
Latitude	24°25′–24°40′ N
Area (km^2)	0.864
Maximum depth (m)	1.28
Minimum depth (m)	0.46
Mean depth (m)	0.87
Mean depth/Maximum depth ratio	0.68
Basin shape	Saucer shaped
Basin slope	Gentle slope
Bottom texture	Sited bottom
Mean maximum temperature (°C)	13.0–32.5
Mean Minimum temperature (°C)	12.0–26.0

For the study of phytosociological characteristics like Frequency, Density and Abundance of the different macrophytes, Quadrat method was used (Curtis, 1959; Misra, 1968). Importance Value Index were then calculated as the sum of relative frequency, relative density and relative abundance (Billore and Vyas, 1982; Misra, 1989). For the detailed study, the phytosociological characters of the different species were investigated at monthly intervals.

Results and Discussion

During the present study a total of 35 macrophytic plant species were found distributed in the lake. Floristic composition of the present study site is shown in Table 49.2. The ranges of frequency, density, abundance, abundance/frequency ratio and the ranges of relative frequency, relative density, relative abundance, importance value index (IVI) are shown in Tables 49.3 and 49.4 respectively.

During the present study, the maximum values of frequency (0–100 per cent) were recorded in *Echinochloa stagnina, Cynodon dactylon* and *Sporobolus diander* followed by *Ceratophyllum demersum* (45–70 per cent), *Alternanthera philoxeroides* (25–70 per cent), *Hydrilla* verticillata (35–60 per cent), and the minimum was exhibited by *Carex cruciata* (0–15 per cent), *Chrysopogon aciculatus* and *Euryale ferox* (0–20 per cent).

The highest value of density were recorded in *Sporobolus diander* (283.00–416.00 plants m^{-2}), *Echinohcloa stagnina* (41.60–370.40 plants m^{-2}), *Alternanthera philoxeroides* (11.20–364.80 plants m^{-2}), *Ceratophyllum demersum* (126.40–253.60 plants m^{-2}) and the minimum value was exhibited by *Euryale ferox* (5.60–12.00 plants m^{-2}).

The peak values for Abundance were exhibited by *Echinochloa stagnina* (194.24–532.00 plants m^{-2}), *Ceratophyllum demersum* (88.00–422.67 plants m^{-2}), *Salvinia* (200–431.12 plants m^{-2}), *Alternanthera philoxeroides* (8.00–416.00 plant m^{-2}), *Sporobolus diander* (283.00–416.00 plant m^{-2}).

The Abundance/Frequency (A/F), ratios of different species varied from 0.00–86. The maximum values of A/F ratio were exhibited by *Salvinia natans* (0.22–0.86), *Alternanthera philoxeroides* (0.02–0.74), *Echinochloa stagnina* (0.04–0.65), and *Hydrilla verticillata* (0.15–0.65).

According to Curtis (1959) if A/F ratios of the different species are less than 0.025, the species are found distributed homogeneously, while the ratios within 0.025 to 0.05 indicate random distribution. When the ratios are higher than 0.05 the same indicates the aggregate distribution of the species. In the present study the different species were found distributed generally in the aggregate manner.

The maximum values of IVI were exhibited by *Echinochloa stagnina* (31.48–122.86), *Sporobolus diander* (96.25–110.60), *Cynodon dactylon* (78.16–96.60), *Alternanthera philoxeroides* (11.26–78.61) and *Ceratophyllum demersum* (27.73–77.60).

The ranges of frequency values (0–100 per cent) which were exhibited by *Echinochloa stagnina, Cynodon dactyl on, Sporobolus diander* are highly comparable with the values reported by Devi, N.B. (1993) in the Phumdiareas of Loktak lake, Manipur (20 per cent to 100 per cent), Devi, K.U. (2002), in the Poiroupat Lake, Manipur; and Devi, L.G. (1993) in the freshwater ponds of Canchipur (26.27 per cent to 100 per cent). Ranges of frequency value of *Ceratophyllum demersum* are in agreement with the values reported by Kumar and Pandit (2005) in Hokarsar wetland, Kashmir. Shah and Abbas (1979) also reported maximum frequency for *Hydrilla verticillata* (80 per cent) *Eichhornia crassipes* (80 per cent) and *Ceratophyllum demersum* (70 per cent) in the river Ganga. Ambasht (1970) and Misra (1989) also reported high percentages of frequency for the macrophytes in Varanasi.

Table 49.2: Floristic Composition of Oksoipat Lake ('+' and '–' Sign Denotes the Presence and Absence of Species Respectively.

Sl.No.	*Name of Species*	*Site-I*	*Site-II*	*Site-III*	*Site-IV*	*Site-V*
1.	*Alternanthera philoxeroides* Griseb	+	+	+	+	+
2.	*Azolla prinnata* R. Br.	+	+	+	+	–
3.	*Carex cruciata.* Wahllenb	–	–	–	–	+
4.	*Centella asiatica* (Linn.) Urban	–	–	–	–	+
5.	*Ceratophyllum demersum.* Linn.	+	+	+	+	
6.	*Chara zeylanica* wild.	+	–	–	–	
7.	*Chrysopogon aciculatus* (Retz). Trin	–	–	–	–	+
8.	*Cynodon dactylon* (Linn.) Pers	–	–	–	–	+
9.	*Echinochloa stagnina* (Retz.) P. Beauv	+	+	+	+	+
10.	*Eichhornia crassipes* (Mart.) Solms.	–	+	+	+	
11.	*Enhydra fluctuans* Lour.	+	+	+	+	
12.	*Euryale ferox* salisb	–	+	+	–	
13.	*Hydrilla verticillata* (Linn. f.) Royle	+	+	+	+	
14.	*Hydrocotyl javanica* Thunb.	–	–	–	–	+
15.	*Hygroryza aristata* (Retz.) Nees	+	+	+	+	
16.	*Ipomoea aquatica* Forsk	+	+	+	+	
17.	*Knoxia mollis.* R.Br.	–	–	–	–	+
18.	*Ludwidgia adscendens* (Linn.) Hara	+	+	+	+	+
19.	*Marsilea quadrifoliata* Linn.	+	+	+	+	+
20.	*Najas minor* Allioni	+	–	–	–	–
21.	*Nymphaea stellata* willd	+	+	+	–	
22.	*Nymphoides cristateum* (Roxb) O. Kuntze	–	+	+	+	
23.	*Oryza minuta presl.* J.S. ex C.B. presl	+	+	+	+	
24.	*Oryza rufipogon* Griff	+	+	+	+	
25.	*Oxalis corniculata* Linn	–	–	–	–	+
26.	*Paspalum scrobiculatum* Linn.	–	–	–	–	+
27.	*Potamogeton crispus* Linn.	+	+	+	+	–
28.	*Pseudognaphalium luteoallbum* spp. Affine	–	–	–	–	+
29.	*Pseudoraphis minuta,* pilger	+	+	+	–	
30.	*Salvinia cucullata* Roxb	+	+	+	+	
31.	*Salvinia natans* Hoffm	+	+	+	+	
32.	*Sporobolus diander* (Retz) P. Beauv	–	–	–	–	+
33.	*Trapa natans* (Linn.) var. bispinosa (Roxb.)	–	+	+	+	
34.	*Utricularia aurea* Lour	+	+	+	+	
35.	*Vallisneria spiralis* Linn.	+	+	+		

Table 49.3: Phytosociological Values of Macrophytes in Oksoipat Lake

Sl.No.	Name of Species	Freq. (per cent)	Density (Plants m^{-2})	Abundance (Plants m^{-2})	A/F
1.	*Alternanthera philoxeroides* Griseb	20–70	11.20–364.80	8.00–416.00	0.02–0.74
2.	*Azolla prinnata* R. Br.	30–60	5.60–105.60	64.00–264.00	0.07–0.53
3.	*Carex cruciata.* Wahllenb	0.00–1 5	4.80–72.00	6.40–48.00	0.03–0.20
4.	*Centella asiatica* (Linn.) Urban	0.00–25	0.00–20.80	0.00–83.20	0.00–0.21
5.	*Ceratophyllum demersum* Linn.	45–70	126.40–253.60	88.00–22.67	0.1 8–0.44
6.	*Chara zeylanica* wild.	30–60	14.40–132.00	43.00–293.28	0.10–0.41
7.	*Chrysopogon aciculatus* (Retz). Trin	0.00–20	11.20–16.80	11.20–84.00	0.04–0.26
8.	*Cynodon dactylon* (Linn.) Pers	0.00–100	210.40–317.60	46.40–317.60	0.03–0.19
9.	*Echinochloa stagnina* (Retz.) P. Beauv	0.00–100	41.60–370.40	1 94.24–532.00	0.04–0.65
10.	*Eichhornia crassipes* (Mart.) Solms.	0.00–30	3.20–28.50	21.28–203.28	0.09–0.26
11.	*Enhydra fluctuans* Lour.	0.00–35	23.00–59.20	91.36–169.14	0.16–0.32
12.	*Euryale ferox* salisb	0.00–25"	5.60–12.00	4.80–48.00	0.09–0.15
13.	*Hydrilla verticillata* (Linn.f.) Royle	35–60	68.00–144.00	135.36–363.43	0.15–0.65
14.	*Hydrocotyl javanica* Thunb.	0.00–25	0.00–22.40	0.00–89.60	0.00–0.22
15.	*Hygroryza aristata* (Retz.) Nees	25–65	30.40–128.80	113.60–269.71	0.08–0.54
16.	*Ipomoea aquatica* Forsk	0.00–30	0.00–24.00	0.00–52.29	0.00–0.22
17.	*Knoxia mollis* R.Br.	30–35	20.00–33.60	11.36–112.00	0.02–0.29
18.	*Ludwidgia adscendens* (Linn.) Hara	0.00–50	12.00–88.00	8.32–372.00	0.00–0.57
19.	*Marsilea quadrifoliata* Linn.	0.0040	14.40–127.20	0.00–176.08	0.00–0.37
20.	*Najas minor* Allioni	25–30	38.40–69.60	153.60–232.00	0.36–0.48
21.	*Nymphaea stellata* willd	0.00–65	5.60–32.80	28.00–70.08	0.05–0.15
22.	*Nymphoides cristateum* (Roxb) O. Kuntze	10–30	5.60–24.80	22.40–99.20	0.06–0.25
23.	*Oryza minuta presl* J.S. ex. C.B. presl	0.00–60	0.00–64.80	0.00–108:00	0.00–0.43
24.	*Oryza rufipogon* Griff	0.00–65	0.00–66.40	0.00–166.00	0.00–0.26
25.	*Oxalis corniculata* Linn	0.00–25	14.40–16.80	57.60–67.20	0.00–0.17
26.	*Paspalum scrobiculatum* Linn.	35–40	27.20–60.00	19.20–150.00	0.03–0.26
27.	*Potamogeton crispus* Linn.	0.00–50	28.80–205.71	82.08–205.71	0.13–0.37
28.	*Pseudognaphalium luteoallbum* spp. Affine	30–35	21.60–42.40	12.80–121.40	0.02–0.24
29.	*Pseudoraphis minuta,* pilger	0.00–35	10.40–71.20	41.60–203.43	0.09–0.36
30.	*Salvinia cucullata* Roxb	25–65	17.60–70.40	60.80–221.76	0.12–0.41
31.	*Salvinia natans* Hoffm	25–55	55.20–227.20	200.00–431.12	0.22–0.86
32.	*Sporobolus diander* (Retz) P. Beauv	0.00–100	283.20–416.00	283.20–416.00	0.04–0.24
33.	*Trapa natans* (Linn.) var. bispinosa (Roxb.)	0.00–35	17.60–33.60	50.24–86.86	0.09–0.15
34.	*Utricularia aurea* Lour	0.00–55	51.20–174.40	124.48–348.50	0.17–0.53
35.	*Vallisneria spiralis* Linn.	0.00–50	19.20–81.60	76.80–2.4.00	0.170.34

Table 49.4: Phytosociological Values of Macrophytes in Oksoipat Lake

Sl.No.	Name of Species	r Freq. per cent	r Density	r Abundance	IVI
1.	*Alternanthera philoxeroides* Griseb	4.76–11.54	1.23–21.08	0.96–14.39	9.88–58.55
2.	*Azolla prinnata* R. Br.	3.51–9.52	3.50–7.98	0.57–8.27	11.22–21.55
3.	*Carex cruciata.* Wahllenb	2.83–3.00	0.49–0.58	2.25–2.71	5.74–6.30
4.	*Centella asiatica* (Linn.) Urban	0.00–4.46	0.00– 1.52	0.00–4.15	0.00–10.13
5.	*Ceratophyllum demersum* Linn.	6.99–12.87	6.31–19.05	0.61–26.42	23.00–46.90
6.	*Chara zeylanica* wild.	5.26–10.91	1.29–10.72	2.07–10.21	0.00–28.94
7.	*Chrysopogon aciculatus* (Retz) Trin	3.77–4.00	1.15–1.37	4.26–4.74	9.28–10.10
8.	*Cynodon dactylon* (Linn.) Pers	17.86–2.47	23.53–6.46	16.29–19.77	56.89–66.89
9.	*Echinochloa stagnina* (Retz.) P. Beauv	4.30–20.00	4.44–33.26	0.94–25.48	12.96–79.72
10.	*Eichhornia crassipes* (Mart.) Solms.	2.59–4.81	0.25– 1.83	0.74–3.94	3.58–11.30
11.	*Enhydra fluctuans* Lour.	3.68–6.31	2.49–3.56	3.33–4.97	9.72–13.41
12.	*Euryale ferox* salisb	2.34–3.73	0.47–0.72	0.88– 1.21	4.07–7.76
13.	*Hydrilla verticillata* (Linn.f.) Royle	4.24–10.35	5.45–11.27	0.67–10.43	15.08–30.61
14.	*Hydrocotyl javanica* Thunb.	0.00–4.46	0.00– 1.64	0.00–4.47	0.00–10.57
15.	*Hygroryza aristata* (Retz.) Nees	3.64–10.19	0.76–9.05	0.61–10.10	9.81–25.90
16.	*Ipomoea aquatica* Forsk	0.00–3.23	0.00–1.39	0.00–2.37	6.06–12.02
17.	*Knoxia mollis.* R.Br.	5.36–6.60	1.34–2.41	4.45–5.57	13.04–13.98
18.	*Ludwidgia adscendens* (Linn.) Hara	3.18–9.26	1.07–7.56	0.41–9.56	10.69–36.59
19.	*Marsilea quadrifoliata* Linn.	0.00–7.69	0.00–5.50	0.00–6.39	9.37–31.98
20.	*Najas minor* Allioni	3.23–4.48	3.50–4.98	0.74–7.60	9.83–17.06
21.	*Nymphaea stellata* willd	2.69–11.82	0.51–2.85	1.24–2.24	4.93–5.18
22.	*Nymphoides cristateum* (Roxb) O. Kuntze	2.99–5.77	0.43–1.60	0.78–2.48	4,66–16.70
23.	*Oryza minuta presl* J.S. ex C.B. presl	0.00–7.20	0.00–3.94	0.00–2.79	6.85–13.55
24.	*Oryza rufipogon* Griff	0.00–7.98	0.00–4.28	0.00–2.80	7.13–15.06
25.	*Oxalis corniculata* Linn	0.00–4.46	0.00–1.23	0.00–3.35	9.04–12.72
26.	*Paspalum scrobiculatum* Linn.	7.00–7.87	3.14–4.34	7.30–8.28	18.36–19.45
27.	*Potamogeton crispus* Linn.	2.92–8.13	1.98–5.70	0.54–9.93	7.94–23.76
28.	*Pseudognaphalium luteoallbum* spp. Affine	6.00–7.78	2.29–3.10	4.99–6.04	13.79–16.19
29.	*Pseudoraphis minuta* Pilger	3.76–6.93	0.95–5.52	0.44–5.95	6.77–17.88
30.	*Salvinia cucullata* Roxb	3.64–11.11	1.32–8.43	0.55–7.82	8.32–27.37
31.	*Salvinia natans* Hoffm	3.51–10.00	5.03–19.74	0.87–16.63	15.56–55.59
32.	*Sporobolus diander* (Retz) P. Beauv.	17.86–2.47	30.7–33.18	21.26–26.61	68.99–82.26
33.	*Trapa natans* (Linn.) var. *bispinosa* (Roxb.)	3.51–5.93	1.51–1.86	1.85–2.55	1.12–9.29
34.	*Utricularia aurea* Lour	4.09–10.19	4.35–9.67	0.71–9.59	14.17–29.01
35.	*Vallisneria spiralis* Linn.	2.92–7.46	1.75–6.31	0.61–6.39	7.94–18.13

The peak density value for *Echinochloa stagnina* (370.40 plants per m^{-2}) is found higher than the maximum values reported by Handoo and Kaul (1982) at the Haokarsar (48 plants m^{-2}), Shalbagh (137 plants m^{-2}) and Kranchu (116 plants m^{-2}), and Devi, Ch. N. (2002) in the Ikop Lake, Manipur (288.00 plants m^{-2}) and Devi, L.G. (1993) in the freshwater ponds of Canchipur. The present finding for *Ceratophyllum demersum* (253.60 plants m^{-2}) is found higher than the maximum values of *Ceratophyllum demersum* (154 plants m^{-2}) reported by Devi, K.I. (1998) in the Utrapat Lake, Manipur.

The present findings of the abundance values for *Echinochloa stagnina* (194.24–532.00 plants m^{-2}), *Ceratophyllum demersum* (88–422.67 plants m^{-2}), *Salvinia natans* (200–431.12 plants m^{-2}), *Alternanthera philoxeroides* (8.00–416.00 plants m^{-2}) and *Sporobolus diander* (283.20–416.00 plants m^{-2}) are highly comparable with the values reported by Devi, Ch.U. (2000) in the various ponds of Canchipur, Devi, Ch.B. (2001) in the Sanapat Lake, Manipur; Devi, L.G. (1993) in the freshwater ponds of Canchipur, as well as Ambasht (1970) for a few dominant species *viz., Azolla* and *Eleocharis*. They reported low values for *Trapa, Utricularia, Paspallum, Vallisnaria* and *Potamogeton*. The values from the present investigation for *Ceratophyllum demersum* (88.00–422.67 plants m^{-2}) is found to be higher than the values observed by Devi, K.I. (1998) in the Utrapat Lake, Manipur. However, the abundance for *Salvinia* (431.12 plants m^{-2}) was found to be lower than that observed by Devi, L.G. (1993) in the freshwater 'ponds of Canchipur.

The ratios of A/F varied from 0.00 to 0.86 and the peak values were found in *Salvina natans* (0.02–0.74), *Echinochloa stagnina* (0.04–0.65), *Hydrilla verticillata* (0.15–0.65). The distribution of the macrophytes in the present study site is found to be in aggregate manner since the, value of A/F ratios have been found higher than 0.05 (Curtis, 1959).

The present aggregate nature of distribution is found to be comparable to the nature of distribution reported earlier from Manipur *viz.*, ponds of Canchipur, Imphal, Manipur, by Devi, L.G. (1993), Devi, N.B. (1993) from Loktak lake, Manipur; Devi, O.I. (1993) from the Waithou Lake, Manipur and Devi, Ch. N. (2002) from the Ikop lake, Manipur.

The present ranges of IVI for species like *Hydrilla verticillata* (15.08–30.61), *Ludwigia adscendens* (10.69–36.59) and *Utricularia* (14.17–29.01) are in agreement with the values reported by Devi, K.I. (1998) in Utrapat Lake, Manipur and Devi, O.I. (1993) from the Waithou Lake, Manipur.

The values reported in the present study are also in conformity with the values reported by Billore and Vyas (1982) in the Pichhola Lake, Udaipur for *Potamogeton* (8.7–42.3). The highest values which were observed in the emergent macrophytes (*Echinochloa stagnina, Alternanthera philoxeroides*) are found much comparable to those ranges (7.4 to 138.7) reported by Sankhla and Vyas (1982) in the Baghela Tank, Udaipur.

Conclusion

In the light of the present investigation, the study site is found to have high values of morphometric and bathymetric characters having saucer-shaped basin with gentle slope and silted bottom and also luxuriant growth of macrophytic plant species with rich species diversity. The successful growth and distribution of the important macrophytes which are distributed regularly throughout the study period may be due to a number of interacting physico-chemical characteristics. The presence of some dominant emergent species like *Echinochloa stagnina, Alternanthera philoxeroides* indicates enhanced eutrophic status of the present study sites. Hence, it is high time to conserve this valuable lake by investigating other ecological characteristics of the lake so as to, protect it from further deterioration.

References

Ambasht, R.S., 1969. *A Textbook of Plant Ecology*. Students Friends and Co., Lanka, Varanasi, India.

Billore, D.K. and Vyas, L.N., 1982. Distribution and production of macrophytes in Pichhola lake, Udaipur, India.

Biswas, K. and Calder, C.C., 1936. *Handbook of Common Water and Marsh Plants of India and Burma*. Government of India Publication, Calcutta.

Curtis, LT., 1959. *The Vegetation of Wisconsin*. University of Wisconsin Press, Madison.

Devi, Ch. Bebika, 2001. Variations in species distribution and primary production of the macrophytes in Sanapat Lake, Manipur. *Ph.D. Thesis*, Manipur University, Manipur.

Devi, Ch. Nivanonee, 2002. Vegetational structure and primary production of the macrophytes in Ikop Lake, Manipur. *Ph.D. Thesis*, Manipur University.

Devi, Ch. Umabati, 2000. Phytosociology and primary production of the macrophytes in the freshwater ecosystems of Canchipur, Manipur. *Ph.D. Thesis*, Manipur University, Manipur.

Devi, K. Indira, 1998. Ecological studies of freshwater macrophytes in Utrapat lake, Manipur. *Ph.D. Thesis*, Manipur University, Manipur.

Devi, Kh. Usha, 2002. Macrophyte ecology of Poiroupat Lake, Manipur. *Ph.D. Thesis*, Manipur University, Manipur.

Devi, L. Geetabali, 1993. Ecological studies of the macrophytes in the freshwater ponds of Canchipur. Imphal. *M. Phil. Dissertation*. Manipur University, Manipur.

Devi, N. Bee-nakumari, 1993. Phytosociology, primary production and nutrient status of macrophytes of Loktak lake, Manipur. *Ph.D. Thesis*, Manipur University, Manipur.

Devi, O. Ibeton, 1993. Distribution, primary production and nutrient status of macrophytes of Waithou Lake, Manipur. *Ph.D. Thesis*, Manipur University, Manipur.

Gopal, B., 1967. Contribution of *Azolla pinnata* R. Br. to the productivity of temporary ponds of Varanasi. *Trop. Ecol.*, 8: 126–130.

Handoo, L.K. and Kaul, V., 1982. Phytosociological and standing crop studies in wetlands of Kashmir. In: *Wetlands Ecology and Management*, (Eds.) Gopal, B., Turner, R.E., Wetzel, R.G. and D.F Whigham. National Institute of Ecology and International Scientific Publications, India, pp. 187–195.

Jha, U.N., 1965. Hydropyhytes of Ranchi. *Trop. Ecol.*, 6: 98–105.

Jha, U.N., 1968. The pond ecosystem. *Ph.D. Thesis*, Banaras Hindu University, Varanasi.

Kumar, R. and Pandit, A.K., 2005. Community architecture of macropytes in Hokarsar wetland, Kashmir. *Ind. J. Environ. and Ecoplan.*, 10(3): 565–573.

Melzer, A., 1981. Veranderungen der Makrophyten vegetation der starn berger Sees und ihre dikatorische Bedeutung. *Limnologica*, 13: 449–458.

Misra, K.C., 1989. *Manual of Plant Ecology*, 3rd edn.. Oxford and IBH, New Delhi.

Misra, R.T., 1968. *Ecology Workbook*. Oxford and IBH Publ. Co., New Delhi.

Pieczynska, E. and Ozimek, T., 1976. Ecological significance of macrophytes. *Int. J. Ecol. Environ. Sci.*, 2: 155–158.

Sankhla, S.K. and Vyas, L.N., 1982. Observations on the moist bank community of Baghela Tank, Udaipur (India), In: *Wetlands Ecology and Management*, (Eds.) Gopal, B., Turner, R.E., Wetzel, R.G. and D.F Whigham. National Institute of Ecology and International Scientific Publications, India, pp. 197–206.

Shah, J.D. and Abbas, S.G., 1979. Seasonal variations in fershwater, density, biomass and rate of production of some aquatic macrophytes of the river Ganges at Bhagalpur (Bihar). *Trop. Ecol.*, 20: 127–134.

Sharma, B.M., 2000. Water resources. In: Environmental Studies, (Ed.) P.S. Yadava. Manipur University, Canchipur, pp. 127–178.

Wetzel, R.G., 1975. *Limnology*. Saunders, W.B., Philadelphia.

Wiegleb, G., 1984. A study of habitat conditions of the macrophytic vegetation in selected river systems in western Lower Sexony (Federal Republic of Germany). *Aquatic Botany*, 18: 313–352.

Chapter 50

Vermicompost: Its Proper and Successful Application in the Cultivation of *Aloe barbadensis*

Jayanta Sinha[1*], Chanchal Kr. Biswas[2], Arup Ghosh[3] and Nazrul Haque[1]

[1]Vermicomposting Unit and Department of Zoology, [2]Department of Botany, [3]Department of Environment and Water Management B.B. College, Asansol – 713 303

ABSTRACT

Today vermicompost is a well established bio-fertilizer and a very important and primal input in sustainable agriculture. It's wide spread application in different fields of agriculture would assist in bringing a second green revolution in India and also can be successfully and properly applied in alternate dry farming of medicinal plants.

Our work aims towards understanding the effect of vermicompost on the overall development and vigour of a very important medicinal plant *Aloe barbadensis* under the family of Liliaceae, as the plant is mostly used for the treatment of various diseases. A dye 'chrysamic acid' is also obtained from it.

The overall experiment is done by selecting morphological data from ten randomly selected plants from two different experimental fields (vermicompost treated and untreated control). These morphological datas are statistically analyzed.

It is surprisingly established that two year regular application of vermicompost has shown a formidable increase in height of plant, in number of leaf, in number of new plantlet, in stalk circumference, in width of the leaf base and a noticeable augmentation of gel content.

* Corresponding Author: Corresponding address: 15, Rupchand Mukherjee Lane, Flat No. 3B, Kolkata – 700 025; E-mail: jayantasinha_02@yahoo.co.in

Above experiment and observation proves that utilization of vermicompost in alternate dry farming of medicinal plant *Aloe barbadensis* is very successful. So such vermicompost boosted organic farming would be promising in the upliftment of agriculture and agro-industry dependent socio-economic condition of Paschimanchal (Western region) in Burdwan District of West Bengal.

Keywords: *Vermicompost, Biofertilizer, Earthworm, Morphometric characters and Aloe barbadensis.*

Introduction

Chemicals were regarded a breakthrough in agricultural productivity when they made an access into India in the identity of Green Revolution during the sixties. They were miraculous components that improved agricultural produce by leaps and bounds. It was only later that it was understood that chemicals were not a boori but a curse. To begin with, chemical fertilisers may appear to result in faster and healthier growth of plants, but they sacrifice the long-term benefits by destroying the soil and the ecosystem. Chemicals can be very damaging to the soil and plants especially when used over and over again. This is where organic farming advents.

Organic farming has the capacity to take care of each of these problem. Besides the obvious immediate and optimistic effects organic or natural farming has on the environment and quality of food, it also greatly supports a farmer to become self-sufficient in his requirements for agro-inputs and reduce his costs. Application of vermicompost is one of the important avenues of organic farming.

Organic waste can be fragmented rapidly by earthworms, which results into stable and nontoxic substances having a potentially high economic value and can be used as a soil conditioner for plant growth. This product is designated as vermicompost. Physically, the substance is finely fragmented peat like material which has a good structure, porosity, aeration and moisture holding capacity. It provide3 suitable mineral balance and enhances the nutrient availability to the plant (Edwards, 1988). Earthworm castings in the home garden often contain 5 to 11 times more nitrogen, phosphorus, and potassium as the surrounding soil. Secretions in the intestinal tracts of earthworms, along with soil passing through the earthworms, make nutrients more concentrated and available for plant uptake, including micronutrients. Redworms in vermicompost act in a similar fashion, breaking down food wastes and other organic residues into nutrient-rich compost. Nutrients in vermicompost are often much higher than traditional garden compost (Dickerson, 2004). It also reduces the population of soil pathogenic microbes (Dominguez *et al.*, 1997). It is known that vermicompost-earthworm-mulch-plant root (VEMP) interaction is important for an effective utilization of soil ecosystem in agricultural practices. Vermicompost also has an important role in nitrogen fixation (Mba, 1987; Tereschenko, 2002). Vermicompost finds the wide application in various types of agricultural practices specially for the horticultural plants. The worm casts are used in production of plants including vegetables and crops (Gajalakshmi, 2004) and mushroom (Dash and Das, 1989).

In chic pea application of vermicompost increased the dry matter accumulation, increase in grain yield and grain protein content (Jat, 2006).

It can also be successfully and properly used in alternate dry farming including various medicinal and aromatic plants including *Aloe barbadensis* (Saha *et al.*, 1988).

Aloe barbadensis is a very important medicinal plant which has an old age usage in treatment of various disorders in human. It is used in the treatment of fevers, liver and spleen enlargement, skin diseases, constipation, jaundice. It can also be applied against intestinal worms in children. The leaves of this plant are a source of a dye known as chrysamic acid.

In India *Aloe sp.* is widely cultivated (DARE/TCAR Annual report, 2003–2004) and used with along history. In this paper we have tried to see the effect of vermicompost on the growth and vigour of *Aloe barbadensis* cultivated in Asansol, which has a barren, rocky and rolling country with a laterite soil that is characterised by a reddish colour, medium to coarse in texture, acidic in reaction, low in nitrogen, calcium, phosphate and other plant nutrients.

Material and Methods

Experimental Plot Preparation

Two experimental plots are selected in the experimental garden of the college measuring 10 × 20 metre. In each plot 60 plantlets of *Aloe barbadensis* are planted under standard soil condition with the spacing 2.5 feet between each plant. During plantation one of the plot are treated with vermicompost (prepared in college vermicompost unit, species of earthworm used *Eisenia foetida*) at the rate of 5 kg/ plantlet and such application are made at six month intervals. The other plot is used as an untreated control. The plants are maintained through standard horticultural procedure.

Data Collection

Various morphometric datas are collected after two year when the plants become fully matured. The morphometric characters includes height of the plant, number of leaves in each plant, number of vegetatively propagated plantlet around each mother plant, stock circumference of each plant, width of the leaf base and gel content of the leaf. In each case 10 randomly selected plants are analysed from the two plots.

Statistical Analysis

The differences between the results obtained from the treated and untreated *Aloe sp.* Plants were analysed statistically by means of two tailed t-test through MS-Excel software.

Results and Discussion

Earthworms live in the soil and feed on decaying organic material. After digestion, the undigested material moves through the alimentary canal of the earthworm, a thin layer of oil is deposited on the castings. This layer erodes over a period of two months. So although the plant nutrients are immediately available, they are slowly released to last longer. The process in the alimentary canal of the earthworm transforms organic waste to natural fertilizer. The chemical changes that organic wastes undergo include deodourizing and neutralizing. This means that the pH of the castings is 7 (neutral) and the castings are odorless. The worm castings also contain bacteria, so the process is continued in the soil, and microbiological activity is promoted. Vermicompost (*i.e.* worm castings and worm tea), act as great additive to soil. It helps to increase soil structure while adding nutrients.

In this paper assessments are being made about the effect of vermicompost on various morphometric characters of *Aloe barbadensis* have been assessed. Effect of vermicompost on new plant height showed that there is an considerable improvement of the character (173.12 per cent increase; Figure 50.1a) when compared to untreated control. While assessing the effect of vermicompost on new leaf generation it is noticed that there has been 142.37 per cent (Figure 50.1b) increase in the number of new leaf generation in comparison the untreated control. The effect of vermicompost on the increment of stock circumference is also remarkable; it showed 172.92 per cent (Figure 50.1c) increase, when compared to the untreated control. Further vermicompost treatment also increases the width of leaf base as reflected

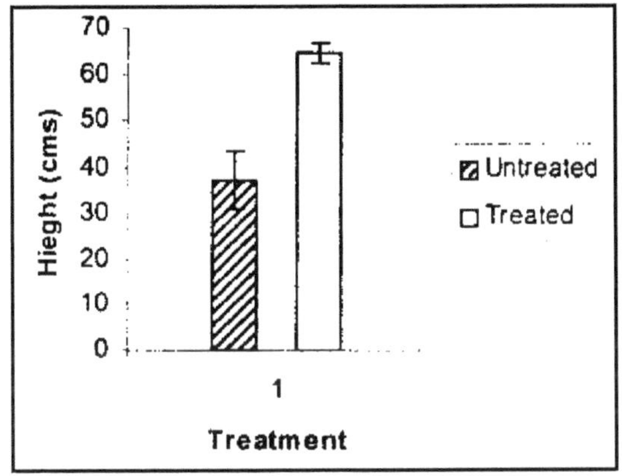

Figure 50.1(a): Effect of Vermicompost on Plant Height

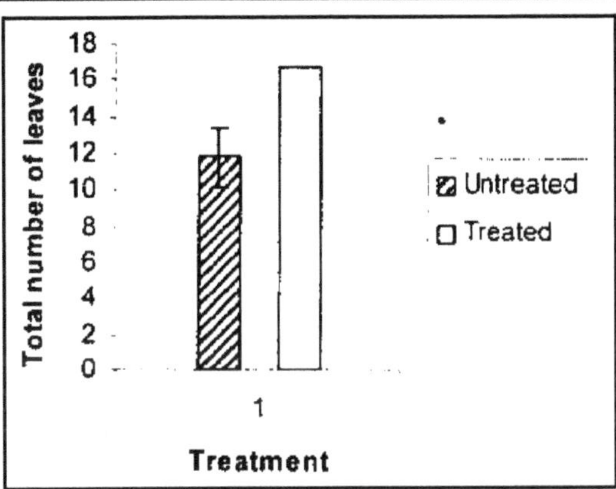

Figure 50.1(b): Effect of Vermicompost on New Leaf Generation

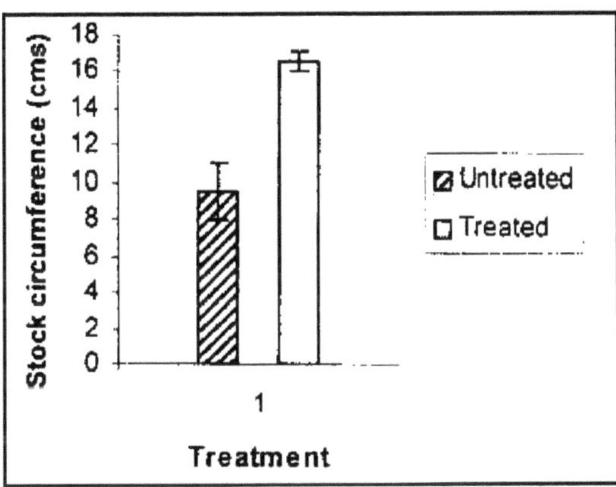

Figure 50.1(c): Effect of Vermicompost on Stock Circumference

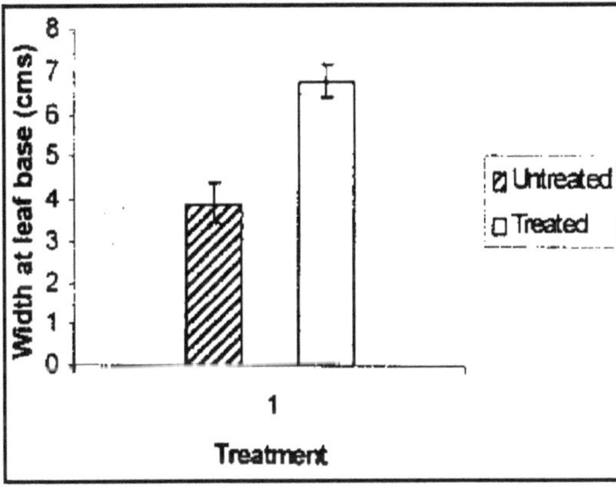

Figure 50.1(d): Effect of Vermicompost on Width at Leaf Base

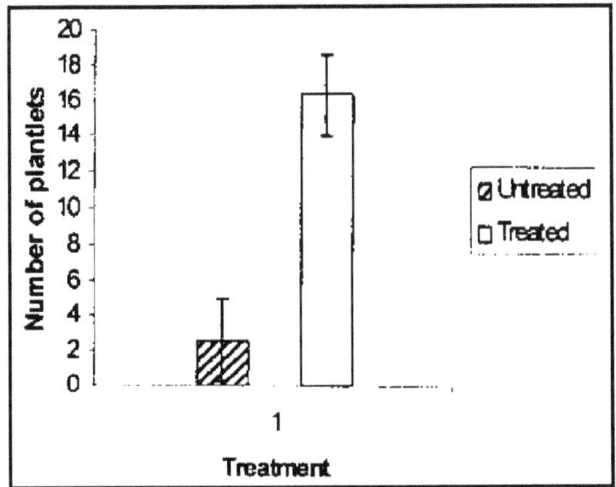

Figure 50.1(e): Effect of Vermicompost on Plantlet Generation

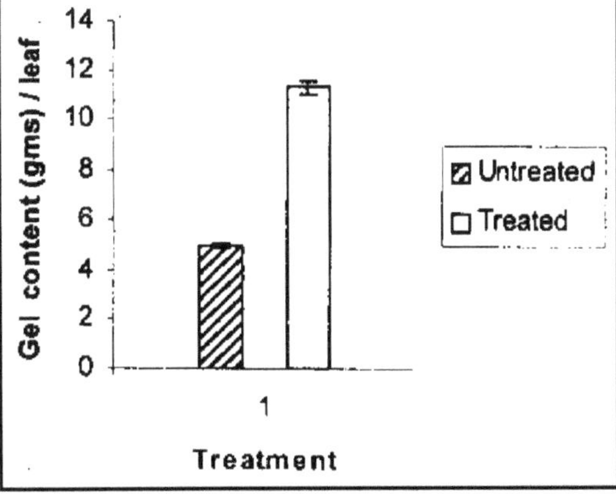

Figure 50.1(f): Effect of Vermicompost on Total Gel Content of Leaf

Figure 50.1(a–f): Effect of Vermicompost Treatment on Various Morphometric Parameters of *Aloe barbadensis* in Comparison to the Untreatred Control. Significantly different form the untreated control at 1 per cent level.

through a rise of 172.46 per cent (Figure 50.1d) when compared to the untreated control. Vermicompost treatment also increases the number of new plantlet by a gigantic rise of 630.77 per cent (Figure 50.1e) when compared to the untreated control. It is also noticed that there is 225.91 per cent increase in the gel content when compared to the untreated control (Figure 50.1f). Vermicompost is also nutrient rich in comparison to normal fertilizer. It contains: nitrogen, phosphorous, potassium, calcium, sodium, magnesium, iron, zinc, manganese, copper, boron, and aluminum. The most important of these being the nitrogen, phosphorous, and potassium that they add to the soil. In vermicomposting there is a decrease in C/N ratio and an increase in mineral N (Bansal, 2000).

Many cropping areas in Asansol are deficient in at least one of these nutrients, which can be supplemented by the use of vermicompost. Thus from the results it can be concluded that vermicompost have got a huge impact upon the growth and vigour of *Aloe barbadensis*. The results are quite encouraging and points towards the organic farming of *Aloe barbadensis* through the application of vermicompost.

Acknowledgement

The authors acknowledge Principal, B.B. College, Asansol, for encouragement and help rendered during the work.

Referneces

Bansal, S. and Kapoor, K.K., 2000.Vermicomposting of crop residues and cattle dung with Eisenia. *Bioresource Technology*, Oxford, U.K., Elsevier Science Limited, 73(2): 93–98.

DARE/ICAR Annual Report 2003–2004. *Improvement and Management of Horticultural Crops*, pp. 5–17.

Dash, M.C. and Das, A.K., 1989. Earhworm cast as a substitute for wheat supplementation in the growth of tropical edible mushroom, *Pleurotus sacorcaju*. *Trop. Agril.*, 66: 179–180.

Dickerson, G.W., 2001. *Vermicomposting Guide* H–164.Cooperative extension service. College of Extension and Home Economics, New Mexico State University, USA, p. 1–4.

Dominguez, J., Edwards, C.A. and Subler, S., 1997. Achieving pathogen stabilization using vermicomposting. *Biocycle*, 38(4): 57–59.

Edwards, C.A. and Neuhauser, E.F., 1988. *Earthworms in Waste and Environmental Management*. SPB Academic Publishing, The Hague, Netherlands, p. 321–328.

Gajalakshmi, S. and Abbasi, S.A., 2004. Earthworms and vermicomposting. *Indian Journal of Biotechnology*, 3: 486–494.

Jat, R.S., 2006. Direct and residual effect of vermicompost, biofertilizers and phosphorus on soil nutrient dynamics and productivity of chickpea-fodder maize sequence. *Journal of Sustainable Agriculture*, 28(1): 41–54.

Mba, C.C., 1987. Vermicomposting and biological N-fixation. In: *Proceedings of the 9th International Symposium on Soil Biology and Conservation of the Biosphere*, (Ed.) J. Szegi. Publication, Budapest, Akademiai Kiado.

Saha, R., Palit, S., Ghosh, B.C. and Mittra, B.N. *ISHS Acta Horticulturae*, 676: III WOCMAP Congress on Medicinal and Aromatic Plants, Volume 2: Conservation, Cultivation and Sustainable Use of Medicinal and Aromatic Plants.

Tereshchenko, N.N. and Naplekova, N.N. 2002. Influence of different ecological hroups of earthworms on the intensity of nitrogen fixation. *Biology Bulletin*, 29(6): 628–632.

Chapter 51

Morphology and Ecology of Charophytes of Two Lakes in Jharkhand, India

Meena Krishnan and Gouri Suresh
Department of Botany, Jamshedpur Women's College,
Jamshedpur – 831 001, Jharkhand, India

ABSTRACT

Jamshedpur is a well known industrial city of India, situated in the district of East Singhbhum in Jharkhand. There are many fresh water bodies in and around the city. Two of the larger water bodies are Sitarampur Lake situated in West Singhbhum and Hudco Lake in East Singhbhum. The former supplies drinking water to many areas in both the districts while the latter has become highly polluted, though it was also constructed mainly for storing water for emergency in summer months. The present paper is based on part of a mammoth work undertaken by the authors to study the algal taxa of these two water bodies in relation to the physico chemical properties of their waters. This paper deals with the morphology and ecology of 3 species each of Chara and *Nitella* collected from these lakes.

Introduction

Some of the chief workers abroad who tried to resolve the taxonomy of Charophyta were Imahori (1954), Round (1963), Wood (1964) etc. Several workers have done a good deal of work on the taxonomy of Charophyta in India (Allen, 1925; Dixit, 1942; Sharma, 1979 etc.). Charophyta of nearby states like UP and West Bengal have also been studied in detail (Iqbal Habib and Pandey, 1990; Mazumdar and Das, 1989 etc.). However, available literature indicates that barring Imahori and Sinha, 1963; Dua, 1982 and Thakur, 1990, very little work has been done on the taxonomy and ecology of Charophyta in

Jharkhand. No work seems to have been done on the Champhyta of either East or West Singhbhum districts at all.

The variations in the taxonomy of Charophytes under various ecological conditions have always imposed problems to taxonomists. In the latest system of classification of charophytes, Wood and Imhori (1965) had recognized 81 species with 395 taxa in the family Characeae. Among the numerous algal taxa that were collected from Sltarampur Lake and Hudco Lake, 3 species were of *Chara* and 3 of *Nitella* during the course of the present study.

Sitarampur Lake

Sitarampur Lake is located in the district of West Singhbhum near Seraikela Kharsawn District in Jharkhand. The lake is a water reservoir spread over 19.2 square km. Here water is collected in a low-lying rain fed catchment area. Water supply from this lake is 50 lakhs gallons per day. Of this, 11 lakhs gallons are supplied to housing colonies, 27 lakhs are supplied to the industries and the rest to non-organized sector in the surroundings. Water is supplied to Jugs alai, Shashtri Nagar, Bagbera, Garnharia, Adityapur etc, spread in two districts of West as well as East Singhbhum. The water is of good quality, as found out by physicochemical analysis. The place is picturesque and is a favourite picnic spot. It is a very neat and clean place, away from city. This dam is unpolluted and is highly populated with many algae.

Common algae that grow here are Chlorococcales, different genera of diatoms and different species of Chara, *Nitella* etc. Blue green algae are comparatively less though *Merismopedia* is rather common. *Microcystis* and *Oscillatoria* are also found.

Hudeo Lake

Hudco is on the south eastern part of Jamshedpur. To the west of the Hudco area, is the giant company, Tata Engineering Works. Construction of the Hudco Lake started in the year 1980 and was completed in 1983. It was constructed for storing water for emergency during summer, to develop the landscape around the lake, for recreation and welfare activities of employees. The catchment area of Hudco Lake is 10.8 acres. It is 90 feet deep and 700 feet long. Water depth is normally up to 60 feet. Storage capacity is 65 million gallons from which 40 MG can easily be drawn and 25 MG is always left undisturbed. Excess water is drained out through an outlet channel to avoid overflow. Pisciculture is being done in this lake.

Reasons for Pollution of Water of the Hudeo Lake

1. Visitors throwing waste materials in and around the lake.
2. Recreation activities like swimming, boating etc.
3. Local tribal people taking bath and washing clothes and utensils using low quality detergents.
4. Immersion of idols of Gods and Goddesses during festivals.
5. Annual picnics.

Materials and Methods

The algal samples were collected in plastic bottles and preserved in 4 per cent Formalin for detailed morphological study. Camera lucida diagrams were made both from live as well as preserved materials. Based on the monograph 'Charophyta' by Pal and Kundu, Sundarlingam and Venkatraman (1962), identification of the taxa was done. Physicochemical properties of the waters of both the lakes were found out regularly on a monthly basis with the help of analytical methods prescribed by APHA, 1998.

Observation and Discussion

While studying the algal flora in relation to different parameters and standards of drinking water stored in the Sitarampur Dam, the authors came across the taxa of Charophyta, the taxonomy of which are described here.

Genus *Chara* Linnaeus

Chara braunii Gmelin

Plant having monoeceous stem 420–700 μ excorticate, spine cell was absent. Stipulodes is in one tier and are well developed one per branch let. Stipulode is 475 μ long and 136 μ wide, acute, bract cells 3–4 small, 330 μ long 75 μ wide, Antheridia 225–445 μ diameter; Oogonia 500–700 μ long and 336–560 μ wide. Convolution 8–12.

Collected from Sitarampur Lake in November.

Chara zevlanica Klein

Plant body monoecious, heavily encrusted axis is moderately stout, 700–800 μ. Internodes are 1.5 to 2 times longer than width. Upper segment is less long than the basal part. Gametangium conjoined at 2–5 lowest nodes of branch lets. Oogonia solitary; 875 μ long; 560 μ broad Wide convolution: 14–15. Coronula 191.5 μ wide at the base. Antheridia: 450 μ in diameter.

Collected from Hudco Lake in June.

Chara globularis Thuill; em. R.D.W.

Plant body slightly encrusted, axis moderately slender 607.98–639.8 μ in diameter. Internodes are 1–2 times longer than branch let. Cortex–3 corticate. Spine cells globular to slightly conical. Stipulodes in two tiers, 2 sets per branch lets. End segment one-celled, reduced, bract cell 4–6. Plant monoecious, gametangia conjoint at 3–4 lowest branches let nodes. Oogania solitary; 700 to 735.94 μ long; 399.9–447.88 μ broad; Convolution 9–12, coronula 157.9–191.9 μ broad striae of 11 low ridges. Antheridia 307.9–335.9 μ in diameter.

Collected from Sitarampur and Hudco Lakes in January.

Genus *Nitella* Agardh (em.Leonh.)

Nitella acuminata A. Braun

Monoecious, plant body internodes 0.5–2 times longer than branch lets. Sterile and fertile branch lets dissimilar. Fertile branch lets with dense heads. Oogonia and antheridia at the same nodes Antheridia solitary, sessile strictly terminal; Globule 312 to 335.16 μ in diameter Convolution 8–10; Nucules–194.04 to 352.8 μ Width of tubular cell–35.28 μ.

Collected from Sitarampur and Hudco Lakes in August.

Nitella terrestris Iyengar

Plant body small with densely crowded branch lets. Axis slender, 129 μ diameter Internodes 0.5–2 times longer than branch lets. Branch lets 5–6, in whorls. Bifurcation of branch lets at the top, tapering gradually, end cell conical and acute 42–56 μ long and 16.2 μ wide at the base. Fertile branch lets conical and end cell conical. Gametangia conjoined at the nodes of fertile branch lets. Oogonia solitary 285 μ long 129 μ wide. Coronula 44.8 μ high 60.2Jl at base. Antheridia solitary 81.6–99.5 μ diameter

Collected from moist soil near Sitarampur Lake in November.

***Nitella psedoflabellala* var. imperialis f wattii (J. Gr.) R.D.W. (N. Wattii J. Gr.) Wood and Imahori (1965)**

Plant monoecious; Axis slender 639 to 959 μ in diameter; Internodes 2 to 4 times longer than branch lets Branch lets fertile 6–7 in whorls. 2,3,(4) furcated covered with slight mucus, Dactyles 45, 2 celled, elongated, penultimate cell with tapering rounded tip. Gametangia solitary at 2nd and 3rd branch let nodes. Oogonia solitary 335 to 399 μ long, 239 μ to 271 μ wide. Convolution 8–9. Oospores dark chestnut brown 335 to 351 μ long, 191 to 220 μ wide Striae of 6–7 prominent flanged ridges, membrane dark brown, strongly vermiferous. Atheridia 191 μ to 223 μ in diameter.

Collected from Hudco Lake in March.

Ecology of Charophytes

It was observed that though *Chara and Nitella* grew in both the lakes, population density was not the same. In October/November, the edges of the Sitarampur Lake were almost filled with *Chara and Nitella.* When the water level started decreasing after rainy season, Charophytes were seen in a continuous zone in the month of October. In Hudco their growth was sparse. They were:

Light

Light is a very important factor for the growth of Charophytes (West, 1905) Leaves of both the genera were found to be poorly developed in the deficiency of light.

Temperature

Temperature does not play any important role. Allen (1925) remarked that "Charophyta is found in every country and climate in the waters at the North and in hot spring of yellow stone"

pH

pH of water of the Sitarampur Lake was mostly neutral as it is supplied for drinking purpose to the local people. A little variation occurs, *i.e.* it varies between 6.9 and 7.1. In the months of May and December, pH was 6.9 while in October–November, it was 7.1.

In the Hudco Lake, pH was 6.3 in December and 7.2 in most of the other months. Fluctuation in pH usually follows changes in sunshine; higher pH values were considered to be indicative of higher rate of productivity.

BOD

Biochemical Oxygen Demand in Hudco Lake was higher, *i.e.* 39 mg/l in the months of August and December and lower, being 21 mg/l in the month of February.

In the Sitarampur Lake, BOD was 26 mg/l in March and lower, being 8 mg/l in the month of June, July, September and December.

DO

Dissolved Oxygen in Hudco Lake was higher, being 7.2 in March and lower, 5.9 in July. In the Sitarampur Lake, it was higher, *i.e.* 7.2 in June, August and September and lower, being 6.8 in the month of December. Solubility of oxygen decreases with increase in temperature.

Calcium

Both the genera grew anchored by means of branching underground rhizoids in the sand and silt of the littoral zone. They formed an association with angiospermic piants. These groups are more frequently populated in calcareous water. Calcium found in Sitarampur was 56 mg/l; this was the peak season for the growth of *Chara* and *Nitella*. Lowest value of 42 was in the month of November. End of November was the dying period for these genera. Values above 25 mg/l indicate calcium rich water.

PO_4

Phosphate in Hudco was 0.46 mg/l in October and 0.05 mg/l, a still lower value in the months of January, March, April and May. In Sitarampur, it was comparatively higher throughout the year, being 0.69 mg/l in April and 1.24 mg/l in August.

Schindler (1971) had suggested that phosphate is the primary limiting nutrients in ponds and lakes.

Chloride

In the Hudco Lake highest value of chloride of 24.42 mg/l was in the month of October while lowest value of 17.16 was in the month of January. In Sitarampur Lake the highest value was 56 mg/l in May and lowest of 42 mg/l was in the month of November.

Thresh *et al.* (1944) had suggested that higher concentration of chloride is due to large quality of organic matter. This view was further supported by Adoni (1985).

Both the genera were found flourishing in the months of August to November. In the rest of the year, plants grew more in Marchi April, before the start of summer when temperatures rose, than in other months. However, in Hudco Lake, species of *Chara* were seen from January to June.

Acknowledgement

The authors are thankful to Dr. Shukla Mahanty, Principal, Jamshedpur Women's College for providing the necessary facilities. Thanks are also due to Mr. Navin Singh and Mr. S. Abid of QA (Quality Assurance) Laboratory, Tata Motors Ltd. for help in physico-chemical analysis of water.

References

Adoni, A.D., 1985. *Workbook on Limnology*. India MDA Committee, Department of Environment, Government of India, pp. 216.

Allen, G.O., 1925. Note on Charophytes from Gonda, U.P. *J. Bombay Nat. Hist. Soc.*, 30: 589–599.

APHA, 1998. *Standard Methods of Examination of Water and Wastewater*, 18th edn. Washington, DC.

Dixit, S.C., 1942. The Charophytes of Bombay Presidency. III. *J. Indian Bo. Soc.*, 21: 355–362.

Dua, B.K., 1982. *Ph.D. Thesis*, Ranchi University.

Imhori, K., 1954. *Ecology, Phytogeography and Taxonomy of Japanese Charophyta*. Ranacawa University, Japan.

Imhori, K. and Sinha, J.P., 1963. Notes on the Asiatic Charophyta IV: Specimens from Ranchi and Nepal. *Adv. Front Pl. Sci.*, 8: 55–68.

Iqbal, Habib and Pandey, U. C., 1990. Freshwater Charophytes of Nakatia river, Bareilley (UP). *India Ad. Plant Sci.*, 3(2): 224–227.

Mazumdar, P.K. and Das, R.N., 1989. Morphotaxonomical studies of Genus *Cham* (Linnaeus) collected from Purulia district of West Bengal. *Res. J. Pl. Env.*, 5(2): 43–47.

Pal, B.P., Kundu, B.C., Sundaralingam, V.S. and Venkataraman, G.S., 1962. *Charophyta*. ICAR, New Delhi.

Round, F.E., 1963. The taxonomy of Charophyta. *Brit. Phyco. Bull.*, 2(4): 224–235.

Schindler, D.L., 1971. Carbon, nitrogen, phosphate and eutrophication in fresh water lakes. *J. Phycol.*, 17: 321–329.

Sharma, Y.S.R.K., 1979. *Two Decades of Algal Cytology from Banaras*. Scholl Presidential address Annual Conference. Soc. Adv. Bot., Ludhiana, p. 1–16.

Thakur, S., 1990. *Ph.D. Thesis*, Ranchi University.

Thresh, J.C., Sacking, E.V. and Beale, J.F., 1944. *The Examination of Water Supplies*, (Ed.) E.W. Taylor.

Wood, R.D., 1964. A synopsis of Characeae. *Ibid*, 91(1): 35–46.

Wood, R.D. and Imhori, K., 1965. The Revision of the Characeae. *Verlag J. Cramer*, Weinheim.

Chapter 52

Water Scarcity Zones of Jamshedpur, Jharkhand

Gouri Suresh[1], Neena Sharma[2], G. Rajalakshmi[2] and Rajni Kumari[2]
[1]Reader in Botany, Jamshedpur Women's College, Jamshedpur
[2]Guest Faculty, Environment and Water Management, Jamshedpur Women's College, Jamshedpur

ABSTRACT

The present article is based on a survey conducted by Jamshedpur Women's College Eco-Club, of which the author is the moderator. The city of Jamshedpur was divided into 9 segments and questionnaires were distributed to households in a random fashion in all the segments. Based on the information collected, a water scarcity map of Jamshedpur was prepared; this map will definitely be of use to the city planners. It was also found that awareness activities are essential in the city.

Keywords: *Water source, Water quality, Awareness, Water scarcity map, Water conservation.*

Introduction

Jamshedpur is a picturesque city situated in the state of Jharkhand, at the confluence of two rivers, Subarnarekha and Kharkhai, and surrounded by the rugged hills of the Chotanagpur plateau. 64 sq.kms in area, the city is located at 532 metres above sea level. The Mapping of Water Scarcity Zones of Jamshedpur was a project, sponsored by the Ministry of Environment and Forests, Government of India, undertaken by the Eco-Club of Jamshedpur Women's College. Eco-Clubs of different schools of Jamshedpur were also involved in this massive venture.

Materials and Methods

Jamshedpur was divided into 9 zones of approximately equal areas based on the main roads separating these zones. Copies of a questionnaire prepared were distributed among the members of the Eco-Clubs of our college and also the participating schools. The members were divided into 9 groups based on the localities they lived in and a leader was selected for each group. Thereafter, the groups fanned out into the 9 zones and collected answers for the questions from the local people. A very large quantity of data was obtained by this method. The questionnaire:

1. Name of the zone:
2. Information collected from:
3. Main source of water–Supply Water/Surface Water/Groundwater
   Any specification:
4. If groundwater source water is available, at what depth:
   Maximum depth Minimum depth
5. Quality of water:
6. Is there any scarcity of water in your area? Yes/No
   If yes, throughout the year/for a few months only
   Specifications, if any:
7. Cleaning procedure practised:
   Self/Community/Company or Municipality Details:
8. Awareness regarding water problems among the local people:
9. Is there any water body in your zone? Lakes/Ponds/River/Any other
10. Any other water related problem faced by the local people of the zone:
11. Suggestions to improve the prevailing condition:
    (*a*) At individual level
    (*b*) At community level
    (*c*) At Government level
12. Suggestions by the Eco-Club members:

Results and Discussion

Based on the collected data, a detailed map of Jamshedpur with regard to its Water scarcity Zones was prepared. This map is sure to help the authorities in chalking out solutions to the problem of water scarcity in parts of Jamshedpur.

From the map of the city, it is clear that Tata Steel Works is situated almost in the centre of the city, towards the southern part and it occupies a considerable area. This area was excluded for the purpose of the study. The 9 zones studied are all outside the company area.

Tata, the biggest company of Jamshedpur, does maintenance of most parts of the city through JUSCO (Jamshedpur Utility Services Company). In these area, there is ample water supply. However, some other areas (about 20 per cent of the total) face water shortage in summer months from April to July. Yet other such areas (about 10 per cent of the total) face the problem throughout the year.

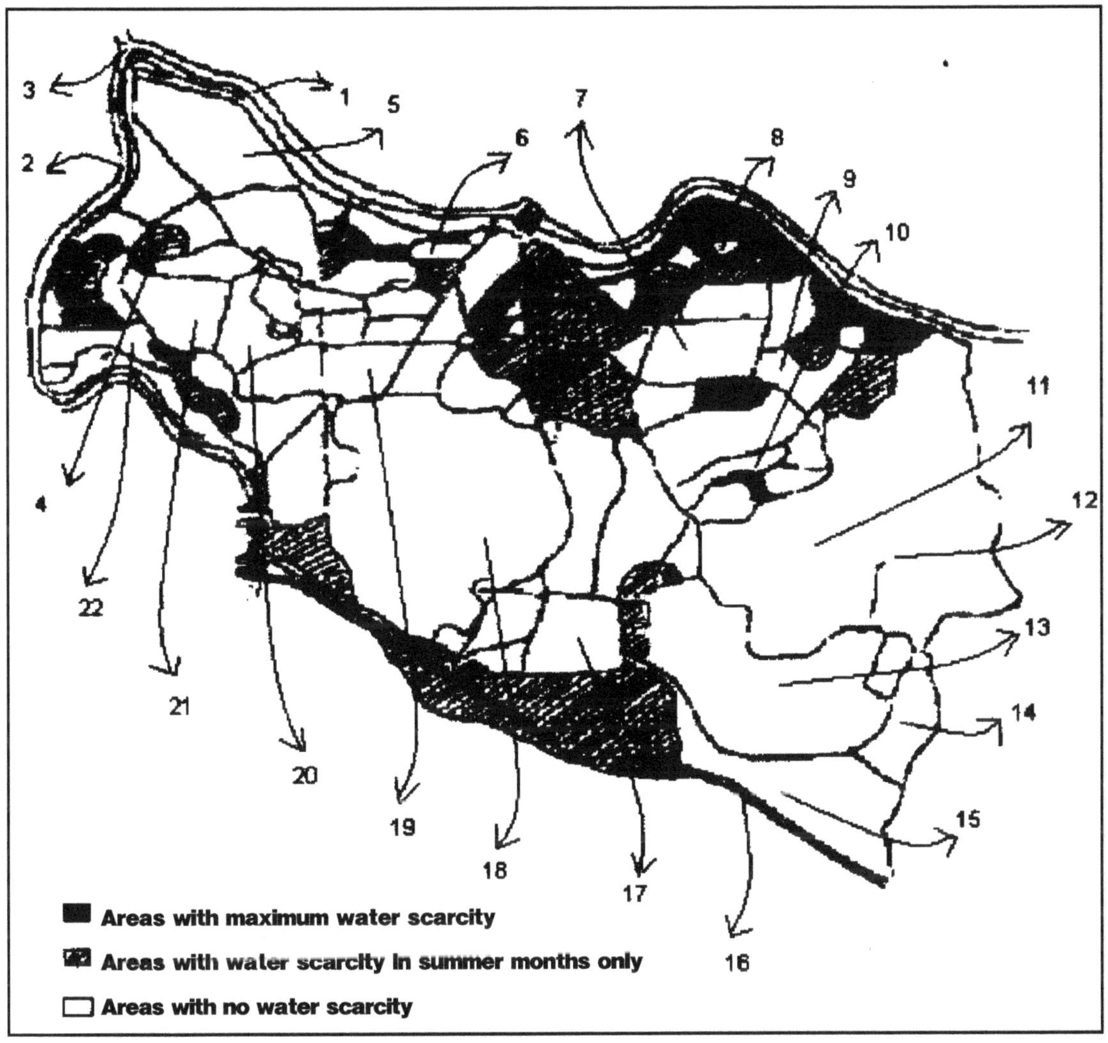

Figure 51.1: Water Scarcity Zones of Jamshedpur, Jharkhand

1: River Subarnarekha; 2: River Kharkhai; 3: Rivers Meet (Do Muhani); 4: Uliyan; 5: Sonari; 6: Jubilee Lake (Jayanti sarovar); 7: Sidgora; 8: Fly Ash Pond; 9: Old Baridih; 10: New Baridih; 11: Telco Colony; 12: TRF Colony; 13: Tata engineering works; 14: Hudco; 15: Lafarge Cement; 16: Railway Line; 17: Tubes Division; 18: Tata Steel Works; 19: Northern Town; 20: B.H. area; 21: Farm Area; 22: Kadma.

Some parts of the city where there is acute water shortage all the year round are located away from water bodies; however, quite contrary to expectations, some such areas are very near to either River Subarnarekha or River Kharkhai. To know the exact reason for this, hydro-geological studies will have to be conducted.

The survey has also thrown up some facts regarding the approach of the public in dealing with water supply and crisis. A majority of the people feel that supplying water is purely the responsibility of the government. It is true that water, like land and forests, is a state subject and it is the duty of the state governments to enforce legislations for surface and subsurface water. Yet people's participation is a must for the redressing of any local scarcity problem. The present survey showed that local people in the affected areas are either non-committal or ignorant about water conservation methods. From this, it is obvious that there is an urgent need of educating the people and creating awareness in them regarding different methods of water conservation.

In many suburban areas of Jamshedpur, where acute water scarcity was seen clusters of residential complexes have been built after draining wetlands. Inland wetlands help to replenish groundwater sources (Miller, Jr., 1996). Therefore draining them is suicidal to water conservation efforts and not many people know this.

It has also been observed that the public is using some water bodies as dumping grounds. These can be cleaned and maintained by the same public–what is needed is motivation. Success stories like that of how such a dump was transformed back to the original 9,000 square meter *jheel* of clear water in Kolkata (Dasgupta, 2003) can be an inspiration.

In areas outside the command of JUSCO, people use mainly hand pumps for all their water needs. But bore wells would yield water only if they are recharged over along period of time (Kumar, 2001). Therefore people must realise that as many types of water harvesting techniques as possible must be implemented. The 3 main methods are roof harvesting, run off harvesting and floodwater harvesting (Suresh, 2004).

Of the several avenues to raising water productivity, the key is pricing water at its market value (Brown, 2001). However, there is a contrary view that privatisation of water resources implies that water is an economic commodity and scarcity is a business opportunity (Jayararnan, 4003). Thus, making the local populace realise the essentiality of practising water conservation measures is the need of the hour. Motivating them to do so is the duty of all those who are aware of the grimness of the situation.

Acknowledgement

The authors are thankful to the Ministry of Environment and Forests, Government of India for sponsoring the project and to Gram Vikas Kendra, Jamshedpur Regional Resource Agency for forwarding the same to the ministry. Thanks are due to all the Eco-Club members who participated in the project. The authors are also grateful to the Principal, Jamshedpur Women's College for providing the facilities needed.

References

Brown, Lester R., 2001. *State of the World*. W.W. Norton and Company, New York, London, pp. 54–55.

Dasgupta, Rajashri, 2003. Kolkata: Clean waters again. *Survey of the Environment*, (Ed.) N. Ravi. The Hindu, pp. 165–170.

Jayaraman, Nityanand, 2003. Privatisation of water. *Survey of the Environment*, (Ed.) N. Ravi. The Hindu, pp. 83–86.

Kumar, Dinesh M., 2001. Drinking water: A scarce resource. In: *Survey of the Environment*, (Ed.) N. Ravi. The Hindu, pp. 93–101.

Miller, G. Tyler, Jr., 1996. *Living in the Environment: Principles, Connections and Solutions*. Wadsworth Publishing Company, New York, pp. 160–162.

Suresh, R., 2004. *Soil and Water Conservation Engineering*. Standard Publishers Distributors, Delhi, pp. 523–527.

Chapter 53

The Allergenicity of Pollen on Environmental Influences

M. Venkateshwarlu, Ch. Sridhar Rao and B. Digamber Rao
Department of Botany, Kakatiya University, Warangal – 9

ABSTRACT

The raising importance of wild relations of crop makes it obligatory for the cryopreservation of pollen of wild species related to cereals, millets, weed plants and other anemophilus agro-crops. The importance parameters of pollen systems in the biomonitoring of phytotoxicants in problems related to environmental health, as enumerated by Rosen (1981). The discussion about the influence of environmental factors, in particular air pollution has not ended yet. If air pollution would have influence on the allergencity of pollen we would have to assume that it influences both *Hypties* and *Eukalyptus* pollen. The increase of sensitizations against *Parthinium* pollen correlates with the increasing number of pollen in the air.

***Keywords**: Allergenicity, Allergenic potency, Air mass, Sensitiation.*

Introduction

The biological materials contained in the air are not only relevant to the allergy problem, but are also of application as bioindicators of air pollution, Air mass is also a repository of organic materials like plant and animal parts. Algal filaments in association with pollen and spores and a medium of life and transport of many organisms, seem or un seen. Another area in which pollen and spores found early application was in coal prospecting consequent to the discovery of sporae dispersae "so named by Van post in (1916). The important parameters of pollen systems in the biomonitoring of phytotoxicants in problems related to environmental health, as enumerated by Rosen (1981). The increase of sensitizations against *Parthinium* pollen correlates with the increasing number of pollen in the air. If air pollution would have influence on the allergenicity of pollen, we would have to assume that it influences both *Hypties* and *Eukalyptus pollen.* The general discussion about the increase of allergic diseases includes the question of environmental influences on the allergenic capacity of pollen.

Materials and Methods

Aerobiology deals not only with different types of organisms but also with their products such as spores of bacteria, fungi, viruses, associated metabolites, allergic pollen of higher plants, fragments and faecal pellets, plant and animal cells during industrial processing; and products of biotechnological processes, mainly enzymes. The impact on other organisms includes infection, allergy and toxicosis in man and animals and infection of plants.

In such cases the pollen is generally coated with nutrients and pigments, which allow the *Parthinium, Hyptis* and *Eucalyptus* pollen grains to adhere to the insect body and serve as an insect-attractant for eventual dispersal. Wind pollinated flowers show adaptations towards reduction of the floral parts, increased pollen production and size of stigma together with a more effective presentation of both organs. In addition, the entomophilous pollen are generally larger, heavier with various types of exine ornamentations. Birds, bats and some small mammals eat pollen and nectar as a major part of their diet, and may act as pollinators. Early speculation as to the origins of disease by Hippocrates and Lucretins and later, the discovery of fungal spores by Micheli, had lid which is aerobiology was first concerned with airborne bacteria and disputes over spontaneous generation extending from the time of Antohine van Leen wenhoek. It is only during the last 40–50 years that more efficient sampling methods have been adopted. New sampling instruments have often originated as instruments used in industrial hygiene, which have been applied first in plant pathology and then in the study of allergens.

Results and Discussion

The data sets for our comparison were annual totals of airborne pollen of *Parthinium, Hypties* and *Eukalyptus.* Thc RAST data have been performed by use of phadabas RAST in the period of July to December, we decided to consider positive results to be those of RAST class 2 and higher or >0.34 PRO, respectively only specimens related to inhalation allergies have been taken into account pollen spores, house dust mite, and animal dander. The total number of included subjects was 2179. For the statistical evaluation, we determined the significance to be on the 5 per cent error probability level (Table 53.1). For the annual total of *Parthinium Hypties* and *Eukalyptus* trends have been calculated by correlation of the totals against the consecutive months. Whilst there was almost no trend obvious for *Parthinium,* a weak trend to ward higher values was observed in *Hypties* pollen, but a significant trend to higher totals was detected in the annual totals of *Eukalyptus* pollen (Figure 53.2) correspondingly, we investigated the trends in the relative amount of sensitizations against those pollen for the same period. The correspondingly, we investigated the trends in the relative amount of sensitisations against those pollen for the same period. The correlation between annual totals of the pollen and RAST sensitivity was not significant, but near to significance for *Grass* correlation in order to prove the response of sensitization rates to increasing pollen numbers taking into consideration a reaction delay of one or two months respectively, also failed to become significant (Figure 53.3).

Table 53.1: Pollen and Percentage of Rast Positive Results Out of All Positive Rast Test on Inhalation Allergens

Month	*Hyptis Pollen (1000)*	*Pos. Rast*	*Eucalyptus Pollens (1000)*	*Pos. Rast*	*Grass Pollen (1000)*	*Pos. Rast*
July	558	40.9%	231	74.4%	182	17.8%
August	120	27.5%	411	66.8%	409	17.2%
September	552	43.8%	298	55.1%	193	23.1%
October	592	31.8%	325	53.6%	801	14.7%
November	554	40.5%	250	64.9%	130	16.5%
December	295	51.1%	262	59.1%	594	15.9%
January	432	46.1%	302	67.1%	389	15.8%
February	591	53.7%	431	66.1%	114	26.9%
March	234	48.7%	338	66.2%	186	24.9%
April	255	33.7%	502	63.9%	131	16.6%

Investigation of Jilek *et al.* (1993) have proved that stressed individuals of birch produce more Bet VI and Bet V II, known as profiline related to the human stress hormone Valenta *et. at.* (1991). Suggest that a number of environmental influences cannot only have influence on human health by itself, but would also influence the pollen grain and its allergenic potency. Schata *et al.* (1993) found no differences in the aggressively of pure pollen and those exposed to high concentrations of diesel exhaust. In their study, there is no need to investigate the influence of air pollution factors, as they should have the same influence on birch pollen as well as on and *Hyptis* pollen (Figure 53.1). A possible explanation

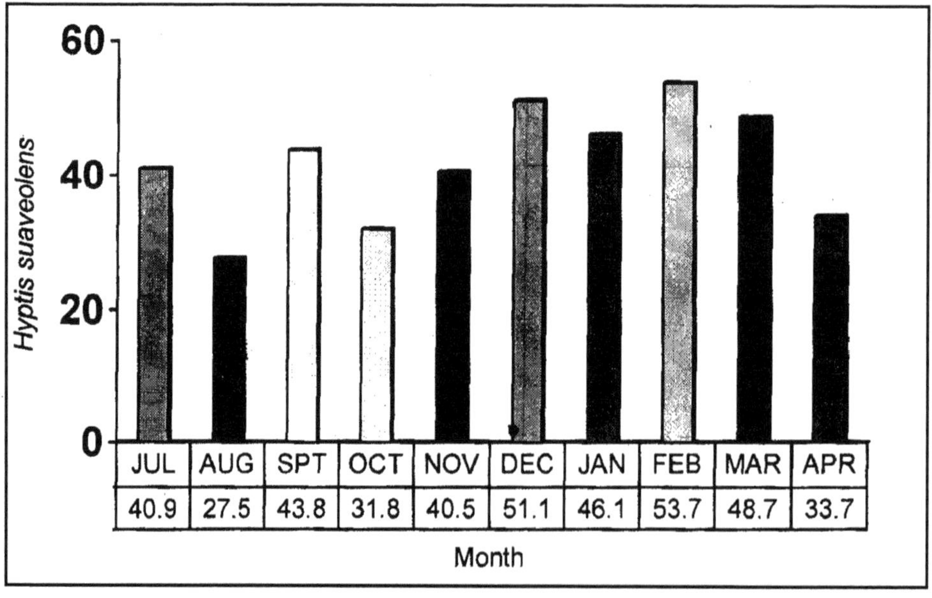

Figure 53.1: Incidence of *Hyptis suaveolens* (July to April)

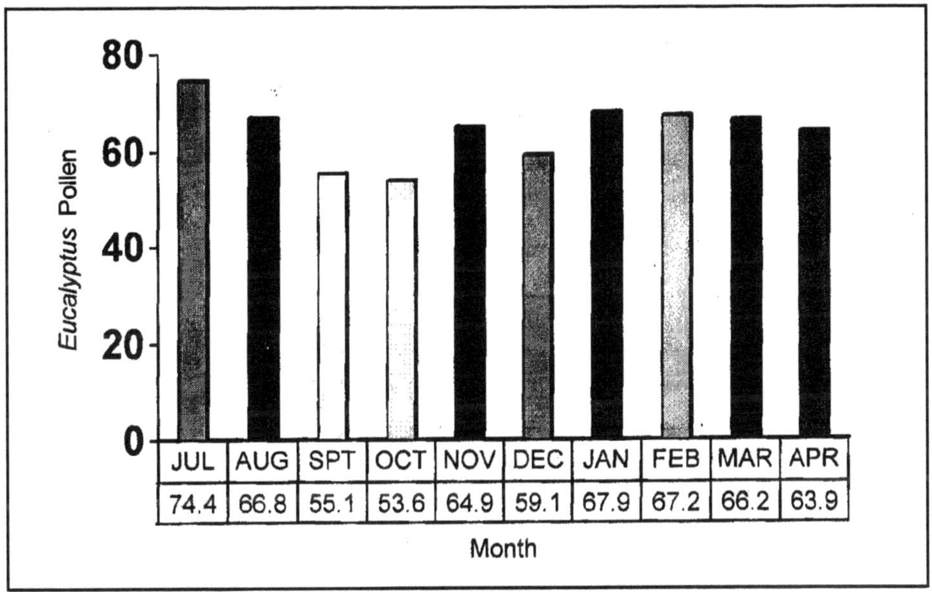

Figure 53.2: Incidence of *Eucalyptus* Pollen (July to April)

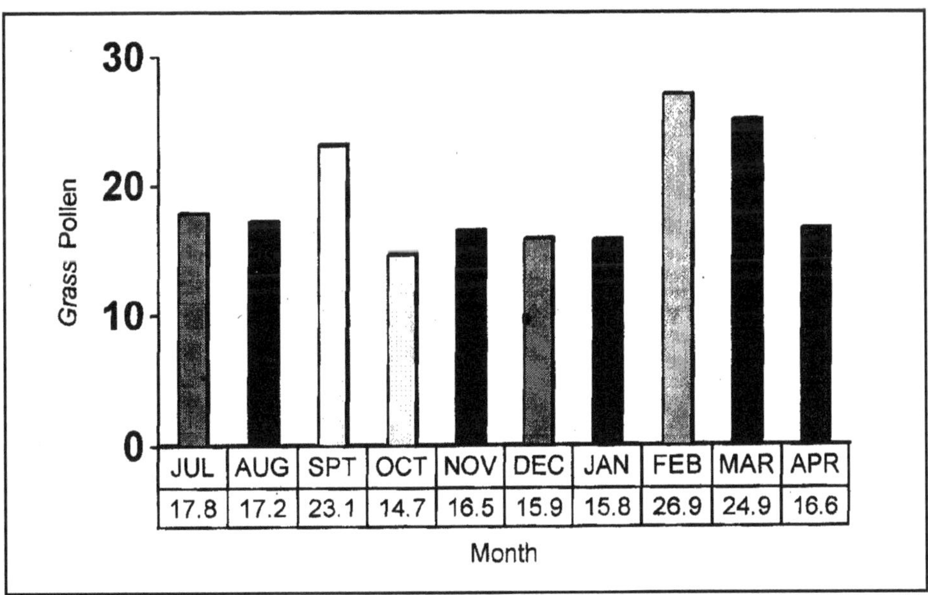

Figure 53.3: Incidence of *Grass* Pollen (July to April)

is, that in the region of the birch, being an element of the cold and moderate climates, is should to the margins of its natural distribution area because of a permanent warning of the atmosphere.

Conclusion

The possibility of using such dispersed aerospora in biomonitoring is the basic factor in applied Aerobiology, and the present review covers only the major aspects of the subject. The raising importance of wild relations of crops makes it obligatory for the cryopreservation of pollen of wild species related to cereals, millets, weed plants and other anemophilus agro-crops. The discussion about the influence of environmental factors, in particular air pollution, has not ended yet. As there is no increase of *Hyptis* pollen sensitizations but a significant increase of, *Eukalyptus* pollen sensitizations, it is not likely that air pollution, which should affect both taxa in the same manner, is the accountable factor.

References

Jilek, A., I. Swoboda, H. Breitenedeve, I. Fogy, F. Ferreria, E., Schmid E., Heberle-Bors, O. Scheiner, H. Rumpold, M., Koller and M. Breitenbach, 1993. Biological functions, isoforms and environmental control in the betvigene family. In: *Molecular Biology and Immunology of Allergens*, (Eds.) D. Kraff and A. Sehon. CRC Press Inc., Boca Raton, Fl., USA, pp. 39–45.

Rosen, G.W., 1981. Pollen systems to detect phytotoxicants in the environment an introduction. *Environ. Hlth. Persp.*, 17: 105.

Schata, M.R., Wahl, H.D., Baner, D., Dahmann and H.W. Stockmann, 1993. Immunological comparison on pure pollen and those exposed to high concentrations of diesel exhaust. *Allergy* 48(Suppl. 16): 184.

Valenta, R., M. Duchece, M. Breitenbach, K. Petteuburgev, I., Koller, H., Rumpold O. Scheiner and D. Krait, 1991. A low molecular weight allergen of white birch (*Betula verrucosa*) is highly homologous to human profiling. *Int. Arch. Allergy Appl. Immunol.*, 94(1–4): 368–370.

Vonpost, L., 1916. Om skogs trad pollen sydsvenska torfmosslagev–folidevr, Geol, foren, stockh, forh., 38: 384–390.

Chapter 54

Nutraceutical and Antioxidant Potentiality Associated with Fruit Development and Maturation in *Capsicum annuum* L. var. Jwalamukhi

V. Sumitha, S.R. Harish, Viann Maria and K. Murogan
Department of Botany, University College, Thiruvananthapuram – 695 034, India

ABSTRACT

Chilli pepper is a popular food additive in many parts of the world. The fruit is economically important as a vegetable, spice and medicine. The objective of this work was to determine the nutritive value in terms of protein, sugar, aminoacids, minerals, carotenoids, ascorbate and tocopherol during the development and ripening. The whole study was focused on capsicum annuum var. Jwalamukhi, a widely used variety in Kerala. Pepper fruit undergo changes in the levels of soluble protein, soluble sugar, β-carotene, ascorbate, tocopherol, and free phenolics during the maturation process. The level of glucose and reducing sugar gradually increase during fruit development and ripening. The content of soluble protein decreases during fruit development but increased in the stage of ripening. Generally the concentration of carotenoid constituents is high in the pepper during maturity. Levels of capxanthin and zeaxanthin increase during maturation where as the level of lutein declined. Endogenous level of tocopherol was quantified by RP-HPLC. α-tocopherol increased from the green to dry red maturity stage while γ-tocopherol content reached its maximum concentration from red succulent fruit and then declined. The large amount of tocopherol formed in fresh ripe fruits, indicate that this vegetable could became an important source of vitamin E in the human diet. Ascorbate content is increased after the green mature stage and peaked in red succulent fruit. Thus chillies are nutritionally balance with respect to the antioxidant vitamins and enzymes, minerals and could become an importance source of vitamin A, vitamin C and vitamin E in the human diet. The basic data presented here are useful to breeders and food scientist working to maximize the nutritional content and carotenoid stability of fresh, processed and dehydrated chilli peppers. In breeding for high nutrient content, selection should be done at the red succulent or red partial dry stages, where the natural antioxidant one present in high levels.

Keywords: *Capsicum annuum, Ripening, Chilli, Ascorbate, β-Carotene, Tocopherol.*

Introduction

Chilli (*Capsicum annuum*) is a dark-green annual or short-lived perennial plant, belongs to the family solanaceae. The fruits are long, cylindrical indehiscent berry. There are not enough studies on the nutritive value and antioxidant nature of chilli fruits. Fruits and vegetables have been conferred on them the status of functional foods (Sajitha *et al.*, 2006). Epidemiological data as well as *in vitro* studies strongly suggest that foods containing phyochemicals with antioxidant potential have strong protective effects against major diseases risks including cancer and cardiovascular diseases (Smirnoff and Wheeler, 1999). Vegetables and fruits are rich sources of antioxidants such as vitamin C, vitamin E and β-carotene, which are suggested to be antiatherogenic in epidemiological studies. Thus, the consumption of dietary antioxidants from these sources is beneficial in preventing cardiovascular diseases, especially atherosclerosis. Phenolic compounds are other type of antioxidant that possesses a strong inhibition effect against lipid oxidation through radical scavenging. The antioxidant activity of phenolics is mainly because of their redox properties which allow them to act as reducing agents, hydrogen donors, singlet oxygen quenchers and metal chelators. In this study we report antioxidant and nutritive properties of the chilli fruit containing high concentrations of these compounds than most of the leafy vegetables and fruits.

Materials and Methods

Plant Material

Fresh seeds of *Capsicum annuum* var. jwalamukhi were collected from the Kerala Agricultural University, Vellayani and were grown in the green house in the department. After anthesis young pepper fruits were collected every 7 days from 14th day (I stage) to 42nd days (V stage). The fruits were subjected for all analytical and biochemical analysis.

Estimation of Total Phenols

Total phenol content of fruit tissues were estimated by the method of Mayr *et al.* (1995). The total phenols/g tissue was calculated from the standard graph.

Amino Acid Analysis

The samples were hydrolyzed in 6 N HCl for 18 hours at 110°C. The hydrolysate was filtered and analyzed by HPLC. Since tryptophan is destroyed during hydrolysis, the method involving the hydrolysis of samples in 5 N NaOH was used for tryptophan (Singhai and Shrivastava, 2006).

Mineral Analysis by Atomic Spectroscopy

The major and minor nutrients were estimated from the fruit tissue by Atomic Spectroscopy (Bharghava and Raghupathy, 1993).

Estimation of Ascorbate

Ascorbate is extracted and quantified as per the methodology of Sadasivam and Subramanian (1987).

β-Carotene

β-carotene of the fruit samples was extracted by soxhlet method using the solvent Hexane. The absorbance was recorded at 429 nm with petroleum ether as blank (IUPAC, 1987).

Fractionation of Carotenoid Pigments and its Quantification

Carotenoids were extracted and separated using the method of Deli *et al.* (1996). Individual carotenoids were quantified using external standards which includes α- and β-carotene, lutein, zeaxanthin, β-cryptoxanthin and capxanthin.

Estimation of Tocopherol by HPLC

Tocopherol was extracted and was quantified by HPLC using isocratic mobile phase (acetonitrile: methanol 85 : 15 v/v). The flow rate was 0.4 ml/min (Picronen *et al.*, 1985). Tocopherols were identified and quantified by comparing retention times, absorption spectra and peak areas with those of authentic standards of α and γ tocopherol obtained from Sigma Chemicals Co (St. Louis, MO).

Results and Discussion

Composition and Physical Changes

Table 54.1 summarizes the physical and compositional characteristics associated with fruit maturity. The pH exhibited slight increases during maturation until the 4th stage, but declined rapidly in the last stage.

Table 54.1: Displays the Physical and Compositional Features Associated with Fruit Maturity

	Stage I	*Stage II*	*Stage III*	*Stage IV*	*Stage V*
pH	5.72	5.8	5.89	6.1	5.3
Glucose (mg/g tissue)	35.4	37.5	40.64	44	46.9
Pentose (mg/g tissue)	229.5	39.1	242	286	516
Hexose (mg/g tissue)	22.9	24.3	27.4	29.5	30.9
Total Protein (mg/g tissue)	0.31	0.33	0.35	0.54	0.66

Soluble proteins decreased during development, then increased during ripening may be due to the increased synthesis of enzyme proteins involved in ripening and senescence of the fruit. Pentose and hexose levels like wise the soluble protein content decreased during fruit development followed by an increase during maturation. The level of glucose continuously increased during the development, ripening of chilli fruits (Kodioglu and Adyin, 2001). The increase in glucose concentration may be due to the rapid growth phase of the fruit.

Mineral Content

Analysis of mineral elements of chilli by AS was exposed in Table 54.2. The data reveals the nutritive value in terms of potassium, phosphorus, magnesium and calcium. The other elements comprises iron, sodium, manganese, copper and zinc are also present in optimum level and. are comparatively higher than other vegetables. Cu, Zn and Mn are essential components of many enzymes that catalyze oxidative reductive reactions.

Table 54.2: Mineral Content in Chilli (mg/g tissue)

Ca	Fe	Mg	P	P	Na	Zn	Cu	Mn	Se
2.67	0.14	2.74	5.3	36.3	0.51	0.1	0.007	0.04	0.158

Amino Acid Profile in Green and Fully Ripe Chilli Fruit

The amino acid profile was relatively higher in green chilli than in the fully ripe stage (Figure 54.1). The amino acids such as tyrosine, proline, cysteine and aspartic acid share a major level than the other amino acids.

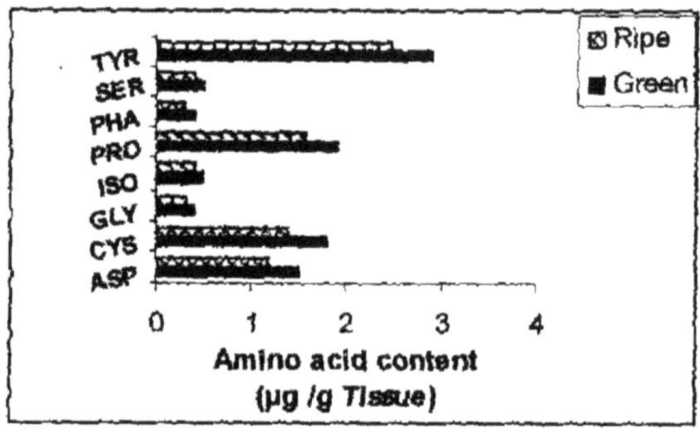

Figure 54.1: Amino Acid Profile of Green and Ripe Chilli

Antioxidants

Carotenoids

The concentration of the chlorophyll and carotenoids pigments during maturation was quantified. The total carotenoid contents were increased; while the chlorophyll content was reduced to non-detectable level (Table 54.3). The concentration of β-cryoptoxanthin, α-carotene, γ-carotene capxanthin and zeaxanthin increased extensively in the pepper variety during maturation, whereas the concentration of lutein decline to zero level. The results were consistent with other studies quantifying pepper carotenoids as a function of maturity (Deli *et al.*, 1996). In the unripe fruits the major carotenoids are lutein and β-carotene. The decrease in the concentration of lutein was coupled with the chlorophyll level. β-Carotene reaches a more or less constant value during ripening (Table 54.4). The other carotenoids like neoxanthin, luteoxanthin and auroxanthin have also been found in unripe stage. The oxygenated carotenoid capxanthin was not detected in the immature fruits and has been associated only with pepper type containing the genetic capacity to synthesize red pigments upon maturation along with zeaxanthin (Biacs *et al.*, 1989).

Vitamin C

Ascorbate content increases with the maturation of fruit and peaked at the red succulent stage (1.48–27.6 mg/g FW). A positive correlation was noticed between the ascorbate and moisture content of the fruit (Figure 54.2). Since ascorbate is a water soluble compound it declines as the fruit dehydrates (Kodioglu and Yavru, 1998). The synthesis of ascorbate in fruits is not fully understood, but the concentration is likely associated with carbohydrate metabolism and senescence (Mozafar, 1994).In green tissues, most ascorbate is located in the chloroplasts were it acts as a free radical scavenger. As such, ascorbate levels in fruits are influenced by diurnal fluctuations temperature, light and glucose, the precursor to ascorbate. Ascorbate can be transported from leaves to fruit along with carbohydrates during fruit ripening. Total chlorophylls and reducing sugars are at maximum levels at red succulent chilli fruit (Moser and Kamellis, 1994). In advanced ripening stages, the respiration rate decreases and oxygen accumulates in chilli fruit and can damage cellular function and hasten senescence. Ascorbic acid is an antioxidant that acts to slow senescence and to maintain. biological integrity during ripening by oxidizing ascorbate to dehydro ascorbate, and several oxides may be involved in ascorbate degradation. Ascorbate play multiroles like inhibiting lipid peroxidation, effective free radical scavenging and regeneration and recycle of vitamin E (Sajitha *et al.*, 2006).

Table 54.3: Summarizes the Pigment Composition in Different Stages of the Fruits

Pigments	*Green*	*Mid ripe*	*Ripe*
Chlorophyll a (mg/g tissue)	0.1258	0.038	ND
Chlorophyll b (mg/g tissue)	0.0511	0.0106	ND
Total Chlorophyll (mg/g tissue)	0.1769	0.0416	ND
Total Carotenoids (mg/g tissue)	0.1062	0.4049	5.84

Table 54.4: Summarizes the Fractionation of Carotenoid Pigments in Different Stages of the Fruits

Pigments	*Green*	*Mid ripe*	*Ripe*
α-Carotene (mg/g tissue)	0,14	1,36	3.30
β-Carotene (mg/g tissue)	0.21	1.49	3.32
Lutein (mg/g tissue)	1.29	0.16	NO
Zeaxanthin (mg/g tissue)	0.17	1.49	4.29
β-cryptoxanthin (mg/g tissue)	0.04	1.01	2.24
Capxanthin (mg/g tissue)	NO	64.25	72.36

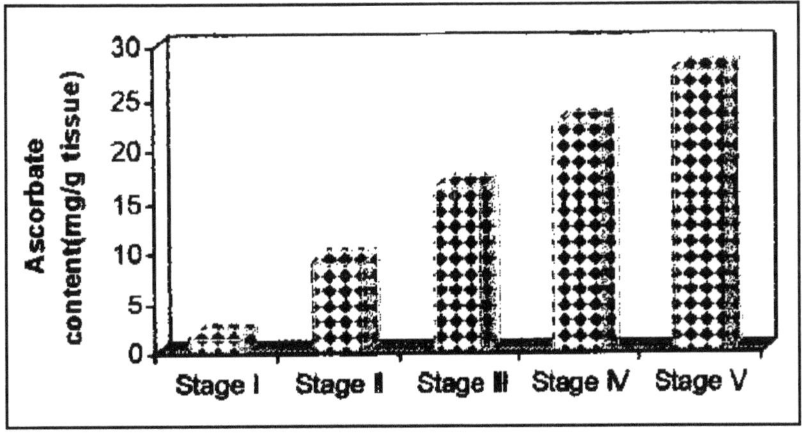

Figure 54.2: Ascorbate Level During Fruit Ripening in Chilli

Vitamin E

The fresh fruit of chili was found to be almost rich in tocopherol similar to spinach and asparagus, which are seems to be the richest sources of vitamin E in vegetables. The level of tocopherol correlates with the maturity of the fruit (Figure 54.2). The amount of tocopherol in the fruits was found to be depending on the lipid content (Kanner *et al.*, 1979). In the study of Biacs *et al.* (1989) a tocopherol increased in chilli fruits from the green to red stages. Our results exhibit a similar trend, but with lower values. Tocopherols are sensitive to oxygen and oxidative reactions, especially in the presence of catalysts such as UV light and metal ions. During storage and processing the fruits are exposed to one or more of these deleterious effects. Tocopherols play many roles like the synthesis of polyunsaturated fatty acids, components of cell membrane and prevent lipid peroxidation.

Total Phenols

Polyphenolics, the largest group of photochemical that stimulate health and minimize the impact of many of the diseases. A negative correlation was observed between the total phenols and development of fruit *i.e.* maximum phenols were observed in the first stage, then decrease with maturation *i.e.* 1/4th at the 5th stage approximately (Figure 54.3). The sink of bound phenolics in the cell may turn over and could thus provide substrates for the capsaicin biosynthetic pathway (Sukrasno and Yeoman, 1993).

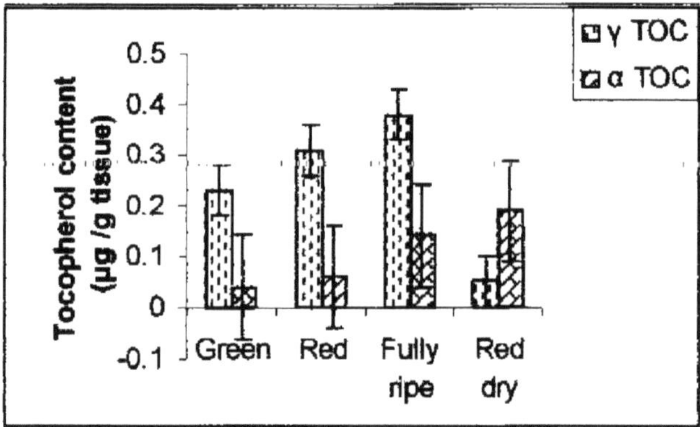

Figure 54.3: Tocopherol Content in Different Stages in Chilli

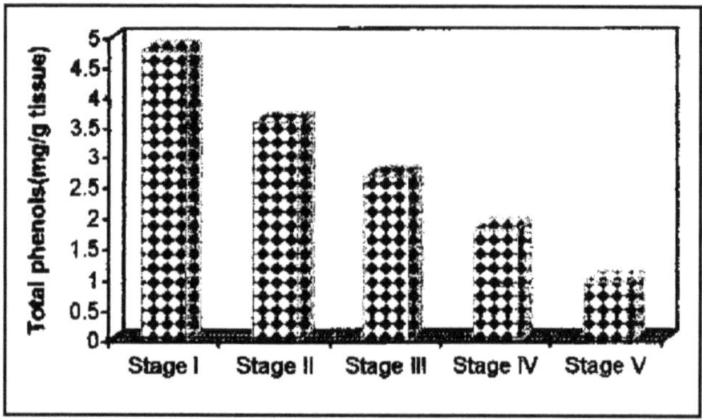

Figure 54.4: Total Phenol Content During Fruit Ripening in Chilli

A pool of phenolic acids (protocatechuic, chlorogenic, coumaric and ferulic acid) was observed by fractionating phenols reveal the potent antioxidant property of the fruit which in turn supports the nutraceutical value of chilli.

Conclusion

Moderate nutrient content and rich tocopherol, ascorbate, provitamin and poyphenolics work synergistically in their antioxidant action strongly suggests that chilli fruits are ideal vegetables and may be categorized as nutraceutical and functional food (NFF). This provides a promise for preventing and ameliorating various diseases.

References

Bharghava, B.S. and Raghupathy, H.B., 1993. Analysis of plant material for macro and micro-nutrients. In: *Methods of Analysis of Soils, Plants, Water and Fertilizer*, (Ed.) H.L.S. Tandon. FDCO, New Delhi.

Biacs, P., Dadood, H. and Huszka, T., 1994. Biochemical and varietal perspective on the colour loss in spice red pepper (paparika). *Hung. Agric. Res.*, 3: 32–37.

Biacs, P., Dadood, H., Pavisa, A. and Haijdu, F., 1989. Studies on the carotenoid pigments of paparika. *J. Agri. Food Chem.*, 37: 350–353.

Deli, J., Matus, Z. and Toth, G., 1996. Carotenoid composition in the fruits of Capsicum annuum cv. Szentesi Ksszarvu during ripening. *J. Agri. Food Chem.*, 44: 711–716.

IUPAC, 1987. *Standard Methods for the Analysis of Oils, Fats and Derivatives*. Blackwell Scientific Publications, Oxford.

Kanner, J., Harel, S., Mendel, H., 1979. Content and stability of a tocopherol in fresh and dehydrated pepper fruits (*capsicum annuum* L.) and processed jalapenos. *J. Agri. Food Chem.*, 27: 1316–1318.

Kodiogiu, A. and Adyin, N., 2001. Changes in the chemical composition, polyphenol oxidase and peroxidase activities during development and ripenening of medlar fruits (*Mespilus germanica* L.). *Bulg. J. Plant Physiol.*, 27: 85–92.

Kodioglu, A. and Yavru, I., 1998. Changes in the chemical content and polyphenol oxidase activity during development and ripening of cherry laurel. *Horn Austria*, 37: 241–251.

Mayr, V., Treutter, D., Santos-Buelga, C., Bauer, H. and Eucht, W., 1995. Developmental changes in the phenol concentration of golden delicious apple fruits and leaves. *J. Agric. Food Chem.*, 38: 1151–1155.

Moser, O. and Kamellis, A.K., 1994. Ascorbate oxidase of *Cucumis melo* L. var. reticulates purification, characterization and antibody production. *J. Exper. Botany*, 45: 717–724.

Mozafar, 1994. *Plant Vitamins: Agronomic, Physiological, and Nutritional Aspects*. CRC Press, Boco Raton, F.L.

Picronen, V., Syvoja, E.L., Varo, P., Salminen, K. and Koivistoinen, E., 1985. Tocopherois and tocotrienols in finnish foods. *Plant Mol. Biol.*, 33: 1215–1218.

Sadasivam, S. and Balasubramanian, T., 1987. *Practical Manual in Biochemistry*, Tamil Nadu Agricultural University, Coimbatore. New Age International, New Delhi.

Sajitha, S. Rajan, Shilpa, L. Satheesh, Kishor Mohan, T.C. and Murugan, K., 2006. Value of ethnic foods in meeting antioxidant needs: The wild plant. *Connection J. Food Sci. and Tech.*, (December Accepted, In Press).

Sajitha, S. Rajan, Shilpa, L. Satheesh. Kishor Mohan, T.C. and Murugan, K., 2006. Nutritive and Antinutritive characteristics of selected ethnic wild plants from south India. *J. Phytol. Res.*, (December Issue, In Press).

Singhai, B. and Shrivastava, S.K., 2006. Nutritive value of new chick pea (*Cicer arietinum*) varieties. *J. Food Qual.*, 4(1): 48–53.

Smirnoff, N. and Wheeler, G.L., 1999. Ascorbic acid metabolism in plants. In: *Plant Carbohydrate Biochemistry*, (Eds.) J.A. Bryant, M.M. Burrell and N.J. Kruger. Bios Scientific Publishers, Oxford, p. 215–229.

Sukrasno, N. and Yeoman, M.M., 1993. Phenylpropanoid metabolism during growth and development of *Capsicum frutescens* fruits. *Phtochem*, 32: 839–844.

Chapter 55

Parametric Variations Due to Metallic Contaminations in SF_6/N_2 Gas Insulated Bus-Duct

Poonam Upadhyay[1], J. Amernath[2], B.P. Singh[3], Pravin Upadhyay[4]
[1]Department of EEE, VNR Vignan Jyothi Institute of Engineering and Technology, Hyderabad, A.P., India
[2]Department of Electrical and Electronics Engineering, Jawaharlal Nehru Technological University, Hyderabad, A.P., India
[3]BHEL Corporate (R&D), Vikas Nagar, Hyderabad, A.P., India
[3]Satyam Computer Services Ltd., Hyderabad, A.P., India

ABSTRACT

Diluted SF_6/N_2 mixtures could be a substitute to pure SF_6 as insulation medium for high voltage gas insulated systems. They promise to have a lower environmental impact than a system insulated with pure SF_6 for the same rating. The main issue concerning the practical use of such mixtures is their behavior in the presence of particle contamination.

In this article, particle movement is determined in gas-insulated bus-duct (GIB) with Sulphur Hexafluoride (SF_6) and Nitrogen (N_2) gas mixtures as insulating medium. In the gas mixture, conducting particle material considered for this study is Aluminium.

The maximum co-axial distance traveled by the particle is calculated using numerical methods. Various parameters like percentage of SF_6 and N_2 in the gas mixture, pressure, radius and length of the particle are changed to obtain the particle movement. It has been observed that 40 per cent to 60 per cent of mixture gives good results when compared to other percentages of gas mixture. It is also observed that the dielectric strenhrth of mixture is similar to that of pure SF_6 gas.

Keywords: GIB, Gas mixture, Protrusions, Contaminations.

Introduction

Due to an exceptional combination of physic-chemical properties, SF_6 has become the only insulating gas used in gas insulated electric power transmission and distribution equipments. However, the fact that SF_6 is one of the strongest man-made greenhouse gases has promised a search for substitute gases with lower or no environmental impact (Harumota *et al.*, 1985; Matsumura *et al.*, 1983).

Diluted SF_6/N_2 mixtures are of particular interest as they promise a reduction of the environmental impacts. Various aspects of the insulation performance of SF_6/N_2 mixtures have already been published.

A lot of basic research work has been done about replacing SF6 with another insulating gas–the so called *'angel gas'*. It had observed in the 1970s at the IEEE Knoxville conference. Christophorou (Christophorou *et al.*, 1995) has presented a summary of work presented in that conference. Many investigations have carried out with CF_4-F_8 and other gas mixtures. Some of the tested gas mixture showed higher insulation capabilities than SF_6, but all of them are more critical in terms of toxicity or degenerating under partial discharge initiated by high electrical field. Strength, investigations have carried out mostly in laboratories and the gas mixture taken was N_2/SF_6, with high Nitrogen content and as low as possible Sulphur-Hexafluoride content to fix the percentage of the SF_6, an optimization process was needed to find the optimum between SF_6 content/gas pressure/enclosure diameter. Authors have mentioned that the basic behavior of N_2/SF_6 gas mixtures shows that with SF_6 content of only 15–25 per cent, an insulating capability of70–80 per cent of pure SF_6 can be reached at the same gas pressure (Koch *et al.*, 1998).

It is well known that the presence of particles in a compressed gas insulated system can deteriorate its electrical performance. These particles may result from mechanical abrasion and damage to the system as well as the enclosure of other contaminants during assembly (Laghori *et al.*, 1981). Numerous reports have been published detailing a large number of experiments involving uniform and coaxial fields. The particles may either be insulating or conducting. They may be free to move under the influence of the applied field or may be fixed on the electrode in the form of a protrusion representing electrode surface roughness.

The aim of this paper is to simulate particle movements in Gas Insulated Bus duct (GIB) with gas mixtures as insulating medium. The conducting particle is considered as Aluminum in the SF_6/N_2 gas mixture respectively. The SF_6 and N_2 content in the mixtures was varied in different percentages. In this paper a method based on particle movement is proposed to determine the particle trajectories in Single Phase GIB is SF_6/N_2 mixture environment In order to determine the movement of the particle in GIB an outer enclosure of diameter 152 mm and inner electrode of diameter 55 mm was considered. Aluminum particle of 10 mm length and 0.25 mm radius were considered to be present in the enclosure surface. The motion of particle was simulated using the charge acquired by particle, the microscopic field at the particle site, the drag coefficient, Reynolds' number and coefficient of restitution. The distance traveled by the particle is calculated. The various parameters like gas mixture, pressure, radius and length of the particle are varied to obtain the particle movement The results have been simulated by using rigorous field calculation method.

The Necessity of this Study

The use of N_2/SF_6 gas mixtures is under considerations as insulation medium especially for gas insulated transmission lines (GIL). Therefore many investigations dealing with the insulating properties of N_2/SF_6 gas mixtures are in progress. The gas N_2, as a natural element of air is the main component of these mixtures. Even admixture of a very low SF_6 content leads to a considerable increase of the dielectric strength. Due to high reliability of the equipment, Gas Insulated Substation (GIS) can be used for longer times without any periodical inspections. Conducting contaminations (*i.e.* Copper, Aluminium and Silver particles) could however, seriously reduce the dielectric strength of gas insulated systems.

Description of GIS System

The presented simulation work for the motion of metallic particles (Al) bus duct of 55 mm/152 mm inner and outer diameter in the presence of gas mixture (SF_6/N_2) insulation is considered.

In the present work the particle is on the surface of the enclosure and the enclosure is earthed. The schematic diagram of the typical compressed Gas insulated bus duct is shown in Figure 55.1.

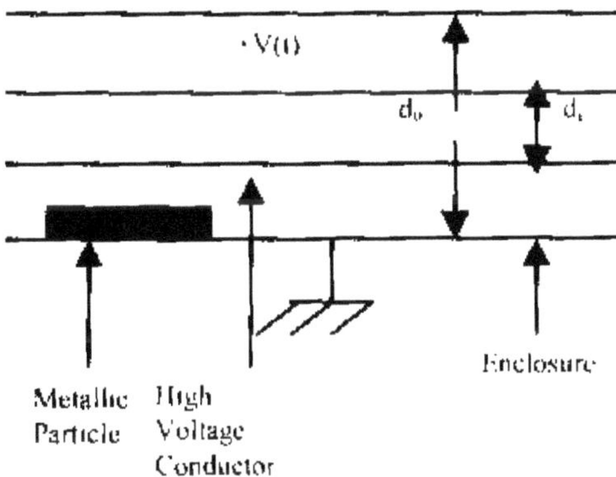

Figure 55.1: Schematic Diagram of a Typical Gas Insulated Bus Duct

Simulation of the Wire Particle Motion

The primary goal of this work is to simulate results to create a satisfactory model of the particle motion in the GIS bus with SF_6/N_2 gas mixture. Following are the major equations used for simulation.

Electrostatic Force

The charge acquired by a vertical particle in contact with necked enclosure can be expressed as:

$$Qnet = \frac{\pi\varepsilon_0 l^2 E_{(to)}}{\ln(2l/r) - 1} \quad (1)$$

where, l is the particle length, r is the particle radius, $E_{(to)}$ is the ambient electrical field at $t = t_o$.

Disregarding the effect of charges on the particle, the electric field in a coaxial electrode system at position of the particle can be written as:

$$E(t) = \frac{V \sin\omega t}{[r_0 - y_{(t)}]\ln(r_0/r_1)} \quad (2)$$

where, V Sinω t is the supply voltage on the inner electrode; r_0 is the enclosure radius; r_1 is the inner conductor radius; $y_{(t)}$ is the position of the particle which is the vertical distance from the surface of the enclosure towards the inner electrode.

The Electrostatic Force

$$Fe = KQ_{net}E_{(t)} \tag{3}$$

where, K is the correction factor smaller than unity and Q_{net} is the net charge acquired by the particle. $E_{(t)}$ is the electric field.

Gravitation Force

The gravitational force is given byL:

$$mg = \pi r^2 l\rho g \tag{4}$$

Drag Force

Drag is a result of energy dissipation in the shock wave near the particle and the skin friction along the surface of the particle. In spherical particles shock wave energy dissipation and in wire particles skin friction is more significant.

The direction of the drag force is always opposed to the direction of motion of particle.

Equation of Motion can be written as:

$$m\frac{d^2y}{dt^2} = F_c - mg - F_d \tag{5}$$

where, F is drag force.

The Reynolds's number is given by

$$Re = \frac{2\,r\rho_g y}{\mu} \tag{6}$$

The viscosity of a mixture of two gases can be approximately calculated from equation (Dean R. Chapman).

$$\mu = \frac{\mu_1}{1+\dfrac{\dfrac{x_2}{x_1}\left[1+\sqrt{\dfrac{\mu}{\mu}}\left(\dfrac{m_2}{m_1}\right)^{1/4}\right]^2}{\dfrac{4}{\sqrt{2}}\left(1+\dfrac{m_1}{m_2}\right)^{1/2}}} + \frac{\mu_2}{1+\dfrac{\dfrac{x_1}{x_2}\left[1+\sqrt{\dfrac{\mu}{\mu}}\left(\dfrac{m_1}{m_2}\right)^{1/4}\right]^2}{\dfrac{4}{\sqrt{2}}\left(1+\dfrac{m_2}{m_1}\right)^{1/2}}} \tag{7}$$

Results and Discussion

The maximum movement for Aluminum particle with variation of percentage of gas mixture is shown in Figure 55.2. As the percentage of nitrogen increases in the gas mixture maximum movement is observed as decreases.

Figure 55.3 shows the maximum movement of Aluminum particle with variation of radius of the particle. As the radius increases maximum movement decreases of any combination of SF_6 and N_2

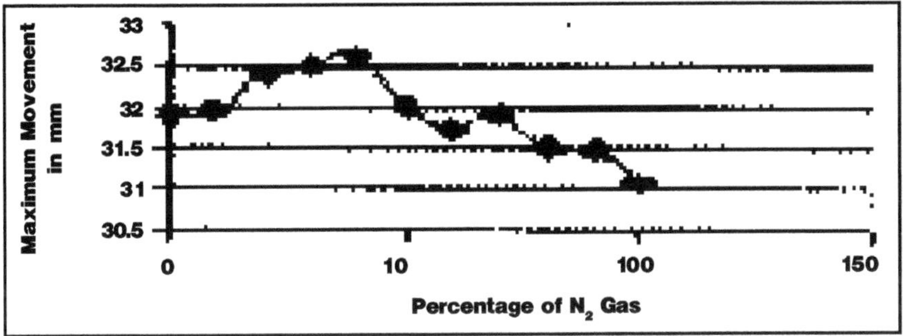

Figure 55.2: Percentage of Gas Mixture Vs Aluminium Particle Movement

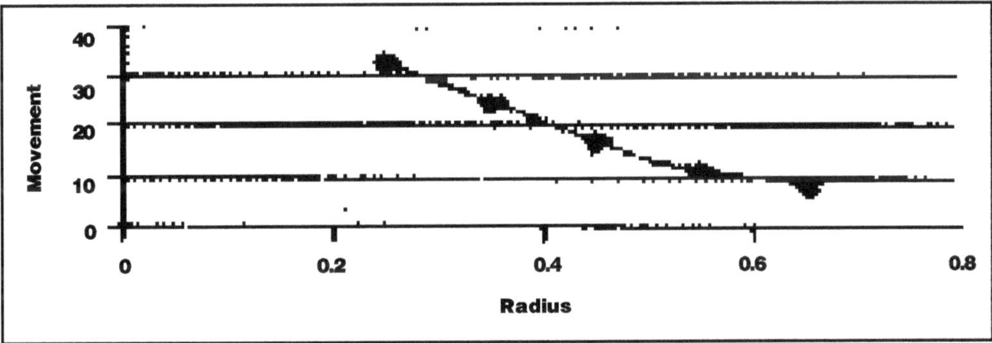

Figure 55.3: Al Particle Radius Vs Maximum Movement

mixture gas. The nature is expected since the charge acquired is proportional to the logarithmic value of radius whereas mass is proportional to the square of the radius. Hence, moving force is a function of radius and the retarding force is proportional to the square of the radius.

Figure 55.4 shows the movement pattern with the increase in the length of the particle. An entirely opposite pattern of movement is seen for the increasing length of particle. This is also expected, since the charge which influences movement of particle, is proportional to the square of the length.

Figures 55.5–55.7 shows the movement with increase in pressure. As the pressure increases there is a marginal change in maximum movement in case of 50 per cent and 60 per cent nitrogen in gas mixture. This reduction may be justified because the maximum movement is a function of product of velocity and density of gas.

Acknowledgement

The authors would like to acknowledge the assistance and considerations from their respective colleges and organization and its head, without which it would have been difficult to achieve and concentrate on the subject. The authors would also like to acknowledge the patience and help of their families, who sacrificed lots of their personal timing due to this study. The authors would like to thanks and the friends, colleagues and others who directly or indirectly helped while preparing the subject.

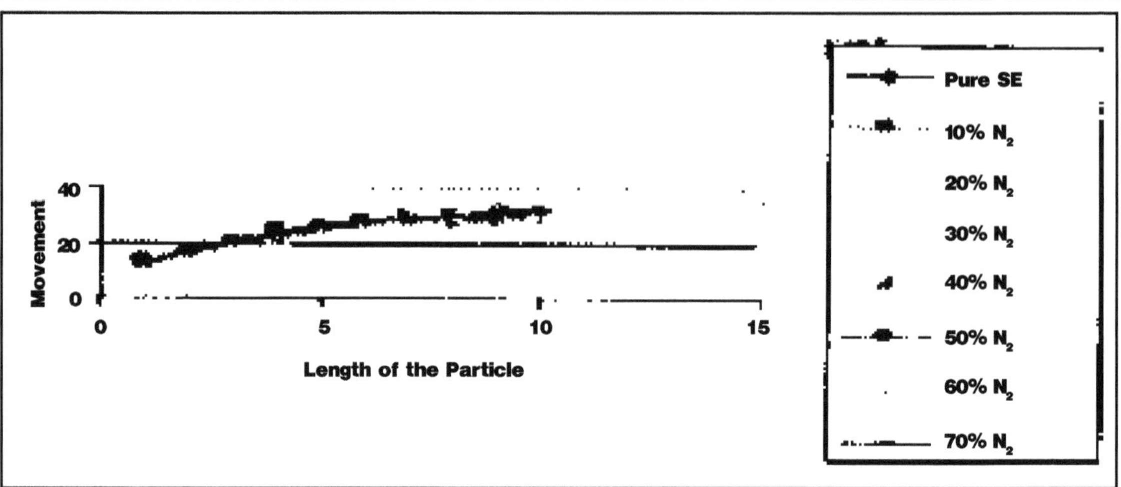

Figure 55.4: Movement of Al Particle with Respect to Length Variations for Different Percentage of N_2

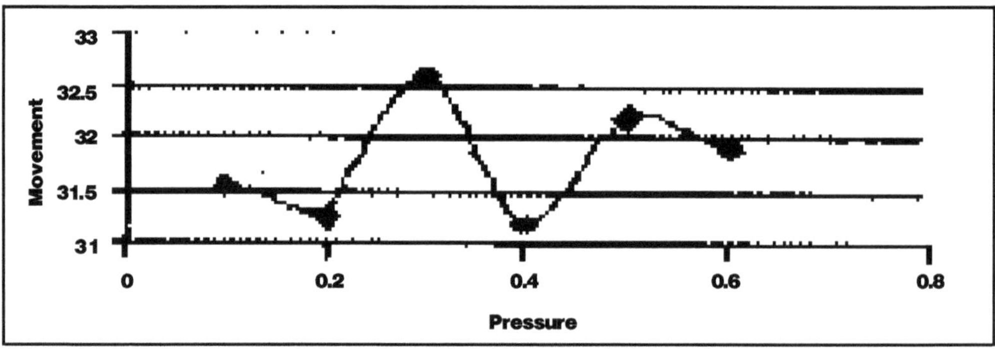

Figure 55.5: Movement of Al Particle with Respect to Pressure Variations for 40 per cent N_2 Gas

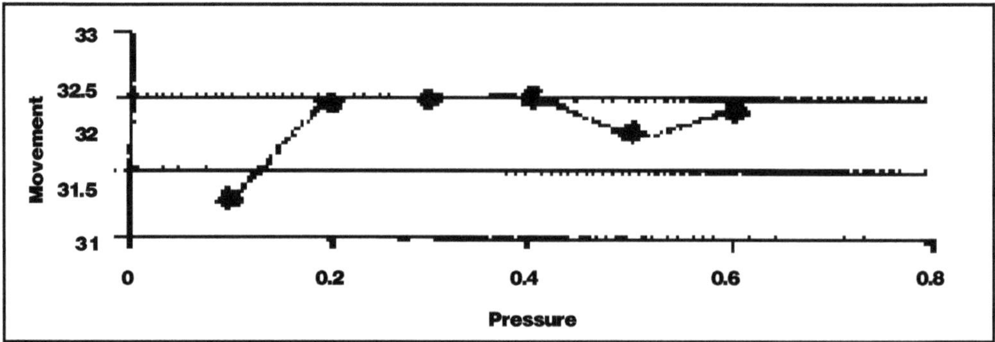

Figure 55.6: Movement of Al Particle with Respect to Pressure Variations for 50 per cent N_2 Gas

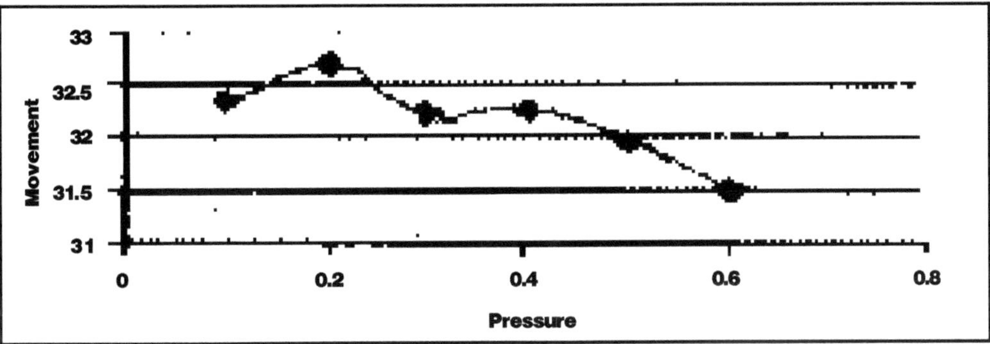

Figure 55.7: Movement of Al Particle with Respect to Pressure Variations for 60 per cent N_2 Gas

References

Amarnath, J, Singh, B.P., Kamakshaiah, S. and Radhakrishna, C., 2001. Influence of power frequency and impulse voltage on motion of metallic particle contamination in GIS-predicted by Monte-Carlo technique. *IEEE Int. High Voltage Workshop*; California.

Amarnath, J., Singh, B.P., Kamakshaiah, S. and Radhakrishna, C., 1999. Determination of particle trajectory in gas insulated bus duct by monte-carlo technique. *IEEE–CEIDP*, Austin, Texas.

Chapman, Dean R. Some possibilities of using gas mixtures other than Air in aerodynamic research;

Christophorou, L.G. and Brunt, R.J., 1995. SF_6/N_2 mixtures, Basic and HV insulation properties. *IEEE Transactions, Dielectrics and Electrical Insulation,* 2(5): 952–1002.

Harumota, Y. *et al.*, 1985. Development of 275Kv EHV class gas insulated power transformer. *IEEE*

Koch, H. and Schuette, A., 1998. Gas insulated transmission lines for high power transmission over long distances. ELSEVIER, *Electric Power System Research,* 44: 69–74.

Laghori, J.R. and Qureshi, A.H., 1981. A review of particle contaminated gas breakdown. *IEEE Transactions on Electrical Insulation*: Vol. EI–16 No. 5.

Matsumura, S. *et al.*, 1983. New approach to SF_6 gas insulated HVDC converter station. *IEEE Transactions on PAS*, Vol. PAS–102: 2871–2880.

Report 1259 National advisory committee for aeronautics, pp. 179–200.

Transactions on PAS, Vol. PAS–104: 2501–2508.

Chapter 56

A Study on the Effect of Commando on the Protein Content of the Freshwater Fish *Labeo rohita*

N. Saradhamani, R. Saraswathi, B. Dhanalakshmi
Post Graduate and Research Department of Zoology, Kongunadu Arts and Science College (Autonomous), Coimbatore – 641 029, Tamil Nadu, India

ABSTRACT

Fishes were exposed to sublethal concentration of detergent Commando (20 mg) for 24, 48, 72 and 96 hrs respectively bio chemical analysis of protein was estimated in liver, kidney, muscles and gills. Protein showed maximum decrease of 93.1 per cent in muscles after 96hrs exposure and minimum of 1.3 per cent in liver tissues after 72 hrs exposure

Keywords: Biochemical effects, Detergents, Labeo rohita.

Introduction

The struggle between Man and Nature began with the rapidly expanding global population which has put a major stress directly or indirectly on land, mineral and vegetation and water resources of the world (Dalela, 1986). Water pollution brings a reduction in diversity of aquatic life and eventually destroys the balance of life and influences the stream and river. Pollution of water is responsible for a very large number of mortalities and incapacitations throughout the world.

Studies were conducted to show toxicity of non-ionic and anionic synthetic detergents on fishes by several authors: (Pittinger and Tennant 1981; Singh and Devi, 1984). Studies on the effect of commercial detergents on freshwater fishes are meager.

A survey of the literature reveals that a lot of work has been carried on various aspects of biochemical changes in various fishes due to herbicide and pesticide pollution but the effect of detergents on the same is les studied. Therefore an attempt has been made to study the possible impact of detergent Commando on some biochemical aspects of a fresh water fish *Labeo rohita.*

Materials and Methods

Labeo rohita, a commonly occurring freshwater major carp fish available in the local ponds and rivers was selected for the experimental studies. Bulk of sample of fishes, (*Labeo rohita*) ranging in weight from 3–4 gms and measuring 3–5 cm in length were procured from Mettur reservoir. Fishes were acclimatized to the laboratory conditions for 2 weeks in large cement tanks (6′ × 4′ × 3′) at 24±3°C. The fishes were fed with ad-libitum and minced boiled egg. Fishes of about the same size (5–8 gms and 7–10 cm) irrespective of sexes were selected for the experiment. Technical grades of Commando a detergent was used in this investigation. Batches of 10 healthy fishes were exposed to different concentrations of detergent Commando to calculate the LC_{50} value. One more set of fishes are maintained as control in tap water. The level of the dissolved oxygen, pH, alkalinity and hardness were monitored and maintained constant. The tanks were continuously aerated with electrically operated aerator. It was found as 200 mg for 24 hrs using probit analysis method, (Finney, 1971).

From this stock solution various sub-lethal concentrations were prepared for bioassay study.

Four groups of fishes were exposed to 20 mg ($1/10^{th}$ of 24 hrs LC_{50} valve) concentration of the detergents Commando for 24, 48, 72 and 96 hrs respectively. Another group was maintained as control. At the end of each exposure period, fishes were sacrificed and tissues such as liver, gill, muscles and kidney were dissected and removed. The tissues (10 mg) were homogenized in 80 per cent methanol, centrifuged at 3500 rpm for 15 minutes and the clear supernatant was used for the analysis of different parameters.

Total proteins concentration was estimated by the methods of Lowry *et al.* (1951).

Results and Discussion

The amount of protein estimated in different tissues of the fish *Labeo rohita.* The amount of protein estimated in different tissues of the fish *Labeo rohita* subjected to different exposures are presented in the Tables 56.1–56.4. Proteins are indispensable constituents of the body and the protein metabolism is also confined to the liver. Kidney is also an important site of protein metabolism. Fall in serum protein may be due to impaired function of kidney or due to reduced protein synthesis owing to liver cirrhosis (Sharma and Maya, 1987 and Garg *et al.,* 1989). The result of the present showed significant decreases in protein content in the tissues studied. The present decreases was found to be greater in 24 hrs (9.9 per cent) in liver, 96 hrs (21.6 per cent) in kidney and muscle (93.1 per cent) and 72 hrs (69 per cent) in gill (Tables 56.1–56.4). Depletion of protein as a result of toxicity stress has already reported by number of workers (James *et al.,* 1979; Natarajan 1983; Swami *et al.,* 1983). A number of workers have been reported depletion in the protein level in different tissues under the stress of various metals, pesticides and chemical, (Rath and Mishra, 1980; Dubale and Awasthi, 1984). Present findings are in good agreement with the above findings by different workers.

Table 56.1: Changes in the Protein Content (mg/g) in the Liver of Fish *Labeo rohita* Exposed to Commando in Different Periods

Exposure Periods	*Exposure Concentration*		
	20 mg		
	Mean±SD	*SE*	*Per cent*
Control	0.502[b]±0.002	0.0012	–
24 hrs	0.452[d]±0.01	0.0058	–9.9
48 hrs	0.516[a]±0.003	0.0017	2.7
72 hrs	0.495[b]±0.005	0.0029	–1.3
96 hrs	0.480[c]±0.01	0.0058	–4.3

Results are mean (±SD) of three observations.

SE: Standard error

%: Percentage increase/decrease over control

In a column, means followed by a common letter are not significantly Different at the 5 per cent level by DMRT

SED	CD (5 per cent)	CD (1 per cent)
0.0056	0.0126	0.0179

Table 56.2: Changes in the Protein Content (mg/g) in the Kidney of Fish *Labeo rohita* Exposed to Commando in Different Periods

Exposure Periods	*Exposure Concentration*		
	20 mg		
	Mean±SD	*SE*	*Per cent*
Control	0.420[a]±0.0040	0.0012	–
24 hrs	0.404[b]±0.004	0.0058	–3.8
48 hrs	0.390[c]±0.0050	0.0017	–7.1
72 hrs	0.347[d]±0.0020	0.0029	–17.3
96 hrs	0.329[e]±0.0010	0.0058	–21.6

Results are mean (±SD) of three observations.

SE: Standard error

%: Percentage increase/decrease over control

In a column, means followed by a common letter are not significantly Different at the 5 per cent level by DMRT

SED	CD (5 per cent)	CD (1 per cent)
0.0029	0.0064	0.0091

The decreased protein content might be due to tissue destruction necrosis and disturbance of cellular fraction and consequent impairment in protein synthetic machinery (Bradbury *et al.*, 1987). The present study clearly supports the hypothesis of Kabeer (1979), and Kumar and Saradhamani (2004) that tissue protein are used to meet the increased energy demand posed by stress.

Table 56.3: Changes in the Protein Content (mg/g) in the Muscle of Fish *Labeo rohita* Exposed to Commando in Different Periods

Exposure Periods	*Exposure Concentration*		
	20 mg		
	Mean±SD	*SE*	*Per cent*
Control	0.290^{a}±0.0035	0.0020	–
24 hrs	0.172^{b}±0.0010	0.0006	–40.6
48 hrs	0.081^{d}±0.001	0.0058	–72
72 hrs	0.090^{c}±0.002	0.0115	–68.9
96 hrs	0.020^{e}±0.001	0.0058	–93.1

Results are mean (±SD) of three observations.

SE: Standard error

%: Percentage increase/decrease over control

In a column, means followed by a common letter are not significantly Different at the 5 per cent level by DMRT

SED	CD (5 per cent)	CD (1 per cent)
0.0016	0.0036	0.0051

Table 56.4: Changes in the Protein Content (mg/g) in the Gills of Fish *Labeo rohita* Exposed to Commando in Different Periods

Exposure Periods	*Exposure Concentration*		
	20 mg		
	Mean±SD	*SE*	*Per cent*
Control	0.252^{a}±0.01	0.0058	–
24 hrs	0.122^{c}±0.002	0.0012	–51.5
48 hrs	0.110^{c}±0.009	0.0052	–56.3
72 hrs	0.0.78^{d}±0.001	0.00058	–69
96 hrs	0.143^{b}±0.009	0.0052	–43.2

Results are mean (±SD) of three observations.

SE: Standard error

%: Percentage increase/decrease over control

In a column, means followed by a common letter are not significantly Different at the 5 per cent level by DMRT

SED	CD (5 per cent)	CD (1 per cent)
0.0060	0.0133	0.0189

The above study clearly indicates the toxic nature of the detergents commando on the biochemical constituents of the fish *Labeo rohita*.

References

Bradbury, S.P., Mckim, J.M. and Coats, J.R., 1987. Physiological responses of rainbow trout *Salmgairdneri* to acute fenvalerate intoxication. *Pest Biochem. Physiol.,* 27: 258

Dalela, R.C., 1986. Need of environmental pollution education and the development of the country. *Environ. Ecotoxicol.,* p. 1–7.

Dubale, M.S. and Awasthi, M., 1984. Biochemical changes in the liver and kidney of a catfish, *Heteropneustes fossilis* exposed to dimethoate. *Camp. Physiol. Ecol.,* 7: 111–114.

Finney, D.J., 1971. *Probit Analysis,* 3rd edn. Cambridge University Press, London, pp. 20.

Garg, V.K., Garg, S.K. and Tyagi, S.K., 1989. Manganese induced haematological and biochemical anomalies in *Heteropneustes fossilis. J. Environ. Biol.,* 10(4): 349–353.

James, J.H., Ziparo, V., Jeppssion, B. and Fischer, J.E., 1979. Hyper ammonemia, plasma amino acid imbalance and blood amino acid transport. *Lancet,* 2: 772–775.

Lowry, A.H., Rose Brough, N.J., Farr, A.L. and Randall, R.J., 1951. Protein measurements with the folin phenol reagents. *J. Biol. Chem.* 193: 265–275.

Natarajan, G.M. (1986. Changes in the carbohydrate metabolites during acute and chronic exposure of air breathing fish *Channa striatus* to oxydementon-methyl (metasystox). *Camp. Physiol. Ecol.,* 14(4): 181.

Pittinger, C.A. and Tennant, D.S., 1981. Critical surface tension of sodium laural sulfate to the fathead minnow *Pimephales promelas. J. Tenn Acad. Sci. Fish Abstr.,* 11: 9633.

Rath, S. and Mishra, B.N., 1980. Changes in nucleic acids and protein content of *Tilapia mossambicus* exposed to dichlorovos (DDVP). *Indian J. Fishery,* 27(1 and 2): 76–81.

Saradhamani, N., 2001. A study on the effect of a paper mill effluent on the water quality of the river Cauvery near Erode and its toxicity to an Indian catfish *Mystius vittatus. Ph.D. Thesis,* Bharathiar University.

Sharma, S.D. and Maya, 1987. Some biochemical alterations in liver and kidney of *Calarias batrachus* in response to arsenic administration. *Himalayan J. Environ. Ecotoxicol.,* p. 49–74.

Swami Rao, K.S., Reddy, K.S.J., Moorthy, K., Murthy, K.S., Sheety, G.L. and Indira, K., 1983. *Indian. J. Camp. Animal. Physiol.,* 1: 95.

Chapter 57

Experimental Study on Waste Heat Recovery for Sustainable Industrial Systems

V. Sivaramakrishnan[1]*, V. Kirubakaran[2], M. Premalatha[3] and P. Subramanian[4]
[1]Research Scholar, [3]Assistant Professor, [4]Professor and Head (Retd)
CEESAT, National Institute of Technology, Tiruchirapalli – 620 015, Tamil Nadu
[2]Lecturer in Energy, Rural Technology Centre,
Gandhigram Rural Institute, Dindugal, Tamil Nadu

ABSTRACT

Drinking water of acceptable quality has become a scarce commodity. In many places of the world only brackish or polluted water is available. Water borne diseases are increasing day by day. Hence it is very essential to have safe drinking water for human beings. This leads to an increasing interest in new technologies with less energy consumption. This paper deals with the production of potable water (boiled) using energy efficient method by employing combined heating and cooling technology. Lot of thermal energy has to be spent to heat the water from room temperature to the required higher temperature. Some of this energy is wasted, while bringing down the water to room temperature. If this waste heat is recovered and used to preheat the raw water, thereby lot of thermal/electrical energy can be saved. This savings of electrical energy gives considerable amount of reduction in CO_2 emission into the atmosphere, when compared with this power generated by burning coal. A novel idea of making the same water to flow through inside and outside heat exchanger pipes in a heat exchanger system is

----

* Corresponding Address: Professor of Mechanical Engineering, Saranathan College of Engineering, Tiruchirapalli – 620 012, Tamil Nadu; E-mail: vsmp1967@yahoo.com; Mobile: +91 98943 88495

employed here. This saves lot of energy and reduces the cost of producing safe drinking water. At laboratory an experimental investigation to this effect has been done by using a shell and tube heat exchanger. It has been observed that more than 50 per cent of the energy can be saved in producing the potable water.

Keywords: *CHP, Fluid flow, Energy conservation, Shell and Tube heat exchanger.*

Introduction

Poor water quality, sanitation and hygiene account for some 1.7 million deaths a year world-wide (3.1 per cent of all deaths), mainly through infectious diarrhea. Nine out of 10 such deaths are in children and virtually all of the deaths are in developing countries (Cho *et al.*, 1996). An *infectious type* is a disease that can be transmitted from one person to another and a major route of transmission of infectious water-related disease is by the faceo-oral route with the pathogen passing from the faces of an infected person and subsequently being ingested through the mouth (Deronzier and Bertolini, 1997). As per doctor's advice it is essential that water has to be boiled before consumption for human beings. To have a very safe drinking water, the raw water may be heated to a higher temperature say 100°C. After boiling the water, it is cooled down to room temperature or slightly higher than the room temperature so that it will be a comfortable temperature to drink, say 30 to 35°C. It is aimed to recover the waste heat energy available during the cooling process of hot water and this recovered thermal energy can be utilized to pre-heat the raw water. This will reduce the energy required to boil the water to 100°C, thereby the process will be energy efficient one. At laboratory a prototype of this system has been tested for the above said advantages and is elaborated in this paper. As per the guidelines (Incropra and DeWitt, 2001) of The World Health Organization (WHO), following are the possible aspects to be considered for manufacture safe drinking water.

1. General Considerations and Principles
2. Microbial Aspects
3. Disinfection
4. Chemical Aspects
5. Radiological Aspects
6. Acceptability Aspects.

As per the same guidelines, the safe drinking water can be produced by employing the following methods:

1. Activated Charcoal
2. Ion Exchange
3. UV Radiation
4. Reverse Osmosis
5. Filtration by Sand Beds
6. Thermal Treatment.

Pasteurization

Contrary to what many people believe, it is not necessary to boil water to make it safe to drink. Heating water to 65°C for 6 minutes, or to a higher temperature for a shorter time, will kill all germs, viruses, and parasites that cause disease in humans, including cholera and hepatitis A and B. This process is called pasteurization. Chlorination, ultra-violet disinfection, and the use of a properly constructed, properly maintained well are other ways of providing clean water. Hence in this work boiling the water to 100°C is not addressed but only heating to a higher temperature is dealt with.

Basis for Energy Calculation

The energy required for this system is as follows. The basis is taken as 2000 liter per day production of safe drinking water (Amofah, 2001).

Mass of water (m) = 2000 kg.

Specific Heat capacity (Cp) = 4.12 kJ/kg °C

Final temperature required = 90°C

Initial temperature of water = 30°C (assumed)

Change in temperature (ΔT) = 90 – 30 = 60°C

Therefore energy required to heat the water,

$$Q = (m) \times (Cp) \times (\Delta T)$$
$$= 2000 \times 4.12 \times 60$$
$$= 494.4 \text{ MJ}$$

If this raw water is preheated by the waste heat energy to say 80°C, then

Mass of water (m) = 2000 kg

Specific Heat capacity (Cp) = 4.12 kJ/kg °C

Temperature of pre heated water = 80°C

Final temperature required = 90°C

Change in temperature (ΔT) = 90 × 80 = 10°C

Therefore energy required to heat the water,

$$Q = (m) \times (Cp) \times (\Delta T)$$
$$= 2000 \times 4.12 \times 10$$
$$= 82.4 \text{ MJ}$$

Therefore energy saving:

$$= 494.4 \text{ MJ} - 824.0 \text{ MJ}$$
$$= 412.0 \text{ MJ}$$

Figure 57.1 shows the energy savings in this waste heat recovery heat exchanger system for various flow rates of water into the system.

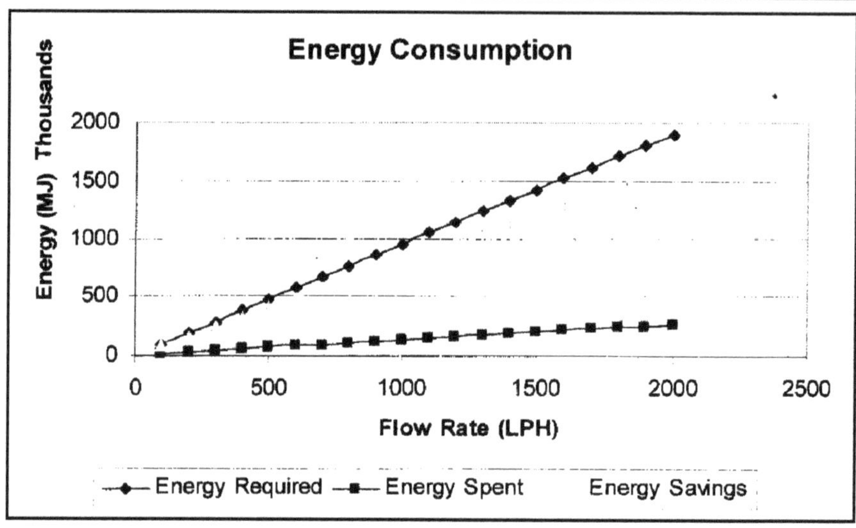

Figure 57.1: Energy Savings

Environmental Impact

Because of the waste heat recovery, the corresponding amount of electrical energy to be spent is reduced. This amount to savings in burning of the coal and hence it reduces the CO_2 emission into atmosphere. For the above case, following is the amount of CO_2 emission, without considering power transmission loss (Jachuck and Ramshaw, 1995)

412 MJ needs, (412 × 1.11)/10	=	45.732 kg of coal
45.732 kg of coal emits		(45.732 × 1.96)/1.11
	=	27.75 kg of CO_2
Total reduction in CO_2 emission	=	27.75 × 10 × 330
	=	91575 kg/year
	=	0.0138 kg I lit of water production

Hence, a plant operating for 10 hours per day and 330 days in a year can prevent 91575 kg of CO_2 emission into the atmosphere. As well, 0.0138 kg of CO_2 per liter of safe water production. Figure 57.2 shows the amount of CO_2 emission that is reduced for a plant operating for 10 hours per day and 330 days in a year.

Experimental Investigation

A waste heat recovery heat exchanger has been designed and experiments were conducted at the laboratory. The observation is provided in the Table 57.2. Where Thi is the inlet temperature of hot water, Tho is the outlet temperature of hot water, Tci is the inlet temperature of cold hot water and Tco is the outlet temperature of cold water. 100 W, 200 W, 300 W, 338 W power to the heater was supplied. The flow rate of water was set at 8, 9, 10, 11 and 12 lit/hour. For a particular experiment shows that raw water entering at 33°C is pre heated to 48°C. Hence it is pre heated to a temperature of 15°C. The fuel or thermal energy corresponding to this change in temperature is saved. This will also amounts to reduction in CO_2 emission to the atmosphere. Thus a combined heating and cooling has been

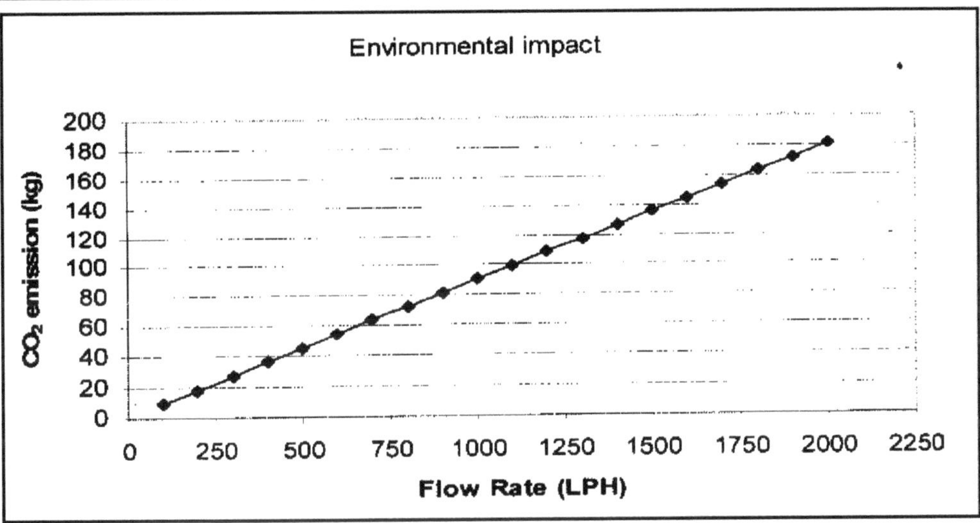

Figure 57.2: Environmental Impact of CO_2 Emission

implemented in this set of experiments by varying the flow rate of water in to the heat exchanger system as well as various power supplies to the heater. Figure 57.3 shows the experimental set up and Table 57.1 shows the corresponding Bill of Material. Thus a waste heat recovery heat exchanger is used in this experimentation (Luzzato *et al.*, 1997; Ashbolt, 2004).

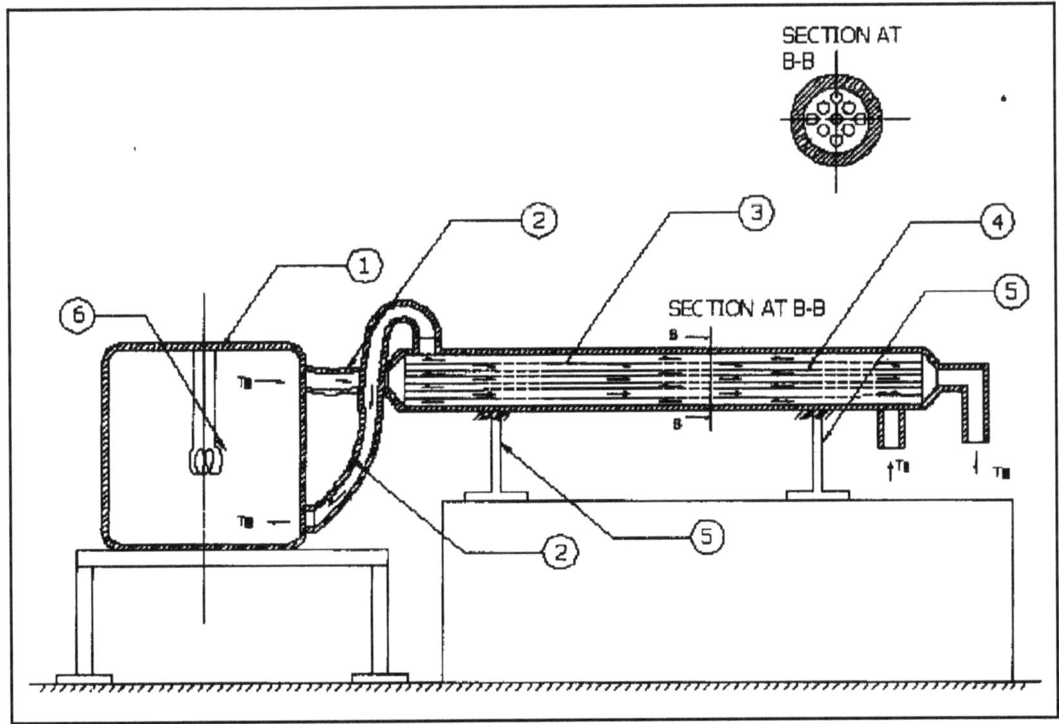

Figure 57.3: Experimental Set Up

Table 57.1: Bill of Material

Sl.No.	Name of the Part	Number	Material
1.	Heating tank	1	Copper
2.	Flexible hose	2	Plastic
3.	Shell	1	PYC
4.	Tube Bundle	1	Aluminum
5.	Stand	2	M.S
6.	Heating coil	1	S.S

Results and Discussion

Table 57.2 shows the experimental observation and efficiency of the combined heating and cooling system. In the Table 57.2, Q required is the energy required to raise the temperature of raw water from room temperature (say 30°C) to a higher temperature (say 90°C). Q supplied is the actual energy supplied by the heater and the efficiency of the system is the ratio of difference between Q required and Q supplied to the Q required.

Table 57.2: Experimental Observation and Results

Flow	Thi	Tho	Tci	Tco	Q Required	Q Supplied	Efficiency
LPH	°C	°C	°C	°C	W	CH&C (W)	Per cent
8	58	41	33	48	228.9	100	56.31
9	57	41	33	48	247.2	100	59.55
10	55	41	33	46	251.8	100	60.28
11	52	41	33	44	239.2	100	58.19
12	51	39	34	44	233.5	100	57.17
8	78	45	33	62	412.0	200	51.46
9	77	45	33	61	453.2	200	55.87
10	76	43	34	61	480.7	200	58.39
11	73	43	33	60	503.6	200	60.28
12	72	42	34	59	521.9	200	61.68
8	95	53	34	79	558.5	300	46.28
9	95	52	34	78	628.3	300	52.25
10	94	51	33	74	698.1	300	57.03
11	93	50	34	73	742.7	300	59.61
12	92	48	33	72	810.3	300	62.98
8	101	58	33	80	622.6	338	45.71
9	100	57	32	78	700.4	338	51.74
10	101	52	32	77	789.7	338	57.20
11	101	51	33	77	856.0	338	60.52
12	100	49	33	76	920.1	338	63.27

Hence the energy saving for this case is as follows:

Mass of water (m) = 8 kg

Specific Heat capacity (Cp) = 4.12 kJ/kg °C

Initial temperature of water = 33°C

Final temperature achieved = 58°C

Change in temperature ΔT = 58 – 33 = 25°C

Therefore energy required to heat the water;

$$Q = (m) \times (Cp) \times (\Delta T)$$
$$= 8 \times 4.12 \times 25$$
$$= 228.9 \text{ kJ}$$

Raw water is preheated by the waste heat energy to 48°C, then

Mass of water (m) = 8 kg

Specific Heat capacity (Cp) = 4.12 kJ/kg °C

Temperature of pre heated water = 48°C

Final temperature required = 58°C

Change in temperature ΔT = 58 – 48 = 10°C

Therefore energy required to boil the water

$$Q = (m) \times (Cp) \times (\Delta T)$$
$$- 8 \times 4.12 \times 10 \times 1000/3600$$
$$= 91.5 \text{ W}$$

Therefore energy saving:

$$= 228.9 \text{ kJ} - 91.5 \text{ W}$$
$$= 137.4 \text{ kJ}$$

The efficiency of the system = $(228.9 - 100) \times 100/228.9$

= 56.31 per cent

Thus, heat Exchangers playa very important role in making these processes economically feasible. Waste heat recovery through heat exchangers plays a major role in the concept called "Sustainable Development". This means "to meet the needs of present generation without compromising the ability of future generation to meet their own needs". It recognizes that economic growth and environmental protection are interlinked. Heat Exchangers have found wide application in pollution control facilities designed to be used in various types of air, water, land or thermal pollution problems. They find a vital role in implementing the concept of pollution prevention. This pollution generation prevention, recovery and recycling, treatment to reduce volume and toxicity and proper disposal of any residual waste. The good and services must be provided in and eco-efficient manner. It should happen progressively reducing the environmental impact and resource intensity throughout the life cycle. In this regard heat exchangers can help to reduce the energy intensity of goods and services. It allows for the utilization of waste heat, process to be integrated to reduce energy requirements and energy to be exchanged between processes (Shah *et al.*, 1999; www.who.int)

Conclusion

The experimental result shows that energy can be saved in producing safe drinking water by employing the waste heat recovery system in an economical way. In this particular experimental investigation it has been estimated that more than 50 per cent of the energy can be saved. Thus the idea of production of safe potable water using an energy efficient method was experimented. This helps in preventing human beings from water borne diseases. It also contributes to the reduction of CO_2 emission to the atmosphere and is an ecofriendly process and sustainable in nature.

Acknowledgements

The first author would like thank Dr. P. Subramanian, Professor and Head, Centre for Energy and Environmental Science And Technology, National Institute of Technology, Tiruchirapalli, for his valuable guidance and support to do this research work. We thank all authors of the reference work, whose contributions are inspiring source to do this work. We thank all the people who have involved in fabricating the experimental set up and their contribution during the course of conducting the experiments.

References

Amofah, George K., 2001. Address on the "*Role of Water in Health; The Public Concern for Improved Water Supply*" Acting Director Public Health at a Public Forum on Water Privatization Process on 16 May at Teachers' Hall, Accra.

Ashbolt, Nicholas John, 2004. Microbial contamination of drinking water and disease outcomes in developing regions. *Toxicology*, 198: 229–238.

Cho, S.M., Seltzer, A.H., Narayanan, T.V., Shah, A.C. and Weddell, J.K., 1996. Design of a continuous ceramic composite heat exchanger for high temperature, high pressure applications (PWE–Vol. 30), Vol. 2, pp. 1–9, ASME, New York

Deronzier, J.C. and Bertolini, G., 1997. Plate heat exchangers in liquid crystal polymers. *Applied Thermal Engineering*, 17: 799–808.

http://www.isodec.org.gh/Papers/DrAmofah_may01.PDF.

http://www.who.int/water_sanitation_health/dwq/gdwq3/en/index.html.

Incropra, Frank P. and DeWitt, David P., 2001. *Introduction to Heat Transfer*, 4th edn. John Wiley and Sons.

Jachuck, R.J. and Ramshaw, C., 1995. Process intensification: spiral polymer film compact heat exchanger, Process Intensification in Practice, BHR Group publication No. 18, Mechanical Engineering Publication, London, pp. 19–26.

Luzzato, C., Morgana, A., Chaudourne, S., O'Doherty, T.O. and Sorbies, G., 1997. A new concept composite heat exchanger to be applied in high temperature industrial processes. *Applied Thermal Engineering*, 17(8–10): 789–797.

Shah, R.K., Thonon, B. and Benforado, D.M., 1999. Opportunities for heat exchanger applications in environmental systems. *Applied Thermal Engineering*, 20: 631–650.

Chapter 58

Screening of Antimutagenic Effects of Green and Black Tea (*Camellia sinensis*) in Reverse Mutation Assay

K.S. Santhy[1], S. Namitha[1], Sherly P. George[1] and P. Arulraj[2]
[1]Department of Life Sciences, Avinashilingam Deemed University, Coimbatore, Tamil Nadu, India
[2]Department of Surgical Oncology, Government Royapettah Hospital, Chennai, Tamil Nadu, India

ABSTRACT

Tea (*Camellia sinensis*) is one of the most consumed beverages worldwide. In the present investigation antimutagenic effects of petroleum ether, chloroform, ethanol and water extracts of green and black tea were evaluated in *Salmonella typhimurium* TA–98 and TA–100 strains. Addition of Sodium azide and Daunomycin, two well known mutagens at a concentration of 10 µl and 6 µl per plate respectively resulted in the induction in the histidine revertant colonies. However addition of 10 µl of petroleum ether, chloroform, ethanol and water extracts of green and black tea to 10 µl of Sodium azide and 6 µl of Daunomycin treated plates resulted in the inhibition in the number of histidine revetant colonies. Further more supplementation with all the four extracts of green and black tea at a concentration of 6 µl and 10 µl per plate respectively in the presence of S9 fraction also led to a significant inhibition in sodium azide and Daunomycin induced colony formation. The antimutagenic activity of ethanolic extract of green and black tea was found to be higher than that of the other extracts. Hence the study revealed that green and black tea has protective efficacy in Sodium azide and Daunomycin induced mutagenicity in the test microbial system.

Keywords: *Green tea, Black tea, Antimutagenic activity, Ames assay.*

Introduction

Plant contains many natural substances that can promote health and alleviate illness. Flavonoids, one of the major constituent of the plants show biological properties that help to reduce the risk of many diseases. Tea plant (*Camellia sinensis*) native to south East Asia is consumed worldwide. Next to water, tea is the most consumed beverage in the world. Green and black tea have many therapeutic uses. It has been found that tea consumption lowers the incidence of skin disorders, guards tooth decay and prevent various diseases.

Flavonoids, a group of phenolic compounds occurring abundantly in vegetables, fruits and green plants attracted special attention as they showed high antioxidant property. The antioxidants are known to prevent cellular damage caused by Reactive Oxygen Species (ROS). Catechins are highly potent flavonoids present in tea and serve perhaps as the best dietary source of natural antioxidants. The fresh tea leaves contains four major catechins as colorless water soluble compounds, Epicatechin (EC), Epicatechin Gallate (ECG), Epigallocatechin (EGC) and Epigallocatechin Gallate (EGCG). Most of the green tea catechins during the manufacture of black tea are oxidized and converted into orange or brown products known as Theaflavins (TF) and Thearubigins (TR). These compounds retain the basic C6-C3-C6 structure and thus are classified as flavonoids.

Reports on potent antimutagenicity of green tea reveals that EGCG is perhaps the most potent antimutagenic agent protecting DNA scissions and non-enzymatic interception of superoxide anions. This leads to the general conclusion that development of cancer is prevented by tea consumption through antimutagenic protection paralleling to their antioxidant efficacy.

In the present investigation an attempt was made to evaluate the antimutagenic effect of green and black tea extracted using four different solvents *viz.*, petroleum ether, chloroform, ethanol and water in Salmonella microsome assay.

Materials and Methods

Green and black tea were procured from a local market. Histidine, biotin, Bactoagar, glucose, sodium ammonium hydrogen phosphate, Daunomycin and Sodum azide were of analytical grade and purity and purchased from Himedia Laboratories Ltd., Mumbai.

Soxhlet Extraction

10 g of the green and black tea powders were weighed using an electrical balance and made into 8 packets using Xerohaze filter paper. Soxhlet extraction of powdered green and black tea were carried out to obtain their extracts. Petroleum ether, chloroform, ethanol and water were used as solvents for soxhlet extraction in the increasing order of polarity. The distillation process was carried out at a low temperature of 40°C. After evaporation of solvents, corresponding residues obtained are stored in the refrigerator for further use. 100 mg of soxhlet extract was dissolved in 2 ml of DMSO and then mixed with 100 ml distilled water and this formed 1000 ppm solution. From this stock solution, solutions of required concentrations were prepared and used in this study.

Bacterial Tester Strains

Salmonella typhimurium tester strains TA-98 and TA-100 were kindly provided by Professor Bruce N. Ames, Berkley, U.S.A. The Strains were checked routinely for Amphicillin resistance, ultraviolet sensitivity and spontaneous revertants.

Preparation of Metabolically Activated Rat Liver S9 Mix

Albino rats (wistar strain) of 200 ± 25 g were obtained from animal laboratory, Food Science and Nutrition Department, Avinashilingam Deemed University, Coimbatore and kept in plastic cages with husk bedding and a stainless steel lid suitable for feeding and watering. The mouse was fed on by standard rodent diet pellet. Mouse was injected with phenobarbitone at a dose of 1 mg/g body weight intra peritoneally for 3 consecutive days. On day six, no food was provided to the mouse for fasting. The mouse was sacrificed on the seventh day for preparation of liver S9 fraction. All the steps were performed at 0–4°C with cold and sterile solutions and glass wares. The liver was excised out after dissecting the animal. The excised liver was then washed in an equal volume of 0.15 M KCl. Then it was mixed in 0.14M KCl and homogenized with a homogenizer. The homogenate was centrifuged for 10 min. at 9000 g and the supernatant which was so collected, was the S9 mix fraction. The freshly prepared S9 fraction was quickly frozen in dry ice and stored at –20°C.

For plate incorporation assay, top agar (2 ml) was distributed into each small test tubes held in a water bath. In different set groups of experiments in above tubes, 10 µl/plate of petroleum ether chloroform, ethanol and water extracts of green tea/10 µl/plate of petroleum ether, chloroform, ethanol and water extracts of black tea plus the mutagen (Daunomycin at the concentration of 6 µl/plate and Sodium azide at the concentration of 10 µl/plate) and 10 µl of metabolically activated S9 mix plus 10 µl of standardised bacterial cultures of TA-98 and TA-100 strains were added to the top agar and then poured into minimal glucose agar plates. The plates were then inverted and placed in an incubator at 37°C for 48 hours and counted for the number of histidine revertant colonies.

Similar experiments were carried out for positive controls (taking Daunomycin and Sodium azide) and negative controls (Untreated groups) for identifying spontaneous culture for both the strains concurrently.

Statistical Analysis

The mean values of number of histidine revertants/plate of different groups were subjected to statistical analysis using student 't' test.

Results

The present investigation depicts the antimutagenic potential of green and black tea extracts in *Salmonella typhimurium* reverse mutation assay (Tables 13.1 and 13.2). The number of spontaneous revertants were found to be 38 and 144 in TA-98 and TA-100 strains respectively. Addition of Daunomycin and Sodium azide to the minimal glucose plates resulted in significant induction in the number of histidine revertants and were found to be 61 and 184 of TA-98 and TA-100 strains respectively. However in plates supplemented with different extracts of black and green tea resulted in the inhibition of induction of histidine revertant colonies either by Daunomycin or Sodium azide.

The percentage inhibition of Daunomycin 26 per cent, 28 per cent, 33 per cent and 19 per cent for green tea and 23 per cent, 26 per cent, 30 per cent and 19 per cent for black tea in petroleum ether, chloroform, ethanol and water extracts respectively in TA-98 strains. The percentage inhibition of different extracts of green and black tea towards Sodium azide induced histidine reversion was found to be 8 per cent, 9 per cent, 10 per cent and 5 per cent and 3 per cent, 8 per cent, 8 per cent and 5 per cent respectively for petroleum ether, chloroform, ethanol, water extracts in TA-100 strains. The number of histidine revertants per plates for TA-98 and TA-100 tester strains in green tea was found to reduce to 45, 44, 41 and 50 and 171, 169, 167 and 175 respectively. The number of histidine revertants per plates for TA-98 and TA-100 tester stains in black tea was found to be reduced to 47, 45, 43 and 50 and 179,

171, 170 and 175 respectively. In the presence of S9 fraction the number of spontaneous revertants for petroleum ether, chloroform, ethanol and water extracts of green and black tea were found to be 106 and 203 in TA-98 and TA-100 strains respectively. On addition of Daunomycin and Sodium azide to the above extracts in TA-98 and TA-100 strains the number of histidine revertant were found to be increased to 155 and 359 respectively. However in plates supplemented with petroleum ether, chloroform, ethanol and water extracts of green and black tea resulted in inhibition of induction of histidine revertant colonies either by Daunomycin or by Sodium azide. The number of histidine revertants/plate of TA-98 and TA-100 tester strains of green tea were found to be reduced to 141, 134, 127 and 136 and 204, 210, 208 and 202 respectively. The number of histidine revertants/plate of TA-98 and TA-100 strains by adding petroleum ether, chloroform, ethanol and water extracts of black tea were found to be reduced to 139, 141, 130 and 138 and 327, 325, 323 and 327 respectively.

Table 58.1: Antimutagenic Effect of Petroleum Ether, Chloroform, Ethanol and Water Extract of Green and Black Tea in *Salmonella typhimurium* TA-98 and TA-100 Strains

Strain	*Treatment*	*Petroleum Ether*	*Chloroform*	*Ethanol*	*Water*
TA-98	SR+GT	39±3.29NS	41±3.23NS	40±2.79NS	37±5.19NS
	SM+GT	45±2.61NS (26%)	44±3.60NS (28%)	41±3.88NS (33%)	50±2.66NS (19%)
	SR+BT	41±3.75**	39±5.01**	39±5.95**	37±2.69**
	SM+BT	47±4.25** (23%)	45±3.57** (26%)	43±4.96** (30%)	50±3.67* (19%)
TA100	SR+GT	146±5.65NS	143±9.08NS	141±10.53NS	145±8.18NS
	SM+GT	171±4.99NS (8%)	169±3.87NS (9%)	167±3.37NS (10%)	175±2.47NS (5%)
	SR+BT	145±7.68**	143±8.48**	139±10.52**	143±6.69**
	SM+BT	179±2.35** (3%)	171±5.23** (8%)	170±3.10** (8%)	175±4.54** (5%)

Results are the average of two independent experiments.

Spontaneous revertant rate for TA-98 was 38±4.1 and TA-100 was 144±3.7.

Standard mutation rate for TA-98 was 61±2.1 and TA-100 was 184±3.2.

NS: Not significant; *: Significant at p = 0.05.

**: Significant at P = 0.01. Per cent inhibition of revertant frequency with the addition of different green and black tea extracts to standard mutagen induced plates is given in the parenthesis.

Discussion

The *Salmonella. typhimurium* reverse mutation assay is most commonly used method to assess mutagenic potential of test chemicals, which may cause base pair and flame shift mutations in the genome of this organism (Maron and Ames, 1983). Its applicability in screening the antimutagenic potential of green and black tea has been performed in the present study. Daunomycin and Sodium azide rule the known genotoxicant in mammalian and microbial test systems. Addition of Dauno mycin and Sodium azide to the minimal glucose plates resulted in the significant induction of histidine revertant colonies. In the present study addition of green and black tea extracts to Sodium azide and

Daunomycin treated plates resulted in the significant inhibition of number of colonies formed in TA-100 arid TA-98 strains respectively.

Table 58.2: Antimutagenic Effect of Petroleum Ether, Chloroform, Ethanol and Water Extracts of Green and Black Tea in *Salmonella typhimurium* TA-98 and TA-100 Strains in the Presence of S9 Fraction

Strain	*Treatment*	*Petroleum Ether*	*Chloroform*	*Ethanol*	*Water*
TA-98	SR+GT	108±8.08NS	105±6.48NS	104±6.55NS	112±8.60 NS
	SM+GT	141±8NS9 (9%)	134±6.11NS (14%)	127±3.15NS (18%)	136±4.05NS (13%)
	SR+BT	102±4.60**	108±4.76**	104±5.63**	102±4.89**
	SM+ BT	139±5.65** (10%)	141±4.12** (9%)	130±3.30** (17%)	138±3.42** (11%)
TA-100	SR+GT	204±8.12NS	210±6.84NS	208±5.86NS	202±3.15NS
	SM+GT	316±7.06NS (12%)	315±5.74NS (13%)	311±6.45NS (14%)	338±7.25NS (6%)
	SR+BT	210+4.79**	207±9.34**	211+9.08**	200±7.87**
	SM+BT	327±13.63** (9%)	325±11.31** (10%)	323±13.11** (11%)	327±9.53** (9%)

Results are the average of two independent experiments.

Spontaneous revertant rate for TA-98 was 106±8.6 and TA-100 was 203±5.8.

Standard mutagen rate for TA-98 was 155±4.9 and TA-100 was 359±5.5.

NS: Not significant; **: Significant at p = 0.01.

Percent inhibition of revertant frequency with the addition of different green and black tea extracts to standard mutagen induced plates is given in the parenthesis.

Catechin component including Epicatechin Gallate (ECG) and Epigallocatechin Gallate (EGCG) provide a significant protection against mutagenicity of Trp-P-2 and N-OH-Trp-P-2 using *Salmonella typhimurium* TA-98 and TA-100 (Hayatsu *et al.*, 1992; Kuroda and Hara, 1999). EGCG also have been reported to provide strong inhibitory effect against mutagenicity of Ba P diol epoxide in TA-100 strain (Hour *et al.*, 1999). Using *Salmonella typhimurium* TA-98 and TA-100, the tea catechins ECG and EGCG have been shown to inhibit the mutagenic activity of direct acting mutagens (Okuda *et al.*, 1984). The extracts of both green and black tea decreased the mutagenic activity of N-methyl-N-nitro-N-nitroso guanidine (MNNG) in *Escherichia coli* WP2 in a desmutagenic manner (Kuroda and Hara, 1999; Jain *et al.*, 1989) (Theaflavins from black tea were found to suppress the mutagenicity of H_2O_2 in *Salmonella typhimurium* (TA 104) (Shiraki *et al.*, 1994). The *Antimutagenic* potential of ethanolic extracts of Rheo discofor in *Salmonella typhimurium* TA 102 pretreated with ROS-generating mutagen nor floxacin in Ames test, protests liver cell structures against diethyl nitrosamine (Avila *et al.*, 2003). Procarcinogens like Benzo (a) pyrene and Aflatoxin B, require metabolic activation by cytochrome P-450 dependent enzymes to manifest their mutagenic/carcinogenic response. The antimutagenic potential of black tea in part, relate to their ability to inhibit cytochrome P-450 dependent metabolic activation of mutagens which inturn results in the inhibition of PAH–DNA binding *(Weisburger et al.,* 1996). Catechins arc competitive inhibitors of NADPH–Cytochrome C reductase enzyme (Hernaez *et al.*, 1998; Wang *et al.*, 1988).

The inhibition of PHlr mutagenicity by black and green tea extracts or polyphenols has been observed in the *Salmonella typhimurium* TA-98 assay containing rat S9 fraction. Green tea extracts were also effective against the mutagenicity of PAH, Benzo (a) pyrene, DMBA with S9 activation (Kuroda and Hara, 1999). The antimutagenic effect of tea involves interaction between the reactive genotoxic species of various promutagens and polyphenolic tea component present in the tea (Kuroda and Hara, 1999; Weisburger, 1999a). The antigenotoxic properties of tea include induction of DNA repair and binding of activated carcinogens (Weisburger, 1999b; Yang *et al.*, 2002).

The antimutagenic activity of aqueous tea poly-phenols and black tea poly-phenols towards Benzo (a) pyrene and cyclophosphamide in *Salmonella typhimurium* tester strain TA-98 and TA-100 (Taneja *et al.*, 2003). The inhibition of Aflatoxin B1-2-amino fluorine and 2-aminoanthracene induced mutagenicity by extracts of *Maytenus ilicifolia* and peltastes peltatus was observed in salmonella microsome assay in the presence of S9 fraction (Horn and Vargas, 2003).

Our findings point to higher antimutagenic activity of ethanolic extracts of green and black tea when compared to other four extracts using petroleum ether, chloroform, ethanol and water. This study throws possibility of reduction of mutagenicity and thereby carcinogenicity in people drinking green tea and black tea regularly.

References

Avila, M.G., Alba, M.A., Delagarza, M., Carmen, M.D., Pretelin, H., Oritz, A.D., Fazenda, S.F. and Trevino, S.V., 2003. Antigenotoxic, antimutagenic and ROS scavenging activities of Rheo dicolor ethanolic crude extract. *Toxicology in vivo*, 17(1): 77–83.

Hayastu, H., Inaba, N., Kakutani, T., 1992. Suppression of genotoxicity of carcinogenesis by(–)epigallocatechin gallate. *Prev. Med.*, 21: 370–376.

Hernaez, F.J., Xu, M. and Dashwood, R.H., 1998. Antimutagenic activity of tea towards 2-hydroxyamino 3-Methilimidazo (4,5-F) quinoline effect of tea concentration and brew time on electrophile scavenging. *Mutat. Research/Fundamental Molecular Mechanism of Mutagenesis*, 902(1–2): 299–306.

Horn, R.C. and Vargas, V.M.F., 2003. Antimutagenic activity of extracts of natural substances in the *Salmonella*/Microsome assay. *Mutagenesis*, 18(2): 113–118.

Hour, T.C., Liang, Y.C., Chu, I.S. and Lin, J.K., 1999. Inhibition of eleven mutagens by various tea extracts epigallocatechin-3-gallate, gallic acid and caffeine. *Food Chem. Toxicol.*, 37: 569–579.

Jain, A.K., Shimoi, K., Nakamura, Y., 1989. Crude tea extracts decrease the mutagenic activity of N-methyl N-nitro-N-nitrosoguanidine *in vitro* and intragastric tract of rats. *Mutat. Res.*, 210: 1–8.

Kuroda, T. and Hara, Y., 1999. Antimutagenicity and carcinogenic activity of tea polyphenols. *Mutat. Res.* 436: 69–97.

Maron, D.M. and Ames, B.N., 1983. Revised methods for *Salmonella* mutagenicity test. *Mutat. Res.*, 113: 173–175.

Okuda, T., Mari, K. and Hayatsu, H., 1984. Inhibitory effect of tanins on direct-acting mutagens. *Chem. Pharm. Bull.*, 32: 3755–3758.

Shikari, M., Hara, Y. and Osawa, T., 1994. Antioxidative and antimutagenic effects of theaflavins from black tea. *Mutat. Res.*, 323: 29–34.

Taneja, P., Arora, A. and Shukla, Y., 2003. Antimutagenic effects of black tea in the *Salmonella typhimurium* reverse mutation assay. *Asian Pacific J. Cancer Prev.*, 4: 193–198.

Wang, Z.Y., Das, M., Brickers, D.R. and Mukthar, H., 1988. Intraction of epicatechins derived from green tea with rat hepatic cytochrome P-450. *Drug Metab. Dispos.*, 16: 98–103.

Weisburger, J.H., 1999a. Mechanisms of action of antioxidants exemplified as in vegetables, tomatoes and tea. *Food Chem. Toxicol.*, 37: 943–948.

Weisburger, J.H., 1999b. Tea and health: The underlying mechanism. *Proc. Soc. Exp. Biol. Med.*, 220: 271–275.

Weisburger, J.H., Hara, Y. and Dolan, L., 1996. Tea polyphenols as inhibitors of mutagenicity of major classes of carcinogens. *Mutat. Res.*, 371: 57–63.

Yang, C.S., Maliakal, P. and Meng, X., 2002. Inhibition of carcinogenesis by tea. *Annu. Rev. Pharmocol. Toxicol.*, 42: 25–54.

Chapter 59

Natural Radioactivity Levels in Sediments Along the East Coast of Region of Potronovo-Pondicherry Coast, East Coast of Tamil Nadu, India

E. Manikandan[1], B. Rajamannan[1*], R. Ravisankar[2], R. Sakthivel[3], M. Arumugam[3], V. Gajendiran[4] and V. Meenakshisundaram[4]

[1]Department of Physics, [3]Department of Earth Science, Annamalai University, Annamalainagar - 608 102, Tamil Nadu, India

[2]Department of Physics, SSN College of Engineering, Chennai - 603 110, Tamil Nadu, India

[4]Radiological Safety Division, IGCAR, Kalpakkam - 603 102, Tamil Nadu, India

ABSTRACT

The activity concentration of primordial radionuclides in beach sediment samples from South East Coast of region of Porto-Novo-Pondicherry Coast of Tamil Nadu, India has been estimated using gamma-ray spectrometer. The specific concentration for ^{232}Th, ^{238}U and ^{40}K are ranged from (7.78–3687.74 $Bq.kg^{-1}$), (BDL –469.21 $Bq.kg^{-1}$) and (BDL-377.11 $Bq.kg^{-1}$) respectively. The measured activity concentrations for these radionuclides were compared to world average activity of soil. These results indicate no radiological anomaly. The data presented in this study will serve as a base line survey for primordial radionuclides concentration in the study area.

***Keywords**: Radionuclides, Beach sediment, Specific activity and Gamma ray.*

* Corresponding Author: E-mail: rajamannan@rediffmail.com.

Introduction

Radiation is present in every environment of the earth's surface and beneath the earth and in the atmosphere. The earth is bombarded by cosmic rays from the space and all matter contains some traces of radioactive substance. Mankind has all long lived in natural radiation environment present due to the various natural sources of terrestrial and extra terrestrial origin. The extra terrestrial sources of radiation are cosmic rays and other radioactive elements produced due to the interaction of cosmic rays with the atmosphere. The sources of terrestrial radiation are radionuclides present in the earth's crust or in the atmosphere. Most of the terrestrial radionuclide originates from ^{238}U and ^{232}Th series and the singly occurring primordial radionuclide ^{40}K. This radionuclide may enter into our food chain and cause internal exposure. The distribution of both the natural and anthropogenic radionuclides is an essential prerequisite for the evaluation and control of public exposures.

Natural radiation level may be enhanced as a result of industrial and technological activities such as phosphate industry, thermal power plants, oil refineries etc. The objective of the work is to collect beach sediment samples from Porto-Novo to Pondicherry, East Coast of Tamil Nadu, India and to study the distribution of these radionuclides in the study area.

Sample Collection and Preparation

Twenty-two sediment samples were collected during low tide. The samples were collected 10 to 20 m away from the high tide when it makes towards the coast. At each sample site, covering a sampling area of 1 m^2, five wet samples were collected, each weighing about one kilogram. Among these five samples from a site, four were from the corners and one from the center. The sediment samples were collected at a depth of 5 cm. The samples were placed in plastic pouches and transported to the laboratory. The study area and sample collection locations are indicated in Figure 59.1.

The five samples collected from a sample site were mixed and weighed after removing stones and other unwanted materials. They were packed in sample containers as detailed in references IAEA, 1989. The samples are stored in the sealed containers for one-month period to attain equilibrium between ^{226}Rn and its daughter products (Mollah *et al.*, 1986). The samples were subjected to gamma spectral analysis with a counting time of 20,000 seconds. A 3″ × 3″ NaI (Tl) detector was employed with adequate lead shielding which reduced the background by a factor of about 95. IAEA standard sources were used to estimate the efficiency factors for the required geometry. The concentrations of various nuclides of interest are determined in $Bq.kg^{-1}$ using the count spectra. The peaks corresponding to 1.46 MeV (^{40}K), 1.76 MeV (^{214}Bi) and 2.614 MeV (^{208}Tl) were considered for arriving at the activity levels. The activity concentrations Cu, C_{Th} and C_K for ^{238}U, ^{232}Th and ^{40}K respectively (in $Bq.kg^{-1}$) were deduced from the spectral data by noting the counts in the respective energy for each location regions of interest (ROI).

Results and Discussion

The distribution of natural radionuclides in the beach sediment samples at 22 sampling sites and the activity ratio are presented in Table 59.1. From the table it is seen that the activities of these vary from 7.78 to 3687.74 $Bq.kg^{-1}$ and the arithmetic mean is 451.26 $Bq.kg^{-1}$. The activity concentration of ^{238}U varies from BDL to 469.21 $Bq.kg^{-1}$ and the arithmetic mean of 62.03 $Bq.kg^{-1}$. Similarly the activity concentration of ^{40}K varies from BDL to 377.11 $Bq.kg^{-1}$ and the arithmetic mean of 157.68 $Bq.kg^{-1}$ these are illustrated in Figure 59.2. The revised world wide median activity counteraction of ^{238}Th, ^{238}U and ^{40}K are 30, 35 and 400 $Bq.kg^{-1}$ is compiled by UNSCEAR (2000).

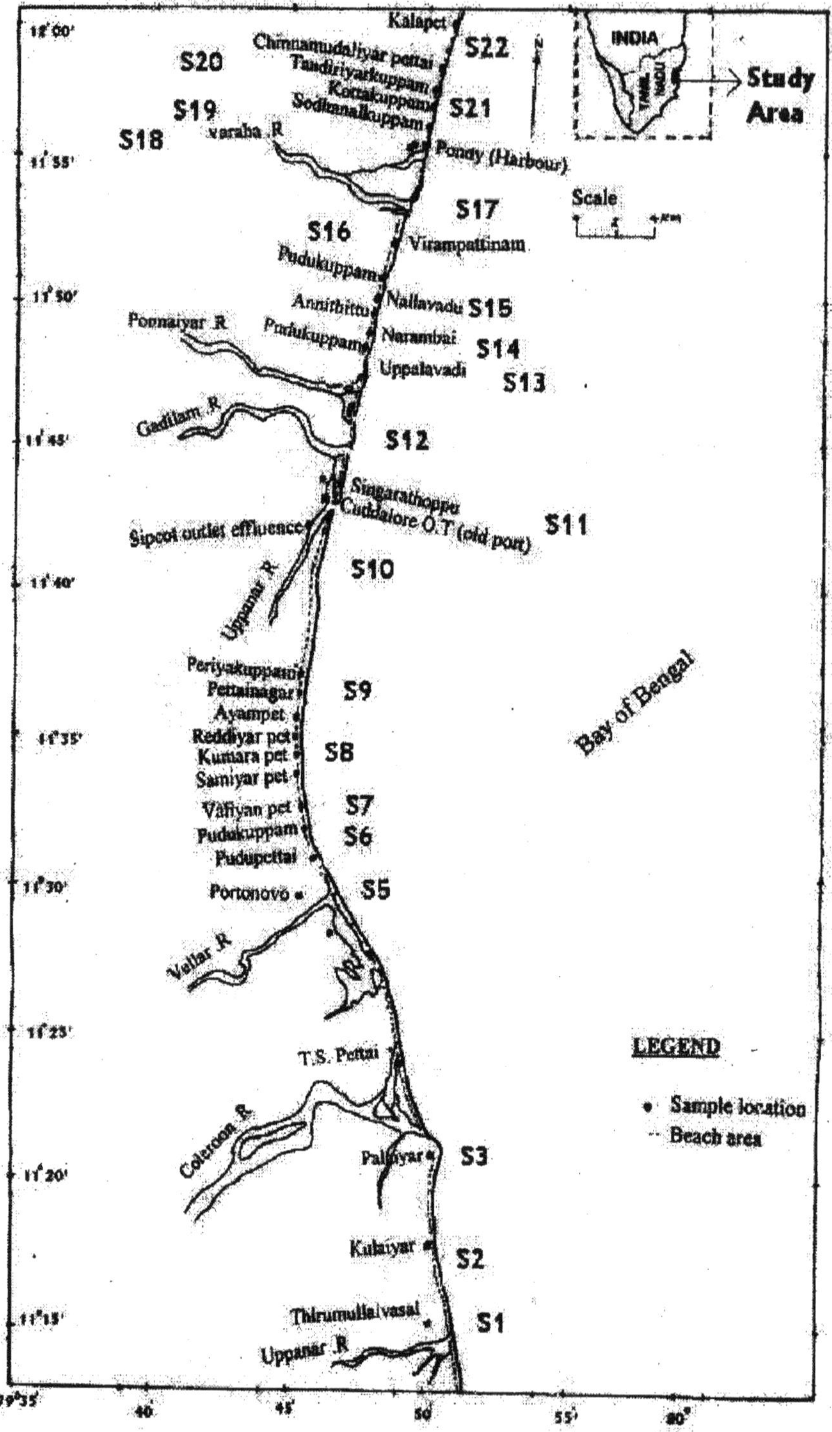

Figure 59.1: Sampling Location Map

Table 59.1: Activity Concentration ^{232}Th, ^{238}U and ^{40}K in Sediment Samples

Sl.No.	*Activity Concentration (Bq.kg^{-1})*			*Activity Ratio's*		
	^{40}K	*^{238}U*	*^{232}Th*	*Th/U*	*K/U*	*K/Th*
1.	246.12	BDL	58.09	–	–	4.23
2.	174	46.05	409.03	8.88	3.77	0.42
3.	237.46	35.08	141.11	4.02	6.76	1.68
4.	222.91	469.21	3687.74	7.85	0.47	0.06
5.	63.32	172.34	1039.27	6.03	0.36	0.06
6.	173.79	44.45	437.54	9.84	3.90	0.39
7.	139.78	42.12	316.13	7.50	3.31	0.44
8.	182.44	14.97	125.81	8.40	12.18	1.45
9.	130.1	62.69	558.84	8.91	2.07	0.23
10.	193.4	14.4	158.47	11.00	13.43	1.22
11.	BDL	132.33	1144	8.64	–	–
12.	145.66	29.61	191.59	6.47	4.91	0.76
13.	94.74	107.15	689.77	6.43	0.88	0.13
14.	BDL	105.73	622.63	5.88	–	–
15.	93.14	40.2	215.81	5.36	2.31	0.43
16.	165.52	BDL	21.93	–	–	7.54
17.	175.75	BDL	14.21	–	–	12.36
18.	155.37	10.65	27.24	2.55	14.58	5.70
19.	160.75	BDL	28.22	–	–	5.69
20.	189.65	BDL	7.78	–	–	24.37
21.	148.14	29.36	8.21	0.27	5.04	18.04
22.	377.11	14.4	24.4	1.69	26.18	15.45
Mean	157.68	62.30	451.26	7.24	2.53	0.34

BDL: Below Detection Limit.

If one compares the activity of the radionuclide to the present study with world average value, ^{232}Th is higher by a factor of 15.04 times while the activity of ^{238}U is higher by a factor of 1.77 and ^{40}K activity is lower by a factor of 0.39. The highest activity of ^{232}Th found in the present study is due to the monazite in the study area as confirmed by the spectral investigations. The variation in the activity concentration values in the present study is attributed to the natural gamma radioactivity values increases with decreasing clay contents. Since most sediment with high clay content is associated with high gamma-ray profiles (Ayes and Theilen, 2000).

Activity Ratio

The activity ratio between radionuclide belonging to same natural radioactive series can provide information about the distribution paths for those radionuclides involved. These results were used to establish the relative mobility of the uranium, thorium and potassium and also the input routes of the sediments. The measurement of activity ratio in sediments may provide clues to their source or supplies

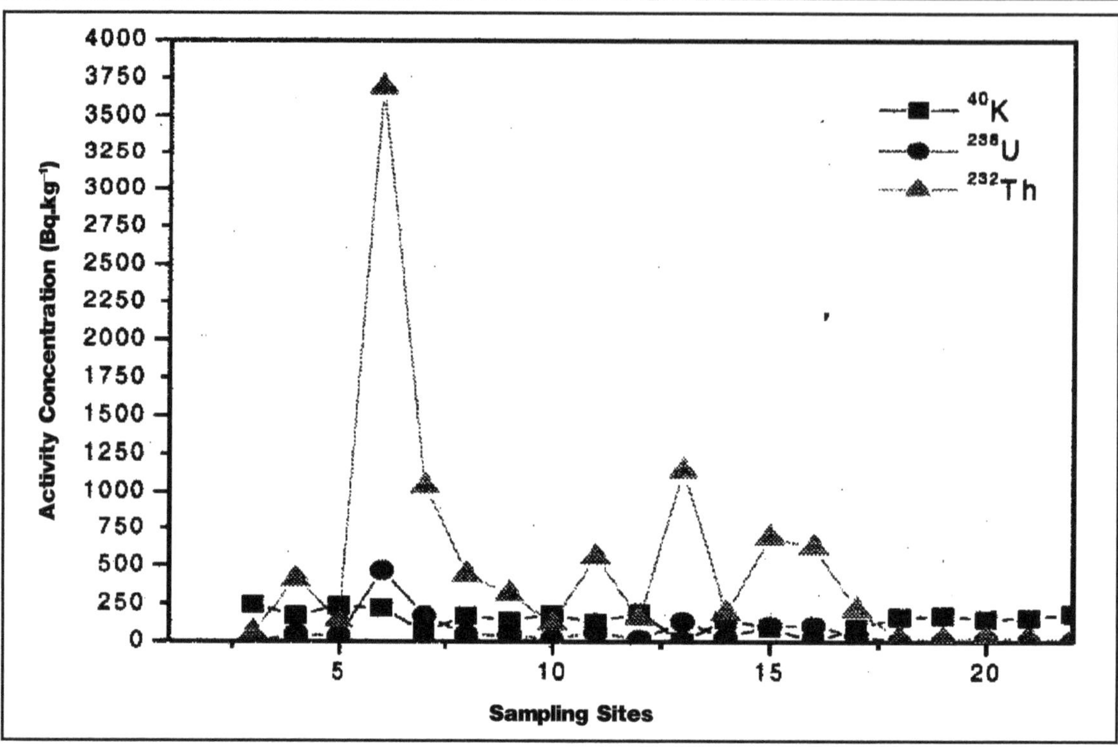

Figure 59.2: Sampling Sites vs Activity Concentration (Bq.kg^{-1})

as well as their pathways in addition to scavenging by particulars (Chung and Chung, 1996). Hence there is a demand to calculate the activity ratio in the present study. The mean activity (Th/U) obtained from the study is found to 7.24.

The theoretically expected Th/U ratio for normal continental crust is about 3.0 while the corresponding value obtained for this ratio is 7.24 which higher than to the expected value. This is due to the different chemical behaviors of Th and U may result in concentration difference another possible reason may be thorium content (and therefore Th/U ratio) of the sediments is much greater in fine size fractions. This concentration of thorium in the fine fraction may reflect a higher content of thorium rich in resistant minerals or relatively a higher content of thorium rich resistant minerals or relatively larger amounts of thorium fixed by clays. The presences of clay minerals in the study area from the spectral (FTIR and XRD) investigations sustain the above statements.

The activity ratios of (Th/U) in almost all the samples are greater than unity, which indicates either a net loss of ^{238}U or an addition of ^{232}Th by weathering processes (Scott, 1968). Further the higher value of the activity ratio (Th/U) supports Th enrichment during weather and transport processes (Balakrishna, 2001).

The Th/U, K/U and K/Th ratio may provide an indication whether relative depletion or enrichment of radioisotopes had occurred. From the Table 59.1, the average values are also calculated and are equal to 7.24, 2.53 and 0.34 respectively. The crystal o rocks have the following ratio Th/U = 3.5, K/U = 1×10^4 and K/Th = 3×10^3, the ratio values in the present are compared to crystal rocks; it is found that, the values of K/Th and K/U is lower. This might reflect a quite different enrichment/depletion of the radionuclides.

This is may be due to the relative mobility of uranium (largely dissolved) and Thorium (largely particulate) depends on the prevailing hydrological regime. The adsorption of uranium by clay minerals insoluble oxides and ox hydroxides and organic matter has significant influence on the transport of uranium in the surfical environment (Plater *et al.*, 1992.)

Statistical Analysis

The statistical data for the corresponding the activities measured for ^{232}Th, ^{238}U and ^{40}K in sediment samples are given in Table 59.2. The Table 59.2 lists the arithmetic mean, standard deviation, skewness and kurtosis. Figure 59.3 gives empirical frequency distributions of the activities of naturally occurring

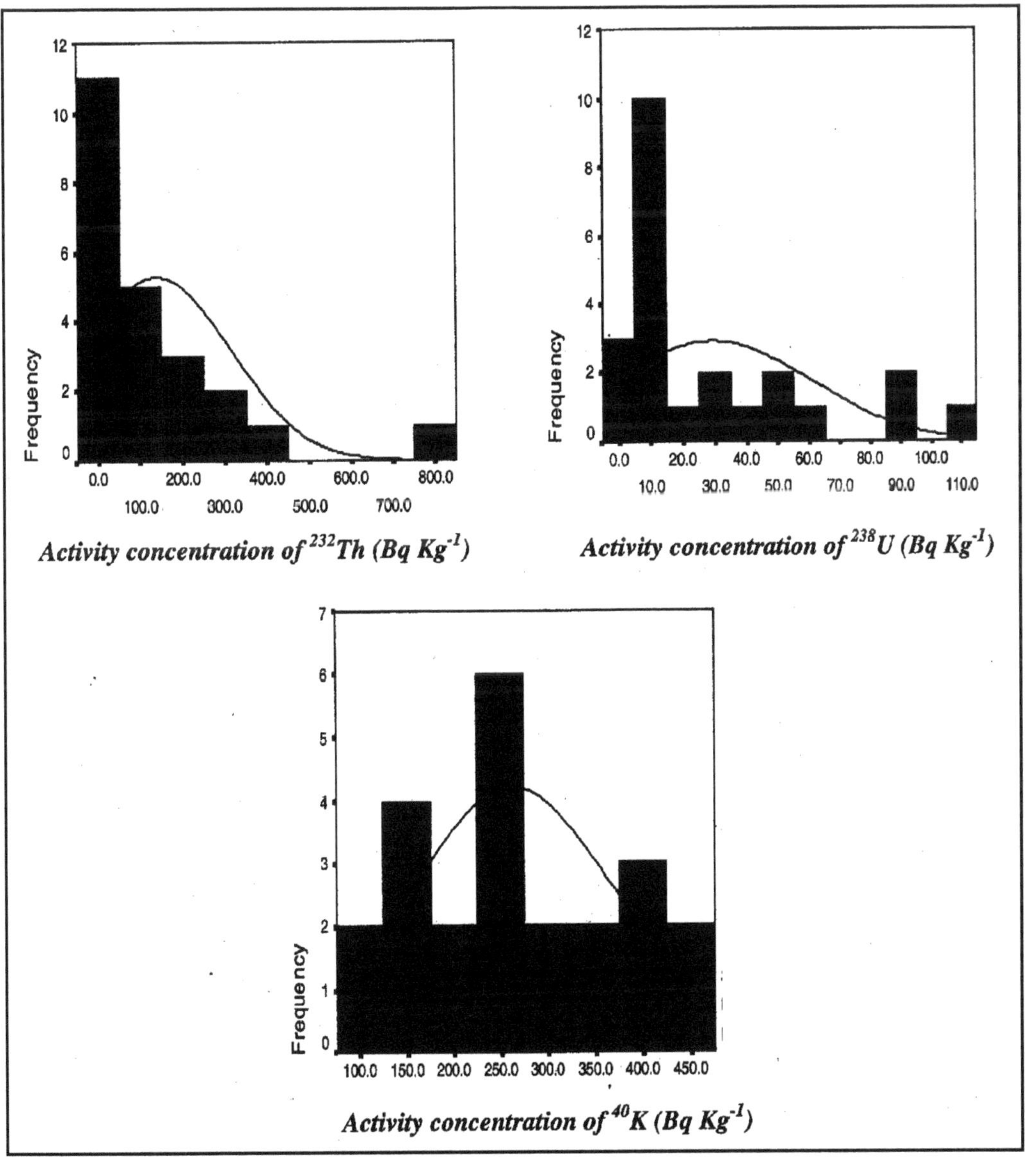

Figure 59.3: Frequency Distribution of Activity Concentration in Bq.kg^{-1}

radionuclides detected in the sediments and their fits plotted considering the skewness and kurtosis coefficients.

Table 59.2: Statistical Data Analysis

Functions	^{40}K	^{238}U	^{232}Th
Mean	157.68	62.31	451.26
Median	163.13	32.34	175.03
Std. Dev.	81.21	102.29	795.62
Skewness	0.28	3.29	3.51
Kurtosis	2.04	12.55	13.98
Freq. Dist.	Normal	Normal	Normal

The positive values of the skewness coefficient indicate that these distributions are asymmetric with right tail longer than left. The very low kurtosis coefficients of ^{40}K suggest that the distribution is close to normal. The positive values of these coefficients for the three radionuclides indicate that the distributions are asymmetrical about the mean and are higher and narrower than would expected for normal distributions. In the correlation between ^{232}Th and ^{238}U, the best fitting relation is of linear type with a correlation coefficient of 0.954, it's illustrated in Figure 59.4.

Conclusion

The analytical results proved that the samples of particular area (S_4) significantly contain three radioactive isotopes (^{232}Th, ^{238}U and ^{40}K) all of which is originated due to the monazite content in this place. This is due to weathering of rocks in Western Ghats and Niligri Hills and also due to the transportation of thorium bearing minerals by long shore current. The activity ratio (Th/U) in the

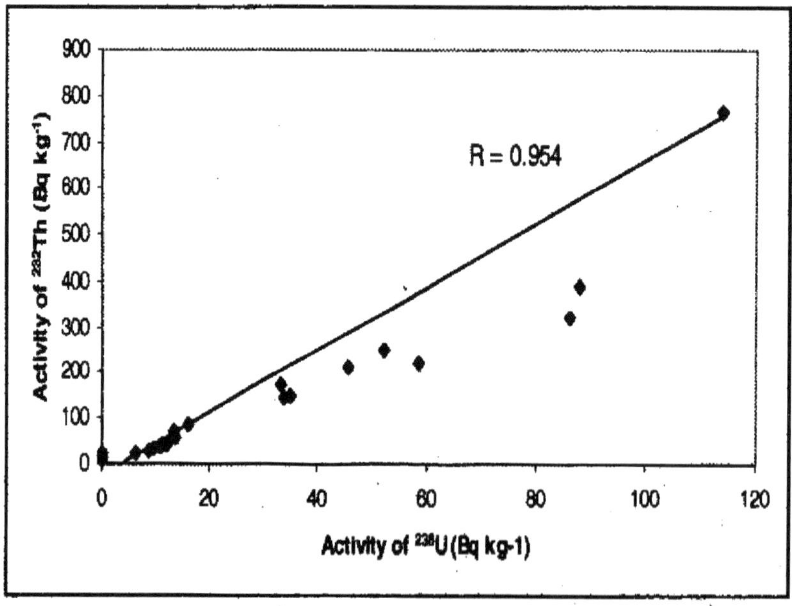

Figure 59.4: Correlation Between ^{238}U and ^{232}Th

present study is greater than unity in all the samples indicate that either a net loss of ^{238}U or ^{230}Th by weathering processes. From the environmental protection point of view, the study area is free from any radiological hazard.

References

Ayres, A. and Theilen, 2001. Natural gamma-ray activity compared to geotechnical and environmental characteristics of near surface marine sediments. *J. App. Geophysics.*, 48: 1.

Balakrishna, K., Shankar, R., Sarin, M.M and Manjunatha, 2001. Distribution of U-Th, nuclides in the riverine and coastal sediments of the tropical South West Coast of India. *J. Env. Radio.*, 57: 21.

Chung, Y. and Chang, W.C., 1996. Uranium and thorium isotopes in marine sediments off Northeastern, Taiwan. *Mar. Geo.*, 133: 89.

IAEA, 1989. Measurements of radionuclides in food and the environment. *Technical Report*, Series No. 295.

Mollah, A.S., Ahmed, G.U., Husain, S.R. and Rahman, M.M., 1968. The natural radioactivity of some building materials used in Bangladesh. *Health Physics*, 50: 849.

Plater, A.J. Ivanovich, M. and Dugdale, R.E., 1992. Uranium series disequilibrium in river sediments and waters; the significance of anamalous activity ratios. *App. Geo.*, 7: 101.

Scott, R., 1968. *Earth and Planet Sci. Letts.*, 4: 245.

United National Scientific Committee on the Effects of Atomic Radiation (UNSCEAR), 2000. *Sources and Risks of Ionizing Radiation*. Report to the general assembly with annexes, New York, United Nations.

Chapter 60

Synthesis and Characterisation of Hydroxyalkylated Dimerised Cardanol-formaldehyde Resins

A. Malar Retna, C.V. Mythili and S. Gopalakrishnan*

Department of Chemistry, Manonmaniam Sundaranar University, Abishekapatti, Tirunelveli – 627 012, Tamil Nadu, India

ABSTRACT

Cardanol the major constituent of CNSL is an excellent source of monomer or starting material for polymer synthesis. It has been dimerised and modified into high ortho multi-nuclear hydroxylated alkylated dimerised cardanol fonnaldehyde resin. The synthesised polyols has been characterized using spectral and physicochemical methods. From the synthesized polyols novel rigid and tough polyurethanes can be synthesized.

Keywords: *Cardanol, Dimerised-cardanol, Dimerised cardanol-formaldehyde resin, Polyol, Polyurethane.*

Introduction

Thermally treated Cashew Nut Shell Liquid contains 86 per cent of cardanol, a meta substituted phenol. One of the unique features of cardanol is its amenability for chemical modification by a variety of means. Cardanol condenses with formaldehyde in the presence of acid catalyst. Dimerisation of

* E-mail: sgkrishrajes@yahoo.co.in.

cardanol have been reported by (Ghatge and Patil, 1980; Trivedi *et al.*, 1989). In the present work dimerised-cardanol was synthesized and treated with formaldehyde in different mole ratios to form novolac resins. The effective utilization of the novolac resin as an active hydrogen nucleophilic compound could be achieved by functionalizing the novo lac resin. The functional group is a hydroxyalkyl group. Hence functionalisation could be carried out by epoxidation followed by hydrolysis of the dimerised cardanol formaldehyde resin. The functionalised resin act as nucleophilic agent. Hydroxyl group containing compounds are the most important nucleophilic reagents for reaction with isocyanates (O. Baeyer, 1947). Hence, an active hydrogen group compound reacts easily with a diisocyanate to form a stable product. Therefore the hydroxyl-alkylated dimerised cardanol-formaldehyde resin synthesized from the novolac resin can effectively be used in the synthesis of polyurethane.

Materials and Methods

Cardanol was obtained from M/s Rishab Resin and chemicals Ltd, Hyderabad. Formaldehyde (40 per cent solution) and methanol was received from M/s BDH (India) Ltd., Adipic acid was received from M/s S.D. fine (India) Ltd, and boron trifloride ethyletherate and epichlorohydrin were received from M/s E. Merck (Germany). The chemicals were used as received.

Synthesis of Dimerised-cardanol

100 g of cardanol was dimerised using 1 gm of borontrifloride ethyletherate catalyst in 500 ml of carbon tetra chloride. The reaction mixture was heated to 50±2°C in a three necked flask fitted with a stirrer and a thermometer for about 1.5 hours. The excess solvent was removed under vacuum.

Synthesis of Dimerised-cardanol Formaldehyde Resin

Dimerised cardanol (I) was treated with formaldehyde in the ratios 1 : 0.9, 1 : 0.8 and 1 : 0.6 in the presence of adipic acid catalyst which was dissolved in methanol by slight warming and refluxed in a three necked flask at 75±5°C for about 5 hours. The reaction resulted in the formation of high ortho multinuclear dimerised cardanol formaldehyde resins [DCRI (II), DCR2 (III), DCR3(IV)]. The initial pH of the reaction mixture was 4 which lowered to 2 on the completion of the reaction. The resins were purified by dissolving in toluene and precipitating with distilled water. Major fraction of the resin was collected and dried using rotary evaporator under vacuum and was analysed (Scheme I-III).

Synthesis of Hydroxy-alkylated Dimerised-cardanol Formaldehyde Resins

Dimerised cardanol formaldehyde resin was treated with 7 moles of epichlorohydrin at 70±5°C for about 1.5 hours. The reaction mixture was cooled to 10–15°C and added with 200 ml of 15 per cent alc, NaOH with vigorous stirring. The reaction mixture was heated once again to 75±5°C for about 3.5 hours. Excess sodium chloride was removed by decantation. The epoxy resin was washed repeatedly with distilled water and extracted with ether. The resins were dried under vacuum. The epoxy resins obtained from dimerised cardanol formaldehyde resins DCR1, DCR2, DCR3 were DCRIE, DCR2E, DCR3E respectively.

The epoxidised resins were hydrolysed using 2N HCl and heated to 120°C for about one hour. The resins were washed repeatedly with distilled water and dried using rotary evaporator. The hydroxy alkylated dimerised cardanol formaldehyde resins were named DCRIEH (V), DCR2EH (VI), DCR3EH (VII) respectively (Scheme IV).

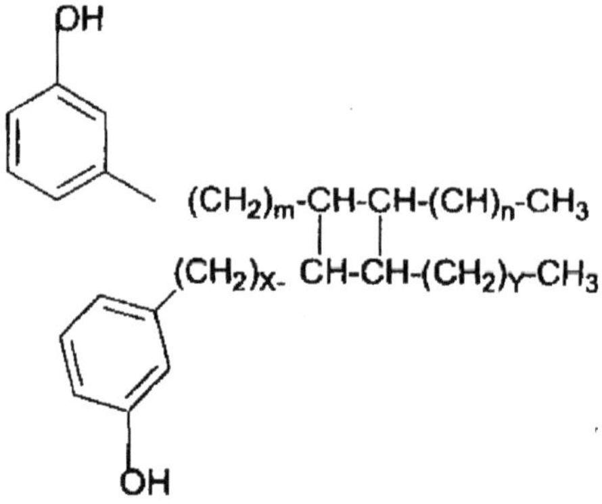

OH

$(CH_2)_m$-CH-CH-$(CH)_n$-CH_3

$(CH_2)_X$- CH-CH-$(CH_2)_Y$-CH_3

OH

(1)

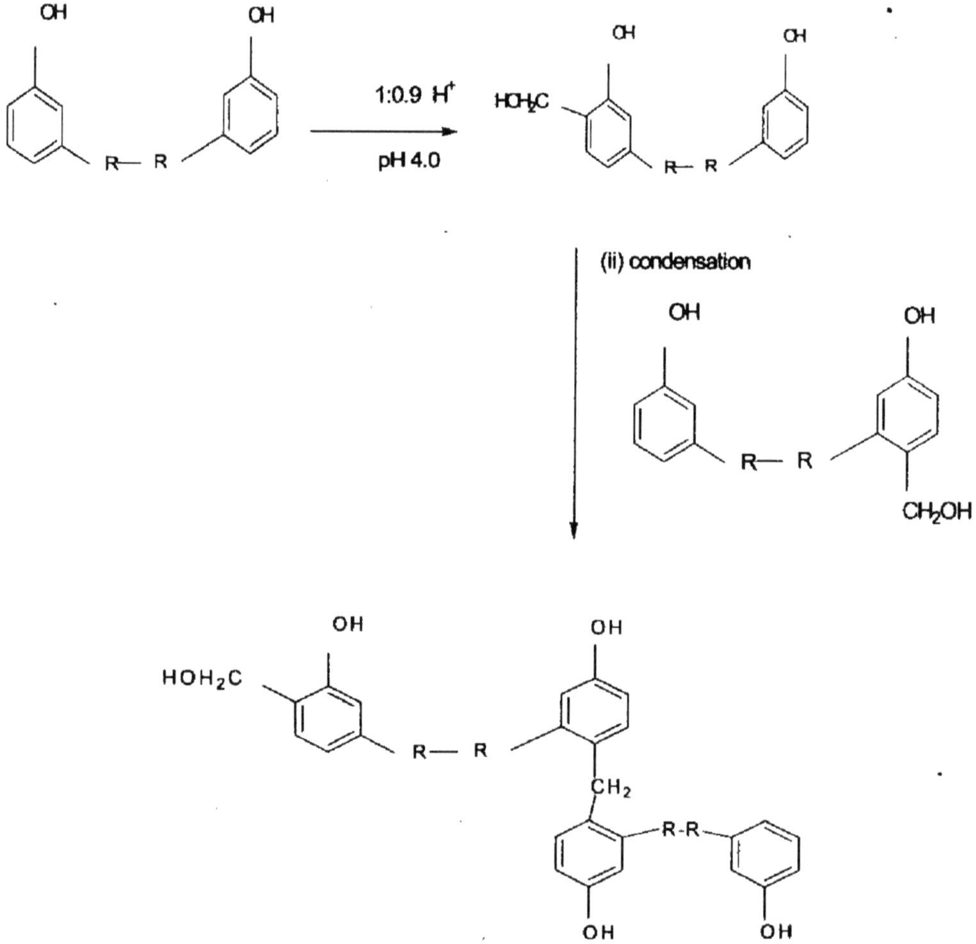

OH OH

R— R

1:0.9 H^+

pH 4.0

OH OH

HOH_2C

R— R

(ii) condensation

OH OH

R— R

CH_2OH

OH OH

HOH_2C

R— R

CH_2

R-R

OH OH

(2)

(Scheme I)

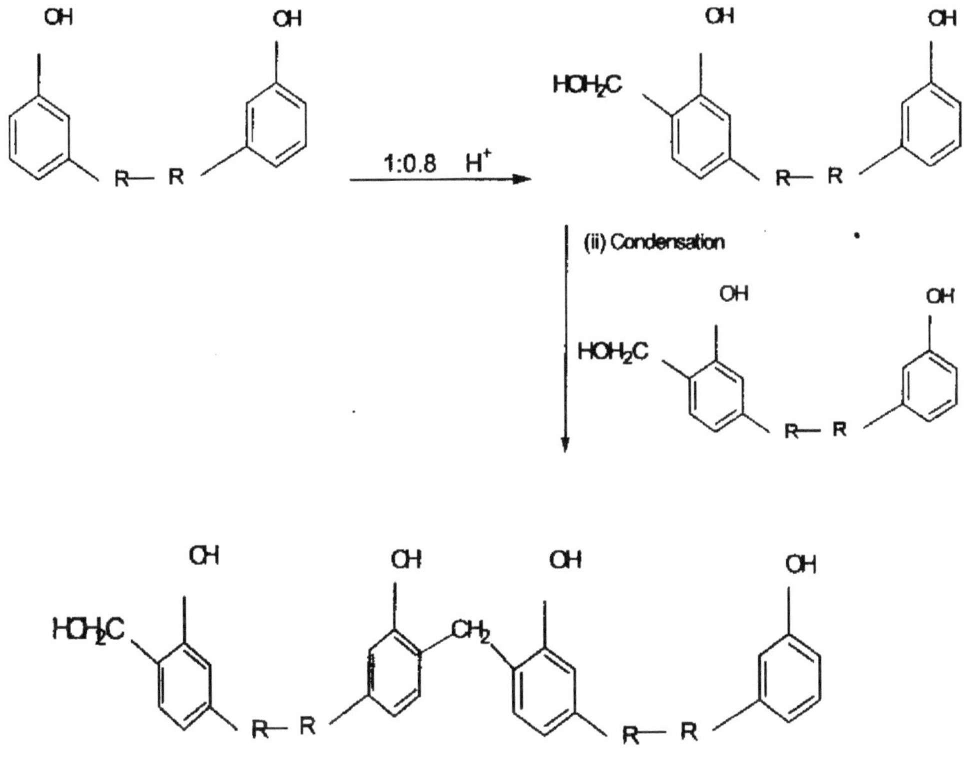

1:0.8 H^+

(ii) Condensation

(3)

SCHEME II

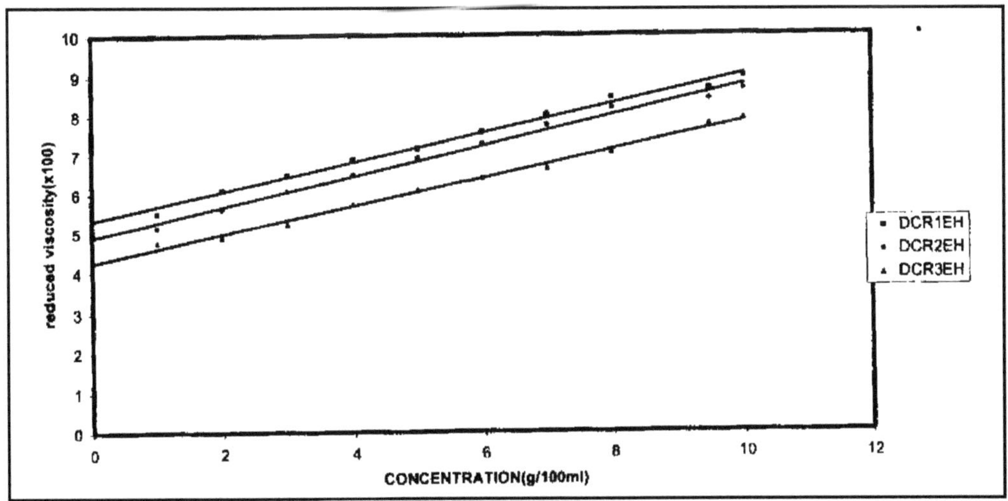

Figure 60.1: Intrinsic Viscosity Curves of Polyols

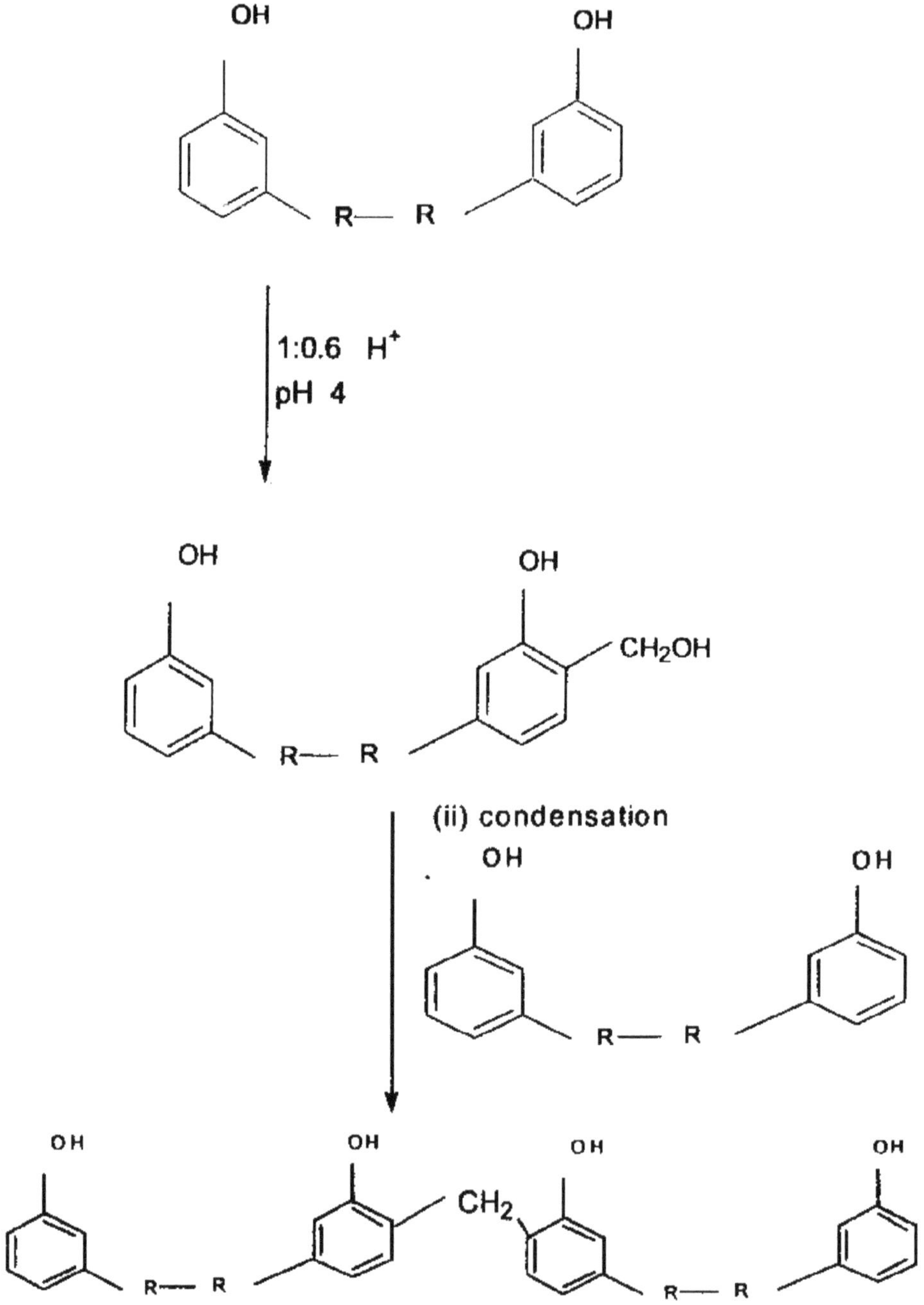

OH
OH
R— R

1:0.6 H^+
pH 4

OH
OH
CH_2OH
R— R

(ii) condensation

OH
OH
R— R

OH
OH
CH_2
OH
OH
R— R
R— R

(4)

Scheme III

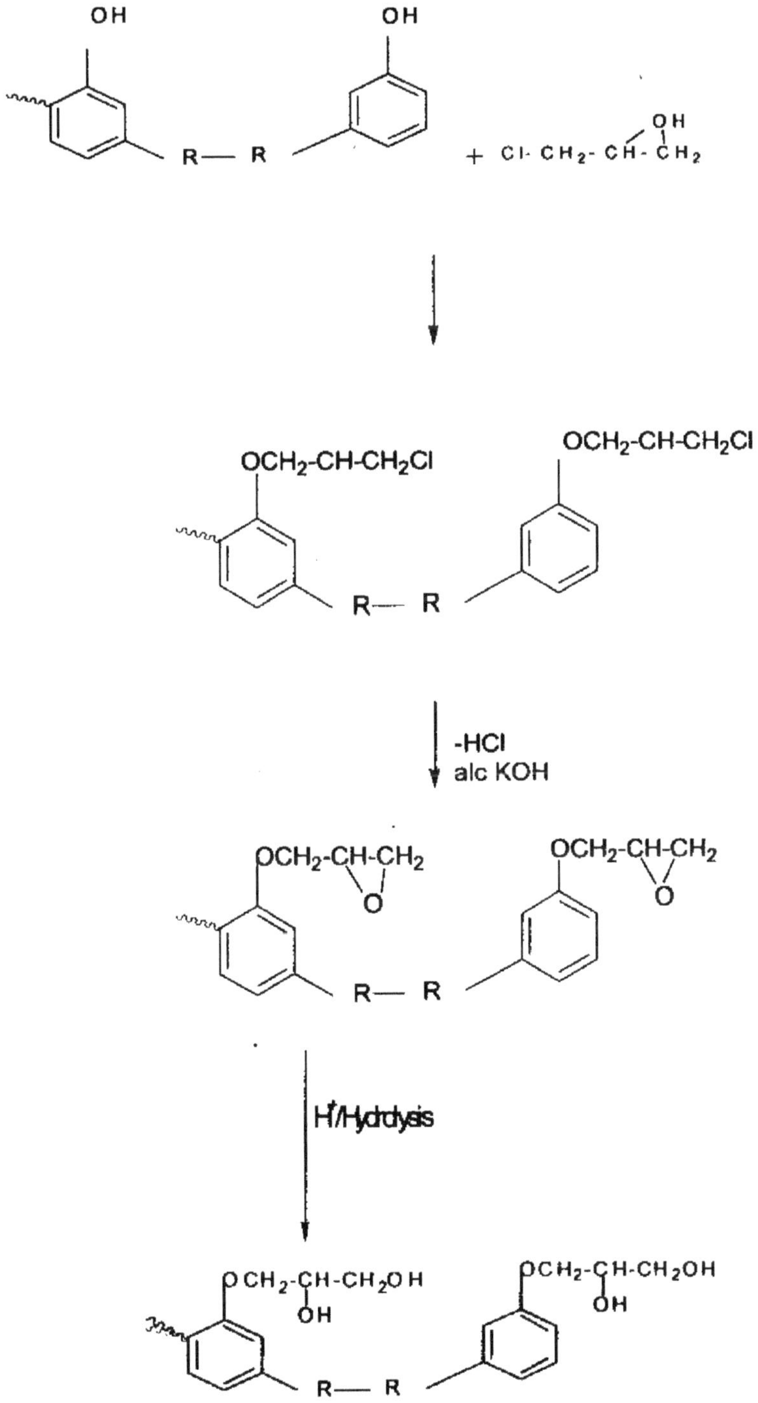

OH OH

R— R + Cl- CH2- CH- CH2 (OH)

OCH2-CH-CH2Cl OCH2-CH-CH2Cl

R— R

-HCl
alc KOH

OCH2-CH-CH2 (O) OCH2-CH-CH2 (O)

R— R

H+/Hydrolysis

OCH2-CH-CH2OH (OH) OCH2-CH-CH2OH (OH)

R— R

SCHEME IV

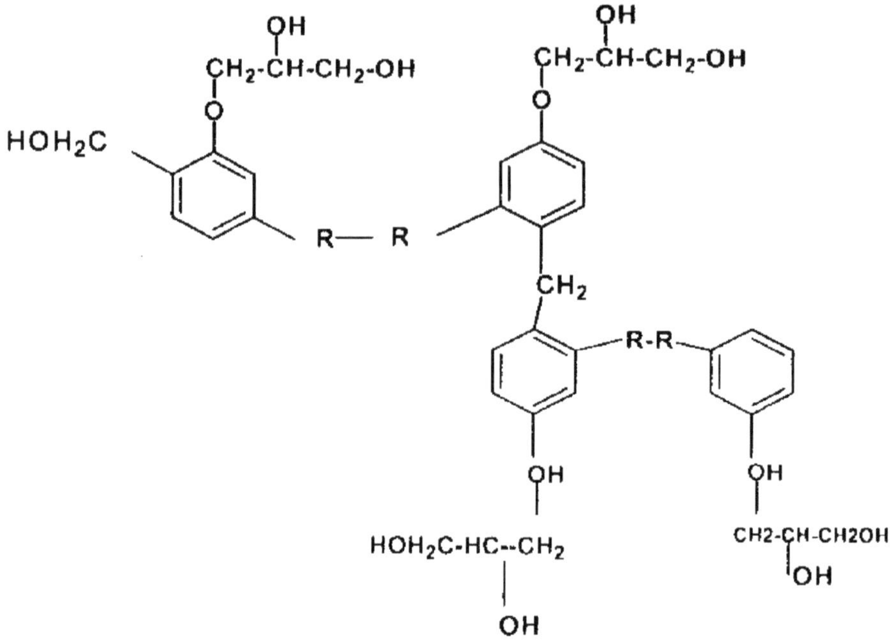
OH
CH2-CH-CH2-OH
O
HOH2C
R— R
CH2
R-R
OH
HOH2C-HC--CH2
OH
CH2-CH-CH2OH
OH

(5)

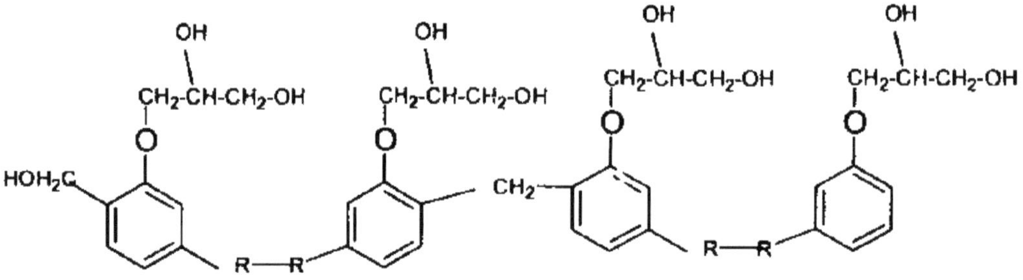
OH
CH2-CH-CH2-OH
O
HOH2C
R—R
CH2
R—R

(6)

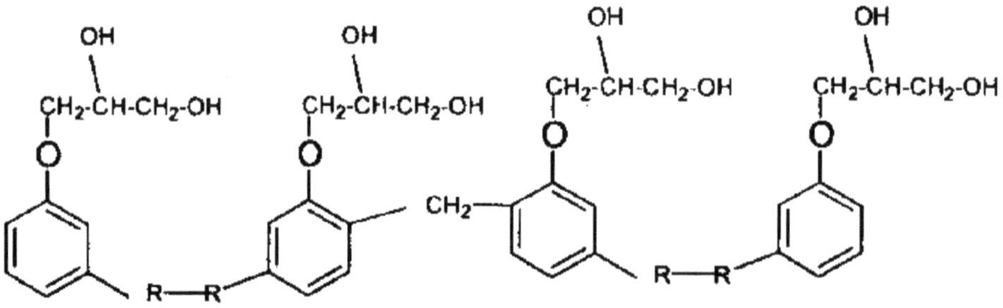
OH
CH2-CH-CH2-OH
O
R—R
CH2
R—R

(7)

Results and Discussion

Thin layer chromatography was studied in three different solvent systems (7 : 3 Pet.ether : diethyl ether, 100 per cent benzene, 1 : 1 Benzene: Chloroform). All the resins showed distinct single spot and different R_f values indicating the purity of the sample and the formation of products (Table 60.1).

Table 60.1: R_f Values of All the Resins

Resin	R_f values		
	Pet. Ether : Diethyl Ether (7 : 3)	Benzene (100 per cent)	Benzene : Chloroform (1 :1)
Cardanol	0.7451	0.8112	0.7839
Dimerised Cadanol	0.7164	0.7826	0.7558
DCR_1	0.6985	0.7550	0.7322
DCR_2	0.6752	0.7635	0.7293
DCR_3	0.6833	0.7568	0.7207
DCR_1E	0.6654	0.7319	0.7142
DCR_2E	0.6433	0.7206	0.7011
DCR_3 E	0.6512	0.715	0.6931
DCR_1EH	0.6235	0.7092	0.6715
DCR_2EH	0.6154	0.6935	0.6843
DCR_3EH	0.6119	0.6861	0.6512

Physico-chemical properties of dimerised-cardanol, dimerised-cardanol, formaldehyde resins, hydroxy-alkylated dimerised cardanol formaldehyde resins are presented in Table 60.2.

Table 60.2: Physico-chemical Properties of the Resins

Sl.No.	Properties	DC	DCR_1	DCR_2	DCR_3	DCR_1EH	DCR_2EH	DCR_3EH
1.	Colour	Brownish black	Brownish black	Brownish black	Brownish black	Brown	Brown	Brown
2.	Odour	Mild phenolic	Mild. phenolic	Mild phenolic	Mild phenolic	Mild phenolic	Mild olic	Mild phenolic
3.	Specific gravity (g/cc) at 30/30°C	0.9332	0.9497	0.9439.	0.9412	0.9642	0.9521	0.9011
4.	Intrinsic viscosity	3.24	3.40	3.5	3.52	5.4	4.9	4.3
5.	Iodine value	135.4	128.1	126.31	124.59	119.14	117.12	115.41
6.	Hydroxyl value (mg KOH/g)	169	197	194	169	269	268	251
7.	Molecular weight	589	1240	1240	1210	1540	1540	1510
8.	Number of hydroxyl groups	2	5	5	4	9	9	8

Both dimerised-cardanol formaldehyde resins and hydroxy-alkylated dimerised cardanol formaldehyde resins have a mild phenolic odour. The iodine values of hydroxy alkylated dimerised cardanol formaldehyde are found to decrease slightly in comparison with that of cardanol. However there is no significant difference in comparison with the parent resins DCR1, DCR2, OCR3. This indicates that there is no change in un saturation at the side chain which may be due to effect of steric hindrance offered by the bulky groups. The specific gravity of the hydroxy-alkylated resins are higher than the respective parent resins. This is due to the intermolecular hydrogen bonding through dihydroxy propyl groups.

The molecular weight was determined by gel permeation chromatography. Hydroxyl value was determined and used for the calculation of number of hydroxyl groups with the molecular weight 1510 indicates the reduced concentration of the hydroxyl alkyl groups in DCR3EH. The intrinsic viscosity is maximum for DCRI EH and minimum for DCR3EH (Figure 60.2). This is in agreement with gel permeation chromatography. The rheological curves of the resins are presented in Figure 60.2. The trend of these curves shows pseudoplastic character. All the resins exhibit sudden decrease of viscosity with increase of spindle speed followed by leveling off at higher speed between 50 and 100 rpm. With increase of spindle speed the secondary bonding forces that hold the aggregates may get dissociated. The leveling off may be due to the buiding up of the broken down structure. Dimerised cardanol shows a peak at 286nm which is attributed to aromatic phenyl ring, whereas dimerised cardanol formaldehyde resins and the polyols shows a peak at 288 nm and 290 nm respectively with increased absorption is due to batho-chromic effect. Knowles 1984. The IR spectrum of dimerised cardanol indicats a new peak at 968 cm^{-1} which is attributed to the cyclobutane ring Marrison (1951). The appearance of a broad band at 1000–1060 cm^{-1} are also due to the cyclobutane ring. The decrease in iodine value of dimerised cardanol when compared to cardanol also indicates the formation of cyclobutane ring. The ^{1}H NMR spectrum of dimerised cardanol indicated the appearance of a small peak at 1.78 adjacent to the main peak at 1.38. The new peak is attributed to the formation of substituted cyclobutane ring in the meta substituted alkyl side chain.

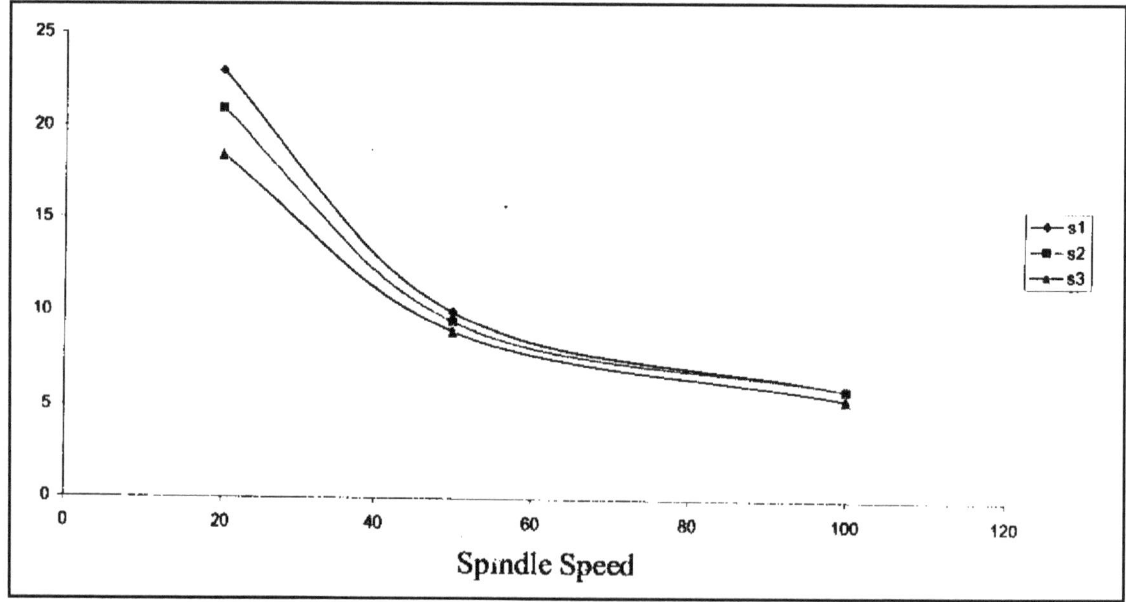

Figure 60.2: Rheological Curves of Polyols

The ^{1}H NMR spectra of DCRI, DCR2, DCR3 show peaks between 6.5 to 7.3 due to aryl protons of benzene nuclei. The peaks around the region 4.9 to 5.2 are due to methylene proton of long alkyl side chain. The strong peak at 1.3 is attributed to the long aliphatic side chain. The ^{1}H NMR spectra of DCRIEH, DCR2EH, DCR3EH shows the appearance of new peaks around 4.2. This is due to methylene protons of dihydroxy propyl groups which are formed in the hydroxyalkylated resins.

The synthesized hydroxyalkylated resins are multifunctional with high molecular weight and nucleophilic character. The resin prepared with higher mole ratio possess high ortho-para linkages and lower ortho-ortho linkages possessing distorted branching while the lower mole ratio resins possess ortk)-ortho linkages possess linear structure. These polyols are one of the best starting materials for the synthesis of novel rigid and tough polyurethanes.

References

Ghatge, N.D. and Patil, D.R., 1980. Epoxy novolac resin based on cardanol and dimerised cardanol. *Indian. J. Tech.*, 18: 203.

Trivedi, M.K., Patni, M.J. and Bindal, L., 1989. Oligomerisation of cardanol. *Indian J. Tech.*, 27: 281.

Baeyer, O., 1947. Synthesis of diisocyanates to manufacture polyurethanes. *Angew Chem.*, 59: 257.

Knowles, A., 1984. *Practical Absorption Spectrometry*, (Eds.) A. Knowles and C. Burgess. Chapman and Hall, London, pp. 10.

Marrison, L.W., 1951. *Journal of Chemical Society*, 1614.

Chapter 61

Moisture Stress and Stomatal Frequency for Preliminary Screening of Mulberry (*Morus alba*)

M. Venkateshwarlu[1], A. Komuraiah[1], K. Sujatha[2] and Ch. Sammaiah[2]
[1]Department of Botany, [2]Department of Zoology (Sericulture Unit), Kakatiya University, Warangal – 506 009, Andhra Pradesh, India

ABSTRACT

The selected genotypes would be evaluated under hot spot condition for getting a sustainable, high yielding, drought resistant mulberry. Ranking on the basis of screening of each genotype separately for individual morpho-anatonical traits showed that three genotypes, No. 50, 4 and 5 were the top scorers with high leaf thickness values and less stomatal size. Preliminary selection of mulberry (*Morus alba*) genotypes for drought tolerance/resistance was done by following a joint scoring technique on multiple parameters like, stomatal frequency, stomatal size, leaf thickness and cuticle thickness. Using this technique a total of twenty one genotypes were selected from a population of forty for further evaluation. Similarly, genotypes with minimum stomatal size, more leaf and cuticle thickness were ranked high. Genotypes exhibiting significant superiority for any three of the four parameters over the check were considered for selection.

Keywords: *Morus alba, Anatomical traits, Multiple parameter's, Moisture stress.*

Introduction

On the other hand, the leaf yield potential of mulberry varieties recommended for rainfed sericulture is still very low (Das *et al.*, 2003). Therefore, a suitable and quick screening technique for preliminary selection of a large gene pool on the basis of drought resistance-tolerance and subsequently evaluating the selected genotypes through multilocational trial is considered for identification of genotype for the regions suffering from chronic water deficit condition. Mulberry (*Morns alba*) is cultivated predominantly as a rainfed crop in India. The recommended high yielding mulberry varieties are however, not performing well under moisture stress condition.

It is fact that stomatal frequency, stomatal size, leaf thickness and cuticle thickness have high correlation with moisture retention capacity of leaves (Susheelamma *et al.*, 1986). Stomata and cuticle retard water loss from leaves and water retention has been proposed as a test of drought resistance (Bhat *et al.*, 1979). Besides, thick cuticled leaves with small stomata are able to retain more moisture and thus can overcome water deficit period by regulating water loss.

The present study was carried out with the objective to select genotypes tolerant/resistant to moisture stress condition based on morpho-anatomical traits and to make decision jointly by following joint score technique for further field trial experiment (Arunachalam *et al.*, 1984; Kebede *et al.*, 1994). An attempt had earlier been made in mulberry to evaluate genotypes on the basis of survival and growth parameters in nursery following this technique. In fact, preliminary selection of genotypes by joint score technique based on morpho-anatonical traits is still lacking in mulberry.

Materials and Methods

The process was repeated three times in each case. Stomatal frequency (mm^{-2}) was determined by using πr^2. The average length and breadth of guard cells was determined on 20 measurements for all the three leaves and repeated five times for final consideration. Mulberry, the only food plant of silkworm (*Bombyx mori*) is a perennial crop and usually multiplied through cuttings. Cuttings survival and leaf yield potential are the most important parameters to be looked into for selection of mulberry genotypes towards mulberry crop improvement programme for commercial exploitation. Subsequently a thin film of quick fix was applied on the lower surface in the central portion of the leaf blade. For stomatal studies 5th, 6th and 7th leaves from 60 days old shots of 11 the genotypes were collected at 9 AM in a beaker containing water. Stomatal impressions were peeled out and the frequency was determined. Twenty microscopic field observations were recorded for each genotype.

Standard procedures were followed for the measurements of leaf and cuticle thickness. Thickness of leaf and cuticle was measured in μm. In all the cases ten readings were taken five times and the mean was calculated.

The process was repeated till all the values of treatments were covered. The groups were numbered as 1, 2, 3......... so on. The score for each genotype was calculated by adding the group values in which genotypes falls and divided by the value obtained by multiplying the number of groups and last group in this calculation, the topmost treatment with the highest value will get the least score. The treatment with the least value will get the highest score. The scores obtained by a genotype for various parameters were added to get the total score. Based on total score, the genotypes were ranked.

Results and Discussion

Stomatal frequency, stomatal size, leaf thickness and cuticle thickness were considered for scoring the genotypes. Significant statistical difference among the genotypes for all the parameters indicated the presence of variability. Stomatal frequency was found maximum in genotype 17 (425 No. mm^{-2}) followed by 11, 21 and 2 in that order. Minimum stomatal size was recorded in genotype 5 (115 μm^2) followed by 37. For this parameter 35 genotypes showed significant difference. Parameter 35 genotypes showed significant difference. Genotype 19 recorded maximum leaf thickness (242 μm) followed by

34, 40, 4 and 39 respectively. Leaf thickness of check genotype was recorded as 124 μm and 35 genotypes were significantly superior over the check for this parameter. Genotype 4 and 22, 20 29, 5, 1, 8, 18 and 40 exhibited statistical significance for cuticle thickness over the check (Table 61.1).

Table 61.1: Ranking of the Mulberry Genotypes

Genotype	*Score Value*				*Total Score*	*Overall Rank*
	Stomatal Frequency	*Stomatal Size*	*Leaf Thickness*	*Cuticle Thickness*		
1.	0.2333	0.3125	0.5789	0.4167	0.3854	5*
2.	0.9667	0.2500	0.7368	1.0000	0.7384	38*
3.	0.5667	1.0000	0.7632	1.0000	0.8325	41
4.	0.5667	0.4375	0.1579	0.1667	0.3322	3*
5.	0.2333	0.1250	0.5263	0.4167	0.3253	2*
6.	0.1333	0.2500	0.8158	1.0000	0.5498	18
7.	0.9000	0.2500	0.3158	0.7500	0.5539	19*
8.	0.6000	0.2500	0.3947	0.4167	0.4154	6*
9.	0.7333	0.5000	0.8421	0.2500	0.5814	22*
10.	0.7667	0.5000	0.7895	0.5833	0.6599	32*
11.	1.0000	0.2500	0.5526	0.8333	0.6590	31*
12.	0.1667	0.6875	0.5789	0.7500	0.5458	17
13.	0.7333	0.2500	0.4211	0.7500	0.5386	16*
14.	0.2333	0.4375	0.8947	0.9167	0.6206	28
15.	0.6667	0.4375	0.8421	1.0000	0.7366	37*
16.	0.3000	0.5625	0.5263	1.0000	0.5972	24
17.	1.0000	0.4375	0.9211	0.9167	0.8188	40
18.	0.5667	0.4375	0.3421	0.4167	0.4407	8*
19.	0.7000	0.2500	0.0526	1.0000	0.5007	13*
20.	0.9667	0.2500	0.5263	0.2500	0.4982	12*
21.	0.9667	0.2500	0.2632	0.7500	0.5575	20*
22.	0.9667	0.3125	0.4737	0.1667	0.4799	10*
23.	0.3667	0.5000	0.2895	0.8333	0.4974	11
24.	0.2333	0.5625	0.6842	0.9167	0.5992	25
25.	0.2333	0.5000	0.2368	0.8333	0.4509	9
26.	0.5667	0.2500	0.9474	1.0000	0.6910	33*
27.	0.5667	0.5000	0.5789	0.8333	0.6197	27*
28.	0.3667	0.4375	0.9211	0.7500	0.6188	26
29.	0.4667	0.5625	0.7895	0.3333	0.5380	15*
30.	0.5667	0.9375	0.2632	0.8333	0.6502	30
31.	0.2333	0.6875	0.8684	1.0000	0.6973	34

Contd...

Table 61.1–Contd...

Genotype	Score Value				Total Score	Overall Rank
	Stomatal Frequency	Stomatal Size	Leaf Thickness	Cuticle Thickness		
32.	0.2333	0.5000	0.7632	1.0000	0.6241	29
33.	0.1000	0.5625	0.6579	0.9167	0.5593	21
34.	0.2333	0.1875	0.0789	1.0000	0.3749	4
35.	0.2667	0.7500	0.9737	0.8333	0.7059	35
36.	0.0667	0.3750	0.8947	1.0000	0.5841	23
37.	0.2333	0.4375	0.3684	1.0000	0.5098	14
38.	0.8333	0.5625	0.6053	0.9167	0.7294	36*
39.	0.1333	0.4375	0.2105	0.9167	0.4245	7
40.	0.2333	0.3125	0.1316	0.4167	0.2735	1*
Check	0.4000	0.8125	1.0000	0.9167	0.7823	39

*: Genotypes selected.

Giving equal importance to each of the traits individually the joint scoring technique was used to selected genotypes performing better over the check. Genotypes exhibiting significant superiority over the check for at least three of total four parameter were considered for selection. Based on total minimum joint score 21 genotypes were selected. Genotype 4, 8, 9, 18, 20 and 22 exhibited significant difference for all the parameters and were secured higher positions in ranking. It has been observed that every genotype which recorded statistical significance for cuticle thickness. was selected and among them genotype 40, 5 and 4 occupied the 1st, 2nd and 3rd rank respectively. Moreover, genotype 40 and 4 also recorded maximum leaf thickness and genotype 5 showed minimum stomatal size. Similar observation on morpho-anatomocal parameters associated with drought resistance had been reported in mulberry (Heichel, 1971a). On the other hand, genotype 17, with maximum stomatal frequency; leaf thickness and cuticle thickness at par with check genotype and could not be selected. This may be because of the role of stomatal size in moisture retention and not stomatal frequency as indicated by several workers (Heicbel, 1971a).

Conclusion

Thus selection of genotypes was done by following joint score technique giving due weightage for all the parameters. Stomatal frequency, stomatal size, leaf thickness and cuticle thickness have significant relationship with drought resistance and higher water use efficiency. Though there are several statistical methods available to rank the genotype on the basis of individual character, the technique used here is the most appropriate one to make decision jointly and select genotypes based on more than one character at a time without giving preference to any individual trait. It if found effective because of easy and quick identification of genotypes prior to conduct the next step of breeding programme. This screening technique ensures drought tolerance among the selected genotypes under moisture stress condition.

References

Arunachalam, V. and Bandyopadhyay, A., 1984. A method of making decision jointly on a number of dependent characters. *Indian J. Genet.*, 44: 419–424.

Bhat, J.G. and Andal, R., 1979. Variation in foliar anatomy in cotton. *Proc. Indian Acad. Sci. Soc.*, 88: 451–453.

Das, C., Misra, A.K., Sengupta, T. and Das, B.K., 2003. Mulberry suits to drought prone red laterite soil of West Bengal. *Indian Farming*, 52: 12–14.

Heichel, G.H., 1971a. Genetic control of epidermal cell and stomatal frequency in maize. *Crop Sci.*, 11: 830–832.

Kebede, H., Martin, B., Nienhuis, J. and King, G., 1994. Leaf anatomy of two *Lycopersicon species* with contrasting gas exchange properties. *Crop Sci.*, 34: 108–113.

Susheelamma, B.N. and Jolly, M.S., 1986. Evaluation of morpho-physiological parameters associated with drought resistance in mulberry. *Indian J. Seric.*, 25: 6–14.

Chapter 62

Role of Mercury in the Food-Chain of Uppanar Estuary, Cuddalore, Southeast Coast of India

R. Rajaram[1*], M. Srinivasan[1] and P. Martin[2]**
[1]CAS in Marine Biology, Annamalai University, Parangipettai – 608 502, Tamil Nadu, India
[2]Department of Chemistry, TBML College, Poriyar

ABSTRACT

Estuarine ecosystems are highly complex, dynamic and subject to many internal and external relationship that are subject to change over time. Uppanar estuary is considered to be one of the highly polluted estuaries in south east coast of India due to industrialization. SIPCOT (Small Industrial Promotion Corporation of Tamil Nadu covering an area of about 520 acres with 44 industries) is located on the bank of Uppanar estuary at Cuddalore. It was established for chemical. petrochemical, pharmaceutical, biocides, fertilizer, fungicides, chlor-alkali and metal processing industries etc. Indiscriminate discharges of waste water from SIPCOT industrial complex into coastal environment affect both biotic and abiotic system and finally cause some ill effects to human beings through food-chain. A detailed study was made on the bioaccumulation of mercury in the food-chain and the results were discussed in this article.

***Keywords:** Mercury, Food-chian, Uppanar estuary.*

---

* E-mail: drrajaram69@rediffmail.com; ** E-mail: sarahprasath@yahoo.com.

Introduction

Coastal zone comprises varied biotopes such as estuaries, backwaters, mangroves, salt marshes, coral reefs, lagoons and near shores. Estuaries are highly protective, sensitive, feeding and breeding ground for fm and shell fishes with rich biodiversity. Realizing the importance of the estuaries, the ever exploding human population exploits not only the biological resources hut also interferes and modifies the basic coastal processes. This results in many environmental episodes such as "Minmatta" and "Itai Itai" diseases.

The coastal zone receives the pollutants principally from three major routes–*viz.*, atmosphere, riverine and glaciers. Of course man also serves as a "Geological Agent" by the way of discharging the effluents through piped outfalls, direct pimping, operation of ships etc. When the pollutants exceed the threshold limits coupled with environmental variables such as salinity, temperature, pH, DO, H_2S etc. give stress to organisms of the environment. So monitoring of estuarine environment needs be carried out periodically to estimate the concentration and distribution of varies pollutant levels to observe the changes in the ecosystem. In southeast coast of India, Uppanar estuary is one of the highly polluted estuaries with 50 industries occupied in 520 acres constitute to form a SIPCOT industrial complex which includes chemical, petro-chemical, pharmaceutical, metal processing, pesticides, fungicides, paints, chlor-alkali paper and pulp manufacturing industries etc. Discharges of waste water from SIPCOT industrial complex estimated to be 56.92 gallons in lakhs/day into the coastal and estuarine environment. The effluents contains the pollutants like carbonates, bicarbonates. oxides of metals, pesticides, hydrocarbon and heavy metals especially more amount of total mercury. So the present studies were, therefore, initiated to understand the mercury distribution pattern in Uppanar estuarine food-chain with regard to effluents from SIPCOT industries.

Sources of Mercury in the Environment

Natural Occurrences

2700–6000 tonnes/year due to earth's crust, emission from volcanoes and evaporation from natural water bodies (Food and Drug Administration).

Anthropogenic Sources

3000 tonnes are released annually into the atmosphere by human activities (Judith, 2000). About 180 tonnes of mercury and its compounds are introduced into Indian coastal environment per annum (Patel and Chandy, 1988) by the following industries:

1. Pharmaceutical and cosmetic industries
    (*a*) Phenyl mercury acetate used in contraceptive vaginal jellies
    (*b*) Various organo-mercury compounds used as diuretics
    (*c*) Organo-mercurial are used as antiseptic products
2. Electrical apparatus industries

    Mercury was used in the manufacture of mercury batteries, mercury pool rectifiers and power tubes, varieties of lamps includes fluorescent, germicidal, photocopying and high intensity arc discharge lamps.
3. Industrial Control Instrument Industries

    It is used in the manufacture of switches, relays, gauges, pump seals and values.

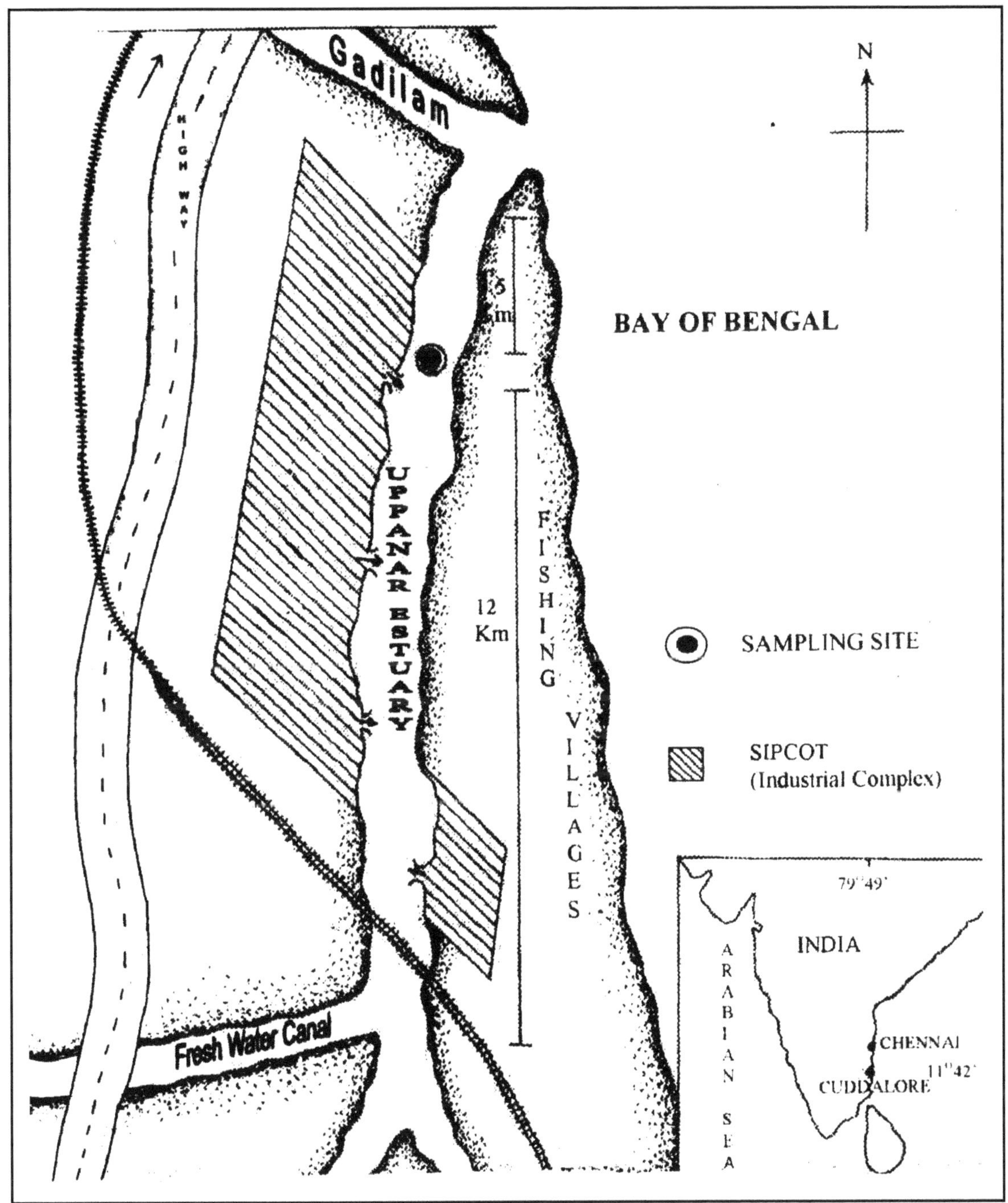

Figure 62.1: Map Showing the Study Area

4. General Laboratory

   Mercury used in experimental equipments such as diffusion pumps, barometers, monometers, thermometers and vibration dampers
5. Chlor-alkali industries uses continuous flow of mercury cathode cell to produce chlorine and caustic soda.
6. Paint Industries
   (*a*) Organo-mercurial compounds: Bactericide-fungicide agents to protect water based paints from bacterial fermentation.
   (*b*) Phenyl mercury derivatives are used as paint preservatives
7. Amalgam alloy is made of silver, tin and copper mixed with mercury to form plastic mass used to fill the tooth cavities
8. Mercury is used to prepare various catalytic salts such as chloride, oxide, sulfate, acetate etc.
9. Various mercury compounds are used as slimicides to control microbial growth in paper and pulp industries.

Materials and Methods

Uppanar estuary (Lat. 11°42′ N; Long. 79°46′ E) is formed by the confluence of Gadilam and Paravanar river, and opens in to the Bay of Bengal near Cuddalore old town on the south east coast of India (Figure 62.1). It forms a potential fishing ground with an annual average landing of about 2000 tonnes. The main water source of the estuary is agricultural drainage and Perumal lake discharge. The tidal effect extends up to a distance of about 6 km and width of the estuary is about 30 m near the mouth and 20 m in the upstream. SIPCOT industrial complex is located on the bank of Uppanar estuary, where the effluents are discharging, mixing and diluting. Apart from the industrial effluents drainage of municipal and domestic sewage from Cuddalore old and new towns and waste from coconut husk retting ground are discharging. The water colour is blackish or brownish ill colour round the year except monsoon months and the bottom soil is characterized by blackish clay in nature. The physico-chemical nature of the environment like salinity varied from 5 to 36 ppt, Surface water temperature ranges from 21.5°C to 34°C, pH showed 7.3 to 8.6. LEC varies from 0.54 to 11.14, dissolved oxygen level varied from 1.34 to 5.2 ml/1, BOD ranges from 1.56 to 6.48 mg^{-1} and nutrients like total phosphorus showed 0.17 to 3.81 µM, inorganic phosphate ranges from 0.34 to 3.69 µM, nitrate and nitrite-level varied from 0.38 to 4.09 µM and 1.03 to 22.29 µM respectively and finally the concentration of reactive silicate showed 7.14 to 140.31 µM (Murugan, 1989; Ananthan, 1991; Murugan and Ayyakkannu, 1991 and 1991a; Gandhiyappan, 1999). Apart from this, the samples like estuarine water and effluent, sediment, phytoplankton, zooplankton, bivalve, fish, were collected from the study areas for the period of one year from June 1998 to May 1999 for the analysis of total mercury.

Water and Effluents

Effluent from SIPCOT industrial discharging site taken carefully and surface water samples were collected in precleaned, acid washed polypropylene bottles and the samples were filtered in Millipore filter paper (Pore size 0.45 µ). The samples were preconcentrated with APDC-MIBK extraction procedure (Brooks *et al.*, 1967)

Sediment

Samples were collected in precleaned, acid washed PVC corers and washed with metal free double distilled water and dried in an oven at 40°C (EPA, 1979) for 5 to 6 hours and ground to powder and redrying the samples from which 500 mg was taken and digested with a mixture of 1 ml of Con. H_2SO_4, 5 ml of Con. HNO_3 and 2 ml of $HClO_4$. A few drops of hydrofluoric acid add to achieve complete digestion and filter the sample to make up 25 ml with metal free double distilled water for mercury analysis (Chester and Hughes, 1967).

Phytoplankton and Zooplankton

Phytoplankton samples were collected from surface water by towing plankton net having 0.35 mouth diameter and 48 µm mesh size (300 µm for zooplankton) for half an hour and dried the sample in an oven at 40°C till the constam weight was obtained. A known weight of sample (500 mg) was digested in Con. NHO_3 and H_2O_2. After centrifugation to remove silica fractions, the, solutions were diluted to 25 ml and analyzed in standard mercury analyzer (Knauer and Martin, 1973).

Bivalve and Fish

The bivalve *Meretrix casta* and fish *Mugil cephalus* were freshly collected from mouth of Uppanar estuary. Stainless steel scalpel steel was used to remove foot and muscle from bivalve and gill, liver, kidney and tissue from fish and the dissected portions were dried in hot air oven at 40°C. The dried samples were powdered and digested in 3 : 1 ratio of HNO_3 and $HclO_4$. The digested samples were diluted to make up 25 ml and stored for mercury analysis (Topping, 1973).

Total mercury was quantified by adopting cold vapour technique in mercury analyzer, model No. 5800D, Serial No. 209, fabricated by Electronic Corporation of India, Bangalore.

Results

Effluents and Water

Monthly variations in the distribution of mercury in effluent during June 1998 to May 1999 were varied from 0.65 to 10.87 µg l^{-1} and in water N.D (Non-detectable) to 1.005 µg l^{-1}. Annual mean concentration in effluents and water were 4.26 µg l^{-1} and 0.55 µg l^{-1} respectively (Figure 62.2).

Sediment, Phytoplankton and Zooplankton

Concentration of mercury in sediment was ranged from 0.79 to 6.71 µg l^{-1} and the annual mean

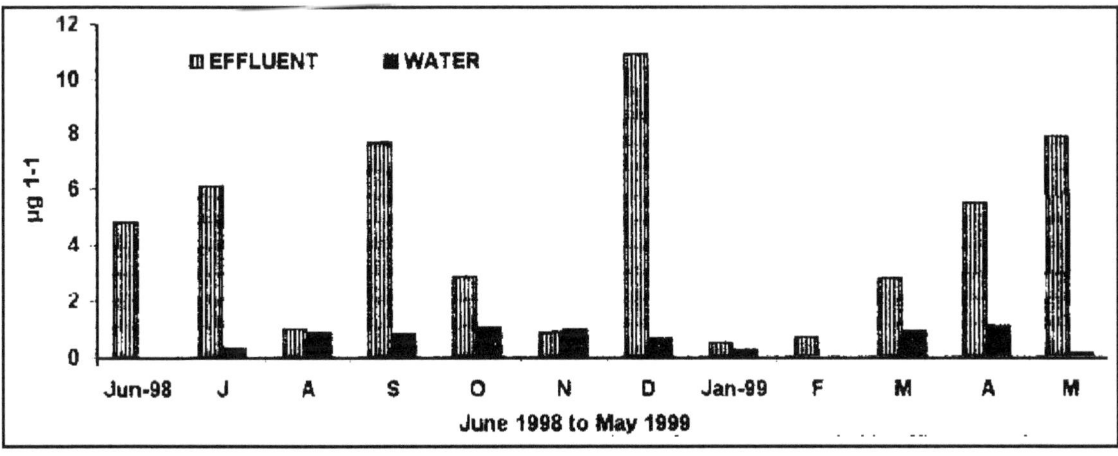

Figure 62.2: Monthly Variations in the Concentrations of Mercury Recorded in Effluents and Water Samples During June 1998 to May 1999

concentration was 2.98 µg l⁻¹. In phytoplankton, it ranges from 0.74 to 6.37 µg l⁻¹ and the actual mean concentration was 2.94 µg l⁻¹. The bioaccumulation of mercury in zooplankton varied from 1.49 to 8.37 µg l⁻¹ and the annual mean concentration was 3.77 µg l⁻¹ (Figure 62.3).

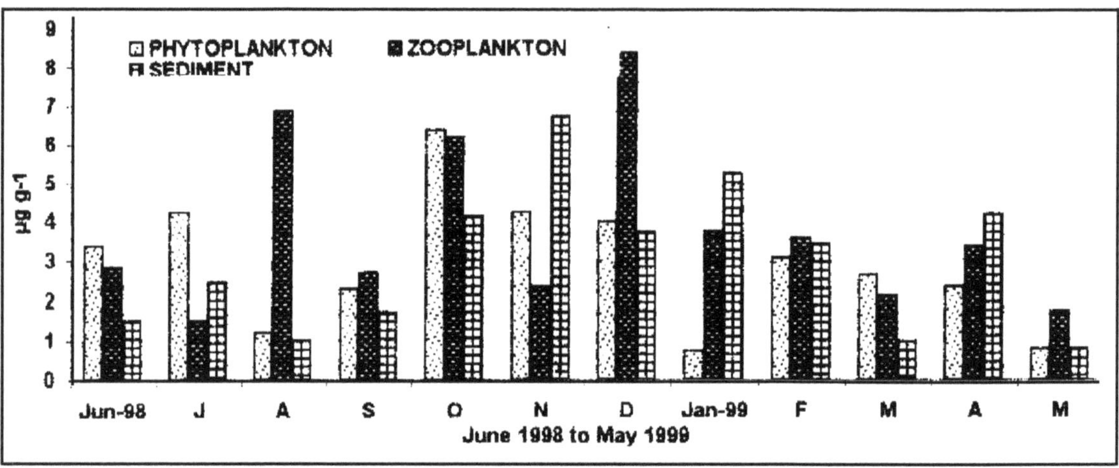

Figure 62.3: Monthly Variations in the Concentrations of Mercury Recorded in Phytoplankton and Zooplankton During June 1998 to May 1999

Bivalve *Merestrix casta*

Concentration of mercury in foot portion of clam showed 0.02 to 5.43 µg l⁻¹ and muscle was ranged between 0.1 to 5.48 µg l⁻¹ (Figure 62.4). Annual mean concentration of mercury in foot and muscle were 2.02 µg l⁻¹ respectively (Figure 62.4).

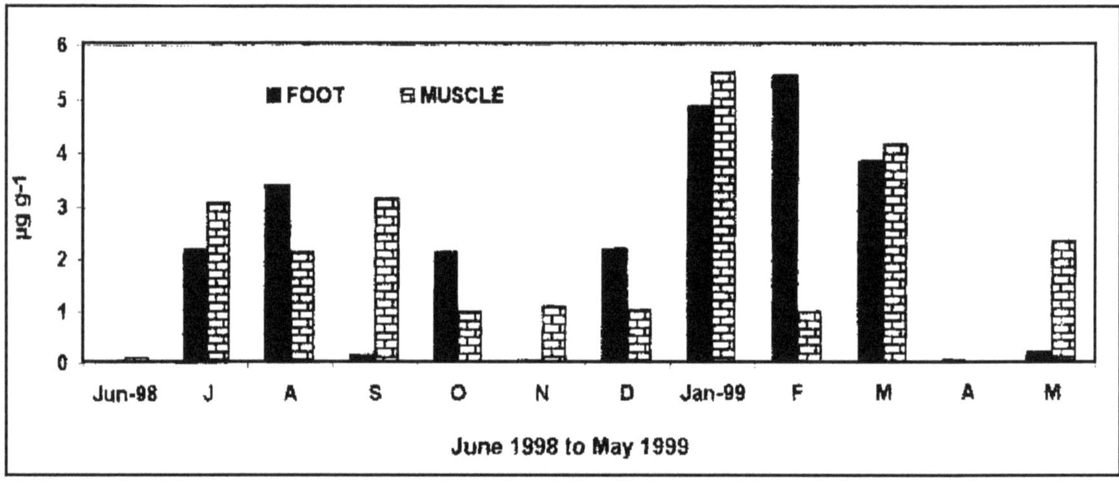

Figure 62.4: Monthly Variations in the Concentrations of Mercury Recorded in Foot and Muscle Portions of Bivalve *Merestrix casta* During June 1998 to May 1999

Fish *Alugil cephalus*

The concentration of mercury in various organs of fish showed, gill (0.29 to 3.1 μg l^{-1}); liver (0.31 to 2.03 μg l^{-1}); Kidney (0.1 to 1.32 μg l^{-1}); and muscle (N.D to 0.14 μg l^{-1}) during the year June 1998 to May 1999. The annual mean concentration of mercury in gill were (1.04 μg l^{-1}); liver (1.15 μg l^{-1}); kidney (0.7 μg l^{-1}) and muscle (0.09 μg l^{-1}) (Figure 62.5).

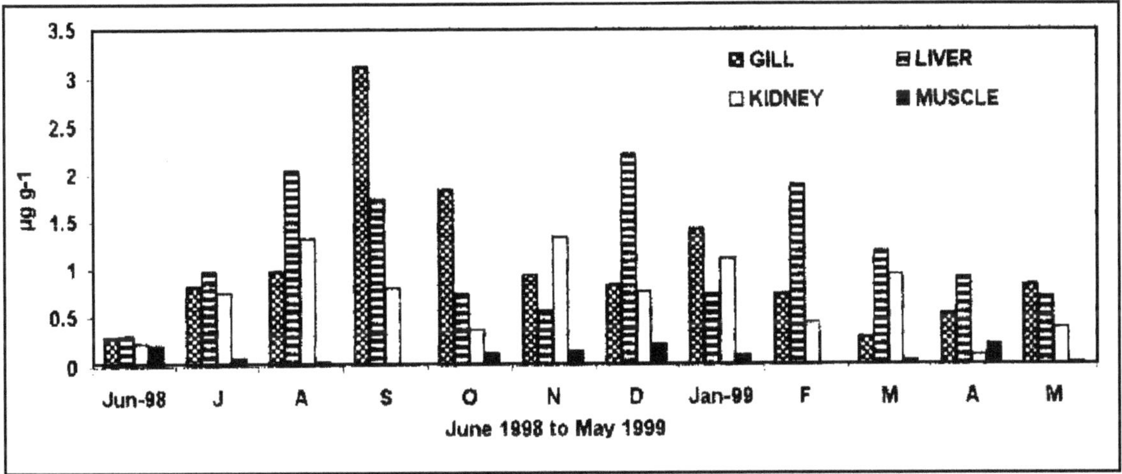

Figure 62.5: Monthly Variations in the Concentrations of Mercury Recorded in Gill, Liver, Kidney and Muscle Portions of Fish *Mugil cephalus* During June 1998 to May 1999

Discussion

Natural sources of mercury in water are through land and river runoff and mechanical and chemical weathering of rocks. The high value of mercury in water due to rainfall causing flooding the stagnant effluent and facilitating higher influence of land derived materials including agricultural runoff and washing of waste coupled with river discharges and direct discharge of effluents. The estuarine sediments may be exchanging part of the exchangeable phase of the metals in water. The high value of mercury in sediment may be due to land drainage, microbial activity, sediment particle type and size and organic content which influence the absorption and accumulation of mercury in sediments.

Many environmental factors influence and affect the uptake of mercury by phytoplankton. This includes physico-chemical. biological and biochemical parameters and also the interaction between the mercury by environment. Phytoplankton species are capable of concentrating large quantities of metals from water. These metals may also he passes on and concentrated at higher levels through the food chain and may pass rapidly back into the water column (Martin, 1970; Fowler, 1977). Phytoplankton should have concentrated more heavy metal by adsorption and absorption mechanisms from the surface waters (Knauer and Martin, 1973; Bryan, 1976). Mercury accumulation by zooplankton is mainly by two pathways *i.e.*, direct uptake from the water through and the assimilation from ingested food and detritus (Davis, 1978). High concentration of mercury in zooplankton is due to land runoff, heavy fresh water flux, industrial effluents and domestic sewage, direct intake of phytoplankton and through water by respiration. Knauer and Martin (1972) stated that zooplankton ingested food (phytoplankton) with high mercury content seems to increase the concentration of mercury in zooplankton. Mainly metal uptake in plankton is in the form of adsorption onto the outside of cells (Flower, 1974).

Accumulation of metals by mussel is influenced by salinity, temperature and concentration of metals in water (Boyden and Romeril, 1974). The size, sex, reproductive condition and season are also factors for the variation. In the present study the accumulation of mercury in *Aferetrix casta* showed N.D. to 5.48 µg gland more amount of mercury accumulation due to filter feeding habit (Paul Raj, 1987).

Fishes are known to concentrate metals in their body tissues in varying proportions depending upon the species, environmental conditions and feeding habits and accumulate high values then their basal environment due to the process of bioamplification (Cooper, 1983). The bioaccumulation of mercury in fish *Mugil cephalus* showed 3.32 µg g^{-1} may be due to behavior of the animal with respect to their environment *i.e.* muddy bottom feeder. The effluent containing organic wastes, exudates, dead and decayed matters are termed as detritus and it is not lost to the ecosystem, rather it serves as the source of mercury flow through detrivores like *Mugil cephalus.* The high level of mercury accumulation (3.32 µg g^{-1}) is mainly due to intake of detritus food absorption by gill and intestine of skin. The excreta of materials present in the bottom sediments are the route of recycling of mercury. Hence, *Mugil cephalus* forms an important item of food for man and it can become in indirect source of mercury entering the body leading to toxic effects. The mercury accumulation in different food-chain organisms was found to be in the decreasing order from the source to water (Effluent > Zooplankton > Phytoplankton > Sediment > Bivalve > Fish). In the present study, the annual mean concentration of mercury in edible part of the muscle in fish showed 0.09 µg g^{-1}, which is lesser then the maximum permissible limit of mercury for human consumption prescribed as 0.5 ppm by USSR, 0.7 ppm by Italy and 0.4 ppm by Japan (FAO, 1983). In the present study the bioaccumulation of mercury found to be safer level but due course the level may increase and cause several problems to the environment. Already there were reports on flourosis in the rural areas near by the industries. Further groundwater turns useless for drinking and bathing, smoke released by the industries cause bad odour and affects the respiratory system also. During monsoon season, the effluents washed from all the industries and discharge in to the Uppanar and hence mass mortality of finfish and shellfish was noted. Therefore, Government of India should take necessary steps to control the pollutants released from al1 the industries in this area.

Acknowledgement

The authors are thankful to the Director, Centre of Advanced Study in Marme Biology, Annamalai University, Parangipeuai for providing the facilities and encouragements during the study.

References

Ananthan, G., 1991. Hydrobiology of Parangipettai and Cuddalore marine environs with special reference to heavy metal pollution. *M. Phill. Dissertation*, Annamalai University, India, 57 pp.

Boyden, C.R and Romeril, M.G., 1974. A trace metal problem in pond oyster culture. *Mar. Pollut. Bull.*, 5: 74–78.

Brooks, R.R., Presley, R.J. and Kalplan, I.R., 1967. APDC–MIBK extraction system for the determination of trace metals in saline water by atomic absorption spectroscopy. *Talanta*, 14: 809–816.

Bryan, G.W., 1976. Some aspects of heavy metal tolerance in aquatic organisms. In: *Effects of Pollutants on Aquatic Organism*, (Ed.) A.P.M. Lockwood. Cambridge University Press, Cambridge, pp. 7–34.

Chester, R. and Hughes, M.J., 1967. A chemical technique for the separation of Ferro-manganese minerals. carbamate and absorbed trace elements from pelagic sediments. *Chem. Geol.*, 2: 249–263.

Cooper, L.T., 1983. Total mercury in fishes and selected biota in Lohotan reservoir. *Navada. Bull. Environ. Contam. Toxicol.*, 31: 9–17.

Davis, A.G., 1978. Pollution status with marine plankton. Part. II. Heavy metals. *Adv. Mar. Biol.*, 15: 381–508.

FAO, 1983. Compilation of legal limits for hazardous substance in fish and fishery products. pp. 102.

Fowler, S.W., 1974. The effect of organism size on the content of certain trace metals in marine zooplankton. *Rapp. Comm. Int. Mer. Medit.*, 22(9): 145–146.

Gandhiyappan, K., 1999. Diversity of phytoplankton and hydrobiological aspects of Paravanar estuary, Cuddalore coast. *M. Phill. Dissertation*, Annamalai University, India, 48 pp.

Judith, E. Foulke, 2000. Mercury in fish; cause for concern. *Sea Food Export J.*, 31(1): 123–134.

Knauer, G.A. and Martin, J.H., 1973. Seasonal variation of cadmium, manganese, lead and zinc in water and phytoplankton in Monterey Bay, California. *Limnol. Oceanogr.*, 18(4): 597–604.

Martin, J.H., 1970. The possible transport of trace metals via molted copepod exoskeletons. *Limnol. Oceanogr.*, 15: 756–761.

Murugan, A. and Ayyakkannu, K., 1991. Ecology of Uppanar backwater, Cuddalore. I. Physico-chemical parameters. *Mahasagar Bull. Nat. Inst. Oceanogr.*, 24(1): 31–38.

Murugan, A. and Ayyakkannu, K., 1991a. Ecology of Uppanar backwater, Cuddalore. II. Nutrients. *Mahasagar Bull. Nat. Inst. Oceanogr.*, 24(2): 103–108.

Murugan, A., 1989. Ecobiology of Cuddalore–Uppanar backwaters, South east coast of India. *Ph.D. Thesis*, Annamalai University, India, 230 pp.

Patel, B and Chandy, J.P., 1988. Mercury in the biotic and abiotic matrices along Bombay coast. *Indian J. Mar. Sci.*, 17: 56–58.

Paul Raj, S., 1986–1987. *Pollution Research*, 5(2): 39–43 and *J. Env. Biol.*, 8(2): 151–155.

Topping, G., 1973. Heavy metals in fish from Scottish water. *Aquaculture*, 1: 373–377.

Chapter 63

Effect of *Trichosanthes dioica* on Blood Glucose and Plasma Antioxidant Status in Streptozotocin Diabetic Rats

G. Sharmila Banu[1], M. Rajasekara Pandian[1] and G. Kumar[2]
[1]Centre for Biotechnology, Muthayammal College of Arts and Science, Rasipuram, Namakkal – 637 408, Tamil Nadu, India
[2]Department of Biochemistry, Selvamm Arts and Science College, Namakkal – 637 030, Tamil Nadu, India

ABSTRACT

Our preliminary study shows that an oral administration of an raw dcseed fruit extract of *Trichosanthes dioica,* an indigenous heat labile plant popularly used in South India, lowers blood glucose level under normal and glucose load conditions, and in streptozotocin (STZ)–induced diabetes in rats. The study was further undertaken to evaluate the antioxidant potential of *Trichosanthes dioica* in STZ diabetic rats. Oral administration of *T. dioica* fruit extract at doses of 25, 50 ml/kg for 21 days resulted in significant reduction in plasma thiobarbituric acid reactive substances (TBARS), hydroperoxide and ceruloplasmin and a significant elevation in plasma reduced glutathione (GSH), ascorbic acid (vitamin C) and tocopherol (vitamin E). The study indicates that *T. dioica* root extract at doses of 25, 50 ml/kg restored all the antioxidant parameters to near normal value.

Keywords: *Antioxidants, Trichosanthes dioica, Streptozotocin, Diabetes mellitus, TRARS*.

Introduction

In the Indian system of medicine (Ayurvedic and Siddha), many plants are used as antidiabetics. In practice, it is being increasingly recognized to be an alternative approach to modern medicine. The World Health Organization (WHO) also recommended that this practice should be encouraged, especially in countries where access to modern treatment of diabetes mellitus is not adequate (WHO, 1980).

Currently oxidative stress is suggested as mechanisms underlying diabetes and diabetic complications (Halliwell and Gutteridge, 1989), which results from an imbalance between radical generating and radical scavenging systems. In diabetes, protein glycation and glucose autooxidation may generate free radicals, which in turn catalyze lipid peroxidation (Baynes, 1991; Mullarkey *et al.*, 1990). Many recent studies showed elevated lipid peroxidation in diabetic animals (Feillet-Coudray *et al.*, 1999), impaired GSH metabolism (Mc Lennan *et al.*, 1991) and decreased ascorbic acid level (Jennings *et al.*, 1987). The level oflipid peroxidation in cell is controlled by various cellular defense mechanisms consisting of enzymatic and non-enzymatic scavenging systems (Halliwell and Gutteridge, 1989). The efficiency of this defense mechanism is altered in diabetes (Wohaieb and Godin, 1987) suggesting that ineffective scavenging of tree radicals may playa crucial role in determining tissue injury.

Trichosanthes dioica Roxb. (Cucurbitaceae) grows as a vegetable in all over India. It is given to improve appetite and digestion (Chopra *et al.*, 1956). The decoction of *T. dioica* is useful as a valuable alternative tonic and febrifuge given in boils and other skin diseases (Kirtikar and Basu, 1980). The leaf juice is applied to patches of alopecia areata (Narayan Das Prajapati *et al.*, 2003). The root is used as hydragogue cathartic tonic and febrifuge (Kirtikar and Basu, 1995). The fruits used as a remedy for spermatorrhoea and the juice of unripe ftuits, tender shoot used as a cooling and laxative (Kirtikar and Basu, 1975). Fruits and seeds show some prospects in the control of some cancer like conditions and hemagglutinating activity (Asolkar *et al.*, 1992). To our knowledge, no detailed investigations have been carried out to shed light the antioxidant property of *Trichosanthes dioica* fruit extract and this paper reports the antidiabetic effect of *Trichosanthes dioica* fruit extract and examines the effect of the test drug on plasma antioxidants.

Materials and Methods

Animals

Male Wistar albino rats (Weighing 160–200 g) were procured from the Animal house, Muthayammal College of Arts and Science, Rasipuram, Tamil Nadu and maintained under standard environmental conditions (12h light/dark cycles at 25–28°C, 60–80 per cent relative humidity); fed with a standard diet (Hindustan Lever, India) and water ad libitum. All studies were conducted in accordance with the National Institute of Health guide (1985).

Plant Material

The fruit of *Trichosanthes dioica* was collected from Palayapalayam, Namakkal District, Tamil Nadu, India and authenticated by Fr. K.M. Matthew, Director, Rapinat Herbarium, St. Joseph's College, Tiruchirapalli. Voucher specimens were deposited in the (collection number 23644) Herbarium for future references.

Preparation of Plant Extract

Fresh raw deseed fruits of *T. dioica* (1 kg) were peeled, washed, cut into small pieces and homogenized in a warring blender with 2 litres of distilled water. The extraction was carried out in a cold room (20°±1°C) with constant stirring overnight. The homogenate was then squeezed through cheese cloth and centrifuged at 2000 rpm for 10 min at 0–4°C. The supernatant being the *T. dioica* fruit extract (yield 210 per cent w/w) was decanted and kept at 4°C until used.

Induction of Diabetes

Rats were made diabetic by single administration of streptozotocin (60 mg/kg/i. p) obtained from Sigma Chemical Co, (St. Louis, MO, USA) dissolved in 0.1 M-citrate buffer, pH 4.5. Forty-eight hour's later, blood samples were collected and glucose levels determined to confirm the development of diabetes. Only those animals which showed hyperglycemia (blood glucose levels > 240 mg/dl) were used in the experiment.

Experiment Design

Assessment of Hypoglycemic Response of Extract in Normal Rats

Preliminary studies were carried out to determine the time necessary to produce peak hypoglycemic effect after oral administration of the test drugs. For each drug preparation (25, 50 ml/kg), 6 rats were used. The drug was given to animals fasted for 12h. In all cases, fasting blood sugar level was determined before oral administrations of the drug, and after drug in take, blood sugar levels were recorded at half hour intervals for a period of 3h. The peak hypoglycemia occurred 3h after the administration of the test preparation.

Assessment of Plant Extract on Glucose Tolerance

Fasted rats were divided into four groups of six animals each. Group I served as control, receiving only vehicle (2 per cent gum acacia, distilled water) and Group II and III received the test drug at the dose level of 25, 50 ml/kg of the fruit extract of *Trichosanthes dioica* suspended in vehicle solution and administered orally. Group IV received Tolbutamide (250 mg/kg). All the animals were given glucose (10 mg/kg) 30 min after dosing, blood samples were collected from the tail vein just prior to 30, 60, 90 min after the glucose loading, and the blood glucose levels measured.

Assessment of Plant Extract on STZ Diabetic Rats

The diabetic rats were divided into 5 groups of 6 animals each. Group I received vehicle alone, served as control. Group II received STZ (60 mg/kg/i.p) dissolved in 0.1 M citrate buffer. Group III and Group IV received the aqueous extract of fruit (25 ml, 50 ml/kg/p.o) suspended in vehicle followed by single intra-peritonial administration of STZ. Group V received Tolbutamide (250 mg/kg/p.o) followed by single intra-peritonial administration of STZ. Blood samples were collected at 30 m, 60 m and 90 m for the estimation of blood glucose and the experiment was conducted further to study the effect of fruit extract on plasma antioxidant status in STZ diabetic rats.

Biochemical Analysis

After 21 days treatment, blood was collected into heparinized tubes. The plasma was separated by centrifugation and the biochemical parameters were determined by following different methods: Blood glucose level by glucose Oxidase method (Triender, 1969), plasma TBARS content by the method of Nichans and Samuelson (Nichans and Samuelson, 1968), oral glucose tolerance test (OGTT) (Du Vigneaud and Karr, 1985) hydroperoxide (Jiang *et al.*, 1992), ceruloplasmin (Ravin, 1961), reduced glutathione (Beutler and Kelley, 1963), α-tocopherol (Baker *et al.*, 1980) and ascorbic acid (Rao and Kuether, 1943). Percent of glycemic changes was calculated as a time function, by applying the formula–

% Glycemia changes = $G_x - G_o / G_o$, *where*, G_o: initial glycemia values; G_x: Glycemia values at x minutes time interval.

Statistical Analysis

Two-way analysis of variance (ANOVA) with interaction effects was employed for analyzing the initial dose response data (Tables 63.1 and 63.2), and one-way analysis of variance was employed for analyzing the antioxidant status and general parameters (Tables 63.3 and 63.4). Inter-group comparisons were done using Duncan's Multiple Range Test (DMRT) with 95 per cent confidence intervals. The SPSS package was used for analysis.

Table 63.1: Effect of Aqueous Extract of Fruit of *Trichosanthes dioica* L. on Blood Glucose Level in Fasted Control Rats

Group	*Treatment (/kg/p.o)*	*Blood Glucose (mg/dl)*			
		Fasting	*1 h*	*2 h*	*3 h*
I	Control	64.3±3.7	64.1±2.1	63.5±3.2	62.2±2.9
II	*Trichosanthes dioica* 25 ml	64.5±5.7	59.5±2.4[a]	54.4±2.6[a]	48.5±4.1[b]
III	*Trichosanthes dioica* 50 ml	67.1±2.9	58.4±2.6	51.5±3.4	47.3±5.8[b]
IV	Tolbutamide 250 mg	65.4±3.6	60.6±1.4	55.2±2.3	48.8±2.2[c]

Values are given as Mean±SD for groups of six animals each. Values not sharing a common superscript differ significantly at P < 0.05, Duncan's Multiple Range Test (DMRT).

Table 63.2: Effect of Aqueous Extract of Fruit of *Trichosanthes dioica* L. on Oral Glucose (10 g/kg) Tolerance

Group	*Treatment (/kg/p.o)*	*Blood Glucose (mg/dl)*			
		Fasting	*30 m*	*60 m*	*90 m*
I	Control + Glucose–10 g	67.3±1.8	142.5±6.9	131.1±6.4	122.5±7.7
II	*Trichosanthes dioica* 25 ml + glucose–10 g	67.8±1.6	126.2±5.2[a]	106.4±4.6 [a]	95.5±5.1
III	*Trichosanthes dioica* 50 ml + glucose–10 g	65.1±1.9	126.8±2.8 [a]	100.5±5.3 [a]	72.3±4.8 [a]
IV	Tolbutamide 250 mg + glucose–10 g	65.4±1.6	127.6±1.4 [a]	108.2±4.3 [a]	78.8±4.2 [a]

Values are given as Mean±SD for groups of six animals each. Values not sharing a common superscript differ significantly at P < 0.05, Duncan's Multiple Range Test (DMRT).

Results

Hypoglycemic Effect

Administration of *Trichosanthes dioica* fruit extract reduced blood glucose level in normal rats (Table 63.1) and maximum reduction 50 ml/kg was noted 2h after the administration of the extract.

Table 63.3: Effect of Aqueous Extract of Fruit of *Trichosanthes dioica* L. on Blood Glucose, Plasma TBARS, Hydroperoxide, and Ceruloplasmin in Control and STZ Diabetic Rats

Group	Treatment (/kg/p.o)	Blood Glucose		TBARS (nmol/ml)	Hydro Peroxide (nmol/ml)	Cerulo Plasmin (mg/dl)
		Initial	Final			
I	Control (2 per cent gum acacia)	66.3±1.8	70.55±2.1^{a}	2.2±0.2^{a}	1.0±0.1^{a}	22.5±4.2a
II	Diabetic control	250.8±3.6	295.8±9.5^{b}	3.4±0.4^{b}	1.8±0.2^{b}	40.6±4.4^{b}
III	Diabetic + *Trichosanthes dioica* 25 ml	248.1±4.9	160.5±8.9^{c}	2.9±0.3^{c}	1.5±0.2^{b}	31.5±4.3^{c}
IV	Diabetic + *Trichosanthes dioica* 50 ml	252.2±5.6	129.8±10.4^{d}	2.1±0.2^{c}	1.2±0.1^{a}	24.8±4.2^{c}
V	Diabetic+ Tolbutamide 250 mg	255.4±5.6	125.6±9.8^{c}	2.2±0.3^{a}	1.1±0.1^{a}	24.2±4.0^{c}

Values are given as Mean±SD for groups of six animals each. Values not sharing a common superscript differ significantly at $P < 0.05$, Duncan's Multiple Range Test (DMRT).

Table 63.4: Effect of Aqueous Extract of Fruit of *Trichosanthes dioica* L. on Plasma GSH, α-tocopherol and Vitamin C in Control and STZ Diabetic Rats

Group	Treatment (/kg/p.o)	GSH (mg/dl)	α-tocopherol (mg/dl)	Vitamin C (mg/dl)
I	Control (2 per cent gum acacia)	23.5±1.5^{a}	1.0±0.4^{a}	2.0±0.3 ^{a}
II	Diabetic control	14.5±1.8^{b}	0.6±0.2^{b}	1.0±0.2^{b}
III	Diabetic+ *Trichosanthes dioica* 25 ml	18.4±2.1^{c}	0.7±0.2^{c}	1.5±0.4^{b}
IV	Diabetic+ *Trichosanthes dioica* 50 ml	22.4±2.5^{a}	1.0±0.1^{a}	1.8±0.3^{a}
V	Diabetic+ Tolbutamide 250 mg	23.2±2.8^{a}	1.0±0.l^{a}	1.9±0.2^{a}

Values are given as Mean±SD for groups of six animals each. Values not sharing a common superscript differ significantly at $P < 0.05$, Duncan's Multiple Range Test (DMRT).

Effect of Oral Glucose Tolerance Test

The effect of glucose tolerance is presented in Table 63.2. In glucose fed rats (10 g/kg), administration of 25, 50 ml/kg of extract significantly increased the tolerance for glucose and maximum tolerance was at 50 ml/kg after glucose loading (3h after drug dosing).

Effect of STZ Induced Diabetic Rats

The fasting blood glucose level in STZ diabetic rats was 240–260 mg/dl. The initial reduction in blood glucose was observed 2h after the administration of fruit extract (Table 63.3). The plasma thiobarbituric acid reactive substance (TBARS), hydroperoxide and ceruloplasmin levels were significantly higher in diabetic rats as compared to controls. Administration of fruit extract (25, 50 ml/kg) or tolbutamide induced significant reduction in plasma hydroperoxide and ceruloplasmin content as compared to diabetic rats (Table 63.3). The plasma GSH, α-tocopherol and ascorbic acid levels were significantly lower in diabetic rats than the normal controls, administration of extract (25 ml, 50 ml/

kg) or tolbutamide (250 mg/kg) increased the glutathione, α-tocopherol and ascorbic acid levels as compared with diabetic rats (Table 63.4).

Discussion

The aim of the present study was to confirm the hypoglycemic effect and to evaluate the antioxidant potential of *Trichosanthes dioica* in STZ diabetic rats. A clinically used tolbutamide (a sulphonylurea drug) is known to lower the blood glucose level by stimulating β-cells to release insulin (Jafri *et al.*, 2000). Among animal models of diabetes, streptozotocin rat was considered especially as the precious tool to study both pathophysiological mechanisms of diabetes mellitus and hypoglycemic activity of medicinal plants (Kedar and Chakrabani, 1982; Obatomi *et al.*, 1994; El Fiky *et al.*, 1996; Jouad *et al.*, 2000). Streptozotocin was recognized as a toxic agent for p-islets of Langerhans (Gunnarsson, 1975; Agarwal, 1980), and widely used for the induction of diabetes with concomitant insulin deficiency (Anderson *et al.*, 1974; Serrada *et al.*, 1989).

Oral administration of fruit extract significantly reduced blood glucose level in control rats and the test drug improved oral glucose tolerance in control rats. The fruit extract (25 ml, 50 ml/kg/p.o) showed antioxidant effect in STZ induced diabetes in rats and an increase in plasma levels of thiobarbituric acid reactive substance (TBARS) which is an index of lipid peroxidation, and hydroperoxide in the STZ diabetic rats. These results confirmed the possibility that the major function of the extract was the protection of vital tissues including liver, kidney, brain and pancreas, thereby reducing the causation of diabetes. Others reported (Halliwell and Gutteridge, 1994; Griesmacher *et al.*, 1995) an increase in the levels of lipid peroxides and TBARS in plasma, which was thought to be the consequence of increased production and liberation of these components in to the circulation.

The antioxidant defense system is significantly altered in diabetes. Ceruloplasmin is a copper containing oxidase, which serves to transport copper in tissues. Ceruloplasmin has been established as chain breaking antioxidant with a potential to scavenge peroxy radicals (Halliwell and Gutteridge, 1984). The level of ceruloplasmin increased in diabetic rats when compared to control rats, which may facilitate the scavenging action on peroxy radicals (Table 63.3). Vitamin C is an excellent water-soluble antioxidant that primarily scavenges oxygen radicals. Vitamin C has been reported to contribute to up to 24 per cent of the total peroxy radical trapping antioxidant activity (TRAP) (Atanasiu *et al.*, 1998). A decreased level of plasma vitamin C was observed in the diabetic rats. The decreased level could be due to increased utilization of vitamin C in deactivation of the increased levels of reactive oxygen or due to the decrease in the GSH level, since the GSH is required for the recycling of vitamin C (Chatterjee and Nandi, 1991; Infers and Sies, 1988).

Glutathione (GSH) is a metabolic regulator and putative indicator of health. We observed lower level of plasma GSH in STZ diabetic rats. It appeared that increased levels of glucose improved utilization of GSH, as reported earlier (Garg *et al.*, 1996; Mitra *et al.*, 1995; Prakasam *et al.*, 2003).

The most important antioxidant in the cell membrane is α-tocopherol; it interrupts the chain reaction of lipid peroxidation by reacting with lipid peroxy radicals, thus protecting the cell structures against damage (Selvam and Anuradha, 1990; Takenaka *et al.*, 1991). The decreased level of α-tocopherol found in the diabetic rats as compared with control could be due to the increased oxidative stress, which accompanied the decrease in the level of antioxidants (Garg *et al.*, 1996). In conclusion this report confirms the hypoglycemic effect of fruit of *Trichosanthes dioica* in normal and STZ treated rats and offers an antioxidant protection in STZ induced diabetes in rats.

References

Agarwal, M., 1980. Streptozotocin: Mechanism of action. *FEBS Lett.*, 120: 1–3.

Anderson, T., Schein, P.S., Mehermanin, M.G. and Couney, D.A., 1974. Streptozotocin diabetes correlation with extent of depression of pancreatic islets nicotinamide adenine dinucleotide. *J. Clin. Invest.*, 54: 672–677.

Asolkar, L.V., Kakkar, K.K. and Chakre, O.J., 1992. *Second Supplement to Glossary of Indian Medicinal Plants with Active Principles*, Part I (A–K) (1965– 1981), pp. 134–135.

Atanasiu, R.L., Stea, D. and Mateescu, M.A., 1998. Direct evidence of ceruloplasmin antioxidant properties. *Mol. Cell. Biochem.*, 189: 127–135.

Baker, H., Frank, O., De Angelis, B. and Feinhold, S., 1980. Plasma tocopherol in man at various time intervals after ingesting free or acetylated tocopherol. *Nutr. Rep. Int.*, 21: 531–536.

Baynes, J.W., 1991. Role of oxidative stress in the development of complications in diabetes. *Diabetes*, 40: 405–412.

Beutler, E. and Kelley, B.M., 1963. The effect of sodium nitrate on red cell glutathione. *Experientia*, 18: 96–97.

Chatterjee, I.B. and Nandi, A., 1991. Ascorbic acid: a scavenger of oxyradicals. *Indian J. Biochem. Biophy.*, 28: 233–236.

Chopra, R.N., Nayar, S.L. and Chopra, I.C., 1956. *Glossary of Indian Medicinal Plants*. CSIR, New Delhi.

Du Vigneaud, V. and Karr, S., 1985. Carbohydrate utilization and disappearance. *J. Biol. Chem.*, 66: 281–300.

El Fiky, F.K., Abou Karam, M.A. and Mify, E.A., 1996. Effect of *Luffa aegyptiaca* (seeds) and *Carissa edulis* (leaves) extracts on blood glucose level of normal and streptozotocin diabetic rats. J. *Ethnopharmacol.*, 50: 43–47.

Feillet-Coudray, C., Rock, E. and Coudray, C., 1999. Lipid peroxidation and antioxidant status in experimental diabetes. *Clin. Chim. Acta*, 284: 31–34.

Garg, M.C., Ojha, S. and Bansal, O.D., 1996. Antioxidant status of streptozotocin diabetic rats. *Ind. J. Exp. Biol.*, 34: 264–266.

Griesmacher, A., Kinder, H.M. and Andert, S., 1995. Enhanced serum levels of thiobarbituric acid reactive substances in Diabetes mellitus. *Amer. J. Med.*, 10: 331–335.

Gunnarsson, R., 1975. Inhibition of insulin biosynthesis by alloxan, streptoztocin and N-nitrosomethylurea. *Mol. Pharmacol.*, 11: 759–765.

Halliwell, B. and Gutteridge, J.M.C., 1984. Oxygen toxicity: Oxygen radical, transition metal and disease. *Biochem. J.*, 219: 1–14.

Halliwell, B. and Gutteridge, J.M.C., 1989. *Free Radicals in Biology and Medicine*, 2nd edn. Carandon Press Publ., Oxford.

Halliwell, B. and Gutteridge, J.M.C., 1994. Lipid peroxidation oxygen radicals, cell damage and antioxidant therapy. *Lancet.*, 1: 1396–1397.

Infers, H. and Sies, H., 1988. The protection by ascorbate and glutathione against microsomal lipid peroxidation is dependent on vitamin E. *Eur. J. Biochem.*, 174: 353–357.

Jatfi, M.A., Aslam, M., Kalim Javad and Singh, Surendar, 2000. Effect of *Punica granatum* L. (flowers) on blood glucose level in normal and alloxan induced diabetic rats. *J. Ethnopharmacol.*, 70: 309–314.

Jennings, P.R., Chirico, S., Jones, A.F., Lunec, I. and Barnett, A.H., 1987. Vitamin C metabolites and microangiopathy in diabetes mellitus. *Diabetes Res.*, 6: 151–154.

Jiang, Z.Y., Hunt, I.V. and Woiff, S.P., 1992. Detection of lipid peroxides using the Fox reagent. *Annl. Biochem.*, 202: 384–389.

Jouad, H., Eddoouks, M., Lacaille-Dubois, M.A. and Lyoussi, R., 2000. Hypoglycemic effect of the water extract of *Spergularia purpurea* in normal and streptozotocin induced diabetic rats. *J. Ethnopharmacol.*, 71: 169–177.

Kedar, P. and Chakrabani, C.H. 1982. Effect of bitter gourd (*Momordica charantia*) seed and glibenclamide in streptozotocin induced diabetes mellitus. *Indian J. Exp. Biol.*, 20: 232–235.

Kirtikar, K.R. and Basu, R.D., 1975. *Indian Medicinal Plants*, 2nd edn., Vol II. Jayyed Press, Allahabad, pp. 1110–1111.

Kirtikar, K.R. and Basu, R.D., 1980. *Indian Medicinal Plants*. Lalit Mohan Basu Publications, Allahabad.

Kirtikar, K.R. Basu, B.D., 1995. *Indian Medicinal Plants*, Vol. 1. International Book Distributors, Dehradun, India.

Mc Lennan, S.V., Heffernen, S. and Wright, L., 1991. Changes in hepatic glutathione metabolism in diabetes. *Diabetes*, 40: 344–348.

Mitra, S.K., Gopumandhavan, S., Muralidhar, S.D. and Sujatha, M.B., 1995. Effect of D-400, a herbomineral preparation on lipid profile, glycated haemoglobin and glucose tolerance in strptozotocin-induced diabetes in rats. *Ind. J. Exp. Biol.*, 33: 798–800.

Mullarkey, C.J., Edelstein, D. and Brownlee, L., 1990. Free radical generation by early glycation products: A mechanism for accelerated atherogeneisis in diabetes. *Biochem. Biophy. Res. Commun.*, 173: 932–939.

National Institute of Health, 1985. *Guide for the Care and Use of Laboratory Animals*. DREW Publication (NIH), Office of Science and Health Reports, DRRINIH, Bethesda, USA.

Nichans, W.G. and Samuelson, B., 1968. Formation of MDA from phospholipid arachidonate during microsomal lipid peroxidation. *Eur. J. Biochem.*, 6: 126–130.

Obatomi, D.K., Bikomo, E.O. and Temple, V.J., 1994. Anti-diabetic properties of the African Mistletoe in streptozotocin induced diabetic rats. *J. Ethnopharmacol.*, 43: 13–17.

Prajapati, Narayan Das, Purohit, S.S., Sharma, Arun K. and Kumar, Tamn, 2003. *A Handbook of Medicinal Plants*. Agrobios (India), Jothpur, pp. 521.

Prakasam, A., Sethupathy, S. and Pugalendi, K.V., 2003. Effect of *Casearia esculanta* root extract on blood glucose and plasma antioxidant status in streptoztocin diabetic rats. *Pol. J. Pharmacol.*, 55: 43–49.

Rao, J.H. and Kuether, C.A., 1943. Detection of ascorbic acid in whole blood and urine through the 2,4-DNPH derivative of dehydroascorbic acid. *J. Boil. Chem.*, 147: 399– 407.

Ravin, H.A., 1961. An improved colorimetric enzymatic assay for ceruloplasmin. *J. Lab. Clin. Med.*, 58: 161.

Selvam, R. and Anuradha, C.V., 1990. Effect of oral metheonine on blood lipid peroxidation and antioxidants in alloxan-induced diabetic rats. *J. Nutr. Biochem.*, 1: 653–658.

Serrada, P., Giroix, M.H and Portha, B., 1989. Evaluation of the pancreatic beta cell function in the rat after prenatal exposure to streptozotocin or N nitrosomethylurea. *Diabetet Metabolisme*, 15: 30–37.

Takenaka, Y., Miki, M., Yasuda, H. and Mino, M., 1991. The effect of α-tocopherol as an antioxidant on the oxidation of membrane protein thiols induced by free radicals generated in different sites. *Arch. Biochem. Biophy.*, 285: 344–350.

Triender, P., 1969. Determination of glucose using glucose oxidase with an alternative oxygen acceptor. *Ann. Clin. Biochem.*, 6: 24–27.

WHO, 1980. The *WHO Expert Committee on Diabetes Mellitus*. Second Report, Technical Report Series, WHO, Geneva, pp. 646.

Wohaieb, S.A. and Godin, D.V., 1987. Alterations in free radical tissue defense mechanism in Streptozotocin-induced diabetes in rats: Effects of insulin treatment. *Diabetes*, 36: 1014–1018.

Chapter 64

Investigation on Sub-Surface Water Quality of Tarikere Taluk with Special Reference to Physico-Chemical Characteristics

K. Harish Babu and E.T. Puttaiah
Department of P.G. Studies and Research in Environmental Science, Kuvempu University, Shankaraghatta – 577 451, Shimoga Distt., Karnataka

ABSTRACT

The research study was carried out in Tarikere taluk, Karnataka (India). This article mainly address the physico-chemical concentration of 56 groundwater sample during September 2004. The results of all the findings are discussed in details which reflect the present status of the groundwater quality of the study area.

Keywords: *Groundwater quality analysis, Tarikere taluk.*

Introduction

Rural India relies a mainly on groundwater for drinking and agriculture practices. Villages once relied on sources like wells, lakes, ponds and streams for their water needs. Contamination of most surface water sources has rendered them unfit for consumption. Also increase in water demand by an increasing population has necessitated resource to tapping groundwater. Unsustainable withdrawal of groundwater has led to the spectra of depleting the problem of water scarcity.

Every human society, be it rural or urban, industrially or technologically advanced, disposal of waste exceeds the limit of natural scavenging or removal process, they are bound to effect the normal functioning of the ecosystems and consequently they bear an adverse effect on the water resources (Miller, 1984).

Supplying inhabitants with safe and clean drinking water is one of the most common problems in developing countries like India, especially in arid and semi-arid regions. Hazardous pathogenic germs and anthropogenic substances do not only contaminate the available water quality but also geogenic substances is adversely affect the water supply of many regions.

The groundwater of Tarikere taluk had many threats such as anthropogenic activities, quality deterioration by agricultural activities and over exploitation. Persistence of continuous drought condition during 2001–2004 had declined the amount of groundwater in the region. With this background an effort has been made to know the quality of groundwater in this region.

Methodology

Study area

Tarikere taluk is located between 75°45′00″–75°52′30″ E and 13°30′00″–13°47′30″ N and it has a geographical area of 1,222 sq. kms. It comprises 45 Gram Panchayats and 245 villages with a total population of 2,54,093 as per 2001 census. As a whole, the region is a flat plain land with an average elevation of about 590.58m msl (Gazetteer of India, 1981).

Meteorology

The climate of Tarikere taluk is semi-arid and enjoys all the three seasons *viz.*, pre-monsoon (February to May), monsoon (June to September) and post-monsoon (October to January). The monthly mean temperature ranges from 8.9 to 38°C. The relative humidity varies between 50 to 84 per cent and it is highest during the month of August and September and lowest humidity observed during the months of January and February.

Precipitation

The precipitation and distribution of rainfall in Tarikere taluk is highly erratic. The annual average rainfall is 980 mm received over 55 rainy days. It varies from as low as 621 mm in the east and as high as 920 mm in west. About 2/3rd of the geological area of Tarikere taluk receives less than 750 mm of annual rainfall. The study area had been receiving rainfall mainly from southwest monsoon and slightly from northeast monsoon.

Geology

Stratigraphically the study area comes under Bababudan belt of Dharwar super group. Bababudan belt is well known for its iron ore (horse shoe shape). Peninsular gneiss, which forms the basement to the younger green stone belts like Shimoga, Chitradurga and Tarikere (Swaminath and Ramakrishnan, 1981). An important gneissic exposure is near Tarikere that transect the structural trend lines of the schist belts while the gneissic foliation is conformable to the schist belt boundary. The tonalitic gneisses occur in Tarikere taluk (Naqvi, 1983).

Pillow breccias and conglomerates are also seen. Metagabbro with titaniferous magnatite is intersheeted with in the metabasites at north of the Tarikere dome of basement gneisses. Serpentinites and talcose schist form a minor component of the Meta volcanic association in the Bababudan group (Fareeddudin *et al.*, 1988). Geological status of Tarikere taluk were represented in Figure 64.1.

Analysis of the Samples

The groundwater samples from the 56 sampling sites were collected and analyzed during September 2004 and Sampling locations have been shown in Figure 64.2.

Water samples from the sampling sites were collected from the bore wells. Initially the water was allowed to run for 15 minutes in order to flush out stationary water. Further, the sample bottles were also flushed with water before the samples were collected. The parameters of water such as dissolved oxygen; total dissolved solids, electrical conductivity and pH were analyzed on the spot with the help

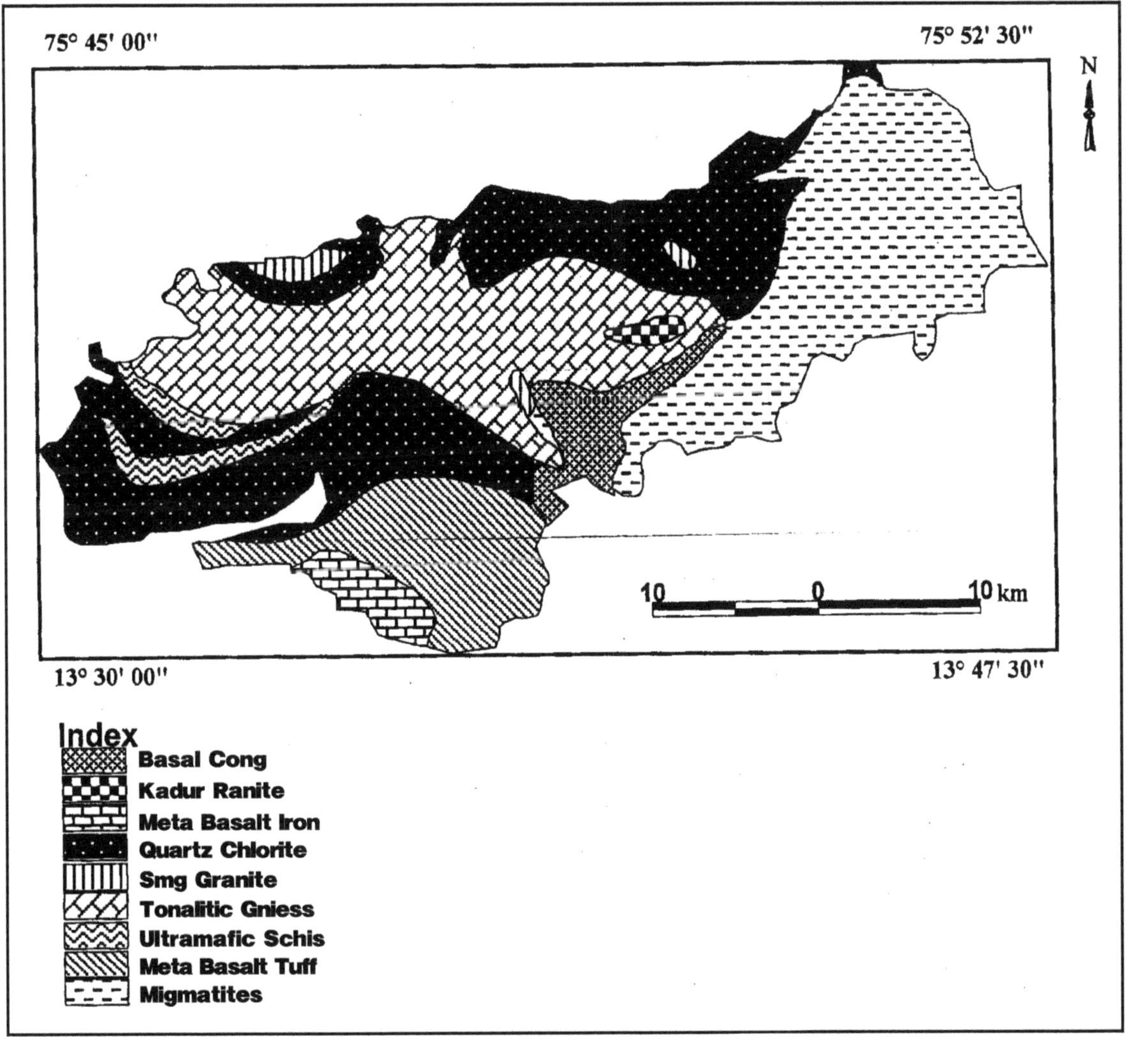

Figure 64.1: The Geological Status of Tarikere Taluk

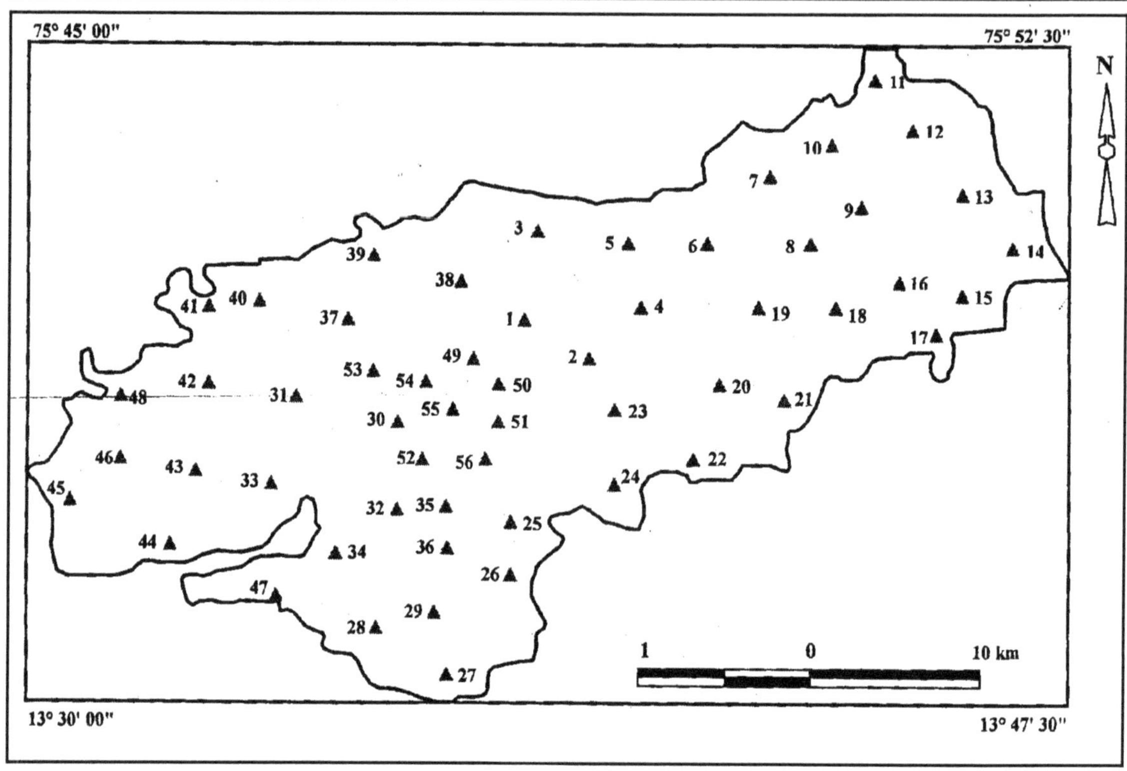

Figure 64.2: Map Showing the Groundwater Locations of the Tarikere Taluk

of water analysis kit (Elico). The remaining parameters were analyzed in the laboratory. Hence, the water was carried to the laboratory in suitable inert bottles. The samples were analyzed using analytical method of APHA (1995) and all analysis was done in triplicate.

Results and Discussion

Physico-chemical Parameters

Water analysis was carried out, by taking 12 parameters, which are very essential to know the water qualities for drinking purpose. The findings of the present investigation are summarized in Table 64.1, and it has been compared with BIS 1998 drinking water standards (Table 64.2) which provides the comprehensive picture of the physico-chemical characteristics of groundwater in the study area.

Turbidity (TUR)

In the present study the turbidity values ranged between 0.15 and 86.1 NTU with a mean value of 9.03. The BIS (1998) acceptable limit for turbidity is 25 NTU. It is proved from the present study, 12.5 per cent of total number of samples cross their permissible limit with reference to the BIS (1998) standards. It is observed from the results that these parameters which crossed their permissible limit are unfit for drinking purposes. It causes health problems like gastro-intestinal disorders, headache and also associated with respiratory diseases (Maiti, 1982; Peter, 1974).

Table 64.1: Water Quality Data of Physico-chemical Parameters of the Study Sites During September 2004

Sam. No.	TUR	pH	EC	TDS	DO	TH	Ca^{2+}	Mg^{2+}	Cl^-	Alk	SO_4^{2-}	F^-
S1	11.5	8.35	1558	975	3.8	1325	327.5	218	678	815.5	170.5	1.8
S2	3.3	8.2	2338	1440	5.7	1775	364	228.5	700	295	135	1.55
S3	6.75	7.95	897.5	560	5.15	560	177.5	34.5	265	162	27.1	1.5
S4	0.15	7.85	1150	750	4.4	970.5	332.5	171	409	190	90.5	1.1
S5	7.6	7.2	1751	1051	5.4	1225	395	205.5	670.5	188.5	120.5	1.4
S6	1.2	6.85	1289	782	4.1	845	371	186.5	350	14.7	101.3	1.6
S7	0.15	7.15	449	276	4.5	350.5	119.5	102.1	75.35	146.5	104	1.25
S8	2.7	7.15	1125	718	5.3	790	130.5	108	281.5	231	125.5	1.4
S9	76.15	7.9	1600	970	5	947.5	164.5	173	515	156.5	170.6	1.55
S10	1.25	8.45	715	458	4	470	130	23.6	650	188.5	78.75	1.15
S11	1.7	7.4	544.5	352	4.7	508	186	27.6	616	120	21.8	1
S12	1.15	8.15	1136	771	4.15	970	191	170.5	451	537.5	126	1
S13	0.75	7.9	1699	1056	6.05	972.5	210.5	104.6	750	532.5	133	1
S14	0.2	7.3	1100	670	4.15	759.5	165	125.5	445	172.5	171	1.1
S15	36.5	8.15	1315	815	4.05	735	234.9	160	425	260	118.5	1.15
S16	41	8.3	633.5	421	4.55	395	165.5	110	330	184.5	143	0.9
S17	69.5	6.7	595	370	4.15	505	68	39.95	565	194	22.7	0.9
S18	2.15	8.7	1125	705	5	800	275	9.45	426.5	615	215.2	0.6
S19	1.3	8.4	955	606	4	695	206	21	163.5	119.5	176	0.7
S20	1.6	7.65	1469	950	3.8	1175	417.5	25.35	650	913	117	0.7
S21	1.4	6.7	101.8	68	4.7	85	26.5	3.5	25.65	116	174.5	0.3
S22	1.85	7.25	210	121	6.04	177	24.5	16.3	33.1	130	115.5	0.55
S23	0.3	7.6	1047	630	4.4	594	124	63	241.5	110	98.5	0.45
S24	0.2	6.8	227	138	5.6	270	45	37.5	110.1	46	98.95	0.6
S25	1.3	7.3	1278	758	4.6	970	220.5	108	553	130	165.1	0.725
S26	10.9	8.05	1373	830	4.8	986	180.5	115.5	367	96	136.3	0.625
S27	0.6	7.3	242	147	5.15	385	103	57.05	346	87	128	0.7
S28	0.25	7.15	805	482	5.35	625	135.5	102	467	110	171.5	0.85
S29	0.65	8.1	1608	995	3.65	960	137	131	547	132	100	0.975
S30	0.15	7.1	603	362	5.2	456	122.5	66.2	183.1	144	140	1
S31	0.9	7.25	466.5	280	3.9	400.5	105.5	32.85	285.5	110	246.5	1.05
S32	2.4	6.75	486.5	286	4.8	460	132	92.35	69.35	168	149.1	0.95
S33	6	6.7	778	467	4.12	623	305.5	128.7	70.2	86	104.1	1.05
S34	1.15	7.75	923	554	4.3	644	47.72	110.5	251	68	231	0.85
S35	3.15	7.25	1037	622	4.4	610	127.5	100.5	206	91.8	130.1	0.95
S36	0.2	7.45	987.5	593	4.63	570	177	91.55	266.5	180	145.1	1.05

Contd...

Table 64.1–Contd...

Sam. No.	*TUR*	*pH*	*EC*	*TDS*	*DO*	*TH*	Ca^{2+}	Mg^{2+}	Cl^-	*Alk*	SO_4^{2-}	F^-
S37	0.2	8.4	840	520	3.75	695	139	84.45	177	82.5	249	1.1
S38	3.45	7.85	545	315	4.05	1700	395.5	229.5	670	145.5	231.5	0.8
S39	1.6	7.95	360	260	3.95	595	83.9	63.35	200.5	79	132	1.55
S40	0.25	8.25	252.5	150	5.1	1160	189	170	735	138.5	215	1.075
S41	0.15	7.7	324	195	4.5	370	76.55	23.05	124.5	44.95	112.1	0.975
S42	10.55	8.35	322.5	178	3.85	1065	181.5	142	535	765.5	171	0.85
S43	1.55	7	452	263	4.6	1125	299.5	119.5	406	74	151.5	0.6
S44	0.5	8.15	784.5	478	4.15	580	279	11.1	285	66.5	141.5	0.95
S45	0.5	7.1	475.5	294	5.1	283	54.45	35.5	156.5	172	161.2	0.8
S46	11.65	7.05	1193	705	4.9	672	234.5	101.5	93	81	139.5	0.9
S47	0.35	6.85	693	425	4	397	108.1	56.7	118	92.1	265.5	1
S48	10.7	7.5	1550	925	3.55	882	212.5	186.2	86.95	75	262	1.05
S49	0.2	7	957	577	5.6	625	78.85	92.5	142	149	152.8	1
S50	1.65	7.05	789	477	4.4	440	91.35	43	175	110	162	1.1
S51	0.6	6.95	900	553	4.4	800	237	45.1	453	126	210.5	1
S52	0.15	7	520.5	315	5.15	640	156.5	56.05	538.5	101	222.5	1.15
S53	86.1	7.05	784.5	478	3.8	1505	520	46.1	652.5	86.7	172	1.1
S54	69.2	7.55	705	416	4.35	219	36.5	25.5	120.5	89	156.4	1.05
S55	1.25	7.9	815	585	4.1	695	215	9.7	685	255	170.3	1.4
S56	7.05	7.8	1200	749	4	695	125	116.5	470	126	57.3	1.2
Mean	9.029	7.542	894.3	551.7	4.552	727.4	185	92.12	361.8	191.1	150.7	1.012
Min	0.15	6.7	101.8	68	3.55	85	24.5	3.5	25.65	14.7	21.8	0.3
Max	86.1	8.7	2338	1440	6.05	1775	520	229.5	750	913	249	1.8

pH

In the present investigation, pH values found to be 6.7 to 8.7 with a mean value of 7.54. The recommended value of pH for drinking purposes is between 6.5 to 8.5 (BIS, 1998). In the present study all the water samples analyzed are all well with in the safer limits except in few sampling stations. However, higher values of pH hasten the scale formation in water heaters and reduce the germicidal potential of chlorine (Mohapathra and Purohit, 2000).

Electrical Conductivity (EC)

The values of electrical conductivity ranged between 101.8 µmhos/cm to 2338 µmhos/cm. Mean values of EC showed 885.31 µmhos/cm. Higher the concentration of acid, base and salts in water, higher will be the EC. The variability of EC could be explained to the natural concentration of ionised substances present in water (Kataria and Jain, 1995). However the higher values of electrical conductivity (>2000 µmhos/cm), may be due to long residence time and lithology. Ballukraya and Ravi (1999) had proved the variation of the conductivity of the water due to the residential times and the geographical features of the sites.

Table 64.2: Comparison of Groundwater Quality Data with Drinking Water Standard (BIS, 1998)

Sl.No.	Parameters	BIS (1998)		Observed Values	
		P	E	Minimum	Maximum
1.	Turbidity	5	25	0.15	86.1
2.	pH	6.5	9.2	6.7	8.7
3.	EC	–	–	101.8	2338
4.	TDS	500	1000	68	1440
5	Dissolved oxygen	–	–	3.55	6.05
6.	Total hardness	300	600	85	775
7.	Calcium	75	200	24.5	520
8.	Magnesium	30	100	3.5	229.5
9.	Chloride	250	1000	25.65	750
10.	Alkalinity	200	600	14.7	913
11.	Sulfate	200	400	21.8	249
12.	Fluoride	0.6–1.2	1.5	0.3	1.8

Note: P: Permissible limit; E: Excessive limit.

All parameters are expressed in mg/L except pH, Turbidity (NTU) and conductivity (μmohs/cm).

Total Dissolved Solids (TDS)

It is justified from the analytical report TDS values ranges from 68 mg/L to 1440 mg/L with a mean value of 551.53 mg/L. However groundwater chemistry changes as the water flows through the subsurface and the increase in geological environment and dissolved solids and major ions. Chebotarev (1985), Ramababu and Somashekara Rao (1986) and Joseph (2001) expressed the dissolution of soil particles containing minerals under slightly alkaline condition; favour the TDS concentration in groundwater. However TDS concentration above the permissible limit (1500 ppm) causes gastrointestinal irritation (Shankar and Muttukrishnan, 1994).

Dissolved Oxygen (DO)

The present research findings revealed for the values of DO varies from 3.55 mg/L to 6.05 mg/L, and observed mean value is 4.55 mg/L respectively.

In the present study, all the samples showed for DO values were well with in the permissible limits for drinking and domestic purposes. However, the presence of oxygen demanding pollutants (like organic wastes) causes rapid depletion of dissolved oxygen from water. Oxidizable inorganic substances like hydrogen, sulphide, ammonia, nitrites, iron etc., decrease the dissolved oxygen concentration (Sawant *et al.*, 2000). And also due to physical, chemical and biological activities in water the level of DO may vary (Jameel, 2002).

Total Hardness (TH)

Total hardness levels varied from 85 mg/L to 775 mg/L, associated with a mean value of 727.36 mg/L. The BIS (1998) acceptable limit for total hardness is 600 mg/L. In the present study, revealed that 60.7 per cent of total number of water samples crosses the permissible limits of BIS (1998) drinking water standards. Owing to fact that higher amount of hardness in the study area comes mainly from the leaching of igneous rock and carbonate rocks (dolomite, calcite and limestone). Water containing the soluble salts of calcium and magnesium such as chlorides, sulphates and bicarbonates are also governs the quality of water (Ramaswamy and Rajaguru, 1991). The adverse effects of total hardness are formation of kidney stone and the heart diseases (Sastry and Rathee, 1998). Nevertheless, groundwater chemistry is controlled by the composition of its recharge components as well as by geological and hydrological variations (Narayana and Suresh, 1989).

Calcium (Ca^{2+})

Present investigation reports stated for calcium values ranged from 24.5 mg/L to 520 mg/L, with a mean value of 184.99 mg/L. BIS (1998) acceptable limit for calcium is 200 mg/L. However, in the present study 30.2 per cent of water samples crosses the permissible limit. Presence of higher amount of calcium in the study area may be due to groundwater receives the calcium minerals leached from the rocks and other deposits like limestone, dolomites, calcite, gypsum, amphiboles, feldspar, and clay minerals leaching or weathering of igneous rocks. Sewage and domestic waste are also important sources of calcium (Mishra and Saxena, 1989).

Magnesium (Mg^{2+})

Investigation report reveals for the magnesium values ranged from 3.5 mg/L to a 229.5 mg/L with a mean value of 92.11 mg/L.

BIS (1998) acceptable limit for magnesium is 100 mg/L and in the present study 48.2 per cent of the water samples crossed the permissible range. Magnesium arises principally from the weathering of rocks contain ferro-magnesium minerals and some carbonate rocks. High concentration of magnesium proves to be diuretic and laxative (Schroeder, 1960).

Chloride (Cl^{-})

Chloride is also one of the important parameter to know the quality of water. Anthropogenic sources of chlorides include fertilizer, road salt, human and animal waste. Concentration of chlorides is considered to be an indicator of organic pollution of animal origin (Kumara, 2002). Here Chloride values ranged from 25.65 mg/L to 750 mg/L, however the mean value observed in the present study is 361.82 mg/L.

The BIS (1998) acceptable limit for chloride is 1000 mg/L. In the present investigation, the values of chloride for all the sampling sites are with in the permissible range as prescribed by BIS (1998) drinking water standards. However, dissolving of the soil constituents had contributed the chloride into the groundwater and also the soil characteristics play an important role in contributing the chloride content in the groundwater (Shivasankaran, 1997).

Total Alkalinity

In the present study total alkalinity values ranged from a 14.7 mg/L to 913 mg/L and a mean value of 189.31 mg/L. The BIS (1998) acceptable limit for total alkalinity is 600mg/L. In the present study, the data revealed that 10.7 per cent of water samples in the study area crossed the permissible limit. When alkalinity of water exceeds the permissible limits, it is likely to produce incrustation sediment deposits, difficulties in chlorination, certain physiological effects on human systems etc. (Raviprakash and Rao, 1989). The constituents of alkalinity result from dissolution of mineral substances in the soil and atmosphere contributes to alkalinity in groundwater (Mittal and Verma, 1997).

Sulphate (SO_4^-)

It is very interesting to note that, sulphate values ranged from 21.8 mg/L to 249 mg/L with a mean value of 150.2 mg/L. The BIS (1998) permissible range for sulphate is 400 mg/L. In the present investigation, the sulphate values for all the samples are within the prescribed limit of BIS (1998) drinking water standards. Sulphate in groundwater takes place the break down of organic substances in the soil. However, geological, hydrological and geomorphologic characteristics showed a remarkable variation in sulphur content (Alexander, 1961).

Fluoride (F^-)

In the present investigation, fluoride values varied from 0.3 mg/L to 1.8 mg/L with a mean value of 1.01 mg/L. The BIS (1998) acceptable limit for fluoride is 1.5 mg/L. In present study 7.14 per cent of total number of water samples in present investigations have crossed the permissible range as prescribed by BIS (1998) drinking water standards.

Degree of weathering and leachable fluoride in terrain is of great significance for the fluoride present in groundwater than the mere presence of fluoride bearing minerals in rocks (Kumar *et al.*, 2000).

Intake of excess fluoride causes dental, skeletal and non-skeletal fluorosis. The non-skeletal fluorosis can be observed such as gastrointestinal complaints, intermittent diarrhoea and flatulence in expectant and lactating mothers hard working young adults and children. Therefore, fluorosis has been considered as one of the incurable diseases. Hence, prevention is the only solution for the disease (Hem, 1985).

Conclusion

Currently carried research investigation should give more precise answer on influence of Geomorphological condition than anthropogenic activities in the examined groundwater samples of the study area. Local geological settings may supports the increasing concentration of physico-chemical characteristics in groundwater. The factors like slow circulation, longer period of contact between aquifer and water, dissolving of minerals at the time of weathering, residential time, drainage pattern and surface water link. Porosity of the soil and rock also alters the characteristics of the groundwater.

The high level contents of the parameters observed may be minimized if the groundwater is recharged with the available water in the rainy season. This not only dilutes the constituents of the groundwater but also raises the groundwater level that depletes due to large-scale exploitation.

Groundwater is extremely important to the future economy and growth of rural India. If the resource is to remain available as high quality water for future generation it is important to protect from possible contamination. Hence it is recommended that suitable water quality management is essential to avoid any further contamination.

References

Alexander. 1961. *Introduction to Soil Microbiology*. Wiley, New York, London, p. 171–172.

APHA. 1995. *Standard Methods for the Examination of Water and Wastewater*; 18th edition, AWWA, WPCF, New York, pp. 1120.

Ballukraya, P.N. and R. Ravi. 1999. Characterization of Groundwater the unconfined aquifers of Chennai city. *Indian Journal of Geological Society of India*, 54: 1–11.

BIS. 1998. *Specifications for Drinking Water*, New Delhi, p. 171–178.

Chebotarev. 1985. Metamorphism of natural waters in the crust of weathering. *Geochem. Cosmochim. Acta*, 8: 22–28.

Fareeddudin, A.S. Janardhan and B. Basavalingu. 1988. Sedimentology, minerology and geochemistry of the Kalasapura conglomerate. *Geol. Soc. India. Mem.*, 9: 65– 82.

Gazetteer of India, Karnataka State, Chikmagalure district. 1981. p. 8–15 and 630–633.

Hem, J.D. 1985. *Study and Interpretation of the Chemical Characteristics of Natural Water*, 3rd ed. U.S. Geological Survey, Water Supply Paper, 225: 263.

Jameel, A. 2002. Physico-chemical studies in Vyakondan channel water of river Cauvery. *Pollution Research*, 17(2): 111–114.

Joseph, W. 2001. Groundwater chemistry in the valley De Yabucoa alluvial aquifer, South-eastern Pnerto Rico. *AWRA 3rd International Symposium on Tropical Hydrology*, San Juan U.S.A.

Kataria, H.C. and O.P. Jain. 1995. Physico-chemical analysis of river Ajhar. *Indian Journal of Environmental Protection*, 5: 569–571.

Kumar, P. Kumari and L.K.R. Singh. 2000. *Environmental Biology*. P.G. Dept. of Zoology, S.K. University, Dumka, Jharkand, p. 180–182.

Kumar. 2002. *Ecology of Polluted Waters*. 1: 144–180.

Maiti, T.C. 1982. *Science Reporter*, 19: 360–361.

Miller, D.R. 1984. Chemicals in the environment. In: *Effects of Pollutant at the Ecosystem Level*. John Wiley and Sons, Chinchester, pp. 7.

Mishra, S.R. and Saxena. 1989. Industrial effluent pollution at Birla Nagar, Gwalior. *Pollution Research*, 8(2): 77–86.

Mittal, S.K and N. Verma. 1997. Critical analysis of groundwater quality parameters. *Indian Journal of Environmental Protection*, 17: 426–429.

Mohapatra, T.K. and K.M. Purohit. 2000. Qualitative aspects of surface and groundwater for drinking purpose in Paradeep area. *Ecology of Polluted Waters*, 1: 144.

Naqvi, S.M. 1983. Geochemistry of gneisses from Hassan district and adjoining areas, Karnataka, India. In: *Precambrian of South India*, (Eds.) Naqvi, S.M. and J.J.W. Rogers. Geological Society of India Mem., 4: 401–416.

Narayana, A.C. and G.C. Suresh. 1989. Chemical quality of groundwater of Mangalore city, Karnataka. *Indian Journal of Environmental Health*, 31: 228–236.

Ramaswamy, V. and P. Rajaguru. 1991. *Indian Journal of Environmental Health*, 33(2): 187–191.

Rambabu, C. and K. Somashekara Rao. 1986. Studies on the quality of bore well water by Nuzuid. *Indian Journal of Environmental Protection*, 16(7).

Raviprakash, S. and K.G. Rao. 1989. The chemistry of groundwater, Parvada area with regard to their suitability for domestic and irrigation purposes. *Indian Journal of Geochem.*, 4: 39–54.

Sastry, K.V. and P. Rathee. 1988. Physico-chemical and microbiological characteristics of water of village Kanneli, (Distt. Rohtak) Haryana. *Proc. Academic, Environmental Biology*, 7(1): 103–108.

Sawant, C.P., Saxena, G.C. and Shrivastava, V.S. 2000. Trace Metals in and around the Industrial belt. *Ecology Environment and Conservation*, 6(6): 135–137.

Schroeder, H.A. 1960. Relation between hardness of water and death rates from certain chronic and degeneration diseases in the US. *J. Chron. Dis.*, 12: 568–573.

Shankar and Muttukrishan. 1994. *In situ* bioremediation of contaminated groundwater. *Proceedings of National Seminar on EPCR–04,* UBDT Engineering College, Davangere, pp. 32.

Shivashankaran, M.A. 1997. Hydrogeochemical assessment and current status of pollutants in groundwater of Pondicherry region, South India. *Ph.D. Thesis*, Anna University, Chennai, p. 80–87.

Swaminath, J. and M. Ramkrishnan. 1981. In: *Early Precambrian Supracrustals of South India*, (Eds.) Swaminath, J. and M. Ramakrishnan. Geol. Surv. Ind. Mem., 3: 23–38.

Chapter 65

Influence of Pollutants of Soil Bacteria

D. Vijaya Lakshmi[1], C. Krishna Kanth[2] and M.A. Singara Charya[3]
[1]Department of Microbiology, [2]Department of Physics
Kavitha Memorial P.G. College, Khammam – 507 001
[3]Department of Microbiology, Kakatiya University, Warangal – 506 009

ABSTRACT

The impact of pollutant generated from Dairy, Sewage, hospital and Automobile site on the soil bacteria was studied during 1996–1999. The minimum bacterial colonies (230×10^6) were recorded during rainy and maximum (606×10^6) during winter months. The dumps of automobiles caused for drastic reduction of bacteria in the soils (59×10^6). No significant relations were drawn between physico-chemical characteristics and bacterial densities in these polluted sites. pH correlated positively with bacterial densities of dairy soils ($r = 0.621$) at 5 per cent significant level. The date revealed the impact of toxic pollutants and the densities and diversities of bacteria and in turn the management of polluted soil systems.

Keywords: *Pollutants, Bacterial population, Degradative enzymes.*

Introduction

Soil is undoubtedly one of the most important need of the environment. Various physical, chemical and biological process are constantly changing soils over geologic time. Soil is a natural, three-dimensional array of vertically differentiated material at the surface of earths crust.

Pollution is often thought of as resulting from chemical contaminant generating from dairy industries, drug factories, sewage, automobiles etc. It is well known that microorganisms play the significant role in dynamics of organic matter by production of enzymes. (Hattori, 1976).

Soil microorganisms in general bacteria in particular reproduce rapidly and adapt to new environmental situations is important to the decomposition and transformation of both natural and anthropogenic products. Some of the functions performed either entirely or in part by bacteria include nutrient cycling, decomposition of organic materials, nitrogen fixation, pesticide, detoxification and oxidation-reduction reactions. Without bacteria to mediate these process is impossible. Soil conditions that affect the growth of bacteria are many. However, the most important factors are the O_2, temperature, moisture, acidity, inorganic and organic nutrient status of the soil environment. Microbes also play an important role in mineralization process. The enzymatic activity of accumulated enzymes and from those in proliferating microorganisms. Accumulated enzymes in soils are regarded as enzymes present and active in a soil in which no microbial proliferation occurs (Kiss *et al.*, 1975). Sources of enzymes present and active in a soil in which no microbial proliferation occurs. Source of enzymes in soil are primarily the microbial biomass, although they can also originate from plant and animal residues. Enzymes can be immobilized on organic and inorganic constituents of soils. It is well known that microorganisms, the most significant sources of soil enzymes, are absorbed by soil organic and inorganic constituents. (Trivedi *et al.*, 2003), Arnaud *et al.* (2004). In view of this to understand this rapid depilation of bacteria at pollution sites. A study was undertaken for three year in different seasons and the facts were established and presented in Tables 15.1–15.3. A correlation was drawn between the physical, chemical parameters and bacterial densities in the sites are increased pollution status were analyzed and presented in Tables 15.4–15.6.

Table 65.1: Bacterial Densities (1 Ч 10^6)/gm of Four Polluted Sites in Khammam During 1996–1999

Polluted Sites	*Jun*	*Jul*	*Aug*	*Sept*	*Oct*	*Nov*	*Dec*	*Jan*	*Feb*	*Mar*	*Apr*	*May*
Dairy	402	500	320	230	570	606	272	398	406	270	368	255
Sewage	270	472	310	310	450	242	292	142	136	142	329	242
Hospital	392	266	276	175	272	190	169	252	190	284	284	271
Automobile	256	370	390	180	182	139	144	59	76	85	163	148

Table 65.2: Bacterial Densities (1 Ч 10^6)/gm of Four Polluted Sites in Khammam During 1997–1998

Polluted Sites	*Jun*	*Jul*	*Aug*	*Sept*	*Oct*	*Nov*	*Dec*	*Jan*	*Feb*	*Mar*	*Apr*	*May*
Dairy	260	300	100	232	317	250	180	62	105	185	272	200
Sewage	205	180	166	300	472	362	309	180	286	192	250	360
Hospital	300	325	380	262	165	326	204	189	300	110	170	220
Automobile	145	172	180	175	170	168	108	72	142	140	148	189

Table 65.3: Bacterial Densities (1 Ч 10^6)/gm of Four Polluted Sites in Khammam During 1998–1999

Polluted Sites	*Jun*	*Jul*	*Aug*	*Sept*	*Oct*	*Nov*	*Dec*	*Jan*	*Feb*	*Mar*	*Apr*	*May*
Dairy	150	320	108	125	310	216	172	187	165	207	216	117
Sewage	230	175	250	256	200	207	162	160	120	272	200	276
Hospital	85	100	95	162	161	150	171	106	162	106	101	106
Automobile	98	176	186	100	121	82	142	197	143	100	65	75

Table 65.4: Correlation Coefficient (r) Between Bacterial Densities and Physico-chemical Characteristics in Four Polluted Sites at Khammam During 1996–1997

Parameters	*Polluted Sites*			
	Dairy	*Sewage*	*Hospital*	*Automobile*
pH	–0.039	0.257	0.455	–0.143
Temperature	–0.069	0.218	0.781***	0.393
Conductivity	0.029	0.443	0.342	0.183
Turbidity	0.009	–0.176	–	–
Dissolved Oxygen	–0.156	0.410	–	–
Biological Oxygen Demand	–0.112	0.295	–	–
Chemical Oxygen Demand	0.284	0.230	–	–
Organic matter	0.135	0.162	–0.024	0.570
Phosphates	0.580	0.408	–0.111	0.118
Sulphates	–0.227	0.465	–0.272	0.423
Nitrates	0.223	0.307	–0.136	0.072

***P < 0.01; –: Not detected.

Table 65.5: Correlation Coefficient (r) Between Bacterial Densities and Physico-chemical Characteristics in Four Polluted Sites at Khammam During 1997–1998

Parameters	*Polluted Sites*			
	Dairy	*Sewage*	*Hospital*	*Automobile*
pH	–0.153	0.010	–0.097	–0.145
Temperature	–0.327	–0.307	–0.098	–0.102
Conductivity	–0.391	–0.164	–0.310	–0.757**
Turbidity	0.022	–0.309	–	–
Dissolved Oxygen	0.042	–0.332	–	–
Biological Oxygen Demand	0.386	0.044	–	–
Chemical Oxygen Demand	0.205	–0.006	–	–
Organic matter	0.224	0.227	–0.413	–0.331
Phosphates	0.005	–0.073	–0.153	–0.576
Sulphates	0.435	0.113	–0.068	0.227
Nitrates	–0.453	–0.496	–0.217	–0.619*

*P < 0.05; **P < 0.01; –: Not detected.

Materials and Methods

The physical Parameters (pH, temperature, electrical conductivity and turbidity) and chemical characteristics (Dissolved Oxygen, BOD, COD, Organic matter, Phosphates, Sulphates and Nitrates) of the soils collected from four polluted sites in Khammam town were analyzed with monthly intervals for a period of three years. The analysis of these parameters were followed as suggested by Bradley

(1999). The serial dilution technique was used to isolate the bacteria from the polluted sites. From different habitats, each 10 gm of soil was suspended or agitated in a known volume of sterile water (90 ml) to make microbial suspension.

Table 65.6: Correlation Coefficient (r) Between Bacterial Densities and Physico-chemical Characteristics in Four Polluted Sites at Khammam During 1998–1999

Parameters	*Polluted Sites*			
	Dairy	*Sewage*	*Hospital*	*Automobile*
pH	0.621*	–0.384	–0.323	–0.041
Temperature	0.102	0.251	–0.589	–0.194
Conductivity	–0.299	0.131	0.393	–0.304
Turbidity	0.432	0.621*	–	–
Dissolved Oxygen	0.256	0.122	–	–
Biological Oxygen Demand	–0.207	–0.058	–	–
Chemical Oxygen Demand	–0.372	0.290	–	–
Organic matter	–0.413	0.056	0.055	0.197 –
Phosphates	0.298	–0.334	–0.474	0.069
Sulphates	0.062	–0.056	–0.291	–0.049
Nitrates	0.437	–0204	–0.179	–0.020

*P < 0.05; –: Not detected.

Serial dilutions were made and desired amount of suspension transferred to culture plates and incubated at 37°C for 24–48 hrs. and analyzed the plates for bacterial count. Standard method used for the analysis of desired samples and two way analysis of variance and correlation coefficient values between physico-chemical and bacterial densities were calculated and recorded in Tables (15.1–15.6).

Results and Discussion

From the Tables, it was clear that the bacterial densities varied significantly with seasons and with the influence of effluents released from different industries. The maximum bacterial colonies were recorded in the month of November, 1996 (606×10^6) while minimum in September, 1996 (230×10^6) in the soils amended with dairy wastes. Remarkable reduction in the bacterial colonies were noticed in the same site during 1997–98, where the maximum bacterial colonies (370×10^6) were observed in the month of October 1997 and the minimum (62×10^6) in the month of January 1998. Subsequently, the bacterial colonies were recovered and stabilized during 1998–99 with maximum and minimum colony count as 320×10^6 and 125×10^6 respectively.

The enumeration of bacterial colonies in the sewage sites of Khammam town revealed that the constituents of the sewage promoted the growth of bacteria where, seasons were not showed their influence on the proliferation of the colonies.

The maximum colonies were 450×10^6 in the month of October while minimum 142×10^6 in the month of March 1997. The fluctuation of bacterial colonies in (1997–98) the site ranged from 166×10^6 (August 1997) to 472×10^6 (October 1997). The moderate densities of bacteria were noticed in all other months. The stability of the bacterial colonies were witnessed during 1998–99 with the range of densities from 120×10^6 (February 1999) to 272×10^6 (March 1999).

The surroundings of the hospital with higher possibilities of contamination with diversified bacterial genera were observed. The soil samples were collected around hospital site (where the waste dumped areas were selected) waste material showed high incidence of bacterial colonies. However, the densities varied with seasonal influence. The maximum colony count (392×10^6) was noticed in the month of June, 1996, while minimum (9×10^6) in December 1996. (The fluctuations in all other months of the year were in between these two values). The bacterial counts were varied from 110×10^6 (March 1998) to 380×10^6 (August 1997) significant decrease was observed in the colony counts of bacteria during 1998–99 (85×10^6 to 162×10^6). It was very difficult to identity the exact reasons and influence of different factors responsible for these bacterial decreases.

The fourth site is amended with used oils, burned or partially burned engine oils and varieties of hydrocarbon related compounds. These soils were shown a variety of bacterial genera with varied densities. The maximum (390×10^6) bacterial colonies were recorded in the month of August, while minimum 59×10^6 in January. In general the months June to October showed moderate to high densities of bacterial population. Similar trend was continued in the enumeration of bacterial colonies. The highest densities of bacteria in the soils contaminated with oils and its related compounds was from May to October (1997–98). The minimum (72×10^6) number of colonies was noticed in the January 1998. The bacterial densities varied from 65×10^6 (April 1999) to 197×10^6 (January 1999). The gradual percolation and attachment of oil globules on the soil moieties could not show any remarkable fluctuations in densities of bacteria.

Similar to our observations many microbiologists (Border *et al.*, 1985), Varada Raj and Ayyappan (1989), Douglas *et al.* (1990), Lindsay and Riley (1994), Jon *et al.* (1999) from different parts of world tried to isolate and identify the different bacterial genera in various polluted environments. They further signified the relationships between the bacteria and environmental conditions, Phyllis (1989).

Akira (1989) characterized the bacterial populations and structure in different degraded environments on the basis of respiratory quinine profiles (Mehta *et al.*, 2002; Datke *et al.*, 2003; Pawar *et al.*, 2004; Amaud *et al.*, 2004; Medina *et al.*, 2005). The effect of pollutant stress on the bacteria flora and relationships between nutritional requirements and presence of available bacteria aerobic and anaerobic systems.

Lalitha Kumari and Singara Charya (1996), they studied on the assessment of industrial and organic pollution and biological diversity of the soils in Warangal-India, observed the high incidence of bacteria in the soils amended by sewage waters. Recently Baranyi and Carmen (1999), isolated variety of bacterial generation in UK with the help of numerical method. Greg *et al.* (1999), observed the dominance of *Pseudomonus aeruginosa* in different decayed environments and elaborated its physical characterization.

The correlation coefficient was determined between various physico-chemical parameters and bacterial densities of four polluted sites were presented (Tables 15.3–15.5). In general no major significant correlation was recorded between physico-chemical and bacterial densities. In most case the significance was positive or negative above 5 per cent level. From the Table 65.4 it was observed that the temperature showed a positive correlation with bacterial densities in soils near to the hospital site ($r = 0.781$) ($P < 0.01$). The correlation during (1997–98) were not in significant level in polluted sites of dairy, sewage and hospital but negative correlation was observed between conductivity and nitrates ($r = -0.757$) ($P < 0.01$) and ($r = -0.619$) ($P < 0.05$) with bacterial densities in the soils of automobile site. The significant correlations were observed in dairy and sewage sites (1998–99). pH correlated positively with bacterial densities of dairy soils ($r = 0.621$) at 5 per cent significant level. In the sewage soils the correlation was significant (r = 0.621) at 5 per cent significant level, and also turbidity and bacterial densities showed a negative correlation at sewage site (r = – 0.621) ($P < 0.05$). Recently, Goovaerts (1999) Mehta *et al.* (2002), Datke *et al.* (2003), Pawar *et al.* (2004), Arnaud *et al.* (2004), Medina *et al.* (2005), reviewed the statistics of soil science, its state of the art and perspective and concluded the close relations between various abiotic factor and their positive role in the bio-productivity of the soils.

Acknowledgement

The authors are thankful to Dr. S. Girisham, Head, Department of Microbiology, Kakatiya University, Warangal for his encouragement. Two of us (DVL, CKK) are grateful to management, Sri Kavitha Educational Society for providing Lab facilities.

References

Akira, H., Kiyoshi, M. and Kitamura, H., 1989. Characterisation of the bacteria Population structure in an anaerobic-aerobic activated sludge system on the basis of respiratory quinone profiles. *Appl. Enviom. Microbiol.*, 55: 897–901.

Arnaud, D., Mathews, R. and Jones, J.E., 2004. Perceptions and assessment of soil fertility by farmer in the mid hills of Nepal. *Agriculture Ecosystem and Environment*, 103: 191–206.

Border, M.M., Firehammer, B.D., Shoop, D.S. and Myer, 1985. Isolation of *Bacteriods fragilis* from the *feces diarrheric* calves and lambs. *J. Clin. Microbiol.*, 21: 472–473.

Baranyi, J. and Carmen, P., 1999. Estimating Bacterial growth parameters by means of detection time. *Appl. Environ. Microbiol.*, 65: 732–736.

Bradley, P.D., 1999. Microbial functional diversity can be influenced by the addition of simple organic substrates to soil. *Soil Biol. Biochem.*, 30: 1981–1988.

Datke, S.B., Minimol, J.S. and Ujjainkar, V.V., 2003. Effect of season on yield and yield components in groundnut. *PKV Res. J.*, 27: 57–60.

Douglas, S., Myers, L. and Julie, B.L., 1990. Enumeration of Enbtero toxigenic *Bactriods fragilis* in Municipal sewage. *Appl. Environ. Microbiol.*, 56: 2243–2244.

Goovaerts, P., 1999. Geostastics in soil science: State of the art and perspectives. *Goederema*, 89: 1–45.

Jon, H., Gatlin, L.E., K. Charles, G.H. Choi and M.S. Hanson, 1999. Identification and characterization of Sir A: An iron regulated protein from *Staphylococcus aureus. J. Bacteriol.*, 181: 1436–1443.

Kiss, S., M.D. Bularda and D. Radulescn, 1975. Nutrient cycling in soil. *Adv. Agron.*, 27: 25–87.

Lalitha Kumari, B., 1996. Assessment of industrial an pollution biological diversity and enzyme productions of the soils in warangal, A.P. *Ph.D. Thesis*, Kakatiya University.

Lindsay, J.A. and T.V. Riley, 1994. Staphylococcus iron requirements siderophore production and iron regulated protein expression. *In fact Immun.*, 62: 2309–2314.

Kumar, A.S., 2005. Effect of salt stress on germination of three varieties of *Vigna radiata. Ind. J. Environ. and Ecoplan.*, 8: 201–203.

Madhuya, D.A., S. Adak and M.K. Purohit, 2003. Assessment of effluents from cement industries: A pollution problem. *Indian J. Environ. and Ecoplan.*, 7: 497–500.

Medina, A., A. Probanza and R. Azcon, 2005. Interactions of arbuscular mycorrhizal fungi and bacilli strains and their effects on plant growth microbial rhizosphere activity. *Applied Soil Ecology*, 22: 15–28.

Pawar, S.N., S.P. Divekar, S.B. Glule and A.S. Kadale, 2004. Effect of mulching on moisture conservation and yield of summer groundnut crop. *J. Sci. Crops,* 14: 410–413.

Phyllis, A., W. Martin and S.T. Rusel, 1989. World wide abundance and distribution of *Bacillus thuringenues* isolates. *Appl. Environ. Microbiol.,* 55: 2437–2442.

Hattori, J., 1976. The physical environment in soil microbiology: An attempt to extend principles of microbiology to soil microorganisms. *CRC Crit. Rev. Microbiol.,* 4: 423–461.

Trivedi, B.S., 2003. Use of farm residues for integrated nutrient management. *Indian J. Agric. Environ. and Biotech,* 1: 75–93.

Varadaraj, P. and Ayyappan, S., 1989. Population and distribution of heterotrophic bacterial flora in freshwater bodies of Bangalore. *Ind. J. Environ. Health,* 31: 162–170.

Chapter 66

Ethnobotany and Biodiversity Conservation in the Niger Delta, Nigeria

Rufus M. Ubom
Department of Botany, University of Uyo, P.M.B. 1017, Uyo, Nigeria

ABSTRACT

A complimentary ethnobotanical uses of forest and homestead garden plant species in the Niger Delta are presented. A total of 339 plant species were identified in the forests and homestead gardens belonging to 88 families; all the species have at least, one reported use. Species with more uses are *Elaeis guineensis* (12), *Raphia hookeri* (12), *Cocas nucifera* (11), *Irvingia gabonensis* (11), *Hevea brasiliensis* (9), *Lonchocarpus cyanescens* (9), *Pterocarpus santalinaides* (9) and *Dacryodes edulis* (8). Ethnobotanical uses with more species include medicine (178/52.5 per cent), fuel (107/31.6 per cent), food/condiment (199/29.4 per cent), timber (72/21.2 per cent), fodder/feed (70/20.6 per cent), commercial (58/17.1 per cent), furniture (55/16.2 per cent), fibre/cordage (55/16.2 per cent), hometool (45/13.2 per cent) and fence (38/11.2 per cent). The presence of commercial species (fruits, flavouring, beverages, thickeners, spices, timber) in the forests allows for the discussion of the importance of the lowland rainforests of Niger Delta and the suggestion that modern management technologies, if applied, could lead to cost effective conservation methods and sustainable production of animal food source, building materials for houses, economic and social benefits of the rural dwellers.

Keywords: *Ethnobotany, Biodiversity, Conservation, Niger Delta.*

Introduction

Biodiversity is a concept that involves different facets of biological variety (Peters, 1991) including inter alia, taxonomic richness, genetic differences in each taxon, communities, ecosystems, landscapes within which organisms inhabit, and the indigenous knowledge of nature possessed by the natives living on the land. Biodiversity is thus a prized variety of interwoven moral, aesthetic, utilitarian and economic reasons. The role of biodiversity for agriculture and medicine is often underestimated. Plant breeding often turns to wild relatives and increased genetic materials from them (Spellberg, 1995). The biodiversity of the Niger Delta is intimately associated with the habitat diversity, each supporting diverse and distinctive plants and animals within their niches, some endemic to the region.

The Niger Delta region of Nigeria is particularly important for its oil exploitation. In addition, its vast plants and animal resources have also been immensely exploited. Niger Delta occupies 37700 km^2 of the 923700 km^2 land area of Nigeria (Figure 66.1). It is situated in the southern most part of Nigeria, and bordered to the south by Atlantic Ocean. It comprises eight states of Nigeria: Akwa Ibom, Bayelsa, Cross-River, Delta, Edo, Imo, Ondo and Rivers. The area is essentially a lowland belt about 30m below sea level. The area supports about 19.21 million people and various densities of livestock. These exert pressure on the resources in addition to oil exploitation by numerous multinational companies. Plants and animal species of economic importance are threatened or endangered. Endemic species of ecological importance are also threatened and need be conserved.

Rural communities, in this region, depend on their interactions with plants for commercial values and thus exploit, albeit, sustainably their environment, including forests. Exploitation of these forests (can) jeopardize certain species existence or even lead them to extinction. Commercial values of the plants are the direct and indirect economic, aesthetic and ethical values. Emphasis on direct commercial value of forest products is placed on a small number of timber species and a short-term profit derived from felled portions of forests for agricultural purposes. Multi-use exploitation and conservation of forest resources, as practiced by these rural communities, have not been adopted as a viable alternative to modern production schemes (Clay, 1998). From ecological perspective rural communities can achieve production in harmony with nature's laws that their production capacity tends to maintain the renewing capacity of the ecosystems, albeit, forests. From this standpoint, the sum of empirical knowledge that the people in the Niger Delta region posses about plants, soils, climate and ecophysiographic units serving their production strategic becomes important in unciphering and understanding traditional economic rationality. Each method of production reveals a strategy of appropriation from nature, a mix of technologies, knowledge about the components of the ecosystems and its interrelations. These depend upon forest compositions, species abundance, appreciation and usage of the species by the native. Due to exploration and exploitation of crude oil much of the area has been polluted and the needed medicinal plants and other useful non-timber species have been going extinct. Little attention has been paid to this seemingly unnoticed but important degradation and elimination of species much averred by the indigenous people of the Niger Delta. Thus, this study attempts to assess the values of forests and their species vis-à-vis the diversity of uses the plants species have been put to and marketed or traded by the natives. An attempt has also been made in eliciting and proffering conservation method(s) the natives should practice in order to protect and preserve these plants for posterity.

Materials and Methods

Study Sites

The Niger Delta, one of the world's largest wetlands in southern Nigeria, is a vast flood plain (over 2000 km^2) built up by the accumulation of sedimentary materials washed down the Niger and Benue Rivers (Pangely, 1994). It comprises four ecological zones; coastal barrier islands, mangrove swamp forests, freshwater swamp forests are low lame rainforests. The later two forests are the most extensive ecological zones in the region. However, anthropogenic activities (timber and fuel wood

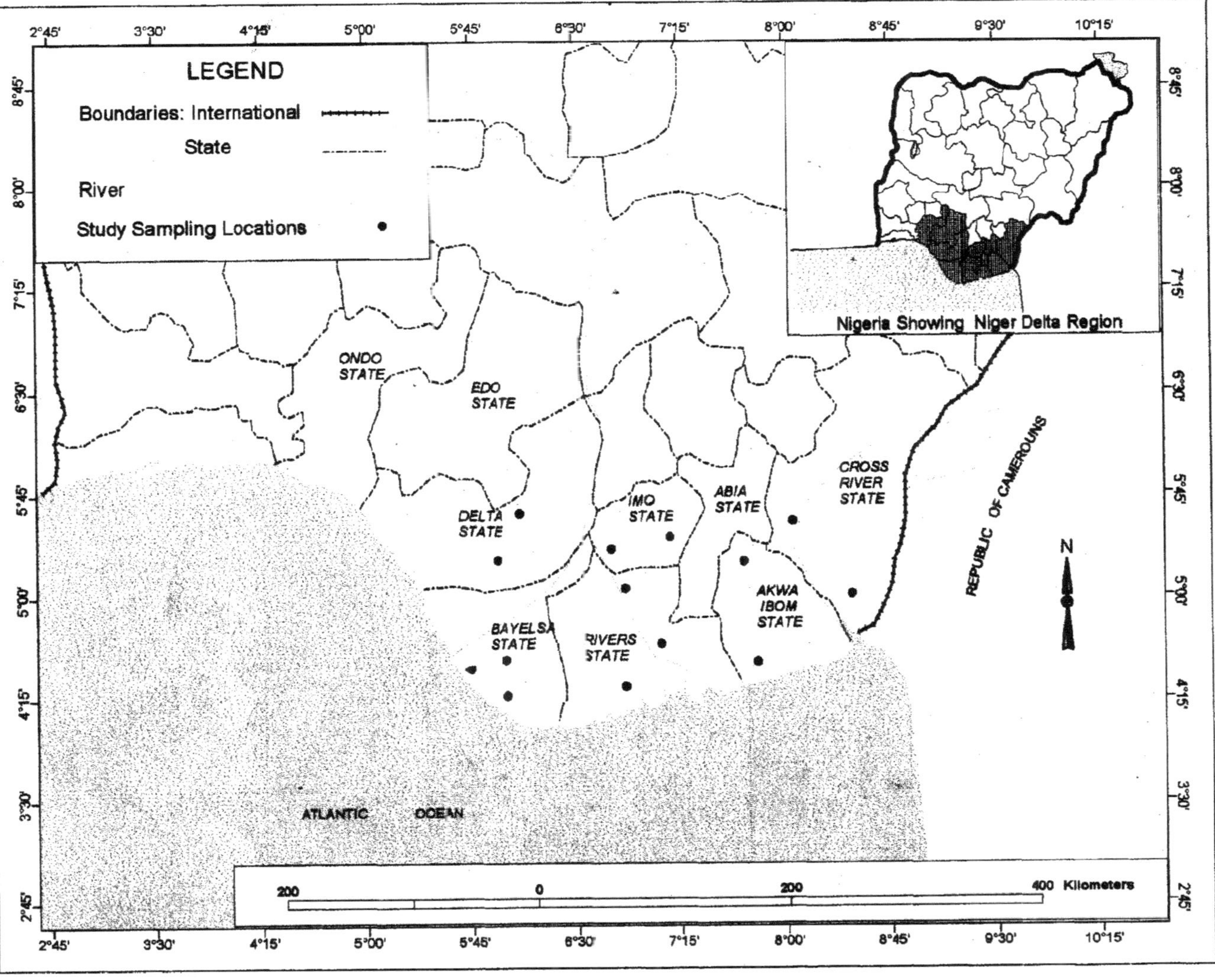

Figure 66.1: Niger Delta Region Showing Study Sites

removal, cultivation, capital project deployment, oil exploration and exploitation) are fast degrading these forests and their biodiversity.

The region is located between latitudes 4°00′ N and 7°′30′ N and longitudes 4°30′ E and 9°15′ E. The area experiences two main season: wet and dry seasons. The wet season is about 7 months in duration from April to October. The dry season lasts about 5 months beginning from November and ending in March. The mean temperature for the area is 25°C; annual rainfall 4921 mm and relative humility 85 per cent. The main crops are cassava, *Manihot esculentus* Krantz; Cocoyam, *Colocasia antiquorum* Schott, yam, *Dioscorea* sp. and oil palm, *Elaeis guineensis.*

Six out of eight states in the region were selected for the study: Akwa-Ibom, Bayelsa, Cross River, Delta, Imo and Rivers. Selection was based on previous knowledge, experience and cumulative work (unpublished) done in these areas by the author. Thirteen study sites (Figure 66.1) were chosen for the study: two in each state except Rivers with three sites.

Sampling and Analysis

Field study was conducted for four years (1999, 2000, 2002, 2004) for both wet and dry seasons. The vegetation (freshwater swamp and lowland rainforests) was sampled systematically within 10 m × 10 m Quadrats along established transects, and randomly throughout the sampling areas. Interviews with different informants were also conducted as to the type of species found in the areas, the use they are being put to and the methods used in conserving the species. The Shannon-Weiner species diversity was used to assess the species quantitatively (Sokal and Rohlf, 1995) according to the formula:

$$H^1 = - \sum_{i=1}^{s} pi \ln pi$$

where,

H^1: Shannon-Weiner index

S: Number of species

pi: Proportion of individuals or abundance of the ith species expressed as a proportion of the total number of individuals of all species.

ln: Log base.

Endangered and endemic species were recorded; identification of species was accomplished by the assistance of Forestry Institute of Nigeria (FRIN) herbarium and texts (Hutchinson and Dalziel, 1954–72; Akobundu and Agyakwa, 1987). Classification of ethnobotanical uses of species involved interviews, complimented by literature (Sofowora, 1982; Kokwara, 1995; Etukudo, 2000, 2003). Diversity of uses was calculated using Shannon-Weiner index ($H^1 = -\Sigma$ pi log pi) *where,* I: uses and pi: proportion of species for the ith use (Rico-Gray *et al.*, 1991). Sorensen index (S.I = 2a/+btc) was used to calculate species similarity between sites; a: number of species common to sites compared, b: number of species in site 1 and c: number of species in site 2.

Results and Discussion

Plant Diversity

Three hundred and thirty nine plant species were encountered in the forests and homestead gardens in the study area (Tables 66.1 and 66.2). Of these, 206 species were recorded for the forests and 79 species for the homestead gardens. All the 339 species belong to 88 families (44 families for forest,

44 for gardens) and 266 genera (205 for forests, 61 for gardens). Fabaccae has the highest number of species (45), followed by Euphorbiaceae (25), Rubiaceae (12) Apocynaceae (11), Asteraceae (11) and Cecropiaceae (10). The species diversity index between the forest and the homestend gardens is 4.89. Species similarity index among the forest is low (47.78 per cent) where 147 species are common to all forests. Similarity index between the homestead gardens is high (49.5 per cent) as 76 species are common to the gardens.

Table 66.1: List of Plant Species Encountered in the Forests of Study Areas. Nomenclature follows Hutchinson and Dalziel (1972) and Akobundu and Agyakwa (1987).

Species	Habit	Family	Ethnobotanical Uses	Plant Parts Used
Abrus precatorius Linn	Climber	Fabaceae	17, 21, 25	Seeds
Abutilon mauritianum (Jacq.) Medic	Shrub	Malvaceae	7, 17	Seeds, bark, leaves
Acanthus montamus (Nees) T. Anders	Herb	Acanthaceae	17, 19	Leaves, whole plant
Achyranthes aspera Linn	Herb	Amaranthaceae	10, 17	Leaves, whole plant
Acioabarteri (Hook f. ex Oliv) Engl.	Shrub	Rosaeceae	1, 2, 3, 5	Stems, fruits
**Aframomum melegueta* K. Schum.	Herb	Zingiberaceae	7, 17, 19, 21, 23, 24	Stems, seeds
**A. sceptrum* (Oliv. and Hanb) K. Schum.	Herb	Zingiberaceae	7, 17, 19, 21, 24	Stem, seeds
**A. strobilaceym* (Sm.) Hepper	Herb	Zingiberaceae	7, 17, 19, 21, 24	Stem, seeds, leaves
Ageratum conizoides Linn.	Herb	Asteraceae	5, 17, 18	Leaves, flowers
Albizia zygia (DC.) J.F. Machr.	Tree	Fabaceae	9, 10, 12, 13, 15, 26, 29	Leaves, stems
Albizia adianthifolia (Schum) W.F. Wight	Tree	Fabaceae	9, 11, 17, 27	Leaves, stem, bark
Alchornea cordifolia (Schum. and Thonn.) Mull. Arg.	Shrub	Euphorbiaceae	3, 6, 7, 9, 11, 30	Stem, bark, leaves, shoots
A. laxiflora (Benth.) Pax and K. Hoffin	Shrub	Euphorbiaceae	3, 6, 7, 11, 17	Stem, bark, leaves
Allanblackia floribunda Oliv.	Tree	Clusiaceae	2, 4, 9, 11, 12, 27, 29	Stems, leaves, fruits
Alstonia boonei De Wild	Tree	Apocynaceae	2, 4, 11, 12, 14, 15, 27	Stems, leaves, bark
A. congensis Engl.	Tree	Apocynaceae	2, 4, 11, 12, 14, 15, 27	Stems, leaves, bark
Alternanthera bettzickiana (Regel) Voss	Herb	Amaranthaceae	9, 19	Leaves, shoots
Alternanthera pungens H.B.K.	Herb	Amaranthaceae	10, 17, 18, 19	Leaves, shoots, stems
A. sesslis (Linn.) R.Br. ex Roth	Herb	Amaranthaceae	10, 17	Leaves, shoots, stems
Anchomanes difformis (Bl.) Engl.	Herb	Araceae	1, 10, 17, 21	Stems tubers, leaves, fruits
**Ancistrophyllum secundiflorum* (P. Beauv.) Wendl.	Shrub	Palmaceae	4, 7, 14, 15, 30	Stems, shoots
Aningeria robusta (A. Chev.) Aubrev. and Pellegr.	Tree	Sapotaceae	4, 11, 12, 15, 27, 29	Bark, stem
Anthocleista vogelii Planch.	Tree	Loganiaceae	4, 9, 11, 12, 17, 27, 29, 30	Stems, bark, leaves
A. dyalonensis A. Chev.	Tree	Loganiaceae	2, 8, 11, 15, 17, 30	Leaves, stems
**Anthostema aubryamum* Baill.	Tree	Euphorbiaceae	11, 15, 17, 27	Bark, leaves, stem
Antiaris africana Engl.	Tree	Moraceae	7, 11, 12, 27, 29	Bark, stems, leaves
Antidesma vogelianum Mull. Arg.	Tree	Euphorbiaceae	2, 12, 15, 27	Stems
***Anthonatha macrophylla* P. Beauv.	Shrub	Fabaceae	6, 11, 13	Stems

Contd...

Table 66.1–Contd...

Species	Habit	Family	Ethnobotanical Uses	Plant Parts Used
Araliopsis soyauxii Engl.	Tree	Rutaceae	3, 11, 15, 27	Stems, roots
Argemone mexicana Linn.	Herb	Papaveraceae	17	Whole plant
Aristolochia albida Duchatre	Herb	Aristolochiaceae	1, 11, 12	Stems
Aspilia africana (Pers.) C.D. Adams	Herb	Asteraceae	5, 9, 17, 30	Leaves, shoots, flowers
Asystacia gangetica (Linn.) T. Anders	Herb	Acanthaceae	9, 17, 30	Leaves
Bacopa creanata (P. Beauv.) Hepper	Herb	Scrophulariaceae	10, 11, 17	Leaves, stem
Bambusa vulgaris Schrad. ex Wendel	Shrub	Poaceae	2, 6, 9, 12, 14	Stem, shoots, leaves
Baphia nitida Lodd.	Shrub	Fabaceae	3, 5, 6, 7, 9, 21, 30	Leaves, stems, bark
Barteria nigritiana Hook. f.	Shrub	Passifloraceae	6, 91 10, 11	Stems, leaves
Begonia microcarpa Warb	Herb	Begoniaceae	17, 21	Leaves, roots
Berlinia grandiflora (Vahl.) Hutch and Dalz.	Tree	Fabaceae	2, 6, 11, 13	Stems
Bixa orellana Linn.	Shrub	Bixaceae	5, 17, 19, 22, 30	Fruits, stems, leaves, whole plant
Blighia sapida Konig	Tree	Sapindaceae	17, 27	Leaves, seeds, stems
Boerhavia erecta Linn.	Herb	Nyctaginaceae	9, 17	Leaves, roots, shoots
Bombax buonopozense P. Beau V.	Tree	Bombacaceae	4, 7, 11, 12, 14, 15	Stems, fruits, leaves
Bosqueia angolensis Ficalho	Tree	Moraceae	11, 12, 14, 17	Stems
Brachystegia eurycoma Harms	Tree	Fabaceae	9, 10, 11, 27	Leaves, stems, seeds
Bridelia grandis Pierre ex Hutch	Tree	Euphorbiaceae	4, 12, 27	Stems
B. micrantha (Hochst.) Baill.	Tree	Euphorbiaceae	4, 12, 14, 27	Stems
Bryophyllum pinnatum (Lan.) Okon	Herb	Crassutaceae	17	Leaves, whole plant
Calamus deerratus Mann and Wendl.	Climber	Palmaceae	4, 7, 14, 15	Stems
Calopogonium mucunoides Desv.	Creeper	Fabaceae	9	Leaves, fruits
Canarium schweinfurthii Linn.	Tree	Burseraceae	4, 10, 11, 17, 24, 29	Stems, fruits, bark
Carpolobia lutea G. Don	Shrub	Polygalaceae	3, 15	Stems
Cassytha filiformis Linn.	Climber	Lauraceae	17	Whole plant
Cathormion altissimum (Hook. f.) Hutch and Dandy	Tree	Fabaceae	11, 12, 27	Stems
Cedrella odorata Linn	Tree	Meliaceae	11, 12, 15, 27	Stems
Ceiba pentandra (L.) Gaertn.	Tree	Bombacaceae	4, 7, 11, 13, 17, 19, 27	Stems, fruits, roots
Celtis zenkeri Engl.	Tree	Ulmaceaee	2, 4, 5, 6, 12	Stems
Centrosoma pubescens Benth.	Creeper	Fabaceae	9	Leaves, fruits
Ceratophylluma demersum Linn.	Herb	Ceratophyllaceae	17	Whole plant
Chromolaena odorata (L.) R.M. King and Robinson	Herb	Asteraceae	17, 30	Leaves, shoots
Chlysobalamus ellipticus Soland. e Sabine	Shrub	Rosaceae	2, 14, 17	Bark, stems
Cissus quadrangularis Linn.	Climber	Ampelidaceae	10, 17	Leaves, stems, fruits
Clappertonia ficifolia (Willdo) Decne	Shrub	Tiliaceae	7, 17	Bark, leaves, flowers
Cleistopholis patens (Benth.) Engl. and Diels	Tree	Annonaceae	2, 11, 27, 29	Stems, leaves

Contd...

Table 66.1–Contd...

Species	*Habit*	*Family*	*Ethnobotanical Uses*	*Plant Parts Used*
Cleome viscosa Linn	Herb	Capparidaceae	9, 17	Leaves, fruits
Clerodendrum splendens G. Don.	Climbing, Shrub	Verbenaceae	5, 17, 19	Leaves, fruits, whole plant
Clitoria ternatea Linn.	Climber	Fabaceae	17, 18	Seeds, roots
Cnestis ferruginea DC.	Shrub	Connaraceae	5, 17	Leaves, fruits, roots
Coelocaryon preusii Warb.	Tree	Myristicaceae	3, 5, 6, 11, 15	Stems, leaves, bark, roots
Coix lachrymal-jobi Linn.	Herb	Poaceae	6, 7, 9	Whole plant
Combretum hispidum Laws	Shrub	Combretaceae	17, 19	Roots
C. racemosum P. Beauv.	Shrub	Combretaceae	17,	Twigs, shoots
***C. zenkeri* Engl. and Diels	Shrub	Combretaceae	5, 7, 17	Leaves, stems
Commelina africana Linn.	Herb	Commelinaceae	17, 19	Whole plant
C. diffusa Burm. f.	Herb	Commelinaceae	17, 19	Whole plant
C. erecta Linn.	Herb	Commelinaceae	17, 19	Whole plant
***Costus afer* Ker-Gawl	Herb	Costaceae	3, 9, 17, 24	Stems, shoots, roots
Costus lucanusianus J. Braun and K. Schum.	Herb	Costaceae	7, 9, 12	Stems, leaves, roots
***C. schlechteri* Winkler	Herb	Costaceae	9, 17, 24	Stems, roots
Crassocephalum togoense C.D. Adams	Herb	Asteraceae	9, 17	Leaves, stems
Crassocephalum crepidinoides S. Moore	Herb	Asteraceae	10, 17	Leaves
Crotalaria retusa Linn.	Herb	Fabaceae	7, 19	Whole plant, bark
Crudia klainei Pierre ex De Wild	Tree	Fabaceae	2, 11	Stems
Cuscuta adstralis R.Br.	Twinner	Cuscutaceae	1, 17	Whole plant
Cynometra vogelii Hook. f	Herb	Fabaceae	11, 29	Stems
Cyperus difformis Linn	Herb	Cypereceae	10, 17	Stems, leaves
Cyperus esculentus Linn	Herb	Cypereceae	7, 10, 17	Tubers, stems, leaves
Dalbergia latifolia Roxb.	Tree	Fabaceae	12, 15, 29	Stems
Daniellia ogea (Harnes) Rolfe ex Holl.	Tree	Fabaceae	4, 91 11, 12, 15, 27	Stems, leaves
Datura stramonium Linn.	Herb	Solanaceae	17, 24, 28	Leaves, seeds, fruit, flowers
Desmodium turtuosum (Sw.) DC.	Herb	Fabaceae	9, 17, 29	Whole plant, roots
Dialium guineense Wild	Shrub	Fabaceae	3, 5, 7, 9, 10, 11	Leaves, stems, bark
Diodia sarmentosa Sw.	Herb	Rubiaceae	17	Leaves
**Diospyros alboflavescens* (Gurke) F. White	Tree	Ebenaceae	4, 8, 11, 12, 15, 27, 29	Stems, leaves
Diplazium sammatii (Kuhn) C. Chr.	Herb	Athyriaceae	10	Shoots
**Dracaena arborea* Link	Tree	Liliaceae	6, 19, 21	Leaves, whole plant
**D. mannii* Bak.	Tree	Uliaceae	6, 19, 21	Leaves, whole plant
Drypetes floribunda (Mull. Arg.) Hutch.	Shrub	Euphorbiaceae	1, 28	Fruits, leaves
Emilia coccinea (Sims) G. Don	Herb	Asteraceae	9, 17	Leaves, shoots

Contd...

Table 66.1–Contd...

Species	Habit	Family	Ethnobotanical Uses	Plant Parts Used
E. sonchifolia (Linn.) DC.	Herb	Asteraceae	9, 17	Leaves, shoots
Erythrina senegalensis DC.	Tree	Fabaceae	6, 27	Stems
Erythrophelm ivorense A. Chev.	Tree	Fabaceae	11, 27, 29	Leaves, stems
Euphorbia heterophylla Linn.	Herb	Euphorbiaceae	13, 17, 19	Leaves, stems, whole plant
Fagara macrophylla Engl.	Tree	Rutaceae	11, 17, 27	Stems, leaves, roots
F. zanthoxyloides Lam	Tree	Rutaceae	3, 6, 11, 17, 27	Stems, leaves, roots
Ficus exasperata Vahl.	Tree	Moraceae	6, 9, 13, 16, 17	Stems, fruits, shoots, leaves
Ficus lepreuri Miq.	Tree	Moraceae	13, 16, 17	Stems, fruits, shoots, leaves
Ambristylis ferruginea (Linn.) Vahl.	Herb	Cyperaceae	17	Fruits
Fimbristylis littoralis Linn.	Herb	Cyperaceae	17	Fruits
Fleurya aestuans (Linn.) Gaudex Miq.	Herb	Urticaceae	17, 28	Leaves, flowers
F. ovalifolia (Schum. and Thonn.) Dandy	Herb	Urticaceae	17	Fruits, roots
Fuirena umbellate Rottb.	Herb	Cyperaceae	17	Fruits
Funtumia elastica (Preuss.) Stapf.	Tree	Apocynaceae	13, 16	Stems, leaves, bark
***Garcinia mannii* Oliv.	Tree	Clusiaceae	3, 13	Stems
Gloriosa superba Linn.	Herb	Liliaceae	17	Whole plant
Glyphaea brevis (Spreng.) Monachino	Shrub	Tiliaceae	3, 5, 6, 7, 10, 11, 17	Stems, leaves, bark
Guarea cedrata (A. Chev.) Pellegr.	Tree	Meliaceae	2, 8, 11, 27, 29	Leaves, stems
G. thompsonii Sprague and Hutch.	Tree	Meliaceae	2, 8, 11, 27, 29	Leaves, stems
Hallea stipulosa (DC) J.F. Leroy	Tree	Rubiaceae	4, 11, 12, 15, 17, 27, 29	Stems, roots, bark
Halopegia azurea K. Schum	Herb	Marantaceae	7, 30	Leaves, stems
Hallea ciliata Aub and Pellegr.	Tree	Rubiaceae	4, 11, 12, 15, 17, 27	Stems, roots, bark
Hannoa klaineana Pierre and Engl.	Tree	Simaroubaceae	2, 4, 6, 11, 12, 27, 29	Leaves, stems
Harungana madagascariensis Lam. ep Poir	Shrub	Hypericaceae	5, 9, 11, 13, 17	Leaves, stems, bark
Heteranthera callifolia Rchb. ex Kunth	Herb	Pontederiaceae	17	Leaves, roots
Heterotis rotundifolia (Sm.) Jac-fet	Herb	Melastomataceae	17	Leaves, fruits, flowers
Hewittia sublobata Linn.	Shrub	Convolvulaceae	17, 19	Flowers, whole plant
**Hibiscus sabdarifa* Linn.	Shrub	Malvaceae	7, 10, 23	Bark, leaves, fruits, flowers
***Hippocratea pallens* Planch. ex Oliv.	Climbing	Hippocrateaceae	11, 17, 25	Fruits, roots shrub
Homalium letestui Pellegr.	Tree	Samydaceae	3, 6, 9	Leaves, stems
Hylodendron gabunense Taub.	Tree	Fabaceae	2, 6, 11, 27	Stems
cacina trichantha Oliv.	Straggling shrub	Icacinaceae	17	Tubers
**Ipomoea aquatica* Forsk.	Twinner	Convolvulaceae	9, 10, 19	Leaves, stems, whole plant
I. involucrata P. Beauv.	Creeper	Convolvulaceae	7, 17	Stems, leaves

Contd...

Table 66.1–Contd...

Species	*Habit*	*Family*	*Ethnobotanical Uses*	*Plant Parts Used*
Irvingia gabonensis Baill.	Tree	Irvingiaceae	9, 10, 11, 18, 23	Fruits, seeds, stems
Justicia flava (Forsk.) Vahl.	Herb	Acanthaceae	17	Whole plant
Khaya grandifoliola C. DC.	Tree	Meliaceae	2, 4, 8, 11, 12, 15, 27, 29	Leaves, stems
K. ivorensis A. Chev.	Tree	Meliaceae	2, 4, 11, 12, 15, 27, 29	Leaves, stems
Klainedoxa gabonensis Pierre ex Engl.	Tree	Irvingiaceae	2, 4, 11, 12, 15, 27, 29	Stems, leaves
Landolphia heudelotll A. DC.	Shrub	Apocynaceae	6, 11, 13, 15, 16	Stems, leaves
L. owariensis P. Beauv.	Liane, shrub	Apocynaceae	6, 11, 13, 15, 16	Stems, leaves
Lannea acida A. Rich	Tree	Anacardiaceae	10, 11, 17	Leaves, stems, fruits
**Leea guineensis* G. Don	Shrub	Leeaceae	17, 19	Leaves, roots, bark
**Lonchocarpus cyanescens* Benth (Schum. and Thonn.)	Shrub	Fabaceae	2, 3, 5, 6, 9, 10, 11, 27, 29	Leaves, stem, roots, seeds
L. griffoneanus (Baill.) Dunn.	Shrub	Fabaceae	1, 2, 6, 11, 21, 27, 29	Leaves, stems, seeds
Lophira alata Barks ex Gaertn. f.	Tree	Ochnaceae	2, 4, 6, 8, 11, 12, 15, 27	Stems
Loranthus aphyllus Sprangue	Parasitic Shrub	Loranthaceae	17	Leaves, fruits
Lovol trichilioides Harms	Tree	Meliaceae	2, 4, 8, 11, 12, 14, 15, 27, 29	Leaves, stems
Ludwigia decurrens Walt.	Herb	Onagraceae	17	Leaves, flowers
Ludwlgia hyssopifolia (G. Don.) Exell	Herb	Onagraceae	17	Leaves, flowers
**Macaranga barteri* Muell. Arg.	Tree	Euphorbiaceae	9, 11, 13, 14, 25	Leaves, stems, fruits, roots
**Maesobotrya barteri* (Baill.) Hutch	Shrub	Euphorbiaceae	3, 9, 10, 11, 14	Stems, leaves, fruits
Mallotus oppositifolius (Geisel) Mull. Arg.	Shrub	Euphorbiaceae	5, 9, 11, 17	Leaves, stems, roots
Manihot glaziovii Mull. Arg.	Shrub	Euphorbiaceae	16, 19, 28	Leaves, tubers
Manniophyton fulvum Mull. Arg.	Shrub	Euphorbiaceae	9	Shoots
Mansonia altissima A. Chev.	Tree	Sterculiaceae	4, 11, 12, 27	Stems
Massularia acuminata (G. Don.) Bull. ex Hoyle	Shrub	Rubiaceae	3, 6, 17	Stems
Maranthochloa cuspidata (Rosc.) Milne-Redh.	Herb	Marantaceae	2, 7, 30	Leaves, stems
Melanthera scandens (Schum. and Thonn.) Roberty	Herb	Asteraceae	17	Leaves, roots
Melastomastrum capitatum (Vahl.) A. and R. Fern	Herb	Melastomataceae	17	Leaves, fruits, flowers
Melochia corchorlfolia Linn.	Herb	Sterculiaceae	7, 10, 17	Bark, leaves, roots
**Microdesmis puberula* Hook. f. ex Planch.	Shrub	Euphorbiaceae	3, 9, 10	Stems, shoots, leaves
Milicia excelsa (Welw.) C.C. Berg.	Tree	Melaiceae	4, 11, 12, 21, 27	Stems

Contd...

Table 66.1–Contd...

Species	Habit	Family	Ethnobotanical Uses	Plant Parts Used
Millettia thonningii (Schum. and Thonn.) Bak	Shrub	Fabaceae	6, 8, 10, 11, 17, 29	Leaves, stems, roots
Mimosa pigra Linn.	Herb	Fabaceae	19, 27	Leaves, whole plant
Mimosa peduca Linn.	Herb	Fabaceae	19, 27	Whole plant
MImosa invisa Mart.	Herb	Fabaceae	4, 11, 12, 17, 27, 29	Leaves, whole plant
Momordica charantia Linn.	Climber	Cucurbitaceae	10, 17	Fruits, leaves
Momodora myristica (Gaertn) Dunal	Climber	Annonaceae	19, 23, 27	Leaves, fruits, whole plant
Musanga cecropioides R.Br.	Tree	Moraceae	9, 11, 16, 17, 30	Stems, leaves, bark, roots
***Mussaenda isertiana* DC	Climbing shrub	Rubiaceae	19, 27	Leaves, whole plant
Myrianthus arboreus P. Beauv.	Tree	Moraceae	2, 11, 12, 15, 17, 27	Stems, leaves, bark, roots
**Napoleona vogelii* Hook. and Planch.	Tree	Lecythidaceae	1, 3, 11, 30	Stems, leaves
Nauclea diderrichii (De Wild. and Th. Dur.) Merrill	Tree	Rubiaceae	6, 17, 27	Stems, fruits
Nauclea latifolia Sm.	Strangling shrub	Rubiaceae	2, 3, 5, 13, 27	Leaves, stems, roots, bark
Nauclea vanderguchtii (De Wild) Petit.	Shrub	Rubiaceae	2, 3, 5, 13	Stems, roots
**Newbouldia laevis* (P. Beauv.) Seemann ex Bureau	Tree	Bignoniaceae	6, 11, 19, 21	Stems, shoots
**Oxystigma mannii* (Baill.) Harms	Tree	Fabaceae	2, 4, 11, 12, 15, 27, 29	Leaves, stem
Pachystela brevipes (Bak.) Baillon ex Engl.	Tree	Sapotaceae	2, 4, 8, 11, 12, 15, 27, 29	Stems
Palisota hirsuta (Thunb.) K. Schum.	Herb	Commelinaceae	9, 17	Leaves, stems
**Pandanus candelabrum* Beauv.	Tree	Pandanaceae	4, 6, 7, 12, 14, 21	Stems
Parkia bicolour A. Chev.	Tree	Fabaceae	4, 9, 10	Leaves, fruits, stems
Parinari excelsa Sabine	Shrub	Chrysobalanaceae	4, 11, 12	Stems
Paspalum vaginatum Sw.	Herb	Poaceae	9	Shoots
Paullinia pinnata Linn.	Climber	Sapindaceae	3, 17	Leaves, stem, roots
**Pentachlethra macrophylla* Benth.	Tree	Fabaceae	4, 9, 10, 11, 18, 21, 23	Leaves, stem, seeds
Pentadesma butryracea Sabine	Tree	Clusiaceae	13, 14, 17	Bark, stems
Peperomia pellucida (L) H.B. and K.	Herb	Piperaceae	17, 28	Whole plant, leaves
Phyllanthus amarus Schum and Thonn.	Herb	Euphorbiaceae	16, 17	Leaves, stem, fruits, flowers
Phyllantus muelleriamus (O. Ktz) Exell	Herb	Euphorbiaceae	17	Flowers, fruits
Physalis angulata Linn.	Herb	Solanaceae	17	Leaves
Piper umbellatum Linn.	Climber	Piperaceae	10, 17, 21	Leaves, roots, whole plant

Contd...

Table 66.1–Contd...

Species	Habit	Family	Ethnobotanical Uses	Plant Parts Used
Plukenetia conophora Mull. Arg.	Climbing shrub	Euphorbiaceae	10, 11, 12, 14, 17, 18	Leaves, seeds, fruits, stems
Poga oleosa Pierre	Tree	Rhizophoraceae	4, 10, 11, 12, 27, 29	Leaves, stems
**Psychotria nigerica* Hepper	Shrub	Rubiaceae	17, 24	Leaves
Pterocarpus mildbraedii Harms	Tree	Fabaceae	5, 6, 9, 11, 15, 17, 29	Leaves, bark, stem
Pterokcarpus santalinoides L'Her Ex DC.	Tree	Fabaceae	5, 6, 9, 10, 12, 14, 15, 17, 26	Bark, leaves, stems
P. osun Craib	Tree	Fabaceae	5, 6, 9, 11, 15, 17, 29	Leaves, bark, stems
Pterygota macrocarpa K. Schum	Tree	Sterculiaceae	12, 27	Stems
Pupalia lappacea (Linn.) Juss.	Herb	Amaranthaceae	17	Shoots
Pycanthus angolensis (Welw.) Warb.	Tree	Myristicaceae	4, 11, 12, 15, 21, 27	Stems, twigs
Rauvolfia vomitoria Afzel.	Shrub	Apocynaceae	9, 11, 16, 17, 28	Leaves, stems, roots
***Raphia vinifera* P. Beauv.	Tree	Palmaceae	2, 7, 11, 25	Leaves, fruits
Ricinodendron heudelotti (Baill) Pierre ex Pax.	Tree	Euphorbiaceae	9, 10, 17, 18, 27	Bark, stems, seeds, leaves
Rothmania hispida (K. Schum.) Fager	Shrub	Rubiaceae	5, 17, 19	Leaves, fruits, whole plant
Rhynchospora corynbosa (Linn.) Britt.	Herb	Cyperaceae	17	Leaves, fruits
***Sacoglottis gabonensis* (Baill.) Urb.	Tree	Humiriaceae	11, 12, 23, 25, 29	Bark, leaves, stems
**Sarcophyrynium braohystachys* (Benth.) K. Schum.	Herb	Marantaceae	17, 21	Leaves, whole plant
Schrankia leptocarpa DC	Straggling herb	Fabaceae	19	Whole plant
Scleria naumanniana Boeck	Herb	Cyperacese	17	Fruits
S. verrucosa Wild.	Herb	Cyperaceae	17	Fruits
Scoparia dulcis Linn.	Shrub	Scrophulariaceae	17	Whole plant
Selaginella nyosurus (Sw) Alston	Creeper	Selaginellaceae	17, 21	Whole plant
S. scandens (P. Beauv.) Spring	Creeper	Selaginellaceae	17, 21	Whole plant
Setaria megaphylla (Steud.) Dur. and Schinz.	Herb	Poaceae	21, 30	Leaves, roots
Senna alata (Linn.) Roxb.	Shrub	Fabaceae	5, 11, 17, 19	Leaves, roots, stems
Senna hirsuta (Linn.) Irwin and Barneby	Shrub	Fabaceae	17	Whole plant
Senna obtusifolia (Linn.) Irwin and Barneby	Shrub	Fabaceae	10, 17, 25	Whole plant
Senna occidentalis (Linn.) Link	Herb	Fabaceae	19	Whole plant
Senna podocarpa (Guill and Perr.) Lock	Shrub	Fabaceae	19	Bark
S. rhombifolia Linn.	Herb	Malvaceae	7, 17	Bark, leaves, stem, roots
Smilax anceps Willd.	Climbing herb	Smilacaceae	7, 17	Roots, tubers, whole plant
Spigelia anthelma Linn.	Herb	Loganiaceae	28	Whole plant
Spilenthes filicaulis (Schum and Thonn.) C.D. Adam	Creeping herb	Asteraceae	17	Flowers, whole plant

Contd...

Table 66.1–Contd...

Species	Habit	Family	Ethnobotanical Uses	Plant Parts Used
Spondiathus preussii Engl.	Tree	Euphorbiaceae	2, 11, 28	Leaves, stems
Stachytarpheta cayenensis (L.C. Rich) Schau.	Shrub	Verbanaceae	17	Whole plant
S. indica (Linn.) Vahl.	Shrub	Verbanaceae	17	Whole plant
Staudtia stipitata Warb	Tree	Myristicaceae	2, 4, 11, 12, 16, 27	Stems, bark
Saterculia tragacantha Lindi	Tree	Sterculiaceae	7, 10, 13, 15, 17	Bark, leaves, stems
Strophanthus sarmentosus DC	Woody climber	Apocynaceae	17, 19	Bark, leaves, seeds
Symphonia globulifera Linn. f.	Tree	Clusiaceae	2, 3, 4, 11, 12, 15, 17, 27	Stems, leaves, roots
***Synsepallum dulcificum* (Schum. and Thonn.) Daniell	Shrub	Sapotaceae	4, 10, 11, 25	Seeds, leaves, stem
Terminalia ivorensis A. Chev.	Tree	Combretaceae	12, 15, 17, 27	Bark, stems
**Tetracera podotricha* Gilg	Shrub	Dilleniaceae	11, 17	Stems, bark
**Tetrapleura tetraptera* (Schum. and Thonn.) Taub.	Tree	Fabaceae	4, 11, 15, 17, 23, 27	Stem, fruits, bark
Thalia welwitschii Ridl	Herb	Marantaceae	19, 30	Leaves, whole plant
Thaumatococcus danielii (Benn.) Benth	Herb	Marantaceae	10, 25, 28, 30	Fruits, leaves
Thevetia peruviana Juss.	Shrub	Apocynaceae	6, 17, 28	Bark, stem, leaves, roots
Treculia africana Decne	Tree	Moraceae	10, 12, 17, 18, 21, 27	Fruits, seeds, stems
Trichilia heudelotii Planch ex Oliv.	Tree, Shrub	Meliaceae	11,12, 27	Stems
Trloax procumbens Linn.	Herb	Asteraceae	9, 17	Leaves, roots
Tristemma incompletum R. Br.	Herb	Melastomataceae	10	Fruits
Triumpheta cordifolia A. Rich	Herb	Tiliaceae	7, 30	Leaves, bark
Uapaca staudtii Pax.	Tree	Euphorbiaceae	2, 4, 11, 17, 27, 30	Stem, leaves, roots
U. togoensis Pax	Tree	Euphorbiaceae	2, 4, 11, 17, 27, 30	Stem, leaves, roots
Urena lobata Linn.	Shrub	Malvaceae	7, 9, 17	Bark, roots, seeds
**Uvaria chamae* P. Beauv	Shrub	Annonaceae	11, 17, 28	Stems, fruits, roots
**U. cristata* R.Br.	Shrub	Annonaceae	11, 17, 28	Stems, fruits, roots
Voacanga africana Stapf.	Tree	Apocynaceae	9, 11, 16, 28	Shoots, stem, leaves
Vossia cuspidata (Roxb.) Griff.	Herb	Poaceae	2, 7, 9	Stems, leaves
Waltheria indica Linn.	Herb	Steruliaceae	7, 21	Bark, whole plant
***Xylopia aethiopica* (Dunal) A. Rich.	Tree	Annonaceae	2, 4, 10, 17, 23, 27	Stems, fruits
***Xylopia villosa* Chipp.	Tree	Annonaceae	14, 15, 17, 27, 29	Stems, seeds

*: Endangered species; **: Endemic species.

Ethnobotanical Uses: 1: Bait; 2: Building; 3: Toothbrush; 4: Commercial; 5: Dyes; 6: Fence; 7: Fibre/Cordage; 8: Field-tool; 9: Fodder, Feed; 10: Food/Condiment; 11: Fuel; 12: Furniture; 13: Gum/Resins; 14: Handicrafts; 15: Home tool; 16: Latex; 17: Medicine; 18: Oil, Oranmental; 20: Plaiting/Weaving; 21: Religious/Cultural; 22: Soap; 23: Spice/Flavouring; 24: Stimulant; 25: Sweetners; 27: Timber; 28: Toxic; 29: Work tool; 30: Wraps/Packing.

Table 66.2: List of Plant Species Present in the Homestead Gardens in the Study Areas. Nomenclature and ethnobotanical uses of the species follow those in Table 66.1.

Species	*Habit*	*Family*	*Ethnobotanical Uses*	*Plant Parts Used*
Abelmoschus esculentus (Linn.) Moech	Herb	Malvaceae	7, 10, 17, 23	Bark, leaves seeds
Amaranthus hybridus Linn.	Herb	Amaranthaceae	7, 10	Bark, leaves
Ananas comosus (Linn.) Merril	Herb	Bromeliaceae	7, 10, 17	Fruits, stems
**Ananas muricata* Linn.	Shrub	Annonaceae	10, 14	Fruits, stems
Arachis hypogea Linn.	Herb	Fabaceae	9, 10, 18, 22, 23	Fruits
**Artocarpus communis* Forest	Tree	Moraceae	10, 12, 13, 14, 30	Fruits, stems, leaves
***A. heterophylla* Lam.	Tree	Moraceae	10, 12, 13, 14, 30	Fruits, stems, leaves
Azadirachta indica A. Juss	Tree	Meliaceae	3, 11, 17, 19	Bark, leaves, stems
Caesalpina pulcherima Swartz	Tree	Fabaceae	11, 12, 19, 27	Flowers, stems
Capsicum annum Linn.	Shrub	Solanaceae	4, 10, 17, 23, 24, 28	Fruits, seeds
C. frutescens Linn.	Shrub	Solanaceae	4, 10, 17, 23, 24, 28	Fruits, seeds
Carica papaya Linn.	Herb	Caricaceae	10, 13, 17, 26	Bark, stems, leaves
**Chrysophyllum albidum* G.Don.	Tree	Sapotaceae	10, 11, 13, 16, 27	Bark, fruits, stems
**Citrullus colosynthis* Schrad	Climber	Cucurbitaceae	10, 17	Fruits, roots
C. vulgaris Linn.	Climber	Cucurbitaceae	10, 17	Fruits, roots
Citrus aurantifolia (Christm.) Swing	Shrub	Rutaceae	17, 24	Fruits
C. limon (L.) Burm. f.	Shrub	Rutaceae	10, 17, 24	Fruits
C. reticulata Blanco	Shrub	Rutaceae	10, 17, 22, 24	Fruits
C. sinensis (Linn.) Osbeck	Shrub	Rutaceae	10, 17, 22, 24	Fruits
Cocos nucifera Linn.	Tree	Palmaceae	2, 4, 7, 10, 14, 15, 17, 18, 22, 24, 27	Fruits, seeds, leaves, whole plant
***Cola argentea* Mast	Tree	Stercuriaceae	2, 4, 11, 24, 27	Fruits, stems
C. nitida (Vent) Schott and Endl.	Tree	Storculiaceae	2, 4, 11, 12, 24, 27	Fruits, stems
***Cofocasia antiquorum* Schott	Herb	Araceae	9, 10, 30	Leaves, corns
C. esculenta (Linn.) Schott.	Herb	Araceae	9, 10, 19, 2.3, 30	Leaves, corns, whole plant
Corchorus olitorius Linn.	Herb	Tiliaceae	7, 10	Bark, leaves
Coula edulis Baill.	Tree	Olacaceae	3, 8, 12, 17, 24, 27	Fruits, stems
Crescentia cujete Linn.	Tree	Bignoniaceae	11, 15, 19, 30	Fruits, stems, whole plant
Cucumis melo Linn.	Herb	Cucurbitaceae	9, 10, 18, 23	Fruits, seeds
Cucumis sativus Linn.	Herb	Cucurbitaceae	10	Fruits, seeds
***Curcubita moschata* Linn.	Herb	Cucurbitaceae	9, 10, 15, 30	Fruits, leaves
Cymbopogon citratus (DC) Stapf.	Herb	Poaceae	17, 24	Leaves, whole plant
***Dacryodes edulis* (G. Don.) H.J. Lam.	Tree	Burseraceae	3,9, 10, 11, 12, 13, 15, 18	Bark, fruits, leaves stems
***Dennettia tripetala* Bak. f.	Shrub	Annonaceae	10, 11, 23, 24	Fruits, seeds, stems, leaves

Contd...

Table 66.2–Contd...

Species	*Habit*	*Family*	*Ethnobotanical Uses*	*Plant Parts Used*
Dioscorea alata Linn.	Climber	Dioscoreaceae	7, 10	Stems, tubers
D. bulbifera Linn.	Climber	Dioscoreaceae	7, 10	Stems, tubers
D. cayenensis Lam	Climber	Dioscoreaceae	7, 10	Stems, tubers
D. dumentorum (Kunth) Pax.	Climber	Dioscoreaceae	7, 9, 10	Stems, tubers
D. esculenta (Lour.) Burkm	Climber	Dioscoreaceae	7, 10	Stems, tubers
**D. rotundata* Poir	Climber	Diosoreaceae	7, 10, 23	Stems, tubers
Elaeis guineensus Jacq	Tree	Palmaceae	2, 7, 9, 11, 12, 14, 15, 17, 18, 20, 22, 23	Fruits, seeds, leaves, bark, stems
Eugenia owariensis P. Beauv.	Tree	Myrtaceae	10, 11, 15	Fruits, stems
Garcinia kola Heckel	Tree	Clusiaceae	3, 8, 11, 12, 17, 24	Fruits, seeds, stems
Gmelina arborea Roxb	Tree	Verbanaceae	4, 7, 11, 15, 29	Stems
***Gnetum africannum* Welw.	Herb	Gnetaceae	4, 7, 10	Leaves, stems
***Gongronema lafiolium* Benth.	Climber	Asclepiadaceae	3, 7, 10, 21, 23	Leaves, stems, bark
***Heinsia crinata* (Afzel) G. Tayl.	Shrub	Rubiaceae	3, 4, 10, 15, 17	Leaves, stems
Hevea braziliensis (Kunth) Muell. Arg.	Tree	Euphorbiaceae	2, 4, 6, 9, 11, 13, 14, 16, 18	Bark, fruits, stems, leaves, seed
***Hyptis suaveolens* Poit	Herb	Lamiaceae	10, 17, 23, 24, 25	Leaves, fruits
**Irvingia gabonensis* (Aubru-Lecomte ex O'Rorke) Baill	Tree	Irvingiaceae	2, 4, 8, 9, 10, 11, 12, 14, 15, 23, 27	Fruits, seeds, stems
L. grandifolia (Engl.) Engl.	Tree	Irvingiaceae	2, 4, 8, 10, 11, 12, 14, 27	Fruits, seeds, stems
***Lasianthera africana* P. Beauv.	Shrub	Icacinaceae	3, 6, 10, 17	Leaves, stems
Lycopersicon esculentum Mill.	Herb	Solanaceae	4, 10	Fruits, seeds
Mangifera indica Linn.	Tree	Anacardiaceae	4, 10, 11, 12, 13, 17, 27, 30	Bark, fruits, stems, leaves
Manihot esculentus Krantz	Shrub	Euphorbiaceae	1, 4, 9, 10, 13, 16, 19	Leaves, stems, tubers, whole plant
Mucuna urens Adans.	Herb	Fabaceae	7, 10, 23	Seeds, stems
Musa paradisiacal Linn.	Herb	Musaceae	7, 10, 17, 22, 30	Leaves, stems, fruits, whole plant
M. paradisiaca subsp. *sapientum* Linn.	Herb	Musaceae	7, 10, 17, 22, 30	Leaves, stems, fruits, whole plants
***Ocimum gratissimum* Linn.	Shrub	Lamiaceae	9, 10, 17, 23, 24	Leaves, fruits
Persea Americana Mill	Tree	Lauraceae	5, 9, 10, 11, 18	Leaves, fruits, stems
Phaseolus lunatus Linn.	Herb	Fabaceae	10, 17	Seeds, leaves, pods
***Piper guineensis* Schurn and Thonn.	Climber	Piperceae	10, 17, 23, 24	Leaves, seeds
Plumeria alba Linn.	Tree	Apocynaceae	16, 17, 19	Bark, stems, whole plant
P. rubra Linn.	Tree	Apocynaceae	16, 17, 19	Bark, stems, whole plant
Psidium guajava Linn.	Shrub	Myrtaceae	3, 9, 10, 11, 15, 17, 26	Fruits, bark, leaves

Contd...

Table 66.2–Contd...

Species	*Habit*	*Family*	*Ethnobotanical Uses*	*Plant Parts Used*
***Raphia hookeri* Mann. and Wendl.	Tree	Palmaceae	2, 4, 6, 7, 11, 12, 13, 14, 16, 20, 24, 28	Fruits, leaves, stems, seed, whole plant
Ricinus communis Linn.	Shrub	Euphorbiaceae	4, 7, 17, 18, 22	Fruits, seeds, stems, bark, leaves
Saccharium officinarum Linn.	Herb	Poaceae	4, 7, 10, 25	Stems
**Sesamum indicam* Linn.	Herb	Pedaliaceae	10, 17, 18	Leaves, fruits, seeds
Solanum macrocarpum Linn.	Shrub	Solanaceae	9, 10	Fruits
**Spondias mombin* Linn.	Tree	Anacardiaceae	6, 9, 10, 12, 27	Fruits, stems
Talinum triangulare (Jacq.) Willd.	Herb	Portulaceaceae	9, 10, 17	Leaves, whole plant
***Telfairia occidentailis* Hook f.	Climber	Cucurbitaceae	9, 10, 17, 23	Fruits, leaves seeds, roots
Terminalia catappa Linn.	Tree	Combretaceae	8, 9, 10, 17, 18, 19, 22, 26, 27	Fruits, seeds, stems, bark, leaves
Theobroma cacao Linn.	Tree	Sterculiaceae	11, 17, 24	Fruits, stems, seeds
Trema guineensis (Schunn. & Thonn.) Ficalho	Tree	Ulmaceae	6, 11, 17	Stems, roots
Vernonia amygdalina Del.	Shrub	Asteraceae	3, 10, 17, 23	Leaves, stems, roots
**Xanthosoma sagittifolium* Schott	Herb	Araceae	10, 23, 30	Leaves, corms, whole plants
Zea mays Linn.	Herb	Poaceae	4, 7, 9, 10, 17, 30	Fruits, stems
Zingiber officinale Rosc.	Herb	Zingiberaceae	4, 17, 10, 23, 24	Stems, rhizomes

*: Endangered species; **: Endemic species.

Diversity of Uses of Plant Species

All plant species have at one reported use. Two hundred and fifty nine forest species and eighty homestead garden species are used for at least a single purpose (Tables 66.1, 66.2). Species with more uses include *Elaeis guineensis* (12), *Raphia hookeri* (12), *Cocos nucifera* (11), *Irvingiagabonensis* (11), *Hevea brasiliensis* (9), *Lonchocarpus cyanescens* (9), *Lovoa trichilioides* (9), *Pterocarpus santalinoides* (9), *Dacryodes edulis* (8), *Khaya grandifoliola* (8), *Lophira alata* (8), *Anthocleista vogelii* (8), *Pachystela brevipes* (8), and *Symphonia globulifera* (8). Ethnobotaincal use with more species include Medicine (178/52.5 per cent), Fuel (107/31.6 per cent), Food/condiment (100/29.4 per cent), Timber (72/21.2 per cent), Food/feed (70/20.6 per cent), Commercial (58/17.1 per cent), Furniture (55/16.2 per cent), Fibre/cordage (55/16.2 per cent), Homestool (45/13.2 per cent) and Fence (38/11.2 per cent). These figures are inappropriate for medicinal and fuel plants as almost every species is used in healing a disease or is used as fuel wood. Diversity index for species ethnobotanical use is high, 7.9 for forest species and 5.0 for homestead garden species. This is indicative of the multiple uses the species are put to.

There is an indication that people from rural areas within the study areas use more forest and homestead garden species than people in towns as the former places are more distant and isolated villages. The surplus forest produce is traded locally. The people in better-communicated towns buy and sell their produce in well-organized markets in and around the towns. Thus, the use of forest and homestead garden produce in the Niger Delta communities is affected by communication systems such as roads and transport.

Economic Importance of the Forests Plants

The type, quality and quantity of resources from the Niger Delta forests have diverse uses (Etukudo, 2000). More important are the timber species (*Antiaris africana, Albizia zygia, Brachystegia enrycoma, Ceiba pentandra, Diospyros alboflavescens, Erythrophleum ivorense, Funtumia elastica, Hallea stipulosa, Hannoa klaineana, Hylodendron gabunense, Khaya grandifoliola, Lannea welwitschii, Lovoa trichilioides, Nauclea diderrichii, Oxystigma mannii, Staudtia stipitata, Tlipochiton scleroxylon* and *Uapaca togoensis*); the fruit species (*Canalium schweinfulrhii, Dialium guineense, Irvingia gabonensis, Maesobotrya balteri, Microdesmis peberula, Momordica charantia, Pentaclethra macrophylla, Synsepallum dulcifilum,* and *Tetracarpidium conophorum*), the spices, flavouring and thickners (*Aframomum melegueta, Irvingia gabonensis, Piper guineense, Tetrapleura tetraptera, Xylopia aethiopica*), the beverages (*Cocos nucifera, Elaeis guineensis, Ananas comosus, Annona muricata, Sacoglottis gabonensis, Synsepalum dulcificum*) and the medical species (Table 66.3). The homestead gardens contain a great number of these species (Table 66.2).

The Forests and Animal Diets

The forests are natural habitats of various animal species, some endemic to Nigeria (*Cercopithecus sclateri, Cercocebus torquatus, Cercopithecus mona, C. niditans*). Some of the animals are important bush meat, source most appreciated by the people of the Niger Delta. The most important venison of these forests are *Cephalophus* sp. and *Thryonomys swinderianus.* The population sizes of these animals have been greatly reduced due to indiscriminate poaching, habitat fragmentation and loss to various forms of anthropogenic activities including oil exploration and exploitation. The plant species used by *cephalophus* sp. and *T. swinderianus.* as food sources include those cultivated in homestead gardens (*e.g. Colocasia* sp., *Corchorus olitorius, Curcubita moschata, Dioscorea dumentorium, Manihot esculentus, Solanum macrocarpum, Telfairia occidentalis, Zea mays*) as well as the fruits of *Elaeis guineensis, Irvingia gabonensis, Persea americana* and *Psidium guajava;* and those found in the forests: *Alchornea cordifolia, Allanblackia floribunda, Centrosoma pubescens, Dialium guineense, Lonchocarpus cyanescens, Macaranga balteri, Maesobotrya balteri, Manniophyton fulvum, Microdesmis puberula, Pterocarpus* sp. and *Voacanga africana.* The monkeys (*Cercopithecus sp., Cercocebus torquatus*) feed on the above fruits species.

The Forests as Sources of Building Materials

The traditional house in the Niger Delta is built from the materials obtained from the forests. This is obvious in the more isolated villages. Materials for the buildings come from species such as *Acioa barteri, Alstonia* sp., *Bambusa vulgaris, Betfinia grandiflora, Cocos nucifera, Elaeis guineensis, Klainedoxa gabonensis, Myrianthus arboreus, Uapaca* sp. and *Raphia hookeri* among others. These traditional houses are fast disappearing and are being replaced by modern ones because over-exploitation of the forests had brought about scarcity of these building materials, the regenerating forests do not contain materials old enough to be used in building; it takes more time to search for and obtain available and needed materials; and since the advent of oil exploration and exploitation some of the rural dwellers who are gainfully employed want to have same standards of living as city dwellers (Rico-Gray *et al.*, 1991). This has resulted in dwindling economic values of these natural materials and their turnover rates. It is suggested that rather than abandoning (the use of) these materials, modern ideas should help improve the production, quality and sustainability of these natural materials, which will enhance their cost benefit status and high turnover rates. These could be used for alternative or different purposes.

Table 66.3: Some Uncultivated Plant Species Identified for Medical Use in the Niger Delta Region, Nigeria

Species	*Family*	*Action*	*Usage*
Abrusprecatorius Linn.	Malvavceae	Nervous disorders treatment	Infusion of seeds
Abutilon mauritianum (Jacq.) Medic	Malvaceae	Alleviates pectoral pains	Mucilage of leaves
Achyranthes aspera L.	Amaranthaceae	Diuretic, anti-pneumonia	Infusion, decoction of whole plant
Aframonum melegueta K. Schum.	Zingiberaceae	Bechic, expectorant, mucolytic	Fresh seeds
Aframomonum sceptrum (Oliv. and Hanb) K. Schum.	Zingiberaceae	Anti-flatulent, calms nervous stomach spasms	Fresh fruits, essence
Aframomum strobilaceum (Sm) Hepper	Zingiberaceae	Anthelmintic	Root decoction
Ageratum conizoides Linn.	Asteraceae	Wound healer, antiseptic	Poultice leaves
Alchornea laxiflora (Benth) Pax. and K. Hoffin	Euphorbiaceae	Gonorrhoea treatment	Leaves and roots decoction
Alternanthera pungens H.B.K	Amaranthaceae	Febrifuge	Infusion of leaves
Althernanthera sessilis (Linn.) R.Br.ex Roth	Amaranthaceae	Febrifuge, snake bite	Infusion, decoction of stems and leaves
Anthonatha macrophylla P. Beauv	Fabaceae	Wound healer	Compress with sap of bark
Argemone mexicana Linn.	Papaveraceae	Skin mycosis, scabies	Lotions, latex
Aspilia africana (Pers.) C.D. Adams	Asteraceae	Alleviates asthma	Leaf decoction
Asystasiagangetica (Linn.) T. Anders	Acanthaceae	Anti-rheumatic, analgesic	Leaf juice and lotion
Bryophyllum pinnatum (Lam.) Oken	Crassulaceae	Coughs, boils treatment	Leaf juice, liniment
Canarium schweinfurthii Linn	Burseraceae	Cough, piles treatment	Bark decoction
Chromolaena odorata (L.) R.M. King and Robinson	Asteraceae	Hepatitis, jaundice healing	Infusion of leaves, decoction of whole plant
Cleome viscosa Linn.	Capparidaceae	Ear troubles treatment	Juice, lotion of leaves, whole plant
Cnestls ferruginea D.C.	Connaraceae	Enema for dysentery, diarrhea,	Root decoction and juice sinusitis, gonorrhea
Combretum hispidium Laws	Combretaceae	Jaundice treatment	Raw root chewed
Combretum zenkeri Engl and Die's	Combretaceae	Depurative, diuretic, sudorific	Leaf extract, decoction
Commelina africana Linn.	Commelinaceae	Laxative, beneficial in leprosy	Infusion, decoction of whole plant
Commelina diffusa Burm f.	Commelinaceae	Soothes skin itching; wound	Poultice plant, compress healer, abscess
Costus afer Ker-Gawl	Zingiberaceae	Rheumatism relief	Leaf decoction
Costus schlechteri Winkler	Zingiberaceae	Wound healer, earaches	Plant extract, decoction

Contd...

Table 66.3–Contd...

Species	*Family*	*Action*	*Usage*
Coula edulis Baill.	Olacaceae	Anaemia, dysentery alleviation	Powered bark decoction
Datura stramonium Linn.	Solanaceae	Anti-rheumatic, asthmatic earache	Poultice of leaves with oil, flower juice
Desmodium turtuosium (Slo.) DC	Fabaceae	Soothing, anti inflammatory in	Leaf extract in enema haemorrhioids
Diodia scandens Sw.	Rubiaceae	Asthma treatment	Leaf extract as emetic
Emilia coccinea (Sims) G. Don.	Asteraceae	Antitussive, calms coughs	Infusion of plant
Emilia sonchifolia (Linn.) DC	Asteraceae	Antitussive, bechic, calms	Infusion of plant tonsillitis
Eramomastox speciosa (Hockst.) Cufod	Acanthaceae	Treatment of diabetes	Leaf juice, lotion
Euphorbia heterophylla Linn.	Euphorbiaceae	Laxative	Decoction of leaves
Euphorbia hirta Linn.	Euphorbiaceae	Sickle cell ailment treatment	Decoction of leaves, roots
Fagara zanthoxyloides Lam.	Rutaceae	Treatment of sickle cell ailment	Decoction, infusion of leaves, bark
Ficus exasperata Vahl	Moraceae	Stops haemorrhage, eye sore	Leaf juice
Aeurya aesatuans (Linn.) Graudex Miq	Urticaceae	Assuages asthma	Air-dried leaves
Fleurya ovalifolia (Schum. and Thorn.) Dandy	Urticaceae	Enema for epilepsy	Leaf decoction
Gloriosa superba Linn.	Liliaceae	Heals leprosy	Root decoction
Glyphaea brevis (Spreug.) Monachino	Tiliaceae	Antiseptic in venereal discharges	Decoction of leaves
Hallea stipulosa (DC) J.F. Lersy	Rubiaceae	Prevents stomach spasms	Boiled root chewed
Heterotis rotundifolia (Sm) Jac-fet	Melastomataceae	Treatment of conjunctivitis	Leaf lotion
Hippocratea pallens Planch. ex Olive	Hippocrateaceae	Calms stomach ache	Fruit pulp
Icacina tirichantha Oliv	Icacinaceae	Treatment of mumps	Tuber juice, lotion
Ipomoea involucrata P. Beauv.	Convolvulaceae	Laxative, analgesic	Leaf infusion
Irvingia gabonensis Bail.	Irvingiaceae	Useful for malaria	Bark decoction
Justicia schimperi (Hochst) Dandy	Acanthaceae	Enema for umbilical pains in	Leaf decoction, infusion children
Loranthes asphyllus Sprauge	Loranthaceae	Antidiabetic	Leaf decoction
Mallotus oppositifolius (Geisel) Mull. Arg.	Euphorbiaceae	Stops bleeding	Leaf juice
Melanthera scandens (Schum and Thonn.) Roberty	Asteraceae	Wound healer, stops bleeding	Leaf juice
Melochia cochorifolia Linn.	Sterculiaceae	Prevents bronchial spasms	Root lotion
Milletia thonningii (Schum. and Thonn.) Bak.	Fabaceae	Prevents abortion, miscarriage in goats	Raw leaves, young branches

Contd...

Table 66.3–Contd...

Species	*Family*	*Action*	*Usage*
Momordica charantia Linn.	Cucurbitaceae	Useful in piles, jaundice, as vermiguge	Fruits, leaves
Musanga cecropioides R.Br.	Cecropiaceae	Indigenous medicine for toothache	Bark infusion
Myrianthus arboreus P. Beauv.	Moraceae	Heals heamaturia	Roots liquid
Nauclea diderrichii (De Wild and Dur.) Merril	Rubiaceae	Enema for indigestion	Bark decoction
Palisota hirsuta (Thunb.) K. Schum	Commelinaceae	Treatment of gonorrhoea	Pounded root decoction with lime juice
Peperoma pellucida (L.) H.B.K.	Piperaceae	Heals earache	Juice of leaf
Phyllanthus amarus Schum and Thonn.	Euphorbiaceae	Eliminates fungi, heals ringworm, scabbies	Poultices of leaves
Physalis angulata Linn.	Solanaceae	Stomachic invigorator	Leaf infusion, decoction
Piper guineense Schum and Thonn.	Piperaceae	Antirheumatic, carminative, insecticidal	Pulverized seeds and fruits
Platostoma africanum P. Beauv.	Lamiaceae	Insecticidal	Pulverized leaves
Pupalia lappacea (Linn.) Juss.	Amaranthaceae	Antitussive, eases cough	Leaf decoction
Rauvolfia vomitoria Afzel	Apocynaceae	Reduces blood pressure	Root decoction, alkaloids
Sida acuta Burm. f.	Malvaceae	Astringent, useful in urinary	Root decoction diseases
Stachytarpheta cayennensis L.C. (Rich) Schau	Verbenaceae	Demulcent, soothes inflammation	Poultice of leaves
Stachytapheta indica (Linn.) Vahl.	Verbenaceae	Indigenous treatment for malaria	Root infusion, decoction
Symphonia globullrera Linn. f.	Clusiaceae	Stomachic invigorator	Bark decoction
Tetrapleura tetraptera (Schum. and Thonn.) Taub.	Fabaceae	Emetic and purgative	Bark decoction
Tridax procumbens Linn.	Asteraceae	Anti-haemorrhage	Plant juice, poultice
Uapaca staudtii Pax	Euphorbiaceae	Useful in cough, antitussive	Raw fruit
Uapaca topensis Pax.	Euphorbiaceae	Useful for whooping cough, antibiotic	Plant and root sap
Urena lobata Linn.	Malvaceae	Anti-rheumatic	Poultice of roots
Uvaria chamae P. Beauv.	Annonaceae	Useful in and stops nose bleeding, diarrhoea	Roots decoction, infusion
Uvaria cristata R.Br.	Annonaceae	Treatment of waist pains	Root decoction
Xylopia aethiopica (Dunal) A. Rich	Annonaceae	Carminative, treatment of kidney and bladder afflictions	Fruit infusion, decoction

Homestead Gardens Produce

A variety of produce is obtained from the homestead gardens: fruits, flavouring, oils, ornamentals, spices, vegetables and thickeners in quantities that are sold or traded locally. Their commercial values can be enhanced if production is co-ordinated and harnessed effectively from the different villages to towns where they are in high demand. These can advance the per capita income of the producers and in turn improve their living standards.

Medical Plants from the Forests

The Niger Delta forests habour a large number of plant species known to be of importance in the medical and pharmaceutical industries. *Alchomea laxinora, Cnestis feniginea, Fagara zanthoxyloides, Icacina trichantha, Piper guineense* and *Uvaria cristata* are widely known and used as species of medicinal utility. *Aspilia africana* contains thiarubine–A in its root, a red oil that kms parasites, viruses, fungi and bacteria. *Smilax kraussiana* contains saponinic glycosides, resin and essential oil that promote elimination of urea, uric acid and other organic wastes as well as decreasing blood cholesterol level. *Strophanthus sarmentosus* is a source of drug called cortisone for the treatment of rheumatoid arthritis. Vernonioside-B_1, is a constituent of *Vernonia amygdalina* which has an antitumoral and antibacterial properties. It is used in treating schistosomiasis. The list of some plants occurring in the Niger Delta forest with their medicinal values is presented in Table 66.3. The information contained therein is based on indigenous knowledge of the local people. The usage of the plants has been, in the main, through extracts or in raw form for ages. However, there is need for scientific analyses of these plants for proven efficacies by biochemists, pharmacognosists, pharmacologists and pharmaceutical industries.

Endemic and Endangered Plants Species

A number of plants species in the Niger Delta forest are endemic and endangered (Tables 66.1 and 63.2). Endemism and a number of sub-species presence in the study area are evidenced in phylogenetic plasticity and the dynamics of ecological and environmental perturbations (Khan *et al.*, 2003).

The loss of plant biodiversity in the Niger Delta is precipitated by anthropogenic factors of over-extraction, unsustainable agriculture and forestry practices, pollution, land use changes, urbanization and probably global climate change. These culminate in biodiversity depletion and diminished productivity. Over extraction target species are those used as fuel wood, in medicine, in building, as timber and as fodder. Observations show that due to extensive use of plant species for these purpose, a large number of plants are endangered.

Conservation Methods

An important element in the relationship between ethnobotany and conservation is the detailed knowledge of the components of biodiversity possessed by the natives over the years and the dynamics of the ecosystem. Through observations, assessments and experimentations, these natives have successfully adopted their production methods towards the characteristics of these biological resources. The possession of this biodiversity-related knowledge should be preserved through capacity building in order to compliment scientific knowledge and provide information on biological system. This will also make it possible for the socio-cultural, economic and religious significance of many plants to be transferred from generation to generation. This requires education strategies that will raise the level of awareness of the natives in the conservation of biodiversity.

The indiscriminate over-exploitation of plants that impinges on food security and health care should be avoided and halted. No wild plant species should be endangered through international trade. There should be the promotion of *in situ* and *ex situ* cultivation of the endemic, medicinal and more obvious economic species. Conservation of the plants should also take the form of systematic protection through breeding. The protected areas in form of groves and nature reserves in the region should be preserved. All these could be achieved through collaborative cooperation of governments, Non-governmental organisations (NGO's), community-based organisations (CBO's), the natives, schools, colleges and other higher institutions.

As biodiversity involves global environmental change, species extirpation or extinction and changes in societal values (Khan *et al.*, 2003), ethnobotany offers an effective approach to plant biodiversity conservation, since this provides a wealth of data on timber and non-timber products which can be rationally exploited. Thus, commodities can be extracted from plant communities with minimal environmental damage and hence provide incentive for conservation. Conservation for diversity and stability, aesthetic, ethical, scientific values and utilitarian purposes is rooted in ethnobotany, which is the most cost-effect method to be practiced.

Conclusion

The economic importance of the tropical lowland rainforests rural dwellers in Niger Delta is emphasized. Though the natural environments of these forests have been modified by activities of man (arable farming, animal and oil exploitation, capital development, forestry production), the establishment of management systems that are flexible, adaptive and experimental would allow for sustainable exploitation while conserving them (Sayer *et al.*, 2000). These management systems should involve immediate economic benefits by integrating components (*e.g.* agroforestry, agrosilviculture) that will promote and complement long-term economic benefits. In addition, enforcement of forests laws, improvement of forest policy and adoption of rules and regulations should be in practice (Li *et al.*, 1999). When these strategies are combined with other vegetation studies and effectively applied, they would provide sustainable means of utilizing the natural resources of these forests and probably bring about compatibility of conservation of nature.

Acknowledgements

The author thanks the Akwa Ibom State Environmental Protection Agency, Uyo; Universal, Scientific and Industrial Company, Uyo and the Prodec-Fugro, Port Harcourt for their financial supports.

References

Akobundu, I.O. and Agyakwa, C.W., 1987. *Guide to West African Weeds.* I.I.T.A. Ibadan, 522 p.

Clay, J.W., 1988. *Indigenous Peoples and Tropical Forests.* Cultural Survival, Inc. Cambridge, M.A.

Etukudo, I., 2000. *Forests: Our Divine Treasure.* Dorand Publishers, Uyo, 194 p.

Etukudo, I., 2003. *Ethnobotany: Conventional and Traditional Uses of Plants.* The Verdict Press, Uyo, 191 p.

Hutchinson, J. and Dalziel, J.M., 1954–72. *Flora of West Tropical Africa,* (Revised by Keay, R.W.J. and Hepper, F.N). Crown Agents for Overseas Governments, London.

Khan, T.I., Dular, A.K. and Solomon, D.M., 2003. Biodiversity conservation in the Thar desert with emphasis on endemic and medicinal plants. *The Environmentalis,* 23: 137–144.

Kokwaro, J.O., 1995. Ethnobotany in Africa. In: *Ethnobotany: Evolution of a Discipline,* (Eds.) Schultes, R.E. and Von Reis, S. Chapman and Hall, London, pp. 216–225.

Li, C., Koskela, J. and Luukkan, O., 1999. Protective forest system in China: Current status, problems and perspective. *Ambio*, 28(4): 341–345.

Pangely, R., 1994. *International River Basin Organisation in Sub Saharan Africa Technical paper* 250, World Bank, Washington DC.

Peters, R.H., 1991. *A Critique for Ecology*. Cambridge University Press, New York.

Rico-Gray, V., Chemas, A. and Mandujano, S., 1991. Uses of tropical deciduous forest species by the Yacatecan Maya. *Agroforestry System*, 14: 149–161.

Sayer, J., Ishwaran, N., Thorsell, J. and Sigaty, T., 2000. Tropical forest biodiversity and the world heritage convention. *Ambio*, 29(6): 302–309.

Soforowa, A., 1982. *Medicinal Plants and Traditional Medicine in Africa*. John Wiley and Sons, Chichester.

Sokal, R.R. and Rohlf, F.J., 1995. *Biometry*. Colt Freeman and Company, New York, 887 p.

Spellerberg, I.F., 1995. Conserving biological diversity. In: *Conservation Biology*, (Ed.) Spellerberg, I.F. Longman, Harlow, pp. 25–35.]

Chapter 67

Groundwater Quality of Bhadravathi Town, Karnataka State

Vijaya Kumara, J. Narayana, K. Harish Babu, Devidas Kamath and E.T. Puttaiah
Department of PG Studies and Research in Environmental Science, Kuvempu University, Shankaraghatta – 577 451, Karnataka

ABSTRACT

A hydro chemical study of the groundwater of Bhadravathi town, Karnataka state, India has been carried out to examine the suitability of water for drinking purposes. Water samples representing the groundwater of the region were collected during pre-monsoon and post-monsoon seasons during the years 2000–2001 and 2001–2002 respectively. A total of 46 water samples collected from different sites of the town were analysed for water quality parameters. The data was analysed with reference to BIS and WHO standards. Ionic relationships were studied. The results of the study provide information needed for groundwater quality management in the region.

Keywords: *Groundwater quality, Bhadravathi town, Physico chemical parameters, Correlation, E. coli.*

Introduction

Water is the life blood of every living creature on earth. Through the wonders of nature, water can take many forms. It is easy to understand the significance of surface water that plays in our lives but it may be difficult to understand the water that exists below the earth's surface. Fresh water, for human consumption is a fragile and finite resource. It is vital for many aspects of economic and social development, for agriculture, energy production, domestic, industrial supply and is a critical component of global environment (Jain *et al.*, 2000). Hydrosphere is the most important factor for the life on earth and its sustainable development. Good quality of water is inadequate even for the normal living and is getting polluted due to industrial discharge including those of paper, textile, pesticides, fertilizers, detergents, oil refineries and photo films. A perusal of the available literature on the groundwater quality assessment, has revealed that, no scientific investigation has been carried out in respect of groundwater source of Bhadravathi town. It is with this background, the present work was taken up.

Materials and Methods

Study Area

Bhadravathi city is a taluk headquarters, situated about 16 kms from Shimoga district, Karnataka state (South India). Bhadravathi is the removed center of industries had a population of 1,63,784. There are 25 major and minor industries situated in the vicinity. The Mysore Paper Milk and Visweswaraya Iron and Steel Limited are the major industries and minor industries cover manufacturing units like rice mills, oil mills, cement pipes, chemicals, machinery parts and poultry farms. The geographical location of Bhadravathi city lies between the latitude of 13°49' to 13°54' north and longitude of 75°40' to 75°45' E m about mid southwestern part of the Karnataka state (Figure 67.1). It is situated at an altitude of 581.55 meters above Mean Sea Level (MSL) and the city covers an area of 67.08 sq. km. The average rainfall of the area is 1029 mm and Temperature varies from 8.9 to 40 °C. Geologically the study area consists of schists and gneisses of Archaean age and forms a part of Dhaewar super group.

Sampling locations in Bhadravathi town have been made using random grid or spatial network method on the basis of geographical ground map of Bhadravathi town. The ground map has been divided into 2 segments and m each segment an average of 23 samples from 23 different localities have been selected so that 46 water samples from 46 localities were collected (Table 67.1). Water samples from different locations were collected as per the guidelines of random sample technique. Borewells fitted with motors for water lifting were allowed to run the water for 5 minutes and the other fitted with hand pump were allowed to run 15 minutes in order to flush out stationary water. As soon as the collection of water, temperature and pH were measured immediately. The physico-chemical analysis were made adopting the standard methods (APHA, 1995)

Results and Discussion

The objective of present study was to assess the quality of fresh water in the study area with respect to physico-chemical parameters, to classify the groundwater on the basis of hydrochemical parameters in order to determine the suitability of water for various uses and to study the statistical correlation analysis and its interpretation, in order to evaluate quality of groundwater.

Colour (Col)

Generally the colour of water is due to degradation of organic matter and oxidation of divalent metal ion species such as Fe^{2+} and Mn^{2+} as were as untreated wastewater percolation or seepage (Sawyer, 1978; APHA, 1985 and Kotaiah, 1994). In the present investigation colour values varied from a minimum of 0 to a maximum of 100 Hazen units in pre-monsoon season (Table 67.3) and minimum of 5 to a maximum of 100 Hazen units in post-monsoon season (Table 67.2). The HIS acceptable limit for colour is 25 Hazen units. In the present study 26 per cent water samples in post-monsoon season and 17.4 per cent water samples in pre-monsoon season cross the BIS (1998) acceptable limit for

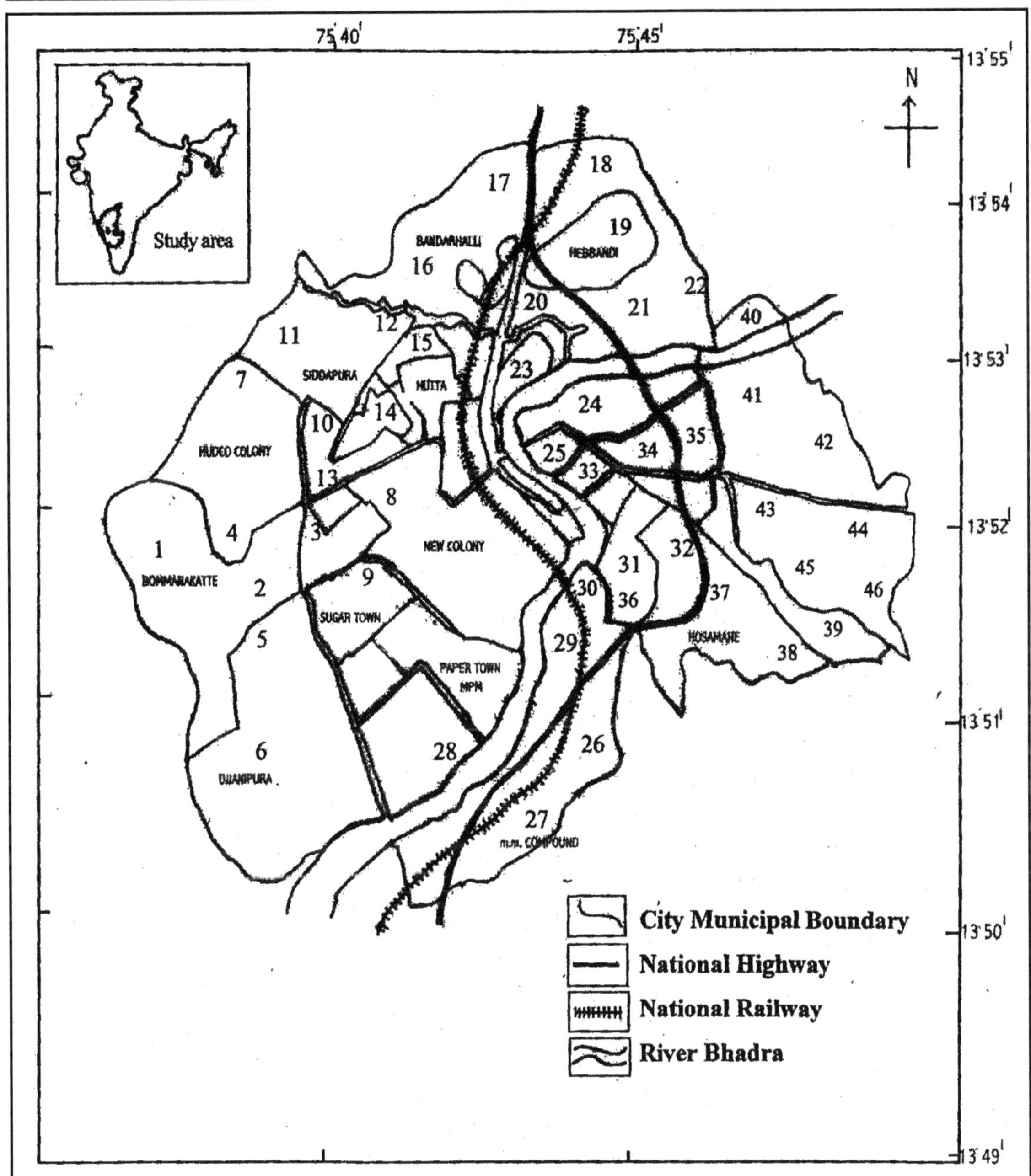

Figure 67.1: Map Showing Groundwater Sampling Locations in Bhadravathi Town

drinking water. Further, the colour has established a positive highly significant correlation with turbidity and iron in both pre and post-monsoon seasons (Tables 67.4 and 67.5). This is in agreement with the findings of Halck *et al.* (1963) and Kotaiah (1994), who made similar observation in their studies.

Table 67.1: Groundwater Sampling Locations of Bhadravathi Town

Sl.No.	Sample No.	Location	Depth of the Borewell in Feet
1.	B1	Bommanakatte, near Baptist Church	225
2.	B2	Bommanakatte, near Bommanakatte pond	180
3.	B3	Cooli block shed, near public toilet	231
4.	B4	Cooli block shed, Slum area, near Vinayaka Balasubramanya temple	250
5.	B5	Coli block, opposite to masjid	281
6.	B6	Ralappa shed, opposite to Manasa fancy center	231
7.	B7	Ganesha colony, opposite to Inchara beauty parlor	287
8.	B8	Jannapura, opposite to Jayashree Kalyana mantapa	293
9.	B9	Jannapura, beside Sridevi nilaya	290
10.	B10	Jannapura, near Karnataka dwaja sthamba	291
11.	B11	Kuvempu Badavane, Jannapura, opposite to Malleswara plastic stores	257
12.	B12	Rosa Siddapura, 8th cross	188
13.	B13	Zinc line, near public toilet and solid waste dumping site	175
14.	B14	Zinc line, slum area, opposite to Sapthagiri nivas	231
15.	B15	Caval gundi, opposite to Keshava nilaya	254
16.	B16	Bandarahathi, beside Shakthi nivas	265
17.	B17	Bandarahathi, opposite to Anganawadi Kendra	241
18.	B18	Kadada Katte, opposite to Nava chetana kannada school	255
19.	B19	Hebbandi Road	245
20.	B20	Caval gundi, opposite to ITI	235
21.	B21	Caval gundi, opposite of Narayanappa stores	268
22.	B22	Bhadravathi, Near bus stand, opposite to Shankara service station	281
23.	B23	Lower hutta, near Tirumala temple	245
24.	B24	Halappa circle, near Shankar film theater	276
25.	B25	Halappa circle, opposite to Viswa Swarupini Mariyamma Temple	282
26.	B26	Nehru Nagar, after railway gate, Tarikere road	255
27.	B27	Nehru Nagar, Tarikere road	268
28.	B28	M.M. compound, Tarikere road	245
29.	B29	Madavanagar, near Agricultural field	256
30.	B30	Bhutanagudi, old town, opposite to Shaneswara devalaya	275
31.	B31	Old town, opposite to Venkateswara nilaya	198
32.	B32	Bhutanagudi, opposite to Venkateswara nilaya	231
33.	B33	Bhutanagudi, near Kiran house	225
34.	B34	Basaweswara circle, OSM road, old town	254
35.	B35	Madavachar Circle, opposite to telephone exchange office	281
36.	B36	Old town, opposite to police station	264

Contd...

Table 67.1–Contd...

Sl.No.	Sample No.	Location	Depth of the Borewell in Feet
37.	B37	Bhrahmanara beedi, opposite to Srinivasa vanijya shale	253
38.	B38	Haladammaa beedi, near Grama devate temple	275
39.	B39	Hosamane, near Harish house	285
40.	B40	Gowdarahalli, oposite to bus stand	265
41.	B41	Gowdarahalli, near agricultural field	231
42.	B42	Hale Seegebagi, opposite to Hindu rudrabhoomi	241
43.	B43	Hale Seegebagi, near Laxmi samil	238
44.	B44	Azad nagar, near Budan Shab house	265
45.	B45	Anwar colony, near Urdu residency school	245
46.	B46	Anwar colony, near masjid	264

Turbidity (Tur)

Most of the suspended matter which may be colloidal or coarse size like clay, silt, organic matter and hydrolyzed metals, together with phytoplankton are responsible for turbidity in water (Knight, 1951). The values of turbidity ranged between 0.1 to 73.45 NTU in post-monsoon season (Table 67.2) and 0.05 to 47.3 NTU in pre-monsoon season (Table 67.3). Further, it is observed that the turbidity values of water samples during post- monsoon season have indicated an increasing trend when compared to pre-monsoon season. However except 3 samples most of the water samples in the present investigation are well within the safe limit of drinking water standards (25 NTU). Turbidity has indicated a significant positive correlation with iron in both pre and post-monsoon seasons (Tables 32.4 and 32.5). Nevertheless, most of other parameter shows insignificant correlation with turbidity.

pH

pH is one of the important parameters of water whose determination facilitates a quick evaluation of acidic or alkaline nature of water. In the present investigation, pH values varied from a minimum of 6.9 to a maximum of 8.35 with a mean value of 7.43 in post-monsoon season (Table 67.2). In pre-monsoon season it ranged between 6.6 to 8.25 with a mean value of 7.25 (Table 67.3). It indicates that the waters under study are alkaline in nature. The recommended value of pH for drinking purposes is between 6.5–8.5 (BIS, 1998). In the present study all the water samples analyzed are all well with in the safer limits. As has been reported here, the water quality of most of the groundwater resources has been found to be alkaline except on a few occasion as per the studies made by Somashekara Rao, 1989 and 1993; Narayana and Suresh, 1989; Govardan, 1990; Gill *et al.*, 1993; Mehta and Trivedi, 1993 and Mittal *et al.*, 1994. pH shows positive correlation with nitrate in pre-monsoon season and with sulfate and potassium in post-monsoon season. However, other parameters studied have exhibited an inverse relationship with pH (Tables 32.4 and 32.5). Observations of present study are in total agreement with the findings of Narayana and Suresh (1989).

Electrical Conductivity (EC)

In the present study, the values of EC ranged between 460–1480 μmhos/cm in pre-monsoon season (Table 67.3) and 456–1566 μmhos/cm in post-monsoon season (Table 67.2). It is observed that the EC values have exhibited an increasing trend in post-monsoon seasons owing to the fact that

Table 67.2: Water Quality Data (Average Values) of Bhadravathi Town During Post-monsoon Season for the Year 2000–2001

Sl.No.	Col.	pH	Tur	EC	TDS	TH	DO	COD	Ca^{2+}	Mg^{2+}	Na^+	K^+	Cl^-	HCO_3^+	SO_4^{2-}	NO_3^-	PO_4^-	F^-	E. coli
B1	75	7.10	37.55	1084	650	328	4.15	7.85	93.47	65	75.60	21.10	317	280	204	5.55	0.01	0.585	–
B2	32.50	7.15	4	1057	634	292	3.55	7	99.90	54.27	27.40	18	344	287	112	26.50	0.17	0.265	–
B3	50	7.35	6.30	916	550	306	3.50	3.25	77.90	43.39	32.05	13.20	244	287	123	3.40	0.03	0.86	–
B4	100	7.50	16.55	640	384	251	3.65	2.85	69.60	42.80	54.95	12.75	114	389	86.75	2.70	0.03	0.395	–
B5	7.50	7.20	3.90	1116	670	462	4.05	5.10	113	70.43	55.05	9.20	200	272	70	0.75	0.035	0.28	10
B6	75	7.20	38.90	789	474	364	4.40	11	125	41.78	56.45	12.85	109	407	82.50	1.05	0.015	0.345	25
B7	5	6.95	0.10	1131	679	450	3.85	12.50	124	63	70.55	16.45	228	520	89	4.50	0.01	0.515	–
B8	2.50	6.90	0.30	1016	610	379	3.70	10.15	131	62.80	116	30.05	223	627	111	3.60	0.125	0.38	–
B9	37.50	8.35	3.20	775	465	320	3.90	14.10	100	51.47	13.90	9.05	107	344	258	3.65	0.02	0.265	–
B10	5	7.40	0.55	535	321	195	4.20	6.10	58.10	25.11	58.30	4.35	108	507	312	6	0.01	0.385	–
B11	5	6.95	0.15	764	459	318	2.95	5.70	66.60	50.43	20.60	4.85	117	422	72.35	2	0.015	0.405	–
B12	2.50	7.25	0.30	787	472	327	4.35	11.70	93.15	80.15	13.75	1.80	121	406	42.50	37.50	0.225	0.60	–
B13	2.50	7.25	0.20	854	512	313	5.35	12.30	49.23	37.14	14.40	3.45	195	264	57.30	50	0.125	0.74	–
B14	2.50	7.25	0.10	907	544	346	5.90	11.55	116	43.21	55.90	6.66	171	467	163	23.50	0.135	0.61	–
B15	10	6.95	2.50	1237	742	465	3.30	11.80	161	64.56	62.30	11.80	245	690	163	23.25	0.15	0.80	–
B16	7.50	8.10	1.65	921	553	364	3.30	9.25	120	63.66	33.65	12.20	171	387	114	5.75	0.045	0.83	–
B17	5	7.50	0.60	1042	625	416	2.90	12.25	126	50.15	60.65	15.30	187	264	109	46.75	0.315	0.91	–
B18	7.50	7.36	0.20	1195	717	495	3.50	7.50	107	78.84	60.70	16.80	202	302	123	4.60	0.135	0.385	–
B19	5	7.45	0.25	970	582	412	4.45	6.80	111	54.91	55.90	11.16	167	267	69.50	12.75	0.01	0.426	–
B20	7.50	7.40	0.70	989	594	360	4.75	5.65	91.98	49.22	49.15	10.76	204	172	90.80	6	0.055	0.435	–
B21	12.50	7.25	2	811	487	309	4.30	3.40	74.90	47.68	32.80	5.30	160	365	96.20	3.35	0.01	0.56	–
B22	12.50	8.25	1.76	602	361	287	3.62	8.05	152	44.38	14.15	43	85.55	272	111	31.80	0.27	0.425	–
B23	5	7.20	0.20	585	351	209	2.90	6.25	60.20	29.15	19.30	3.70	83.40	387	154	45.45	1.7	0.36	–
B24	2.50	7.20	0.10	702	421	320	5.Z5	7.25	70	48.04	13.65	2.30	67.84	362	72.80	2.70	0.145	0.575	–
B25	2.50	7.40	0.15	774	464	372	3.70	4	63.50	66.33	25.30	6.15	81.18	475	80.30	2.15	0.035	0.575	–

Contd...

Table 67.2–Contd...

Sl.No.	Col.	pH	Tur	EC	TDS	TH	DO	COD	Ca^{2+}	Mg^{2+}	Na^+	K^+	Cl^-	HCO_3^+	SO_4^{2-}	NO_3^-	PO_4^-	F^-	E. coli
B26	32.50	7.10	0.70	1076	646	442	4.35	3	87.68	71.66	50.80	5.40	208	387	67.50	0.95	0.075	0.91	–
B27	15	7.95	2.50	1073	644	425	5.10	3.25	50.65	85.12	67.80	6.70	154	367	92.05	1.70	0.025	0.33	–
B28	5	7.30	0.15	949	569	383	3.85	6.40	70.21	63.60	35.30	7.50	149	362	82.30	2.16	0.015	0.285	–
B29	12.50	7.70	1.06	456	274	201	3.55	4.45	45.05	29.70	64.95	8.30	54.16	215	96.10	33	0.145	0.615	–
B30	100	7.35	73.75	874	525	233	5.35	6.50	33.37	43.51	43.40	11.60	57.67	282	117	3.85	0.025	0.80	–
B31	10	7.55	1.30	1006	604	463	2.90	11.90	122	61.41	35.30	4.20	195	257	154	6	0.01	0.76	–
B32	5	7.40	0.20	807	489	346	3.40	8.90	52.50	64.10	44.80	12.40	158	387	171	8.50	0.03	0.90	–
B33	45	7.50	9	1037	622	531	3.40	7	52.70	95.64	60.20	17.50	172	297	232	12.80	0.795	0.59	–
B34	20	7.95	3.90	986	592	382	4	10.90	80.51	61.31	41.90	13.80	205	364	302	11	0.045	0.585	–
B35	0.25	8.25	0.10	1024	615	431	4.10	6.95	60.28	98.33	38.30	16.10	183	327	280	42.50	0.70	0.515	–
B36	0.25	7.20	0.10	1566	940	752	3.70	6.55	127	113	60.80	12.40	205	562	101	5.65	0	0.52	–
B37	0.50	7.65	2.55	1012	607	472	3.60	7.45	92.09	74.37	70.30	13	140	457	70.15	13	0.055	0.59	–
B38	50	7.40	13.40	989	593	458	2.85	12	67.38	88.50	114	15.30	169	337	80.10	12.85	0.11	0.445	–
B39	32.50	7.50	9.40	904	543	342	3.25	5.30	116	32.25	13	13	207	427	77.35	14.8	0.10	0.575	–
B40	12.50	7.40	3.55	971	583	357	4.30	6.95	69.06	58.74	13.80	11	102	272	86.45	5.70	0.015	0.59	–
B41	10	7.05	0.55	608	365	286	4.20	7.65	56.84	47.64	26.50	8.35	65.58	280	69.15	5.65	0.01	0.56	–
B42	2.50	7	0.45	631	379	290	3.70	7.10	57.30	47.18	34.95	10.65	75.70	262	140	6.35	0.01	0.685	–
B43	5	7.45	0.40	601	361	210	5.90	11	87.40	19.67	46.35	12.75	122	252	58.70	31.10	0.03	1.07	–
B44	32.50	7.45	3.15	711	425	299	3.35	5.25	117	49.67	35.75	14	127	347	43.85	45.10	0.225	0.63	–
B45	50	7.90	19.16	976	586	379	3.55	5.60	130	64.55	34.85	5.55	175	376	38.95	2.65	0.055	0.47	–
B46	25	8.25	10.16	811	487	342	4.15	8.20	100	40.34	70	19.65	152	262	80.40	3.10	0.105	0.71	–
Min	5	6.90	0.10	456	274	195	2.85	2.85	33.37	19.67	13	2.30	54.16	172	38.95	0.75	0	0.265	10
Max	100	8.35	73.75	1566	940	752	5.90	12.50	152	113	116	30.05	344	690	312	46.75	1.70	1.07	25
Mean	20.56	7.43	6.02	897	408	362	3.96	7.73	89.93	57.36	45.56	11.77	159	359	115	13.55	0.139	0.566	–
SD	25.65	0.37	13.26	211	137	100	0.775	2.99	31.08	19.60	23.95	7.28	64	106	69	14.70	0.284	0.195	–

Note: All parameters are expressed in mg/l (ppm) except pH, colour (Hazen units), Turbidity (NTU) and conductivity (μmohs/cm).

Table 67.3: Groundwater Quality Data (Average Values) of Bhadravathi Town During Pre-monsoon Season for the Year 2001–2002

Sl.No.	Col.	pH	Tur	EC	TDS	TH	DO	COD	Ca^{2+}	Mg^{2+}	Na^{+}	K^{+}	Cl^{-}	HCO_3^{+}	SO_4^{2-}	NO_3^{-}	PO_4^{-}	F^{-}	E. coli
B1	32.50	7	11.25	1064	638	307	3.50	7	92.50	52.40	60.65	11.50	291	207	160	5.75	0.05	0.50	–
B2	12.50	7.25	2.95	954	572	278	3	5.75	78.85	48.51	27.70	7.45	315	299	102	17.50	0.12	0.35	–
B3	50	7.55	5.10	774	464	287	3.55	2.40	62.60	54.65	30.80	12.80	212	303	123	7.50	0.001	0.74	–
B4	100	7.05	47.30	622	373	239	3.65	2.95	67	41.79	26.80	5.40	113	355	95.90	3.45	0.056	0.29	–
B5	5	7.15	0.10	986	591	402	4.35	3.80	73	80.06	37	16.35	170	258	67	0.65	0.01	0.245	10
B6	45	6.90	21.25	863	518	357	4.35	14	123	56.86	34.95	11.50	126	394	79	1.40	0.065	0.325	25
B7	2.50	7.25	0.15	1124	674	421	3.90	13	116	74.23	63.30	7.75	226	492	87	3	0.005	0.57	–
B8	5	6.95	0.30	960	576	366	3.40	9.85	158	50.60	113	37.25	202	632	102	3	0.08	0.475	–
B9	25	7.25	1.65	851	510	361	4.15	11	99.60	63.64	12.45	12	113	336	173	3.85	0.015	0.32	–
B10	5	7.15	0.80	580	348	227	3.85	5.90	47.25	43.64	36.30	2	107	515	246	6.55	0.005	0.34	–
B11	5	6.70	0.30	687	412	312	2.70	5.25	114	48.23	25.85	7.75	99.60	407	69.50	1	0.005	0.56	–
B12	5	7.15	0.40	734	441	341	6.40	15	118	54.18	13	12.50	112	362	23.60	40	0.155	0.57	–
B13	2.50	7.15	0.15	972	584	311	3.50	12	96.75	52.06	14.40	1.25	167	247	47.30	51.50	0.095	0.68	–
B14	7.50	7.55	0.15	915	549	324	6.35	16	71.80	61.40	43.20	1.65	261	437	157	15.75	0.095	0.43	–
B15	5	7.55	0.25	1144	686	404	4.35	14	112	70.77	35.30	2.40	231	637	123	21.90	0.10	0.605	–
B16	5	7.25	0.25	881	529	358	3.40	9.10	126	56.37	44.15	3.60	199	322	107	0.50	0.10	0.87	–
B17	5	7.15	0.50	1004	603	404	3.70	4	123	68.23	30.50	5.45	194	262	102	51	0.32	0.90	–
B18	5	6.95	0.10	1110	666	437	3.85	6.40	103	81.11	60.30	5.05	172	302	118	30.50	0.06	0.30	–
B19	5	6.85	0.25	801	481	381	3.55	5.45	101	68.15	31.80	4.05	204	255	86.50	2	0.30	0.51	–
B20	32.50	7.20	5.45	836	502	352	3.95	4.85	86.90	64.28	26.10	4.25	176	168	90.25	1.25	0.075	0.52	–
B21	12.50	7.15	4.05	782	469	299	4.05	4.20	75.65	59.25	27	7	146	314	86.40	3.40	0.01	0.67	–
B22	5	6.95	1.50	611	367	281	3.60	8.25	108	40.54	7.85	3.40	70.80	305	96.40	30.40	0.05	0.40	–
B23	0	8.10	0.25	537	322	261	3.40	5.45	129	31.95	15.30	2.80	67.65	380	110	66.50	0.36	0.445	–
B24	0	7,20	0.10	699	420	319	5.75	6.35	69.50	60.74	16	1.45	70.95	359	66.80	0.60	0.275	0.65	–
B25	7.50	7.15	0.95	762	457	331	3.85	2.50	82.15	60.58	25.60	5	82	435	73.80	1.50	1.175	0.615	–

Contd...

Table 67.3–Contd...

Sl.No.	Col.	pH	Tur	EC	TDS	TH	DO	COD	Ca^{2+}	Mg^{2+}	Na^{+}	K^{+}	Cl^{-}	HCO_3^{+}	SO_4^{2-}	NO_3^{-}	PO_4^{-}	F^{-}	E. coli
B26	15	6.95	0.20	1148	689	419	3.95	3.15	57	87.96	44.30	11.30	216	366	67.35	1	0.02	0.50	–
B27	5	7.15	1.50	1058	635	408	3.85	3.45	191	52.66	80.30	14	174	364	87.80	2.75	0.085	0.46	–
B28	2.50	8.50	0.60	916	550	382	4.15	6.50	73	75.07	36.30	6.50	171	364	52.80	11.50	0.015	0.33	–
B29	10	7.20	4.45	460	276	204	3.40	4.40	73	31.82	46.15	5.85	40.40	188	29.30	43.15	0.01	0.65	–
B30	47.50	7.30	27.75	901	541	270	5.05	5.65	45.40	54.66	43.45	10.30	57.90	266	106	8.80	0.01	0.15	–
B31	5	7.10	3.15	1004	602	505	3.20	10	120	93.51	37.30	3.25	193	267	92.65	6	0.02	0.75	–
B32	15	7.20	0.08	812	487	329	2.75	5.65	54.95	66.59	48	9.50	140	375	99	3	0.01	0.9	–
B33	22.50	7.30	0.10	1070	642	479	3.85	5.80	65	100	41.20	10.50	159	317	193	6.85	0.025	0.59	–
B34	12.50	7.25	1.25	926	556	317	3.65	14	70.30	60.06	40.70	1	202	347	219	7.70	0.005	0.34	–
B35	2.25	8.10	5	1021	613	404	4.55	5.75	63.30	82.66	32.50	15	177	331	241	52	0.41	0.61	–
B36	7.50	7	0.68	1480	888	652	3.90	4.90	179	114	47.50	12.20	204	550	116	6.70	0.05	0.595	–
B37	5	7.25	0.40	1049	630	450	3.70	6	117	80.85	76.30	6.50	135	454	75.35	11.50	0.10	0.45	–
B38	40	7.25	16.80	999	599	405	3.35	17	69.50	81.60	117	52.50	159	317	108	12.60	0.01	0.47	–
B39	25	7.35	9.50	883	530	336	3.55	5.25	129	50.06	1.20	0.70	187	400	66.40	10.30	0.25	0.54	–
B40	5	7.25	1.60	947	568	315	4.55	6.65	66.10	60,47	15.30	1.85	96.30	256	97.15	6.85	0.15	0.55	–
B41	5	7	0.10	595	357	290	4.25	8.05	66.20	54.38	25.50	8.65	50.40	259	58.25	3.65	0.01	0.63	–
B42	5	7.05	0.15	602	361	270	3.85	5.45	53.90	52.50	26	6.15	66.20	236	122	5.40	0.01	0.68	–
B43	0.25	6.95	0.10	562	337	215	3.75	5.35	60	38.27	26.50	10	123	259	69	32.65	0.05	1.11	–
B44	22.50	7.20	2.70	683	410	330	3.65	6.35	63.80	64.80	39.20	12.10	112	307	110	42.70	0.10	0.55	–
B45	40	7.65	11.95	944	566	373	3.65	3.6	136	57.46	33.40	4.35	147	352	46.70	1.70	0.05	0.64	–
B46	17.50	8.25	2.85	816	490	330	3.55	8.7	104	54.69	75.95	15.30	140	261	26.70	3.25	0.01	0.73	–
Min	0	6.70	0.05	460	276	204	2.70	2.4	45.40	31.82	1.20	0.70	40.40	168	23.60	0.50	0.001	0.245	10
Max	100	8.25	47.30	1480	888	652	6.40	17	191	114	117	37.25	315	632	246	66.5	1.175	1.11	25
Mean	15.50	7.25	4.26	872	523	349	3.91	7.41	93.50	62.16	38.99	8.89	154.2	344	101	14.14	0.109	0.558	–
SD	18.70	0.36	8.72	200	120	81.44	0.75	4	34.04	17.16	23.89	9.02	63	102	50.26	17.34	0.19	0.182	–

Note: All parameters are expressed in mg/l (ppm) except pH, colour (Hazen units), Turbidity (NTU) and conductivity (μmohs/cm).

Table 67.4: Correlation Matrix Between Different Physico-chemical Parameters During Post-monsoon Season for the Year 2000–2001

Sl.No.	Col.	pH	Tur	EC	TDS	TH	DO	COD	Ca^{2+}	Mg^{2+}	Na^+	K^+	Cl^-	HCO_3^+	SO_4^{2-}	NO_3^-	PO_4^-	F^-	Fe^{2+}
Col.	1																		
pH	0.051	1																	
Tur	**0.826**	–0.024	1																
EC	–0.045	–0.133	0.013	1															
TDS	–0.046	–0.133	0.012	**0.999**	1														
TH	–0.046	–0.037	–0.170	**0.869**	**0.869**	1													
DO	0.003	–0.036	0.183	–0.129	–0.129	**–0.238**	1												
COD	–0.169	0.036	–0.040	0.114	0.114	0.120	0.033	1											
Ca^{2+}	–0.101	0.036	–0.122	**0.411**	**0.410**	0.120	**–0.217**	0.356	1										
Mg^{2+}	–0.096	0.056	–0.095	**0.733**	**0.734**	**0.838**	**–0.231**	0.008	0.092	1									
Na^+	0.136	–0.141	0.141	**0.362**	**0.362**	**0.352**	–0.089	0.128	0.150	**0.306**	1								
K^+	0.148	**0.234**	0.115	0.129	0.128	0.095	–0.208	0.135	**0.414**	0.092	**0.334**	1							
Cl^-	0.047	–0.174	–0.060	**0.708**	**0.708**	**0.419**	–0.166	0.127	**0.417**	**0.327**	**0.308**	**0.247**	1						
HCO_3^-	–0.153	–0.301	–0.125	**0.285**	**0.285**	**0.303**	–0.158	0.135	**0.343**	**0.235**	**0.258**	–0.022	0.133	1					
SO_4^{-2}	0.015	**0.281**	0.003	0.023	0.024	–0.01	–0.108	**0.208**	–0.151	0.091	0.057	0.113	0.136	0.101	1				
NO_3^-	–0.252	0.106	–0.227	**–0.210**	**–0.211**	–0.223	0.007	0.276	0.038	–0.174	**–0.197**	0.075	–0.020	–0.148	0.013	1			
PO_4^-	–0.102	0.063	–0.114	–0.140	–0.140	–0.087	**–0.246**	–0.026	–0.144	0.012	–0.127	0.022	–0.107	–0.024	**0.241**	**0.545**	1		
F^-	–0.040	–0.028	0.039	–0.038	–0.037	–9.064	0.133	0.163	–0.038	–0.154	–0.053	–0.078	0.009	–0.108	–0.074	**0.213**	–0.108	1	
F^{2+}	**0.356**	–0.152	**0.249**	**0.194**	**0.194**	–0.063	0.133	–0.110	0.080	0.075	0.048	0.027	**0.462**	0.121	–0.041	–0.076	**–0.211**	–0.068	1

Note: Bolded values shows significant correlation.

Table 67.5: Correlation Matrix Between Different Physico-chemical Parameters During Pre-monsoon Season for the Year 2001–2002

Sl.No.	Col.	pH	Tur	EC	TDS	TH	DO	COD	Ca^{2+}	Mg^{2+}	Na^+	K^+	Cl^-	HCO_3^+	SO_4^{2-}	NO_3^-	PO_4^-	F^-	Fe^{2+}
Col.	1																		
pH	–0.041	1																	
Tur	**0.898**	–0.051	1																
EC	–0.074	0.023	–0.101	1															
TDS	–0.074	0.023	–0.101	**0.999**	1														
TH	–0.182	–0.007	**–0.223**	**0.840**	**0.840**	1													
DO	–0.083	0.134	0.002	0.026	0.026	0.005	1												
COD	–0.100	0.029	–0.046	0.163	0.162	0.111	**0.325**	1											
Ca^{2+}	–0.184	–0.066	–0.139	**0.398**	**0.398**	**0.502**	–0.150	0.095	1										
Mg^{2+}	–0.122	0.021	–0.190	**0.775**	**0.775**	**0.906**	0.080	0.077	0.092	1									
Na^+	0.041	0.003	0.045	**0.390**	**0.390**	**0.323**	**–0.214**	**0.214**	**0.227**	**0.262**	1								
K^+	0.179	–0.034	0.152	0.192.	0.191	0.196	–0.117	0.245	0.068	0.193	**0.724**	1							
Cl^-	–0.014	0.055	–0.126	**0.664**	**0.664**	**0.401**	–0.112	0.193	**0.224**	**0.357**	**0.280**	0.086	1						
HCO_3^-	–0.142	0.035	–0.108	**0.316**	**0.316**	**0.317**	0.090	**0.249**	**0.393**	0.175	**0.260**	0.121	0.184	1					
SO_4^{-2}	0.027	0.063	–0.024	0.159	0.159	0.060	0.017	0.122	**–0.244**	0.185	0.043	–0.001	**0.238**	0.182	1				
NO_3^-	**–0.264**	**0.255**	–0.154	0.159	–0.188	–0.173	0.051	0.071	0.011	**–0.209**	**–0.219**	–0.092	–0.138	–0.147	0.030	1			
PO_4^-	–0.180	0.090	–0.103	–0.078	–0.078	–0.006	0.094	**–0.214**	0.064	–0.041	**–0.224**	–0.151	–0.122	0.117	–0.023	0.185	1		
F^-	–0.167	–0.009	–0.175	–0.119	–0.119	–0.079	–0.106	–0.161	–0.066	–0.051	–0.074	–0.070	–0.117	**–0.210**	**–0.229**	0.137	0.076	1	
Fe^{2+}	**0.325**	–0.018	**0.226**	0.176	0.176	–0.113	0.089	0.064	0.084	–0.157	0.012	0.123	**0.361**	0.072	–0.031	–0.144	**–0.237**	–0.164	1

Note: Bolded values shows significant correlation.

during post-monsoon season the dissolution of salts, minerals and other soil constituents increases due to increase in the groundwater table (Shivasankaran, 1997).

Most of the inorganic salts such as sodium chloride, potassium sulfate and potassium nitrate are responsible for increasing the EC values of groundwater systems. Sharma *et al.* (1995) classified the groundwaters based on electrical conductivity values as indicated below:

Excellent	000–333
Good	333–500
Permissible	500–1000
Brackish	1000–1500
Saline	1500–10,000

The results obtained revealed that more than 65 per cent of water belongs to permissible, 30.43 per cent of water belongs to brackish and 2.1 per cent belongs to excellent and saline category.

The statistical correlation of conductivity with other parameters is presented in correlation matrix (Tables 67.4 and 67.5). Electrical conductivity showed a significant positive correlation with total dissolved solids, total hardness, calcium, magnesium, sodium, chloride and bicarbonates. The correlations established for electrical conductivity in the present work are similar to those observed by Kaza Somashekara Rao (1994) and Basavarajappa (2002).

Total Dissolved Solids (TDS)

The term solid refers to the matters either filterable or non filterable that remain as a residue in water *i.e.* TDS of water includes all soluble materials in solutions whether ionized or non-ionized. It does not include suspended sediments, colloids or dissolved gasses. TDS values are estimated by pursuing the empirical relationship (USSLS, 1954; Hem, 1985; Kotaiah *et al.*, 1994; Rambabu *et al.*, 1996 and Chandankeri, 1996).

In the present study TDS values ranged from a minimum of 276 mg/l to a maximum of 888 mg/l in pre-monsoon season (Table 67.3) and a minimum of 274 mg/l to a maximum of 940 mg/l in post-monsoon season (Table 67.2). It is observed that the its values have exhibited an increasing trend in post-monsoon seasons owing to the fact that during post-monsoon season dissolution of more quantity of constituents of soil particles as groundwater table increases during post-monsoon season. In the present study almost all water samples are well with in the limit of drinking water standards.

Carrol, 1962 classified fresh water based on TDS values as indicated below.

Fresh	0–1000 ppm
Brackish	1000–10000 ppm
Saline	10000–100000 ppm
Brine	Above 100000 ppm

The results obtained revealed that water in the study area belongs to the fresh category.

Total dissolved solids showed positive correlation with total hardness, calcium, magnesium, sodium, chloride and bicarbonates in both the seasons (Tables 67.4 and 67.5). Narayana and Suresh (1989) also observed similar correlation in their study.

Total Hardness (TH)

In the present investigation, total hardness values varied from a minimum of 204 mg/l to a maximum of 652 mg/l in pre-monsoon season (Table 67.3) and a minimum of 195.5 mg/l to a maximum of 752.5 mg/l in post-monsoon season (Table 67.2).

Das Gupta (2000) and Shivasankaran (1997) are of the opinion that the concentration of hardness increases towards the post-monsoon season and decreases towards pre-monsoon season. In the present study also a similar behavior of hardness is noticed.

The degree of hardness in ppm has been classified in terms of equivalents of calcium carbonate concentration (APHA, 1995; Kotaiah, 1994) as:

Soft	0–50 mg/l
Medium	50–150 mg/l
Hard	150–300 mg/l
Very Hard	Greater than 300 mg/l

The results obtained revealed that more than 75 per cent of water samples belongs to the very hard category.

The statistical correlation of total hardness shows a highly significant correlation with electrical conductivity, total dissolved solids, calcium, magnesium, chloride and bicarbonates (Tables 63.4 and 63.5). This is conformity with the findings of Hussain (2001).

Chemical Oxygen Demand (COD)

In the present investigation COD values ranged from a minimum of 2.4 mg/l to a maximum of 17.5 mg/l in pre-monsoon season (Table 67.3) and a minimum of 2.85 mg/l to a maximum of 12.5 mg/l in post-monsoon season (Table 67.2). The WHO permissible limit for COD is 10 mg/l. In the present study the sample numbers 8, 9, 12, 13, 14, 16, 31 and 34 in post-monsoon season and sample numbers 6, 7, 9, 12, 14, 15, 31, 34 and 38 in pre-monsoon season showed higher concentration of COD than the standards prescribed for drinking water.

The statistical correlation of chemical oxygen demand shows a positive correlation with dissolved oxygen, sodium, potassium, bicarbonates and negative correlation with phosphate in pre-monsoon season (Table 67.5).

Calcium (Ca^{2+}) and Magnesium (Mg^{2+})

In the present study, calcium values ranged from a minimum of 33.37 mg/l to a maximum of 152.6 mg/l in post-monsoon season (Table 67.2) and a minimum of 45.4 mg/l to a maximum of 191.25 in pre-monsoon season (Table 67.3). However, the magnesium values ranged from a minimum of 31.82 mg/l to a maximum of 114.7 mg/l in pre-monsoon season (Table 67.3) and a minimum of 19.67 to a maximum of 113.08 mg/l in post-monsoon season (Table 67.2). In the present investigation the calcium and magnesium values are were with in the maximum permissible limit of drinking water standards.

Further, it is important to note that both calcium and magnesium have indicated strong significant correlation with total hardness, total dissolved solids, electrical conductivity, sodium, chloride but inverse correlation is found with sulfate and nitrate (Tables 67.4 and 67.5). Garg *et al.* (1988) has also expressed a similar opinion on correlation of calcium and magnesium.

Sodium (Na^+) and Potassium (K^+)

In the present investigation the sodium values are ranged from a minimum of 1.2 mg/l to a maximum of 117 mg/l in pre-monsoon season (Table 67.3) and a minimum of 13 mg/l to a maximum of 116 mg/l in post-monsoon season (Table 67.2). However, the potassium ranged from a minimum of 0.7 to a maximum of 37.25 mg/l in pre-monsoon season (Table 67.3) and a minimum of 2.3 mg/l to a maximum of 30.05 in post-monsoon season (Table 67.2). The values of sodium and potassium are well with in the drinking and domestic water standards prescribed by BIS.

Sodium shows positive correlation with electrical conductivity, total dissolved solids, chemical oxygen demand, calcium, magnesium, potassium, chloride, bicarbonates and inverse correlation with dissolved oxygen, nitrate and phosphate (Tables 67.4 and 67.5). The above correlations are in total agreement with those observed by Hussain *et al.* (2001).

Chloride (Cl)

Chloride ion is generally present in natural waters and its presence can be attributed to the dissolution of salt deposits, discharge of effluents from chemical industries, oil were operations, sewage discharges, irrigation drainage, contamination from refuge leachates. The salty taste produced by chloride ion depends on chemical composition of the water (Vijaya Kumara *et al.*, 2002). In the present study, chloride values ranged from a minimum of 54.16 mg/l to a maximum of 344.3 mg/l in post-monsoon season (Table 67.2) and a minimum of 40.4 mg/l to a maximum of 315.5 mg/l in pre-monsoon season (Table 67.3). It is observed that the concentration of chloride ions has shown increased trend during post-monsoon season. This may due dissolution of chloride containing substances in soil during infiltration or percolation of rain water (Shivasankaran, 1997).

The correlation matrix indicates that chlorides have a strong positive correlation with electrical conductivity, total dissolved solids, calcium, magnesium, sodium, sulfate and iron. However, it shows insignificant correlation with nitrate and fluoride (Tables 67.4 and 67.5).

Bicarbonate (HCO_3^-)

In the present study, the bicarbonate values ranged from a minimum of 172.5 mg/l to a maximum of 690 mg/l in post-monsoon season (Table 67.2) and a minimum of 168.5 mg/l to a maximum of 632.5 mg/l in pre-monsoon Season (Table 67.3). The results revealed that all the water samples are well with in the standards prescribed for drinking water.

Bicarbonates establish a positive correlation with electrical conductivity, total dissolved solids, total hardness, chemical oxygen demand, calcium, sodium and inverse correlation with fluoride (Tables 67.4 and 67.5). However, Narayana and Suresh (1989) have indicated that bicarbonate shows positive correlation with total dissolved solids, total hardness, calcium, sodium and also showed inverse correlation with dissolved oxygen.

Sulfate (SO_4^-)

The sulfate content of natural water is an important parameter in determining the suitability of water for residential use or public use. Higher concentration of sulfate (> 250 ppm) cause cathartic action and miss functioning of alimentary canal and gastrointestinal irrigation (ISI, 1982) in human beings. Hence, determination of sulfate in fresh water becomes essential. In the present investigation, sulfate values ranged from a minimum of 38.95 mg/l to a maximum of 312.5 mg/l in post-monsoon season (Table 67.5) and 23.6 mg/l to a maximum of 246 mg/l in pre-monsoon season (Table 67.3). The study reveals that the concentration of sulfate is high during post-monsoon season as compared to pre-monsoon season. Alexander (1961) and Miller (1979) are of the opinion that the break-down of organic substances in soil, leachable sulfates present in fertilizers and other human interference's are the expected causes for high concentration of sulfate.

The statistical correlation revealed that the sulfate is negatively correlated with fluoride, colour, magnesium, sodium and positive correlation with pH (Tables 67.4 and 67.5).

Fluoride (F)

In the present study, the fluoride values ranged from a minimum of 0.245 mg/l to a maximum of 1.11 mg/l in pre-monsoon season (Table 67.3) and a minimum of 0.265 mg/l to a maximum of 1.07 mg/l in post-monsoon season (Table 67.2). In the study area the fluoride concentration is well with in the limit of drinking water standards.

Nitrate (NO_3)

In the present study, the nitrate values ranged from a minimum of 0.75 mg/l to a maximum of 46.75 mg/l in pre-monsoon season (Table 67.3) and a minimum of 0.50 mg/l to a maximum of 66.50 mg/l in post-monsoon season (Table 67.2).

Laksmanan *et al.* (1986) reported that human wastes also contains nitrogen for about 5 kg/ person/year (WHO, 1984). Further, fertilizers, animal wastes, municipal and industrial untreated wastes are considered as important sources of nitrate contamination in groundwater (WHO, 1984 and Handa, 1983). Even in the absence of fertilizer applications and geological deposits, the high level nitrates were observed in urban regions of Hyderabad and Secunderabad in some groundwaters owing to the lack of good sanitary systems and proper drainage management (Davina *et al.*, 1999). Percolation of sewage increases the concentration of nitrate in groundwater (Vijaya Kumara, *et al.*, 2002). The same opinion can also be considered in some sampling locations of the study area. Therefore the nitrate levels in the sample number 13, 17, 23 and 44 revealed that the waters of the particular location are unsuitable for drinking purposes with respect to nitrate concentration in the water.

Phosphate (PO_4^-)

In the present study, the phosphate values ranged from a minimum of 0.001 mg/l to a maximum of 1.17 mg/l in pre-monsoon season (Table 67.3) and a minimum of 0 mg/l to a maximum of 1.07 mg/l in post-monsoon season (Table 67.2). Rajmohan *et al.* (2000) have reported phosphorous concentration at a minimum of 0.03 ppm to a maximum of 0.70 ppm in their study area owing to the intensively irrigated area of Kanchipuram Taluk of Tamil Nadu.

E. Coli

The examination of water samples for *E. coli* contamination has revealed that the bacterial count was minimum of zero count in the almost all the samples except sample nos. 5 and 6 where the *E. coli* number was recorded to be 10/100 ml and 25/100 ml respectively (Tables 67.2 and 67.3). This is attributed to the percolation of untreated municipal sewage and solid waste leachate in to the borewell water.

Conclusion

Following conclusions could be drawn from this study:

1. From the results of the seasonal samples, it is observed that the concentration of major physico-chemical parameters have exhibited an increasing trend during post-monsoon

season compared to the pre-monsoon season. This is attributed to the dissolution of salts and minerals in soil through the recharge of groundwater by rainfall and rising the groundwater table during post-monsoon season.

2. It is observed that 30 per cent of water samples investigated are non-potable and 70 per cent of the samples are potable and it is in accordance with that of BIS for drinking water.
3. The present study has revealed that the fluoride content in the study area is well with in the prescribed limits.
4. In five sampling locations, the nitrate content is higher than the prescribed standards of BIS for drinking water. This is attributed to the percolation of untreated municipal sewage and industrial effluents.
5. The examination of water samples for *E. coli* contamination has revealed that the bacterial count was minimum of zero count in almost all the samples except in water sample numbers 5 and 6 where the *E. coli* number is 10/100 ml and 25/100 ml respectively. This is attributed to the percolation of untreated municipal sewage and solid waste leachate.
6. In the study area, the total hardness content of 75 per cent of water samples belongs to the very hard category.
7. The cations and anions concentration in ppm is in the order of relative dominance in the following sequences. $Ca^{2+} > Mg^{2+} > Na^{+} > K^{+}$, $HCO_3 > Cl > SO_4^{2-} > NO_3 > F > PO_4^{3-}$.
8. The nitrate concentration is in the order of higher magnitude than phosphate.

Further this study has given the criteria that help in providing important information for implementation of water quality control practices. It is expected that the criteria should serve the public interest at large and it is based upon scientific facts.

References

Alexander, M., 1961. *Introduction to Soil Microbiology*. Wiley, New York, London, pp. 472.

APHA, 1985. *Standard Methods for the Examination of Water and Wastewater*, 17th edn. American Public Health Association, AWWA, WPCF, Washington D.C., New York.

APHA, 1995. *Standard Methods for the Examination of Water and Wastewater*, 18th edn. American Public Health Association, Washington D.C., New York, pp. 208.

Basavarajappa, B.E., 2002. Studies on the impact of environmental pollution on groundwater quality in and around Davanagere. *Ph.D. Thesis*, Kuvempu University, Jnanasahyadri, Shankaraghatta.

Black, 1963. *Water Pollution Technology*. Reston, Virginia.

Bureau of Indian Standards, 1998. *Specification for Drinking Water*, New Delhi.

Chandankeri, G.G., 1996. An integrated approach to groundwater studies of Kumudavathi River lower Basin. *Ph.D. Thesis*, Karnataka University, Dharwad.

Davina, V. Gonsalves and Joe D'Souza, 1999. Impact of the tourism industry on groundwater in Calangute of Goa. *Ecology, Environment and Conservation*, 5(1): 19–24.

Garg, V.K., Dahiya, Sudhir, Chaowdhary, Arti and Deepshika, 1998. Fluoride distribution in underground of Gind district of Haryana (India). *Ecology, Environment and Conservation.*

Gill, S.K., Sahota, S.K., Sahota G.P.S., Sahota, B.K. and Sahota, H.S., 1993. Comparisons of physico-chemical parameters of groundwater from shallow aquifer near 2 thermal power plants in Punjab. *Indian Journal of Environmental Protection*, 13: 584–587.

Govardan, V., 1990. Groundwater pollution hazardous to human life: A case study of Nalgonda district. *Indian Journal of Environmental Protection*, 10: 54–61.

Handa, B.K., 1983. Effect of fertilizers use on groundwater quality in India. In: *Groundwater in Water Resource Planning*, UNESCO, IAH-IAHS, Koblenz.

Hem, J.D., 1985. *Study and Interpretation of the Chemical Characteristics of Natural Water*, 3rd edition. U.S. Geological Survey Water Supply Paper, 2254: 263.

Hussain, J., Sharma, K.C., Hussain, I. and Ojha, K.G., 2001. Physico-chemical characteristics of water from borewells of an industrial town Bhilwara, Rajasthan: A correlation study. *Asian J. Chemistry*, 13(2): 470–476.

Jain, C.K., Sharma, M.K., Bhatia, K.K.S. and Seth, S.M., 2000. Groundwater pollution: Endemic flurosis. *Pollution Research*, 4(19): 505–509.

Kaza, Somashekara Rao, Raju, V.A., Singanan, M., Someshwara Rao, B., Sheshagiri Rao, P.V. and Chakravarti, K.R., 1994. Studies on the quality of water supplied by the municipality of Kakinanda and groundwater of Kakinanada town. *Indian Journal of Environmental Protection*, 14(3): 167–169.

Knight, A.G., 1951. The photometric estimation of colour in turbid waters. *Journal of Institution of Water Engineering*, 5: 623.

Kotaiah, B. and Kumaraswamy, N., 1994. *Environmental Engineering Laboratory Manual*. Charotar Publishing House, Anand–388 001 (India).

Lakshmanan, A.R., Krishnarao, T. and Viswanathan, S., 1986. Nitrate and fluoride levels in drinking waters in the twin cities of Hyderabad and Secunderabad. *Indian Journal of Environmental Health*, 28(1): 39–47.

Mehta, S.B. and Trivedi, V.H., 1993. Groundwater contamination. *Indian Journal of Environmental Protection*, 13: 577–579.

Miller, J.C., 1979. Nitrate contamination of the water table aquifer by septic tank systems in the coastal plain of Dilaware. *Rural Environmental Conference*, Warren V.P.

Mittal, S.K., Rao, A.L.J., Singh, S. and Kumar, R., 1994. Groundwater quality of some areas in Patiala city. *Indian Journal of Environmental Health*, 31: 228–236.

Narayana, A.C. and Suresh, G.C., 1989. Chemical quality of groundwater of Mangalore city in Karnataka. *Indian Journal of Environmental Health*, 31(3): 228–236.

Rajamohan, N.L., Elango, S., Ramachandran and Natarajan, M., 2000. Major ion correlation in groundwater of Kancheepuram region (South India). *Indian Journal of Environmental Protection*, 20(3): 188–193.

Rambabu, C. and Somashekara Rao, K., 1996. Studies on the quality of borewell water of Nuzvid. *Indian Journal of Environmental Protection*, 16(7): 41–47.

Sawyer, N.C. and MeCarty, P.L., 1978. *Chemistry for Environmental Engineers*, 3rd Edition, McGraw-Hill Book Company.

Sharma, D.K. *et al.*, 1998. Studies in quality of water in and around Jaipur. *IWWA (Indian Waste Water Analysis)*, 20(3).

Sharma, Sanjay, Jain, P. C. and Mathur, R., 1995. Quality assessment of groundwater in municipal and fringe areas near Gwalior. *Indian Journal of Environmental Protection*, 15(7): 534–538.

Shivashankaran, M.A., 1997. Hydrological assessment and current status of pollutants in groundwater of Pondicherry region of South India. *Ph.D. Thesis*, Anna University, Chennai.

Somashekara Rao, K., 1989. Quality and assessment of groundwaters. *Indian Journal of Environmental Protection*, 6: 483–444.

Somashekara Rao, K., 1993. Correlations among water quality parameters of groundwater of Nuzvid town and Nuzvid Mandalam. *Indian Journal of Environmental Protection*, 4: 261–266.

USSLS, 1954. Diagnosis and improvement of saline alkali soils. U.S. Salinity Laboratory Staff, U.S. Department of Agriculture Handbook, 60: 164.

Vijaya Kumara, Narayana, J. and Puttaiah, E.T., 2002. Assessment of groundwater quality of Bhadravathi town, Karnataka. *Geobios*, 29: 217–220.

Vijaya Kumara, Narayana, J. and Puttaiah, E.T., 2002. Evaluation of groundwater quality in urban and rural regions of Bhadravathi taluk, Shimoga district, (Karnataka). *Science Journal*, Kuvempu University, 2(1): 61–66.

WHO, 1984. *Guidelines for Drinking Water Quality*, WHO recommendations, Vol. 2.

Chapter 68

Saccharification Studies by a Thermophilic Fungus *Sporotrichum thermophile* Isolated from Agriculture Waste of Bhopal

Rajhans Singh[1], Namita Singh[2], Anil Prakash[3] and Sarika Poonia[2]
[1]Maharana Partap Degree College, Kalor Road, Bhopal – 462 016
[2]Department of Biotechnology, Guru Jambheshwar University, Hisar – 125 001, Haryana
[3]Department of Biotechnology, Barkatullah University, Bhopal – 462 026, M.P., India

ABSTRACT

Thermophilic fungus has been reported from various habitat *viz.* municipal waste, natural compost, coal refuse, oil palm, and stored grains etc. Thermophilic fungus has been exploited for number of unique properties, like enzymes, antibiotic, amino acids production.

Cellulose is available in nature as plant residues. The cellulosic wastes cause environmental pollution, conservation of these wastes into useful products through microbes is in progress. Number of thermophilic fungi produces extra cellular enzymes, which diffuse and break the cellulose chain by saccharification.

In the present study attempt has been made to characterize and standardize the production of enzymes from *Sporotrichum thermophile* fungus isolated from agriculture waste. Following observations were recorded. High enzymatic activity was recorded between 12–15 day of incubation and a sharp decline in the activity after 15th days. 45°C temperatures shows better growth and production of enzyme and pH 6.8 showed better production of cellulose. Hence *S. thermophile* can be utilized for minimizing the environmental pollution by conservation of cellulosic waste into the useful products.

Keywords: *Cellulose, Endo-β-glucanase, Exo-β-glucanase, Sporotrichum thermophile.*

Introduction

Cellulose is most abundantly produced biopolymer in terrestrial environments. Each year fixation of CO_2 yields more than 10^{11} tons of dry plant material all over world (Schlesinger, 1991) and almost half of this consist of cellulose (Eriksson *et al.*, 1990). Cellulose is one of the major components of municipal waste and agro-wastes which cause potent sources of environmental pollution. This potentially valuable source of food energy is largely unavailable to monogastric animals because of resistance of cellulose to digestive enzymes (Coutts and Smith, 1976). Cellulose almost never occurs alone in nature but is usually associated with plant substances. This association affects its natural degradation. Bacteria and fungal degradation of cellulose occur exocellularly either in association with outer cell envelope layer or extracellularly. Their enzymes help in conversion of cellulose into simpler saccharides. Previously mesophilic microorganisms were known to produce cellulose enzyme (Coutts and Smith, 1976) but now-a-days thermophilic microorganisms have received considerable attention for production of more amount (Singh *et al.*, 1990; Wase *et al.*, 1985) and variety of enzymes. This is largely due to the fact that enzyme produced by thermophiles exhibit optimal activity at high temperature that their mesophilic counterparts (Maheshwari *et al.*, 2000; Feldman *et al.*, 1988; Grajek, 1987; Hagerdal *et al.*, 1980).

The thermophilic microorganisms have adopted themselves to a wide array and ecological niche such as comport, soil, birds nest, wood chips, seed grams, volcano and hot springs etc. (Maheshwari *et al.*, 1987). The resistance to survive at high temperature has modified the metabolism of thermophiles in favour of production of enzymes and other secondary metabolites of industrial importance (Feldman *et al.*, 1988; Bhat and Maheshwari, 1987).

Thermophilic fungi have become an experimental system for the basic and applied research due to their unique properties (Maheshwari *et al.*, 1987, 2000). These include growth and temperature optima, habitat relationship, enzymatic abilities, degradation and deterioration characteristics as well as possible allergenic reactions. They produce important extracellular enzymes such as xylanase, amylase, protease, lipase and cellulase. The production of thermostable enzymes at elevated temperature by thermophilic fungi is of economic importance (Feldman *et al.*, 1988). Some of these have already been exploited commercially (Chaudhuri *et al.*, 1988).

The existence of cellulolytic microorganism is well documented. Several efforts have been made by many workers to develop preproductive cellulose mutants of thermophilic fungi (Hayashida and Mo, 1986; Sen and Chakarbarty, 1984). Cellulose is a prominent agrowaste in each country and creating too much environmental pollution. To minimize the environmental pollution these cellulosic materials should be converted into useful product. Among thermophilic fungi cellulose system has been examined exhaustively. However, *Sporotrichum thermophile* has been studied in the most detail as it synthesizes a complete set of enzymes necessary for the break down of crystalline cellulose. Therefore, *Sporotrichum thermophile* appeared to be a useful source of cellulases. Keeping this in view, *Sporotrichum thermophile* was selected to study the cellulolytic potential.

The present study was carried with the objective to identify different environmental conditions for optimum production of cellulose enzymes by locally isolated thermophilic fungi *S. thermophile.*

Materials and Methods

Organism and Culture Conditions

The organism, *Sporotrichum thermophile,* a thermophilic fungi used for the enzymatic saccharification of cellulose, was isolated from local agriculture waste (Bhopal, M.P., India). Culture of

Sporotrichum thermophile was maintained on basal medium (Pattersson *et al.*, 1963). Slants were kept at 4°C in cold room. Experiments were performed undersubmerged culture conditions.

Growth Parameters

The growth of *Sporotrichum thermophile* was estimated by measuring the extra cellular protein and mycelial protein. The extra cellular protein was measured by using Lowry's method (1951). Bovine Serum albumin was used as a standard mycelial protein was estimated by microbiuret method using Benedict's reagent (Herbert *et al.*, 1971). For studying growth behavior under different environmental conditions samples were taken after 3, 6, 9, 12, 15, 18 and 21 days incubation under different environmental conditions.

Enzymes Assay of Cellulose Enzymes

Cellulose activity was assayed viscometrically using Ostwald Fenske Viscometer (Reese *et al.*, 1950). The percentage loss in viscosity was calculated according to the following formula:

$$\%\text{ loss in viscosity} = \frac{(ET_o - ET_t)}{(ET_o - ET_w)} \times 100$$

where,

ET_o: Efflux time at zero hour

ET_t: Efflux time at time t

ET_w: Efflux time of distilled water

Endo-β-glucanase (EC. 3.2.1.4)

The Endo-β-glucanase was asseyed by the method of Sumner and Somer (1944) in terms of reducing sugar (R.S.) released from the substrate carboxyl methyl cellulose (CMC). 1 ml of cold CMC reagent was added to 1 ml of enzyme sample incubated at 50°C for 30 mins and then cooled to 4°C. Then amount of reducing sugar produced assayed by DNS reagent (Miller, 1959).

Exo-β-glucanase (E.C. 3.2.1.9.1)

The Exo-β-glucanase was assayed by the method suggested by Mendels and Weber (1969) by analyzing reducing sugars (R.S.) with Dinitrosalicyclic acid reagent 50 mg or cotton fibers in 1 mM citrate phosphate buffer pH-5 and 1 ml of crude enzyme preparation were incubated at 45°C for 24 hrs. Then reducing sugar was assayed by DNS reagent (Miller, 1959).

Results and Discussion

It is a well known fact that incubation period greatly influences the growth and production of the organism. To find optimum period of incubation cultures were kept for 3 to 15 days under submerged conditions. The optimum conditions of incubation for growth vary from organism to organism (Bhat and Maheshwari, 1987). The maximum growth in our experiment in terms of extracellular protein (105.4 μg/ml) and mycelial protein (98.2 μg/ml) was observed on 15th day of incubation after which the growth sharply declines (Figure 68.1). This finding was in contrast to Bhat and Maheshwari (1987) who reported maximum growth of organism on 4th day and Coutts and Smith (1976) reported that growth of organism increases with increase in incubation period.

The mycelial dry weight maximum was on 18th day of incubation (0.92 gm/70 ml) (Figure 68.1). It increases with increase in period of incubation and than remained approx stationary with culture age. Similar observation was reported by Bhat and Maheshwari (1987).

Bhat and Maheshwari (1987) reported that growth of *Trichoderma ressei* (mesophilic strain) was excellent in comparison to *S. thermophile* (thermophilic fungus). This was in contrast to our results that *S. thermophile* grow very well and to result of Maheshwari *et al.* (2000) who found that *S. thermophile* grew at five times the rate of *T. ressei* and to Tansey (1971) results.

Cellulolytic system mainly consists of three components, exoglucanase, endoglucanase and β-glucosidase that act in synergy to degrade cellulose (Ryu and Mandels, 1980; Grajek, 1987; Beguin, 1990; Gilbert and Hazlewood, 1993; Nidetzky and Steiner, 1993). Endoglucanase, exogluctnase and β-glucosidase may form ternary complex with substrate (Moloney, 1985). It is hypothesized that in *S. thermophile*, β-glucosidase physically associates with endo and exoglucanase to form a catalytically more active cellulose complex (Maheshwari *et al.*, 2000). The different level of celluloses produced by *S. thermophile* could be because of difference in induction by same substrate (Macris and Galiotou-Panayotou, 1986). Bhat and Maheshwari (1987) and Bagga *et al.* (1990) have reported that cellulases were not synthesized at the same time and same rate. Cellulose productivity also varies among strains of *S. thermophile* (Maheshwari *et al.*, 2000; Oberson *et al.*, 1992). The maximum production of endo-β-glucanase (40.6 μg R.S./ml) and exo-β-glucanase (31.2 μg R.S./ml) was on the same 12th day of incubation. These findings are not agreeable with the earlier concept of Bhat and Maheshwari (1987). Coutts and Smith (1976) reported that endo-β-glucanase appeared in early stages of growth. Similar observations were made by Neudoerffer and Smith (1970). However present study the production of both enzymes are on the same day. Production of endo and exoglucanase enzyme increased with increase in incubation period upto 12th day after which it declines. But the production of endo-β-glucanase is higher than the exoglucanase. This result was agreeable to some extent with Coutts and Smith (1976).

The maximum growth was on 15th day and maximum enzyme produced was on 12th day which is in actively growing period. This result was same to that of Maheshwari *et al.* (2000) that in *S. thermophile* endo and exoglucanase were typically formed during active growth. The percentage viscosity loss was maximum (67.5 per cent) on 12th day of incubation (Figure 68.4).

Growth and the production of enzymes by *Sporotrichum thermophile* was greatly affected by the hydrogen ion concentration of the synthetic medium. Exo-β-glucanase and endo-β-glucanase can tolerate a wide range of pH, the maximum production was at 6.5 pH (41.25 μg R.S./ml and 31.5 μg R.S./ml) (Figure 68.5). The percentage viscosity loss is maximum at pH 6.5 (69.5 per cent) and it declines to approx 31 per cent on moving from pH 6.5 to 8.5. Both exo and endoglucanase activity decline with increase in pH from 6.5. In contrast to our result many author reported that the cellulose production was observed at pH 6.0 by *Sporotrichum thermophile* (Bhat and Maheshwari, 1987). Singh *et al.* (1990) have also reported pH 6.0 as optimum for cellulase production by immobilized cells of *Sporotrichum thermophile.* Grajek (1987) has reported that the optimum pH for extracellular β-glucosidase production for his isolate of *Sporotrichum thermophile* from wood waste material was around 6.5 pH. In contrast to our results Coutts and Smith (1976) reported that his organism *Sporotrichum thermophile* produces exo and endoglucanases more efficiently at acidic pH. Similar results were observed by Rao (1977) that initial pH of 5.6 was found to be most favourable for obtaining high yield of cellulase.

For better growth of organism the basal medium was adjusted in the pH range from 3.5–8.5. The maximum growth was found when pH of the medium was adjusted to 6.5 prior to autoclaving

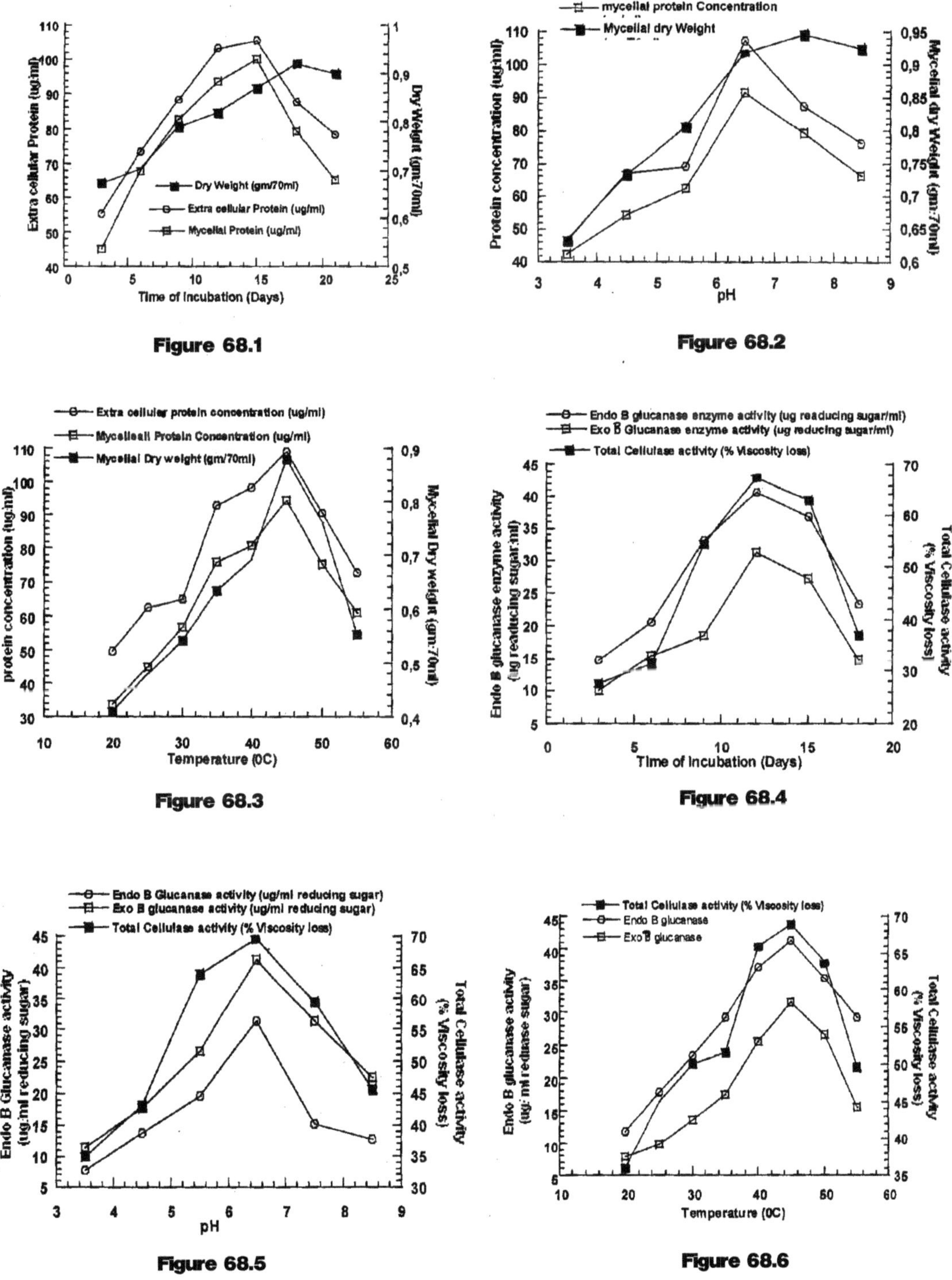

Figure 68.1

Figure 68.2

Figure 68.3

Figure 68.4

Figure 68.5

Figure 68.6

Figure 68.1–68.6

(extracellular protein concentration– 107.42 µg/ml and mycelial protein concentration–91.5 µg/ml) (Figure 68.2). pH 3.5–4.5 was found to be too acidic for growth of *S. thermophile*. However enhancement in pH from 6.5–8.5 show poor growth of organism. The maximum mycelial dry weight (0.9468 gm/70 ml) was found at 7.5 pH. This finding is in contrast with Coutts and Smith (1976), Bhatt and Maheshwari (1987), Maheshwari *et al.* (1987), Singh *et al.* (1990). Cossar and Canevascini (1986) reported that the growth of thermophilic fungus *Thermomyces lanuginosus* was at 6.5 pH. Optimum pH range for enzymatic activity was 4.6 to 5.2 and no activity was observed at 8.5 pH. Jensen and Olsen (1992) reported that the growth of the fungus at pH 6.5 was in accordance with our result. While about enzymatic activity the results are in contrast. Maheshwari *et al.* (2000) reported growth of thermophilic fungi at pH range 5.5–7.0.

From the results it is also apparent that percentage viscosity loss was also higher at 6.5 pH and low at acidic pH. From the above results it may be concluded that proper pH control is necessary to obtain high enzyme yield with the substrate.

Temperature plays an important role in production of enzymes. To optimize ideal temperature for growth and enzyme production flasks were kept at temperature ranging from 20–55°C. The production of cellulases by *Sporotrichum thermophile* was maximum at 45°C *i.e.* endo-β-glucanase, 41.25 µg R.S./ml and exo-β-glucanase 31.5 µg R.S./ml. This temperature was good or both, fungal growth as well as higher yield of cellulases. In the case of mesophilic moulds the optimum temperature for cellulase is in the range 25–37°C (Rao, 1977; Johnson *et al.*, 1982; Wase *et al.*, 1985; Macris and Galiotou-Panayotou, 1986), but in the case of thermophiles an optimum of 45–60°C has earlier been reported (Prakash, 1984; Bhat and Maheshwari, 1987). Maheshwari *et al.* (2000) reported temperature optimum for *S. thermophile* at 45–50°C. Coutts and Smith (1976) suggested that greater enzyme production is at 45°C. Romanelli *et al.* (1975) suggest that maximum cellulose utilization was at 40°C. Grajek (1987) suggested that maximum glucosidase production by *S. thermophile* at 40°C.

From the technological point of view, the temperature effect on the enzymatic hydrolysis of cellulose would be of considerable interest. Thermophilic microorganisms have been reported to be more advantageous over the mesophilic microorganisms because thermophiles hydrolyze cellulose faster than mesophilic counterpart, (Coutts and Smith, 1976; Maheshwari *et al.*, 2000).

The growth of the organism always depends on tire temperature. Maheshwari *et al.* (2000) reported that for thermophiles the optimum temperature is 35°C. The effect of temperature on their growth has been evaluated taking into consideration either radial mycelial growth on solid media (Johri and Satyanarayana, 1986) or dry mycelial weight in liquid media (Emersson, 1968). Similar observation was also made in the present study that at low temperature the growth of *Sporotrichum thermophile* was low and with the increase in temperature the growth increases. The maximum extracellular protein concentration was 108.8 µg/ml and mycelial protein concentration 94.4 µg/ml at 45°C. The mycelial dry weight was also maximum (0.88 gm/70 ml) at 45°C (Figure 68.3).

Fergus and Amelung (1971), Satyanarayana and Johri (1981) have reported that thermophilic fungi differ in their ability to survive after exposure to temperature ranging from 60°C to 70°C to different length of time. However, in the present case *Sporotrichum thermophile* survive upto 45°C. Grajek (1987) reported best growth temperature was 43°C but maximum cellulose production at 40°C. Romanelli *et al.* (1975) reported that his strain of *Sporotrichum thermophile* grow at 40°C and degrade more cellulose. The results of Romanelli *et al.* (1975) was in contrast to our results because *Sporotrichum thermophile* in present case grown at 45°C and produced maximum enzyme at 45°C. Similar observation was also reported by Singh *et al.* (1990) and Coutts and Smith (1976). Therefore with the above discussion it may be concluded that temperature plays a very important role in growth as well as in the production of metabolites.

The results obtained in this experiment were very useful and encouraging, *Sporotrichum thermophile* is one of the most important organism which produce thermostable enzymes in nature.

References

Bagga, P.S., Sandhu, D.K. and Sharma, S., 1990. Purification and characterization of cellulolytic enzymes produced by *Aspergillus nidulans*. *Journal of Applied Bacteriology*, 68: 61–68.

Beguin, P., 1990. Molecular biology of cellulose degradation. *Ann. Rev. Microbiol.*, 44: 219–228.

Bhat, K.M. and Maheshwari, R., 1987. *Sporotrichum thermophile:* Growth, cellulose degradation, and cellulase activity. *Appl. Environ. Microbiol.*, 53(9): 2175–2182.

Chaudhuri, S., Thakur, I.S., Goel, R. and Johri, B.N., 1988. Purification and characterization of two thermostable xylanases from melanocarbus albomyces. *Biochem. Int.*, 17: 563–575.

Cossar, D. and Canevascini, G., 1986. Cellulose enzyme production during continuous culture growth of *Sporotricum* (*Chrysoporium*) *thermophile*. *Appl. Microbiology Biotechnology*, 24: 306–310.

Coutts, A.D. and Smith, R.E., 1976. Factors influencing the production of celluloses by *Sporotrichum thermophile*. *Appl. and Environ. Microbiol.*, 31: 819–825.

Emersson, R., 1968. Thermophiles. In: *The Fungi*, (Eds.) Ainsworth, G.C. and Sursman, A.S. Academic Press, New York, 3: 105–128.

Eriksson, K.E.L., Blanchette, R.A. and Ander, P., 1990. *Microbial and Enzymatic Degradation of Wood and Wood Components*. Springer-Verlag, Berlin, 407 pp.

Feldman, K.A., Lovett, J.S. and Tsao, G.T., 1988. Isolation of the cellulose enzymes from the thermophilic fungus *Thermoascus aurantiacus* and regulation of enzyme production. *Enzyme Microb. Technol.*, 10: 202–271.

Fergus, C.L. and Amelung, R.M., 1971. The heat resistance of some thermophilic fungi on mushroom compost. *Ibid*, 68: 675–679.

Gilbert, H.J. and Hazlewood, G.P., 1993. Bacterial celluloses and xylanases. *Journal of General Microbiology*, 139: 187–194.

Grajek, W., 1987. Comparative studies on the production of cellulases by thermophilic fungi in submerged and solid state fermentation. *Appl. Microbiol. Biotechnol.*, 26: 126–129.

Hagerdal, B., Ferchak, J.O. and Pye, E.K., 1930. Saccharification of cellulase by cellulolytic enzyme system of *Thermomonospora* sp. I. stability of cellulolytic activities with respect to time, temperature and pH. *Biotech. Bioeng.*, 22: 1515–1526.

Hayashida, S. and Mo, K., 1986. Production and characteristics of avicel-disintegrating endoglucanase from a protease-negative *Humicola grisea* var. *thermoidea* mutant. *Applied and Environ. Microbiol.*, 51(5): 1041–1046.

Herbert, D., Phipps, P.J. and Strange, R.E., 1971. Chemical analysis of microbial cells. In: *Method in Microbiology*, 5B: 209–344.

Jensen, B. and Olsen, J., 1992. Physio-chemical properties of a purified α-amylase from the thermophilic fungus *Thermamyces lanugenuses*. *Enzyme Microb. Technol.*, 14: 112–116.

Johnson, Eric A., Saka, J.M., Halliwell, G., Madia, A. and Demain, A.L., 1982. Saccharification of complex cellulosic substrate by the cellulose system from *Clostridium thermocelium*. *Appl. Environment Microbiol.*, 43(5): 1125–1132.

Johri, B.N. and Satyanarayana, T., 1986. Thermophilic moulds perspective in basic and applied research. *Indian Rev. Life Sci.*, 6: 75–100.

Lowry, O.H., Rosebrough, N.J., Farr, A.L. and Randall, R.J., 1951. Protein measurement with the folin phenol reagent. *Journal of Biological Chemistry*, 193: 265–275.

Macris, B.J. and Galiotou-Panayotou, M., 1986. Enhanced cellobiohydrolase production from *Aspergillus ustus* and *Trichoderma harzianum*. *Enzyme Microbiol Technology*, 8: 141–144.

Maheshwari, R., Bharadwaj, G. and Bhat, M.K., 2000. Thermophilic fungi: Their physiology and enzymes. *Microbiology and Molecular Biology Reviews*, 64(3): 461–488.

Maheshwari, R., Kamalam, P.T. and Balasubramanyam, P.V., 1987. The biogeography of thermophilic fungi. *Curr. Sci.*, 56: 151–155.

Mendels, M. and Weber, J., 1969. The production of cellulases. In: *Cellulases and their Applications.* Advances in chemistry series, (Ed.) F. Gould. American Chemical Society, Washington, 95: 391–413.

Miller, G.L., 1959. Use of dinitrosalicilic reagent for determination of reducing groups. *Anal. Chem.*, 31: 426–428.

Moloney, A.P., McCrae, S.I., Wood, T.M. and Coughlan, M.P., 1985. Isolation and characterization of 1,4-β-D-glucan glucanohydrolases of *Talaromyces emersonii*. *Biochem. J.*, 225: 365–374.

Neudoerffer, T.S. and Smith, R.E., 1970. An evaluation of fungal enzymes for the solubilization of wheat bean constituents. *Can J. Microbiol.*, 16: 139–146.

Nidetzky, B. and Steiner, W., 1993. A new approach for modeling cellulase cellulose adsorption and the kinetics of the enzymatic hydrolysis of microcystalline cellulose. *Biotechnology and Bioengineering*, 42: 469–479.

Oberson, J., Binz, T., Fracheboud, D. and Canevascini, G., 1992. Comparative investigation of cellulose-degrading enzyme systems produced by different strains of *Myceliophthora thermophila* (Apinis) v. Oorschot. *Enzyme. Microb. Technol.*, 14: 303–312.

Pettersson, B., Cowling, E.B. and Porath, J., 1963. Studies on cellulolytic enzymes. I. Isolation of a low molecular weight cellulose from *Polypores versicolor*. *Biochem. Biphys. Acta*, 67: 1–8.

Prakash, A., 1984. Antagonistic attributes of thermophilic fungi and thermophilism. *Ph.D. Thesis*, Bhopal University, Bhopal.

Rao, R.R., 1977. Studies on cellulose production by an Aspergillus strain. *Indian Journal of Microbiology*, 17(3): 123–128.

Reese, E.T., Siu, R.G.H. and Levinson, H.S., 1950. The biological degradation of soluble cellulose derivatives and its relationship to the mechanism of cellulose hydrolysis. *Journal of Bacteriology*, 59: 485–497.

Romanelli, R.A., Houston, C.W. and Barnett, S.M., 1975. Studies on thermophilic cellulolytic fungi. *Appl. Microbiology*, 30: 276–281.

Ryu, D.D.Y. and Mandels, M., 1980. Cellulases: Biosynthesis and application. *Enzyme Microb. Technol.*, 2: 91–102.

Satyanarayana, T. and Johri, B.N., 1981. Lipolytic activity of thermophilic fungi of paddy straw compost. *Current Sci.*, 50: 682–688.

Schlesinger, W.H., 1991. *Biogeochemistry: An Analysis of Global Change*. Academic, San Diego, 443 pp.

Sen, S. and Chakraborthy, S.L., 1984. Isolation of cellulose hyperproductive mutants of *Myceliophthora thermophila.* In: *Proc. 7th In. Biotechnol. Symp.*, New Delhi, 2: 604–605.

Singh, A., Goel, R. and Johri, B.N., 1990. Production of cellulolytic enzymes by immobilized *Sporotrichum thermophile. Enzyme Microb. Technol.*, 12: 464–468.

Sumner, J.B. and Somer, G.F., 1944. *Laboratory Experiments in Biological Chemistry*. Academic Press, Inc., New York.

Tansey, M.R., 1971. Agar-diffusion assay of cellulolytic ability of thermophilic fungi. *Arch. Microbiol.*, 77: 1–11.

Wase, D.A. J., McManamey, W.J., Raymahasay, S. and Vaid, A.K., 1985. Comparison between cellulose production by *Aspergillus fumigatus* IMI 255901. *Enzyme Microb. Technol.*, 7: 225–229.

Chapter 69

Impact of Open Cast Coal Mining on Soil Environment and its Management

Indu Gupta[1], R.S. Singh[2] and A.C. Gorai[3]
[1]Department of Zoology, S.S.L.N.T. Mahila Mahavidyalaya, Dhanbad, Jharkhand
[2]Scientist, Central Mining Research Institute, Dhanbad, Jharkhand
[3]Reader, R.S.P. College, Jharia, Jharkhand

ABSTRACT

Opencast mining causes land degradation, land fragmentation, soil disruption, soil contamination, erosion and soil quality degradation. The contribution of opencast mining to total coal production around 350 million tones in India during 2003–04 was more than 75 per cent. Mine spoils created due to open-cast mining not fit for plant growth due to their high concentration of heavy metals and destruction of flora and fauna. The study describes the procedures to improve the damage to the soil and improve the ability to support plant growth in long term. It also includes the practices of mine rehabilitation with emphasis on re-vegetation. The same can also be adapted in rejuvenation of mined out areas.

Introduction

In India, there are over 500 coal mines spread over an area of 16,650 sq.km. Mining activity till date are confined to 36 million hectares including the state of Assam, West Bengal, M.P., A.P., Maharashtra and Orissa. Major coal reserves are confined in eastern states of India, known as Jharkhand is full of shrubs, forests and green covers. The entire coal reserves in Indian Sub-continent is riverine in nature. The river valleys in being lust green and a store of biodiversity (Singh, 1995).

Table 69.1: Distribution of Riverine Coal Deposits in India (Singh, 1995)

River Basin	*Percentage of Reserve*
Damodar river	64.85
Sone and its tributaries	11.60
Mahanadi and tributaries	8.80
Godavari and tributaries	10.29
Others	4.00

With the nationalization of coal industry in the year 1971/72, the opencast mining became extremely popular. Initially, it was introduced as a packet mining. In 1994 the share of opencast mining was 68 per cent of the total coal production. The nature increased to 75 per cent by 2004, however, opencast mining is responsible for extensive damage to the environment and ecology which has caused havoc in the fragile ecosystem.

Table 69.2: Pre-dominantly Opencast Mining Belt (Singh, 1995)

Coal-Field	*Coal in mt/sq.km.*	*Percentage Contribution to Total Surface Production*
North Karanpura	11.16	7.30 High concentration
Singrauli	4.95	22.45
I.B. Valley	14.74	6.10 High concentration
Korba	9.69	13.00 High concentration
Wardha	1.23	10.20
Talcher	14.43	8.70 High concentration

Due to opencast mining nearly 60 sq.km. areas per year was estimated to be directly upturned and an additional 150 sq.km. area was likely to be degraded every year due to overburden dumping. Over 70 per cent of coal blocks earmarked for opencast mining during the 8th and 9th Five Year Plans were within the declared forest land. The irrigated land also affected due to opencast mining. The average human population was below 100 sq.km., while the biodiversity was extremely rich with rare species and dense forest.

According to plan of Eastern Coalfields Limited (Anan, 1988), at least 14,000 ha of land in Raniganj Coal field would be damaged due to mining by that time. In Jharia Coal-field, the total of 426.16 ha was identified as degraded area (CMPDI Report, 1986).

During opencast mining, the overlying soil is removed and the fragmented rock is heaped in the form of overburden. This overburden becomes mine spoil. Mining disrupts soil components such as soil horizon and structure, soil microbial population and nutrient cycles that are crucial to sustaining a healthy ecosystem. This disrupted mine spoil are very reluctant and are generally not favorable for soil microbes, invertebrates and vegetational growth.

Mine Spoil

Mine Spoils refers to any overburden material not specifically salvaged in the pre-mining phase (Jha *et al.*, 1992). These are acidic in nature with many scarce organic matter contents. Mine spoils are generally not fit for plant growth. They possess very compacted texture resulting into poor drainage. These are unfit for microbial population and microbes as well as plant growth. Mine spoils are devoid of soil microbes, organic matter, soil moisture and plant available nutrients. They consist of excess amount of heavy metal with little amount of macro nutrients. (Singh *et al.*, 2004)

Chemical Aspects of Mine Spoil

Mine spoils vary in their total heavy metal concentration, pH, cation exchange capacity, organic carbon content and plant available nutrient concentration. Metals in soil solution can be present in the form of free ions; soluble complexes with organic or inorganic may be associated with mobile colloidal materials. Several mine spoils contain toxic levels of soluble elements such as Fe, Zn, Al, Mn, and Cu and inadequate supply of mg, Ca, N and P (Singh *et al.*, 2004). An integrated biotechnological approach has been studied by reclamation of manganese mine spoil dumps (Juwarkar *et al.*, 2002).

Mine spoil vary in their heavy metal concentration, pH, cat-ion exchange capacity, organic carbon and plant available nutrient concentration (Singh *et al.*, 2004). It is generally devoid of macronutrients with substantial amount of micronutrients and heavy metals which are toxic to the microbial population.

Microbial Aspects of Mine Spoil

The high concentration of heavy metal in contaminated mine spoils may inhibit microbial activity (Pierzynski *et al.*, 1994). Soil microbe make the soil structure stable, release required nutrient in inorganic forms by mineralization, produce growth, regulating substance and immobilize heavy metal in soil. Microbial number and microbially mediated soil such as nitrification, denitrification and decomposition of organic matter (Chang *et al.*, 1981; Nordgren, 1988; Bollag *et al.*, 1984; Doelman *et al.*, 1979) are also adversely affected, microbial immobilization of heavy metals in the root zone results into reduced availability and uptake by plants (Pierzynski *et al.*, 1994; Jha, Singh, 1992; Pederson *et al.*, 1980) found that spoils have high bulk density as low porosity. Most of the spoils retained only 25 per cent as much plant available water as the water of natural soil.

Effect of Mining on Soil Environment

1. The clearance of forests for mining results in the loss of genetic resources (Hussain, 1998). Reduce the productivity of soil and create a dust nuisance. Erosion from the overburden and tailing dumps create silt blanketing over agricultural lands and destroy cultivation.
2. Flora-Mining cause disappearance of vegetation and regionally rare plant species.
3. Fauna-on the soil Annelids like earthworm, (*Pheretima posthuma, Genus-Lumbricus*). Arthropods likes termites, white ants, centipede, flies, beetles, julus are very common on soil. Some are useful for the vegetation. These fauna are destroyed due to mining. Loss of fauna including direct losses through clearing and the indirect effects on species through reduction and fragmentation of habitat.
4. Release of toxic elements from overburden: From the environmental point of view the tailings have an impact over the quality of air, water and flora and fauna from the aesthetic aspect. Without doubt the tailings of a processing plant when in current disposed pollute the adjacent area and give prejudice and these terrains for use as agricultural or pastoral destination. (Gergorio *et al.*, 1975).
5. Damage to soils including salination, acidification, pollution and loss of structure.
6. Damage to heritage site: Damaged soil by mining can improve by the following method so that vegetation can grow.

Techniques for Re-vegetation on Mine Spoil

Physical Techniques, Landform Design and Reconstruction

Essential aspects of rehabilitation is reshaping and grading of overburden dump and mined out areas. Following factors should be considered for the shaping of land.

1. *Stability*: The erosion potential of the different materials on site needs to be assessed and a geotechnical investigation may be required. Terraced land form with short steep (angle of repose) stopes and gently sloping terraces (C5 per cent) may be stable and have a higher land capability than a conventional land form of around 15–18 per cent slope.
2. *Drainage Density*: Gradient of slopes increase the drainage density and can change the nature of surface material.
3. *Erosion Control*: Before vegetation, it is important that soil should be controlled by erosion. Soil particle can be lost by three ways *i.e.* blown away by winds, washed away, whole surface may slip away or slump.

Soil having size change from 01–0.05 per cent are effected by wind and 40 per cent are not effected.

Wind erosion can be controlled by three means:

1. Protect the soil surface with a mulch of natural or manufactured material.
2. Maintain the soil in erosion resistant condition *i.e.* must, or with a compacted surface.
3. Reduce the velocity of wind by wind breaks.

On disturbed area water erosion can be controlled by:

1. Slow down of the water flow across the soil surface.
2. Reduce the impact of rain drops on soil surface.
3. Maintain the soil in an erosion resistant condition.

Mulches can be used to protect the soil from raindrops impact. Suitable material for mulches are brush matting, stubble mulch, hay mulch, saw mill wastes and other chemical stabilizers. These mulches are effective only on a small area.

Top soil is the important part for the growth of plant. If the Top soil contains large number of seeds of undesirable species, then it may be better to use the sub-soil as a substrate for rehabilitation. The Top soil contains the majority of the seeds and other plant propagules (such as rhizomes, lignotubers, roots etc), soil microorganism, organic matter, plant nutrients.

The top soil is the upper layer and usually darker than the under lying soils because of the accumulation of organic matter. A void stripping deeper soil horizons which may have poor structure or high clay contents. When the A1 horizon is not obvious, the top 100–300 mm of soil should be recovered. When the aim is to restore the native flora, because most of the seeds of flora stored in top-layer and its removal and return as a thin layer on the surface will maximize the contribution of these seeds to the restoration of micro flora and fauna.

Timing of Top Soil Stripping

Soil should not be stripped or replaced when they are too wet or too dry, as this can lead to compaction, loss of structure and loss of viability of seeds and mycorrhizal innoculum for handling the soil, moisture content vary in different soil local language and experience will help for handling the soil without damage. After setting of seeds clearing and soil stripping should take place. Maximise the store of seed in the soil after setting of seed.

If we want the option is to store other plant (*e.g.* corms bulbs, rhizomes and roots) then the optimum time to strip the soil is when the soil is cool and wet.

Soil Microorganism

Symbiotic microorganisms: Plants form beneficial symbiotic association with a number of soil micro-organisms including fungi, bacteria and actinomycites (single cell plants found usually in soil).

Mycorhiza are a natural component of the ecosystem. Majority of plant species form association with vesicular mycorhiza (VAM) and extomycorrhial fungi. These fungi are effective in increasing the uptake of phosphorous by plants growing in phosphorous deficient soil.

To conserve mycorhizal inoculum Top soil should be:

1. Direct return, wherever possible
2. The pile should be low and revegetated as soon as possible.

Nirtrogen fixation by legumes forms a symbiotic association between the plant and bacteria, known as Rhizobia. This Rhizobia is more tolerant to disturbance and stockpiling than mycorrhizal fungi.

An actinomycetes of the genus Frankia form nodules with the capacity to fix nitrogen on the roots of species from the genera Casuarina and Allocasuarina. These genera are often included to revegetation of Frankia to soil disturbance and stockpiling.

Plant Pathogen: When the soil is infected strictly separate infected soil from uninfected control the movement of infected soil into uninfected area and use of fungicide.

Alternate to Top Soil

If the top soil contains high level of weed soil or plant pathogen then it is unsuitable for rehabilitation, then sub-soil, waste rock or similar material must be used as a substrate for revegetation. To increase chemical properties of such soil, maintain the pH, the physical and chemical properties of proposed substrate should be thoroughly investigated before use in rehabilitation.

The following techniques can improve the plant growth in the long term:

1. Application of animal manure, sewage sludge or other waste.
2. Use of gypsum to reduce the pH of higher alkaline substances.
3. Use of lime to raise the pH of acid substances.
4. Inorganic fertilizer
5. Growing green manure
6. To grow nitrogen fixing species such as legumes.
7. Applying mulch.

Re-vegetation

To restore a native ecosystem, the initial re-vegetation effort is identical to original vegetation. The initial vegetation effort must be establish the building blocks for a self sustaining system so that desired vegetation can grow.

Best time for vegetation is the rainy season. All the preparatory works are completed before the seeds are put in the soil, so that seeds can generate properly.

Species Selection

The species selection depends upon the future land use of the area and the soil condition of the area. If the objective is to restore the native vegetation and fauna then the species are pre-determined. Some species can not grow in the soil where the conditions are different after mining. In this case species should be from outside the mining area. One of the major approaches in terms of establishment of the native ecosystem include search of the local areas for natural analogues of post mining landscape and mine soil which may be used as a model for the past mining ecosystem.

Establishment

Plant species can be established on rehabilitated areas from:

1. Propagules (seeds, lignotubers, corm, bulb, rhizomes and roots) stored in the top soil.
2. Sowing seeds.
3. Spreading harvested plants with bradysporous seeds (seed retained on the plant in persistent woody capsules) on to areas being rehabilitated.
4. Planting nursery–raised seedling.
5. Transplants of individuals from natural areas.
6. Habitat transfer–the transfer of substantial amounts (around 1 m^2 or more in area and 200–300 mm depth) of relatively undisturbed soil with its vegetation intact from natural areas.
7. Invasion from surrounding areas through vectors including birds, animals and wind.

Top Soil

Top soil play an important role in the establishment of native species. It is most important aspects of restoring the full suit of pre-mining species to rehabilitated areas.

Seedling

Sowing seed is an economical and reliable method for establishing some species. Seeding results in a more random distribution of plants than planting the seedlings which leads to more natural looking vegetation. The best species are those which have high germination and survival rates in the field. (Singh *et al.*, 1997) Re-vegetated the overburden mine soil with native species *viz.*, are very successful–Sisum (*Dalbergia Sisoo*), Neem (*Azadirchta indica*), Siris (*Albizia procera*), Gulmohar (*Delonix regia*), Babool (*Acacia nilotica*) these are found to be useful as early colonizer plant species.

Planted Seedling

It is more economical than to seeding. Plant taken from the nursery and nursery raised seeding is more appropriate than to seeding in the top soil. Seedling should always be well adapted before being planted in the field.

Transplanting and Habitat Transfer

Direct transplanting of species is possible by transferring slices or front end loader bucket of soil with the vegetation intact. It is an expensive method. Transplanting is best carried out in wet seasons so that there is maximum survival of plant species.

Seed Collection and Treatment

Seed should be cleaned before storage. Drying the capsules or pods in the sun or in an oven fleshy fruits should be soaked in water to release the seeds.

Clean seeds should be stored in dry containers. The containers should be clearly labeled with details of the species, date collected and collection location. Storage areas should be fumigated regularly. Seeds will be treated with insecticide and fungicide. Seeds must be stored at low humidity (<10 per cent relative humidity) and low temperature. Temperature varies for temperate and tropical plant species.

Seeds of some species require pre-sowing treatment. Germination of most native legumes and other species is enhanced by heat treatment. These species are commonly immersed in boiling water for 30 seconds to five minutes before sowing.

Seed Bed Preparation

Seed bed is a suitable place where seed can germinate. In seed bed soil must be well aerated and seed must be in good contact with soil. In the soil, around the seed, soil must be loose, so that root and seed can grow properly. Seed bed should be free from weeds. In the soil moisture level should be adequate.

Seeding Method

The seeding technique chosen will depend on local factors such as topography, size of the area being rehabilitated and the type of seed various mechanics have been developed for sowing seeds but by hand is still the best options. Care must be taken not to bury the seeds too deep for successful establishment.

Timing of seeding is an important factor for successful revegetation. Seed should be sown immediately after reliable rains or after the break of season. Native seeds require specific moisture and temperature conditions for germination. So that they establish at the optimum time of the year for survival. When seeds are sown early, then they may be eaten by ants, birds, and small mammals, but however, in native vegetation this may be prevented by disturbing the fauna.

Weed Control

Spread of weed is an important consideration in rehabilitation. Care must be taken that weeds are not introduced to the area in manure or as contaminated in seed of desirable species.

Herbicides can be applied selectively using wicks application in some cases, when the weeds are much taller than the desirable species.

Fertilizers and Soil Amendments

The type of fertilizers and the application rate will vary according to the site, soil type and post mining land use. Instead of fertilizer mycorrhizal fungi can be used for the growth of plant. Care should be taken when applying fertilizers to plants. In India, high rates of nitrogen and phosphorous are more favourable for plant growth. If nitrogen is close to roots, then it damages them. High rates of phosphorous and low rates of nitrogen fertilizer will favour the growth of legumes (EPA, 1995).

Calcium carbonate increase the pH of acid soil. Application of lime are in the range of 2–5 tonnes/ha but will vary according to soil type, initial pH and source of lime. Gypsum improve the structure of poorly structured sodic soil. If exchangeable Na is above 6 per cent then soil at around 5–10 tonnes/ha.

Various organic wastes (*e.g.* Animal manures, sewage, sludge and blood and bone) can have value as both fertilizer and soil amendments. However, it is difficult to spread for large-scale rehabilitation.

Nutrient Accumulation

During the mining, vegetation are removed from the soil which lead to the loss of plant nutrients. It is necessary to improve the nutrient in the soil before vegetation. Best method to increase the nitrogen contents in the soil is by establishment of nitrogen fixing plants like legumes, which increase the nitrogen level in ecosystem.

Fauna

To restore the natural ecosystem, the fundamental part is to restore the fauna at the place of mining. Some invertebrates species are introduced if fresh Top soil is placed on the area, but most of the fauna re-colonise from the surrounding area. The total population of the mesofauna (0.1 to 10 mm) varies greatly between sites but can exceed 2000 million per hectare in old grassland soils (Alan Wild, 1996). Some fauna spreads quickly colonise if the areas is fit for their food, shelter and breeding places. If in the surrounding area, population of fauna species is low, then a captive breeding and release program or re-introduction programme may be appropriate. Animals, like invertebrates are important in many ecological processes such as nutrient cycling, litter decomposition, soil aeration, seed dispersal, seed predation and pollination. Invertebrates also form the food of other fauna group. The abundance and diversity of invertebrates evaluate the success of restoration program.

Maintenance

It includes:

1. Replanting failed or unsatisfactory areas.
2. Repairing of any erosion problem.
3. Watering plants in dry areas especially at the establishment phase.
4. Fire management.
5. Pest and weed control.
6. Control of feral and native population, including fencing.
7. Fertilizer Application.
8. Application of lime or gypsum to control pH and improve soil structure.

Site Monitoring

It is important that after rehabilitation its success should be monitored to assess the extent programme is successful. The success can be defining as to what constitutes a viable ecosystem. What is an acceptable level of species diversity, how indistinguishing should be rehabilitated areas from neighbouring untouched areas and so on.

According components of success criteria could include (EPA, 1995).

1. Physical (stability, resistance to erosion, re-establishment of drainage).
2. Biological (species richness, plant density, seed production, fauna return, weed control, productivity, establishment of nutrient cycle).

3. Water quality standards for drainage water, and
4. Public safety issue.

Conclusion

Rehabilitation of mined out land is not a operation but it is a part of effective environmental management through all phases of resource development, from exploration to construction, operation and closure and soil is considered to be the medium in which plant life sustain.

Restoration of ecosystem is multi-disciplinary science with complicated life forms ecology even though human have been disturbing the land for many centuries. In biodiversity–there is a technique by which mine site can be developed in to economic improvement of the society. They can achieve this by careful attention to all aspects of rehabilitation and revegetation from initial planning. Cleaning, soil removal, storage and replacement through species selection and re-establishment of vegetation with its associated organism to maintenance of soil micro flora and microfauna for the sustainable development of soil structure and fertility status of mining industry.

References

Chang, F.H. and Broadbent, F.E., 1981. *Soi.*, 132: 461–421.

CMPDI, 1986. Environmental restoration of mined out areas broad estimates: A report, Ranchi.

Doelman, P. and Haanstra, L., 1979. *Soi. Biol. Biochem.*, 11: 487–491.

EPA, 1995. *Overview of Best Practice Environment Management in Mining*. The Environmental Protection Agency of the Australian Environmental Department, Commonwealth of Australia, pp. 30.

EPA, 1995. *Rehabilitation and Revegetation: Best Practice Environmental Management in Mining*. The Environmental Protection Agency of the Australian Environmental Department, Commonwealth of Australia, pp. 30.

Hussain, A., 1998. *Surface Mine Environment and Reclamation*. Samya, Ambikapur, India.

Jha, A.K. and Singh, J.S., 1992. *Rehabilitation of Mine Spoils: Restoration of Degraded of Land Concepts of Strategies*, pp. 211–253.

Juwarkar, A., Khobragade, R., Nimje, M., Dubey, K., Singh, S.K., Gandhi, P.K. and Singh, R.N., 2002. Restoration of biodiversity in coal mine dump. In: *Proceeding of National Seminar on Indian Mining: Problems and Prospects*, 20–21st December. Mining Engineers Association of India, Nagpur, pp. 177–188.

Loveson, V.J. and Dhar, B.B., 1995. *Satellite Remote Sensing Survey Over Some Environmentally Critical Areas in Coalfields of Damodar Basin, India*, pp. 228.

Pierzynski, G.M., Schnoor, J.L., Banks, M.K., Tracie, J.C., Litch, L.A. and Erickson, L.E., 1994. *Vegetative Remediation at Superfund Site, Mining and its Environmental Impact*. Royal Society of Chemistry, pp. 49–69.

Singh, R.S. and Tripathi, N., Singh, S.K., Paul, D. and Gupta, A., 2004. Bioreclamation of contaminated mine spoils. In: *Proceedings International Conference on Soil and Groundwater Contamination Enviro.*, India, pp. 149–150.

Singh, R.S., Tewary, B.K. and Dhar, B.B., 1997. Reclamation of coal mine overburden dumps in India. In: *Engineering Geology and Environment*, (Eds.) Marinos, Koukis, Tsiambos and Stumaras, Balkema, Rotterdam.

Singh, T.N., 1995. Impact of coal mining on environment and ecology. In: *Proceeding of First World Mining Environmental Congress*, New Delhi, December, pp. 70–72.

Wild, Alan, 1995. Soils and the environment: An introduction. Ch. 5: *Organisms and Soil Processes*, pp. 68–88.

Chapter 70

Limnology of Lentic Freshwater Systems in North Cachar Hills, Assam, India

T. N. Majumdar[1], A. Gupta[2] and M. Daolagupu[1]
Department of Zoology, Haflong Government College, Haflong – 788 819, Assam
[1]Department of Ecology and Environmental Science, Assam University, Silchar – 788 011, Assam, India
E-mail: agecol@rediffmail.com

ABSTRACT

A study was carried out on the limnology of five lentic systems, at different altitudes in North Cachar Hills district of Assam. The influence of altitude on temperature is evident from the results obtained. Marked seasonal variations were observed in dissolved oxygen and free carbon dioxide content which also showed variations among the different lentic systems. Total alkalinity and conductivity of water were found to be low. Comparatively higher levels of total alkalinity, conductivity, phosphates and nitrates were recorded in the dry season. A strong correlation between primary productivity and phosphates, nitrates and total alkalinity could be observed. Statistical analysis using ANOVA and Duncan's Long range Test revealed significant difference among the different lentic systems in several physico-chemical properties such as dissolved oxygen, total alkalinity, phosphates, nitrates and conductivity of water.

Keywords: *Lentic system, Physico-chemical properties, North Cachar Hills, Assam.*

Introduction

The North Cachar Hills District is one of the two hill districts of Assam. It lies between *c* 25° to 26° North latitude and *c* 92° to 94° East longitude. The district consists of undulating hills, valleys and small patches of plain lands. The total geographical area of North Cachar Hills is approximately 4888 sq. km. On its eastern side are the states of Nagaland and Manipur, while Cachar district of Assam lies towards its south. On the western side is the state of Meghalaya, while Nowgaon and Karbianglong district of Assam are located towards its north (Figures 70.1). The main mountain range of the district is the Borail spreading from east to west in the southern part of the district.

Although extensive work has been done on the limnology of lentic systems in other parts of India (Singh and Ray, 1995; Tiwari, 1999; Sahat *et al.*, 2001; Mariappan and Vasudevan, 2002), few such investigations have been carried out in the lentic bodies of North-East India (Das *et al.*, 2002; Dutta Gupta *et al.*, 2004) with no published account on those in North Cachar Hills district. This article, therefore, attempts to generate base line information on the limnology of five lentic systems at different altitudes in this district. Besides physico-chemical variables, the vegetation near the systems along with aquatic macrophytes were also studied, since they have a pronounced effect on ecosystem processes (Carpenter and Lodge, 1986) and can be used to monitor the quality of water (Dewanji and Matai, 2000).

Materials and Methods

Study Area

For the present investigation five water bodies were selected at three different places in N.C. Hills District. The Fishery Tank, the Haflong Lake and the Upper Lake are situated in Haflong, (altitude

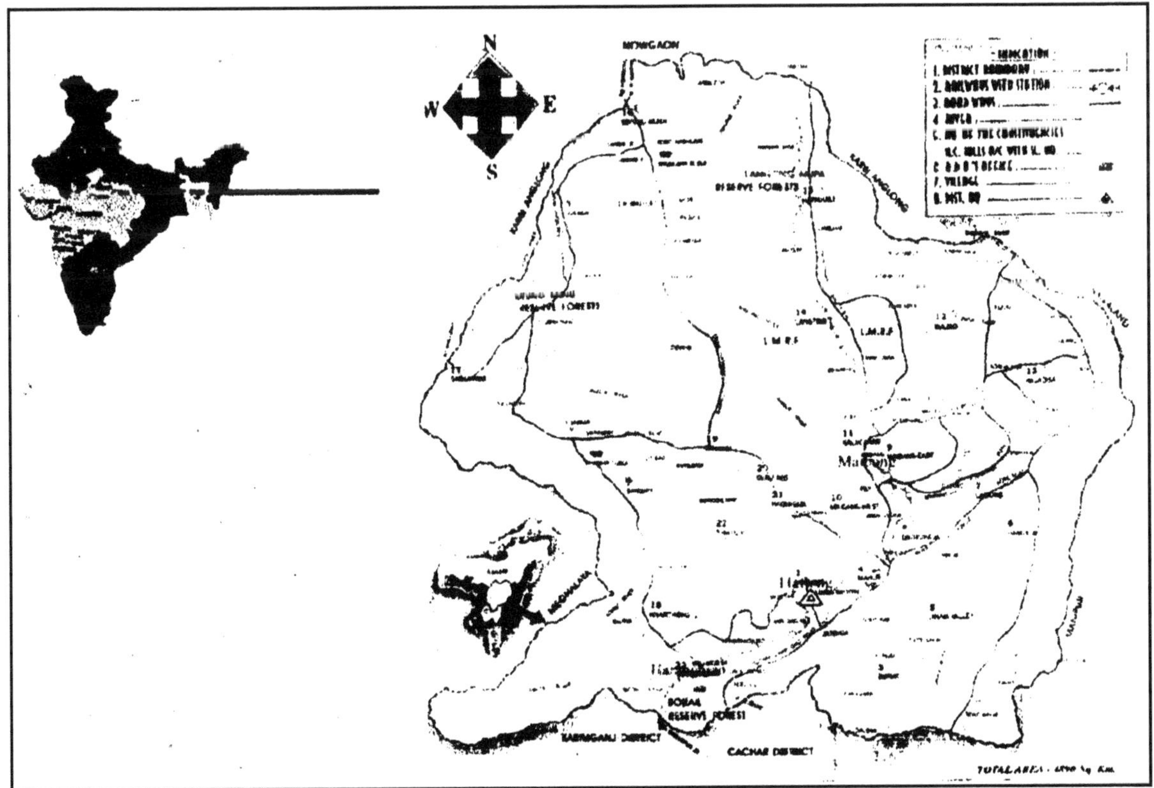

Figure 70.1: Map of N.C. Hills District of Assam, Showing Study Sites

c 935.10 msl.), the headquarter of N.C. Hills District, while a pond each in Maibong (altitude *c* 365,87 msl), a small town in the northern part of the district, and Harangajao (altitude *c* 240.79 msl) in the south, were investigated (Figures 70.1). The Fishery Tank is *c* 250 m long and *c* 50 m wide with an average depth of 2 metres. It is surrounded by dense human habitation, public buildings and roads. The Haflong Lake is an artificial lake *c* 1.0 km long and *c* 135 m wide with an average depth of 2 m. It is situated in a low lying area and receives sewage from a large part of Haflong town. The Upper Lake is located about 100 m above the level of the other two waterbodies in Haflong. It lies between two steep hills and is *c* 250 m long, *c* 30.0 m wide, and *c* 12.0 m deep. Human habitation around the lake is very thinly dispersed. The pond in Maibong, located near the railway station, has a length of *c* 80 m and width and depth of 40 m and 2 m, respectively, and is surrounded by paddy fields and human habitation. The pond in Harangajao is a small waterbody located amidst paddy fields near the Haflong-Silchar Road about 6.0 km from Harangajao town. It is *c* 45 m long with an average width and depth of 18 m and 2.1 m, respectively.

Climate and Vegetation

Information on the, climate and vegetation of North Cachar Hills District was collected from different secondary sources (Champion and Seth, 1968; Parashar, 1994). Primary data on vegetation in and around the waterbodies were also collected.

Physico-chemical Properties of Water

Physico-chemical properties of water were studied in three lentic system at Haflong either monthly (Fishery Tank and Haflong Lake) or seasonally (Upper Lake) from August, 1999 to July, 2000. These systems along with the two ponds at Maibong and Harangajao were also investigated in three seasons (pre-monsoon, monsoon and winter) during 2001–2002. Air and water temperature were measured with a mercury bulb thermometer, pH and conductivity with a pH and conductivity meter (Merck-WTH, Germany), respectively, and total alkalinity, dissolved oxygen and free carbon dioxide by standard methods (Michael, 1984). Nitrate and phosphate were measured using a spectrophotometer (Thermo Spectronics) by phenol disulphonic acid digestion method and stannous chloride-ammonium molybdate method respectively. Primary productivity was estimated by the light and dark bottle method (Michael, 1984).

Results and Discussion

Climate and Vegetation

The climate in North Cachar Hills District is moderate with maximum temperature ranging between 24°C and 30°C and minimum between 14°C to 19°C. Annual rainfall ranges from 2200 mm to 3000 mm. Distribution of rainfall from August, 1999 to July 2000 and from March, 2001 to February, 2002 is depicted in Figures 70.2a and 70.2b, which show that there was no precipitation from December to February in the first year of study and in the second year, November, December and February were totally devoid of any precipitation. The highest precipitation was recorded in the month of August (475.00 and 569.00 mm respectively) in both the years. Total rainfall was higher in the second study year (2889.3 mm) than that in the first (2278.75 mm).

The hills of the Borail range have thick vegetation. The forest is basically moist Semi-Evergreen type with patches of east Himalayan Moist-Mixed Deciduous and pure Bamboo forest. Trees, shrubs and herbs around each water body and the aquatic macrophytes in them were studied in the present investigation (Table 70.1). Relatively fewer trees species were present in the catchment of Upper Lake

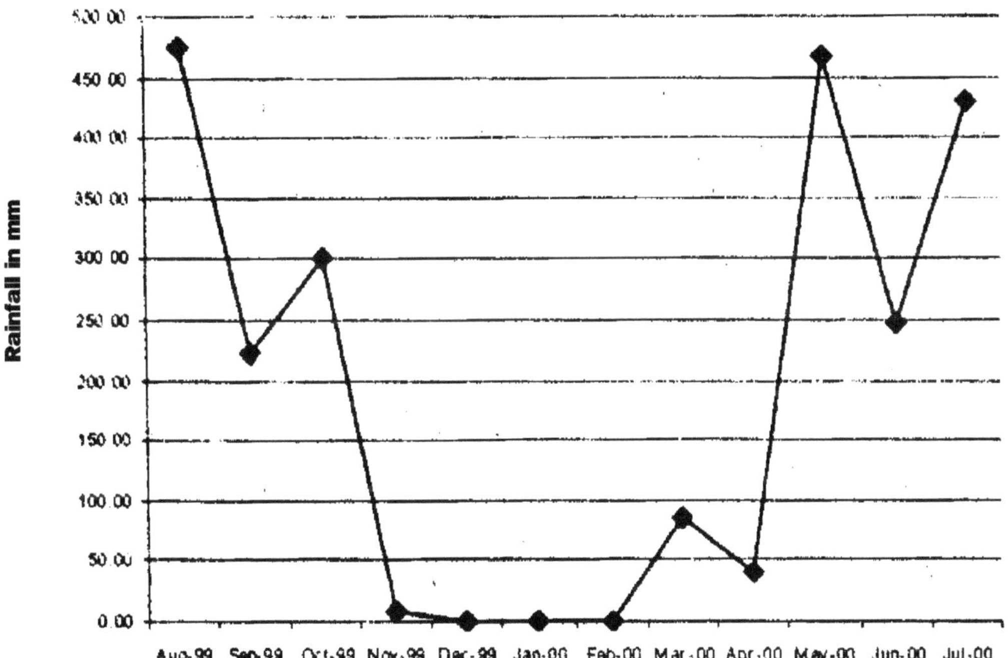

Figure 70.2a: Rainfall in N.C. Hills District from August 1999 to July 2000

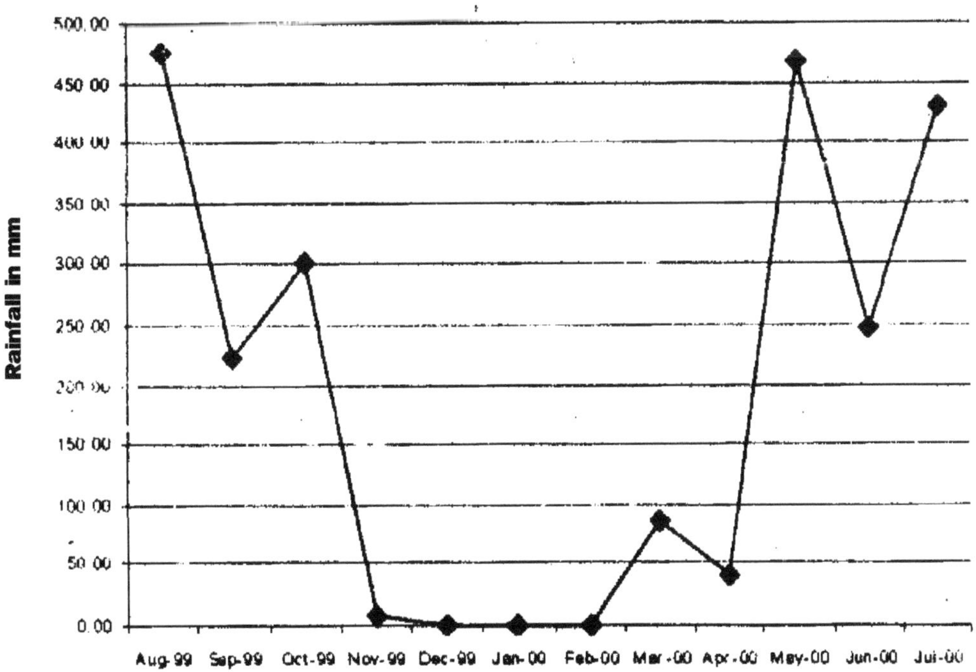

Figure 70.2b: Rainfall in N.C. Hills District from March 2001 to February, 2002

Table 70.1: Vegetation in and Around the Water Bodies of North Cachar Hills, Assam

Types of Vegetation	*Fishery Tank*	*Haflong Lake*	*Upper Lake*	*Maibong Pond*	*Harahgajao Pond*
Marginal					
Trees/Shrub	*Mangifera indica*	*Pinus* sp.	*B. tulda*	*Tectona grandis*	*T. grandis*
	Areea catechu	*Margo* sp.	*M. paradisiaca*	*Ficus carica*	*B. tulda*
	Dioscorea bulbifera	*Acacia* sp.		*Dioscorea bulbifera*	
	Anthocephalus cadamba	*Cassia* sp.		*Acacia nilotica*	
	Musa paradisiaca	*Delonix regia*		*Gmelina arborea*	
	Bambusa tulbla	*Eucalyptus* sp.			
		Ficus benghalensis			
Herbs/Small shrubs	*Lantana camara*	*Digitaria sanguinalis*	*Eupatorium odoratum*	*Chrysopogon aciculatus*	*Digitaria sanguinalis*
	Chrysopogon aciculatus	*C. aciculatus*	*Hedyotis lineata*	*Cynodon dactylon*	*Imperata cylindrica*
	Cynodon dactylon	*C. benghalensis*	*C. aciculatus*	*Digitaria sanguinalis*	*Chrysopogon aciculatus*
	Digitaria sanguinalis	*A. conyzoides*	*Achyranthes aspara*	*Commelina benghalensis*	*Ageratum conyzoides*
	Commelina benghalensis	*Parthenium* sp.	*C. benghatensis*	*Ageratum conyzoides*	*Lantana camara*
	Ageratum conyzoides	*A. spinosus*	*Bambusa subtilis*	*Lantana camara*	*Achyranthes aspara*
	Amaranthus spinosus	*Centella asiatica*	*Blachnum* sp.	*Eupatorium odoratum*	
	Amaranthus viridis		(Fern)		
	Blainvillea latifolia				

Contd....

Table 70.1–Contd...

Types of Vegetation	*Fishery Tank*	*Haflong Lake*	*Upper Lake*	*Maibong Pond*	*Harahgajao Pond*
Aquatic Macrophytes					
Marginal	*Axonopus compressus*	*Axonopus compressus*	*Axonopus compressus*	*Axonopus compressus*	*Axonopus compressus*
	Polygonum perfoliatum	*Sagittaria* sp.	*Alternanthera sessilis*	*Monochoria hastaefolia*	*Sagittaria* sp.
	Polygonum hydropiper	*Cardanthera triflora*	*Amaranthus* sp.	*Sagittaria* sp.	*Amaranthus* sp.
	Ipomea aquatica	*Limnophila heterophylla*	*Polygonum perfoliatum*	*Amaranthus* sp.	*Monochoria hastaefolia*
	Phylanthus nirurii	*Monochoria hastaefolia*	*Polygonum hydropiper*	*Limnophila heterophylla*	*Limnophila heterophylla*
	Amaranthus sp.	*Monochoria vaginalis*			*Ipomea aquatica*
		Amaranthus sp.			
Floating	*Eichhorrnia crassipes*	*Trapa natans*	*Trapa natans*	*Trapa natans*	*Azolla pinnata*
	Azolla pinnata	*Eichhornia crassipes*		*Pistia stratiotes*	*Utricularia* sp.
	Pistia stratiotes	*Pistia stratiotes*		*Azolla pinnata*	*Lemna* sp.
	Utricularia sp.	*Azolla pinnata*		*Lemna* sp.	*Hydrilla* sp.
		Utricularia sp.		*Utricularia* sp.	*Eichhornia*
		Lemna sp.		*Eichhornia crassipes*	*crassipes*
Submerged	NIL	*Hydrocharis* sp.	*Salvinia* sp.	*Salvinia* sp.	NIL
		Salvinia sp.		*Vallisneria* sp.	
		Vallisneria sp.		*Hydrilla* sp.	
		Hydrilla sp.		*Chara* sp.	
Emergent	NIL	*Nelumbo* sp.	NIL	*Nelumbo* sp.	NIL
		Nymphaea sp.		*Nymphaea* sp.	
		Oryza sp.		*Oryza* sp.	

and Harangajao Pond when compared with those around the other waterbodies studied. Emergent type of aquatic macrophyte was totally absent in Fishery Tank, Upper Lake and Harangajao Pond. Submerged aquatic macrophyte was absent in both Fishery Tank and Harangajao Pond whereas in Upper Lake only one species of submerged (*Salvinia* sp.) and one species of floating macrophyte (*Trapa natans.*) were found covering a very small area of the waterbody. Several species of marginal macrophytes were present in all the lentic systems, studied.

Physico-chemical Properties of Water

The physico-chemical properties of water in differet lentic systems studied are depicted in Tables 70.2 and 70.3. The influence of altitude on temperature is evidenced by the lowest temperature regime in Upper Lake, Haflong, and the highest in Harangajao. It is a well-known fact that no other single factor has so profound influence on aquatic ecosystem than temperature (Welch, 1952). Hydrogen ion concentration (pH) is another important environmental factor influencing the aquatic ecosystem (Welch, 1952). In the present investigation, the Fishery Tank at Haflong showed higher pH values (6.5 to 8.35) than the other systems during the first annual cycle (1999–2000). This probably reflects the effect of periodic liming and manuring of the tank by the Department of Fisheries during that period. In the absence of liming during 2001–2002, the pH regime became much lower (6.5–7.5). Lentic systems located at a lower altitude (< 100 msl) in both Brahmaputra and Barak valleys in Assam exhibited a somewhat higher pH range (Das and Baruah, 2002; Dutta Gupta *et al.*, 2004). In contrast, the pH values recorded in certain lentic systems of Shillong, Meghalaya, which are located at a higher altitude (1495–1900 msl) than those in the present study, were somewhat lower (Gupta *et al.*, 1992). Thus the lentic systems of North Cachar Hills, located at an intermediate altitude, exhibit a pH regime lower than that in the valleys, but higher than those at a higher altitude. Conductivity was found to range from 17.6–78.1 $\mu S\ cm^{-1}$, which is lower than those reported from the other regions of India (Habib *et al.*, 1997). The entry of nutrient-rich sewage into Haflong Lake and liming/ manuring during the first annual cycle in Fishery Tank is reflected in the relatively high conductivity in these two systems. The conductivity in the latter system declined in the second annual cycle when liming was discontinued. The range of dissolved oxygen is wide in small waterbodies and may vary greatly from one waterbody to the other in the same area (Kaushik *et al.*, 1991). In the present study, dissolved oxygen was found to vary from 1.6 mg l^{-1} to 10.4 mg l^{-1} in the different study sites. Relatively higher dissolved oxygen values were recorded in Upper lake and Haranbrajao pond (6.2–10 mg l^{-1} and 5.2–10.4 mg l^{-1} respectively) when compared to those in Fishery tank, Haflong Lake and Maibong Pond (2.7–10 mg l^{-1}, 1.6–6.4 mg l^{-1} and 2.8–6.4 mg l^{-1} respectively) as a result of the entry of large amount of organic matter in the form of manure and sewage in the latter. Again, in Haflong Lake lower oxygen values were recorded in the first annual cycle (1.6–6.4 mg l^{-1}) than that in second annual cycle (4.8–6.4 mg l^{-1}). This might be due to the accidental drainage of water from this system in the first annual cycle when the water level was much lower. In most of the water bodies higher oxygen values were recorded during winter. Higher values of dissolved oxygen in colder months even in tropical systems were reported by some workers (Kauashik *et al.*, 1989; Prasad, 1990). Higher level of free carbon dioxide (6–13 mg l^{-1}) could be observed during post monsoon in the first annual cycle (1999–2000) in Fishery Tank, Haflong Lake and Upper Lake. In contrast, in second annual cycle (2001–2002), Fishery Tank, Haflong Lake and Harangajao Pond showed the highest level of free CO_2 (15.6 mg l^{-1}, 11 mg l^{-1}, and 8 mg l^{-1} respectively) during monsoon whereas it was higher during pre-monsoon in Upper Lake and Maibong Pond (5.2 mg l^{-1} and 9 mg l^{-1} respectively). The rain absorbs small amount of CO_2 from the atmosphere and delivers it to the water bodies (Unni, 1972). In the second study year rainfall was considerably heavier in the monsoon months. This might be the cause of higher CO_2 during Monsoon.

Table 70.2: Physico-chemical Properties of Water in the Lentic Systems of North Cachar Hills, Assam, in the First Annual Cycle (1999–2000)

Month	Water Body	Air Temp. °C	Water Temp. °C	pH	Conductivity μScm^{-1}	Dissolved Oxygen mgl^{-1}	Free Carbon dioxide mgl^{-1}	Total Alkalinity mgl^{-1}	Phosphates μmgl^{-1}	Nitrates mgl^{-1} μgml^{-1}	Gross Primary Productivity $mgC/m^3/24$ hrs (Range)	Respiration $mgC/m^3/24$ hrs (Range)	Net Primary Productivity $mgC/m^3/24$ hrs (Range)
Aug'99	Fishery Tank	27	28	7.5	73.9	5.8	8	68	26.6	0.94	1576.5±523.53 (975.9–1951.9)	925.9±114.66 (825.8–900.9)	650.6±458.72 (150.1–1051)
	Haflong Lake	27	28	7.5	52	4.2	10	64	15.3	1.2	1764.9 (1576.5–1951.9)	1351.3	412.9 (225.23–600.58)
	Upper Lake	24	26	7.24	20.9	8.6	3	48	5.1	1.8	2402.3 (1726.66–3077.45)	2026.9 (1501.4–2552.4)	375.36 (225.2–525.5)
Sep'99	Fishery Tank	26	28.5	6.75		4.8	12.4	64					
	Haflong Lake	28	29	6.9		3.2	10.5	64					
	Upper Lake												
Oct'99	Fishery Tank	24	25.5	6.5		3.2	13	104					
	Haflong Lake	25.5	26	6.7		3.2	14	68					
Nov'99	Fishery Tank	21	19.5	7.15	78.1	5.6	6	60	127	4.6	1701.63±312.55 (1351.3–1951.9)	925.10±144.66 (825.80–1051.0)	775.73±263.64 (525.5–1051)
	Haflong Lake	20	19	6.8	64.2	4.0	8	64	12.7	4.6	1351.3 (1051–1651.6)	488 (150.14±825.8)	863.33 (150.14–1776.5)
	Upper Lake	19	18	6.5	30.1	6.8	6	30	18.1	5.9	1426.4 (1351.3–1501.4)	825.8 (750.7–900.9)	750.7
Dec'99	Fishery Tank	17	17	6.55		5.6	6	80					
	Haflong Lake	17	16	6.6		3.8	6.8	92					
	Upper Lake												
Jan'00	Fishery Tank	16	14.5	6.78	73.9	3.2	12	92	30	1.6	625.60±188.94 (450.41±825.8)	425.4±263.69 (150.1–675.7)	200.2±86.69 (150.1–300.3)
	Haflong Lake	16	14	6.5	59	3.6	6.8	94	10	2.8	750.7 (450.4–1051)	375.36 (300.3–450.4)	375.36 (150.14–600.58)
	Upper Lake	19	17.5	6.95	24.1	7	2.8	36	12.5	6.0	1126.1	825.8 (750.7–900.9)	300.3 (325.2–375.36)

Contd...

Table 70.2–Contd...

Month	Water Body	Air Temp. °C	Water Temp. °C	pH	Conductivity μScm^{-1}	Dissolved Oxygen mgl^{-1}	Free Carbon dioxide mgl^{-1}	Total Alkalinity mgl^{-1}	Phosphates μmgl^{-1}	Nitrates mgl^{-1} μgml^{-1}	Gross Primary Productivity $mgC/m^3/24$ hrs (Range)	Respiration $mgC/m^3/24$ hrs (Range)	Net Primary Productivity $mgC/m^3/24$ hrs (Range)
Feb'00	Fishery Tank	20	19	6.95		2.8	13	64					
	Haflong Lake	20	19	7.0		1.6	9	80					
	Upper Lake												
Mar'00	Fishery Tank	23	22.5	7.6		3.6	11.5	60					
	Haflong Lake	22	23	7.1		5.0	8.4	68					
	Upper Lake												
Apr'00	Fishery Tank	24	21	8.2		5.8	6.4	66					
	Haflong Lake	24	20	7.7		6.4	9.5	64					
	Upper Lake												
May'00	Fishery Tank	24.5	25	7.8	64.2	2.7	13	68	17	0.48	1676.63±991.23 (600.6–2552.2)	1151.1±740.63 (300.3–1501.4)	525.53±456.7 (225.2–1051.1)
	Haflong Lake	25	26	7.4	48	4.2	10	66	9.5	0.38	2552.4 (1651.6–3453.3)	1764.8 (1051–2477.4)	825.8 (600.58–975.9)
	Upper Lake	22	21	7.01	22.9	6.2	5.8	40	5.0	2.5	825.8 (450.43–1201.15)	337.8 (75.1–600.6)	487 (375.36–600)
Jun'00	Fishery Tank	25	22	8.2		4.6	12.4	68					
	Haflong Lake	24.5	21	7.3		4.8	11	66					
	Upper Lake												
Jul'00	Fishery Tank	28	29	8.35		4.8	12.4	66					
	Haflong Lake	28.5	29	7.6		4.6	11	68					
	Upper Lake												

Table 70.3: Physico-chemical Properties of Water in the Lentic Systems of North Cachar Hills, Assam, in the Second Annual Cycle (2001–2002)

Season	Waterbody	Air Temp. °C	Water Temp. °C	pH	Conductivity μ Scm^{-1}	Dissolved Oxygen (mgl^{-1})	Free Carbon Dioxide (mgl^{-1})	Total Alkalinity (mgl^{-1})	Phosphates $\mu g\ ml^{-1}$	Nitrates mgl^{-1}
Premonsoon (May'01)	Fishery Tank	24	25.2	7.5	23.6	6	8	84	1.08	5.8
	Haflong Lake	23	24	7.6	55.8	5.6	8.4	80	21.2	45.9
	Upper Lake	20	20.5	7.2	19.2	8	5.2	20	–	2.8
	Maibong Pond	26	28	6.52	55.4	2.8	9	56	2.4	2.2
	Harangajao Pond	24	22.5	7	20.2	6.4	5.2	56	1.02	0.76
Monsoon (Aug'01)	Fishery Tank	26	29	6.5	20.2	5.6	15.6	76	0.24	6.7
	Haflong Lake	28	29	7.3	63	4.8	20	68	24.6	23
	Upper Lake	24	26	6.45	18.9	9.6	4.8	20	–	2.05
	Mailbong Pond	22	23	7	59	5.6	7	46	11.68	19.7
	Harangajao Pond	30	32	6.05	18.2	5.2	8	48	1.33	0.88
Winter (Jan'02)	Fishery Tank	20	18	7.15	28.6	6	6	84	4.62	4.85
	Haflong Lake	16	14.5	6.8	74.6	6.4	8.4	94	21.3	45.9
	Upper Lake	18	17	7	20.1	10	3.4	26	–	2.8
	Maibong Pond	20.5	19	6.2	61	6.4	5.6	68	5.35	4.6
	Harangajao Pond	20	18	6.85	17.6	10.4	2.4	20	0.72	0.68

Table 70.4: Comparison Among Physico-chemical Properties of Water in Different Waterbodies Using One Way ANOVA and Duncan's Long Range Test

Period	*Physico-Chemical Parameters*	*F Value*	*df*	*Level of Significance*	*Duncan's LRT*
First Annual Cycle	Dissolved Oxygen	9.8	9,2	P ≤ 0.01	UL > HL, FT; HL ≅ FT
	Free Carbon Dioxide	0.13	9,2		
	Total alkalinity	5.016	9,2	P ≤ 0.05	FT, HL < UL
	pH	0.88	9,2		
	Air Temp.	0.09	9,2		
	Water Temp.	0.05	9,2		
	Conductivity of Water	69.63	9,2	P ≤ 0.001	FT > HL > UL
	Phosphates	0.67	9,2		
	Nitrates	1.34	9,2		
	Gross Primary Productivity	30.07	9,2	P ≤ 0.001	
	Respiration	0.076	9,2		
	Net Primary Productivity	0.36	9,2		
Second Annual Cycle	Dissolved Oxygen	3.398	10,4		
	Free Carbon Dioxide	2.15	10,4		
	Total Alkalinity	14.36	10,4	P ≤ 0.001	FT, HL > MP, HL, UL; MP > UL, FT ≅ HL; MP ≅ HP; HP ≅ UL
	Phosphates	41.30	10,4	P ≤ 0.001	HL > MP, FT, HP, UL; MP > HP, UL
	Nitrates	13.42	10,4	P < 0.001	HL > MP, FT, UL, HP
	pH	1.17	10,4		
	Conductivity	64.35	10,4	P ≤ 0.001	HL > MP > FT, UL, HP
	Air Temp.	0.36	10,4		

Higher total alkalinity could be observed in Fishery Tank (60 mg l^{-1}–104 mg l^{-1}) and Haflong Lake (64 mg l^{-1}–94 mg l^{-1}) followed by Maibong (46–68 mg l^{-1}) and Harangajao Ponds (20–56 mg l^{-1}), while the lowest was in Upper Lake (20–48 mg l^{-1}). Alikunhi (1957) stated that in highly productive waters, alkalinity ought to be more than 100 mg l^{-1}. Alkalinity greater than 100 mg l^{-1} was recorded only once (October 1999 in Fishery Tank) during the present study, although concentrations of 80–94 mg l^{-1} were also recorded in Fishery Tank and Haflong Lake in the post monsoon and winter seasons, while in all the other systems, alkalinity values were consistently much lower. Phosphate levels were higher in Fishery Tank during first annual cycle (26.6–127 µg l^{-1}), which again may be due to manuring and management. Nitrates were found to fluctuate widely between 0.38 mg l^{-1} and 45.9 mg l^{-1}. The least concentration of nitrate was recorded from Harangajao Pond. During the second annual cycle, the range of nitrate content was the highest (23.0–45.9 mg l^{-1}) in Haflong Lake among all the waterbodies studied. This might be due to the sewage disposal into the lake. However, the fluctuations were very narrow during the first annual cycle, when there was a stress in the lake ecosystem due to the accidental draining out of its water. Conductivity, total alkalinity, phosphates and nitrates were relatively high in all the systems during dry seasons which might be due to evaporation of water (Furch and Junk, 1993). In the present study, higher phosphate concentrations were recorded in Haflong Lake, Maibong Pond and Harangajao Pohd during monsoon. Both Maibong and Harangajao Ponds being situated in the vicinity of paddy fields received nutrient enriched surface runoff during monsoon, while the bulk disposal of sewage into Haflong Lake during monsoon might be the reason behind the observed high level of phosphate in these waterbodies during this season. Fluctuations in both gross primary productivity and net primary productivity could be observed in all the waterbodies. The highest GPP was observed in Haflong Lake (2552.4 mgC/m^3/24 hrs) in pre-monsoon, although highest NPP was recorded in the same system during post-monsoon (863.33 mgC/m^3/24 hrs). A strong correlation between primary productivity and phosphates, nitrates and total alkalinity contents of the lentic systems was apparent.

Statistical Analysis

Statistical analysis of data for physico-chemical properties of water in different waterbodies with ANOVA and Duncan's Long Range Test is shown in Table 70.4. The table revealed a significant difference among the lentic systems only in dissolved oxygen, total alkalinity and conductivity during the first annual cycle and in total alkalinity, phosphates, nitrates and conductivity during the second annual cycle. During the first annual cycle, Upper Lake showed significantly higher dissolved oxygen content than Haflong Lake and Fishery Tank, while both Fishery Tank and Haflong Lake showed significantly higher total alkalinity than Upper Lake. Fishery Tank had higher conductivity of water than Hatlong Lake, which in its turn had higher conductivity of water than Upper Lake. During the second annual cycle, significantly higher total alkalinity was observed in both Fishery Tank and Haflong Lake than that in the other systems. Halfong Lake also showed higher concentrations of phosphates, nitrates and conductivity compared to the others.

Acknowledgements

TNM is thankful to Mr. GK. Das, Principal, Haflong Government College, for his help and cooperation. Help rendered by Sri Arun Chandra Das in collection of samples is also acknowledged.

References

Alikunhi, K. H., 1957. Fish culture in India. *Farm Bulletin*, ICAR, New Delhi, 20: 1–6.

Carpenter, S.R. and Lodge, D.M., 1986. Effects of submerged macrophytes on ecosystem processes. *Aquat. Bot.* 26: 341–370.

Champion, H.G. and Seth, S.K., (1968. *Forest Types of India*. Forest Research Institute, Dehradun.

Das, A C., Baruah, B.K., Baruah, D. and Sengupta, S., 2002. Study on wetlands of Guwahati city-1. Water quality of ponds and beets. *Poll. Res.*, 21(4): 511–513.

Dewanji, A. and Matai, S., 2000. Aquatic weeds as indicators of water quality. In: *Environmental Stress: Indication Mitigation and Eco-Conservation*, (Eds.) Yunus, M., Sing, N. and Dekok, L.J. Kluwer Academic Publishers, Boston, p. 251–258.

Dutta Gupta, S., Gupta, S. and. Gupta, A., 2004. Euglenoid blooms in the flood plain wetlands of Barak Valley, Assam, Northeastern India. *J. Environ. Biol.*, 25(3): 369–373.

Furch, K. and Junk, W.J., 1993. Seasonal nutrient dynamics in an Amazonian Lake Arch. *Hydrobiol.*, 128: 277–295.

Gupta, S., Michael, R.G. and Gupta, A., 1992. A preliminary investigation on physico-chemical factors, periphyton and invertebrate communities in a protected water works of Shillong, Meghalaya. In: *Proc. Nat. Acad. Sci.*, India, 62(B)III: 333–337.

Habib, O.A, Tripett, R. and Murphy, K.J., 1997. Seasonal changes in phytoplankton community structure in relation to physico-chemical factors in Lack Lomond, Scotland. *Hydrobiologia*, 350: 63–79.

Kaushik, S., Agarkar, M.S. and Saxena, D.N., 1991. Water quality and periodicity of phytoplantonic algae in Chambal tank, Gwalior, Madhya Pradesh. *Bionature*, 11: 87–94.

Kaushik, S., Sharma, S. and Saksena, M.N., 1989. Physico-chemical factors and the aquatic insect density of a pond receiving cotton mill effluent at Gwalior. *Indian J. Ecol.*, 16(1): 64–67.

Mariappan, P., and Vasudevan, T., 2002. Correlation coefficients of some physico-chemical parameters of drinking water ponds in eastern part of Sivagangai District, Tamil Nadu. *Poll. Res.*, 21(4): 403–407.

Michael, P., 1984. *Ecological Methods for Field and Laboratory Investigations*. Tata McGraw-Hill Publishing Company, Limited, New Delhi.

Parashar, D. (Ed.), 1994. *N.C. Hills: The Enchanting Land and the Bird Mystery of Jatinga*. North Cachar Hills Autonomous District Council, Haflong in Association with Konark Publishers. Pvt. Ltd., New Delhi.

Prasad, D.Y., 1990. Primary productivity and energy flow in Upper Lake, Bhopal (India). *Indian J. Environ. Health*, 32(2): 132–199.

Sahat, T., Manna, N.K., Som Majumdar, S. and Bhattacharjya, I.N., 2001. Primary productivity of the Subhas Sarobar Lake in East Calcutta in relation to some selected physicochemical parameters. *Poll. Res.*, 20(1): 47–52.

Singh, J.P. and Roy, S.P., 1995. Limnobiotic investigation of Kawar Lake, Begusarai, Bihar. *Env. Eco.*, 13(2): 330–335.

Tiwari, D.R., 1999. Physicochemical studies of the Upper Lake Water, Bhopal, Madhya Pradesh, India. *Poll. Res.*, 18(3): 323–326.

Unni, K.S., 1972. An ecological study of the macrophytic vegetation of Doodhadhari lake, Raipur (M.P.). 3. Chemical factors. *Hydrobiologia*, 40: 25–36.

Welch, P.S., 1952. *Limnology*. Mc-Graw Hill Book Company, New York.

Index

www.ingramcontent.com/pod-product-compliance
Ingram Content Group UK Ltd.
Pitfield, Milton Keynes, MK11 3LW, UK
UKHW052224270726
14059UKWH00003B/109